T0201955

ELECTROMAGNETIC RADIATION

Electromagnetic Radiation

Richard Freeman, James King,
and Gregory Lafyatis

OXFORD
UNIVERSITY PRESS

OXFORD
UNIVERSITY PRESS

Great Clarendon Street, Oxford, OX2 6DP,
United Kingdom

Oxford University Press is a department of the University of Oxford.
It furthers the University's objective of excellence in research, scholarship,
and education by publishing worldwide. Oxford is a registered trade mark of
Oxford University Press in the UK and in certain other countries

First published 2019
First published in paperback 2024

Published in the United States of America by Oxford University Press
198 Madison Avenue, New York, NY 10016, United States of America

British Library Cataloguing in Publication Data
Data available

Library of Congress Cataloging in Publication Data
Data available

ISBN 978–0–19–872650–0 (Hbk.)
ISBN 978–0–19–889968–6 (Pbk.)

DOI: 10.1093/oso/9780198726500.001.0001

Printed and bound by
CPI Group (UK) Ltd, Croydon, CR0 4YY

Contents

Part III Electromagnetism and Special Relativity

Part I

Introductory Foundations

Essentials of Electricity and Magnetism

1

- Review of Maxwell's steady-state equations in vacuum
- Modifications of Maxwell's steady-state equations in the presence of matter: electric and magnetic polarization
- Generalization of Maxwell's equations in the presence of time varying sources leading to a causal unification of fields in the form of additional sources
- Origin of electromagnetic radiation directly from time-dependent Maxwell's equations and the response of materials to electromagnetic radiation
- Electromagnetic conservation laws, including electromagnetic energy, momentum and angular momentum

1.1 Maxwell's static equations in vacuum

Maxwell's equations are the foundational equations of classical electromagnetic phenomena. They are comprised of four 1st order linear partial differential equations and are essentially statements that define the electric and magnetic vector fields (e.g., specify their divergence and curl) in terms of specific boundary conditions and electric charge and current distributions. The mathematical origins of Maxwell's equations can be found in the basic inverse square laws of electrostatics and magnetostatics, which were mainly formulated in the late eighteenth through early nineteenth centuries, but far from exclusively, through the observations and work of Coulomb, Ampere, Biot, and Savart. When coupled with Faraday's concept of a field and the general mathematical theorems of Gauss, Laplace, and Poisson, we begin to see the formal modern description of electric and magnetic phenomena–at least for steady-state conditions in vacuum.

Electromagnetic Radiation. Richard Freeman, James King, Gregory Lafyatis,
Oxford University Press (2019).
DOI: 10.1093/oso/9780198726500.001.0001

1.1.1 Electrostatic equations

The integral form of the law of electrostatics or Coulomb's law is:

$$\vec{E}(\vec{r}_o) = \frac{1}{4\pi\varepsilon_o}\int \frac{\rho(\vec{r})(\vec{r}_o - \vec{r})}{|\vec{r}_o - \vec{r}|^3} dV \tag{1.1}$$

where the position vectors $\vec{r}$ and $\vec{r}_o$ refer to the source and observer locations, respectively, and ρ is the (static) charge density. We note from this equation a number of important features: first, the static electric field falls off as the inverse square of the distance to the observer and is proportional to the charge density; second, a contribution to the field at $\vec{r}_o$ due to an element of charge $\rho(\vec{r})\,dV$ will point along $\vec{R} = \vec{r}_o - \vec{r}$, the direction from source to observer, with a polarity dependent on the charge sign; and third, the field $\vec{E}(\vec{r}_o)$ is a linear vector superposition of contributions from charge elements integrated over all space, independent of time. If we now look at the divergence taken with respect to r_o of this field,

$$\nabla\cdot\vec{E}(\vec{r}_o) = \frac{1}{\varepsilon_o}\int \rho(\vec{r})\,\delta(\vec{r}_o - \vec{r})\,dV = \frac{\rho(\vec{r}_o)}{\varepsilon_o} \tag{1.2}$$

where we have used $\nabla\cdot(\hat{R}/R^2) = 4\pi\,\delta(\vec{R})$. This is the differential form of what is known as Gauss' law and is equivalent to Coulomb's law. It is the first of Maxwell's equations. In this form, we see that any divergence in the field is local to and proportional to the charge density. In its integral form, which can be directly obtained from Eq. 1.2 using the divergence theorem, it states that the integral of the $\vec{E}$ field over an arbitrary closed surface is equal to the charge enclosed within that surface divided by ε_o. Noting that $\vec{R}/R^3 = -\nabla(1/R)$, Eq. 1.1 can be rewritten as a gradient

$$\vec{E}(\vec{r}_o) = \frac{-1}{4\pi\varepsilon_o}\nabla\int \frac{\rho(\vec{r})}{|\vec{r}_o - \vec{r}|} dV = -\nabla\phi(\vec{r}_o) \tag{1.3}$$

where $\phi(\vec{r}_o)$ is the scalar potential. Because in general the curl of a gradient vanishes, it follows from Eq. 1.3 that

$$\nabla\times\vec{E}(\vec{r}_o) = 0 \tag{1.4}$$

which is the second of the two electrostatic Maxwell's equations and states that electrostatic fields are irrotational (curlless) everywhere. The integral form of Eq. 1.4, which can be directly obtained using Stokes's theorem, states that the line integral of the $\vec{E}$ field around an arbitrary closed curve is zero, thus confirming its status as a gradient.

Because the curl of a gradient is always zero, Eqs. 1.2 and 1.4 can be more compactly expressed in terms of ϕ as $\nabla^2\phi = \rho/\varepsilon_o$ (Poisson's equation) or $\nabla^2\phi = 0$ (Laplace's equation) in charge free regions.

1.1.2 Magnetostatic equations

The integral form of the law of magnetostatics or the Biot–Savart law is:

$$\vec{B}(\vec{r}_o) = \frac{\mu_o}{4\pi}\int \frac{\vec{\jmath}(\vec{r}) \times (\vec{r}_o - \vec{r})}{|\vec{r}_o - \vec{r}|^3} dV \tag{1.5}$$

where, as before, the distance and direction from a source element to the observation point is represented by $\vec{R} = \vec{r}_o - \vec{r}$ but the source is now a distribution of steady-state current density elements, $\vec{\jmath}(\vec{r})\, dV$, each contributing to $\vec{B}(\vec{r}_o)$ an amount proportional to $\vec{\jmath}(\vec{r})$, in the direction, $\vec{\jmath}(\vec{r}) \times (\vec{r}_o - \vec{r})$, given by the right-hand rule. Also, as with the electric field, the static magnetic field (due to each current element) falls off as the inverse square of the distance. Following, analogously, the electrostatic development of Section 1.1.1 to obtain a differential form, we consider the curl of $\vec{B}$

$$\nabla \times \vec{B}(\vec{r}_o) = \frac{\mu_o}{4\pi}\int \nabla \times \left(\vec{\jmath}(\vec{r}) \times \frac{\vec{r}_o - \vec{r}}{|\vec{r}_o - \vec{r}|^3}\right) dV \tag{1.6}$$

which, with some manipulation (see **Discussion 1.1**), can be written

$$\nabla \times \vec{B}(\vec{r}_o) = \mu_o \int \vec{\jmath}(\vec{r})\,\delta(\vec{r}_o - \vec{r})\, dV = \mu_o \vec{\jmath}(\vec{r}) \tag{1.7}$$

where, again, we have used $\nabla \cdot (\hat{R}/R^2) = 4\pi\,\delta(\vec{R})$. This is the differential form of what is known as Ampere's law and is equivalent to the Biot–Savart law. It is the third of Maxwell's equations. The integral form of Eq. 1.7, which can be directly obtained using Stokes's theorem, states that the line integral of the $\vec{B}$ field around an arbitrary closed curve is equal to the current enclosed by that curve multiplied by μ_o.

To obtain the fourth differential form of Maxwell's equations under steady-state conditions in the absence of matter, we take the divergence of Eq. 1.5 to obtain[1]

$$\nabla \cdot \vec{B}(\vec{r}_o) = 0 \tag{1.8}$$

which is the second of the two magnetostatic Maxwell's equations and states that magnetostatic fields are solenoidal (divergenceless) everywhere. The integral form of Eq. 1.8, which can be directly

[1] Expand the divergence of the integrand of Eq. 1.6

$$\nabla \cdot \left(\vec{\jmath}(\vec{r}) \times \frac{\vec{r}_o - \vec{r}}{|\vec{r}_o - \vec{r}|^3}\right) = \frac{\vec{r}_o - \vec{r}}{|\vec{r}_o - \vec{r}|^3} \cdot \left(\nabla \times \vec{\jmath}(\vec{r})\right) - \vec{\jmath}(\vec{r}) \cdot \left(\nabla \times \frac{\vec{r}_o - \vec{r}}{|\vec{r}_o - \vec{r}|^3}\right)$$

where the first term vanishes because the $\vec{\jmath}$ is not a function of the observer coordinates. For the second term, we again note that $\frac{\vec{R}}{R^3} = -\nabla\frac{1}{R}$ and the curl of a gradient vanishes.

obtained using the divergence theorem, states that the surface integral of the normal component of $\vec{B}$ field around an arbitrary closed surface is zero. Because $\vec{B}$ has no divergence value, it can be written as the curl of another field, $\vec{A}$. This vector field, known as the "vector potential", is analogous to the scalar potential, ϕ, encountered in electrostatics. So, continuing in close analogy with electrostatics, we are tempted to write Eqs. 1.7 and 1.8 in terms of a single second order differential equation of the potential such as the Poisson or Laplace equations. Thus, we note that much like writing $\vec{E}$ as $-\nabla\phi$ automatically satisfies $\nabla \times \vec{E} = 0$ and turns $\nabla \cdot \vec{E} = \rho/\varepsilon_o$ into the Poisson equation, writing $\vec{B}$ as $\nabla \times \vec{A}$ automatically satisfies $\nabla \cdot \vec{B} = 0$ and turns $\nabla \times \vec{B} = \mu_o \vec{J}$ into $\nabla \times \left(\nabla \times \vec{A}\right) = \nabla\left(\nabla \cdot \vec{A}\right) - \nabla^2 \vec{A} = \mu_o \vec{J}$. This is a more compact way of expressing Eqs. 1.7 and 1.8. In summary, Maxwell's equations for steady state and in the absence of matter are:

$$\nabla \cdot \vec{E} = \rho/\varepsilon_o \tag{1.9}$$

$$\nabla \times \vec{E} = 0 \tag{1.10}$$

$$\nabla \times \vec{B} = \mu_o \vec{J} \tag{1.11}$$

$$\nabla \cdot \vec{B} = 0 \tag{1.12}$$

1.1.3 Lorentz force

The effects of magnetic and electric fields on a charge q were given their modern form by Lorentz in 1892, building on the work of Thomson's (1881)[2] and Heaviside's (1889)[3] extrapolations of Maxwell's exposition of his equations (1865):

$$\vec{F} = q[\vec{E} + \vec{v} \times \vec{B}] \tag{1.13}$$

This description of the total force on a charge in the presence of external fields $\vec{E}$ and $\vec{B}$ has been so well verified experimentally, even for charge velocities approaching the speed of light, that it is used as an empirical definition of $\vec{E}$ and $\vec{B}$ at any space-time point when $q, \vec{F}, \vec{v}$ are known.[4]

1.2 Maxwell's static equations in matter

Within matter, where there are charges that respond to external fields by moving freely, or charges bound to other charged objects that orient or displace in response to external fields, Maxwell's equations become exceedingly difficult to solve exactly. It is useful then, when

[2] Thomson, J. J. On the electric and magnetic effects produced by the motion of electrified bodies. *Philosophical Magazine*, 11, 229–249, https://doi.org/10.1080/14786448108627008 (1881).

[3] Heaviside, Oliver. On the electromagnetic effects due to the motion of electrification through a dielectric. *Philosophical Magazine* 324 (April 1889).

[4] The relativistic formulation of Eq. 1.13 is the same, with the proviso that the force is related to the velocity by

$$\vec{F} = \frac{d}{dt}(\gamma m\vec{v})$$

working with fields in matter, to divide the problem conceptually into microscopic and macroscopic fields with the microscopic fields, in a sense, being the true yet practically intractable fields in all their grainy detail, while the macroscopic fields are spatial and temporal averages of the micro-fields over regions and times that are microscopically large yet macroscopically small. In this subsection it will be shown that the response of matter to applied fields generally results in so-called "bound" sources of charge and current density and for materials with a component of free electrons, an additional source of "free" current. While this will modify the two inhomogenous Maxwell's equations, it will, in the steady-state case, leave unaffected the two homogeneous equations. As a consequence of this, we can immediately see that the macroscopic versions of the two homogeneous equations will be identical to the microscopic versions. That is,

$$\nabla \times \vec{E} = 0 \textit{ and } \nabla \cdot \vec{B} = 0 \tag{1.14}$$

1.2.1 Response of material to fields

Polarization, either electric $\vec{P}(\vec{r})$ or magnetic $\vec{M}(\vec{r})$, is defined macroscopically as dipole moment per unit volume and its existence within a material is a result of the local alignment of atomic or molecular electric $\vec{p}$ or magnetic $\vec{m}$ dipole moments within a macroscopically small but microscopically large volume about the evaluation point, $\vec{r}$. This alignment can be permanently frozen into the material as in the case of ferromagnets and the less often encountered electric analogs known as electrets. Alignment of dipoles resulting in polarization is, however, more commonly a response to the presence of electric and magnetic fields. Two basic types of dipole response have been found: Either pre-existing dipole moments are rotated into alignment by the fields or dipole moments are induced by the applied fields within the material. A well known example of the first type of response to electric fields occurs within water because the positive and negative charge centers of the "polar" H_2O molecule are intrinsically separate. Similarly, the pre-existing magnetic atomic dipoles (due to unpaired electrons) within paramagnetic materials will align with an applied magnetic field. The second type of response in which dipole moments are induced occurs in all materials but is most noticeable within "non-polar" materials devoid of pre-existing dipoles. A classical picture of such a material response to an electric field is that of neutral atoms with initially overlapping positive (nuclear) and negative (electron) charge centers that, upon application of the field, get stretched in opposite directions to the mechanical limits of their bonds, thus forming electric dipole moments. The induction of magnetic dipole moments by a magnetic field is known as diamagnetism. In this case, there is no stretching but rather currents within atomic or molecular structures are induced via Faradays law (Section 1.4.1) resulting in an anti-alignment of the dipoles to the field. Macroscopic polarization in terms of the microscopic electric dipole moments is given as:

$$\vec{P}(\vec{r}) = N(\vec{r}) < \vec{p}(\vec{r}) > \tag{1.15}$$

where $< \vec{p}(\vec{r}) >$ is the average of all the electric dipole moments in a macroscopically small but microscopically large volume centered at the location $\vec{r}$, and $N(\vec{r})$ is the number of such objects per unit volume. In the case of magnetically active materials, whether there is an orientation of magnetic objects, or induced currents centered around atoms or molecules, the corresponding expression is the generation of a macroscopic magnetic polarization or "magnetization":

$$\vec{M}(\vec{r}) = N(\vec{r}) < \vec{m}(\vec{r}) > \tag{1.16}$$

where in the same way $< \vec{m}(\vec{r}) >$ is the average magnetic dipole centered at the location $\vec{r}$ and $N(\vec{r})$ is the number of objects per unit volume at that location.

The most general instance of material response to a field is not linear, isotropic or homogeneous and therefore requires a non-linear, spatially dependent tensor for its mathematical description. In the present case, we will initially assume a simpler material that is linear and isotropic but not necessarily homogeneous. In this case, for example, the average dipole moment and the polarizing electric field $\vec{E}_p$ within the material are related by a coefficient, α, known as the polarizability:

$$< \vec{p}(\vec{r}) > = \varepsilon_o \alpha(\vec{r}) \vec{E}_p \tag{1.17}$$

where ε_o is generally included for later convenience. Combining Eq. 1.17 with Eq. 1.15 gives us an expression for the polarization in terms of the polarizing field:

$$\vec{P}(\vec{r}) = \varepsilon_o N(\vec{r}) \alpha(\vec{r}) \vec{E}_p \tag{1.18}$$

Now, it is an easily overlooked but important point that the macroscopic applied field amplitude, $\vec{E}$, within the material is not necessarily the average amplitude of the polarizing field, $\vec{E}_p$, felt by the atoms and molecules in the matter. Indeed, both field amplitudes are average values, however, while the applied field results from macroscopically averaged surface and volume charge densities external to the material, the polarizing field refers specifically to volumes local to the atoms and molecules and so additionally takes into account the fields of all nearby dipoles. The source of these additional local fields are represented in the form of bound charges that, along with bound currents, are discussed next.

1.2.2 Bound charges and currents

The presence of polarization or magnetization in dielectric and magnetically active materials is characterized by the existence of "bound" charges and currents. For the case of polarized material, this can be shown by considering the potential Φ at a point $\vec{r}_o$ due to all the electric dipole moments within a non-uniformly polarized material in a volume, V

$$\Phi(\vec{r}_o) = \frac{1}{4\pi\,\varepsilon_o}\int_V \frac{\hat{R}\cdot\vec{P}(\vec{r})}{R^2}dV \tag{1.19}$$

where $\vec{P}(\vec{r})\,dV = d\vec{p}$ is a macroscopically small but microscopically large element of dipole moment within the material and as usual, $\vec{r}$ is a source point and $\vec{R} = \vec{r}_o - \vec{r}$ is the vector pointing from the source to the observation point. Through the use of integration by parts and the divergence theorem, this equation can be re-expressed as[5]

$$\Phi(\vec{r}_o) = \frac{1}{4\pi\,\varepsilon_o}\oint_S \frac{\vec{P}(\vec{r})\cdot\hat{n}}{R}da + \frac{1}{4\pi\,\varepsilon_o}\int_V \frac{-\nabla\cdot\vec{P}(\vec{r})}{R}dV \tag{1.20}$$

where S is the surface bounding the material and $\hat{n}$ is the outward surface normal at points of integral evaluation. The numerators times their respective differentials have the form and units of elements of charge so that we equate them to surface charge and volume charge densities. That is,

$$\vec{P}\cdot\hat{n} = \sigma_b \quad and \quad -\nabla\cdot\vec{P} = \rho_b$$

Physically, this is not hard to visualize. For example, to see the latter equivalence, consider a small, yet still macroscopic volume within the non-uniformly polarized material and in this region let there be generally positive divergence. Consider just the x direction. Because a larger polarization means more positive charge displaced in the positive x direction and more negative charge displaced in the negative x direction, a positive divergence means that more positive charges are pushed out of the right bounding surface than negative charges are pushed out of the left boundary, thus leaving a net negative charge density.

[5] Noting $\frac{\hat{R}}{R^2} = \nabla\frac{1}{R}$ and $\vec{P}\cdot\nabla\left(\frac{1}{R}\right) = \nabla\cdot\left(\frac{\vec{P}}{R}\right) - \frac{\nabla\cdot\vec{P}}{R}$, we have

$$\Phi(\vec{r}_o) = \frac{1}{4\pi\,\varepsilon_o}\int_V \nabla\cdot\frac{\vec{P}(\vec{r})}{R}dV + \frac{1}{4\pi\,\varepsilon_o}\int_V \frac{-\nabla\cdot\vec{P}(\vec{r})}{R}dV$$

and then use the divergence theorem on the first term.

By similar considerations of the vector potential at a point $\vec{r}_o$ due to all the magnetic dipole moments within a non-uniformly magnetized material it can be shown that

$$\vec{M}(\vec{r}) \times \hat{n} = \vec{K}_b \quad and \quad \nabla \times \vec{M}(\vec{r}) = \vec{J}_b$$

which is to say that the curl of the magnetization $\nabla \times \vec{M}(\vec{r})$ in a magnetizable material can be identified with a real, yet bound, current density, $\vec{J}_b$, within the volume and $\vec{M}(\vec{r}) \times \hat{n}$ can be identified as a bound surface current density, $\vec{K}_b$, on the surface.[6]

The physical interpretation of this can be seen by imagining a uniformly magnetized material in which each little magnetic dipole has an associated current loop. Because of the uniformity, all the neighboring dipole current loops cancel out. However, at surfaces not perpendicular to $\vec{M}$, or equivalently, where $\vec{M}(\vec{r}) \times \hat{n} \neq 0$, there are missing neighbors and so there is net bound current on the surface. On the other hand, if the magnetization is not uniform in such a way that $\nabla \times \vec{M}(\vec{r}) \neq 0$ within the volume, then the magnitude of $\vec{M}$ varies in a direction perpendicular to $\vec{M}$ so neighboring dipoles do not completely cancel and again there is a net current in the direction of $\nabla \times \vec{M}(\vec{r})$.

1.2.3 Macroscopic fields

We have seen that the presence of polarization and magnetization within matter is equivalent to a distribution of "bound" charge and current densities as given by

$$\rho_b = -\nabla \cdot \vec{P} \tag{1.21}$$

$$\vec{J}_b = \nabla \times \vec{M} \tag{1.22}$$

In terms of the total (free plus bound) charge and current densities, the two inhomogeneous Maxwell's equations in a material possessing both a polarization and a magnetization can now be written:

$$\varepsilon_o \nabla \cdot \vec{E} = \rho_f + \rho_b = \rho_f - \nabla \cdot \vec{P} \tag{1.23}$$

$$\frac{1}{\mu_o}(\nabla \times \vec{B}) = \vec{J}_f + \vec{J}_b = \vec{J}_f + \nabla \times \vec{M} \tag{1.24}$$

If we now combine divergence and curl terms to get

$$\nabla \cdot \left(\varepsilon_o \vec{E} + \vec{P}\right) = \rho_f \tag{1.25}$$

$$\nabla \times \left(\frac{1}{\mu_o}\vec{B} - \vec{M}\right) = \vec{J}_f \tag{1.26}$$

[6] The vector potential, at a field point $\vec{r}_o$, resulting from a superposition of all the little magnetic dipole moments within a volume is:

$$\vec{A}(\vec{r}_o) = \frac{1}{4\pi\varepsilon_o c^2}\int_v \frac{\vec{M}(\vec{r}) \times \hat{R}}{R} dV$$

Using the identity $\frac{\hat{R}}{R^2} = \nabla \frac{1}{R}$, and integrating by parts,

$$\vec{A}(\vec{r}_o) = \frac{\mu_o}{4\pi}\oint_s \frac{\vec{M}(\vec{r}) \times \hat{n}}{R} da + \frac{\mu_o}{4\pi}\int_v \frac{\nabla \times \vec{M}(\vec{r})}{R} dV$$

we can then define two new macroscopic fields

$$\vec{D} = \varepsilon_o \vec{E} + \vec{P} \tag{1.27}$$

$$\vec{H} = \frac{1}{\mu_o}\vec{B} - \vec{M} \tag{1.28}$$

So that in terms of these new fields, which represent the fundamental fields plus polarization and magnetization effects due to the material, the steady-state macroscopic Maxwell's equations then simplify to:

$$\nabla \cdot \vec{D} = \rho_f \tag{1.29}$$

$$\nabla \times \vec{E} = 0 \tag{1.30}$$

$$\nabla \cdot \vec{B} = 0 \tag{1.31}$$

$$\nabla \times \vec{H} = \vec{\mathcal{J}}_f \tag{1.32}$$

It is important to emphasize that while both $\vec{D}$ and $\vec{H}$ have only the free charges and currents as sources, both of these quantities are just convenient constructs introduced to permit a compact method of accounting for the response of the material to the fundamental fields, $\vec{E}$ and $\vec{B}$. It is also important to keep in mind that while $\vec{P}$ and $\vec{M}$, like $\vec{E}$ and $\vec{B}$, are macroscopically averaged vector fields within matter (i.e., they are the same type of mathematical object), they differ substantially in that $\vec{E}$ and $\vec{B}$ represent the fundamental fields in the purest ethereal sense as envisioned by Faraday, while $\vec{P}$ and $\vec{M}$ are essentially representations of charge and current distributions within matter. From this perspective, we can see the conceptual difference between, for example, the two equations $\nabla \cdot \vec{E} = \rho/\varepsilon_o$ and $-\nabla \cdot \vec{P} = \rho_b$: We read the first equation as "a collection of charge (any charge) acts as a source of the electric field" while the second equation reads "a collection of charge (bound charge) results from a distortion of charge distribution in a material" Implicit in these statements is that in the first case the source somehow "causes" or at least accompanies the field but the two things are not physically the same whereas in the second case the charge distribution is equivalent to a distortion in $\vec{P}$. Finally, $\vec{D}$ and $\vec{H}$, while often treated more like fields akin to $\vec{E}$ and $\vec{B}$, are composite vector fields that are part pure field and part material response.

1.2.4 Polarizability and Susceptibility

Earlier, in our discussion of the response of matter to electric fields, we obtained an expression (Eq. 1.18) for the polarization $\vec{P}$ in terms of the polarizability α and the polarizing field, $\vec{E}_p$. We further noted that the polarizing field, $\vec{E}_p$, felt by the atoms and molecules in the matter, was not generally of the same amplitude as the macroscopic applied field amplitude, $\vec{E}$, within a material and this was said to be due to the specific accounting, by $\vec{E}_p$, of fields from other nearby dipoles. As can be imagined, this difference is density dependent and in fact, for the case of gases, the material is tenuous enough that we can approximate $\vec{E}_p \simeq \vec{E}$. However, this is not the case for denser liquids and solids and we would therefore like to find the relation between these two field amplitudes so we can then write an expression relating the macroscopic applied field amplitude, $\vec{E}$, to the polarization $\vec{P}$ in terms of the microscopic polarizability, α. This constant of proportionality is known as the (DC) electric susceptibility, $\chi_e = \chi_e(\alpha)$, and can be seen as the macroscopic equivalent to the microscopic polarizability, α. The electric susceptibility is thus defined by

$$\vec{P}(\vec{r}) = \varepsilon_o \chi_e \vec{E} \tag{1.33}$$

where, for example, in the case of gases, this connection is trivial: $\chi_e = N(\vec{r})\,\alpha(\vec{r})$. An analogous relation exists that expresses the magnetization response, $\vec{M}$, of a magnetic material to the macroscopic field, $\vec{H}$. The magnetic susceptibility, χ_m, is thus similarly defined as,

$$\vec{M}(\vec{r}) = \chi_m \vec{H} \tag{1.34}$$

For the electric field case, to see how the difference between the applied and polarizing fields ($\vec{E}$ and $\vec{E}_p$) comes about, we divide our treatment of the material into two regions: (a) a macroscopically small but microscopically large spherical cavity, centered on the point in question, in which we must account for the specific charge configurations of the surrounding atoms and molecules and (b) all the rest of the material outside the cavity that we can safely treat as smooth and macroscopically averaged. Let us express the relation between the two average fields as $\vec{E}_p = \vec{E} + \Delta\vec{E}$. We first note that if region (a) were to be treated like (b), smooth and macroscopically averaged, then we would essentially be eliminating any reference to specific fields, which is required for the evaluation of $\vec{E}_p$, and our result would yield $\Delta\vec{E} = 0$. So, $\Delta\vec{E}$ is what we get when we replace the field resulting from a smooth and macroscopically averaged treatment of region (a) with a more detailed and accurate treatment of the region. Two results are important: (a) It is a well-known result from electrostatics in matter[7] that the electric field within a dielectric sphere of uniform polarization $\vec{P}$ is also uniform and given by $\vec{E} = -\vec{P}/3\varepsilon_o$ and (b) it can be shown that in material lattices of sufficient symmetry, the total contribution to the electric field at a given lattice point due to atoms at all nearby lattice points (i.e., within the small cavity region in our problem)

[7] Jackson, J. D., *Classical Electrodynamics, 3rd edition*, Wiley, New York (1999).

goes to zero.[8] Now, with the additional assumptions that within the small cavity region (a) the polarization is constant and the material is sufficiently symmetric, these results can be used to write $\Delta\vec{E} = \vec{P}/3\varepsilon_o$. Then letting

$$\vec{E}_p = \vec{E} + \vec{P}/3\varepsilon_o \tag{1.35}$$

the substitution of this result into Eq. 1.18, with some rearranging, yields the polarization in terms of the applied field:

$$\vec{P}(\vec{r}) = \varepsilon_o \frac{N\alpha(\vec{r})}{\left(1 - \frac{1}{3}N(\vec{r})\alpha(\vec{r})\right)}\vec{E}(\vec{r}) = \varepsilon_o\chi_e(\vec{r})\vec{E}(\vec{r}) \tag{1.36}$$

which, as mentioned, gives the macroscopic susceptibility (χ_e) in terms of the microscopic polarizabilty (α) for dense materials. This is known as the Clausius–Mossotti equation.

1.2.5 The canonical constitutive relations

Note that Eqs. 1.27 and 1.28 make no assumption about whether the polarization or magnetization is frozen into the material or is, for example, a linear response to an applied field. If we consider the latter case, then following from $\vec{P} = \varepsilon_o\chi_e\vec{E}$ of the previous section along with the analogous result of $\vec{M} = \chi_m\vec{H}$ for the macroscopic magnetization response, Eqs. 1.27 and 1.28 lead to[9]

$$\vec{D} = \varepsilon\vec{E} \tag{1.37}$$

$$\vec{B} = \mu\vec{H} \tag{1.38}$$

in which $\varepsilon = \varepsilon(\vec{r}) = \varepsilon_o(1+\chi_e)$ and $\mu = \mu(\vec{r}) = \mu_o(1+\chi_m)$ are the permittivity and permeability, respectively, of the material (see **Discussion 1.2**).

1.2.6 Electric fields and free charges in materials

If a material has free charges there is a further potential relation between the free current densities discussed before and the applied electric field:

$$\vec{J}_f = \sigma\vec{E} \tag{1.39}$$

where σ is the conductivity of the material. The introduction of conductivity here is properly a "constitutive" relation, because the concept

[8] Purcell, E.M. and Morin, D.J. *Electricity and Magnetism*, 3rd edition, Cambridge University Press, 2013.

[9] The macroscopic derived fields are given by Eqs. 1.27 and 1.28 as

$$\vec{D} = \varepsilon_o\vec{E} + \vec{P} \text{ and } \vec{H} = \frac{1}{\mu_o}\vec{B} - \vec{M}$$

For linear responses ($\vec{P} = \varepsilon_o\chi_e\vec{E}$ and $\vec{M} = \chi_m\vec{H}$) to an applied field, these become

$$\vec{D} = \varepsilon_o(1+\chi_e)\vec{E} = \varepsilon\vec{E}$$
$$\vec{B} = \mu_o(1+\chi_m)\vec{H} = \mu\vec{H}$$

is inherently macroscopic. The conductivity, as expressed in Eq. 1.39, is a relationship that essentially says that in a macroscopic region of the material under consideration, a free current is associated with an applied electric field, and the magnitude of the current depends upon macroscopic parameters of the material, in this case the resistivity ρ. For materials with inherently large resistivity, the conductivity ($\sigma = 1/\rho$) is small enough that an applied field can exist in the material with no excitation of a free current. On the other hand, if the resistivity is extremely low (in a superconductor, e.g.,), there can be no equilibrium applied field for then there would be extremely large current flow. This idea is perhaps best understood by considering the free current to be the flow of individual free charges. In a material with extremely low resistivity, the free charges will move to cancel out the applied field; that is, the current described by Eq. 1.39 will be extremely high until the free electrons have arranged themselves to electrostatically cancel the applied field.

1.3 Energy of static charge and current configurations

1.3.1 Electrostatic field energy

The simplest starting point for the calculation of the field energy arising from a static placement of charges is to consider a collection of charges, q_i. The electrostatic energy of the ith charge is given by

$$\mathscr{E}_i^s = q_i \varphi_i^s$$

where $\varphi_i^s = \sum_j^n \varphi_{ij}^s$ is the summed electrostatic potential from all the other charges, q_j, $(j \neq i)$, evaluated at the position of q_i. The total energy of n assembled charges (so-called "configuration energy") is then

$$\mathscr{E}^s = \frac{1}{2} \sum_i^n q_i \varphi_i^s \tag{1.40}$$

where the factor of $1/2$ arises in this summation because we have essentially counted the potential energy from each charge pair twice. Also, we have been careful to specify that Eq. 1.40 is only the energy of assembling the charges (relative to their being infinitely separated); that is, the energy to assemble the individual charges themselves is not included. To calculate the total electrostatic energy of a charge distribution, we proceed formally by restating Eq. 1.40 for a continuous charge distribution

$$\mathscr{E}_c^s = \frac{1}{2}\int \rho\,\varphi_c^s\,dV \tag{1.41}$$

If we note that the charge density is given at each location by $\rho = \varepsilon_o \nabla \cdot \vec{E}$ and that the electric field within the volume is given at each location by $\vec{E} = -\nabla\varphi_c^s$, then with some manipulation,[10] Eq. 1.40 can be written

$$\mathscr{E}_c^s = \frac{\varepsilon_o}{2}\int E^2\,dV \tag{1.42}$$

To make contact with the expression in Eq. 1.40, now imagine the continuous charge distribution to be made up of a large collection of individual charges and that at any location in the volume the total electric field is given by[11]

$$\vec{E}(\vec{r}) = \sum_j \vec{E}_j(\vec{r}) \tag{1.43}$$

where the vectors $\vec{E}_j(\vec{r})$ are the Coulomb fields of each individual charge evaluated at the field point of interest.[12] Then, Eq. 1.42 can be written

$$\mathscr{E}_c^s = \mathscr{E}_o^s + \mathscr{E}_I^s$$

where the first term does not depend upon the relative position of the charges under consideration and is given by[13]:

$$\mathscr{E}_o^s = \frac{\varepsilon_o}{2}\int \sum_i E_i^2(\vec{r})\,dV \tag{1.44}$$

while the second one does, and is given by:

$$\mathscr{E}_I^s = \frac{\varepsilon_o}{2}\int \sum_{i,j}(1-\delta_{i,j})\vec{E}_i(\vec{r})\cdot\vec{E}_j(\vec{r})\,dV \tag{1.45}$$

which clearly vanishes for $j = i$. We can analyze $\mathscr{E}_I^s$ further by using the relation

$$\vec{E}_j = -\nabla\varphi_j^s(\vec{r})$$

[10] Express ρ in terms of $\vec{E}$ in Eq. 1.41 then use the vector identity $\left(\nabla\cdot\vec{E}\right)\varphi_c^s = -\vec{E}\cdot\left(\nabla\varphi_c^s\right) + \nabla\cdot\left(\varphi_c^s\vec{E}\right)$ and then $\nabla\varphi_c^s = -\vec{E}$; use the divergence theorem and note that the product of the electric field and the potential tends to zero faster than $1/R^2$.

[11] This argument follows that of Panofsky and Phillips (Panofsky, W. K. H. and Phillips, M., *Classical Electricity and Magnetism. 2nd edition*, Addison-Wesley, 1962.).

[12] Here and in what follows, the number of individual charges is considered to be so very large so that the idea of an approximate continuous charge distribution is reasonable.

[13] If the sum of the fields, Eq. 1.43, is inserted into Eq. 1.42, we have

$$\mathscr{E}_c^s = \frac{\varepsilon_o}{2}\int\left(\sum_i \vec{E}_i\right)\cdot\left(\sum_j \vec{E}_j\right)dV$$

which can be separated into a sum in which $i = j$,

$$\mathscr{E}_o^s = \frac{\varepsilon_o}{2}\int\sum_j E_j^2\,dV$$

and a sum in which $i \neq j$,

$$\mathscr{E}_I^s = \frac{\varepsilon_o}{2}\int\sum_{i,j}(1-\delta_{ij})\vec{E}_i\cdot\vec{E}_j\,dV.$$

where $\varphi_j^s(\vec{r})$ is the potential at the field point of interest due to the jth charge. Then Eq. 1.45 can be written[14]

$$\mathscr{E}_I^s = -\frac{\varepsilon_o}{2}\int\left[\sum_{i,j}(1-\delta_{ij})(\nabla\cdot(\vec{E}_i\cdot\varphi_j^s)-\varphi_j^s\nabla\cdot\vec{E}_i)\right]dV \qquad (1.46)$$

Now the volume is occupied by point charges, $\rho(\vec{r}) = \sum_i q_i\delta(\vec{r}-\vec{r}_i)$ so that $\nabla\cdot\vec{E}_i(\vec{r}) = (q_i/\varepsilon_o)\delta(\vec{r}-\vec{r}_i)$. Inserting this into Eq. 1.46 and using the divergence theorem while noting that the product of the field times the potential goes to zero faster than R^{-2}, yields

$$\mathscr{E}_I^s = \frac{1}{2}\sum_{i,j}(1-\delta_{ij})q_i\int\varphi_j^s(\vec{r})\delta(\vec{r}-\vec{r}_i)dV$$

If we note that $\int\varphi_j^s(\vec{r})\delta(\vec{r}-\vec{r}_i)dV = \varphi_{ij}^s$, then the portion of the electrostatic energy that depends upon the arrangement of the finite point charges expressed in terms of the fields in Eq. 1.42 is given by

$$\mathscr{E}_I^s = \frac{1}{2}\sum_{i,j}(1-\delta_{ij})q_i\varphi_{ij}^s = \frac{1}{2}\sum_i^n q_i\varphi_i^s$$

which is in agreement with the "configuration energy" we noted in Eq. 1.40. The meaning of this result is that when the total electrostatic energy of a collection of charges is calculated by the sum of the electrostatic field energies, there is a term that does not depend on the relative positions and one that does. The one that does is exactly equal to the energy we would have calculated by assuming the energy was contained in the charges as they are brought together. The portion of the electrostatic energy calculated in Eq. 1.44 is evidently the energy associated with creating the finite charges themselves. It is not possible to determine in any meaningful manner whether the electrostatic energy is "in the fields" or "inherent in the charges."

1.3.2 Magnetic field energy

In this section, we will show that in the same way the electrostatic energy of a system of charges can be represented as a volume integral of the product of charge density and electric potential (Eq. 1.41), the magnetic energy of a system of currents can be represented by a volume integral of the scalar product of current density and the vector potential. In the case of currents, unlike for charges and electrostatic energy, the reversible work of assembling the system does

[14] Use the vector identity $\nabla\cdot(f\vec{V}) = f\nabla\cdot\vec{V}+\vec{V}\cdot\nabla f$.

not include bringing the components in from infinity. Rather, the final magnetic energy of the system can be obtained by starting at an initial situation in which all the currents are zero and ramping up to the final system values.

As we saw, for a system of charges the potential at the location of charge i due to all the other charges is $\phi_i^s = \sum_j^n \phi_{ij}^s$ where φ_{ij}^s is linearly proportional to the value of charge q_j. Likewise, for a system of current loops, the magnetic flux passing through the ith current loop due to all the other loops is $\phi_i^s = \sum_j^n \phi_{ij}^s$ where ϕ_{ij}^s is linearly proportional to current in the jth loop, I_j. If the jth current increases by dI_j in time dt, then the associated flux through the ith current loop increases and a back emf, $V_{ij} = -d\phi_{ij}^s/dt$, is induced by Faraday's law of induction. To maintain the current during this time, the external source must then provide an equal and opposite emf, $-V_{ij}$. If an amount of charge dq_i has passed through the source during this time then the work done by the external source for the ith current loop as a result of a current change in the jth loop is,

$$dW_{ij} = -V_{ij}dq_i = I_i d\phi_{ij}^s$$

And if the currents are now all increased to their final values and we sum over j all the flux contributions to the ith current loop, we find that the magnetic energy of the ith current loop is given by

$$\mathscr{B}_i^s = \sum_j^n W_{ij} = I_i \phi_i^s$$

where $\phi_i^s = \sum_j^n \phi_{ij}^s$ is the magnetic flux due to all the other currents, I_j, $(j \neq i)$, passing through the current loop of I_i. The total energy of the n fully energized current loops is then

$$\mathscr{B}_i^s = \frac{1}{2}\sum_i^n I_i \phi_i^s \tag{1.47}$$

where, again, the factor of $1/2$ arises in this summation because we have essentially counted the energy from each current pair twice. To calculate the total magnetic energy of a current distribution, we proceed by rewriting the total flux in the ith loop, ϕ_i^s, in terms of the vector potential $\vec{A}(\vec{r}_i) = \vec{A}_i$. We know that

$$\phi_i^s = \int_S \left(\nabla \times \vec{A}_i\right) \cdot d\vec{a} = \oint_C \vec{A}_i \cdot d\vec{s}$$

and thus Eq. 1.47 can be written,

$$\mathscr{B}_i^s = \frac{1}{2}\sum_i^n \oint_C \vec{A}_i \cdot I_i d\vec{s}$$

so the closed line integral sums the contributions from all the current elements of the ith loop and the sum over i then combines the contributions from all current loops. Finally, if we re-express the current element in terms of current density and a volume element, $Id\vec{s} \Rightarrow \vec{\mathcal{J}} dV$, we can formally restate Eq. 1.47 for a continuous current distribution,

$$\mathscr{B}_c^s = \frac{1}{2}\int \vec{\mathcal{J}} \cdot \vec{A}\, dV \tag{1.48}$$

which is the magnetic analogue to Eq. 1.41 for electrostatic energy. In the forms of Eqs. 1.41 and 1.48, the configuration energy is emphasized to reside at the location of the charges and currents. However, it is often useful to express electric and magnetic configuration energies in forms that emphasize the fields themselves as carriers of this energy. We have already obtained this form for electrostatic energy (Eq. 1.42). We now obtain this "field" form for magnetic energy in an analogous way.

Writing the current density at each location as $\vec{\mathcal{J}} = 1/\mu_o\left(\nabla \times \vec{B}\right)$ and noting that the magnetic field within the volume is given at each location by $\vec{B} = \nabla \times \vec{A}$, we obtain–in much the same way as for the electrostatic case,[15]

$$\mathscr{B}_c^s = \frac{1}{2\mu_o}\int B^2\, dV$$

which is the magnetic energy counterpart to Eq. 1.42 for electrostatic energy of continuous distributions.

1.4 Maxwell's dynamic equations in vacuum

To this point, we have limited our consideration to steady-state sources of charge and current. In the context of radiation, for which the sources are necessarily changing in time, this consideration is mostly peripheral but has been included to serve as a review. Thus, in a practical

[15] Express $\vec{\mathcal{J}}$ in terms of $\vec{B}$ in Eq. 1.48; use the vector identity $\left(\nabla \times \vec{B}\right)\cdot\vec{A} = \left(\nabla \times \vec{A}\right)\cdot\vec{B} - \nabla\cdot\left(\vec{A}\times\vec{B}\right)$ and $\nabla \times \vec{A} = \vec{B}$; use the divergence theorem and note that the cross product of the magnetic induction and the vector potential tends to zero faster than $1/R^2$.

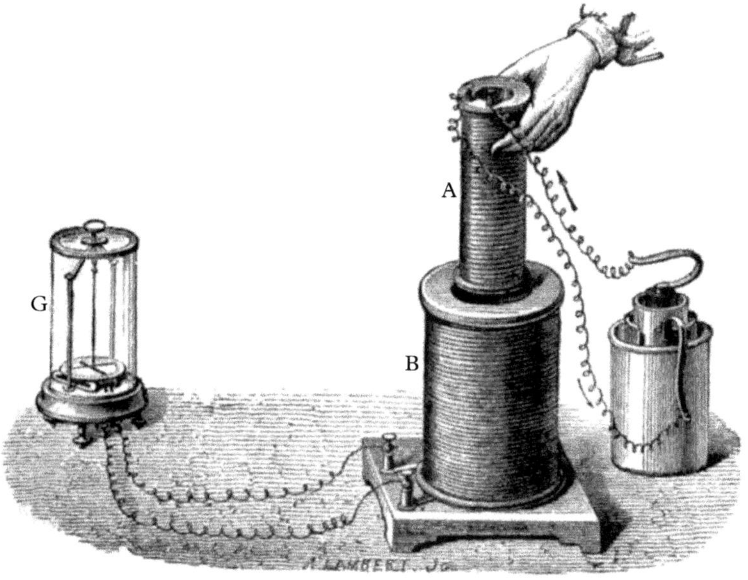

Fig. 1.1 *Faraday's apparatus showing electromagnetic induction from 1892 text book on Magnetism and Electricity by A.W. Poyser. Battery on the right powers the coil that is inserted into or removed from the standing coil. During the insertion or removal, the galvanometer connected to the standing coil responds. (By J. Lambert [Public domain], via Wikimedia Commons).*

sense, this section and the next on the time-dependent Maxwell's equations is the starting point for our discussion of radiation.[16]

1.4.1 Faraday's contribution

Historically, it was a series of experiments carried out and reported by Faraday (1831) that provided the first quantitative results on the relation between time-varying electric and magnetic fields. Specifically, he studied circuits placed in temporally and/or spatially varying magnetic fields and noticed that a current was induced when the magnetic flux Φ enclosed by the circuit changed in time. He observed three distinct cases:

(1) Change in flux due to a circuit moving through a spatially varying magnetic field.

(2) Change in flux due to a spatially varying magnetic field moving across a stationary circuit (see Fig. 1.1).

(3) Change in flux due to a spatially constant, temporally varying magnetic field within a stationary circuit.

where, with the hindsight of special relativity and its consequences for the transformation of $\vec{E}$ and $\vec{B}$ fields, we can immediately see the

[16] Generally, in this book, when discussing the fields and sources in vacuum, we will omit the subscripts, "*b*" and "*f*", for bound and free, since all charges are necessarily free and there is no need for a distinction.

equivalence of cases 1 and 2. However, prior to special relativity, case 1 current, which was understood to result from the $\vec{J} \times \vec{B}$ magnetic force, was distinct from cases 2 and 3 current for which there was no moving charge and therefore no magnetic force. So, it was the great insight of Faraday to identify the shared experience of a changing flux within the three cases and then to conclude that this changing flux induced an electric field that acted on the charges to generate current in the wire– a first step toward both the causal (classical) and actual (relativistic) unification of the $\vec{E}$ and $\vec{B}$ fields. In differential form, Faraday's law is expressed (see **Discussion 1.3**):

$$\nabla \times \vec{E} = -\frac{\partial \vec{B}}{\partial t} \tag{1.49}$$

which we see is the first time-dependent modification to the steady-state vacuum Maxwell's equations. Note that the induced electric field is independent of the existence of a circuit and that the current in the wire is circumstantial in comparison to Eq. 1.49.

1.4.2 Conservation of charge and the continuity equation

Electric charge is locally conserved, and in a volume V where the current inflow and outflow are not equal, this conservation of charge requires there to be a build-up or or depletion of charge (assuming no additional charge sources or sinks). More specifically, the total current, I, flowing through the surface of the volume is exactly the negative of the rate of change of the total charge Q within a volume,

$$I = -\frac{dQ}{dt} \tag{1.50}$$

Noting that $I = \int \vec{J} \cdot d\vec{A}$ and $Q = \int \rho dV$, where A is the surface of volume V, the divergence theorem then immediately yields the continuity equation for charge and current densities

$$\nabla \cdot \vec{J} + \frac{\partial \rho}{\partial t} = 0 \tag{1.51}$$

For example, an overall outflow of current in a region is represented by a positive value for $\nabla \cdot \vec{J}$ and this term is exactly balanced by a corresponding depletion of charge within that region expressed as a negative value for $\partial \rho / \partial t$.

1.4.3 Maxwell's contribution

For the final time-dependent modification to Maxwell's equations, the stage is set by considering an inconsistency that arises within them when dealing with time-varying charge and current densities. As we know from the mathematics of vector calculus, the divergence of the curl of a vector field vanishes. In particular, within Maxwell's equations, the divergence of the right-hand side terms of the curl equations must vanish. In electrostatics, this requirement is satisfied trivially since $\nabla \times \vec{E} = 0$, and in magnetostatics the currents are steady state by definition so that Ampere's law is satisfied: $\nabla \cdot (\nabla \times \vec{B}) = \mu_o \nabla \cdot \vec{J} = 0$. Neither is there a problem with Faraday's law (Eq. 1.49) since $\nabla \cdot (\partial \vec{B}/\partial t) = \partial/\partial t(\nabla \cdot \vec{B}) = 0$. On the other hand, if we consider the case of a time-varying (non-steady-state) current then there will be regions in which the rates of inflow and outflow of charge are not equal and therefore we will have a situation where $\nabla \cdot \vec{J} \neq 0$ and thus an inconsistency within Maxwell's equations: $\nabla \cdot (\nabla \times \vec{B}) = 0 \neq \mu_o \nabla \cdot \vec{J}$. It was this inconsistency in Ampere's law that was so brilliantly resolved in 1865 by Maxwell and for this revolutionary work, which essentially completed the classical theory of electrodynamics, the set of equations were named in honor of Maxwell.

So, how did Maxwell resolve the inconsistency? It has to do with local conservation of charge and the resulting continuity equation (Eq. 1.51) as discussed previously. With the use of Gauss' law or Maxwell's 1st equation, Eq. 1.9, the charge density can be expressed as a divergence and Eq. 1.51 can be rewritten as,

$$\nabla \cdot \left(\vec{J} + \varepsilon_o \frac{\partial \vec{E}}{\partial t} \right) = 0 \tag{1.52}$$

And it was Maxwell's realization that the zero divergence term in parentheses, which properly reduces to $\nabla \cdot \vec{J} = 0$ for steady-state conditions, is the correct term to be equated to the curl of $\vec{B}$ for the general case of time-varying charges and currents. Ampere's law is thus modified to

$$\nabla \times \vec{B} = \mu_o \left(\vec{J} + \varepsilon_o \frac{\partial \vec{E}}{\partial t} \right) = \mu_o \vec{J} + \mu_o \vec{J}_d \tag{1.53}$$

So now we have, for time dependent cases, an additional source for the $\vec{B}$ field - the time rate of change of the $\vec{E}$ field. Maxwell named this term the "displacement current". However, it hardly resembles any kind of traditional current, and, in fact, although it was derived for a situation in which it represented a changing charge density necessary to compensate for the divergence of an actual current, its existence within the equation is more general than that. Indeed, it need not accompany any actual current and it alone can act as a source of $\vec{B}$. It is this point that greatly adds to the symmetry of the Maxwell's equations–Faradays law says that a changing magnetic field induces an electric field and now the addition of Maxwell's displacement current says that a changing electric field induces a magnetic field. The (now complete) Maxwell's equations are

$$\nabla \cdot \vec{E} = \frac{\rho}{\varepsilon_o} \tag{1.54}$$

$$\nabla \times \vec{E} = -\frac{\partial \vec{B}}{\partial t} \tag{1.55}$$

$$\nabla \cdot \vec{B} = 0 \tag{1.56}$$

$$\nabla \times \vec{B} = \mu_o \vec{J} + \frac{1}{c^2} \frac{\partial \vec{E}}{\partial t} \tag{1.57}$$

where $c = (\sqrt{\varepsilon_o \mu_o})^{-1}$. Faraday's and Maxwell's contributions have shown us that the fields can result not only from charge and current densities but also from changing fields. The implications of this are profound. We will see that this result led not only to the unification of electric and magnetic phenomena but also to an explanation of light and radiation as self-propagating waves of electric and magnetic fields. Furthermore, as will be detailed in Chapter 5, it was the prediction of an absolute speed of light in the face of the intuitive notion of absolute space and time, which led to the development of special relativity. These equations are universally valid both in vacuum and in matter and–for reasons that will be explained in the next section–are known as the "microscopic" Maxwell's equations.

1.5 Maxwell's dynamic equations in matter

1.5.1 Origin of material currents

Within matter, the time-dependent Maxwell's equations (Eqs. 1.54–1.57) are microscopically correct but, as described in Section 1.2.2, a macroscopic description is more useful. In that section, we saw that polarization and magnetization (macroscopic concepts) within matter are equivalent to distributions of "bound" charge and current densities as given by Eqs. 1.21 and 1.22. We then expressed the two steady-state inhomogeneous Maxwell's equations within a material in terms of these "bound" contributions to the charge and current densities (Eqs. 1.23 and 1.24). Now, in the time-dependent case, we expect that the polarization and magnetization and thus the bound charge and current densities, in so much as they are responding to the time-varying fields, are also varying in time. That is, we now expect that $-\partial/\partial t(\nabla \cdot \vec{P}) = \partial \rho_b/\partial t \neq 0$ and $\partial/\partial t(\nabla \times \vec{M}) = \partial \vec{\mathcal{J}}_b/\partial t \neq 0$. So, at first sight, if we allow Eqs. 1.23 and 1.24 to vary with time, there appears to be no change to the equations–the sources vary and the fields follow. Closer inspection, however, reveals that a time varying polarization also represents another type of bound current density (in addition to the previously discussed $\vec{\mathcal{J}}_b = \nabla \times \vec{M}$) within the material. This general "polarization current" source term is written as (see **Discussion 1.4**):

$$\vec{\mathcal{J}}_p = \frac{\partial \vec{P}}{\partial t} \tag{1.58}$$

Thus, combining the current density source terms due to the material responses of magnetization and polarization (Eqs. 1.22 and 1.58) with the Maxwell time-dependent equation in vacuo (Eq. 1.57), we obtain the 4th Maxwell time-dependent equation including material response,

$$\nabla \times \vec{B} = \mu_o \left(\vec{\mathcal{J}}_f + \nabla \times \vec{M} + \frac{\partial \vec{P}}{\partial t} + \varepsilon_o \frac{\partial \vec{E}}{\partial t} \right) \tag{1.59}$$

where we can define a total material current density, $\vec{\mathcal{J}}_t$: (see **Discussion 1.5**)

$$\vec{\mathcal{J}}_t = \vec{\mathcal{J}}_f + \nabla \times \vec{M} + \frac{\partial \vec{P}}{\partial t} \tag{1.60}$$

so that Eq. 1.59 can be written more compactly as

$$\nabla \times \vec{B} = \mu_o \left(\vec{\mathcal{J}}_t + \varepsilon_o \frac{\partial \vec{E}}{\partial t} \right)$$

Alternatively, through the use of the derived fields $\vec{D} = \varepsilon_o \vec{E} + \vec{P}$ and $\vec{H} = \vec{B}/\mu_o - \vec{M}$ as defined before, Eq. 1.59 can be arranged into the macroscopic form

$$\nabla \times \vec{H} = \vec{\mathcal{J}}_f + \frac{\partial \vec{D}}{\partial t} \tag{1.61}$$

so that the remaining equations may now be expressed through the collection of Eqs. 1.29, 1.55, 1.56, and 1.61.

$$\nabla \cdot \vec{D} = \rho_f \tag{1.62}$$

$$\nabla \times \vec{E} = -\frac{\partial \vec{B}}{\partial t} \tag{1.63}$$

$$\nabla \cdot \vec{B} = 0 \tag{1.64}$$

$$\nabla \times \vec{H} = \vec{\mathcal{J}}_f + \frac{\partial \vec{D}}{\partial t} \tag{1.65}$$

which is a macroscopic representation of the fields including the effects of material response alone ($\nabla \cdot \vec{P}$ and $\nabla \times \vec{M}$), the effects of time dependence alone ($\partial \vec{B}/\partial t$ and $\varepsilon_o \partial \vec{E}/\partial t$) and the combined effect of material response and time dependence ($\partial \vec{P}/\partial t$).

1.6 Plane wave propagation in vacuum

In the absence of charge and current ($\rho = 0$ and $\vec{\mathcal{J}} = 0$), the time dependent Maxwell's equations for a vacuum, Eqs. 1.54–1.57, are given as

$$\nabla \cdot \vec{E} = 0 \tag{1.66}$$

$$\nabla \times \vec{E} = -\frac{\partial \vec{B}}{\partial t} \tag{1.67}$$

$$\nabla \cdot \vec{B} = 0 \tag{1.68}$$

$$\nabla \times \vec{B} = \frac{1}{c^2}\frac{\partial \vec{E}}{\partial t} \tag{1.69}$$

If we take the curl of Eq. 1.67,

$$\nabla \times (\nabla \times \vec{E}) = -\frac{\partial}{\partial t}(\nabla \times \vec{B}) \tag{1.70}$$

then expansion of the left-hand side and use of Eq. 1.66 yields

$$\nabla(\nabla \cdot \vec{E}) - \nabla^2 \vec{E} = -\nabla^2 \vec{E} = -\frac{\partial}{\partial t}(\nabla \times \vec{B}) \tag{1.71}$$

Finally, substitution of Eq. 1.69 results in a wave equation for the electric field,

$$\nabla^2 \vec{E} - \frac{1}{c^2}\frac{\partial^2 \vec{E}}{\partial t^2} = 0 \tag{1.72}$$

And, in the exact same way, Eqs. 1.69, 1.68, and 1.67 can be used to obtain an identical wave equation for the magnetic field, $\vec{B}$. The solution to such an equation, a plane wave, is of infinite extent and of constant value along the planes transverse to its propagation and so is not practically realizable (A more practical solution obtained by imposing a transverse cylindrical constraint will be discussed later in Chapter 12). In the direction of propagation (say, z), the general solution to Eq. 1.72 has the form

$$\vec{E}(z,t) = \vec{E}_+(z - ct) + \vec{E}_-(z + ct) \tag{1.73}$$

where both $\vec{E}_+(t=0)$ and $\vec{E}_-(t=0)$ can be any general function of z. For $t > 0$, $\vec{E}_+$ propagates toward positive z and $\vec{E}_-$ propagates toward negative z both with a speed $c = (\varepsilon_0 \mu_0)^{-1/2}$. In contrast to this general solution, the simplest (and in some sense the most fundamental) solution to Eq. 1.72 are the harmonic (monochromatic) transverse electric plane waves. Indeed, not surprisingly, they form the basis set by which any finite electric transverse wave, of the general form given by Eq. 1.73, can be represented. Such harmonic solutions are, by definition, of the form[17] $\vec{F}(\vec{r}_o, t) = \vec{F}(\vec{r}_o)exp(-i\omega t)$ so that

$$\frac{\partial \vec{E}}{\partial t} = -i\omega\vec{E} \quad and \quad \frac{\partial \vec{B}}{\partial t} = -i\omega\vec{B} \tag{1.74}$$

while the vacuum plane wave equations for $\vec{E}$ (Eq. 1.72) and $\vec{B}$ reduce to the Helmholtz equations

$$\left(\nabla^2 + \frac{\omega^2}{c^2}\right)\vec{E}(\vec{r}_o) = 0 \quad and \quad \left(\nabla^2 + \frac{\omega^2}{c^2}\right)\vec{B}(\vec{r}_o) = 0 \tag{1.75}$$

to which the solutions are of the form $\vec{F}(\vec{r}_o) = \vec{F}_o^+ exp(i\frac{\omega}{c}\hat{n}\cdot\vec{r}_o) + \vec{F}_o^- exp(-i\frac{\omega}{c}\hat{n}\cdot\vec{r}_o)$. Thus, solutions to Eq. 1.75 have required that we set $\vec{E}$ and $\vec{B}$ spatial dependence proportional to $exp(\pm i\frac{\omega}{c}\hat{n}\cdot\vec{r}_o)$ and so, for example, a harmonic solution of Eq. 1.72 can be represented as

$$\vec{E}(\vec{r}_o, t) = \vec{E}(\vec{r}_o)e^{-i\omega t} = \vec{E}_o^+ exp[i(k\hat{n}\cdot\vec{r}_o - \omega t)] + \vec{E}_o^- exp[-i(k\hat{n}\cdot\vec{r}_o + \omega t)] \tag{1.76}$$

in which we have introduced the wave vector $\vec{k} = k\hat{n} = (\omega/c)\hat{n}$ that points along the wave propagation direction. The physical field is the real part of the expression in Eq. 1.76. Notice in Eq. 1.76 that if we pull k out of the parentheses, we are left with the arguments $(\hat{n}\cdot\vec{r}_o \pm \frac{\omega}{k}t) = (\hat{n}\cdot\vec{r}_o \pm ct)$ thus obtaining the form of Eq. 1.73 with $\vec{E}_o^+$ moving in the direction of $\hat{n}$ and $\vec{E}_o^-$ moving opposite the direction of $\hat{n}$.

[17] The physical fields are represented by the real parts of these wave functions.

If we consider only the positive-going solutions, we can express the harmonic field solutions as

$$\vec{E}(\vec{r}_o,t) = \vec{E}_o exp[i(\vec{k}\cdot\vec{r}_o - \omega t)] \tag{1.77}$$

$$\vec{B}(\vec{r}_o,t) = \vec{B}_o exp[i(\vec{k}\cdot\vec{r}_o - \omega t)] \tag{1.78}$$

Inserting these harmonic solutions, the two Maxwell curl equations (Eqs. 1.67 and 1.69) become[18]

$$\vec{k}\times\vec{E} = \omega\vec{B} \;\; and \;\; \vec{k}\times\vec{B} = -\omega\varepsilon_0\mu_0\vec{E}. \tag{1.79}$$

both of which individually indicate that $\vec{E}$ and $\vec{B}$ are orthogonal and, when taken together, indicate that $\vec{k}$ is orthogonal to both fields. And so the curl equations tell us $\vec{E}$, $\vec{B}$, and $\vec{k}$ are mutually orthogonal.[19] Finally, putting Eqs. 1.79 together with Eqs. 1.77 and 1.78,

$$\vec{E}(\vec{r}_o,t) = -c(\hat{n}\times\vec{B}) = \vec{E}_o\, exp(i\vec{k}\cdot\vec{r}_o - i\omega t) \tag{1.82}$$

$$\vec{B}(\vec{r}_o,t) = \frac{1}{c}(\hat{n}\times\vec{E}) = \vec{B}_o\, exp(i\vec{k}\cdot\vec{r}_o - i\omega t) \tag{1.83}$$

where we have used $\vec{k} = k\hat{n} = (\omega/c)\hat{n}$ and $\varepsilon_0\mu_0 = c^{-2}$. And so, from these equations we see that the field amplitudes are related by $|\vec{E}| = c|\vec{B}|$. In summary, the transport of radiation in vacuum is a transverse plane wave, moving at the speed of light in the direction $\vec{k}$ with $\vec{E}$, $\vec{B}$ and $\vec{k}$ mutually orthogonal and with the field amplitudes related by $|\vec{E}| = c|\vec{B}|$. Figure 1.2 is a representation of the properties of a plane wave.

1.6.1 Polarization of plane waves

According to the discussion before, $\vec{E}$ is subject to the condition that it must be perpendicular to the direction of travel ($\hat{n}\cdot\vec{E} = 0$). A moment's reflection on this reveals that there must be two independent solutions for $\vec{E}$ corresponding to the two dimensions of the transverse plane. That is, given any arbitrarily chosen orthogonal pair of directions $\hat{e}_1$ and $\hat{e}_2 = \hat{n}\times\hat{e}_1$ within a plane defined by this condition, two independent solutions for the wave equation would then be two electric vectors, $\vec{E}_1$ and $\vec{E}_2$, of arbitrary magnitude pointed along $\hat{e}_1$ and $\hat{e}_2$, respectively. For example, for a wave propagating in the positive z direction, $\hat{e}_1$ and $\hat{e}_2$ could be taken to be in the x and y directions. Specifically, the two solutions would read

[18] For example, inserting 1.77 and 1.78 into Eq. 1.67,

$$\nabla\times\left(\vec{E}_o exp[i(\vec{k}\cdot\vec{r}_o - \omega t)]\right)$$
$$= -\frac{\partial}{\partial t}\left(\vec{B}_o exp[i(\vec{k}\cdot\vec{r}_o - \omega t)]\right)$$

noting that $\vec{k}\cdot\vec{r}_o = k_x x_o + k_y y_o + k_z z_o$, we consider just the x component of the left side

$$\left[\nabla\times\vec{E}\right]_x = \frac{\partial E_z}{\partial y_o} - \frac{\partial E_y}{\partial z_o}$$
$$= ik_y E_z - ik_z E_y$$
$$= i\left[\vec{k}\times\vec{E}\right]_x$$

combining with the x component of the right side

$$i\left[\vec{k}\times\vec{E}\right]_x = i\omega B_x$$

which for all components

$$\vec{k}\times\vec{E} = \omega\vec{B}$$

and in a like manner

$$\vec{k}\times\vec{B} = -\omega\varepsilon_0\mu_0\vec{E}.$$

[19] Further confirmation that both $\vec{E}$ and $\vec{B}$ must be perpendicular to $\vec{k}$, the direction of wave travel, is found by inserting the harmonic solutions (Eqs. 1.77 and 1.78) into the divergence equations (Eqs. 1.66 and 1.68). We get

$$\nabla\cdot\vec{E} = \vec{k}\cdot\vec{E} = 0 \tag{1.80}$$

$$\nabla\cdot\vec{B} = \vec{k}\cdot\vec{B} = 0 \tag{1.81}$$

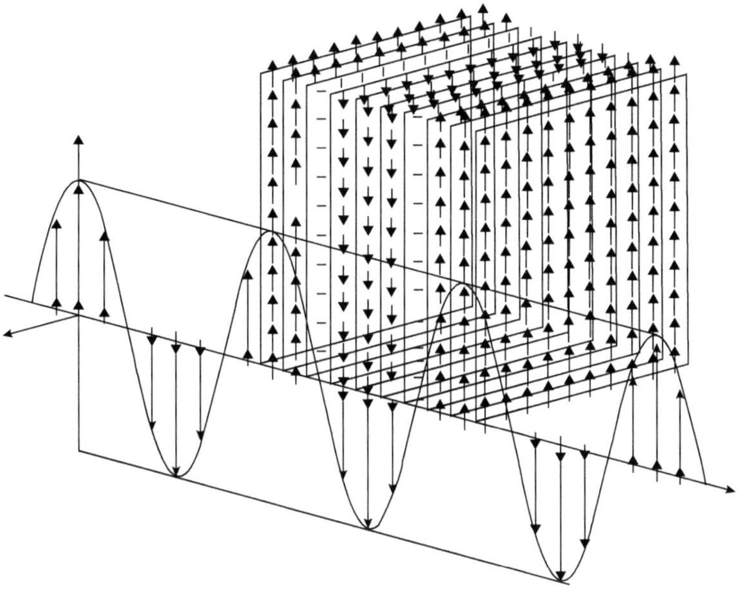

Fig. 1.2 *Representation of a plane wave. The electric field vectors are shown. The magnetic field (not shown) is perpendicular to the electric field and the direction of propagation. The "sheets" of constant phase extend infinitely in the directions transverse to the propagation ([Public domain], via Wikimedia Commons).*

$$\vec{E}_1 = E_1\, exp(i\vec{k}\cdot\vec{r}_o - i\omega t - i\alpha)\,\hat{e}_1 \tag{1.84}$$

$$\vec{E}_2 = E_2\, exp(i\vec{k}\cdot\vec{r}_o - i\omega t - i\alpha_2)\,\hat{e}_2 \tag{1.85}$$

where α and α_2 are independent arbitrary phases set by the initial conditions. In what follows, E_1 and E_2 are taken to be real and α is assumed to be zero (if, necessary, this may be accomplished by redefining the time origin). The total electric vector $\vec{E}\,(\vec{r}_o,t)$ is then the vector sum of 1.84 and 1.85:

$$\vec{E}\,(\vec{r}_o,t) = E_1\, exp(i\vec{k}\cdot\vec{r}_o - i\omega t)\,\hat{e}_1 + E_2\, exp(i\vec{k}\cdot\vec{r}_o - i\omega t - i\alpha_2)\,\hat{e}_2 \tag{1.86}$$

Defining $|\vec{E}| \to E_o$, since $E_0^2 = E_1^2 + E_2^2$ we can parametrize the amplitudes by the normalization of the field:

$$\cos\theta = \frac{E_1}{E_o} \quad and \quad \sin\theta = \frac{E_2}{E_o} \tag{1.87}$$

Then Eq. 1.86, may be rewritten:

$$\vec{E}(\vec{r}_o,t) = E_o\, exp(i\vec{k}\cdot\vec{r}_o - i\omega t)\,(\cos\theta\hat{e}_1 + \sin\theta e^{-i\alpha_2}\hat{e}_2) \tag{1.88}$$

we define the unit polarization vector for $\vec{E}$ as $\hat{e}_\perp$:

$$\hat{e}_\perp = \cos\theta\hat{e}_1 + \sin\theta e^{-i\alpha_2}\hat{e}_2 \tag{1.89}$$

so that the wave is represented by

$$\vec{E}(\vec{r}_o, t) = E_o exp(i\vec{k}\cdot\vec{r}_o - i\omega t)\,\hat{e}_\perp \tag{1.90}$$

The physical field is found by taking the real part of this expression. The direction of the electric field is given by the direction of

$$\text{Re}\left[\left(\cos\theta\hat{e}_1 + \sin\theta e^{-i\alpha_2}\hat{e}_2\right) exp(i\vec{k}\cdot\vec{r}_o - i\omega t)\right].$$

We consider three cases:

Case 1: If the two components are in phase, $\alpha_2 = 0$. Then, for a particular position and time:

$$\text{Re}\left[\left(\cos\theta\hat{e}_1 + \sin\theta e^{-i\alpha_2}\hat{e}_2\right) exp(i\vec{k}\cdot\vec{r}_o - i\omega t)\right] = \left(\cos\theta\hat{e}_1 + \sin\theta\hat{e}_2\right)\cos(i\vec{k}\cdot\vec{r}_o - i\omega t) \tag{1.91}$$

The field is linearly polarized at an angle θ with respect to $\hat{e}_1$ and its amplitude varies in time and space.

Case 2: If the two components are equal in magnitude, $\theta = 45^o$, and are out of phase by 90°, $e^{-i\alpha_2} = i$, the wave has circular polarization:

$$\text{Re}\left[\left(\cos\theta\hat{e}_1 + \sin\theta e^{-i\alpha_2}\hat{e}_2\right) exp(i\vec{k}\cdot\vec{r}_o - i\omega t)\right] = \frac{1}{\sqrt{2}}\left(\cos(\vec{k}\cdot\vec{r}_o - \omega t)\hat{e}_1 \pm \sin(\vec{k}\cdot\vec{r}_o - \omega t)\hat{e}_2\right) \tag{1.92}$$

The magnitude of the field is everywhere the same but it rotates in time at a fixed point in space and in space for a fixed time. For example, taking the plus sign in the right-hand side of Eq. 1.90, for $t = 0$, the electric field's direction starting at the origin for increasing distance along the wave's direction of propagation, $\vec{k}\cdot\vec{r}_o$, increases and the electric field direction rotates from $\hat{e}_1$ (at the origin) toward $\hat{e}_2$ and beyond. This is "right circular polarization." See Fig. 1.3. Note that a person standing at a fixed location would observe a field rotating clockwise at the wave frequency as the wave passed. Similarly, taking the minus sign in the right-hand side expression produces left circular polarization.

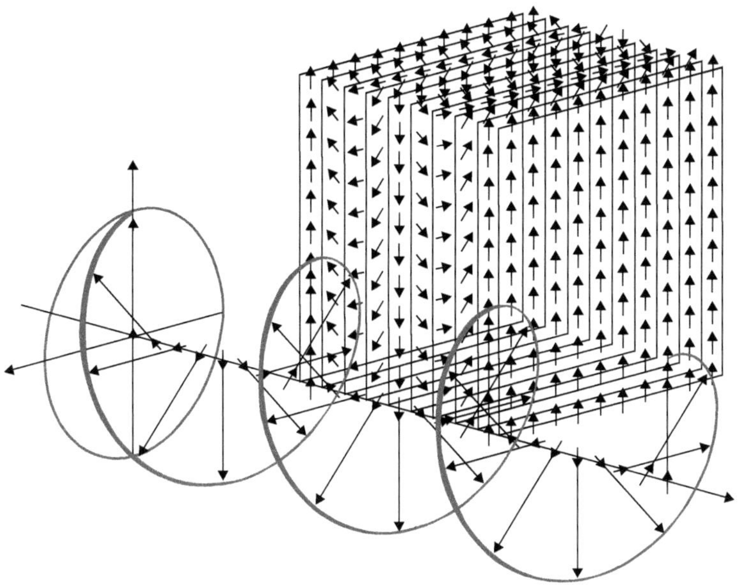

Fig. 1.3 *The orientation of the total perpendicular $\vec{E}$ field as it propagates for circular polarization. This particular wave appears to be left handed to the observer, right handed as viewed from the source ([Public domain]. via Wikimedia Commons).*

Case 3: The general case is when neither of these two conditions hold and this leads to elliptical polarization. Here, the electric field's direction rotates and its amplitude oscillates at the wave frequency.

1.7 E&M propagation within simple media

As shown in Section 1.6 for a vacuum, the microscopic time dependent Maxwell's equations can be used to obtain wave equations for $\vec{E}$ and $\vec{B}$. In this section we will show that the macroscopic time dependent Maxwell's equations, Eqs. 1.62–1.65, are similarly used to obtain the general equations for electromagnetic propagation in material. In the following, we will assume the polarization $\vec{P}$, magnetization $\vec{M}$, and current density $\vec{J}$ all have the following simplifying properties:

(1) They respond linearly to applied fields (linearity)
(2) Their response is independent of the direction of the applied fields (isotropy)
(3) Their response is independent of position and time (homogeneity)

If we make the further assumption that ε, μ and σ are independent of frequency, then we can use the constitutive relationships $\vec{D}=\varepsilon\vec{E}$ and $\vec{B}=\mu\vec{H}$ (Eqs. 1.37 and 1.38) and Ohm's law $\vec{\mathfrak{J}}_f=\sigma\vec{E}$ (Eq. 1.39) for general time-dependent fields (see **Discussion 1.6**) Finally, if we assume that our medium has no regions of non-zero charge density,[20] then the usual manipulation of Eqs. 1.62–1.65 yields:

$$\nabla^2\vec{E}=\mu\varepsilon\frac{\partial^2\vec{E}}{\partial t^2}+\mu\sigma\frac{\partial\vec{E}}{\partial t} \tag{1.93}$$

which is the general time-dependent, second order partial differential equation for electric fields within linear, isotropic, homogeneous, and frequency-independent material in the absence of free charge. In a completely analogous way, we can obtain the identical equation for $\vec{B}$,

$$\nabla^2\vec{B}=\mu\varepsilon\frac{\partial^2\vec{B}}{\partial t^2}+\mu\sigma\frac{\partial\vec{B}}{\partial t} \tag{1.94}$$

In Chapter 9, we will revisit electromagnetic propagation in matter more systematically and in greater detail.

1.8 Electromagnetic conservation laws

1.8.1 Energy density

Maxwell's equations give a prescription for how the total energy in the electromagnetic field can change. If we add together the results of taking the dot product of $\vec{H}$ with both sides of Eq. 1.63 and the dot product of $\vec{E}$ with both sides of Eq. 1.65[21]

$$\vec{E}\cdot(\nabla\times\vec{H})-\vec{H}\cdot(\nabla\times\vec{E})=\vec{E}\cdot\vec{\mathfrak{J}}_f+\vec{E}\cdot\dot{\vec{D}}+\vec{H}\cdot\dot{\vec{B}} \tag{1.95}$$

$$\nabla\cdot(\vec{E}\times\vec{H})=-\vec{E}\cdot\vec{\mathfrak{J}}_f-\vec{E}\cdot\dot{\vec{D}}-\vec{H}\cdot\dot{\vec{B}} \tag{1.96}$$

In the following, we continue to restrict ourselves to materials in which the susceptibility properties are all linear, isotropic and homogeneous in the applied fields, and crucially, are also independent of the frequency, at least in the neighborhood of the frequency spread in the applied field. Then, if the surface element of the surface A surrounding the volume V is designated $d\vec{A}$, and we use the divergence theorem,

$$\int_S(\vec{E}\times\vec{H})\cdot d\vec{A}=-\int\vec{E}\cdot\vec{\mathfrak{J}}_f\,dV-\frac{\partial}{\partial t}\left(\int\frac{1}{2}[\vec{E}\cdot\vec{D}+\vec{H}\cdot\vec{B}]dV\right) \tag{1.97}$$

[20] Note that the conductivity σ of the material is independent of the existence of ρ_f so that we have here a situation in which $\rho_f=0$ and $\vec{\mathfrak{J}}_f=\sigma\vec{E}\neq 0$.

[21] Make use of the vector identity: $\nabla\cdot(\vec{a}\times\vec{b})=\vec{b}\cdot(\nabla\times\vec{a})-\vec{a}\cdot(\nabla\times\vec{b})$.

If we now make the reasonable physical identifications of rates:

$$\int_S (\vec{E} \times \vec{H}) \cdot d\vec{A} =$$

energy flow through surface A of volume V

$$\int \vec{E} \cdot \vec{\mathcal{J}}_f \, dV =$$

electric field work on charges in volume V

$$\frac{\partial}{\partial t}\left(\int \frac{1}{2}[\vec{E} \cdot \vec{D} + \vec{H} \cdot \vec{B}] dV \right) =$$

electromagnetic energy change in volume V

With these identifications, Eq. 1.97 expresses the conservation of energy: In words, the energy flow into the volume through the surface "A" is equal to the work done by the fields on the sources within the volume, plus the increase in the field energy within the volume.

1.8.2 Poynting's Theorem

Taking the differentials of Eq. 1.97 yields Poynting's Theorem, a point by point relationship between the field energy density, flow of energy, and the work performed by the fields on charges:

$$\frac{\partial}{\partial t}\mathscr{U} + \nabla \cdot \vec{S} + \vec{E} \cdot \vec{\mathcal{J}}_f = 0 \qquad (1.98)$$

where $\mathscr{U} = \frac{1}{2}[\vec{E} \cdot \vec{D} + \vec{H} \cdot \vec{B}]$ is the total electromagnetic field energy density. Here, $\vec{S} = (\vec{E} \times \vec{H})$, known as the Poynting vector, is the flow of electromagnetic energy per unit area (i.e., intensity), and $\vec{E} \cdot \vec{\mathcal{J}}_f$ is the work per unit volume the electric field does on the free charges in the volume. Equations 1.97 and 1.98 express the macroscopic and microscopic conservation of energy, respectively, Just as with Maxwell's equations, Eq. 1.98 applies to each point in space, at any given time. That is, the change of the energy density is constrained at each space-time point by this microscopic relationship (see **Discussion 1.7**).

1.8.3 Linear momentum density

We seek to construct an equation that expresses the conservation of momentum in analogy to the conservation of energy. From our discussion before, we have noted that the term $\vec{E} \cdot \vec{\mathcal{J}}_{free}$ is the rate of work done by the electromagnetic fields on the free charges, and is thus the rate of change of the mechanical energy of the free charges in the medium. This statement is a concept readily adopted by considering the current to be the sum of individual particles of charge q moving with velocity $\vec{v}$. Then, writing E as the total energy in the volume (electromagnetic fields plus mechanical movement) (see **Discussion 1.8**):

$$\frac{d}{dt}E = -\int_S \vec{S}\cdot d\vec{A} \tag{1.99}$$

Physically, this equation states that when the influx of energy $|\vec{S}|$ is inward on the surface A, the total energy, mechanical plus field, within the volume surrounded by A increases with time in a prescribed manner. Likewise, the analogous expression in the case of momentum is expected to be of the form,

$$\frac{d}{dt}\mathscr{L}_{total} = \int_S \mathscr{P}_{fields}\cdot d\vec{A}$$

where $\mathscr{L}_{total}$ is the total linear momentum of the fields and the charges, and $\mathscr{P}_{fields}$ is the flow of electromagnetic momentum density per unit area. Here there are two momenta to consider: One is the momentum associated with the fields, and the other is the mechanical momentum of the charges. The expression for the mechanical momentum is obtained, assuming only forces from electromagnetic fields on a single charge, from

$$\begin{aligned}\frac{d\vec{p}_m}{dt} &= \text{total electromagnetic force on the charged particle}\\ &= q(\vec{E}+\vec{v}\times\vec{B})\end{aligned} \tag{1.100}$$

Note that while the $\vec{B}$ field cannot do work on a charge, it can affect a charge's mechanical momentum, depending on the direction of the charge's velocity with respect to the field.

We anticipate the expression for the momentum of the fields by noting the expression for the electromagnetic field energy density:

$$\mathscr{U} = \frac{1}{2}[\vec{E}\cdot\vec{D}+\vec{H}\cdot\vec{B}] = \frac{1}{c}(\vec{E}\times\vec{H})\cdot\hat{k}$$

where $\hat{k}$ is perpendicular to both $\vec{E}$ and $\vec{B}$ and points in the direction of energy flow. Because classical time-varying fields are a representation of photons, they must share their energy/momentum relation, that is, $P_{photon} = (E_{photon})/c$, so the electromagnetic field momentum density is

$$\mathscr{P} = \frac{\mathscr{U}}{c} = \frac{1}{c^2}(\vec{E}\times\vec{H})\cdot\hat{k} \tag{1.101}$$

from which we obtain the field momentum in a volume, V (see **Discussion 1.9**)

$$\vec{p}_f = \frac{1}{c^2}\int (\vec{E}\times\vec{H})dV \tag{1.102}$$

1.8.4 Maxwell stress tensor

The task is to find a single expression that relates the total momentum (mechanical and field) to the fields themselves. We are seeking a relationship of the form:

$$\begin{aligned}\vec{\mathscr{F}} &= \frac{d}{dt}\left(\vec{p}_m + \vec{p}_f\right)\\ &= \vartheta(\vec{E},\vec{B})\end{aligned}$$

where $\vec{\mathscr{F}}$ is the total force on the charge, q. We seek the form of $\vartheta(\vec{E},\vec{B})$. This derivation will lead us to the concept of the Maxwell stress tensor, a powerful idea that the total force acting on charges within a volume can be expressed in terms of a second-rank tensor describing the flow of per unit area of momentum through the surface surrounding the volume.

Start with the generalization of 1.100 for charge and current distributions $\rho(\vec{r})$ and $\vec{\mathcal{J}}(\vec{r})$[22]

$$\frac{d\vec{p}_m}{dt} = \int (\rho\vec{E} + \vec{\mathcal{J}}\times\vec{B})\,dV \tag{1.103}$$

Use Eqs. 1.54 and 1.57 to eliminate ρ and $\vec{\mathcal{J}}$, then use Eq. 1.55 together with the time derivative of Eq. 1.102, to obtain the total force $\vec{\mathscr{F}}$:[23]

$$\begin{aligned}\vec{\mathscr{F}} = \frac{d}{dt}\left(\vec{p}_m + \vec{p}_f\right) = \int\Bigg(\varepsilon\Big[(\nabla\cdot\vec{E})\vec{E} - \vec{E}\times(\nabla\times\vec{E})\Big]\\ + \frac{1}{\mu}\Big[(\nabla\cdot\vec{B})\vec{B} - \vec{B}\times(\nabla\times\vec{B})\Big]\Bigg)dV\end{aligned} \tag{1.104}$$

Applying the vector identity expanding the gradient of the dot product of two vectors, we obtain[24]

$$\begin{aligned}\vec{\mathscr{F}} = \int\Bigg(\varepsilon[(\nabla\cdot\vec{E})\vec{E} + (\vec{E}\cdot\nabla)\vec{E} - \nabla E^2/2]\\ + \frac{1}{\mu}[(\nabla\cdot\vec{B})\vec{B} + (\vec{B}\cdot\nabla)\vec{B} - \nabla B^2/2]\Bigg)dV\end{aligned} \tag{1.105}$$

[22] This canonical derivation is found in nearly all texts at the advanced undergraduate and introductory graduate level. Here we follow the notation given in Griffiths.

[23] After eliminating ρ and $\vec{\mathcal{J}}$ using Maxwell's equations, use Faraday's law to write

$$\frac{\partial\vec{E}}{\partial t}\times\vec{B} = \frac{\partial}{\partial t}(\vec{E}\times\vec{B}) + \vec{E}\times(\nabla\times\vec{E})$$

To gain the symmetric form of Eq.1.104, add the term $(\nabla\cdot\vec{B})\vec{B}$ $(=0)$ to the result.

[24]
$$\nabla(\vec{A}\cdot\vec{B}) = \vec{A}\times(\nabla\times\vec{B}) + \vec{B}\times(\nabla\times\vec{A}) + (\vec{A}\cdot\nabla)\vec{B} + (\vec{B}\cdot\nabla)\vec{A}$$

We are seeking a form of the total force expression that can be written in the form of a surface integral. Inspection of Eq. 1.105 suggests that if the integrand of the volume integral could be expressed as a divergence of some tensor entity, the application of the general divergence theorem would produce the desired result. That is, we seek

$$\vec{\mathscr{F}} = \int \nabla \cdot \overleftrightarrow{T}\, dV = \int_S \overleftrightarrow{T} \cdot d\vec{A} \tag{1.106}$$

The second rank tensor that satisfies the form of Eq. 1.106 and is consistent with Eq. 1.105 is the Maxwell Stress Tensor (using Cartesian coordinates)[25]

$$T_{ij} = \varepsilon\,[E_iE_j + c^2B_iB_j - \frac{1}{2}\delta_{ij}(E^2 + c^2B^2)] \tag{1.107}$$

The total force can then be expressed as a function of $\overleftrightarrow{T}$ as

$$\begin{aligned} \vec{\mathscr{F}} &= \sum_i \hat{x}_i \int \left(\sum_\alpha \frac{\partial}{\partial x_\alpha} T_{i\alpha} \right) dV \\ &= \sum_i \hat{x}_i \int_S \left(\sum_\alpha \hat{n}_\alpha T_{i\alpha} \right) dA \end{aligned} \tag{1.108}$$

where the vector $\hat{n}$ is the unit vector normal to the surface element dA of the surface A that surrounds the volume V. The tensor expressed in Eq. 1.107 carries the name Maxwell Stress Tensor; Eq. 1.108 is understood to mean that $\sum_\alpha \hat{n}_\alpha T_{i\alpha}$ is the ith component of force (per unit area) flowing through the surface A that surrounds the volume containing the charges. The total forces acting on charges enclosed by a surface A is given by summing the fields over this surface in accordance with Eq. 1.108.

1.9 Radiation in vacuum

In the most general sense, radiation is the transport of energy away from a charge and/or current density within a well-defined source volume by the E and B fields of the that source; and this energy is generally measured (conceptually at least) by its effect on a test charge placed some distance, r, removed from the source. Our discussion of radiation will include the transport of energy from charges or currents that are changing within the source, or charges and currents fixed

[25] In general, the dot product of a vector and a second rank tensor is a vector. Here the divergence of a second rank tensor is

$$\nabla \cdot \overleftrightarrow{T} = \sum_\alpha \sum_\beta \hat{x}_\alpha \left(\frac{\partial}{\partial x_\beta} T_{\alpha\beta}\right)$$

relative to a source point moving relative to the test charge. We exclude from this discussion, however, the "normal" energy exchange between a test charge and a fixed charge/current element not moving relative to the test charge. To be clear, a stationary charge can impart momentum and energy to a test charge, and a static current can change the momentum of a moving test charge, with the subsequent recoil of the source in accordance to Newton's laws. Our discussion here on radiation centers on the reaction of a test charge to movement and/or rearrangement of the constituents of a source.

1.9.1 Field amplitude as a function of distance from the source

Within our discussion, there is room for a finer differentiation of forms of radiation. Since the E fields associated with Coulomb's force law, and the B fields associated with the Biot–Savart law have an R^{-2} dependence on the distance between the source and the test charge, and since the energy density scales as the square of either the B or E field, we can conclude that these fields cannot give rise to transport of energy to large distances away from the source even though they are clearly changing in time, driven by the changes within the source and/or the source position. The reason is energy conservation: at sufficiently large distances from the source, the total energy passing through a spherical surface centered at the source would scale as $R^2(E^2) \sim R^{-2}$; that is, the total energy recorded as radiated from the source by these fields would asymptotically approach zero as the measurement sphere was made larger. This unphysical result suggests we must look elsewhere for the form of the source fields that are to be associated with free energy transport away from the source.

1.9.2 Decoupling of radiation fields from the source

In Chapter 3, we will find in our derivation of fields arising from moving charges or changing currents that when there is an acceleration within the source, or the source position accelerates, terms that scale with distance as R^{-1} necessarily develop in the E and B fields emanating from the source. These fields clearly satisfy the requirement that the recorded total energy emitted by the source is independent of the distance from the source (for large enough distances). What is more subtle, but perhaps just as physically relevant, is that field components from the moving sources that scale as R^{-2} couple any test charge used to measure the field to the source in the sense that any motion

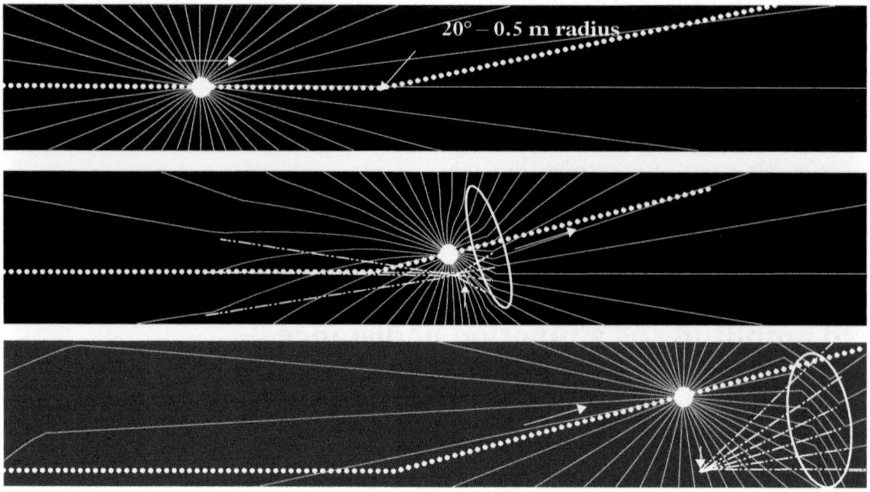

Fig. 1.4 *The output from a computer program that calculates and displays the electric field lines from a charge undergoing prescribed motion. In this simulation, the charge moves with a constant speed of $v = c/2$. In the upper screen the charge is moving uniformly along the dotted line. The middle screen shows the charge just after it has undergone the acceleration that changes its direction by 20°, by making a turn with a radius of curvature of 0.5 m. In the bottom screen, the charge is moving uniformly along a new straight line. The white circles in the middle and bottom panels indicate the location of the so-called acceleration front; that is, where the direction of the field lines change - indicating the acceleration at the 20° curve. Note that for all space beyond the acceleration front, the field lines point to where the charge* would have been *had acceleration not taken place.*

induced in the test charge by these fields will be accompanied by a change in motion of the source itself. On the other hand, the field components that scale as R^{-1}, and (necessarily) arise from accelerations of or within the source, decouple themselves from the source, so that any subsequent interaction of these fields with a test charge is completely independent of the source. We refer to this transport as the "free" radiation transport of energy from the source.

1.9.3 Illustration of coupled and decoupled fields from an accelerated charge

An illustration of this "decoupling", shown in Fig. 1.4, shows the results from a computer program[26] that calculates and displays the electric field lines from a moving charge that begins in uniform motion, executes a small bend at a constant velocity, then continues in uniform motion at an angle relative to its initial velocity. We will investigate the fields of a point charge moving at relativistic velocities in Chapter 6; here we use movement and acceleration of a point charge

[26] Shintake, T., *Nuc. Inst. & Method. in Physics Research*, 507(1), 89 (2003).

to illustrate the origin and differences between the "coupled" fields, and the "decoupled" ones. In the top panel, the charge is shown moving at a constant speed ($v = c/2$) along the yellow line. It presumably has been doing so for enough time that the information that the charge is moving without acceleration has permeated (at the speed of c) all of the space encompassed in the simulation. The strength of the field at any point in space is indicated roughly by the density of the field lines at that point: The field strengths associated with this constant motion are then seen to be highest in the direction perpendicular to the motion - a relativistic effect discussed in Chapter 6. For all space in this panel, a test charge would experience a force pointed directly away from the charge. Depending upon the properties of the test charge (e.g., charge, mass, velocity), kinetic energy and momentum will be gained by the test charge with the necessary reduction in kinetic energy and (generally) change in direction of the moving point charge. This is, then, a case of the fields emanating from the moving charge being capable of transmitting energy while remaining coupled: The behavior of the source (its vector velocity) is influenced by the properties of the very test charge used to measure these fields.[27]

In the middle panel, the charge is shown, still with a speed of $v = c/2$, just after it has completed the turn of 20° relative to its initial direction. For regions of space beyond approximately the location of the white oval, there is no available information that the charge has executed the turn: regions of space are "informed" only after a time $t = R/c$, where R is the distance from the turning point (the location of the acceleration) to the region of interest. Inside of this sphere, the fields at this time are in the process of "adjusting" their orientation to reflect the new position and direction. Note that outside the sphere, *the field lines point to the extrapolated position where the charge would have been at this time had there been no acceleration* (white vertical arrow): This somewhat non-intuitive result is the inescapable result of the finite maximum speed of information transmission, made manifest by the charge's speed approaching c.

In the bottom panel, the charge has settled again into a constant, no acceleration condition. For regions of space within the sphere of radius $R = ct_2$, where t_2 is the time when the constant acceleration at the curve ends, the field lines are the same relative to the charge's new position as they were in the top panel. For regions outside of the sphere of a larger radius $R' = ct_1$, where t_1 is the time when the constant acceleration at the curve begins, the field lines all point to the now well-separated extrapolated position the charge would have had were there no acceleration (white vertical arrow). The region between these curves contain both longitudinal field components (pointed toward the acceleration position) and transverse fields components

[27] Feynman, R. P., Leighton, R. B., and Sands, M., *The Feynman Lectures on Physics, Vol. 1* chapter 28, eq., 28.3, Addison-Wesley (1964).

(perpendicular to the longitudinal components). This region between these two ever increasing radii continues to propagate independently of the subsequent behavior of the point charge: that is, this region of mixed longitudinal and transverse fields moves outward in a sphere centered at the acceleration point, and *any interaction of the fields in this region with a test charge will have no influence on the subsequent movement of the point charge*. This is precisely what is meant by "decoupled". We show in Chapter 3 that the transverse component in this region has a radial dependence of R^{-1}.

In the chapters that follow we derive the forms of the fields from a source described locally by a charge current density and its time behavior. In the course of this derivation we will come across the fundamental restrictions and conditions imposed by the maximum speed of "information" transmission being c. We will explore how this condition, along with causality, plays out mathematically in the evaluation of integrals and spatial and time differentiation. We will see that the mathematics will lead us not only to forms of radiation usually associated with light propagation in free space, but to radiation from charges undergoing arbitrary accelerations that give rise to special radiation conditions, for example, Bremsstrahlung and synchrotron radiation.

Exercises

(1.1) Find the magnetic field inside a sphere of uniform magnetization, $\vec{M}$.

(1.2) A magnetic monopole generates a magnetic field diverging from a point source exactly analogous to the Coulomb field of an electric charge. These have been searched for but never found. Rewrite Maxwell's equations including the new terms required, should they be found in nature.

(1.3) Fill in the steps needed to derive Eq. 1.99.

(1.4) Find the plane wave solutions of Eq. 1.99.

(1.5) Find the polarizability of a classical hydrogen atom: a charge $(-e)$ moving in a circular orbit of radius a_o about a charge of $+e$, for an electric field perpendicular to the orbital plane.

(1.6) Find E and D inside a thin disk of a dielectric with permittivity ε for an applied field that is oriented perpendicularly to the disk.

(1.7) Find the electrostatic energy of a sphere with uniform charge density throughout its volume (relative to the charges being separated to infinity). Find B and H for the case where the half space, $z < 0$, is filled with a material of permeability μ, $z > 0$ is vacuum, and a thin wire located at a height z_o carries current I along the y-axis.

(1.8) Determine the force on a wall due to the partial reflection, reflectivity R, of a light wave with electric field E_o, and angle θ with respect to the perpendicular to the wall.

(1.9) Consider a uniformly charged hemisphere of charge density ρ and radius r. Using Maxwell's stress tensor find the electrical force on the planar part of the hemisphere.

(1.10) A diode consists of an electron emitting cathode held at ground $(V = 0)$ potential and an anode of voltage V_o, toward which the electrons are accelerated. Assume a planar geometry, a cathode-anode distance, d, and that the initial velocity of the electrons is 0 at the cathode. For low currents, the field is essentially uniform. But as the current is increased, the space-charge field created by the electrons affects their motion. Derive the Child–Langmuir law for the space-charge limited maximum current: $I_max = kV^{3/2}$ and find the constant k.

(1.11) Find the charge distribution that produces a so-called screened Coulomb potential: $V(r) = \frac{q}{r}exp[-ar]$.

1.10 Discussions

Discussion 1.1

Equation 1.7 is obtained from Eq. 1.6 by making use of the vector identity for the curl of the outer product of two vectors and noting that ∇ is an operation in the observer's coordinates, then the integrand in Eq. 1.6 reduces to:

$$\nabla \times \left(\vec{J} \times \frac{\vec{R}}{R^3}\right) = -\left[\left(\vec{J} \cdot \nabla\right)\frac{\vec{R}}{R^3} - \vec{J}\left(\nabla \cdot \frac{\vec{R}}{R^3}\right)\right]$$

The first term here also vanishes: note that $-\left(\vec{J} \cdot \nabla\right)\frac{\vec{R}}{R^3} = \left(\vec{J} \cdot \nabla'\right)\frac{\vec{R}}{R^3}$. Then, considering just the x component,

$$\int \vec{J} \cdot \nabla'\left(\frac{x_o - x}{R^3}\right) dV = \int \nabla' \cdot \left(\frac{(x_o - x)}{R^3}\vec{J}\right) dV - \int \frac{(x_o - x)}{R^3}\left(\nabla' \cdot \vec{J}\right) dV$$

The second term on the right vanishes because $\vec{J}$ is steady state, and using the divergence theorem the first term on the right can be expressed, via the divergence theorem, as

$$\int \nabla' \cdot \left(\frac{(x_o - x)}{R^3}\vec{J}\right) dV = \oint \left(\frac{(x_o - x)}{R^3}\vec{J}\right) \cdot \hat{n}da$$

and selecting our volume of integration to be large enough that the current on the surface of integration is zero, this term also vanishes.

Discussion 1.2

It is tempting (and more intuitive, by comparison with the relationship $\vec{P} = \varepsilon_o \chi_e \vec{E}$) to define the magnetic susceptibility as the material response to the $\vec{B}$ field, $\vec{M} = \frac{1}{\mu_o}\chi_m \vec{B}$, such that from Eq. 1.28 we would have $\mu = \mu_o(1 - \chi_m)^{-1}$ within Eq. 1.38. However, convention dictates otherwise. Furthermore, and more importantly, if we ignore the fact that the "microscopic" $\vec{B}$ is more fundamental than the "macroscopic" $\vec{H}$ and we consider these definitions of susceptibility as relating practical, measurable quantities to the polarizations, then immediately we see that $\vec{P} = \varepsilon_0 \chi_e \vec{E}$ is appropriate because the $\vec{E}$ field within a material can easily be measured using a voltmeter. And likewise, the definition of magnetization, $\vec{M} = \chi_m \vec{H}$, is more appropriate because $\vec{H}$ is easily obtained from the measurable quantity of free current $\vec{j}_f$ through the macroscopic Maxwell Equation of 1.32. The relations 1.37 and 1.38 are known as the constitutive relations and can be used when the material is isotropic and responds linearly to the fields. In general, the constitutive relations, which are not necessarily homogeneous or linear and isotropic as they are here, relate the macroscopic derived fields $\vec{D}$ and $\vec{H}$ to the real fields $\vec{E}$ and $\vec{B}$ within a material.

Discussion 1.3

Faraday's conclusion was that the temporal change in flux passing through a closed circuit induces an electric field coincident with the circuit. Mathematically, it is stated that the line integral of the induced electric field around the circuit (or the electromotive force, *emf*) is equal to the negative change rate of enclosed flux:

$$emf = \oint \vec{E} \cdot d\vec{s} = -\frac{d\Phi}{dt} = -\int \frac{\partial \vec{B}}{\partial t} \cdot \hat{n} da$$

where in the last term we have identified the enclosed flux as the integral of the field over a surface bounded by the circuit. Invoking Stokes's theorem, we obtain the differential form of Eq. 1.49.

Discussion 1.4

This result is readily visualized by considering a cylinder of material uniformly polarized along its axis so that the only resulting bound charges are oppositely signed surface charges on the ends given by $Q_p = \sigma_b A = (\vec{P} \cdot \hat{n}) A = PA$, where A is the cross sectional area of the cylinder. Then, if the polarization increases in a time dt, the amount of separated bound charge also increases resulting in a current, $I_p = dQ_p/dt$, where I_p is called the "polarization" current. Dividing by the area, we get Eq. 1.58.

Discussion 1.5

If Maxwell's displacement current source term, $\varepsilon_0 \frac{\partial \vec{E}}{\partial t}$, is treated as such, then we can lump all sources into one generalized current density term as

$$\nabla \times \vec{B} = \mu_0 \vec{J}_g$$

where $\vec{\mathcal{J}}_g = \vec{\mathcal{J}}_f + \nabla \times \vec{M} + \frac{\partial \vec{P}}{\partial t} + \varepsilon_o \frac{\partial \vec{E}}{\partial t}$. Unlike the other terms, however, the displacement term is not technically charge in motion but rather a temporally changing field so it turns out that its more appropriate to consider this field source term as separate from the total material current, $\vec{\mathcal{J}}_t = \vec{\mathcal{J}}_f + \nabla \times \vec{M} + \frac{\partial \vec{P}}{\partial t}$.

Discussion 1.6

The relationships $\vec{D} = \varepsilon \vec{E}$ and $\vec{B} = \mu \vec{H}$ were originally derived in Section 1.2.5 for the steady-state (DC) cases. However, for time-dependent fields, it turns out that the relationship $\vec{P}(t) = \varepsilon_o \chi_e(\omega) \vec{E}(t)$ is only valid for harmonic fields, $\vec{E}(t) = \vec{E}_o e^{-i\omega t}$, since if the susceptibility is frequency-dependent then the linear material response to a general time dependent field is a superposed manifold of response components over a range of frequencies and amplitudes. Thus, extending this argument to frequency-dependent conductivity $\sigma(\omega)$ and magnetic susceptibility $\chi_m(\omega)$, we can see that statements such as $\vec{D}(t) = \varepsilon(\omega) \vec{E}(t)$, $\vec{B}(t) = \mu(\omega) \vec{H}(t)$, and $\vec{\mathcal{J}}_{free}(t) = \sigma(\omega) \vec{E}(t)$ have no meaning for general time-dependent fields. Alternatively, if ε, μ, and σ are frequency independent, these statements are completely valid.

Discussion 1.7

As we indicated before, Eq. 1.95 was derived under the assumption that the material had no absorption, nor frequency dependence in the susceptibilities. This assumption, while providing a compelling conservation equation, is clearly of limited validity. When there is dispersion in the electric susceptibility, for example, there is necessarily a time delay between the application of $\vec{E}$ and the establishment of $\vec{D}$ (and similarly in the magnetic case between the application of $\vec{B}$ and the resulting $\vec{H}$; (see Section 9.7), invalidating the step simplifying the $[\vec{E} \cdot \dot{\vec{D}} + \vec{H} \cdot \dot{\vec{B}}]$ term in going from Eq. 1.95 to Eq. 1.97. In the case of material dispersion at a frequency in the neighborhood of the applied $\vec{E}$ and $\vec{B}$ fields there will be absorption because any variation in the real part of the susceptibility with frequency will be reflected in a corresponding imaginary term as well (see Section 9.8). This means that in addition to the work done on the free charges, the fields can deposit energy in the material through absorption. Consider for example a collection of neutral (cold) hydrogen atoms: for frequencies near the energy spacings of the $n = 1$ to $n = 2$ quantum levels (divided by h) there is a probability that an electromagnetic wave will cause the electron in the hydrogen atom to move from the ground state to an excited one. There is no work done by the fields on any free charges in this case, nevertheless the electromagnetic wave will have deposited energy into the collection. To see the consequences, consider a large enough volume of these hydrogen atoms; if the density of the atoms and the volume are sufficiently large, then there will be no energy transferred across the surface and no work done by the fields on charges. However, the fields will decay as their energy is transferred to the material. In practice this means that Poynting's theorem (Eq. 1.98) for practical materials needs to be modified to an effective form:

$$\frac{\partial}{\partial t}\mathcal{U} + \nabla \cdot \vec{S} + \vec{E} \cdot \vec{\jmath}_{free} + \mathcal{R} = 0$$

where $\mathcal{R}$ is the rate of absorption (non-Ohmic) in the material. The absorption can be readily calculated in the (simplified) approximation of Drude materials (see Chapter 10).

Discussion 1.8

As noted in **Discussion 1.7** here again we have assumed that there is no non-Ohmic absorption within the material. If there were, then we would have to account for energy contained in the material that is neither mechanical nor field. Such a case involves the constituents of the material excited to a higher internal energy, or those constituents giving energy back to the system when they return to their ground state; in the case of de-excitation this internal energy can express itself as mechanical motion (heat) or as radiation (emission). The more general form of Eq. 1.99 is then

$$\frac{d}{dt}E = -\int_S \vec{S} \cdot d\vec{A} \pm \mathcal{R}$$

where the – sign denotes absorption, and the + sign emission.

Discussion 1.9

The argument that the electromagnetic momentum density is simply related to the electromagnetic energy density by the photon energy/momentum relationship is, of course, only rigorously valid for the case of classical radiation, that is, when $\vec{E}$ and $\vec{H}$ represent the fields of a traveling wave within the medium. However, if we proceed with Eq. 1.100 we end up with the Eq. 1.104 valid for all configurations of $\vec{E}$ and $\vec{H}$ by making the identification that the momentum density is, in fact, given by Eq. 1.101 for both static and time-varying fields.

The Potentials

2

- The scalar potential $V(\vec{r},t)$ and the vector potential $\vec{A}(\vec{r},t)$ such that

$$\vec{E}(\vec{r},t) = -\nabla V(\vec{r},t) - \frac{\partial \vec{A}(\vec{r},t)}{\partial t} \text{ and } \vec{B}(\vec{r},t) = \nabla \times \vec{A}(\vec{r},t)$$

- Requirements on the choices of $V(\vec{r},t)$ and $\vec{A}(\vec{r},t)$ through gauge considerations, specifically introducing the Lorenz Gauge, the most applicable for radiation:

$$\nabla \cdot \vec{A} + \frac{1}{c^2}\frac{\partial V}{\partial t} = 0$$

- The wave equations prescribing the potentials in terms of the source using the Lorenz Gauge
- Modification of the scalar and vector potentials to account for the speed of light and causality: "retarded time"
- The potentials expressed in terms of moments of the source along with the concepts of "near", "intermediate," and "far" zones to facilitate the mathematics of deriving approximate expressions for the potentials evaluated at appropriate distances from the source
- Expressions for the vector potential in terms of the electric and magnetic dipole, and electric quadrupole moments of the source in the approximation zones

2.1 The magnetic and electric fields in terms of potentials

We start with the time-dependent Maxwell's equations written in a form to emphasize the sources:

Electromagnetic Radiation. Richard Freeman, James King, Gregory Lafyatis,
Oxford University Press (2019).

DOI: 10.1093/oso/9780198726500.001.0001

$$\varepsilon_o \nabla \cdot \vec{E} = \rho \quad (2.1)$$

$$\nabla \cdot \vec{B} = 0 \quad (2.2)$$

$$\nabla \times \vec{E} + \frac{\partial \vec{B}}{\partial t} = 0 \quad (2.3)$$

$$\frac{1}{\mu_o}(\nabla \times \vec{B} - \frac{1}{c^2}\frac{\partial \vec{E}}{\partial t}) = \vec{J} \quad (2.4)$$

Mathematically, if a vector field has a curl of zero, that field may be written as the gradient of a scalar potential field. Similarly, if a vector field has a divergence of zero, that field may be expressed as the curl of a vector potential field. Thus, if $\partial \vec{B}/\partial t \neq 0$, then $\nabla \times \vec{E} \neq 0$ so that in general $\vec{E}(r,t) \neq \nabla V(r,t)$. However, because $\nabla \cdot \vec{B} = 0$, we have $\vec{B} = \nabla \times \vec{A}$, where $\vec{A}$ is the so-called vector potential; substituting for $\vec{B}$ in terms of $\vec{A}$ in Eq. 2.3 yields:

$$\nabla \times (\vec{E} + \frac{\partial \vec{A}}{\partial t}) = 0 \quad (2.5)$$

Although $\vec{E}$ cannot by itself be derived from a scalar potential, the vector $\vec{E} + \partial \vec{A}/\partial t$ can, and we have the generalization of the static case:

$$\vec{E} + \frac{\partial \vec{A}}{\partial t} = -\nabla V$$

And so by considering the two homogeneous Maxwell's equations, we can state the defining equations for E and B in terms of V and A as:

$$\vec{E}(\vec{r},t) = -\nabla V(\vec{r},t) - \frac{\partial \vec{A}(\vec{r},t)}{\partial t} \quad \textit{and} \quad \vec{B}(\vec{r},t) = \nabla \times \vec{A}(\vec{r},t) \quad (2.6)$$

2.2 Gauge considerations

Since we seek the expressions for the E and B fields, and because these quantities do not depend on the values of the potentials but rather upon spatial and temporal derivatives of the potentials it is evident that for a given $\vec{E}$ or $\vec{B}$ field there is not a unique choice of V or A. Indeed, since $\vec{B} = \nabla \times \vec{A}$, whatever the value of $\vec{A}$, the vector $\vec{A}' = \vec{A} + \nabla \psi$, where ψ is any continuously differentiable scalar function, will result in the same $\vec{B}$ field. However, with this new $\vec{A}'$, $\vec{E}$ is given by

$$\vec{E} = -\nabla V - \frac{\partial}{\partial t}\vec{A}' = -\nabla V - \frac{\partial}{\partial t}(\vec{A} + \nabla \psi)$$

so that to leave $\vec{E}$ unchanged, we must simultaneously change V to V' where

$$V' = V - \frac{\partial \psi}{\partial t}$$

This is the transformation of $\vec{A}$ and V together that leaves $\vec{E}$ and $\vec{B}$ unchanged, the so-called gauge transformation:

$$\vec{A} \rightarrow \vec{A} + \nabla \psi \tag{2.7}$$

$$V \rightarrow V - \frac{\partial \psi}{\partial t} \tag{2.8}$$

If we restrict our choice of $\vec{A}$ and V by the condition

$$\nabla \cdot \vec{A} + \frac{1}{c^2}\frac{\partial V}{\partial t} = 0 \tag{2.9}$$

then it is readily shown that this relationship is invariant under the gauge transformation of Eqs. 2.7 and 2.8 provided that ψ is chosen to satisfy

$$\nabla^2 \psi - \frac{1}{c^2}\frac{\partial^2}{\partial t^2}\psi = 0$$

The constraining condition of Eq. 2.9 is called the Lorenz gauge. We will make immediate use of this gauge to simplify and separate the wave equations for V and $\vec{A}$.

2.3 The wave equations prescribing the potentials using the Lorenz gauge

The task is to derive equations relating V and $\vec{A}$ to the source terms ρ and $\vec{J}$. The first of these relations is obtained from Eq. 2.1 by using $\vec{E}$ given in Eq. 2.6

$$\varepsilon_o \left[\nabla^2 V + \frac{\partial}{\partial t}(\nabla \cdot \vec{A}) \right] = -\rho \tag{2.10}$$

and the second is obtained from Eq. 2.4 by inserting the definitions of $\vec{E}$ and $\vec{B}$ from Eq. 2.6

$$\frac{1}{\mu_o}\left[\nabla^2 \vec{A} - \frac{1}{c^2}\frac{\partial^2 \vec{A}}{\partial t^2} - \nabla\left(\nabla \cdot \vec{A} + \frac{1}{c^2}\frac{\partial V}{\partial t} \right) \right] = -\vec{J} \tag{2.11}$$

Applying the Lorenz gauge (Eq. 2.9) immediately decouples the differential equations for $\vec{A}$ and V. Defining the symbol operator:

$$\Box \equiv \nabla^2 - \frac{1}{c^2}\frac{\partial^2}{\partial t^2} \tag{2.12}$$

Eqs. 2.10 and 2.11 can be written in this compact form in which $V$and $\vec{A}$ are seen to be solutions to separate inhomogeneous wave equations with source terms determined by the source charge and current densities. The equivalent of Maxwell's equations can then be written in terms of only the potentials and the source parameters:

$$\varepsilon_o \Box V = -\rho \tag{2.13}$$

$$\frac{1}{\mu_o}\Box \vec{A} = -\vec{\jmath} \tag{2.14}$$

$$\nabla \cdot \vec{A} = -\frac{1}{c^2}\frac{\partial V}{\partial t} \tag{2.15}$$

Here, we have chosen to elevate the Lorenz gauge condition (Eq. 2.15) to the level of the two inhomogeneous wave equations (Eqs. 2.13 and 2.14) to emphasize that although the V and $\vec{A}$ satisfy their own, separate, wave equations, they are explicitly not independent, and must simultaneously satisfy Eq. 2.15. Mathematically, it is often less cumbersome to solve for the potentials of a given charge and current distribution than deal directly with the Maxwell's equations for the fields. The fields are retrieved from the potentials by application of the expressions given in Eq. 2.6.

2.4 Retarded time

The basic tenet of relativity, that information can travel no faster than the speed of light, requires a firm departure from thinking of fields as simply a mathematical construct from which one calculates the force on a test charge at a distance r removed from the source. In the concept of "action at a distance," the field lines are thought of as rigid connections to their originating source. When this source moves, or appears and/or disappears, the entire field line reflects this change immediately. Because "c" is so large compared to most other velocities encountered in nature, the ideas associated with action at a distance are often satisfactory. However, when source conditions change in a time δt, where $\delta t \lesssim R/c$, and R is the distance of the observation point from the source, then action at a distance fails, for the information about the displacements of the source, or any changes within the source, can be transmitted to the observer no faster than the speed of light. This leads to a definition of the concept of "retarded time," τ: A **change** in the charge density ρ or current density $\vec{J}$ in the source at location $\vec{r}$ at time τ is **observed** at the field point located at $\vec{r}_o$ at a time $t = \tau + R/c$, where $R = |\vec{R}| = |\vec{r}_o - \vec{r}|$. A consequence of this fact is that fields from charges and currents take on a true physical reality: they can affect a test charge located in space at a given point and time even when their originating charge or current source no longer actually exists.

This concept is familiar in the context of observations in astronomy in which a major change takes place in the luminosity of an object located at a great distance, say R, from the Earth. The change we see on Earth will then have taken place a time R/c ago. Indeed, we measure distances in astronomy in light-years, the distance light travels in one year. By definition, a change in a stellar object at a distance of 1 light year, would take at least a year before being recorded on Earth. It is certainly the case that some stars that we "see" today may in fact no longer exist. Nonetheless, the light waves arriving from the star when it did exist have a measurable effect, and appear to be coming from a location in space corresponding to where the object existed at a time R/c ago.

The consequences of accounting for retarded time, and for the finite speed of information gathering, are profound and give rise mathematically to many of the radiation phenomena we observe from rapidly changing conditions of a source. Again, it is worth re-emphasizing what is meant by "rapidly changing": If the timescale of changes in a source, δt, occur faster than, or on the order of R/c, where R is the distance between the observer and the source, then retarded time effects will be important. This means, of course, that for a source changing in time there will be regions, delineated by distance from the source, in which the retarded time effects are trivial, and others in which they dominate the radiation effects.

Figure 2.1 shows the coordinate system and the time it takes for information to move from the source to the observer. Anything the observer experiences at time t has originated at the source at an earlier time, $t - R/c$. In other words, with t as the time of observation, the observer will experience potentials and fields associated with values

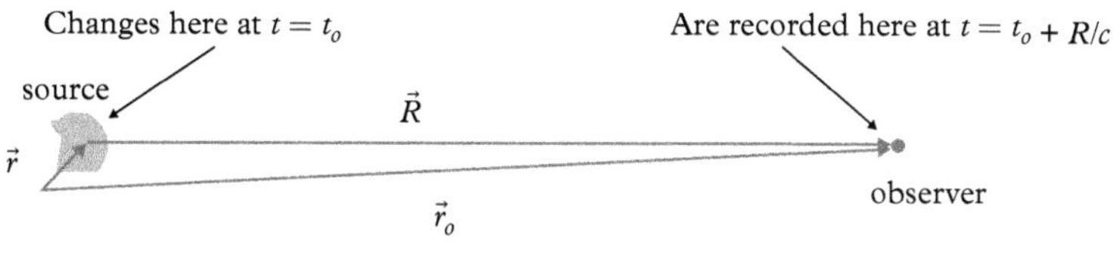

Fig. 2.1 *Relationship between the source coordinates and the observers.*

and changes in ρ and $\vec{J}$ that occurred in the source at the "retarded time," τ, where evidently

$$\tau = t - R/c \tag{2.16}$$

2.4.1 Potentials with retarded time

We seek the solutions to 2.13 and 2.14 when ρ and $\vec{J}$ are time dependent. From the previous discussion, we are faced with the fact that while we need the potentials at the field point at time t, the sources are located at a distance $R = |\vec{r}_o - \vec{r}|$ from this field point. This means that the potentials at the field point at time t must reflect the conditions at the source at the earlier time $\tau = t - R/c$. The static solutions for $V(\vec{r}_o)$ and $\vec{A}(\vec{r}_o)$ are well known:

$$V(\vec{r}_o) = \frac{1}{4\pi\varepsilon_o}\int \frac{\rho(\vec{r})}{R}\, d^3r \tag{2.17}$$

$$\vec{A}(\vec{r}_o) = \frac{\mu_o}{4\pi}\int \frac{\vec{J}(\vec{r})}{R}\, d^3r \tag{2.18}$$

An informed conjecture is that the solutions for $V$and $\vec{A}$ in the presence of temporal variations in ρ and $\vec{J}$ are given by

$$V(\vec{r}_o,t) = \frac{1}{4\pi\varepsilon_o}\int \frac{\rho(\vec{r},\tau)}{R}\, d^3r = \frac{1}{4\pi\varepsilon_o}\int \frac{\{\rho(\vec{r},t)\}_{ret}}{R}\, d^3r \tag{2.19}$$

$$\vec{A}(\vec{r}_o,t) = \frac{\mu_o}{4\pi}\int \frac{\vec{J}(\vec{r},\tau)}{R}\, d^3r = \frac{\mu_o}{4\pi}\int \frac{\{\vec{J}(\vec{r},t)\}_{ret}}{R}\, d^3r \tag{2.20}$$

where the $\{\}_{ret}$ is an alternate way to indicate that a quantity is to be evaluated at the retarded time, $\tau = t - R/c$.

The conjecture conveyed by Eqs. 2.19 and 2.20 is physically appealing: The only obvious difference between a large distance from the source and a closer distance is that "information" about changes in the source should take longer to appear at the far distance, something that is satisfied by the use of the retarded time. The expressions given in Eqs. 2.19 and 2.20 are referred to as "Retarded Potentials."

In order for this conjecture to be correct, the forms of V and $\vec{A}$ as given in Eqs. 2.19 and 2.20 must satisfy the wave equations in Eqs. 2.13 and 2.14, as well as satisfy the Lorenz (or radiation) gauge condition, $\nabla \cdot \vec{A} = -(1/c^2)\partial V/\partial t$. The first requirement is shown by recognizing that τ is a function of both t and $R = |\vec{r} - \vec{r}_o|$ (see **Discussion 2.1**) while the second is shown by expressing $\nabla \cdot \vec{A}$ in terms of the divergence of $\vec{J}$ within the source, which can in turn be related to $\dot{\rho}$ through the continuity equation (see **Discussion 2.2**).

2.5 Moments of the retarded potential

Up to this point we have described the source by a general charge density $\rho(\vec{r}, t)$, and a current density $\vec{\mathcal{J}}(\vec{r}, t)$ and have discussed how to evaluate the potentials arising from such a source. Here, we parse the general potentials in order to find simpler, more explicit expressions relating to the various charge moments of the source. This is by far the most effective means of description, provided that the expansion is adequately described by the first few source moment terms. (see **Discussion 2.3**) The vector potential is given by Eq. 2.20

$$\vec{A}(\vec{r}_o, t) = \frac{\mu_o}{4\pi} \int \frac{\vec{\mathcal{J}}(\vec{r}, \tau)}{R} d^3r \tag{2.21}$$

Take the time dependence to be sinusoidal, then for a given frequency, ω, $\vec{\mathcal{J}}(\vec{r}, \tau) = \vec{\mathcal{J}}(\vec{r})e^{-i\omega\tau}$. Noting that $\tau = t - R/c$, we have

$$\vec{A}(\vec{r}_o, t) = \left[\frac{\mu_o}{4\pi} \int \frac{\vec{\mathcal{J}}(\vec{r})}{R} e^{ikR} d^3r \right] e^{-i\omega t} \tag{2.22}$$

where $k = \omega/c = 2\pi/\lambda$.

2.5.1 Potential zones

If, for all regions in our discussion, we consider only the cases in which both R and λ are much greater than the characteristic size of the source region, D, then Eq. 2.22 invites a comparison of $R = |\vec{r}_o - \vec{r}|$ to λ. In the literature, there are usually three zones called out comparing R to λ:

(1) $R << \lambda$, the near zone
(2) $R \sim \lambda$, the intermediate zone
(3) $R >> \lambda$, the radiation or far zone

Since the definition of zones depends upon the ratio of the distance of the measurement (observer) position to the wavelength of the radiation, a source undergoing arbitrary motion and radiating a wide range of frequencies will produce a situation in which a given distance from the source may be characterized as "near" or "far," depending on the portion of the emitted spectrum that is of concern. Mathematically, the three zones are defined within Eq. 2.22 according to which of the two factors, $1/R$ or e^{ikR}, provides the dominant variation in $\vec{A}$ for a given variation in R. In the near zone, the change in $1/R$ dominates.

In the far zone, it is the change in e^{ikR} that dominates. Finally, in the intermediate zone, the changes in the two factors contribute approximately equally.

Near zone potential

In the near zone, the exponential in the integrand of Eq. 2.22 becomes $exp[i\frac{2\pi}{\lambda}R] \approx 1$ so in this region there are negligible retarded effects or phase delays, and the potential has the same form as the static case but with the potential oscillating in phase with the source:

$$\vec{A}(\vec{r}_o, t) \simeq \left[\frac{\mu_o}{4\pi}\int \frac{\vec{\jmath}(\vec{r})}{R} d^3r\right] e^{-i\omega t} \tag{2.23}$$

where we note that in this zone, for which e^{ikR} is approximated as constant, a spatial derivative of $\vec{A}$ will yield only the term that is proportional to $1/R^2$.

Intermediate zone potential

In the intermediate zone, unlike for the near zone, we cannot approximate the exponential of Eq. 2.22 as unity. However, since R is in all cases presumed to be very much larger than any dimension of the source, we can approximate $R \simeq r_o - r\cos\theta$ within Eq. 2.22 where θ is the angle between the direction of $\vec{r}_o$ and $\vec{r}$. Then Eq. 2.22 becomes:[1]

$$\vec{A}(\vec{r}_o, t) \simeq \frac{\mu_o}{4\pi}\frac{e^{ikr_o}}{r_o}\left[\int \vec{\jmath}(\vec{r})\left(1 + \frac{r}{r_o}\cos\theta\right) e^{-ikr\cos\theta} d^3r\right] e^{-i\omega t} \tag{2.24}$$

Next, writing the exponential within the integral as a series expansion,

$$exp(-ikr\cos\theta) = \sum_{n=0}^{\infty} \frac{(-i)^n}{n!}(kr\cos\theta)^n$$

Equation 2.24 becomes

$$\vec{A}(\vec{r}_o, t) \simeq \frac{\mu_o}{4\pi}\frac{e^{ikr_o}}{r_o}\left[\sum_{n=0}^{\infty}\frac{(-i)^n}{n!}\int \vec{\jmath}(\vec{r})\left(1 + \frac{r}{r_o}\cos\theta\right)(kr\cos\theta)^n d^3r\right] e^{-i\omega t} \tag{2.25}$$

where, because we have restricted our discussions to source sizes much smaller than the wavelength of any emitted radiation ($kr \ll 1$), this expansion will converge in the first few terms. Expanding out to the $n = 0$ and $n = 1$ terms, we get

[1] Approximating $R \simeq r_o - r\cos\theta$ within Eq. 2.22,

$$\simeq \frac{\mu_o}{4\pi}\int \vec{\jmath}(\vec{r})\frac{exp(ikr_o - ikr\cos\theta)}{r_o - r\cos\theta} d^3r$$

$$\simeq \frac{\mu_o}{4\pi}\frac{e^{ikr_o}}{r_o}\int \vec{\jmath}(\vec{r})\left(1 - \frac{r}{r_o}\cos\theta\right)^{-1} \times exp(-ikr\cos\theta)\, d^3r$$

$$\simeq \frac{\mu_o}{4\pi}\frac{e^{ikr_o}}{r_o}\int \vec{\jmath}(\vec{r})\left(1 + \frac{r}{r_o}\cos\theta\right) \times e^{-ikr\cos\theta} d^3r.$$

$$\vec{A}(\vec{r}_o, t) \simeq \frac{\mu_o}{4\pi}\frac{e^{ikr_o}}{r_o}\left[\int \vec{\jmath}(\vec{r})\left((1+\frac{r}{r_o}cos\theta)(1-ikr\,cos\theta)\right)d^3r\right]e^{-i\omega t} \tag{2.26}$$

While there is little literature available on closed form expansions in this region, and currently most computation is handled by computer calculation, we know that in this region the $\vec{E}$ and $\vec{B}$ fields associated with the curl, divergence and time derivative of $\vec{A}$ migrate from being strongly "coupled" for $R \lesssim \lambda$ to completely "decoupled" as R becomes much larger than λ. Further, as we have already discussed, it is in this zone that the $\vec{E}$ fields transition from being completely radial and proportional to R^{-2} in the near zone to having transverse components proportional to R^{-1} (along with radial components proportional to R^{-2}) in the radiation zone. Indeed, in this zone the change in $\vec{A}$ for a given δr_o has significant contributions from both e^{ikR} and $1/R$ factors in Eq. 2.22. A spatial derivative of Eq. 2.22 in this zone will therefore include both terms: one that is proportional to R^{-2} and one proportional to $(\lambda R)^{-1}$.

Far (radiation) zone potential

In the far zone, we expect the radiation fields, $\vec{E}$ and $\vec{B}$, to be transverse and scale as R^{-1}. We start by considering the vector potential in the intermediate zone, Eq. 2.26, which assumes $R \gg r$, $\lambda \gg r$, and $R \simeq \lambda$. Note that this form is valid in all three zones. If we move out to the far zone in which $R \simeq r_o \gg \lambda$, then the change in $\vec{A}$ for a given δr_o is dominated by the factor e^{ikr_o} with $1/r_o$ remaining relatively constant. A spatial derivative of $\vec{A}$ in the far zone can therefore be approximated as having only one term that is proportional to $(\lambda R)^{-1}$. With the radiation zone condition, $r_o \gg \lambda \gg r$, Eq. 2.26 can be approximated as,

$$\vec{A}(\vec{r}_o, t) \simeq \frac{\mu_o}{4\pi}\frac{e^{ikr_o}}{r_o}\left[\int \vec{\jmath}(\vec{r})\,(1-ikr\,cos\theta)\,d^3r\right]e^{-i\omega t} \tag{2.27}$$

which, as the vector potential expanded out to $n=1$, is an accurate representation in the radiation zone.

2.5.2 General expansion of the retarded potential

Electric dipole term

Expanding Eq. 2.25 and keeping only terms with no $cos\theta$ dependence yields the zeroth approximation to the vector potential,

$$\left[\vec{A}(\vec{r}_o,t)\right]_{zero\ order} = \frac{\mu_0}{4\pi}\frac{e^{ikr_o}}{r_o}\left[\int \vec{J}(\vec{r})\,d^3r\right]e^{-i\omega t} \tag{2.28}$$

Integrating 2.28 by parts, and using the continuity equation[2] yields the electric dipole term of the expansion,

$$\vec{A}(\vec{r}_o,t)|_{ed} = \frac{\mu_0}{4\pi}\frac{e^{ikr_o}}{r_o}(-i\omega)\vec{p}_o e^{-i\omega t} = \frac{\mu_0}{4\pi}\frac{e^{ikr_o}}{r_o}\dot{\vec{p}}(t) \tag{2.29}$$

with

$$\vec{p}_o \equiv \int \vec{r}\rho(\vec{r})\,d^3r$$

The form of Eq. 2.29 is satisfying: The electric dipole contribution to the vector potential is a vector field pointed in the direction of the time derivative of the dependent dipole, with a magnitude proportional to this derivative, and oscillating as a spherical wave with a R^{-1} dependence on the distance from the dipole to the field point. It might be noted that as an approximation to the full vector potential, the zero order term is valid only at large distances while as an expression for the electric dipole term, it is valid in all zones. Finally, in Eq. 2.29, we specifically cast the largest term of the vector potential expansion in terms of the charge distribution, but there are circumstances under which the field of a given current distribution is desired.

Magnetic dipole and electric quadrupole terms

Expanding Eq. 2.25 and keeping only terms linear in $\cos\theta$ yields the 1st order contribution:

$$\left[\vec{A}(\vec{r}_o,t)\right]_{1st\ order} = \left[\frac{\mu_0}{4\pi}\frac{e^{ikr_o}}{r_o}\left(\frac{1}{r_o} - ik\right)\int \vec{J}(\vec{r})\,r\cos\theta\,d^3r\right]e^{-i\omega t} \tag{2.30}$$

The task is to manipulate this expression to display a form that is amenable to the usual multipole expansion. The canonical approach is to write the integrand here as $\vec{J}[\vec{r}\cdot\hat{r}_o]$ and then express this term as a sum of symmetric and its anti-symmetric parts under exchange of $\vec{J}$ and $\vec{r}$:

$$\vec{J}(\vec{r})r\cos\theta = (\vec{r}\cdot\hat{r}_o)\vec{J} = \frac{1}{2}\Big((\vec{r}\cdot\hat{r}_o)\vec{J} + \vec{r}(\vec{J}\cdot\hat{r}_o)\Big) + \frac{1}{2}\Big((\vec{r}\cdot\hat{r}_o)\vec{J} - \vec{r}(\vec{J}\cdot\hat{r}_o)\Big)$$

[2] Integration by parts of Eq. 2.28 yields (since the current is bounded by the finite source region) so that

$$\vec{A}(\vec{r}_o,t) = -\frac{\mu_0}{4\pi}\frac{e^{ikr_o}}{r_o}\left[\int \vec{r}[\nabla'\cdot\vec{J}(\vec{r})]\,d^3r\right]e^{-i\omega t}$$

the continuity equation provides:

$$\nabla'\cdot\vec{J} = -\frac{\partial}{\partial t}\rho(\vec{r}) = i\omega\rho(\vec{r})$$

then use the definition of the electric dipole moment of a charge distribution

$$\vec{p} = \int \vec{r}\rho(\vec{r})\,d^3r.$$

and replace the last term using the triple cross product vector identity,

$$(\vec{r}\cdot\hat{r}_o)\vec{\jmath} = \frac{1}{2}\left((\vec{r}\cdot\hat{r}_o)\vec{\jmath} + \vec{r}\,(\vec{\jmath}\cdot\hat{r}_o)\right) - \frac{1}{2}\left(\hat{r}_o \times (\vec{r}\times\vec{\jmath})\right) \tag{2.31}$$

Noting the standard vector form for the magnetic dipole of the source:

$$\vec{m}_o \equiv \frac{1}{2}\int\left(\vec{r}\times\vec{\jmath}(\vec{r})\right)d^3r$$

and defining

$$\vec{A}'\,(\vec{r}_o) = \frac{\mu_o}{4\pi}\frac{e^{ikr_o}}{r_o}\left(\frac{1}{r_o} - ik\right)\frac{1}{2}\int\left((\vec{r}\cdot\hat{r}_o)\vec{\jmath} + \vec{r}\,(\vec{\jmath}\cdot\hat{r}_o)\right)d^3r \tag{2.32}$$

we can write the 1st order contribution to $\vec{A}$ in Eq. 2.30 as:

$$\vec{A}\,(\vec{r}_o) = \left[\vec{A}'\,(\vec{r}_o) - \frac{\mu_o}{4\pi}\frac{e^{ikr_o}}{r_o}\left(\frac{1}{r_o} - ik\right)\left(\hat{r}_o \times \vec{m}_o\right)\right] \tag{2.33}$$

The integral of $\vec{A}'\,(\vec{r}_o)$ is completed using an integration by parts and exchanging the divergence of $\vec{\jmath}$ with the negative time derivative of the charge density to yield (see **Discussion 2.4**)

$$\vec{A}'\,(\vec{r}_o) = -\frac{\mu_o}{4\pi}\frac{e^{ikr_o}}{r_o}\left(\frac{1}{r_o} - ik\right)\left(\frac{i\omega}{2}\right)\int\vec{r}\,(\hat{r}_o\cdot\vec{r})\;\rho(\vec{r})\,d^3r \tag{2.34}$$

Using the definition of the quadrupole tensor $\overleftrightarrow{Q}$ elements,

$$Q_{ij} \equiv \int \rho(\vec{r})\,(3r_ir_j - r^2\delta_{ij})\,d^3r \tag{2.35}$$

this can be written as (Exercise 2.6)

$$\vec{A}'\,(\vec{r}_o) = -\frac{\mu_o}{4\pi}\frac{e^{ikr_o}}{r_o}\left(\frac{1}{r_o} - ik\right)\frac{i\omega}{6}\left[\overleftrightarrow{Q}\cdot\hat{r}_o + \left(\int r^2\rho\,d^3r\right)\hat{r}_o\right] \tag{2.36}$$

Substituting this into Eq. 2.33, we obtain

$$\begin{aligned}\vec{A}\,(\vec{r}_o) = &-\frac{\mu_o}{4\pi}\frac{e^{ikr_o}}{r_o}\left(\frac{1}{r_o} - ik\right)\\ &\times\left\{\frac{i\omega}{6}\left[\overleftrightarrow{Q}\cdot\hat{r}_o + \left(\int r^2\rho\,d^3r\right)\hat{r}_o\right] + (\hat{r}_o\times\vec{m}_o)\right\}\end{aligned} \tag{2.37}$$

which is a general form of the 1st order contribution to the vector potential and is therefore valid in all zones.

In the radiation zone, Eq. 2.37 simplifies: Anticipating that the fields in the radiation zone are proportional to $\vec{A} \times \hat{r}_o$, we drop the $\hat{r}_o$ term. And because we are here considering the radiation zone ($kr_o \gg 1$), we also drop the terms proportional to $1/r_o^2$. We have, in summary, for the radiation zone:

$$\left[\vec{A}(\vec{r}_o, t)\right]_{1st\ order} = ik\frac{\mu_o}{4\pi}\frac{e^{ikr_o}}{r_o}\left[\frac{i\omega}{6}\overleftrightarrow{Q}\cdot\hat{r}_o + \left(\hat{r}_o \times \vec{m}_o\right)\right]e^{-i\omega t} \quad (2.38)$$

The two terms in the brackets are identified as the electric quadrupole and magnetic dipole terms in the multipole expansion.

Transverse fields in the radiation zone

Using the Lorenz gauge, it is possible to express both the $\vec{E}$ and $\vec{B}$ fields in terms of $\vec{A}$. The expression for $\vec{B}$ in terms of $\vec{A}$ remains $\vec{B} = \nabla \times \vec{A}$. For each frequency ω, the Lorenz condition gives:

$$\nabla\cdot\vec{A} + \frac{1}{c^2}\frac{\partial}{\partial t}V = \nabla\cdot\vec{A} - \frac{ik}{c}V = 0 \Rightarrow V = -\frac{ic}{k}\nabla\cdot\vec{A} \quad (2.39)$$

then the gradient of the potential in terms of the vector potential is:

$$\nabla V = -\frac{ic}{k}\nabla(\nabla\cdot A) \quad (2.40)$$

which makes it possible to write $\vec{E}$ completely in terms of $\vec{A}$:

$$\vec{E} = -\nabla V - \frac{\partial}{\partial t}\vec{A} = \frac{ic}{k}\nabla(\nabla\cdot\vec{A}) + i\omega\vec{A} \quad (2.41)$$

If we substitute into this equation the far zone approximation for $\vec{A}$, Eq. 2.27, and drop the resulting r_o^{-2} terms (since in the radiation zone, $(\lambda r_o)^{-1} \gg r_o^{-2}$), we obtain[3]

$$\vec{E} = ick\left[\vec{A} - (\hat{r}_o\cdot\vec{A})\,\hat{r}_o\right] = i\omega\vec{A}_T \quad (2.42)$$

where we have resolved the $\vec{A}$ vector into its components that are parallel ($\vec{A}_P = (\hat{r}_o\cdot\vec{A})\,\hat{r}_o$) and transverse ($\vec{A}_T = \vec{A} - \vec{A}_P$) to the vector between the source and field point ($\vec{R} \simeq \vec{r}_o$). Now, because $\vec{A}$ is linearly proportional to $\vec{J}$, we are left with the useful statement that the observed radiated field in the far or radiation zone arises

[3] The radiation zone vector potential expanded out to $n = 1$, Eq. 2.27,

$$\vec{A}(\vec{r}_o, t) = \vec{A}(\vec{r}_o)\,e^{-i\omega t} = \left[\frac{\mu_o}{4\pi}\frac{e^{ikr_o}}{r_o}\int \vec{J}(\vec{r}) \times (1 - ikr\cos\theta)\,d^3r\right]e^{-i\omega t}$$

We note that the integral is a vector and is not a function of r_o so that,

$$\nabla\cdot\vec{A}(\vec{r}_o) = \frac{\mu_o}{4\pi}\left[ik\frac{e^{ikr_o}}{r_o} - \frac{e^{ikr_o}}{r_o^2}\right]\hat{r}_o \cdot\left(\int \vec{J}(\vec{r})(1 - ikr\cos\theta)\,d^3r\right)$$

Dropping the r_o^{-2} term yields

$$\nabla\cdot\vec{A}(\vec{r}_o) = ik\,\hat{r}_o\cdot\left(\frac{\mu_o}{4\pi}\frac{e^{ikr_o}}{r_o}\int \vec{J}(\vec{r}) \times (1 - ikr\cos\theta)\,d^3r\right) \rightarrow ik(\hat{r}_o\cdot\vec{A})$$

And, in a like manner, to the same order in kr_o, we find

$$\nabla(\nabla\cdot\vec{A}) = -k^2(\hat{r}_o\cdot\vec{A})\,\hat{r}_o$$

solely from the source current transverse to the direction to the observer.

Exercises

(2.1) Following Eq. 2.23, the claim is made that a spatial derivative of the vector potential will yield only the term that is proportional to $1/R^2$. Explain this statement. What approximations are made? Presumably a time derivative of this potential–as is required to find the electric field–will yield a term $\sim 1/R$. Discuss this term.

(2.2) Rewrite Maxwell's equations using the scalar and vector potentials in place of the fields. The Coulomb gauge condition is $\nabla \cdot \vec{A} = 0$. For a source-free region of space, write the electric and magnetic fields using the Coulomb gauge potentials.

(2.3) The Coulomb gauge condition is $\nabla \cdot \vec{A} = 0$. Given a field configuration specified by potentials, $\Phi_L, \vec{A}_L$, that satisfy the Lorenz gauge condition, find the gauge transformation that transforms these into potentials that satisfy the Coulomb gauge condition and return the same electric and magnetic fields.

(2.4) Consider a charge, q, rotating about the origin in the x–y plane at a radius, a, with angular frequency ω. Find the electric dipole, electric quadrupole, and magnetic dipole vector potentials in the radiation zone for this charge distribution. Describe the radiation pattern for each multipole and compare the magnitude of the vector potential for the three multipoles evaluated at a common (large) distance.

(2.5) Using arguments similar to those leading up to Eq. 2.42, find $\vec{B}$ in the radiation zone and show that it is orthogonal to $\vec{E}$.

(2.6) Show that

$$\vec{V} = \int \vec{r}\,(\hat{r}_o \cdot \vec{r})\,\rho(\vec{r})\,d^3r$$
$$= \frac{1}{3}\left[\overleftrightarrow{Q} \cdot \hat{r}_o + \left(\int r^2 \rho\, d^3r\right)\hat{r}_o\right]$$

2.6 Discussions

Discussion 2.1

We start by expanding the gradient (operating only on r_o). The divergence of the gradient is then:

$$4\pi\varepsilon_o \nabla V = \nabla\left[\int \frac{\{\rho\}}{R} d^3r\right] = \int \left[\{\rho\}\nabla(\frac{1}{R}) + \frac{1}{R}\nabla\{\rho\}\right] d^3r = \int \left[-\frac{\{\rho\}}{R^2}\hat{R} - \frac{\{\dot{\rho}\}}{cR}\hat{R}\right] d^3r$$

$$4\pi\varepsilon_o \nabla^2 V = \int \left[-4\pi\{\rho\}\delta(\vec{R}) - \nabla\{\rho\}\cdot\frac{\hat{R}}{R^2} - \frac{\{\dot{\rho}\}}{c}\left(\nabla\cdot\frac{\hat{R}}{R}\right) - \frac{\nabla\{\dot{\rho}\}}{c}\cdot\frac{\hat{R}}{R}\right] d^3r \tag{2.43}$$

Next, noting the common relation (∇ operating only on r_o)

$$\nabla f(r,\tau) = \frac{\partial f(\tau)}{\partial \tau}\nabla\tau = \dot{f}(\tau)\nabla(t - \frac{R}{c}) = -\dot{f}(\tau)\frac{\hat{R}}{c}$$

in which

$$\dot{f}(\tau) \equiv \frac{\partial f(\tau)}{\partial t} = \frac{\partial f(\tau)}{\partial \tau}\frac{\partial \tau}{\partial t} = \frac{\partial f(\tau)}{\partial \tau}$$

and the relation

$$\nabla \cdot \frac{\hat{R}}{R} = \frac{1}{R^2}$$

the two middle terms in Eq. 2.43 cancel and we can evaluate $\nabla^2 V$ as

$$\nabla^2 V = -\frac{\rho}{\varepsilon_o} + \frac{1}{4\pi\varepsilon_o}\int \frac{\{\ddot{\rho}\}}{c^2 R} d^3 r$$

The second time derivative of the potential can be written,

$$\left(\frac{1}{c^2}\right)\frac{\partial^2}{\partial t^2}V = \frac{1}{4\pi\varepsilon_o}\int \frac{\partial^2}{\partial t^2}\left(\frac{\{\rho\}}{c^2 R}\right)d^3 r = \frac{1}{4\pi\varepsilon_o}\int \frac{\{\ddot{\rho}\}}{c^2 R} d^3 r \tag{2.44}$$

Subtracting Eq. 2.44 then yields

$$\nabla^2 V - \left(\frac{1}{c^2}\right)\frac{\partial^2}{\partial t^2}V = -\frac{\rho}{\varepsilon_o} \tag{2.45}$$

thus showing that the wave equation is satisfied. A similar treatment for the wave equation in $\vec{A}$ gives the same result.

Discussion 2.2

Using the notation ∇ to mean differentiation with respect to r_o (observer coordinate) while holding r (source coordinate) fixed, and ∇' to mean differentiation with respect to r while holding r_o fixed and noting that $\vec{R} = \vec{r}_o - \vec{r}$, then evidently, $\nabla\left(\frac{1}{R}\right) = -\nabla'\left(\frac{1}{R}\right)$. Further, because $\vec{J}$ is a function of τ, which is in turn a function of R, we have

$$\nabla \cdot \vec{J} = \left(-\frac{1}{c}\hat{R} \cdot \dot{\vec{J}}\right)$$

while differentiation with respect to the source coordinates picks up two terms:

$$\nabla' \cdot \vec{\mathscr{J}} = (\nabla')^\tau \cdot \vec{\mathscr{J}} + \left(\frac{1}{c}\hat{R}\cdot\dot{\vec{\mathscr{J}}}\right) = -\dot{\rho} + \left(\frac{1}{c}\hat{R}\cdot\dot{\vec{\mathscr{J}}}\right)$$

where the τ superscript indicates holding τ (r) constant while differentiating $\vec{\mathscr{J}}(\vec{r},\tau)$ with respect to $\vec{r}$ and where we have used the continuity equation with coordinates of the source. Then applying the chain rule

$$\nabla \cdot \left(\frac{\vec{\mathscr{J}}}{R}\right) = \frac{1}{R}\nabla\cdot\vec{\mathscr{J}} + \vec{\mathscr{J}}\cdot\nabla\left(\frac{1}{R}\right) = \frac{1}{R}\nabla\cdot\vec{\mathscr{J}} - \vec{\mathscr{J}}\cdot\nabla'\left(\frac{1}{R}\right) = \frac{1}{R}\nabla\cdot\vec{\mathscr{J}} + \left[\frac{1}{R}\nabla'\cdot\vec{\mathscr{J}} - \nabla'\cdot\left(\frac{\vec{\mathscr{J}}}{R}\right)\right]$$

$$= \frac{1}{R}\left(-\frac{1}{c}\hat{R}\cdot\dot{\vec{\mathscr{J}}}\right) + \frac{1}{R}\left(-\dot{\rho} + \frac{1}{c}\hat{R}\cdot\dot{\vec{\mathscr{J}}}\right) - \nabla'\cdot\left(\frac{\vec{\mathscr{J}}}{R}\right) = -\frac{1}{R}\dot{\rho} - \nabla'\cdot\left(\frac{\vec{\mathscr{J}}}{R}\right)$$

Using these results, the divergence of $\vec{A}(\vec{r}_o,t)$ as given in Eq. 2.19 becomes

$$\nabla\cdot\vec{A} = \frac{\mu_o}{4\pi}\nabla\cdot\left[\int\frac{\vec{\mathscr{J}}(\vec{r},\tau)}{R}d^3r\right] = \frac{1}{4\pi\varepsilon_o c^2}\int\left[-\frac{1}{R}\dot{\rho} - \nabla'\cdot\left(\frac{\vec{\mathscr{J}}}{R}\right)\right]d^3r = -\frac{1}{c^2}\frac{\partial}{\partial t}V$$

where the term

$$\int\nabla'\cdot\left(\frac{\vec{\mathscr{J}}}{R}\right)d^3r = \oint\left(\frac{\vec{\mathscr{J}}}{R}\right)\cdot d\vec{a} \to 0$$

using the divergence theorem and noting that the surface is taken to be outside of all the charges and currents of the source so there is no contribution from the surface integral.

Discussion 2.3

In this example, we motivate the expansion of the dynamic retarded vector potential in the text by the simpler expansion of the electrostatic potential into spherical harmonics. We then show that the first few associated multipole moments of the charge distribution can be represented by linear combinations of the Cartesian monopole, dipole and quadrupole moments. Our coordinate system is that $\vec{r}$ is the coordinate of a charge element $[\rho(\vec{r})\,d^3r]$, $\vec{r}_o$ is the coordinate of the observation (field) point, and $\vec{R} = \vec{r}_o - \vec{r}$ is the vector from the charge element to the field point. We make use of the identity:

$$\frac{1}{R} = \frac{1}{|\vec{r}-\vec{r}_o|} = 4\pi\sum_{l=0}^{\infty}\sum_{m=-l}^{l}\frac{1}{2l+1}\frac{r^l}{(r_o)^{l+1}}Y_{lm}^*(\hat{r})\,Y_{lm}(\hat{r}_o)$$

in order to separate the source coordinates from the field ones. Then

$$V(\vec{r}_o) = \frac{1}{4\pi\varepsilon_o}\int \frac{\rho(\vec{r})}{R}d^3r = \frac{1}{\varepsilon_o}\sum_{l=0}^{\infty}\sum_{m=-l}^{l}\frac{1}{2l+1}\frac{1}{(r_o)^{l+1}}Y_{lm}(\hat{r}_o)\left[\int Y^*_{lm}(\hat{r})\,r^l\rho(\vec{r})\,d^3r\right] \tag{2.46}$$

where the term in brackets is the lm^{th} moment of the charge distribution, often written in the literature as $q_{lm}[\rho(\vec{r})]$. If the origin is chosen within the distribution, where $|\vec{r}|$ is assumed to be much less than $|\vec{r}_o|$, then only the first few source moment terms are significant. Writing the Cartesian monopole, dipole and quadrupole moments, respectively, as $q = \int \rho(\vec{r})\,d^3r$; $\vec{p} = \int \vec{r}\rho(\vec{r})\,d^3r$ and $Q_{ij} = \int \rho(\vec{r})(3r_ir_j - r^2\delta_{ij})d^3r$, the q_{lm} can be written as in Jackson: (Landau and Lifshitz use a different normalization for their corresponding q's and Q's).

(1) $q_{00} = \frac{1}{\sqrt{4\pi}}q$

(2) $q_{10} = \sqrt{\frac{3}{4\pi}}\,p_z$

(3) $q_{1\pm1} = \mp\sqrt{\frac{3}{8\pi}}\,(p_x \mp ip_y)$

(4) $q_{20} = \frac{1}{2}\sqrt{\frac{5}{4\pi}}\,Q_{zz}$

(5) $q_{2\pm1} = \mp\frac{1}{3}\sqrt{\frac{15}{8\pi}}\,(Q_{xz} \mp iQ_{yz})$

(6) $q_{2\pm2} = \frac{1}{12}\sqrt{\frac{15}{2\pi}}\,(Q_{xx} \mp 2iQ_{xy} - Q_{yy})$

The corresponding expression for $\vec{A}(\vec{r}_o)$ is more involved as it requires the use of vector spherical harmonics, as does the expansion for $\vec{E}$ and $\vec{B}$. The negative spatial derivative of $V(\vec{r}_o)$ can be formally obtained as a sum of spatial derivatives for each moment term, where the three spherical components for each moment can be written as linear combinations of Y_{lm}'s. However, this procedure is mathematically involved for the individual components do not necessarily have zero divergence.

Discussion 2.4

The result, Eq. 2.34, can be written more simply as

$$\vec{A}'(\vec{r}_o) = \xi\int \vec{r}(\hat{r}_o\cdot\vec{r})(-i\omega)\,\rho(\vec{r})\,d^3r$$

where

$$\xi = \frac{\mu_o}{4\pi}\frac{e^{ikr_o}}{r_o}\left(\frac{1}{r_o} - ik\right)\left(\frac{1}{2}\right)$$

Next, using the continuity equation, we obtain (∇' operates only on r)

$$\vec{A}'(\vec{r}_o) = -\xi\int \vec{r}[(\hat{r}_o\cdot\vec{r})\nabla'\cdot\vec{J}]d^3r$$

Defining $\vec{\eta} = -\xi\,\vec{r}(\hat{r}_o \cdot \vec{r})$ and considering just the x component, we can integrate by parts

$$A'_x = \int \eta_x \left(\nabla' \cdot \vec{\mathcal{J}}\right) d^3r = \int \nabla' \cdot \left(\eta_x \vec{\mathcal{J}}\right) d^3r - \int \left(\nabla' \eta_x \cdot \vec{\mathcal{J}}\right) d^3r = -\int \left(\nabla' \eta_x \cdot \vec{\mathcal{J}}\right) d^3r$$

where the first term in the result vanishes due to the fact that the source is finite in spatial extent. Direct evaluation of the right-hand side then yields,

$$A'_x = \xi \int \left[\left(2\hat{r}_{ox} r_x + \hat{r}_{oy} r_y + \hat{r}_{oz} r_z\right)\mathcal{J}_x + \left(r_x \hat{r}_{oy}\right)\mathcal{J}_y + \left(r_x \hat{r}_{oz}\right)\mathcal{J}_z\right] d^3r$$

which can be generalized for the ith component as

$$A'_i = \xi \int \left[\sum_j [\delta_{ij}(\hat{r}_o \cdot \vec{r}) + r_i \hat{r}_{o_j}]\mathcal{J}_j\right] d^3r$$

and can be more compactly written as

$$A'_i = \xi \int \left[\mathcal{J}_i(\vec{r} \cdot \hat{r}_o) + r_i(\vec{\mathcal{J}} \cdot \hat{r}_o)\right] d^3r$$

so that

$$\vec{A}'(\vec{r}_o) = \xi \int \left[\vec{\mathcal{J}}(\vec{r} \cdot \hat{r}_o) + \vec{r}(\vec{\mathcal{J}} \cdot \hat{r}_o)\right] d^3r$$

or

$$\vec{A}'(\vec{r}_o) = \frac{\mu_o}{4\pi}\frac{e^{ikr_o}}{r_o}\left(\frac{1}{r_o} - ik\right)\left(\frac{1}{2}\right) \int \left[\vec{\mathcal{J}}(\vec{r} \cdot \hat{r}_o) + \vec{r}(\vec{\mathcal{J}} \cdot \hat{r}_o)\right] d^3r$$

and so we have shown that Eq. 2.32 is equivalent to Eq. 2.34.

Part II

Origins of Radiation Fields

General Relations between Fields and Sources

3

- Effects of the finite speed of light on the forms of the electric and magnetic fields at a field point removed from the source location: retarded time
- Spatial derivatives of the potential at a field point removed from the source necessarily contain the temporal derivative of the source variations
- Expressions for the electric and magnetic fields at a field point removed from the source in terms of the charge and current distributions, and their corresponding derivatives, all evaluated at the retarded time: "Jefimenko's equations"
- A graphical argument that demonstrates the transverse nature of radiation fields, based on electric field lines envisioned by Faraday
- Jefimenko's equations without reference to retarded potentials using the Green Function formalism

3.1 Relating retarded potentials to observable fields

As we have discussed in Section 2.4, the potentials V and $\vec{A}$, arising from a source at $\vec{r}$ and experienced at a field point, $\vec{r}_o$, reflect the conditions at the source at a time τ, where τ is just that time earlier by $|\vec{R}|/c = |(\vec{r}_o - \vec{r})|/c$ than the current time at the field point, t. We proved through direct substitution into the wave equations specifying the potentials that the expressions for the potentials at a field point, $\vec{r}_o$, in terms of parameters of the rapidly changing source are identical to those for a static source, provided the characteristics of the source at the appropriate retarded time are used.

Electromagnetic Radiation. Richard Freeman, James King, Gregory Lafyatis,
Oxford University Press (2019). © Richard Freeman, James King, Gregory Lafyatis.
DOI: 10.1093/oso/9780198726500.001.0001

This simple, intuitive result is not expected to carry over into our investigation of how to relate the electric and magnetic fields to the source conditions when there are rapid changes in the source. There are several reasons for anticipating that the forms of the fields will be more complicated than just calculating the equivalent static solution for the fields at the field point, and evaluating the source at the appropriate retarded time: First, simple inspection of Coulomb's expression for the electric field (or Ampere–Biot–Savart expression for the magnetic field) shows that such a procedure would predict that while both the electric and magnetic fields would be time varying, they would both necessarily possess only terms that fall off proportional to r_o^{-2} for $r_o >> r$. In anticipation of rapidly varying source conditions producing fields that carry energy away from the source, the product of E and B $(= E/c)$ must fall exactly proportional to r_o^{-2} for $r_o >> r$ in order to conserve energy in free space. That is, while we might expect the correct forms of the fields for a rapidly changing source to contain some terms that are simple retarded time adjustments of the static solutions, there must be radiation terms for both the magnetic and the electrical fields that are proportional to r_o^{-1} (for $r_o >> r$) as well. Simply using the retarded time in the otherwise static solutions cannot produce fields with r_o^{-1} terms. The second reason that we expect more complicated forms in the field expressions is the fact that the fields are related to the potentials by spatial derivatives evaluated at the field points. Such derivatives represent the express evaluation of the potential at two nearby field points at the same local time, t; but the source time $\tau = t - |\vec{r} - \vec{r}_o|/c$ that produced these potential values is not the same. Thus the values of these spatial derivatives will depend on the time derivatives of the source at time τ. As we show next, this mixes time and spatial derivatives in a manner that produces additional field terms, in contrast to the case of the potentials.

We derive the fields using the general relations relating the fields to the potentials, namely,

$$\vec{E}(\vec{r}_o, t) = -\nabla V(\vec{r}_o, t) - \frac{\partial \vec{A}(\vec{r}_o, t)}{\partial t} \tag{3.1}$$

$$\vec{B}(r_o, t) = \nabla \times \vec{A}(\vec{r}_o, t) \tag{3.2}$$

for which ∇ is the gradient to be taken with respect to the position of the field observation point, holding the time, t, constant. Equations 3.1 and 3.2 mean that the fields at $(\vec{r}_o, t)$ are determined by taking the spatial and temporal derivatives of $V$and $\vec{A}$ evaluated at $(\vec{r}_o, t)$; with these potentials, in turn, having been determined from the charge and current characteristics over an integrated range of source locations at their respective retarded times, τ (see Eqs. 2.19 and 2.20).

3.1.1 Spatial derivatives of retarded potentials

We can use the concept of an information gathering sphere (IGS)[1] to create a visual mnemonic of how the spatial derivatives of the retarded potentials at the observation (field) point depend upon the time derivatives of the source.[2] Figure 3.1 shows the contribution from a point-like source (for simplicity, here located near the origin: $\vec{R} \simeq \vec{r}_o$) to two field points near the observer that are in line with the source and separated by $|\Delta\vec{r}_o| = \Delta R << c \cdot \delta t_{char}$, where δt_{char} is the time scale for a significant change in the source. Spatial differentiation is the process of comparing the values of a function at two points separated by an infinitesimal distance at the same local time, t. The complication is that a temporal change in a non-local (source) can translate into a local (observer) spatial variation. This is illustrated in Fig. 3.1, which shows a snapshot in time in which the IG spheres for the two simultaneous observer points have just passed through and "gathered information" about the state of the source at two different times separated by $\Delta t = |\Delta\vec{r}_o|/c$. These two IG spheres, as shown, will deliver their information to the two infinitesimally separated field points at the same local time, t: Clearly when the IG spheres converge on the two points, at the same local time t, point 2 contains information about the source at a slightly earlier time (evaluated at the source) than point 1. Thus a spatial derivative of the potential at the field points (points 1 and 2) at local time t will contain a time derivative of the source variation at the earlier time $\tau = t - r_o/c$. To see how quantitatively to treat the spatial derivative for a field created by the retarded potential of an extended time-varying source, we first consider the field of a time-varying point-like source, $\rho(\vec{r}, t) \cdot \delta V$, such as in Fig. 3.1. Referring to Fig. 3.1 and noting our simplifying assumption that the compact source is localized near the origin ($\vec{R} \simeq \vec{r}_o$), we can see that the resulting scalar potential, which is spherically symmetric in R, is likewise approximated as $V(R) \simeq V(r_o)$. Thus, the gradient of the potential at the observation point can be approximated by

$$\nabla V(\vec{r}_o, t) \simeq \hat{r}_o \frac{\partial}{\partial r_o} V(r_o) = lim_{\Delta r_o \to 0} \left[\frac{V(r_o, t) - V(r_o - \Delta r_o, t)}{\Delta r_o} \right] \hat{r}_o \tag{3.3}$$

Here, for the source to appear point-like to the two observation locations, we require the linear size of the source, $\sim \delta r_{char}$ to satisfy $\delta r_{char} \ll \Delta r_o \ll c \cdot \delta t_{char}$. Noting that Δr_o is uniquely related to the time interval, $\Delta\tau$, between the two observation points[3] we have in terms of the source,

[1] See Panofsky, W.K.H. and Phillips, M., *Classical Electricity and Magnetism*, 2nd edition, Addison-Wesley, 1962.; An Information Gathering Sphere is essentially an imaginary space-time sphere collapsing in from infinity, at the speed of light, thereby collecting information from all space-time events which are causally connected to the observation event at its ultimate point of convergence.

[2] In what follows it will be most convenient to consider point-like sources, although the argument applies to extended sources as well.

[3] For point 2, the more distant point from the source, we have $\tau \simeq t - \frac{r_o}{c}$ while for point 1, the closer point to the source, we have $\tau' \simeq t - \frac{(r_o - \Delta r_o)}{c}$. Thus, we note that

$$\tau' \simeq t - \frac{r_o}{c} + \frac{\Delta r_o}{c} \simeq \tau + \frac{\Delta r_o}{c}$$

or

$$\Delta\tau = \tau' - \tau \simeq \frac{\Delta r_o}{c}$$

so the closer point 1 sees the source at a later retarded time than point 2.

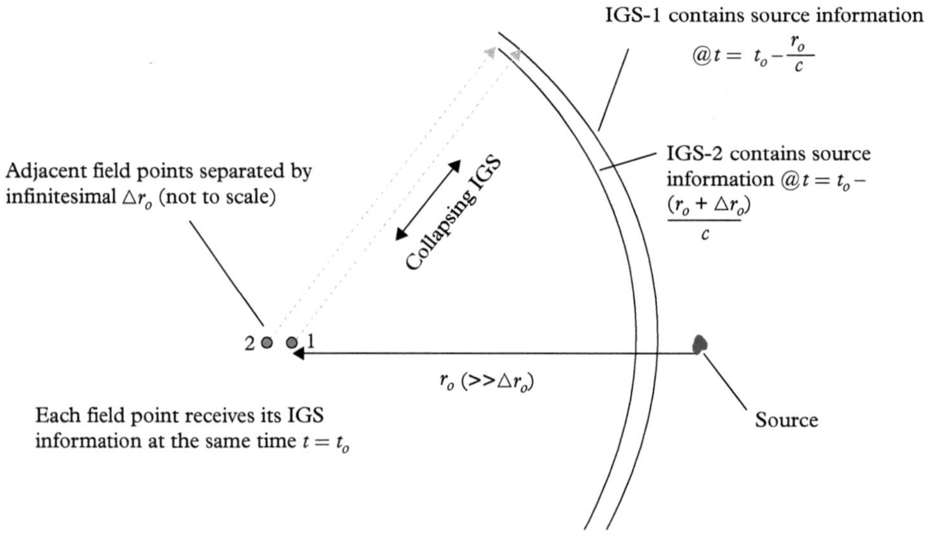

Fig. 3.1 *At time, t, because of the finite speed of light, the information observed at a field point about a particular source (its magnitude, sign, position, etc.) has been evaluated at an earlier time, τ, uniquely determined by the distance between that source and the specific field point. Here the contribution at a particular time t of the marked source to the potential at field point 2 (IGS-2) is determined a time $\Delta r_o/c << \delta t_{char}$ earlier than a contribution of the same source at the same time t to the potential at field point 1 (IGS-1).*

$$\nabla V(\vec{r}_o, t) \simeq \left(\frac{\partial}{\partial r_o} V(\vec{r}_o)\right)\hat{r}_o = \frac{\hat{r}_o}{4\pi\varepsilon_0}\frac{\partial}{\partial r_o}\left(\frac{\rho(\vec{r},\tau)\cdot\delta V}{r_o}\right)$$
$$= \frac{\hat{r}_o}{4\pi\varepsilon_0}\left[-\frac{\rho(\vec{r},\tau)\cdot\delta V}{r_o^2}\right.$$
$$\left. + \, lim_{\Delta r_o\to 0}\left(\frac{\rho(\vec{r},\tau) - \rho(\vec{r},\tau+\Delta r_o/c)}{r_o\Delta r_o}\right)\cdot\delta V\right] \tag{3.4}$$

where, within the limit term, τ and $\tau + \Delta r_o/c$ refer to points 2 and 1, respectively. The numerator of the limit term is expanded as[4]

$$\rho(\vec{r},\tau) - \rho(\vec{r},\tau+\Delta r_o/c) = \rho(\vec{r},\tau) - \left[\rho(\vec{r},\tau) + \left(\frac{\Delta r_o}{c}\right)\frac{\partial}{\partial t}\rho(\vec{r},\tau)\right]$$
$$= -\frac{\dot{\rho}(\vec{r},\tau)}{c}\Delta r_o \tag{3.5}$$

Substituting this result into Eq. 3.4 gives the spatial gradient of the electrical potential for a time varying point-like source near the origin:

[4] Note that for a retarded time, $\tau = (t - \frac{R}{c})$, in which R is time-independent, we have $\frac{\partial\tau}{\partial t} = 1$. Furthermore, we consider non-relativistic motion so that an interval of time is the same for the source and the observer. Mathematically this amounts to the statement that $\frac{\partial\tau}{\partial t} = 1$ in all cases.

$$\nabla V(\vec{r}_o, t) \simeq -\frac{\hat{r}_o}{4\pi\varepsilon_o}\left(\frac{\rho(\vec{r},\tau)}{r_o^2} + \frac{\dot{\rho}(\vec{r},\tau)}{cr_o}\right)\cdot \delta V \qquad (3.6)$$

In the general case of an extended but localized region of charges near the origin, this expression may be integrated over the charged region:

$$\nabla V(\vec{r}_o, t) \simeq -\frac{1}{4\pi\varepsilon_o}\int\left[\frac{\rho(\vec{r},\tau)}{r_o^2} + \frac{\dot{\rho}(\vec{r},\tau)}{cr_o}\right]\hat{r}_o\, dV \qquad (3.7)$$

As we had anticipated, the finite speed of light necessarily injects information concerning the rate of change of the source conditions into the value of the gradient of the potential at the field point for a rapidly varying source. If the speed of light, or more correctly, the speed with which electromagnetic field information is transmitted, were to become arbitrarily large (i.e., $c \to \infty$), then the difference between t and τ would vanish, and this term would vanish as well. The term containing the source temporal information has a spatial distribution such that it behaves as r_o^{-1} for $r_o >> r$, and we anticipate that it will contribute to terms in the electric field capable of transmitting energy out to infinity.[5]

3.2 Jefimenko's equations from the retarded potentials

The evaluation of the $\vec{E}$ field from the retarded potential proceeds by the use of Eq. 3.1 with the result of the last section (Eq. 3.7), and by dropping the restriction that $\vec{r}_o >> \vec{r}$:

$$\begin{aligned}
\vec{E}(\vec{r}_o, t) &= -\nabla V - \frac{\partial \vec{A}}{\partial t} \\
&= \frac{1}{4\pi\varepsilon_o}\int\left[\frac{\rho(\vec{r},\tau)}{R^2} + \frac{\dot{\rho}(\vec{r},\tau)}{cR}\right]\hat{R}\, dV - \frac{\mu_o}{4\pi}\int\left[\frac{\partial}{\partial t}\left(\frac{\vec{\mathcal{J}}(\vec{r},\tau)}{R}\right)\right]dV \\
&= \frac{1}{4\pi\varepsilon_o}\int\left[\frac{\{\rho\}\hat{R}}{R^2} + \frac{\{\dot{\rho}\}\hat{R}}{cR} - \frac{\{\dot{\vec{\mathcal{J}}}\}}{c^2 R}\right]dV \qquad (3.8)
\end{aligned}$$

Arguments similar to those which gave rise to the source terms in the integral of 3.8 yield an expression for the $\vec{B}$ field (**see Discussion 3.1**)

$$\begin{aligned}
\vec{B}(\vec{r}_o, t) &= \nabla \times \vec{A} = \frac{\mu_o}{4\pi}\int\left[\nabla \times \left(\frac{\vec{\mathcal{J}}(\vec{r},\tau)}{R}\right)\right]dV \\
&= \frac{\mu_o}{4\pi}\int\left[\frac{1}{R^2}\left(\{\vec{\mathcal{J}}\}\times\hat{R}\right) + \frac{1}{cR}\left(\{\dot{\vec{\mathcal{J}}}\}\times\hat{R}\right)\right]dV
\end{aligned}$$

[5] We can also see, using this method of limits argument, that

$$\begin{aligned}
\nabla\rho(\vec{r},\tau) &= lim_{\Delta R\to 0}\left(\frac{\rho(\vec{r},\tau)-\rho(\vec{r},\tau+\Delta R/c)}{\Delta R}\right)\hat{R} \\
&= -\frac{\dot{\rho}(\vec{r},\tau)}{c}\hat{R}
\end{aligned}$$

And we note that a similar argument gives a similar result for the divergence of a vector dependent upon retarded time, e.g.,

$$\nabla\cdot\vec{\mathcal{J}}(\vec{r},\tau) = -\frac{\dot{\vec{\mathcal{J}}}(\vec{r},\tau)}{c}\cdot\hat{R}$$

$$= -\frac{1}{4\pi\varepsilon_0 c^2}\int\left[\hat{R}\times\left(\frac{\{\vec{\mathcal{J}}\}}{R^2}+\frac{\{\dot{\vec{\mathcal{J}}}\}}{cR}\right)\right]dV \tag{3.9}$$

So, as predicted, for both $\vec{E}$ and $\vec{B}$ we have picked up additional radiation field terms all arising from the time rate of change of charge or current density and all of which fall off as R^{-1}. Summarizing, we have what is now generally referred to as "Jefimenko's equations":

$$\vec{E} = \vec{E}(R^{-2})_{near} + \vec{E}(R^{-1})_{radiation} \tag{3.10}$$

$$\vec{E}(\vec{r}_o, t) = \frac{1}{4\pi\varepsilon_0}\int\left[\frac{\{\rho\}\hat{R}}{R^2}+\frac{\{\dot{\rho}\}\hat{R}}{cR}-\frac{\{\dot{\vec{\mathcal{J}}}\}}{c^2R}\right]dV \tag{3.11}$$

$$\vec{B} = \vec{B}(R^{-2})_{near} + \vec{B}(R^{-1})_{radiation} \tag{3.12}$$

$$\vec{B}(\vec{r}_o, t) = -\frac{1}{4\pi\varepsilon_0 c^2}\int\left[\hat{R}\times\left(\frac{\{\vec{\mathcal{J}}\}}{R^2}\right)-\frac{\{\dot{\vec{\mathcal{J}}}\}\times\hat{R}}{cR}\right]dV \tag{3.13}$$

The value of Jefimenko's formulation of the electric and magnetic fields is their explicit dependence on charge and current densities (and their time derivatives of the source) at the retarded time. If we know these source characteristics at a time τ then these equations give an explicit procedure for calculating the fields at a position r_o at a time t later, where $t = \tau + |\vec{r}_o - \vec{r}|/c$. When the field distance is not large compared to source dimension (the so-called "near field"), the calculation of the observed fields using this formulation can be tedious, usually no less difficult than calculating the values of the (retarded) potentials at the field point and taking the necessary time and spatial derivatives.

Examination of these expressions (Eqs. 3.11 and 3.13) reveals a clear delineation between the closely coupled "DC" fields and the radiation portion of the field that dominates at larger distances. Both the electric and magnetic fields have a term that is just the static solution for a given charge and current distribution of the source, only evaluated at the retarded time. As expected, this term has a radial dependence proportional to R^{-2} with the polarization of the electric field along the direction between the charge and the observer. In addition, however, the expressions given in Eqs. 3.11 and 3.13 also have terms that depend explicitly on the time derivatives of the current and charge density within the source evaluated at the retarded time, all with a radial dependence of R^{-1}. These terms are generated, as we have seen, by correctly accounting for the retarded time within the derivatives relating the retarded potentials to the value of the fields at the observation point.

We now consider these expressions in more detail: From fundamental considerations we expect (and require) that valid expressions for the radiation portion of these fields demonstrate a polarization transverse to the direction of their propagation and have source terms that represent charge acceleration. Equation 3.13 for the magnetic radiation field explicitly fulfills these requirements with the radiation portion dependent on $\{\dot{\vec{J}}\} \times \hat{R}$ that indicates a magnetic field transverse to $\vec{R}$ and proportional to the transverse component of the time derivative of the source current density. However, when we consider Jefimenko's expression for the radiation electric field given in Eq. 3.11, it does not transparently yield the necessary form. Note that the radiation term involving the time derivative of the charge density is neither transverse nor does it represent charge acceleration. Evidently then, for this expression to be a valid representation of the radiation electric field, this term containing the time derivative of the charge density must exactly cancel out the radial part of the time derivative of the current density. This is seen by using the continuity equation to write $\dot{\rho}$ in terms of $\vec{J}$ (**see Discussion 3.2**) and rearranging Eq. 3.11 to be

$$\vec{E}(\vec{r}_o, t) = \frac{1}{4\pi\varepsilon_o}\int\left[\frac{\{\rho\}}{R^2}\hat{R} + \frac{(\{\vec{J}\}\times\hat{R})\times\hat{R} + (\{\vec{J}\}\cdot\hat{R})\hat{R}}{cR^2} + \frac{(\{\dot{\vec{J}}\}\times\hat{R})\times\hat{R}}{c^2R}\right]dV \tag{3.14}$$

Written in this form, the electric field radiation term is now seen (**see Discussion 3.3**) to be proportional to the time derivative of the transverse component of the current density, and transverse to the direction of propagation. In addition, Eq. 3.14 together with Eq. 3.13 explicitly demonstrate that $\vec{E}_{rad} = c\vec{B}_{rad} \times \hat{R}$. On the other hand, this expression has within its form the potential for confusion, because as written it suggests that the electric field would depend upon the current even in steady state. Proving that the second term in Eq. 3.14 does not contribute to the static electric field can be algebraically tedious for a specific current distribution, but a general proof is obtained by writing this term in a form that explicitly depends on the time derivatives of the charge density and the current density. This essentially amounts to retracing our derivation back to Eq. 3.11.[6]

[6] This proof is taken from: *The Relation Between Expressions for Time-Dependent Electromagnetic Fields Given by Jefimenko and by Panofsky and Phillips* by Kirk T. McDonald Joseph Henry Laboratories, Princeton University, Princeton, NJ 08544 (Dec. 5, 1996; updated June 28, 2013)": In Discussion 3.3, we showed that

$$-\int\frac{(\nabla\cdot\{\vec{J}\})\hat{R}}{cR}dV = \int\frac{(\{\vec{J}\}\times\hat{R})\times\hat{R} + (\{\vec{J}\}\cdot\hat{R})\hat{R}}{cR^2}dV$$

in addition, we have the relation from Discussion 3.2

$$\nabla\cdot\{\vec{J}\} = -\{\dot{\rho}\} + \{\dot{\vec{J}}\}\cdot\left(\frac{\hat{R}}{c}\right) \tag{3.15}$$

comparing Eqs. 3.14 and 3.15 we see that for steady state, that is, when the time derivative of the current and charge density are both zero, then the middle term in Eq. 3.14 is identically zero. These equations basically retrace out derivation back to the form of Eq. 3.11 which is seen to properly reduce to the static solution for the steady state condition.

3.3 Graphical representation of transverse fields arising from acceleration

Although we have seen a mathematical argument of how changes in a source involving the acceleration of charges (or time dependent

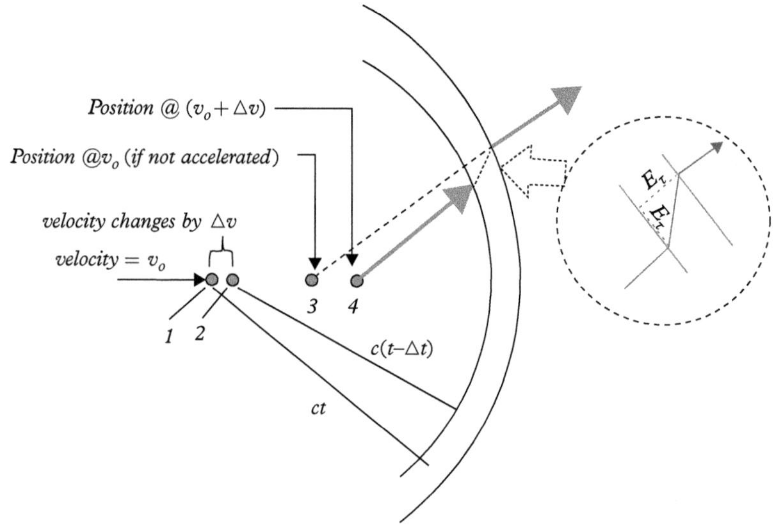

Fig. 3.2 *A charge is moving at velocity v_o for all times before $t = 0$. At $t = 0$ the charge reaches Point 1 where it undergoes a constant acceleration $\dot{v} = c\dot{\beta}$ for a time Δt resulting in an increase of velocity Δv when it reaches Point 2. Subsequently, the charge moves with constant velocity $v = v_o + \Delta v = v_o + c\dot{\beta}\,\Delta t$. Point 3 is the location the charge would have reached by time t if there had been no change in velocity (no acceleration between points 1 and 2). Point 4 is the position at time t assuming the acceleration took place as indicated. The solid arrows indicate the direction of the "dc" longitudinal electric fields.*

currents) can give rise to terms in the E and B fields that have an R^{-1} spatial dependence, the generation of a propagating disturbance with an R^{-1} dependence in an otherwise static field with an R^{-2} dependence can appear physically strange. A visual mnemonic can be generated by using the concept of field lines as envisaged by Faraday; which have a physical reality (quite apart from their environment), and are continuous. Recall that for a uniformly moving charge, the field lines are such that they appear to be projected back to the present location of the charge in the laboratory frame, not to a putative retarded position. As a continuation of the discussion in Section 1.9, where we considered the momentary acceleration of a point charge perpendicular to its initial velocity, we now consider, in Fig. 3.2, acceleration of a point charge in the direction of the velocity. Beyond the "event horizon" at a distance $R = ct$ from the original position (Point 1) of the charge, the field lines all appear to originate from Point 3; that is, to all observers beyond this horizon the charge appears at the position associated with constant velocity. Within the second event horizon (associated with the end of acceleration), at a distance $c(t - \Delta t)$ from Point 2, field lines appear to emanate from a charge at

Point 4. Since the field lines are presumed to be continuous, in the region between the ct horizon and the $c(t-\Delta t)$ horizon the lines are bent. In this transition region, each line has two components: the radial part and the transverse part. The value of the radial part is determined simply from Gauss' law

$$E_r = \frac{q}{4\pi\varepsilon_o}\frac{1}{R^2} \tag{3.16}$$

where R is the distance from the transition region to the region between Points 3 and 4. For $\beta << 1$, we can make the approximation that the transverse field is the ratio of the lengths of the two components of E (evaluated for small Δt) times E_r[7]

$$\begin{aligned} E_t &= lim_{\Delta t\to 0}\left[\frac{(c\dot{\beta}\,\Delta t)\,t\,sin\theta}{c\Delta t}\right]E_r = \left(\frac{\dot{\beta}}{c}R\,sin\theta\right)\left(\frac{q}{4\pi\varepsilon_o}\frac{1}{R^2}\right) \\ &= \frac{q\,a\,sin\theta}{4\pi\varepsilon_o c^2 R} \end{aligned} \tag{3.17}$$

where θ is the angle measured from the direction of acceleration,[8] $\vec{a}$. So we see that E_t and E_r are moving outward along R, inside the transition region, at the speed of light. And our geometrical argument has confirmed that E_t, with its R^{-1} dependence, will dominate E_r, with its R^{-2} dependence, at large R. If we add that $\vec{E}_t$ is orthogonal to $\hat{R}$, the direction from the original position of the charge at the time of the acceleration, and can thus be seen as proportional to the (negative) projection of $\vec{a}$ onto the surface of the event horizon, we can write the vector version of Eq. 3.17:

$$\vec{E}_t = \frac{q}{4\pi\varepsilon_o c^2 R}\left(\left(\vec{a}\times\hat{R}\right)\times\hat{R}\right)$$

which is identical to the radiation zone component of Jefimenko's equation for $\vec{E}$ if we let $\dot{\vec{J}} = \rho\vec{a} = q\delta(r)\,\vec{a}$ within Eq. 3.14.

3.4 Jefimenko's equations without regard to retarded potentials: Green Functions

Here, we re-derive the expressions for $\vec{E}$ and $\vec{B}$, but this time introducing, then demonstrating, and employing, the power of the Green function formalism. In this development we will obtain the $\vec{E}$ and $\vec{B}$

[7] The discussion presented here is valid only for $\beta << 1$ but can be extended to all $\beta < 1$. Use the form of the radial E field from a charge moving at arbitrary velocities and the analogous argument to obtain the ratio of the transverse to longitudinal ratio, which due to the relativistic transform of the time becomes

$$\begin{aligned} |E_t| &= \frac{q}{4\pi\varepsilon_o}\left|\left(\frac{\gamma^2\dot{\beta}}{c}R\,sin\theta\right)\right. \\ &\quad\times\left.\left(\frac{(\hat{R}-\vec{\beta})}{\gamma^2(1-\beta\cos\theta)^3R^2}\right)\right| \\ &= \left(\frac{q}{4\pi\varepsilon_o c}\right)\frac{|(\hat{R}-\vec{\beta})|\dot{\beta}\,sin\theta}{(1-\beta\cos\theta)^3 R} \end{aligned}$$

[8] We recall that $c\dot{\beta}\tau$ is the difference in velocity Δv and so the numerator $(c\dot{\beta}\,\tau)t\,sin\theta$ represents the horizontal displacement of the inner event horizon field lines with the outer event horizon field lines (weighted by $\sin\theta$ relative to the direction of acceleration) and the denominator $c\,\Delta t$ is the radial separation of the inner and outer event horizons. If the field lines are continuous, then this ratio is the lengths of the two components of E in the transition region.

fields directly, by-passing the retarded potentials. The Green Function technique sweeps up all the relevant physics into the initial setup; the physical concept of retarded time will be seen to fall out of standard mathematical manipulations. Our development uses the Fourier transforms of Maxwell's equations, primarily to reduce the complexity by eliminating the time derivatives. Taking the Fourier transform with respect to time[9] of each of the four Maxwell's equations generates a set of similar equations in which $\vec{E}, \vec{B}, \vec{\mathcal{J}}$, and ρ are replaced by their Fourier transforms $\vec{E}_F, \vec{B}_F, \vec{\mathcal{J}}_F$, and ρ_F

$$\varepsilon_o \nabla \cdot \vec{E}_F = \rho_F$$

$$\nabla \cdot \vec{B}_F = 0$$

$$\nabla \times \vec{E}_F - i\omega \vec{B}_F = 0$$

$$\frac{1}{\mu_o}(\nabla \times \vec{B}_F + \frac{i\omega}{c^2}\vec{E}_F) = \vec{\mathcal{J}}_F$$

These equations are combined to generate vector-valued inhomogeneous Helmholtz equations[10]

$$\nabla^2 \vec{B}_F + k^2 \vec{B}_F = -\mu_o \nabla \times \vec{\mathcal{J}}_F \qquad (3.18)$$

$$\nabla^2 \vec{E}_F + k^2 \vec{E}_F = \frac{1}{\varepsilon_o}\nabla \rho_F - i\omega \mu_o \vec{\mathcal{J}}_F \qquad (3.19)$$

We now seek solutions for $\vec{B}_F$ and $\vec{E}_F$ using a Green function. The Green function technique is a general method for solving non-homogeneous linear partial differential equations of the form:

$$\nabla \cdot (p(\vec{r}) \nabla) \Phi(\vec{r}) + q(\vec{r}) \Phi(\vec{r}) = -f(\vec{r})$$

subject to Dirichlet or Neumann boundary conditions, $\Phi = constant$ or $\frac{\partial \Phi}{\partial n} = 0$, respectively, on a surface of the volume under study. Here we are specifically interested in solving the non-homogeneous Helmholtz equation,

$$\left(\nabla^2 + k^2\right)\Phi(\vec{r}) = -f(\vec{r}) \qquad (3.20)$$

subject to the boundary condition that at infinity, $\Phi \to 0$. The source or driving term, $f(\vec{r})$, is assumed confined to a limited region in space. The technique is based on Green's Theorem[11], an equation that relates the values of arbitrary functions $\varphi(\vec{r})$ and $\psi(\vec{r})$ and their second derivatives within a volume V to their values and their normal derivatives over the surface S containing the volume, that is:

---

[9] Our conventions for temporal Fourier transform pairs are as follows. Given a function of time, $f(t)$, its Fourier transform is given by:

$$f_F(\omega) = \frac{1}{\sqrt{2\pi}}\int_{-\infty}^{\infty} f(t)\, e^{i\omega t}\, dt$$

With the inverse transform:

$$f(t) = \frac{1}{\sqrt{2\pi}}\int_{-\infty}^{\infty} f_F(\omega)\, e^{-i\omega t}\, d\omega.$$

[10] To obtain the inhomogeneous Helmholtz equation in $\vec{E}_F$, take the curl of the third equation using the expansion of the "curl of the curl"

$$\nabla \times \nabla \times E = \nabla(\nabla \cdot E) - \nabla^2 E = i\omega \nabla \times \vec{B}_F$$

where $\nabla^2 E$

$$= \sum_{i,j} \frac{\partial^2 E_{x_i}}{\partial x_j^2}\hat{x}_i \quad (\text{for } x_i \text{ or } x_j = x, y, z)$$

then substitute the first and fourth equations. Similarly, for the equation in $\vec{B}_F$, take the curl of fourth equation using the same expansion for the curl of the curl and substitute in the second and the third equations.

[11] Eq. 3.21 is obtained directly from divergence theorem for a volume V with a closed surface S bounding it ($\hat{n}$ is normal to the surface)

$$\int dV \nabla \cdot \vec{D} = \oint dS \vec{D} \cdot \hat{n}$$

Take the vector $\vec{D} = \varphi(r) \nabla \psi(r)$; then substitute $\nabla \cdot \vec{D} = \varphi(r) \nabla^2 \psi(r) + \nabla\varphi(r) \cdot \nabla\psi(r)$ and $\vec{D} \cdot \hat{n} = \varphi(r)\frac{\partial \psi(r)}{\partial n}$; do the same with $\varphi(r)$ and $\psi(r)$ interchanged and subtract the two resultant divergence theorem integral relations to obtain Eq. 3.21. The restrictions on the functions and their derivatives are general and usually easily satisfied. (For our current application, $\psi(r)$ and $\varphi(r)$ must vanish at $r = \infty$ and their derivatives must be well defined over the volume and the surface.

$$\int dV[\varphi(\vec{r})\nabla^2\psi(\vec{r}) - \psi(\vec{r})\nabla^2\varphi(\vec{r})]$$
$$= \oint dS[\varphi(\vec{r})\frac{\partial}{\partial n}\psi(\vec{r}) - \psi(\vec{r})\frac{\partial}{\partial n}\varphi(\vec{r})] \qquad (3.21)$$

Here we will carefully select our functions $\varphi(\vec{r})$ and $\psi(\vec{r})$ (with malice of forethought). For $\varphi(\vec{r})$, we choose it to be $\Phi(\vec{r})$, the solution to Eq. 3.20 that we are seeking. For $\psi(\vec{r})$, we choose it to be the solution of:

$$(\nabla^2 + k^2)\psi(\vec{R}) = -\delta(\vec{R}) \qquad (3.22)$$

subject to the problem's boundary condition. $\psi\left(\vec{R}\right) = G_F(\vec{R})$ is thus identified as the Green Function for the problem, in this case, the Helmholtz equation. As usual, $\vec{R} = \vec{r}_o - \vec{r}$ is the vector pointing from source to observer. Using the Eqs. 3.20 and 3.22 to eliminate the Laplacians in the volume integrals on the left-hand side of Eq. 3.21 and noting that since the surface integral on the right-hand side of Eq. 3.21 can be evaluated at $r = \infty$, so that the integrand in the surface integral is zero, we have[12]

$$\Phi(\vec{r}_o) = \int dV\left[f(\vec{r})\right]\psi(\vec{R}) = \int dV\left[f(\vec{r})\right]G_F(\vec{R}) \qquad (3.23)$$

where we note here that the integration is with respect to the source coordinates and $\vec{R} = \vec{r}_o - \vec{r}$. The Green function $G_F(\vec{R})$ for the Helmholtz equation (Eq. 3.22) is the canonical normalized spherical wave (**see Discussion 3.4**)

$$G_F(\vec{R}) = \frac{e^{ikR}}{4\pi R}$$

Each component of Eqs. 3.18 and 3.19 is an inhomogeneous Helmholtz equation sourced by the rhs for the corresponding component. Our vector solution for $\vec{E}_F$ is then found by letting $f(\vec{r}) \to \frac{1}{\varepsilon_o}\nabla'\rho_F - i\omega\mu_o\vec{\mathfrak{J}}_F$ in Eq. 3.23[13]

$$\vec{E}_F(\vec{r}_o,\omega) = -\int dV\left[\frac{1}{\varepsilon_o}\nabla'\rho_F - i\omega\mu_o\vec{\mathfrak{J}}_F\right]\frac{e^{ikR}}{4\pi R} \qquad (3.24)$$

Here the primed gradient is taken with respect to the source coordinates. Our solution follows the standard prescription of superposition that is at the core of the Green function technique: namely, integration of the Green Function over the specific source distribution. We integrate the first term in Eq. 3.24 by parts to remove the gradient, obtaining (**see Discussion 3.5**)

[12] In the left-hand side of Eq. 3.21 insert the Laplacians obtained in Eqs. 3.20 and 3.22, and set the surface integral to zero, then Eq. 3.21 gives

$$\int\left[\varphi(\vec{r})[-\delta(\vec{r}_o - \vec{r}) - k^2\psi(\vec{r}_o - \vec{r})] - \psi(\vec{r}_o - \vec{r})\right.$$
$$\left.[-\frac{1}{\varepsilon_o}\frac{\partial}{\partial x_i}\rho(\vec{r}) - i\omega\mu_o\mathfrak{J}_i(\vec{r}) - k^2\varphi(\vec{r})]\right]dV = 0$$

the first term yields $-\varphi(r_0)$, and the second and fifth terms cancel.

[13] In Eq. 3.24 with $\vec{R} = \vec{r}_o - \vec{r}$ for cases in which $r_o >> r$, $R = (\vec{R}\cdot\vec{R})^{1/2} \approx r_o - \vec{r}\cdot\hat{r}_o$

$$\vec{E}_F(r_o,\omega) \approx -\frac{e^{ikr_o}}{4\pi r_o}$$
$$\times \int\left[\frac{1}{\varepsilon_o}\nabla\rho_F - i\omega\mu_o\vec{\mathfrak{J}}_F\right]e^{-ik(\vec{r}\cdot\hat{r}_o)}\,dV.$$

Now, when r is additionally always very much smaller than the wavelength of the radiation–that is, the source dimensions are such that there is essentially no phase difference across the source as seen by the observer–then $e^{-ik(\vec{r}\cdot\hat{r}_o)} \approx 1$ and the Fourier transform of the E field is a simple spherical wave with an extended source given by $\int\left[\frac{1}{\varepsilon_o}\nabla\rho_F - i\omega\mu_o\vec{\mathfrak{J}}_F\right]dV$.

$$\vec{E}_F(\vec{r}_o,\omega) = -\frac{1}{4\pi\varepsilon_0}\int dV\left[\frac{i\omega}{cR}-\frac{1}{R^2}\right]\hat{R}\,\rho_F\,e^{i\omega R/c} + \frac{1}{4\pi\varepsilon_0}\int dV\,\frac{i\omega}{c^2R}\vec{\jmath}_F\,e^{i\omega R/c} \tag{3.25}$$

In returning this expression to the time domain we identify term of the form $-i\omega f_F$ as the Fourier transform of the time derivative of the original function $\dot{f}(t)$. Similarly, $e^{i\omega R/c}f_F$ is the Fourier transform of a function with a temporal offset, $f\left(t-\frac{R}{c}\right)$. Thus: $\vec{E}(\vec{r}_o,t)$ is obtained from Eq. 3.25 via the inverse Fourier transform (as in **Discussion 3.6**)

$$\vec{E}(\vec{r}_o,t) = \frac{1}{4\pi\varepsilon_0}\int dV\left[\frac{\{\dot{\rho}\}\hat{R}}{cR}-\frac{\{\dot{\vec{\jmath}}\}}{c^2R}+\frac{\{\rho\}\hat{R}}{R^2}\right] \tag{3.26}$$

$$\vec{B}(\vec{r}_o,t) = \frac{-\mu_0}{4\pi}\int dV\left[\hat{R}\times\left(\frac{\{\vec{\jmath}\}}{R^2}+\frac{\{\dot{\vec{\jmath}}\}}{cR}\right)\right] \tag{3.27}$$

If we were to start with Eq. 3.18 instead of Eq. 3.19, we would find the corresponding expression for the B field (**see Discussion 3.6**). As expected, Eqs. 3.26 and 3.27 are equivalent to Jefimenko's equations given by Eqs. 3.11 and 3.13.

3.4.1 Field characteristics

The expressions, given by Eqs. 3.11, 3.13, and 3.14 for the electric and magnetic field amplitudes and directions at the observer space-time location ($\vec{r}_o$, t) exhibit several expected features:

- The observed field at time t is determined by the volume integrals of the source charge and current distributions and their time derivatives, all at the earlier (τ) time.
- These expressions give the expected static solutions when there is no time dependence of either ρ or $\vec{\jmath}$. That is, for $\dot{\rho}=0$ and $\dot{\vec{\jmath}}=0$, they properly reduce to:

$$\vec{E}_{static} = \frac{1}{4\pi\varepsilon_0}\int_{-\infty}^{\infty}\frac{\rho\hat{R}}{R^2}dV; \qquad \vec{B}_{static} = -\frac{\mu_0}{4\pi}\int_{-\infty}^{\infty}\frac{\hat{R}\times\vec{\jmath}}{R^2}dV$$

- As anticipated, these results show that the radiation fields arising from time-varying currents and charges are not in general obtained from the static field solutions by the substitution $\rho\to\rho(\tau)$ and $\vec{\jmath}\to\vec{\jmath}(\tau)$. Rather, the radiation terms depend explicitly on the rate of change of the charge and current density within the source, all at the earlier (τ) time.

- In the radiation zone, the field amplitudes fall off as $1/R$ and the energy flux falls off as $1/R^2$, as required by energy conservation.
- In the radiation zone, the fields are transverse to the observation direction, $\vec{R} \simeq \vec{r}_o$.

3.4.2 Example: fields directly from Jefimenko's equations

Jefimenko's equations give a recipe for the calculation of the electric and magnetic fields at a field point $\vec{r}_o$ at time t in terms of the properties of the source located at $\vec{r}$ at the time $\tau = t - |\vec{R}|/c$ where $\vec{R} = \vec{r}_o - \vec{r}$. Most authors, while acknowledging the pedagogical usefulness of Jefimenko's formulation, point out that most field calculations are easier using the retarded potentials at the field point, then performing the necessary spatial and time derivatives. This canonical method, however, requires handling the derivatives of the retarded potentials, which can be intricate, often obscuring the physics of the results. Here, we give an example for which the field calculations via Jefimenko's equations are not only easier mathematically, but more physically transparent as well.[14]

Consider an infinite neutral wire extended in the y direction at $x = z = 0$. For $t < 0$, there is no current in the wire; at $t = 0$ the current jumps to I_o; at $t = t_o$ the current suddenly falls back to zero (See Fig. 3.3). Our goal is to determine the electric and magnetic fields at a distance d in the $+\hat{x}$ direction as a function of time.

Examining Eqs. 3.27 and 3.26 and noting both ρ and $\dot{\rho}$ are zero we have (For this exercise, assume the current is uniformly the same everywhere in the wire, and $t_o >> d/c$)

$$\vec{E}(\vec{r}_o, t) = \frac{-1}{4\pi\varepsilon_o} \int \frac{\{\dot{\vec{\mathcal{J}}}\}}{c^2 R} dV \tag{3.28}$$

$$\vec{B}(\vec{r}_o, t) = \frac{-\mu_o}{4\pi} \int dV \left[\hat{R} \times \left(\frac{\{\vec{\mathcal{J}}\}}{R^2} + \frac{\{\dot{\vec{\mathcal{J}}}\}}{cR} \right) \right] \tag{3.29}$$

From the problem description, the current can be described mathematically using the Heaviside function, and its time derivative.[15]

$$\{\vec{\mathcal{J}}\} = I_o \delta(x) \delta(z) \Big[\theta(t - R/c)(1 - \theta(t - t_o - R/c)) \Big] \hat{y} \tag{3.30}$$

$$\{\dot{\vec{\mathcal{J}}}\} = I_o \delta(x) \delta(z) \Big[\delta(t - R/c) - \delta(t - t_o - R/c) \Big] \hat{y} \tag{3.31}$$

[14] The problem idea taken from D.J. Griffiths' *Introduction to Electrodynamics*, 3rd edition, Example 10.2 (pp. 425-426) and Problem 10.9 (p. 426); Aspects of this problem are treated by Chung, J. M. *JKPS*, 49, 1339 (2006). Griffiths analyzes this problem using retarded potentials.

[15] Here we use the definition of the Heaviside function as the integral of the Dirac delta function:

$$H(x) \equiv \int_{-\infty}^{x} \delta(s)\, ds = \theta(x).$$

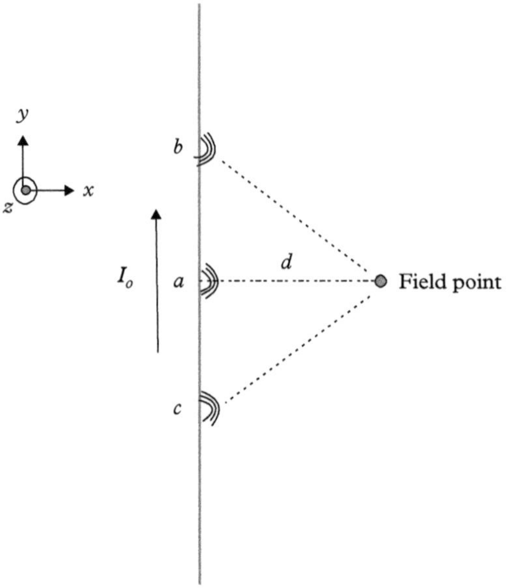

Fig. 3.3 *A $t = 0$, a current I_o starts flowing uniformly in the wire. This "information" leaves the wire at all points at $t = 0$, but it arrives at the field point from "a" first, and with a larger contribution to the observed fields than any other part of the wire. No fields can be detected at the Field Point until a time d/c has passed since the turn on of the current.*

The integrals resolve to simple one dimensional integrations along y. Substituting Eq. 3.31 into Eq. 3.28 yields:

$$\vec{E}(\vec{r}_o, t) = \frac{-I_o}{2\pi\varepsilon_o}\left[\int_0^\infty \frac{\delta(t - R/c)}{c^2\sqrt{y^2 + d^2}}\,dy - \int_0^\infty \frac{\delta(t - t_o - R/c)}{c^2\sqrt{y^2 + d^2}}\,dy\right]\hat{y} \tag{3.32}$$

A change of variables and a canonical manipulation of the delta function form changes Eq. 3.28 to

$$\vec{E}(\vec{r}_o, t) = \frac{-I_o}{2\pi\varepsilon_o c}\left[\int_d^\infty \frac{\delta(R - ct)}{\sqrt{R^2 - d^2}}\,dR - \int_d^\infty \frac{\delta(R - c(t - t_o))}{\sqrt{R^2 - d^2}}\,dR\right]\hat{y}$$

$$\vec{E}(\vec{r}_o, t) = \frac{-I_o}{2\pi\varepsilon_o c}\left[\frac{1}{\sqrt{(ct)^2 - d^2}} - \frac{1}{\sqrt{c^2(t - t_o)^2 - d^2}}\right]\hat{y} \tag{3.33}$$

where the first term is zero until time $t > d/c$ and the second is zero until time $t > d/c + t_o$.[16]

The evaluation of the $\vec{B}$ field requires the evaluation of terms involving both the current and the derivative of the current. For $0 \leq t < t_o$ the current is I_0, for all other times, the current is zero. The derivative of the current contributes to the magnetic field in a similar manner, at the time of turn on and turn off, as to the calculation of the electric field. The contributions to $\vec{B}$ arising from the turn on of the current at $t = 0$ and turn off at $t = t_o$ are given in analogy to Eq. 3.32[17]

[16] Note the scaling property:

$$\delta(\alpha x) = \frac{\delta(x)}{|\alpha|}$$

to write

$$\delta(t - R/c) = \delta(c^{-1}(ct - R)) = c\delta(R - ct)$$

In the integrals in 3.32, we change variables:

$$R = \sqrt{y^2 + d^2}$$
$$y = \sqrt{R^2 - d^2}$$
$$dy = \left(\frac{R}{\sqrt{R^2 - d^2}}\right)dR$$

and note that when $y = 0$; $z = d$

[17] Note that the cross product of $\vec{R}$ with the current and its derivative introduces a $\sin\theta$ term into the integral, which in this case is simply d/R.

$$\vec{B}(\vec{r}_o,t)\Big| = \frac{-I_o\mu_o d}{2\pi}\left[\int_d^{\infty}\frac{\delta(R-ct)}{R\sqrt{R^2-d^2}}\,dR - \int_d^{\infty}\frac{\delta(R-c(t-t_o))}{R\sqrt{R^2-d^2}}\,dR\right]\hat{z}$$

$$= \frac{-I_o\mu_o d}{2\pi}\left[\frac{1}{(ct)\sqrt{(ct)^2-d^2}} - \frac{1}{(c(t-t_o))\sqrt{c^2(t-t_o)^2-d^2}}\right]\hat{z} \qquad (3.34)$$

where we have the same restrictions on the two terms as in Eq. 3.33. The contribution to the magnetic field due to the steady current occurs only for the time $0 \le t < t_o$. This involves an integral over the wire, taking into account the length of wire that can contribute to the $\vec{B}$ field at the source as a function of time.[18] Thus the time dependence of the contribution from the constant current (for $0 \le t < t_o$) is

$$\vec{B}(\vec{r}_o,t)\Big|_I = \frac{-I_o\mu_o d}{2\pi}\left[\int_{\sqrt{c(t-t_o)^2-d^2}}^{\sqrt{(ct)^2-d^2}}\left(\frac{dy}{[y^2+d^2]^{3/2}}\right)\right]\hat{z}$$

$$= \frac{-I_o\mu_o}{2\pi d}\left[\frac{\sqrt{(ct)^2-d^2}}{ct} - \frac{\sqrt{c^2(t-t_o)^2-d^2}}{c(t-t_o)}\right]\hat{z} \qquad (3.35)$$

Combining the results of Eqs. 3.35, 3.34, and 3.33, we have the solution for all times:

For times $t < d/c$,

$$\vec{E} = 0$$
$$\vec{B} = 0$$

For times $d/c \le t < (t_o + d/c)$,

$$\vec{E}(\vec{r}_o,t) = \frac{-I_o}{2\pi\varepsilon_o c}\frac{1}{\sqrt{(ct)^2-d^2}}\hat{y}$$
$$\vec{B}(\vec{r}_o,t) = \frac{-I_o\mu_o}{2\pi d}\frac{ct}{\sqrt{(ct)^2-d^2}}\hat{z}$$

For times $t \ge t_o + d/c$,

$$\vec{E}(\vec{r}_o,t) = \frac{-I_o}{2\pi\varepsilon_o c}\left[\frac{1}{\sqrt{(ct)^2-d^2}} - \frac{1}{\sqrt{c^2(t-t_o)^2-d^2}}\right]\hat{y}$$
$$\vec{B}(\vec{r}_o,t) = \frac{-I_o\mu_o}{2\pi d}\left[\frac{ct}{\sqrt{(ct)^2-d^2}} - \frac{c(t-t_o)}{\sqrt{c^2(t-t_o)^2-d^2}}\right]\hat{z}$$

[18] As noted in Fig. 3.3, the "information" that the current is non-zero reaches the source point from the portion of the wire at $y = 0$ first, at a time delay d/c. Subsequently, the only portions of the wire that can contribute are between $y = 0$ and $y = \sqrt{(ct)^2 - d^2}$. As before, the cross product of $\vec{R}$ with the current and its derivative introduces a $sin\theta$ term into the integral, which in this case is simply $d/R = d/\sqrt{y^2+d^2}$.

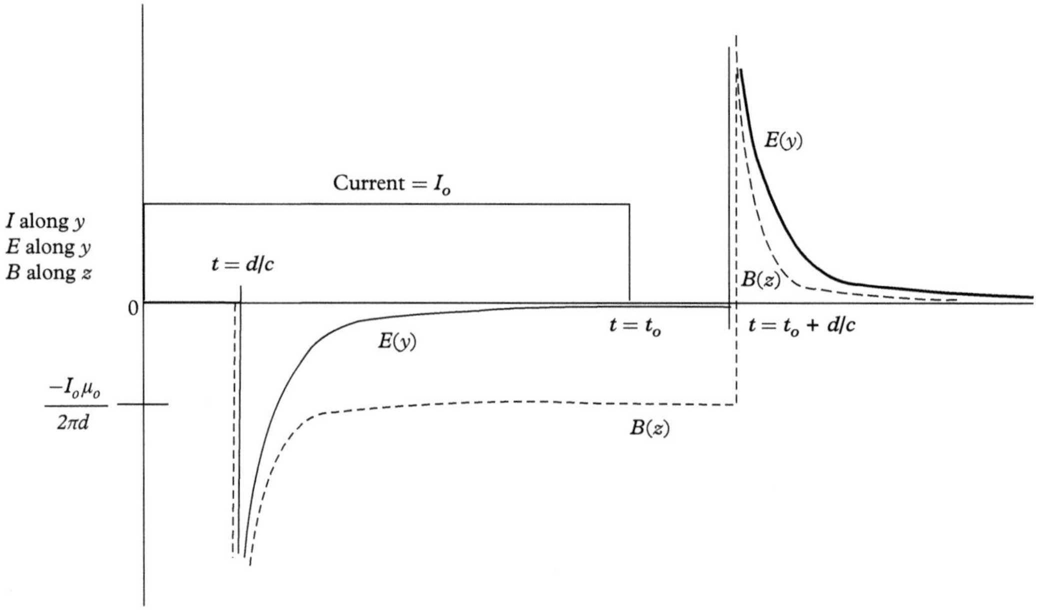

Fig. 3.4 *The current distribution in the wire and the observed electric and magnetic fields at the field point a distance d away from the wire (Fig. 3.3). The fields from portions of the wire for $|y| > 0$ arrive at the field point at increasingly later times, and contribute progressively less because of their increasing distance from the field point. The directions of the $\vec{E}$ and $\vec{B}$ fields are consistent with Lenz's and Ampere's laws.*

These results are sketched in Fig. 3.4, where the fields are shown as recorded at the field point. (The current is shown as measured at the wire.) The current, electric the magnetic fields are plotted in arbitrary units on the y-axis. The value of the magnetic field for a constant current is also indicated. The horizontal axis is time, where at $t = 0$ the current is turned on, and at $t = t_o$ the current is turned off. Because of the sudden nature of the turn on of current, an electric field develops (mathematically in Jefimenko's formulation, there is a contribution to $\vec{E}$ from $\dot{\vec{\mathcal{J}}}$). In accordance with Lenz's law, the electric field is in the direction to oppose the current, and the effects of retardation are immediately evident from the time delay for the fields to arrive at the observer. The magnetic field also has a sudden spike at the turn on, then settles toward its static value. At the turn off (when the current suddenly returns to zero at $t = t_o$), the transient fields are reversed in sign, again in accordance with Lenz's law. The trailing edge nature of the fields as observed at the field point is due to the finite time for the full effects of the current properties in the entire wire to be felt at the observation point.

Exercises

(3.1) Show explicitly that $\vec{E}(\vec{r}_o, t)$ and $\vec{B}(\vec{r}_o, t)$ given in Eqs. 3.11 and 3.13 satisfy

$$\nabla \times \vec{E}(\vec{r}_o, t) = -\frac{\partial}{\partial t}\vec{B}(\vec{r}_o, t) \text{ and}$$

$$\nabla \times \vec{B}(\vec{r}_o, t) = \frac{1}{c^2}\frac{\partial}{\partial t}\vec{E}(\vec{r}_o, t)$$

(3.2) Show that when $\tau = t - \frac{|\vec{R}|}{c}$, with $\vec{R} = \vec{r}_o - \vec{r}$, then $\nabla \cdot \vec{\mathcal{J}}(\vec{r}, \tau) = -\frac{\partial}{\partial t}\vec{\mathcal{J}} \cdot \frac{\hat{R}}{c}$

(3.3) Show that Eqs. 3.11 and 3.13 may be written in terms of derivatives of the retarded source parameters as:

$$\vec{B}(\vec{r}_o, t) = \frac{\mu_o}{4\pi}\int \frac{\nabla' \times \{\vec{\mathcal{J}}\}_r}{R} dV$$

$$\vec{E}(\vec{r}_o, t) = \frac{1}{4\pi}\int \frac{[\nabla'\{\rho\}_r + \frac{\partial}{\partial t}\{\vec{\mathcal{J}}\}_r]}{R} dV$$

where the derivatives are taken on the source variable r.

(3.4) Show, using Eqs. 3.11 and 3.13, that if the current in the source is independent of time, then Coulomb's law holds in its non-retarded form, that is

$$\vec{E}(\vec{r}_o, t) = \frac{1}{4\pi\varepsilon_o}\int \frac{\rho(\vec{r}, t)\hat{R}}{R^2} dV$$

(3.5) Consider a wire in the $\hat{y}$ direction with a current that is oscillating in magnitude according to $\vec{I}(y, t) = \hat{y}I_o sin\omega t$. Derive the electric and magnetic fields measured.

(3.6) Setting up Jefimenko's equations to describe the fields of a moving point charge #1: Consider a line of charge of length L and charge density λ_o, moving at $v/c << 1$ along the y-axis such that at $t = t'$, the "plug of charge" just passes the origin with a velocity $\overline{v} = v\hat{y}$. Using the Heaviside step function (and its derivative) derive the expressions for the charge and current density at position $\vec{r}$ and time t' as well as their respective time derivatives.

(3.7) Setting up Jefimenko #2: Write the integral Jefimenko expressions for the electric and magnetic fields in terms of your answers from Problem 5. Discuss the mathematical difficulty involved in solving these equations as written.

(3.8) Setting up Jefimenko #3: Rewrite your expressions from Problem 5 in terms of integrals over delta functions of argument $t' - t_{ret}$. That is, derive the companion expressions $\dot{\rho}, \vec{\mathcal{J}}, \dot{\vec{\mathcal{J}}}$ corresponding to (Θ is the Heaviside function):

$$\rho(\vec{r}, t_{ret}) = \lambda_o\delta(x)\delta(z)$$

$$\int \Theta(y - vt')[1 - \Theta(y - vt' - L)]\delta(t' - t_{ret})dt'$$

3.5 Discussions

Discussion 3.1

Equation 3.8 is obtained by starting with

$$\vec{B}(\vec{r}_o, t_o) = \nabla_o \times \vec{A} = \frac{\mu_o}{4\pi} \int \left[\nabla_o \times \left(\frac{\vec{\mathcal{J}}(\vec{r}, t_{ret})}{R} \right) \right] dV$$

and noting that in general

$$\nabla_o \times \left(\frac{\vec{\mathcal{J}}}{R} \right) = \frac{1}{R^2} \left(\vec{\mathcal{J}} \times \hat{R} \right) + \frac{1}{R} \left(\nabla_o \times \vec{\mathcal{J}} \right)$$

we further consider, for the second term, that the $\hat{x}$ component of

$$\nabla_o \times \{\vec{\mathcal{J}}\} = \left[\frac{\partial \{\mathcal{J}_z\}}{\partial t_{ret}} \left(\frac{\partial t_{ret}}{\partial y_o} \right) - \frac{\partial \{\mathcal{J}_y\}}{\partial t_{ret}} \left(\frac{\partial t_{ret}}{\partial z_o} \right) \right] = \frac{1}{c} \left[\{\dot{\vec{\mathcal{J}}}\} \times \hat{R} \right]_x$$

so that

$$\nabla_o \times \left(\frac{\{\vec{\mathcal{J}}\}}{R} \right) = \frac{1}{R^2} \left(\{\vec{\mathcal{J}}\} \times \hat{R} \right) + \frac{1}{cR} \left(\{\dot{\vec{\mathcal{J}}}\} \times \hat{R} \right)$$

and we conclude

$$\vec{B}(\vec{r}_o, t_o) = \frac{\mu_o}{4\pi} \int \left[\frac{1}{R^2} \left(\{\vec{\mathcal{J}}\} \times \hat{R} \right) + \frac{1}{cR} \left(\{\dot{\vec{\mathcal{J}}}\} \times \hat{R} \right) \right] dV$$

Discussion 3.2

Start with

$$\vec{E}(\vec{r}_o, t) = -\frac{1}{4\pi\varepsilon_o} \int \nabla \left(\frac{\rho(\vec{r}, \tau)}{R} \right) dV - \frac{\mu_o}{4\pi} \int \left[\frac{\partial}{\partial t} \left(\frac{\vec{\mathcal{J}}(\vec{r}, \tau)}{R} \right) \right] dV \qquad (3.36)$$

remembering that ∇ is a field variable operator. Then the first integrand,

$$\nabla \left(\frac{\rho(\vec{r}, \tau)}{R} \right) = -\frac{\rho(\vec{r}, \tau)\hat{R}}{R^2} - \dot{\rho}(\vec{r}, \tau) \frac{\hat{R}}{cR}$$

At this point we introduce the continuity equation applied to the source coordinates at the retarded time. That is, we use the source variable operator ∇' that, by definition, differentiates with respect to the source variables r, θ, and ϕ and we evaluate at the retarded time because the continuity equation relates a local divergence to a local time variation. Thus, employing the {} notation to indicate retarded time and ∇' to denote the gradient with respect to the source coordinate r,

$$\{\nabla' \cdot \vec{\mathcal{J}} + \dot{\rho}\} = 0 \Longrightarrow -\{\dot{\rho}\} = \{\nabla' \cdot \vec{\mathcal{J}}\}$$

Next we note the difference between $\{\nabla' \cdot \vec{\mathcal{J}}\}$ and $\nabla' \cdot \{\vec{\mathcal{J}}\}$,

$$\nabla' \cdot \{\vec{\mathcal{J}}\} = \{\nabla' \cdot \vec{\mathcal{J}}\} + \{\dot{\vec{\mathcal{J}}}\} \cdot \nabla' \tau$$

where our notation is to be read that the first term on the right represents $\nabla' \cdot \{\vec{\mathcal{J}}\}$ taken with respect to the explicit r (with the implicit r within τ held constant), and the second term represents $\nabla' \cdot \{\vec{\mathcal{J}}\}$, taken with respect to the implicit r (with the explicit r held constant). Now, making use of the continuity equation and noting that $\nabla' \tau = -\nabla \tau = \frac{\hat{R}}{c}$, we get

$$\nabla' \cdot \{\vec{\mathcal{J}}\} = -\{\dot{\rho}\} + \{\dot{\vec{\mathcal{J}}}\} \cdot (\frac{\hat{R}}{c}) \tag{3.37}$$

and with these considerations, Eq. 3.36 becomes:

$$\vec{E}(\vec{r}_o, t) = -\frac{1}{4\pi \varepsilon_o} \int \left[-\frac{\{\rho\}\hat{R}}{R^2} + \left(\nabla' \cdot \{\vec{\mathcal{J}}\} - \{\dot{\vec{\mathcal{J}}}\} \cdot (\frac{\hat{R}}{c}) \right) \frac{\hat{R}}{cR} + \frac{1}{c^2} \frac{\{\dot{\vec{\mathcal{J}}}\}}{R} \right] dV \tag{3.38}$$

Discussion 3.3

The second of the four terms on the right in Eq. 3.38 is most easily evaluated by writing out components. The ith component of the second term is (where ∇' denotes the gradient with respect to the source coordinate r),

$$\frac{R_i}{cR^2} \nabla' \cdot \{\vec{\mathcal{J}}\} = \nabla' \cdot \left(\frac{R_i \{\vec{\mathcal{J}}\}}{cR^2} \right) - \nabla' \left(\frac{R_i}{cR^2} \right) \cdot \{\vec{\mathcal{J}}\} \tag{3.39}$$

In Eq. 3.39 the first of the two terms on the right is of the form such that, using the divergence theorem, the integral over each of the ith components is zero because the current distribution is assumed to be confined to some finite region of space. We write the last term in Eq. 3.39 in the canonical form so that it becomes

$$\frac{R_i}{cR^2}\nabla'\cdot\{\vec{\mathcal{J}}\} = -\nabla'\left(\frac{R_i}{cR^2}\right)\cdot\{\vec{\mathcal{J}}\} = \frac{1}{c}\left[\frac{\{\mathcal{J}_i\}}{R^2} - 2\left(\{\vec{\mathcal{J}}\}\cdot\hat{R}\right)\frac{R_i}{R^4}\right] \tag{3.40}$$

Next, noting that $\hat{R} = \sum_i \frac{R_i}{R}\hat{i}$ and $\{\vec{\mathcal{J}}\} = \sum_i \{\mathcal{J}_i\}\hat{i}$, we substitute Eq. 3.40 for the second term on the right in Eq. 3.38 and rearrange

$$\vec{E}(\vec{r}_o, t) = \frac{1}{4\pi\varepsilon_o}\int\left[\frac{\{\rho\}}{R^2}\hat{R} + \frac{2(\{\vec{\mathcal{J}}\}\cdot\hat{R})\hat{R} - \{\vec{\mathcal{J}}\}}{cR^2} + \frac{(\{\dot{\vec{\mathcal{J}}}\}\cdot\hat{R})\hat{R} - \{\dot{\vec{\mathcal{J}}}\}}{c^2R}\right]dV$$

Finally, using the triple cross product vector formula, $(\vec{A}\times\vec{B})\times\vec{C} = (\vec{C}\cdot\vec{A})\vec{B} - (\vec{C}\cdot\vec{B})\vec{A}$ for both the second and third terms, this reduces further to

$$\vec{E}(\vec{r}_o, t) = \frac{1}{4\pi\varepsilon_o}\int\left[\frac{\{\rho\}}{R^2}\hat{R} + \frac{(\{\vec{\mathcal{J}}\}\times\hat{R})\times\hat{R} + (\{\vec{\mathcal{J}}\}\cdot\hat{R})\hat{R}}{cR^2} + \frac{(\{\dot{\vec{\mathcal{J}}}\}\times\hat{R})\times\hat{R}}{c^2R}\right]dV$$

Discussion 3.4

From Eq. 3.22 we note that because G_F is the response to a point charge at $\vec{r}$ with no additional boundary conditions, it is spherically symmetric (i.e., a function only of R). With such spherical symmetry, the angular terms of ∇^2 of Eq. 3.22 vanish so that over a space in which $R \neq 0$, Eq. 3.22 simplifies to an equation which yields the solution $G_F = \frac{Ae^{ikR}}{R}$. Consideration of the volume surrounding $\vec{R} = 0$ determines the value of A. Taking the volume integral of Eq. 3.22 with $\psi\left(\vec{R}\right) = G_F\left(\vec{R}\right) = \frac{Ae^{ikR}}{R}$ and making use of the divergence theorem yields

$$-\oint\frac{Ae^{ikR}}{R^2}\hat{R}\cdot\hat{R}dS + ik\oint\frac{Ae^{ikR}}{R}\hat{R}\cdot\hat{R}dS + k^2\int\frac{Ae^{ikR}}{R}dV = -1$$

which simplifies to

$$-A\oint e^{ikR}d\Omega + ikA\oint e^{ikR}Rd\Omega + k^2A\int e^{ikR}Rd\Omega dR = -1$$

Because Eq. 3.22 vanishes for all points but $\vec{R} = 0$, we can make the integration volume arbitrarily small so that $e^{ikR} \simeq 1$ and the last two terms on the left become negligible and thus

$$-A\oint d\Omega = -4\pi A = -1 \rightarrow A = 1/4\pi$$

and A is determined

$$G(R) = \frac{e^{ikR}}{4\pi R}$$

Discussion 3.5

(∇' denotes the gradient with respect to the source coordinate r) Integrate $\int \nabla' \rho_F \frac{e^{ikR}}{R} dV$ by parts using the form $\int_a^b u\nabla v = uv|_a^b - \int_a^b v\nabla u$. Set $u = \frac{e^{i\omega R/c}}{R}$ and $\nabla v = \nabla' \rho_F$ then $\nabla u = \nabla' \left(\frac{e^{i\omega R/c}}{R}\right) = -[\frac{i\omega}{cR} - \frac{1}{R^2}]\hat{R} e^{i\omega R/c}$ and $v = \rho_F$ where the last negative sign on ∇u arises due to differentiation with respect to the source coordinates. Because the charge density ρ_F is bounded and the volume integral is arbitrarily large, the first term, $uv|_a^b$, vanishes and the first term in Eq. 3.24 is expressed

$$-\frac{1}{4\pi\varepsilon_0}\int \nabla' \rho_F \frac{e^{ikR}}{R} dV = -\frac{1}{4\pi\varepsilon_0}\int dV \left[\frac{i\omega}{cR} - \frac{1}{R^2}\right]\hat{R}\rho_F e^{i\omega R/c}$$

Fourier Transform: Writing

$$\left(\frac{\partial f}{\partial t}\right)_F = \frac{1}{\sqrt{2\pi}}\int_{-\infty}^{\infty}\left(\frac{\partial f}{\partial t}\right)e^{i\omega t'}dt'$$

Integrating by parts and throwing away the terms at the limits (the function is assumed to go to zero at $\pm\infty$), gives:

$$\left(\frac{\partial f}{\partial t}\right)_F = -\frac{1}{\sqrt{2\pi}}\int_{-\infty}^{\infty} i\omega f e^{i\omega t'}dt' = -i\omega f_F$$

and

$$(f(t - R/c))_F = \frac{1}{\sqrt{2\pi}}\int_{-\infty}^{\infty} f(t' - R/c)\, e^{i\omega t'}dt' = \frac{1}{\sqrt{2\pi}}\int_{-\infty}^{\infty} f(t')\, e^{i\omega t' + i\omega R/c}dt' = e^{i\omega r/c} f_F$$

Inverse Fourier Transform: The required integral operation is

$$\int dV \int_{-\infty}^{\infty} dt' \left[\frac{1}{cR}[\hat{R}\dot{\rho}(t') - \frac{1}{c}\dot{\mathcal{J}}(t')] + \frac{\hat{R}}{R^2}\rho(t')\right]\frac{1}{\sqrt{2\pi}}\int_{-\infty}^{\infty} e^{i\omega(t' + \frac{R}{c})}e^{-i\omega t}d\omega$$

The integrals are performed in order starting with ω. The integral over ω yields a delta function in t', $2\pi\delta(t' + \frac{R}{c} - t) = 2\pi\delta\left[t' - (t - \frac{R}{c})\right]$. That is, t' can only have the value $t' = (t - \frac{R}{c})$ in the subsequent integral over t'. We have then

$$\frac{1}{\sqrt{2\pi}}\int dV \int_{-\infty}^{\infty} dt' \left[\frac{1}{cR}[\hat{R}\dot{\rho}(t') - \frac{1}{c}\dot{\mathcal{J}}(t')] + \frac{\hat{R}}{R^2}\rho(t')\right]2\pi\delta\left[t' - (t - \frac{R}{c})\right]$$

and finally carrying out the integral over t', and including the ignored factor,

$$\vec{E}(\vec{r}_o,t) = \frac{\sqrt{2\pi}}{\varepsilon_o\sqrt{32\pi^3}}\int dV\left[\frac{1}{cR}[\hat{R}\dot{\rho}(t-\frac{R}{c}) - \frac{1}{c^2R}\dot{\vec{\mathcal{J}}}(t-\frac{R}{c})] + \frac{\hat{R}}{R^2}\rho(t-\frac{R}{c})\right]$$

In Eq. 3.26, recall that the notation $\left\{ret\right\}$ indicates that all variables $(\rho, \dot{\rho}, \vec{\mathcal{J}}, \dot{\vec{\mathcal{J}}})$ within the curly brackets are evaluated at $\tau = (t - \frac{R}{c})$ to obtain the value of the E field at time t.

Discussion 3.6

Instead of Eq. 3.24 we would have:

$$\vec{B}_F(\vec{r}_o,\omega) = \int [\mu_o\nabla' \times \vec{\mathcal{J}}_F(\vec{r})]\frac{e^{ikR}}{4\pi R}dV$$

then following the development given before for $\vec{E}$ in which we integrated by parts and assumed a bounded source, we get,

$$\vec{B}_F(\vec{r}_o,\omega) = \frac{\mu_o}{4\pi}\int dV\left[\frac{i\omega}{cR} - \frac{1}{R^2}\right][\hat{R}\times\vec{\mathcal{J}}_F(\vec{r})]\,e^{i\omega R/c}$$

where we have used the vector formula $\int(\nabla\times\vec{A})\,dV = -\oint(\vec{A}\times\hat{n})\,dS$ and as usual $\vec{R} = \vec{r}_0 - \vec{r}$. The rest of the derivation follows essentially line/line with that presented for the $\vec{E}$ field. For the specific case in which $r_0 >> r$, we can approximate $R = (\vec{R}\cdot\vec{R})^{1/2} \approx r_0(1 - (\vec{r}_o\cdot\vec{r})/r_0^2) \approx r_o$. Then, in the radiation zone, where the $1/R^2$ term is negligible, we can write

$$\vec{B}_F^{rad}(\vec{r}_o,\omega) \cong \frac{i\omega\mu_o}{4\pi\, cr_o}e^{(i\omega r_0/c)}\left[\hat{r}_o\times\int dV\,\vec{\mathcal{J}}_F(\vec{r})\,e^{-i\omega(\hat{r}_o\cdot\vec{r}/c)}\right]$$

Substituting $\vec{\mathcal{J}}_F = \frac{1}{\sqrt{2\pi}}\int_{-\infty}^{\infty}\vec{\mathcal{J}}(t')\,e^{i\omega t'}dt'$ and integrating by parts over t':

$$\vec{B}_F^{rad}(\vec{r}_o,\omega) \cong \frac{-\mu_o}{c\sqrt{32\pi^3}}\int_{-\infty}^{\infty}dt'\,e^{i\omega(t'+r_0/c)}\frac{1}{r_o}\left[\hat{r}_o\times\int dV\,e^{-i\omega(\hat{r}_o\cdot\vec{r}/c)}\frac{\partial\vec{\mathcal{J}}(\vec{r},t')}{\partial t'}\right]$$

and taking the inverse Fourier transform yields the expression for the B field in the radiation zone:

$$\vec{B}_{rad}(\vec{r}_o,t) = \frac{-\mu_o}{4\pi}\frac{\hat{r}_o}{cr_o}\times\int dV\,\dot{\vec{\mathcal{J}}}(\vec{r},\tau + \hat{r}_o\cdot\vec{r}/c)$$

where $\tau = t - r_o/c$.

Fields in Terms of the Multipole Moments of the Source

4

- Evaluation of Jefimenko's equations for the electric and magnetic fields in the radiation zone by expansion of the charge and current distributions in their lowest order moments
- Electric and magnetic fields in terms of the moments of the charge and current distributions: electric and magnetic dipole, and the electric quadrupole
- Electric and magnetic fields for all regions of space relative to the source location (near, intermediate, and radiation) using the multipole expansion of the vector potential in terms of the charge and current distributions of the source
- Expressions for the power radiated by the multipole moments of the source

4.1 Multipole radiation using Jefimenko's equations

4.1.1 Approximate spatial dependence

In general, the evaluation of $\vec{E}(\vec{r}_o, t)$ and $\vec{B}(\vec{r}_o, t)$ using Eqs. 3.26 and 3.27 requires detailed knowledge of the source at the retarded time, specifically the time derivatives of the charge and current density distributions with the source. We can, however, see how these Jefimenko expressions give rise to physically reasonable expressions for the radiation fields by considering the simple case in which the source current density is oscillating sinusoidally in time at a frequency ω: then $\vec{\mathcal{J}}(\vec{r},\tau) = \vec{\mathcal{J}}(\vec{r})e^{i\omega\tau}$ and $\dot{\vec{\mathcal{J}}}(\vec{r},\tau) = i\omega\vec{\mathcal{J}}(\vec{r})e^{i\omega\tau}$. From the last term in Eq. 3.14 we have the expression for the radiation field:

Electromagnetic Radiation. Richard Freeman, James King, Gregory Lafyatis,

DOI: 10.1093/oso/9780198726500.001.0001

$$\vec{E}(\vec{r}_o, t)_{rad} = \frac{1}{4\pi\varepsilon_o}\int\left[\frac{(\vec{\mathcal{J}}(\vec{r})\times\hat{R})\times\hat{R}}{c^2R}i\omega e^{i\omega\tau}\right]dV \tag{4.1}$$

If we next take $\frac{r}{r_o} << 1$, then we can make the approximations $R = (\vec{R}\cdot\vec{R})^{1/2} \simeq r_o - \vec{r}\cdot\hat{r}_o \simeq r_o$ and we can write $i\omega\tau \simeq i\omega t - ikr_o + ik\vec{r}\cdot\hat{r}_o$ and $R \simeq r_o$. When r is additionally always very much smaller than the wavelength of the radiation (i.e., the source dimensions are such that there is essentially no phase difference across the source as seen by the observer) then $k\vec{r}\cdot\hat{r}_o \simeq 0$ and Eq. 4.1 can be expressed in the radiation zone as

$$\vec{E}(\vec{r}_o, t)_{rad} \simeq \frac{i\omega e^{i(\omega t - kr_o)}}{4\pi\varepsilon_o c^2 r_o}\left[\int\left[(\vec{\mathcal{J}}(\vec{r})\times\hat{r}_o)\times\hat{r}_o\right]dV\right] \tag{4.2}$$

which is just a spherical wave traveling away from a source that has been effectively approximated, with regard to direction, distance, and phase, as a point source at the origin. Thus, in this single frequency point source (zeroth order) approximation, each field point is related to every other field point at time t by the phase factor $e^{-ikr_o} = e^{-i\frac{\omega}{c}r_o}$. That is, the effect of accounting for the retarded time in this approximation of the radiation term in Jefimenko's equations is the establishment of the phase relation between field points located far from the source.

We next consider the more general case in which the source time dependence is not necessarily sinusoidal and the source has some extent. One of the useful features of expressing the fields in terms of the retarded source parameters is the ability to find radiation fields from the moments of the current distributions within the source. This process gives expressions for the radiation fields directly from the various moments of the source, without having to construct the expansion of the potential first and then compute the fields from various derivatives. This development follows from the dependence of the current density time derivative on the first order correction $(\vec{r}\cdot\hat{r}_o/c)$ to the zeroth order approximation of τ.[1] Using the last (radiation zone) terms in Eqs. 3.13 and 3.14,

$$\vec{E}(\vec{r}_o, t)_{rad} = \frac{1}{4\pi\varepsilon_o c^2}\int\frac{(\dot{\vec{\mathcal{J}}}(\vec{r},\tau)\times\hat{R})\times\hat{R}}{R}dV \tag{4.3}$$

$$= c\vec{B}(\vec{r}_o, t)_{rad}\times\hat{R} \tag{4.4}$$

$$\vec{B}(\vec{r}_o, t)_{rad} = \frac{1}{4\pi\varepsilon_o c^3}\int\frac{(\dot{\vec{\mathcal{J}}}(\vec{r},\tau)\times\hat{R})}{R}dV \tag{4.5}$$

[1] This section follows that given in Melo e Souza, R. de, Cougo-Pinto, M. V., Farina, C., and Moriconi, M. Multipole radiation fields from the Jefimenko equation for the magnetic field and the Panofsky-Phillips equation for the electric field *Am. J. Phys.* 77, 67 (2009).

we consider a confined source, and field points that satisfy $\frac{r}{r_o} \ll 1$ such that we can approximate $R \simeq r_o - \vec{r} \cdot \hat{r}_o$, and subsequently, $\tau \backsimeq t - \frac{r_o}{c} + \frac{\vec{r} \cdot \hat{r}_o}{c}$. Noting that $t - \frac{r_o}{c}$ would be the exact retarded time if the source were a point with no extent, we label this time as $\tau^o = t - \dfrac{r_o}{c}$. The time derivative of the current density in the source can then be expanded in first order as

$$\dot{\vec{\mathcal{J}}}(\vec{r}, \tau) \backsimeq \dot{\vec{\mathcal{J}}}(\vec{r}, \tau^o) + (\frac{\vec{r} \cdot \hat{r}_o}{c}) \frac{\partial}{\partial t} \dot{\vec{\mathcal{J}}}(\vec{r}, \tau^o) \tag{4.6}$$

Substituting Eq. 4.6 into expression Eq. 4.5 for $\vec{B}(\vec{r}_o, t)$, and pulling R and $\hat{R}$ (approximated as r_o and $\hat{r}_o$) out of the integral,

$$\vec{B}(\vec{r}_o, t)_{rad} \backsimeq \frac{1}{4\pi \varepsilon_o c^3 r_o} \int \left[\dot{\vec{\mathcal{J}}}(\vec{r}, \tau^o) + (\frac{\vec{r} \cdot \hat{r}_o}{c}) \frac{\partial}{\partial t} \dot{\vec{\mathcal{J}}}(\vec{r}, \tau^o) \right] dV \times \hat{r}_o \tag{4.7}$$

4.1.2 Radiation from zeroth order moments

The zeroth order term in the $\vec{B}_{rad}$ field expansion of Eq. 4.7 is the electric dipole (*ed*) term, while the second term represents a mixture of the magnetic dipole and electric quadrupole. The first term can be written as[2]

$$\vec{B}(\vec{r}_o, t)^{ed}_{rad} = \frac{1}{4\pi \varepsilon_o c^3 r_o} \int \left[\sum_i \{\nabla \cdot (x_i \dot{\vec{\mathcal{J}}})\} \hat{x}_i - \{\vec{r} \nabla \cdot \dot{\vec{\mathcal{J}}}\} \right] dV \times \hat{r}_o \tag{4.8}$$

Equation 4.8 can be evaluated, using Gauss' theorem and the continuity equation,[3] to be

$$\begin{aligned} \vec{B}(\vec{r}_o, t)^{ed}_{rad} &= \frac{1}{4\pi \varepsilon_o c^3 r_o} \left[\frac{\partial^2}{\partial t^2} \left(\int \rho(\vec{r}, \tau^o) \vec{r} \, dV \right) \right] \times \hat{r}_o \\ &= \frac{\ddot{\vec{p}}(\tau^o) \times \hat{r}_o}{4\pi \varepsilon_o c^3 r_o} \end{aligned} \tag{4.9}$$

where we have identified the electric dipole moment of the charge distribution in the source as

$$\vec{p}(\tau^o) = \int \rho(\vec{r}, \tau^o) \vec{r} \, dV$$

From Eq. 4.4 we obtain the corresponding expression for the electric field:

[2] The first term in Eq. 4.7 is the electric dipole term

$$\vec{B}(\vec{r}_o, t)^{ed}_{rad} \backsimeq \frac{1}{4\pi \varepsilon_o c^3 r_o} \left[\int \dot{\vec{\mathcal{J}}}(\vec{r}, \tau^o) dV \right] \times \hat{r}_o$$

but $\dot{\vec{\mathcal{J}}}(\vec{r}, \tau^o) = \{\dot{\vec{\mathcal{J}}}\}$ can be expressed as

$$\{\dot{\vec{\mathcal{J}}}\} = \sum_i \{\dot{\vec{\mathcal{J}}} \cdot \hat{x}_i\} \hat{x}_i = \sum_i \{\dot{\vec{\mathcal{J}}} \cdot \nabla x_i\} \hat{x}_i$$

Integrating this by parts we get

$$\begin{aligned} &\sum_i \int \{\dot{\vec{\mathcal{J}}} \cdot \nabla x_i\} \hat{x}_i dV \\ &= \sum_i \int \left(\{\nabla \cdot (x_i \dot{\vec{\mathcal{J}}})\} - \{x_i \nabla \cdot \dot{\vec{\mathcal{J}}}\} \right) \hat{x}_i dV \end{aligned}$$

[3] The integral part of Eq. 4.8 is

$$\int \left[\sum_i \{\nabla \cdot (x_i \dot{\vec{\mathcal{J}}})\} \hat{x}_i - \{\vec{r} \nabla \cdot \dot{\vec{\mathcal{J}}}\} \right] dV$$

Using Gauss' theorem on the first term

$$\sum_i \hat{x}_i \int \{\nabla \cdot (x_i \dot{\vec{\mathcal{J}}})\} dV = \sum_i \hat{x}_i \oint \{x_i \dot{\vec{\mathcal{J}}} \cdot \hat{r}\} dS = 0$$

so we see this term vanishes because the surface of the volume is beyond the bounded distribution. Using the time derivative of the continuity equation, $\{\nabla \cdot \dot{\vec{\mathcal{J}}}\} = -\{\ddot{\rho}\}$, the second term becomes

$$\begin{aligned} -\int \{\vec{r} \nabla \cdot \dot{\vec{\mathcal{J}}}\} dV &= \int \{\ddot{\rho} \vec{r}\} \, dV \\ &= \frac{\partial^2}{\partial t^2} \left(\int \rho(\vec{r}, \tau^o) \, \vec{r} \, dV \right) \end{aligned}$$

where we have used $d\tau^o = dt$.

$$\vec{E}(\vec{r}_o, t)^{ed}_{rad} \backsimeq \frac{(\ddot{\vec{p}}(\tau^o) \times \hat{r}_o) \times \hat{r}_o}{4\pi\varepsilon_o c^2 r_o} \tag{4.10}$$

The Poynting vector associated with the electric dipole moment is then

$$\vec{S}(\vec{r}_o, t)^{ed} \backsimeq \frac{1}{\mu_o}\left(\vec{E}(\vec{r}_o, t)^{ed}_{rad} \times \vec{B}(\vec{r}_o, t)^{ed}_{rad}\right) \tag{4.11}$$

$$\backsimeq \frac{\hat{r}_o}{16\pi^2\varepsilon_o c^3 r_o^2}\left(\ddot{\vec{p}}(\tau^o)\cdot\ddot{\vec{p}}(\tau^o) - [\ddot{\vec{p}}(\tau^o)\cdot\hat{r}_o]^2\right) \tag{4.12}$$

With a spherical polar coordinate system oriented such that the direction of the dipole moment is along the z-axis, Eq. 4.12 becomes

$$\vec{S}(\vec{r}_o, t)^{ed} \backsimeq \frac{\hat{r}_o[\ddot{\vec{p}}(\tau^o)]^2}{16\pi^2\varepsilon_o c^3 r_o^2} sin^2\theta \tag{4.13}$$

and the instantaneous electric dipole contribution to the total radiated power is

$$P(t)^{ed} = \int \vec{S}(\vec{r}_o, t)^{ed} \cdot d\vec{a} = \frac{[\ddot{\vec{p}}(\tau^o)]^2}{6\pi\varepsilon_o c^3} \tag{4.14}$$

Physical consequences of τ^o

The physical consequences of τ^o in the expressions in Eqs. 4.9–4.14 are most clearly seen by considering the dipole to have a single frequency sinusoidal time dependence:

$$\vec{p}(\tau^o) = \frac{\vec{p}_o}{2}\left[e^{-i\omega\tau^o} + e^{i\omega\tau^o}\right] = \vec{p}_o\, cos\omega\tau^o$$

$$\ddot{\vec{p}}(\tau^o) = -\omega^2\vec{p}(\tau^o) = -\omega^2\vec{p}_o\, cos\left(\omega t - \frac{\omega r_o}{c}\right)$$

$$\left[\ddot{\vec{p}}(\tau^o)\right]^2 = \omega^4 p_o^2\, cos^2\left(\omega t - \frac{\omega r_o}{c}\right) = \frac{\omega^4 p_o^2}{2}\left(1 + cos\left(2\omega t - \frac{2\omega r_o}{c}\right)\right)$$

We see that the effect of correctly including the retarded time in the zeroth order approximation of Jefimenko's equations gives the physically reasonable result that the radiated fields measured at a distance r_o removed from a dipole oscillating at frequency ω have the same form as those at a distance r'_o, differing only by a phase factor $e^{\frac{i\omega(r'_o - r_o)}{c}}$, and an overall amplitude scaling as $\frac{r'_o}{r_o}$. This is because, in the zeroth order approximation, the source dimensions are such that there is essentially no distance, direction or phase difference across the source as seen by an observer at time t and position $\vec{r}_o$. The measured instantaneous total power oscillates at 2ω and depends upon the distance from the source according to the phase factor $e^{\frac{2i\omega(r'_o - r_o)}{c}}$.

4.1.3 Radiation from first order moments

We now turn to the second term in the expression for the magnetic radiation field built on the first order correction of the retarded current density time derivative. This term takes into account the non-zero extension of the source in the direction of the observer, $\hat{r}_o$. From Eq. 4.7

$$\vec{B}(\vec{r}_o, t)^{1st}_{rad} \backsimeq \frac{1}{4\pi\varepsilon_o c^4 r_o}\left[\int (\vec{r}\cdot\hat{r}_o)\dddot{\vec{\mathcal{J}}}(\vec{r}, \tau^o)\, dV\right] \times \hat{r}_o \tag{4.15}$$

We anticipate that this first order field is a combination of "magnetic dipole" and "electric quadrupole" terms, which we untangle in the canonical manner by splitting the integral into symmetric and anti-symmetric (under the exchange of $\dddot{\vec{\mathcal{J}}}$ and $\vec{r}$) parts and employing the triple cross product expansion.[4] The first order magnetic field is then

$$\vec{B}(\vec{r}_o, t)^{1st}_{rad} \backsimeq \frac{1}{4\pi\varepsilon_o c^4 r_o}\left[\frac{1}{2}\int \left\{\left((\vec{r}\cdot\hat{r}_o)\dddot{\vec{\mathcal{J}}} + (\dddot{\vec{\mathcal{J}}}\cdot\hat{r}_o)\vec{r}\right)\right.\right.$$
$$\left.\left. + \left((\vec{r}\times\dddot{\vec{\mathcal{J}}})\times\hat{r}_o\right)\right\} dV\right] \times \hat{r}_o \tag{4.16}$$

where $\{\ldots\}$ indicates evaluation at the retarded time $\tau^o = t - r_o/c$.

Magnetic dipole

Considering just the second (antisymmetric) term of Eq. 4.16

$$\vec{B}(\vec{r}_o, t)^{anti-sym}_{rad} = \frac{1}{4\pi\varepsilon_o c^4 r_o}\left[\left(\frac{1}{2}\int \vec{r}\times\dddot{\vec{\mathcal{J}}}(\vec{r}, \tau^o)dV\right)\times\hat{r}_o\right]\times\hat{r}_o \tag{4.17}$$

and identifying the magnetic dipole moment of the current distribution in the source as

$$\vec{m}(\tau^o) = \frac{1}{2}\int \vec{r}\times\vec{\mathcal{J}}(\vec{r}, \tau^o)dV$$

we can write the expression for the magnetic dipole contribution to the radiation fields as (using Eq. 4.4 to obtain $\vec{E}_{rad}$ from $\vec{B}_{rad}$)

[4] Write the integrand in Eq. 4.15 as the simple sum of two terms, each equal to 1/2 of the original integrand. Then add and subtract $\frac{1}{2}(\dddot{\vec{\mathcal{J}}}\cdot\hat{r}_o)\vec{r}$, resulting in four terms:

$$(\vec{r}\cdot\hat{r}_o)\dddot{\vec{\mathcal{J}}} = \frac{1}{2}\left((\vec{r}\cdot\hat{r}_o)\dddot{\vec{\mathcal{J}}} + (\dddot{\vec{\mathcal{J}}}\cdot\hat{r}_o)\vec{r}\right) + \frac{1}{2}\left((\vec{r}\cdot\hat{r}_o)\dddot{\vec{\mathcal{J}}} - (\dddot{\vec{\mathcal{J}}}\cdot\hat{r}_o)\vec{r}\right)$$

Identifying the second (antisymmetric) term as the standard expansion of the triple cross product $(\vec{A}\times\vec{B})\times\vec{C} = (\vec{A}\cdot\vec{C})\vec{B} - (\vec{B}\cdot\vec{C})\vec{A}$, we then have the required result:

$$(\vec{r}\cdot\hat{r}_o)\dddot{\vec{\mathcal{J}}} = \frac{1}{2}\left((\vec{r}\cdot\hat{r}_o)\dddot{\vec{\mathcal{J}}} + (\dddot{\vec{\mathcal{J}}}\cdot\hat{r}_o)\vec{r}\right) + \frac{1}{2}\left((\vec{r}\times\dddot{\vec{\mathcal{J}}})\times\hat{r}_o\right)$$

$$\vec{B}(\vec{r}_o, t)^{md}_{rad} = \frac{(\dddot{\vec{m}}(\tau^o) \times \hat{r}_o) \times \hat{r}_o}{4\pi\varepsilon_o c^4 r_o} \tag{4.18}$$

$$\vec{E}(\vec{r}_o, t)^{md}_{rad} = \frac{-\dddot{\vec{m}}(\tau^o) \times \hat{r}_o}{4\pi\varepsilon_o c^3 r_o} \tag{4.19}$$

Comparing the magnetic dipole fields (Eqs. 4.18 and 4.19) with the electric dipole fields (Eqs. 4.9 and 4.10), we have

$$\frac{1}{c}\vec{E}(\vec{r}_o, t)^{md}_{rad} = \left[\vec{B}(\vec{r}_o, t)^{ed}_{rad}\right] \text{ with the substitution of } \dddot{\vec{p}} \text{ by} - \frac{\dddot{\vec{m}}}{c} \tag{4.20}$$

$$\vec{B}(\vec{r}_o, t)^{md}_{rad} = \left[\frac{1}{c}\vec{E}(\vec{r}_o, t)^{ed}_{rad}\right] \text{ with the substitution of } \dddot{\vec{p}} \text{ by } \frac{\dddot{\vec{m}}}{c} \tag{4.21}$$

The Poynting vector and the total instantaneous power for magnetic dipole radiation are readily obtained with the help of Eqs. 4.20 and 4.21 and, in a manner identical to that which determined the same quantities (4.13 and 4.14), for the electric dipole radiation

$$\vec{S}(\vec{r}_o, t)^{md} \backsimeq \frac{\hat{r}_o[\dddot{\vec{m}}(\tau^o)]^2}{16\pi^2\varepsilon_o c^5 r_o^2} sin^2\theta \tag{4.22}$$

$$P(t)^{md} = \int \vec{S}(\vec{r}_o, t)^{md} \cdot d\vec{a} = \frac{[\dddot{\vec{m}}(\tau^o)]^2}{6\pi\varepsilon_o c^5} \tag{4.23}$$

Electric quadrupole

Consider now the symmetric part of $\vec{B}(\vec{r}_o, t)^{1st}_{rad}$ in Eq. 4.16:

$$\vec{B}(\vec{r}_o, t)^{sym}_{rad} = \frac{1}{4\pi\varepsilon_o c^4 r_o}\int \frac{1}{2}\left[(\hat{r}_o \cdot \vec{r})\ddot{\vec{\mathcal{J}}}(\vec{r}, \tau^o) + \left(\ddot{\vec{\mathcal{J}}}(\vec{r}, \tau^o) \cdot \hat{r}_o\right)\vec{r}\right]dV \times \hat{r}_o \tag{4.24}$$

and write the first of these terms using the vector relation $\vec{A} = \sum(\vec{A} \cdot \hat{x}_i)\hat{x}_i = \sum(\vec{A} \cdot \nabla x_i)\hat{x}_i$, then integrate the first term by parts, and use the continuity equation to get (**see Discussion 4.1**)

$$\int [(\hat{r}_o \cdot \vec{r})\ddot{\vec{\mathcal{J}}}(\vec{r}, \tau^o)]dV = \int \left(\frac{\partial^3}{\partial t^3}[(\hat{r}_o \cdot \vec{r})\rho(\vec{r}, \tau^o)]\vec{r} - [\ddot{\vec{\mathcal{J}}}(\vec{r}, \tau^o) \cdot \hat{r}_o]\vec{r}\right)dV$$

then interchanging the order of integration and time differentiation and noting the cancellation of $(\ddot{\vec{\mathcal{J}}} \cdot \hat{r}_o)\vec{r}$ terms, Eq. 4.24 can be written,

$$\vec{B}(\vec{r}_o, t)^{sym}_{rad} = \frac{1}{8\pi\varepsilon_o c^4 r_o}\frac{\partial^3}{\partial t^3}\left[\int [(\hat{r}_o \cdot \vec{r})\rho(\vec{r}, \tau^o)\vec{r}]\right]dV \times \hat{r}_o. \quad (4.25)$$

Here, again we follow the canonical development by subtracting a term that contributes nothing to the magnetic field ($\int r^2\rho(\vec{r}, \tau^o)\hat{r}_o dV$) from the integrand in Eq. 4.25 (since $\hat{r}_o \times \hat{r}_o = 0$)

$$\vec{B}(\vec{r}_o, t)^{sym}_{rad} = \frac{1}{24\pi\varepsilon_o c^4 r_o}\frac{\partial^3}{\partial t^3}\left[\int [3\vec{r}(\vec{r} \cdot \hat{r}_o) - r^2\hat{r}_o]\rho(\vec{r}, \tau^o)\right]dV \times \hat{r}_o$$

Noting the definition of the elements, $Q_{ij} = \int [3r_i r_j - r^2\delta_{ij}]\rho(\vec{r})dV$, of an electric quadrupole moment tensor $\overleftrightarrow{Q}$, for a charge distribution $\rho(\vec{r})$, we identify the electric quadrupole contribution to $\vec{B}_{rad}$ (**see Discussion 4.2**)

$$\vec{B}(\vec{r}_o, t)^{eq}_{rad} = \frac{\dddot{\vec{Q}}(\hat{r}_o, \tau^o) \times \hat{r}_o}{24\pi\varepsilon_o c^4 r_o}$$

where $\vec{Q}(\hat{r}_o, \tau^o) = \overleftrightarrow{Q} \cdot \hat{r}_o$ and we note that, like $\vec{p}$ and $\vec{m}$, $\vec{Q}$ depends on the retarded time but unlike for $\vec{p}$ and $\vec{m}$, $\vec{Q}$ also depends, in magnitude and direction, on the direction of observation. This is because this term additionally takes into account the non-zero extension of the source in the direction of the observer, which varies with the direction of the observer. The electric quadrupole contribution to $\vec{E}_{rad}$ is again obtained using Eq. 4.4

$$\vec{E}(\vec{r}_o, t)^{eq}_{rad} = \frac{\left(\dddot{\vec{Q}}(\hat{r}_o, \tau^o) \times \hat{r}_o\right) \times \hat{r}_o}{24\pi\varepsilon_o c^3 r_o}$$

4.2 Multipole radiation from the scalar expansion of the vector potential

Because we considered just the radiation part of Jefimenko's equations in the previous treatment, the fields arising from the zeroth and first order moments of the charge distribution are restricted to the radiation zone in which $kr_o \gg 1$. The reason for using Jefimenko's formulation was the relative transparency and directness of the derivation of the radiation fields. There is another, canonical, approach to deriving these multipole fields, one that yields expressions for these fields for all three zones, $kr_o \ll 1$, $kr_o \sim 1$, and $kr_o \gg 1$. This approach involves applying the gradient of the retarded potentials at the field point, while carefully accounting for the coupling to the source through the retarded time.

4.2.1 Fields from an electric dipole moment

Equation 2.26 is the single frequency vector potential expanded out to $n = 1$. The zeroth order term in this expansion, given by Eq. 2.28, is the electric dipole term, and is valid as such in all zones. Using $\tau^o = t - r_o/c$, the retarded time of a point source approximation, it can be written as,

$$\left[\vec{A}(\vec{r}_o, t)\right]^{ed}_{zero\ order} = \frac{\mu_o}{4\pi}\frac{e^{ikr_o}}{r_o}(-i\omega)\vec{p}_o e^{-i\omega t} = \frac{\mu_o}{4\pi}\frac{\dot{\vec{p}}(\tau^o)}{r_o} \tag{4.26}$$

with $dt = d\tau^o$ and $\vec{p}(t) = \vec{p}_o e^{-i\omega t} = \left(\int \vec{r}\rho(\vec{r})\, dV\right) e^{-i\omega t}$. The $\vec{B}$ field is the curl of the potential,

$$\vec{B}(\vec{r}_o, t)^{ed} = \nabla \times \vec{A}(\vec{r}_o, t)^{ed} = \frac{\mu_o}{4\pi}\nabla \times \frac{\dot{\vec{p}}(\tau^o)}{r_o}$$

and because $\vec{A}$ depends on r_o through retarded time τ^o, a careful differentiation yields (**see Discussion 4.3**)

$$\vec{B}(\vec{r}_o, t)^{ed} = -\frac{\mu_o}{4\pi r_o}\hat{r}_o \times \left[\frac{\dot{\vec{p}}(\tau^o)}{r_o} + \frac{\ddot{\vec{p}}(\tau^o)}{c}\right] \tag{4.27}$$

$$= i\omega\frac{\mu_o}{4\pi}\frac{e^{ikr_o}}{r_o}\left(\frac{1}{r_o} - ik\right)\left(\hat{r}_o \times \vec{p}(t)\right) \tag{4.28}$$

that, using Eq. 4.26, can also be expressed in terms of the vector potential as

$$\vec{B}(\vec{r}_o, t)^{ed} = \left(\frac{\vec{A}(\vec{r}_o, t)^{ed}}{r_o} + \frac{\dot{\vec{A}}(\vec{r}_o, t)^{ed}}{c}\right) \times \hat{r}_o \tag{4.29}$$

$$= \left(\frac{\vec{A}(\vec{r}_o, t)^{ed}}{kr_o} + \frac{\dot{\vec{A}}(\vec{r}_o, t)^{ed}}{\omega}\right) \times k\hat{r}_o \tag{4.30}$$

The $\vec{B}$ field from the electric dipole moment is then approximated in the "near" and "radiation" zones as:

$$\vec{B}(\vec{r}_o, t)^{ed}_{near} = \vec{A}(\vec{r}_o, t)^{ed} \times \frac{\hat{r}_o}{r_o}$$
$$= ick \frac{\mu_o}{4\pi} \frac{e^{ikr_o}}{r_o^2} \left(\hat{r}_o \times \vec{p}(t)\right) \quad (kr_o \ll 1) \tag{4.31}$$
$$\vec{B}(\vec{r}_o, t)^{ed}_{rad} = \dot{\vec{A}}(\vec{r}_o, t)^{ed} \times \frac{\hat{r}_o}{c}$$
$$= ck^2 \frac{\mu_o}{4\pi} \frac{e^{ikr_o}}{r_o} \left(\hat{r}_o \times \vec{p}(t)\right) \quad (kr_o \gg 1) \tag{4.32}$$

Having found the electric dipole expressions for $\vec{B}$ (Eqs. 4.27–4.30), the $\vec{E}$ field is then determined directly from Ampere's law for harmonically varying fields,

$$\vec{E} = \frac{ic^2}{\omega} \nabla \times \vec{B} \tag{4.33}$$

Using Eq. 4.30, we can write Eq. 4.33 as

$$\vec{E}^{ed} = \frac{ic^2}{\omega} \nabla \times \left[\left(\frac{\vec{A}^{ed}}{kr_o} + \frac{\dot{\vec{A}}^{ed}}{\omega} \right) \times k\hat{r}_o \right] \tag{4.34}$$

Alternatively, using Eq. 4.27, we can write Eq. 4.33 as

$$\vec{E}^{ed} = \frac{ik}{4\pi\varepsilon_o} \nabla \times \left(\frac{e^{ikr_o}}{r_o} (1 - \frac{1}{ikr_o}) \right) \left(\hat{r}_o \times \vec{p}(t)\right) \tag{4.35}$$

which expands to (**see Discussion 4.4**)

$$\vec{E}^{ed} = \frac{e^{ikr_o}}{4\pi\varepsilon_o r_o} \left[\left(\frac{1}{r_o^2} - \frac{ik}{r_o} \right) \left(3\hat{r}_o(\hat{r}_o \cdot \vec{p}) - \vec{p} \right) + k^2 \left((\hat{r}_o \times \vec{p}) \times \hat{r}_o \right) \right] \tag{4.36}$$
$$= \left[\vec{E}(r_o^{-3}) + \vec{E}(r_o^{-2}) + \vec{E}(r_o^{-1}) \right]_{ed} \tag{4.37}$$

where

$$\vec{E}(r_o^{-3}) = \vec{E}^{ed}_{near} = \frac{1}{4\pi\varepsilon_o} \frac{e^{ikr_o}}{r_o^3} \left[3\hat{r}_o(\hat{r}_o \cdot \vec{p}(t)) - \vec{p}(t) \right] (kr_o \ll 1) \tag{4.38}$$
$$\vec{E}(r_o^{-2}) = \frac{-ik}{4\pi\varepsilon_o} \frac{e^{ikr_o}}{r_o^2} \left[3\hat{r}_o(\hat{r}_o \cdot \vec{p}(t)) - \vec{p}(t) \right] (kr_o \sim 1) \tag{4.39}$$
$$\vec{E}(r_o^{-1}) = \vec{E}^{ed}_{rad} = \frac{k^2}{4\pi\varepsilon_o} \frac{e^{ikr_o}}{r_o} \left[(\hat{r}_o \times \vec{p}(t)) \times \hat{r}_o \right] (kr_o \gg 1) \tag{4.40}$$

where a comparison of the last term, the $\vec{E}$ field radiation term, with Eq. 4.32, the B field radiation term, shows $\vec{E}^{ed}_{rad} = c\vec{B}^{ed}_{rad} \times \hat{r}_0$. That is, in the radiation zone the electric and magnetic fields are perpendicular.

There are several notable characteristics for these fields of the $n = 0$ vector potential expansion:

(1) The expressions for the electric and magnetic radiation fields (for $kr_o \gg 1$) are identical to those obtained from the Jefimenko approach (e.g., compare Eqs. 4.40 and 4.10 or Eqs. 4.32 and 4.9, using the definition of the retarded time).

(2) In all zones, which is to say, for all values of r_o, $\vec{B}$ is transverse to the direction to the field point.

(3) In general, the spatial dependence of $\vec{E}$ is considerably more complex than that of $\vec{B}$: at the smallest values of kr_o, it starts as the quasi-static classic dipole field, undergoing a $\pi/2$ phase shift as r_o becomes larger, then finally assuming the form of a transverse spherical wave at the largest values of kr_o.

(4) The directions of $\vec{E}$, $\vec{B}$, and $\hat{r}_0$ are mutually orthogonal only when $kr_o \gg 1$ (i.e., in the radiation zone). At such points, the dipole fields are spherical waves with $sin\theta$ amplitude dependence relative to the dipole direction, and a phase factor, e^{ikr_o}, relative to the dipole oscillation.

Figure 4.1 shows a slice of the electric field lines at an instant of time, in the r_o, θ coordinate plane, from an oscillating electric dipole: The regions shown are characterized by the "near" and "intermediate" fields. For larger distances beyond the figure, the electric field becomes entirely transverse with spherical phase fronts. In the near field, the electric fields are nearly the "classic" static fields of an electric dipole oscillating at the frequency of the dipole oscillation, with a very slight phase delay of e^{ikr_o}. In the intermediate region the fields are a mixture of radial and transverse and are seen to be more and more transverse at increasing distances from the dipole.

4.2.2 Fields from magnetic dipole moment

The first order term in the expansion of Eq. 2.26, given in Eq. 2.30, includes both the magnetic dipole and the electric quadrupole terms, as seen explicitly in Eq. 2.37. Considering just the magnetic dipole term of Eq. 2.37, we have:

$$\left[\vec{A}(\vec{r}_o, t)\right]^{md}_{first\ order} = -\frac{\mu_o}{4\pi}\frac{e^{ikr_o}}{r_o}\left(\frac{1}{r_o} - ik\right)\left(\hat{r}_o \times \vec{m}(t)\right) \tag{4.41}$$

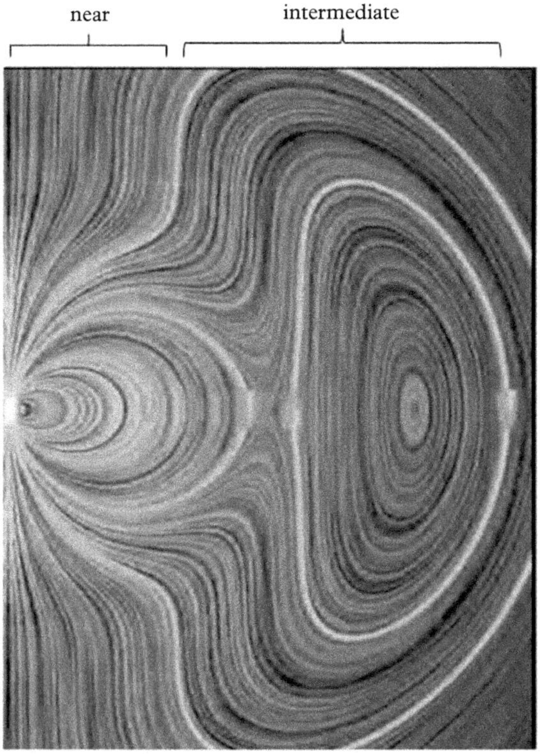

Fig. 4.1 *The Electric fields of an oscillating dipole at time t_o. The electric field near the dipole is a (time delayed) oscillating "static-like" dipole field with both transverse and longitudinal components. For larger r_o, the fields become more and more transverse to the direction from the dipole to the field point. At all locations, the associated B fields are perpendicular to the vector between the dipole location and the field point.*

with the magnetic dipole moment of the current distribution, $\vec{m}(t) = \vec{m}_o e^{-i\omega t}$ and,

$$\vec{m}_o = \frac{1}{2}\int \left(\vec{r} \times \vec{\mathcal{J}}(r)\right) dV$$

Previously we found the expression for the $\vec{B}$ field from the electric dipole expansion, Eq. 4.28,

$$\vec{B}(\vec{r}_o, t)^{ed} = i\omega \frac{\mu_o}{4\pi} \frac{e^{ikr_o}}{r_o} \left(\frac{1}{r_o} - ik\right) \left(\hat{r}_o \times \vec{p}(t)\right) \tag{4.42}$$

Comparison of Eqs. 4.41 and 4.42 shows that $\vec{A}(\vec{r}_o, t)^{md}$ and $\vec{B}(\vec{r}_o t)^{ed}$ have the same radial dependence and orientation relative to their respective dipole moments. Letting that form be represented by

$$f\left(\vec{V}\right) \equiv \frac{\mu_o}{4\pi}\frac{e^{ikr_o}}{r_o}(\frac{1}{r_o} - ik)\left(\hat{r}_o \times \vec{V}(t)\right)$$

so that Eqs. 4.41 and 4.42 are expressed as $\vec{A}(\vec{r}_o, t)^{md} = -f(\vec{m})$ and $\vec{B}(\vec{r}_o, t)^{ed} = i\omega f(\vec{p})$, and noting that

$$\vec{B}(\vec{r}_o t)^{md} = \nabla \times \vec{A}(\vec{r}_o, t)^{md} = -\nabla \times f(\vec{m}) \tag{4.43}$$

$$\frac{1}{c}\vec{E}(\vec{r}_o t)^{ed} = \frac{i}{k}\nabla \times \vec{B}(\vec{r}_o, t)^{ed} = -\nabla \times f(c\vec{p}) \tag{4.44}$$

we see that replacing $\vec{p}$ with $\vec{m}/c$ in $\frac{1}{c}\vec{E}(\vec{r}_o t)^{ed}$ yields $\vec{B}(\vec{r}_o t)^{md}$,

$$\vec{B}(\vec{r}_o t)^{md} = \frac{1}{c}\vec{E}(\vec{r}_o t)^{ed}|_{\vec{p}\to\vec{m}/c} \tag{4.45}$$

which is consistent with our earlier observations in Eq. 4.21. Separately, using Faraday's and Ampere's laws, we find that replacing $\vec{p}$ with $\vec{m}/c$ in $\vec{B}(\vec{r}_o t)^{ed}$ yields $\frac{1}{c}\vec{E}(\vec{r}_o t)^{md}$,[5]

$$\frac{1}{c}\vec{E}(\vec{r}_o t)^{md} = -\vec{B}(\vec{r}_o t)^{ed}|_{\vec{p}\to\vec{m}/c} \tag{4.46}$$

And so, using Eqs. 4.42 and 4.46 for $\frac{1}{c}\vec{E}(\vec{r}_o t)^{md}$ and Eqs. 4.36 and 4.45 for $\vec{B}(\vec{r}_o t)^{md}$, we obtain the electric and magnetic fields arising from the magnetic dipole part of the 1st order term of the vector potential expansion:

$$\vec{E}(\vec{r}_o t)^{md} = -c\vec{B}(\vec{r}_o t)^{ed}|_{\vec{p}\to\vec{m}/c} = \frac{-i\omega}{4\pi\varepsilon_o c^2}\frac{e^{ikr_o}}{r_o}(\frac{1}{r_o} - ik)\left(\hat{r}_o \times \vec{m}(t)\right) \tag{4.47}$$

$$\vec{B}(\vec{r}_o t)^{md} = \frac{e^{ikr_o}\mu_o}{4\pi\varepsilon_o r_o}\left[\left(\frac{1}{r_o^2} - \frac{ik}{r_o}\right)\left(3\hat{r}_o(\hat{r}_o\cdot\vec{m}) - \vec{m}\right) + k^2\left((\hat{r}_o \times \vec{m}) \times \hat{r}_o\right)\right] \tag{4.48}$$

A comparison of the $1/r_o$ radiation field terms in these equations with those found earlier directly from a field expansion, Eqs. 4.19 and 4.18, shows the forms are in agreement. The directions of the electric and magnetic fields relative to their generating dipole are shown in Fig. 4.2. Note that for the case of a charge distribution with an electric dipole moment aligned with its magnetic dipole moment, the total electric field is no longer in the plane of the dipoles. This "depolarization" depends on the details of the relative field strengths.

[5] Again, starting with equations 4.41 and 4.42,

$$\frac{1}{c}\vec{A}^{md} = -\frac{1}{i\omega}\vec{B}^{ed}|_{\vec{p}\to\vec{m}/c}$$

and applying the curl twice to each side of the equation,

$$\begin{aligned}
\frac{1}{c}\nabla \times \vec{A}^{md} &= -\frac{1}{i\omega}\nabla \times \vec{B}^{ed}|_{\vec{p}\to\vec{m}/c} \\
\vec{B}^{md} &= \frac{1}{c}\vec{E}^{ed}|_{\vec{p}\to\vec{m}/c} \\
\nabla \times \vec{B}^{md} &= \frac{1}{c}\nabla \times \vec{E}^{ed}|_{\vec{p}\to\vec{m}/c} \\
-\frac{i\omega}{c^2}\vec{E}^{md} &= \frac{i\omega}{c}\vec{B}^{ed}|_{\vec{p}\to\vec{m}/c} \\
\frac{1}{c}\vec{E}^{md} &= -\vec{B}^{ed}|_{\vec{p}\to\vec{m}/c}.
\end{aligned}$$

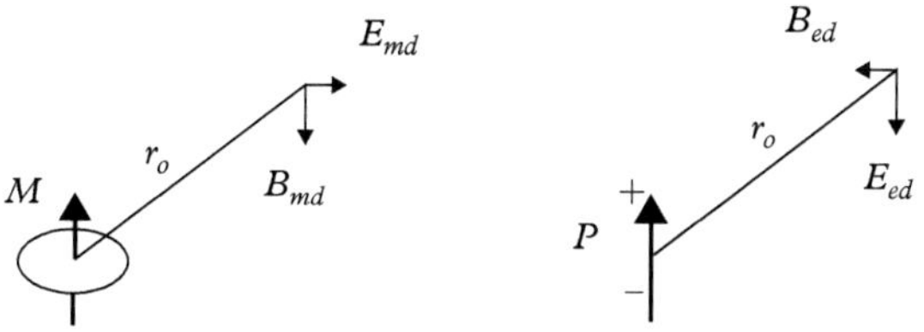

Fig. 4.2 *Electric and magnetic field directions for electric dipole and magnetic dipole.*

Because of the effective replacement $\vec{p} \rightarrow \vec{m}/c$, the magnitudes of the $\vec{E}/c$ and $\vec{B}$ fields from the time varying magnetic dipole are reduced compared to those from the time varying electric dipole by the factor $\frac{m}{cp}$. That is,

$$\frac{\left(E^{md}/c\right)}{B^{ed}} = \frac{B^{md}}{\left(E^{ed}/c\right)} = \frac{(m/c)}{p}$$

and, specifically, in the radiation zone where $E/c = B$, this becomes

$$\frac{B^{md}}{B^{ed}} = \frac{E^{md}}{E^{ed}} = \frac{(m/c)}{p}$$

where m is the strength of the magnetic dipole, and p is the strength of the electric dipole. We expect this factor to be small, if for no other reason than the nature of the charge expansion formalism we used to derive the terms. Physically, we also expect this factor to be small because of the nature of the moments: Consider a charge distribution centered at the origin that has a characteristic dimension D with a charge q that can be distributed over D. Then the size of the electric dipole moment will be on the order of

$$Electric\ Dipole = p \sim qD$$

For a frequency ω, the characteristic current will be on the order $I = q\omega$, and the size of the magnetic dipole moment will be on the order of

$$Magnetic\ Dipole = m \simeq I(area) \simeq q\omega D^2$$

the ratio of magnetic and electric dipole field sizes for a source of size D and radiation wavelength of λ is then:

$$\left(Magnetic/Electric\right)_{dipole fields} = \frac{(m/c)}{p} \simeq \frac{\omega D}{c} \simeq \frac{D}{\lambda} \tag{4.49}$$

So as long as the wavelength of the radiation is much larger than the size of the oscillating charge distribution, the magnetic dipole fields are small compared to the electric dipole fields. This was one of the restrictions we placed upon our analysis from the beginning.

4.2.3 Fields from electric quadrupole moment

The first order term in the expansion of Eq. 2.26, given by Eq. 2.30, includes both the magnetic dipole and electric quadrupole parts, as seen explicitly in Eq. 2.37. Considering just the electric quadrupole part of Eq. 2.37, we have,

$$\vec{A}(\vec{r}_o) = -\frac{\mu_o}{4\pi}\frac{e^{ikr_o}}{r_o}\left(\frac{1}{r_o} - ik\right)\left\{\frac{i\omega}{6}\left[\overleftrightarrow{Q}(t)\cdot\hat{r}_o + \left(\int r^2\rho(t)\,dV\right)\hat{r}_o\right]\right\}$$

which, as before, is valid in all zones and where $\overleftrightarrow{Q}(t) = \overleftrightarrow{Q}(0)e^{-i\omega t}$ and $\rho(t) = \rho(0)e^{-i\omega t}$. However, because of the complexity of the formalism for anything other than the radiation zone, we now consider only radiation fields. The simplified equation for the electric quadrupole contribution in the radiation zone is given by the first term of Eq. 2.38.

$$\left[\vec{A}(r_o,t)\right]^{eq}_{n=1} = -\frac{\mu_o\omega k}{24\pi}\frac{e^{ikr_o}}{r_o}\overleftrightarrow{Q}\cdot\hat{r}_o e^{-i\omega t}$$

where the expressions for the $\vec{E}$ and $\vec{B}$ fields are simply given by:

$$\vec{B}^{eq} = \nabla\times\vec{A}^{eq} = ik\left(\hat{r}_o\times\vec{A}^{eq}\right)$$
$$\vec{E}^{eq} = \frac{ic^2}{\omega}\nabla\times\vec{B}^{eq} = -i\omega\hat{r}_o\times\left(\hat{r}_o\times\vec{A}^{eq}\right)$$

then we have explicitly:

$$\vec{B}^{eq} = \frac{-i\omega^3}{24\pi\varepsilon_o c^4}\frac{e^{ikr_o}}{r_o}(\hat{r}_o\times\overleftrightarrow{Q}\cdot\hat{r}_o) \quad kr_o \gg 1 \tag{4.50}$$

$$\vec{E}^{eq} = \frac{i\omega^3}{24\pi\varepsilon_o c^3}\frac{e^{ikr_o}}{r_o}\hat{r}_o\times(\hat{r}_o\times\overleftrightarrow{Q}\cdot\hat{r}_o) \quad kr_o \gg 1 \tag{4.51}$$

where the components of the electric quadrupole moment tensor $\overleftrightarrow{Q}$ are given by $Q_{ij} = \int(3r_ir_j - \delta_{ij}r^2)\rho(r)dV$. Comparison of the form of the fields from the dipole expansions (either electric or magnetic) with the form of the fields from the electric quadrupole shows that while the spatial dependence of the radiation from dipoles depends only on the sine of the angle between the moment and the direction to the observer, the quadrupole fields have an additional angular dependence because the source vector ($\overleftrightarrow{Q} \cdot \hat{r}_o$) is not in a fixed direction, but rather depends upon the direction to the observer. This extra angular dependence makes the general forms of the distribution of $[\vec{E}]_{eq}$ and $[\vec{B}]_{eq}$ quite complicated and not readily useful. Here, we follow the literature and restrict ourselves to calculating the power distribution, and the total power, first in general terms, and for several symmetric charge configurations.

4.3 Power radiated in terms of multipole moments of the source

4.3.1 Power radiated by electric dipole moment

The radiated energy per unit area per unit time is given by the time average of the Poynting vector:

$$\vec{S} = Real\left[\frac{1}{\mu_o}\left(\vec{E} \times \vec{B}^*\right)\right] \tag{4.52}$$

and the associated radiant intensity is given by

$$\left\langle \frac{d|\vec{P}|}{d\Omega} \right\rangle = Real\left[\frac{\varepsilon_o c^2}{2}\left(\vec{E} \times \vec{B}^*\right) \cdot \hat{r}_o r_o^2\right] \tag{4.53}$$

Using E and B given by Eqs. 4.9 and 4.10, we obtain the radiant intensity of the electric dipole in the radiation zone:

$$\left\langle \frac{d|\vec{P}|}{d\Omega} \right\rangle_{ed}^{rad} = \omega^4 \frac{|\vec{p}|^2 sin^2\theta}{32\pi^2\varepsilon_o c^3} \tag{4.54}$$

where θ is the angle between the dipole direction and $\hat{r}_o$. The total power radiated is obtained by integrating the expression in Eq. 4.54 over the total solid angle

$$\left\langle |\vec{P}| \right\rangle = \int_0^{2\pi} \int_o^{\pi} \left\langle \frac{d|\vec{P}|}{d\Omega} \right\rangle sin\theta \, d\theta \, d\varphi. \tag{4.55}$$

The ω^4 dependence is characteristic of dipole radiation. The radiant intensity and total power radiated from an electric dipole (zeroth order in the expansion of Eq. 2.25) are then

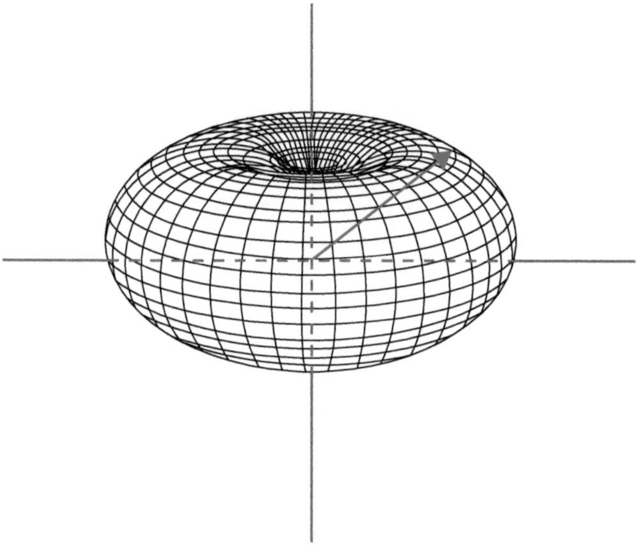

Fig. 4.3 *A schematic of the radiation intensity $[\mathcal{J}/(sec - m^2)]$ for the fields in the radiation zone of an electric dipole as a function of angle (at a fixed radius from the dipole). The red arrow length is proportional to the intensity at that polar angle. The power density is maximum in directions perpendicular to the dipole axis, and there is no energy radiated along the axis of the dipole.*

$$\left\langle \frac{d|\vec{P}|}{d\Omega} \right\rangle_{ed} = \omega^4 \frac{|\vec{p}|^2 sin^2\theta}{32\pi^2 \varepsilon_o c^3} \tag{4.56}$$

$$\left\langle |\vec{P}| \right\rangle_{ed} = \omega^4 \frac{|\vec{p}|^2}{12\pi \varepsilon_o c^3} \tag{4.57}$$

The radiated power density for the radiation from an electric dipole in the radiation zone is shown in Fig. 4.3. The distance from the origin (red arrow) is proportional to the power density in that polar angle. The power density is maximum in the direction orthogonal to the dipole and there is no energy radiated along the axis of the dipole.

4.3.2 Power radiated by magnetic dipole moment

The fields from the magnetic dipole in the radiation zone are given by the $1/r_o$ terms in Eqs. 4.47 and 4.48. Following the discussion of the power radiated for an electric dipole (Eqs. 4.52–4.57), we can similarly find the power radiated for a magnetic dipole by using the expressions for the $1/r_o$ radiation zone terms in Eqs. 4.47 and 4.48:

$$\begin{aligned}\left\langle \frac{d|\vec{P}|}{d\Omega} \right\rangle_{md} &= Real\left[\frac{1}{2\mu_o}\left(\vec{E}^{md} \times \vec{B}^{md*}\right) \cdot \hat{r}_o r_o^2\right] \\ &= \omega^4 \frac{\mu_o sin^2\theta}{32\pi^2 c}\left(\frac{m}{c}\right)^2\end{aligned}$$

where the angle θ is now measured relative to the direction of the magnetic dipole $\vec{M}$. The total power is the integral over the total solid angle:

$$\left\langle |\vec{P}| \right\rangle_{md} = \omega^4 \frac{\mu_o}{12\pi c} \left(\frac{m}{c} \right)^2$$

As expected, the radiated power results are the same as for the electrical dipole case, only with

$$p \rightarrow \left(\frac{m}{c} \right)$$

The total power radiated by a magnetic dipole will be reduced relative to an electric dipole moment by an amount approximately the square of the field ratio given in Eq. 4.49 so that if the characteristic size of the charge distribution is D

$$\textit{Total Radiated Power Ratio} \left(\frac{\textit{Magnetic Dipole}}{\textit{Electric Dipole}} \right) \sim (kD)^2 \qquad (4.58)$$

This result strongly suggests that in virtually all instances of radiation from a charge distribution, the power radiated will be completely dominated by the electric dipole fields: Only in the case where the electric dipole moment of the charge distribution is zero due to spatial charge distribution symmetries will the magnetic dipole fields be important.

4.3.3 Power radiated by electric quadrupole moment

The energy radiated per unit time per unit solid angle by the electric quadrupole term of the charge distribution is given by the usual expression:

$$\left\langle \frac{d|\vec{P}|}{d\Omega} \right\rangle_{eq} = Real \left[\frac{1}{2\mu_o} \left(\vec{E}^{eq}_{eq} \times \vec{B}^{eq*} \right) \cdot \hat{r}_o r_o^2 \right]$$

$$= \frac{\omega^6}{1152\pi^2 \varepsilon_o c^5} \left[(\hat{r}_o \times \overleftrightarrow{Q} \cdot \hat{r}_o) \right.$$

$$\left. \times (\hat{r}_o \times (\hat{r}_o \times \overleftrightarrow{Q} \cdot \hat{r}_o) \right] \cdot \hat{r}_o \qquad (4.59)$$

which can be written

$$\left\langle \frac{d|\vec{P}|}{d\Omega} \right\rangle_{eq} = \frac{\omega^6}{1152\pi^2 \varepsilon_o c^5} \left[(\overleftrightarrow{Q} \cdot \hat{r}_o) \cdot (\overleftrightarrow{Q} \cdot \hat{r}_o) - (\hat{r}_o \cdot \overleftrightarrow{Q} \cdot \hat{r}_o)^2 \right] \tag{4.60}$$

and, written in terms of the components, is:

$$\left\langle \frac{d|\vec{P}|}{d\Omega} \right\rangle_{eq} = \frac{\omega^6}{1152\pi^2 \varepsilon_o c^5 r_o^2} \left[\sum_{ijk} (Q_{ij} r_{oj})(Q_{ik} r_{ok}) - \frac{1}{r_o^2} \sum_{ijkl} (r_{oi} Q_{ij} r_{oj})(r_{ok} Q_{kl} r_{ol}) \right] \tag{4.61}$$

Integration of Eq. 4.61 over the solid angle gives the total radiated power of the electric quadrupole distribution of a charge configuration (**see Discussion 4.5**)

$$\left\langle |\vec{P}| \right\rangle_{eq} = \frac{\omega^6}{1440\pi \varepsilon_o c^5} \sum_{ij} |Q_{ij}|^2 \tag{4.62}$$

In general, the expression given in Eq. 4.61 is complicated and seldom used; however, a common simplification of axial symmetry about the z-axis (cylindrical symmetry) is often encountered. For this symmetry all $Q_{\alpha\beta} = 0$ for $\alpha \neq \beta$ and we adopt the convention that the "quadrupole moment of the cylindrically symmetric charge distribution" Q_o is given by

$$Q_o \equiv Q_{33} = \int (3z^2 - r^2)\rho(r)\, dV$$

Then, because of the cylindrical symmetry and the fact that the trace of the quadrupole tensor is zero,

$$Q_{11} = Q_{22} = -\frac{Q_{33}}{2} \tag{4.63}$$

and Eq. 4.61 reduces for cylindrical symmetry to[6]

$$\left\langle \frac{d|\vec{P}|}{d\Omega} \right\rangle_{eq} = \frac{\omega^6 Q_o^2 \cos^2\theta}{512\pi^2 \varepsilon_o c^5} \left[1 - \cos^2\theta \right] \tag{4.64}$$

$$= \frac{\omega^6 Q_o^2}{512\pi^2 \varepsilon_o c^5} \cos^2\theta \sin^2\theta \tag{4.65}$$

Using these results and integrating over the solid angle, we obtain the power radiated by a cylindrically symmetric electric quadrupole moment:

[6] Starting with Eq. 4.61 setting all off-diagonal terms to zero, and using Eq. 4.63 we have

$$\left\langle \frac{d|\vec{P}|}{d\Omega} \right\rangle_{eq} = \frac{\omega^6}{1152\pi^2 \varepsilon_o c^5 r_o^2} (\frac{1}{4}) \times \sum_i \left[Q_o^2 (r_{oi}^2 + 3r_{o3}^2) - \frac{1}{r_o^2} Q_o^2 (r_{oi}^2 - 3r_o^2)^2 \right]$$

and taking the angle θ to be the angle between direction of the $z-component$ (Q_o) and the observation direction:

$$(r_o^2 + 3z^2) = r_o^2 (1 + 3\cos^2\theta)$$
$$(r_o^2 - 3z^2)^2 = r_o^4 (1 - 3\cos^2\theta)^2$$

we recover Eq. 4.65

$$\left\langle |\vec{P}| \right\rangle_{eq} = \frac{\omega^6 Q_o^2}{960\pi\,\varepsilon_o c^5} \tag{4.66}$$

with

$$Q_o = \int (3z^2 - r^2)\rho(r)\,dV$$

The ω^6 dependence is characteristic of quadrupole radiation. We expect the strength of the quadrupole radiation to be similar to the strength of the magnetic dipole and much less than the power radiated by the electric dipole term, because of the relative order of the expansion of the vector potential. The relative powers are (using Eqs. 4.57 and 4.65)

$$\textit{Relative Power}\left(\frac{\textit{Electric Quad}}{\textit{Electric Dipole}}\right) \sim \frac{\omega^2 Q_o^2}{c^2 p^2}$$

For a characteristic charge distribution size of D, the ratio of Q_o to p scales as $\frac{(q_o D^2)}{(q_o D)}$, so the relative powers scale as:

$$\textit{Relative Power} \sim (kD)^2$$

which is the same as the ratio of the magnetic dipole to electric dipole radiated powers (Eq. 4.58). For quantum charge distributions that are characterized either by extremely energetic transitions (usually in gamma-ray emissions from nuclei) and/or when the dipole moment radiation at a given frequency is absent due to symmetry reasons, this power ratio is no longer negligible and the quadrupole radiation can become significant.

Exercises

(4.1) In the text, the fields in the radiation zone due to an electric dipole were obtained in the forms of Eqs. 4.9 and 4.10,

$$\vec{B}^{ed}_{rad} = \frac{\ddot{\vec{p}}\left(\tau^0\right) \times \hat{r}_o}{4\pi\,\varepsilon_o c^3 r_o} \quad \text{and} \quad \vec{E}^{ed}_{rad} = \frac{\left(\ddot{\vec{p}}\left(\tau^0\right) \times \hat{r}_o\right) \times \hat{r}_o}{4\pi\,\varepsilon_o c^2 r_o}$$

respectively, where τ^0 is the retarded time for a source approximated as a point with no extent. Show that these electric dipole fields can be written in terms of the associated current

$$\vec{B}^{ed}_{rad} = \frac{\dot{I}\left(\tau^0\right) d\vec{s} \times \hat{r}_o}{4\pi\varepsilon_o c^3 r_o} \quad \text{and}$$

$$\vec{E}^{ed}_{rad} = \frac{\left(\dot{I}\left(\tau^0\right) d\vec{s} \times \hat{r}_o\right) \times \hat{r}_o}{4\pi\varepsilon_o c^2 r_o}$$

(4.2) Find the radiation fields, $\vec{E}_{rad}$ and $\vec{B}_{rad}$, for an electric dipole from Eqs. 4.9 and 4.10, assuming $\vec{p} = p_o \cos\left(\omega\tau^0\right)\hat{z}$. Next find $\vec{S}$ and show that this result for $\vec{S}$ can be put in the form of the more general Eq. 4.13.

(4.3) Show that Eq. 4.12 can be obtained from Eq. 4.11 through twice applying the triple cross product vector formula, $(\vec{A} \times \vec{B}) \times \vec{C} = (\vec{C} \cdot \vec{A})\vec{B} - (\vec{C} \cdot \vec{B})\vec{A}$

(4.4) Using an example of the duality property of electromagnetic fields in which,

$$p_o \rightarrow \frac{m_o}{c} \quad \vec{E}^{ed}_{rad} \rightarrow c\vec{B}^{md}_{rad} \quad \vec{B}^{ed}_{rad} \rightarrow -\frac{\vec{E}^{md}_{rad}}{c}$$

confirm the form of the magnetic dipole fields in the radiation zone given by Eqs. 4.18 and 4.19,

$$\vec{B}^{md}_{rad} = \frac{1}{4\pi\varepsilon_o c^4 r_o}\left(\ddot{\vec{m}}\left(\tau^0\right) \times \hat{r}_o\right) \times \hat{r}_o$$

$$\vec{E}^{md}_{rad} = -\frac{1}{4\pi\varepsilon_o c^3 r_o}\ddot{\vec{m}}\left(\tau^0\right) \times \hat{r}_o$$

(4.5) The quadrupole moment tensor $Q_{ab} = \int [3r_a r_b - r^2\delta_{ab}]\rho(\vec{r})dV$ in full matrix form is

$$\overleftrightarrow{Q} = \int \begin{pmatrix} 3x^2 - r^2 & 3xy & 3xz \\ 3yx & 3y^2 - r^2 & 3yz \\ 3zx & 3zy & 3z^2 - r^2 \end{pmatrix} \rho(\vec{r})\, dV$$

Show that its inner product with the unit vector in the observation direction, $\hat{r}_o$, can be expressed as the vector

$$\vec{Q}(\hat{r}_o) = \overleftrightarrow{Q} \cdot \hat{r}_o = \int [3\vec{r}(\vec{r} \cdot \hat{r}_o) - r^2\hat{r}_o]\rho(\vec{r})dV$$

(4.6) Find the radiation fields, $\vec{E}_{rad}$ and $\vec{B}_{rad}$, for an electric quadrupole from the two unlabeled equations before Eq. 4.26, with the quadrupole moment tensor of a charge distribution defined by $Q_{ab} = \int [3r_a r_b - r^2\delta_{ab}]\rho(\vec{r})dV$ and assuming a harmonic charge distribution, $\rho\left(\tau^0\right) = \rho_o \cos\left(\omega\tau^0\right)$.

(4.7) a) Starting with Eq. 4.27, the general magnetic field arising from the electric dipole moment of a distribution, and assuming a harmonically varying electric dipole moment along the z-axis: $\vec{p} = p_o \exp(-i\omega\tau^0)\hat{z}$, where p_o is assumed to be real, re-express this field (in spherical coordinates) as

$$\vec{B}(\vec{r}_o, t)^{ed} = -\frac{\mu_o\omega k^2 p_o}{4\pi}\left[\frac{1}{kr_o} + \frac{i}{(kr_o)^2}\right] \times \sin\theta \exp\left(-i\omega\tau^0\right)\hat{\phi}$$

b) Using Maxwell's equation, $\nabla \times \vec{B} = \mu_o\varepsilon_o\frac{\partial\vec{E}}{\partial t}$, show that the associated general electric field, $\vec{E}^{ed}$, is given by

$$\vec{E}(\vec{r}_o, t)^{ed} = -\frac{k^3 p_o}{4\pi\varepsilon_o}\left\{\left[\frac{2i}{(kr_o)^2} - \frac{2}{(kr_o)^3}\right]\cos\theta\hat{r}_o + \left[\frac{1}{kr_o} + \frac{i}{(kr_o)^2} - \frac{1}{(kr_o)^3}\right]\sin\theta\hat{\theta}\right\}\exp\left(-i\omega\tau^0\right)$$

c) Using parts (a) and (b), find $\vec{S}$ and show that, in general, it has oscillatory components along $\hat{r}_o$ and $\hat{\theta}$.

d) From the general equations of parts (a) and (b), show that the fields in the radiation zone ($kr_o \gg 1$) due to a harmonically varying electric dipole moment can be written as:

$$\vec{B}(\vec{r}_o, t)^{ed}_{rad} = -\frac{\mu_o\omega k p_o}{4\pi r_o}\sin\theta \exp\left(-i\omega\tau^0\right)\hat{\phi}$$

$$\vec{E}(\vec{r}_o, t)^{ed}_{rad} = -\frac{k^2 p_o}{4\pi\varepsilon_o r_o}\sin\theta \exp\left(-i\omega\tau^0\right)\hat{\theta}$$

(4.8) Use the electric field given in Problem 4.7, part (d) and superposition to find the fields in the radiation zone arising from two oppositely directed electric dipoles which have amplitudes $p_o = \mp q_o b$ and which are located on the z axis at $z = \pm b/2$, respectively. This is a good approximation to the fields in the radiation zone of a linear electric quadrupole.

(4.9) a) Using the substitutions in Problem 4.4 (expressions of the duality property of electromagnetic fields) show that the fields in the radiation zone ($kr_o \gg 1$) due to a magnetic dipole moment harmonically varying along the z-axis can be written (in spherical coordinates) as

$$\vec{B}(\vec{r}_o,t)^{md}_{rad} = -\frac{k^2 m_o}{4\pi\varepsilon_o c^2 r_o}\sin\theta\exp\left(-i\omega\tau^o\right)\hat{\theta}$$
$$\vec{E}(\vec{r}_o,t)^{md}_{rad} = -\frac{\mu_o\omega k m_o}{4\pi r_o}\sin\theta\exp\left(-i\omega\tau^o\right)\hat{\phi}$$

b) Confirm that these results are consistent with the radiation terms of Eqs. 4.47 and 4.48.

(4.10) a) The electric and magnetic fields arising from the magnetic dipole part of the vector potential expansion are given by Eqs. 4.47 and 4.48. Express the radiation parts of these equations in terms of the vector potential given in Eq. 4.41.
b) The radiation part of the electric and magnetic fields arising from the electric dipole part of the vector potential expansion are given, respectively, by Eqs. 4.40 and 4.32. Express Eq. 4.40 in terms of the undifferentiated vector potential using Eq. 4.32.
c) Using the results of parts (a) and (b), along with the equations for the electric and magnetic fields arising from the electric quadrupole part of the vector potential expansion (two Eqs. just before Eq. 4.50), show that all the fields in the radiation zone can be expressed in terms of their associated vector potential as,

$$\vec{E}_{rad}(\vec{r}_o,t) = \left(\dot{\vec{A}}(\vec{r}_o,t)\times\hat{r}_o\right)\times\hat{r}_o$$
$$\vec{B}_{rad}(\vec{r}_o,t) = \dot{\vec{A}}(\vec{r}_o,t)\times\frac{\hat{r}_o}{c}$$

regardless of whether they result from electric dipole, magnetic dipole or electric quadrupole moments.

(4.11) The general fields due to an harmonically varying electric dipole moment are given in Problem 4.7, part (d). Compute the associated:
a) Instantaneous Poynting vector (the radiated power per unit area), $\vec{S} = \frac{1}{\mu_o}\vec{E}\times\vec{B}$
b) Time averaged Poynting vector, $\left\langle\vec{S}\right\rangle = \frac{1}{2\mu_o}Re\left(\vec{E}\times\vec{B}^*\right)$
c) Time averaged angular power (the radiated power per unit solid angle), $\left\langle\frac{dP}{d\Omega}\right\rangle = \left\langle\vec{S}\right\rangle\cdot r_o^2\hat{r}_o = \frac{1}{2\mu_o}Re\left(\vec{E}\times\vec{B}^*\right)\cdot r_o^2\hat{r}_o$
d) Time averaged total power, $\langle P\rangle = \iint\left\langle\frac{dP}{d\Omega}\right\rangle\sin\theta\, d\theta\, d\phi$

(4.12) a) Using the general fields from an harmonically oscillating electric dipole (Problem 4.7, part d), calculate the general $\vec{E}$ and $\vec{B}$ fields arising from a permanent electric dipole rotating in the x–y plane at constant angular frequency, ω. (Note that this can be seen as the sum of two orthogonal oscillating electric dipoles (along x and along y) that are 90^oout of phase.
b) As in Problem 4.11, calculate from the general $\vec{E}$ and $\vec{B}$ fields: $\vec{S}$, $\left\langle\vec{S}\right\rangle$ $\left\langle\frac{dP}{d\Omega}\right\rangle$, and $\langle P\rangle$.

(4.13) The general fields due to an harmonically varying magnetic dipole moment are given in Problem 4.9, part (a). Repeat the computations done in Problem 4.11 for these general magnetic dipole fields.

4.4 Discussions

Discussion 4.1

The first integral in Eq. 4.24 can be written

$$\int [(\hat{r}_o \cdot \vec{r})\overset{\dots}{\vec{\mathcal{J}}}(\vec{r},\tau^o)]dV \tag{4.67}$$

but, as before, $\overset{\dots}{\vec{\mathcal{J}}}(\vec{r},\tau^o) = \{\overset{\dots}{\vec{\mathcal{J}}}\}$ can be expressed as

$$\{\overset{\dots}{\vec{\mathcal{J}}}\} = \sum_i \hat{x}_i\{\overset{\dots}{\vec{\mathcal{J}}} \cdot \hat{x}_i\} = \sum_i \hat{x}_i\{\overset{\dots}{\vec{\mathcal{J}}} \cdot \nabla x_i\}$$

substituting this into the integral and integrating by parts

$$\sum_i \hat{x}_i \int \{(\hat{r}_o \cdot \vec{r})\overset{\dots}{\vec{\mathcal{J}}} \cdot \nabla x_i\}dV = \sum_i \hat{x}_i \int \left(\{\nabla \cdot x_i[(\hat{r}_o \cdot \vec{r})\overset{\dots}{\vec{\mathcal{J}}}]\} - \{x_i \nabla \cdot [(\hat{r}_o \cdot \vec{r})\overset{\dots}{\vec{\mathcal{J}}}]\}\right) dV$$

Using Gauss' theorem on the first term on the right

$$\sum_i \hat{x}_i \int \{\nabla \cdot [x_i(\hat{r}_o \cdot \vec{r})\overset{\dots}{\vec{\mathcal{J}}}]\}dV = \sum_i \hat{x}_i \oint \{x_i(\hat{r}_o \cdot \vec{r})\overset{\dots}{\vec{\mathcal{J}}} \cdot \hat{r}\}dS = 0$$

so this term vanishes because the surface of the volume is beyond the bounded distribution. In the remaining term, we expand the divergence

$$\nabla \cdot [(\hat{r}_o \cdot \vec{r})\overset{\dots}{\vec{\mathcal{J}}}] = (\hat{r}_o \cdot \vec{r})\nabla \cdot \overset{\dots}{\vec{\mathcal{J}}} + \overset{\dots}{\vec{\mathcal{J}}} \cdot \nabla(\hat{r}_o \cdot \vec{r})$$

and thus the top integral (Eq. 4.67) becomes

$$\begin{aligned}\int [(\hat{r}_o \cdot \vec{r})\overset{\dots}{\vec{\mathcal{J}}}(\vec{r},\tau^o)]dV &= -\sum_i \hat{x}_i \int \left(\{x_i(\hat{r}_o \cdot \vec{r})\nabla \cdot \overset{\dots}{\vec{\mathcal{J}}}\} + \{x_i\overset{\dots}{\vec{\mathcal{J}}} \cdot \nabla(\hat{r}_o \cdot \vec{r})\}\right) dV \\ &= -\int \left(\{\vec{r}(\hat{r}_o \cdot \vec{r})\nabla \cdot \overset{\dots}{\vec{\mathcal{J}}}\} + \{\vec{r}\overset{\dots}{\vec{\mathcal{J}}} \cdot \nabla(\hat{r}_o \cdot \vec{r})\}\right) dV\end{aligned}$$

Using the second time derivative of the continuity equation, $\nabla \cdot \overset{\dots}{\vec{\mathcal{J}}} = -\dddot{\rho}$, for the first term on the right and $\nabla(\hat{r}_o \cdot \vec{r}) = \hat{r}_o$ for the second term

$$\int [(\hat{r}_o \cdot \vec{r})\overset{\dots}{\vec{\mathcal{J}}}(\vec{r},\tau^o)]dV = \int \left(\{(\hat{r}_o \cdot \vec{r})\dddot{\rho}\,\vec{r}\} - \{(\overset{\dots}{\vec{\mathcal{J}}} \cdot \hat{r}_o)\vec{r}\}\right) dV$$

Discussion 4.2

The quadrupole moment tensor $Q_{ij} = \int [3r_i r_j - r^2\delta_{ij}]\rho(\vec{r})dV$ in full matrix form is

$$\overleftrightarrow{Q} = \int \begin{pmatrix} 3x^2 - r^2 & 3xy & 3xz \\ 3yx & 3y^2 - r^2 & 3yz \\ 3zx & 3zy & 3z^2 - r^2 \end{pmatrix} \rho(\vec{r})\, dV$$

and its inner product with the unit vector in the observation direction, $\hat{r}_o$, can thus be expressed as the vector

$$\begin{aligned} \vec{Q}(\hat{r}_o) = \overleftrightarrow{Q} \cdot \hat{r}_o &= \int \begin{pmatrix} 3x^2 - r^2 & 3xy & 3xz \\ 3yx & 3y^2 - r^2 & 3yz \\ 3zx & 3zy & 3z^2 - r^2 \end{pmatrix} \begin{pmatrix} x_o \\ y_o \\ z_o \end{pmatrix} \frac{\rho(\vec{r})}{r_o} dV \\ &= \int \left[\begin{pmatrix} 3x(xx_o) + 3x(yy_o) + 3x(zz_o) \\ 3y(xx_o) + 3y(yy_o) + 3y(zz_o) \\ 3z(xx_o) + 3z(yy_o) + 3z(zz_o) \end{pmatrix} - r^2 \begin{pmatrix} x_o \\ y_o \\ z_o \end{pmatrix} \right] \frac{\rho(\vec{r})}{r_o} dV \\ &= \int [3\vec{r}(\vec{r}\cdot\hat{r}_o) - r^2\hat{r}_o]\rho(\vec{r})dV \end{aligned}$$

Discussion 4.3

Using the vector identity $\nabla \times f\vec{V} = f\left(\nabla \times \vec{V}\right) - \left(\vec{V} \times \nabla f\right)$,

$$\begin{aligned} \nabla \times \frac{\dot{\vec{p}}(\tau^o)}{r_o} &= \frac{\nabla \times \dot{\vec{p}}(\tau^o)}{r_o} - \dot{\vec{p}}(\tau^o) \times \nabla \frac{1}{r_o} \\ &= \frac{\nabla \times \dot{\vec{p}}(\tau^o)}{r_o} + \frac{\dot{\vec{p}}(\tau^o) \times \hat{r}_o}{r_o^2} \end{aligned}$$

and evaluating the numerator of the first term,

$$\nabla \times \dot{\vec{p}}(\tau^o) = \nabla \times \left(-i\omega\vec{p}_o e^{-i\omega\tau^o}\right) = -i\omega e^{-i\omega t} \nabla \times \left(e^{ikr_o}\vec{p}_o\right)$$

and again using this vector identity

$$\nabla \times \left(e^{ikr_o}\vec{p}_o\right) = e^{ikr_o}\nabla \times \vec{p}_o - \vec{p}_o \times \nabla e^{ikr_o} = -\vec{p}_o \times \nabla e^{ikr_o}$$

where, since $\vec{p}_o$ is a constant, the first term has vanished. Next, noting that $\nabla e^{ikr_o} = ike^{ikr_o}\hat{r}_o$, we have

$$\nabla \times \dot{\vec{p}}(\tau^o) = -i\omega e^{-i\omega t}\left(-\vec{p}_o \times ike^{ikr_o}\hat{r}_o\right)$$
$$= -\frac{\omega^2}{c}\vec{p}_o e^{-i\omega\tau^o} \times \hat{r}_o$$
$$= \frac{\ddot{\vec{p}}(\tau^o)}{c} \times \hat{r}_o$$

so that finally,

$$\nabla \times \frac{\dot{\vec{p}}(\tau^o)}{r_o} = \frac{\ddot{\vec{p}}(\tau^o) \times \hat{r}_o}{cr_o} + \frac{\dot{\vec{p}}(\tau^o) \times \hat{r}_o}{r_o^2}$$
$$= -\frac{\hat{r}_o}{r_o} \times \left[\frac{\ddot{\vec{p}}(\tau^o)}{c} + \frac{\dot{\vec{p}}(\tau^o)}{r_o}\right]$$

Discussion 4.4

Rewriting Eq. 4.35 with

$$f(r_o) = \frac{e^{ikr_o}}{r_o}\left(1 - \frac{1}{ikr_o}\right) \quad \text{and} \quad \vec{V} = \hat{r}_o \times \vec{p}(t)$$

we obtain

$$\vec{E}_{ed} = \frac{ik}{4\pi\varepsilon_o}\nabla \times f(r_o)\vec{V}$$

Using the vector identity $\nabla \times f\vec{V} = f\left(\nabla \times \vec{V}\right) - \left(\vec{V} \times \nabla f\right)$, along with the gradient,

$$\nabla f(r_o) = \left(\frac{ik}{r_o} - \frac{2}{r_o^2} + \frac{2}{ikr_o^3}\right)e^{ikr_o}\hat{r}_o$$

we obtain

$$\vec{E}_{ed} = \frac{ike^{ikr_o}}{4\pi\varepsilon_o}\left[\left(\frac{1}{r_o} - \frac{1}{ikr_o^2}\right)\left(\nabla \times \vec{V}\right) + \left(-\frac{ik}{r_o} + \frac{2}{r_o^2} - \frac{2}{ikr_o^3}\right)\left(\vec{V} \times \hat{r}_o\right)\right] \tag{4.68}$$

For the first term of Eq. 4.68, we note

$$\nabla \times \vec{V} = \nabla \times (\hat{r}_o \times \vec{p}) = (\vec{p} \cdot \nabla)\hat{r}_o - (\nabla \cdot \hat{r}_o)\vec{p} \tag{4.69}$$

where

$$(\vec{p} \cdot \nabla)\hat{r}_o = \frac{1}{r_o}\Big(\vec{p} - (\hat{r}_o \cdot \vec{p})\hat{r}_o\Big) \quad \text{and} \quad (\nabla \cdot \hat{r}_o)\vec{p} = \frac{2}{r_o}\vec{p} \tag{4.70}$$

so that

$$\nabla \times \vec{V} = \nabla \times (\hat{r}_o \times \vec{p}) = -\frac{1}{r_o}\Big(\vec{p} + (\hat{r}_o \cdot \vec{p})\hat{r}_o\Big) \tag{4.71}$$

Meanwhile, for the second term of Eq. 4.68,

$$\vec{V} \times \hat{r}_o = (\hat{r}_o \times \vec{p}) \times \hat{r}_o = \vec{p} - (\hat{r}_o \cdot \vec{p})\hat{r}_o \tag{4.72}$$

So, plugging in Eqs. 4.71 and 4.72, we can rewrite Eq. 4.68 as

$$\begin{aligned}
\vec{E}_{ed} &= \frac{ike^{ikr_o}}{4\pi\varepsilon_o}\left[\left(-\frac{1}{r_o^2} + \frac{1}{ikr_o^3}\right)\Big(\vec{p} + (\hat{r}_o \cdot \vec{p})\hat{r}_o\Big) + \left(\frac{2}{r_o^2} - \frac{2}{ikr_o^3}\right)(\vec{p} - (\hat{r}_o \cdot \vec{p})\hat{r}_o) - \frac{ik}{r_o}(\hat{r}_o \times \vec{p}) \times \hat{r}_o\right] \\
&= \frac{e^{ikr_o}}{4\pi\varepsilon_o r_o}\left[\left(\frac{1}{r_o^2} - \frac{ik}{r_o}\right)\left[\Big(\vec{p} + (\hat{r}_o \cdot \vec{p})\hat{r}_o\Big) - 2\Big(\vec{p} - (\hat{r}_o \cdot \vec{p})\hat{r}_o\Big)\right] + k^2\left[(\hat{r}_o \times \vec{p}) \times \hat{r}_o\right]\right] \\
&= \frac{e^{ikr_o}}{4\pi\varepsilon_o r_o}\left[\left(\frac{1}{r_o^2} - \frac{ik}{r_o}\right)\left[3\hat{r}_o(\hat{r}_o \cdot \vec{p}) - \vec{p}\right] + k^2\left[(\hat{r}_o \times \vec{p}) \times \hat{r}_o\right]\right]
\end{aligned}$$

Discussion 4.5

Integration of Eq. 4.61 over the total solid angle is straightforward upon noting that the $Q_{\alpha\beta}$ in 4.61 are constants of the integration, and the integration reduces to:

$$\left\langle |\vec{P}| \right\rangle_{eq} = \frac{\omega^6}{1152\pi^2\varepsilon_o c^5 r_o^2}\left[\sum_{ijk}(Q_{ij})(Q_{ik}) \int (r_{oj})(r_{ok})\, d\Omega - \frac{1}{r_o^2}\sum_{ijkl}(Q_{ij})(Q_{kl}) \int (r_{oi}\, r_{oj}\, r_{ok}\, r_{ol})\, d\Omega\right] \tag{4.73}$$

Because the first integral in Eq. 4.73 is odd under the reversal of direction of the $r_{o\alpha}$:

$$\int (r_{oj})(r_{ok})\, d\Omega = K\delta_{jk}$$

and K is determined from

$$\int\left[\sum_j r_{oj}^2\right]d\Omega = 4\pi r_o^2 = K\sum_j \delta_{jj} = 3K$$

where all three K's must be equal by symmetry of the coordinates, then

$$K = \frac{4\pi r_o^2}{3}$$

In a similar manner, the second integral is determined by noting that the integral

$$\int (r_{oi}\, r_{oj}\, r_{ok}\, r_{ol})\, d\Omega \tag{4.74}$$

is completely symmetric under interchange of the indices, and further, the integral only has a nonzero value if the product within the integral occurs in pairs of like coordinates, that is:

$$\int (r_{0i}\, r_{oj}\, r_{ok}\, r_{ol})\, d\Omega = \frac{4\pi r_o^4}{15}\left(\delta_{ij}\delta_{kl} + \delta_{ik}\delta_{jl} + \delta_{il}\delta_{jk}\right)$$

Insertion of these results, along with the additional use of the fact that the trace of $\overleftrightarrow{Q}$ is zero in the second integral of Eq. 4.73 yields Eq. 4.62.

Part III

Electromagnetism and Special Relativity

Introduction to Special Relativity

5

- Historical overview of the development of concepts of special relativity
- Einstein's postulates of relativity, and the relativity of simultaneity
- The Lorentz transformation and principles of covariance among inertial frames
- Geometry of space-time: Minkowski space-time diagrams
- Two examples of the physical consequences of special relativity: the twin paradox and the connected rockets
- 4-vectors in space-time: contravariant and covariant. The metric tensor, 4-vector gradient, 4-vector velocity, 4-vector momentum, 4-vector force, 4-wave vector, 4-current density, and 4-potential.
- Manifest covariance of Maxwell's equations. The electromagnetic field tensor
- Examples of electromagnetic field transformations: static electric and magnetic fields parallel and transverse to the velocity relating two inertial frames. Transformation of fields from a charge moving at relativistic velocities
- The Einstein stress-energy tensor

The views of space and time that I wish to lay before you have sprung from the soil of experimental physics, and therein lies their strength. They are radical. Henceforth, space by itself and time by itself, are doomed to fade away into mere shadows, and only a kind of union of the two will preserve an independent reality.

H. Minkowski, *Space and Time*

Electromagnetic Radiation. Richard Freeman, James King, Gregory Lafyatis,
Oxford University Press (2019). © Richard Freeman, James King, Gregory Lafyatis.
DOI: 10.1093/oso/9780198726500.001.0001

Einstein's special relativity theory is a turning point in our understanding of the physical universe. The birth of special relativity dates to two papers: Lorentz's (1904) *Electromagnetic Phenomena in a System Moving with any Velocity Smaller than that of Light*[1] and Einstein's (1905) *On the Electrodynamics of Moving Bodies*. The focus of this chapter is that of Einstein's 1905 work. That paper has two parts: first Einstein shows how the two postulates of special relativity tie together space and time and lead to the new kinematics Minkowsi is referring to here. The second part of the 1905 paper lays out how electromagnetism as contained in Maxwell's equations fits into this new world view. It is important to realize that Maxwell's equations are already relativistically correct, though in the way that the equations are usually presented, their relativistic covariance is not immediately evident. Indeed, the strongest impact of this new concept of "space-time" was felt not as much within Maxwell's theory of electromagnetism as it was in the classical mechanics of Newton, which were decidedly not relativistically covariant. Here, we begin with just enough history to explain the experimental observations that required Lorentz and Einstein to go beyond the common sense kinematics of classical physics. Einstein's 1905 paper is, arguably, the clearest explanation of the essential physics behind his theory. Accordingly, we quote extensively from that paper in finding the Lorentz transformation equations. The novel kinematics of relativity are largely summarized by the phenomena of "Lorentz contraction" and "time dilation." Frequently, ambiguous claims are made of the sort that "moving rulers appear shorter" or "a moving clock appears to run slowly."[2] To see that these phenomena are real in the sense that they violently conflict with our classical intuition, we look at two thought experiments: the "rocket problem" that wonderfully demonstrates a practical consequence of length contraction, and the "twin paradox" that explores time dilation by considering the round trip to a distant star taken by a space-traveling twin. For the description and explanation of the twin paradox, we introduce a helpful visualization of special relativity for one spatial dimension: Minkowski diagrams. 4-vectors and 4-tensors constitute a mathematical apparatus generally used for working through all but the simplest special relativity problems. We develop 4-vector mathematics by comparing and contrasting them with regular 3-vectors. The 4-vector, 4-tensor notation is used to express and study Maxwell's equations and to show that they are indeed, as written, already consistent with special relativity. Finally, we show how Maxwell's 3D stress tensor is generalized as Einstein's stress-energy tensor.

[1] https://www.lorentz.leidenuniv.nl/IL-publications/Lorentz.html

[2] In fact, if one uses light, for example, to observe an object, then, unless care is taken, the observed "appearance" of many moving systems can lead to a confusing entanglement of Doppler effects with the relativistic phenomena of interest.

5.1 Historical introduction–1666 to 1905

5.1.1 The nature of space and time

That things feel the same regardless of our motion is a common intuition built upon our accumulation of experience. In physics, this observation has long been stated as a principle, as early as Galileo. This principle of relativity, or more precisely, the equivalence of inertial frames for doing physics, is incorporated in Newtonian physics, where it is expressed mathematically as an invariance of the equations of physics under Galilean transformations. Starting with Newton and Leibniz, the philosophical underpinnings of this relativity concept evolved dramatically over the course of time. It reached its apogee in the writings of Einstein, who promoted it to one of the two foundational postulates of the special theory of relativity.

Newton was insistent in his belief that absolute space and time served as a separately existing stage upon which material objects evolved according to physical laws (**see Discussion 5.1**). Samuel Clarke, writing on Newton's behalf, further elaborated and defended this view in the Leibniz–Clarke correspondence. There, Leibniz contested this Newtonian notion of absolute space claiming instead that space and time are not containers, in which objects are located and move, but are actually defined by the relations, such as distance, velocity and acceleration, between the objects. This relational point of view thus effectively denied the objective existence of absolute space and time in favor of space and time abstracted from the relations between material objects–a view summarized by the statement that "all is relative." Despite this disagreement over the ontological nature of space and time, both Newton and Leibniz agreed that adding to all objects an additional (identical) velocity–described as giving the system a "boost" to a different inertial frame–should make no difference to the physics. However, their differences point up what is special about the special theory of relativity. In particular, although special relativity rejects the Newtonian notion of an absolute space and time, it does not go so far as to claim, as Leibniz did, that there are only relative positions and times between objects and events and that space-time has no existence apart from the material within it. Indeed, the kinematic covariance of special relativity is limited to inertial frames–frames that are neither accelerating or rotating. And among these there is no preferred absolute stage for the universe. The fact that the physics of differently accelerating or rotating frames is not covariant and the associated question of "accelerating or rotating with respect to what" is not addressed by special relativity but is rather treated in general relativity.

Minkowski, writing after special relativity had become generally accepted, noted that, classically, the two major symmetries of relativity, symmetry under rotations and symmetry with respect to boosts, are on very different conceptual footings. On the one hand, that physics is covariant under rotations is inherent in the very nature of space–space is isotropic. On the other hand, in the Newtonian world, in which the "stage" of space defines a special state of rest, one can ask whether experiments done on systems identical but for a relative boost give results differing only by the relative boost. And, indeed, for Maxwell's electromagnetism, the answer is in the negative–electromagnetism is not covariant to Galilean boosts. That we feel some compulsion to ask "Is there a preferred frame among comoving frames?" and little to ask "Is there a preferred orientation?" likely stems from our natural experiential constraint in the former and not the latter; that is, while we can fully explore all rotations, we have little or no every day experience with velocities approaching the speed of light. Thus, covariance of physics for different inertial frames, related by a boost must be checked on a case by case basis- (**see Discussion 5.2**).

For what follows we will focus on the boosts by consistently comparing descriptions of physical phenomena in two representative inertial frames that we define now. These are a "laboratory frame" that we will alternatively call the "unprimed frame" or "rest frame," and designate, K, and the "primed frame," K', with a relative velocity V in the $+x$ direction that we refer to as the "moving frame." We use a capital V for the relative velocity between two frames. Lower case v's are particle velocities or other velocities in a problem. Physical laws must not depend on how we choose our temporal and spatial origins. We generally use this flexibility to choose coordinates such that the spatial origins of the two frames coincide at $t' = t = 0$. For a classical boost, an object in the rest frame at time t with spatial coordinates x, y, z will have moving frame coordinates given by a so-called "Galilean" transformation:

$$t' = t \qquad x' = x - Vt \qquad y' = y \qquad z' = z \tag{5.1}$$

for which classical "Newtonian" kinematics is covariant. In the following, we will explore a series of experimental and theoretical developments which ultimately led to a reconsideration of the "absolute" space/time structure foundations of classical mechanics, culminating in a relativistic analog to Eq. 5.1–the Lorentz transformation.

5.1.2 The nature of light

Any reading of the considerable body of experimental work leading up to the work of Lorentz and Einstein leaves one breathless at the skill and cleverness of those experimentalists operating without modern optical and electronic technology.[3] Here, we describe a selection of key experiments that, taken together, point up the inadequacy of the world view of Newtonian classical physics for describing the motion of light.

On the theoretical side, in the early eighteenth century, Newton's *Opticks* (1704) explained reflection, refraction, and related phenomena using a model where light is taken to be a stream of particles which he called corpuscles. Alternatively, Christiaan Huygens, in his *Treatise on Light* (1690), proposed an important competing view in which descriptions of optical phenomena were based on taking light to be undulatory motion in an extremely rarefied medium–the ether. This view immediately raises questions such as "Does this ether define a preferred frame, an absolute rest?" and "How does ether interact with regular matter?"

Chronologically, the first experiment here considered is James Bradley's observation of a *New Discovered Motion of the Fixed Stars* (1727)–what we now call "the aberration of star light" (**see Discussion 5.3**). The origin of the motion is readily understood by considering Fig. 5.1. If we imagine ourselves to be located on the Earth's surface such that the starlight is coming in overhead, we should not orient our telescope perfectly vertically, because in the time it takes the light to travel from the top of the telescope to the bottom, the Earth will have moved slightly in the horizontal direction. One needs to tilt the telescope to compensate. In different parts of the Earth's orbit, the direction of the Earth's velocity is different and so is the direction of the necessary tilt (Fig. 5.1a). That annual variation in the tilt is what Bradley measured (Fig. 5.1b) and this is called the stellar aberration. Bradley's measurements were consistent with both Newton's particle theory and a version of Huygen's wave theory in which the ether frame was that of the fixed stars.

In 1802, Young carried out his double slit experiment decisively establishing that light is a wave and that Newton's particle picture was wrong. Now, as a wave, it seemed natural that, like all waves previously studied, light would have a preferred frame - the frame of the medium through which it traveled. However, identifying such a preferred frame would jeopardize the "principle of relativity," which presumes an equivalence of inertial frames in describing physical phenomena. Alternatively, the nature of this medium could be otherwise entirely compatible with the classical picture–as is the case of a medium carrying sound waves.

[3] An excellent review of the highlights of the history of special relativity is given by Oliver Darrigol, *Seminaire Poincare* 1, 1–22 (2005).

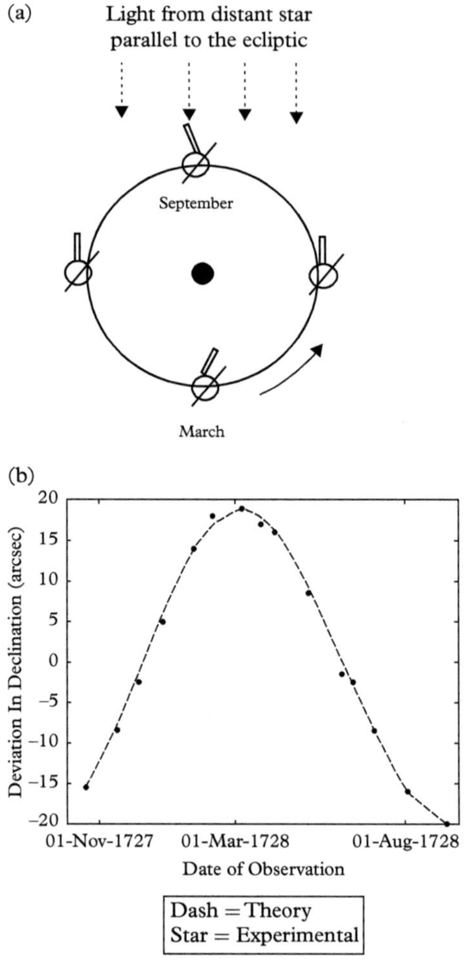

Fig. 5.1 *Stellar aberration refers to the change in the apparent location of a star in the sky due to the Earth's motion. (a), The view from the perspective of the solar system shows how the aberration varies at different times of the year for a star on the ecliptic. (b) Bradley's measurements of the relative angular deviations of Gamma Draconis taken over a year compared with the predictions from the stellar aberration hypothesis.*

In 1865, finally cementing the idea of light as a wave, Maxwell published his *Dynamical Theory of the Electromagnetic Field* making the study of light, or optics, a branch of electromagnetism (**see Discussion 5.4**). More specifically, the source-free Maxwell's equations are readily manipulated to give wave equations for the fields so identification of light as an electromagnetic wave–a major unification at the time–enabled detailed quantification of light phenomena. For instance, a y-polarized light wave traveling in the $+x$ direction is described by the partial differential wave equation:

$$\frac{\partial^2 E_y}{\partial x^2} = \frac{1}{c^2}\frac{\partial^2 E_y}{\partial t^2} \tag{5.2}$$

where c is the wave velocity. But again, this immediately raises the question "velocity with respect to what?" Three possibilities were suggested that are generally consistent with the classical view of space and time:

(1) The c from Maxwell's equations is the speed of light relative to the frame of the "luminiferous ether" a tenuous medium through which the Earth passed unimpeded and at rest with respect to the fixed stars. The ether was stiff at optical frequencies and thus allowed transmission of electromagnetic waves.
(2) Maxwell's c is to be measured with respect to an ether similar to that in 1, but which was entrained or "completely dragged" within and in the vicinity of matter passing through it.
(3) Maxwell's c is to be measured relative to the source of the electromagnetic wave. This is referred to as the "emission theory" and is the way that Newtonian matter particles behave.

Let us now consider each of these possibilities in more detail:

Possibility 1: The existence of a non-interacting, universal, tenuous ether, is consistent with Bradley's stellar aberration since it preserves rectilinear motion for light rays. However, such a preferred frame for the speed of light would violate the principle of relativity. A variation of this picture, in which the ether interacts and is partially dragged within moving matter, was proposed by Augustin-Jean Fresnel in 1818 to reconcile the results of Arago's 1810 experiment in which stellar aberration was measured through a prism (**see Discussion 5.5**).

Best known of the ether dragging experiments was that of Fizeau in 1851. Fizeau set up an optical apparatus that today we would call a Sagnac interferometer. Light, entering the interferometer is split into two beams–one circulating clockwise and the other counterclockwise–and recombined so that an interference is set up at the initial beam splitter. Within a segment of the interferometer common to both beam paths, the two light beams pass, in opposite directions, through a glass tube of controllable flowing water. Thus, any dragging of the ether by the water would act to increase the speed of one beam (the CCW one in this picture) and decrease the speed of the other. The measurement consisted of observing any changes in the interference pattern as the water flow increased. Experimentally, there is something to be said for actually seeing and measuring an effect–in contrast to experiments such as the water telescope where you see nothing happening when you expect to see something. Fizeau did see the interferometer fringes shift as he increased the water velocity. And the fringe shifts were exactly those expected based on Fresnel's partial dragging hypothesis,

$\Delta v_{light} = \left(1 - 1/n^2\right) v_{water}$. While a triumph for the Fresnel assumption, this result was extremely distasteful to many including Einstein. Optical dispersion results from a material's index of refraction being a non-constant function of wavelength. Fizeau's result and Fresnel's explanation would indicate that there is a different ether for each color of light and these are individually dragged differently. While this does not seem to be impossible, it detracts enormously from the simplicity of the theory.

Possibility 2: Another plausible hypothesis, proposed by George Stokes in 1845 and consistent with Maxwell's understanding of his equations, is that the ether within and in the vicinity a body, such as the Earth, which is passing through it is completely dragged by that body. Because this could be developed into a theory for light waves similar to that for sound waves and does not distinguish an absolute rest frame, it does not conflict with the principle of relativity. Bradley's earlier measurements of stellar aberration, however, effectively rule out this possibility (**see Discussion 5.6**).

Possibility 3: Though it refers to electromagnetic waves, it is consistent with the way massive classical particles act and is an approach that Einstein initially pursued but abandoned. Indeed, in the "speed of light" postulate, to be discussed later, Einstein pointedly states that the speed of light is independent of its source's speed. Following Lorentz and Einstein's publications and challenging special relativity, in an ultimately futile attempt to maintain classical notions of space and time, Walther Ritz took this "emission theory" as a starting point and by gently modifying Maxwell's equations was able to develop a theory of electromagnetism with virtues of being consistent with both stellar aberration measurements and the Michelson–Morley experiment (to be discussed momentarily); it allowed light to satisfy Galilean covariance and kept a classical view of space and time. This possibility was eliminated by consideration of light from a "beacon" star orbiting a much more massive star–the so-called "double star observations": As the speed of the light is modulated at the orbital frequency, the fast light emitted at a later time, when the star is moving toward the observer can overtake and pass slower light from an earlier time, when the star was moving away. This should lead to an apparent blinking of the rotating star when viewed from a sufficient distance. This effect was distinctly not observed in the survey conducted by Willem de Sitter in 1913 (**see Discussion 5.7**).

Thus, elimination of possibilities 2 and 3 by experimental evidence suggested that it would be necessary to give up the principle of relativity: the idea of universal covariance of physics among inertial frames. Specifically, it appeared as if the rest frame of the fixed stars, the ether frame, assumes a privileged position among inertial frames. Practically, such a frame, for example, could be discovered by a physicist in a windowless train car by measuring and comparing the speeds of light beams moving in various directions within the car. Mathematically, lack of covariance meant that Maxwell's equations would hold rigorously only in this "rest" frame. An example of this can be seen by noting that upon a Galilean boost of coordinates to a moving frame and the substitutions of $t \to t'$ and $x \to x' + Vt'$ in Eq. 5.2, the result returned in the new coordinates is not the same wave equation. That is, Maxwell's equations as written are not covariant under Galilean transformations.

5.1.3 Michelson–Morley experiments

To address the validity of possibility 1, it remained to measure the relative motion between the Earth and the ether. In 1887, Michelson, aided by Morley, performed a suitable "windowless train car measurement," their famous interferometry experiment, in which any motion of the earth relative to the ether would be detected as a displacement of the interference bands produced by their interferometer (**see Discussion 5.8**). Famously, their results were negative. They were consistently unable to detect any motion relative to the ether or the ether wind. How was this to be interpreted? For one, it could not be argued that the motion of the Earth happened to match that of the ether during their multiple measurements since the Earth's motion, simultaneously spinning on its axis and orbiting the sun, is constantly changing. The only other conclusion to be drawn from these null results, it seemed, was that the ether hypothesis was wrong. They concluded:

> The interpretation of these results is that there is no displacement of the interference bands. The result of the hypothesis of a stationary ether is thus shown to be incorrect and the necessary conclusion follows that the hypothesis is erroneous. This conclusion directly contradicts the explanation of the phenomenon of aberration [of starlight], which presupposes that the earth moves through the ether, the latter remaining at rest.

5.2 Einstein and the Lorentz transformation

Einstein and Lorentz both provided explanations for the failure of the Michelson-Morley experiment to measure Earth's velocity through the ether. Although their assumptions about the underlying space/time geometry differed significantly, both started with Maxwell's equations, which are already relativistically covariant, so the two explanations are mathematically equivalent and correctly predict experimental results testing special relativity. Lorentz, aligning firmly with an ether-filled Galilean space/time and noting that the forces binding molecules together to make up an interferometer arm were likely electromagnetic in origin, concluded that just as the wave equation expression for an electromagnetic wave was frame-dependent, so too must be these binding forces. More to the point, any motion relative to the ether could affect these forces and in turn affect the length of the interferometer arm. Indeed, by solving Maxwell's equations for a moving point charge, Lorentz argued quantitatively that an interferometer arm oriented parallel to its motion through the ether would contract in length an amount exactly enough to compensate for the expected wave transit time difference of the two orthogonal arms and thereby account for Michelson–Morley's null measurement of fringe shifts. This gave the Lorentz–Fitzgerald contraction a physical justification. Thus, we see that Lorentz regarded the Lorentz equations as mathematical artifices to enable calculations for objects at rest within non-ether frames. In contrast, Einstein, holding rigidly to the principle of relativity and finding no need for an ether, radically revised the concepts of time and space such that the Lorentz equations were to be viewed not as mere artifices but rather as transformation equations, which effectively supplant the classical Galilean boosts of Newton. Lorentz, for his part, continued to believe in the existence of the ether long after it was shown by Einstein to be superfluous.

5.2.1 Einstein's approach

Einstein started on his path to special relativity by holding firmly to both the principle of relativity and Maxwell's electromagnetism, which, through its apparent dependence upon a preferred reference frame, was in stark disagreement with the former. Einstein notes, in particular, the following curious observed fact regarding the current induced due to motion between a wire and a magnet: It is only the relative motion between the two that determines the induced current whereas the classical electromagnetic interpretation depends on whether the wire or the magnet is in "motion." In his famous 1905 article, he wrote (**see Discussion 5.9**)

> It is known that Maxwell's electrodynamics–as usually understood at the present time–when applied to moving bodies, leads to asymmetries which do not appear to be inherent in

> the phenomena. Take, for example, the reciprocal electrodynamic action of a magnet and a conductor. The observable phenomenon here depends only on the relative motion of the conductor and the magnet, whereas the customary view draws a sharp distinction between the two cases in which either the one or the other of these bodies is in motion. For if the magnet is in motion and the conductor at rest, there arises in the neighborhood of the magnet an electric field with a certain definite energy, producing a current at the places where parts of the conductor are situated. But if the magnet is stationary and the conductor in motion, no electric field arises in the neighborhood of the magnet. In the conductor, however, we find an electromotive force, to which in itself there is no corresponding energy, but which gives rise–assuming equality of relative motion in the two cases discussed–to electric currents of the same path and intensity as those produced by the electric forces in the former case... Examples of this sort, together with the unsuccessful attempts to discover any motion of the earth relatively to the "light medium," suggest that the phenomena of electrodynamics as well as of mechanics possess no properties corresponding to the idea of absolute rest.

Thus, in addition to the postulate of relativity, which abolished absolute space and time and granted equality to all inertial frames, Einstein specifically recognized the validity of Maxwell's equations and, by extension, the reality of Lorentz transformations and the spacetime it suggested. This equally important recognition of Maxwell's equations was summarized in Einstein's second postulate. Unlike Lorentz, Einstein extended a single transformation covariance (the Lorentz type) to *all physical phenomenon*, and in particular mechanics. Bottom line, the first postulate stated that for both light and matter, the notion of a preferred inertial frame–ether-based or otherwise–is wrong and the second postulate, in combination with the first, stated that Maxwell's equations are correct, or more precisely, the correct transformations between inertial frames are Lorentz transformations. In Einstein's words, the two postulates of relativity are:

(1) **The Principle of Relativity**: *the same laws of electrodynamics and optics will be valid for all frames of reference for which the equations of mechanics hold good.*

(2) **The Invariance of Light Speed**: *light is always propagated in empty space with a definite velocity c which is independent of the state of motion of the emitting body.*

These two postulates of relativity are fundamental principles that constrain all physical systems and are thus analogous to the universal laws of thermodynamics which, unlike electromagnetism and mechanics, for example, transcend any one particular domain. While it was light and Maxwell's equations that pointed up what would ultimately be identified as problems with Newtonian mechanics and basic concepts of space and time, the broader significance of the second postulate is *for all inertial frames there is a maximum velocity, c, for which signals may be transmitted* (**see Discussion 5.10**).

Relativity of simultaneity

Einstein's key insight was the need for a re-examination of the nature of time. His recognition, encapsulated by his two postulates, of the connection between signaling and simultaneity necessitated a redefining of the latter in terms of the former. Previously, in the Newtonian world, two events would be judged simultaneous if they occurred at the same tick of the cosmic clock specifying absolute time. Previously, in practice, given an infinitely fast method of signaling, all events could be instantly linked and ordered in time and simultaneity established. Einstein writes in part 1, section 1, the "Definition of Simultaneity:"

> Now we must bear carefully in mind that a mathematical description of this kind has no physical meaning unless we are quite clear as to what we understand by "time." We have to take into account that all our judgments in which time plays a part are always judgments of simultaneous events. If, for instance, I say, "That train arrives here at 7 o'clock," I mean something like this: "The pointing of the small hand of my watch to 7 and the arrival of the train are simultaneous events."

However, as Poincare noted five years earlier, simultaneity and the time ordering of events at different, separate locations is a matter of convention and "must be defined so that the expression of the laws of physics should be the simplest possible."[4] Einstein's two postulates point to the only consistent redefinition of simultaneity but at the cost of a new found frame dependence. This is seen in the famous thought experiment involving two observers moving relative to one another. The first observer is in a moving train car which contains a flash bulb at the exact center while the second observer stands on the embankment. When the bulb flashes, the pulse of light spreads out in all direction with speed *c*, regardless of inertial frame, according to Einstein's postulates. Thus the observer on the train will record that the pulse, traveling equal distances must have taken equal times and

[4] Poincare, Henri La mesure du temps, *Revue de mÃl'taphysique et de morale* 6, 1–13 (1898).

will reach the front and back of the car simultaneously. The observer on the ground, however, sees the front of the car moving away from the advancing pulse and the back of the car moving toward it. The pulse reaching the back of the car travels the shorter distance and therefore he claims it must arrive before the light reaching the front. Einstein's postulates lead to a universe in which time can no longer be considered absolute. Even the order of two events which are separated in space depends on the inertial frame of the observer.

Synchronizing clocks at different locations

First, keeping Einstein's postulates in mind, lets consider how we might systematically compare times at points separated in space. In particular, it is crucial to note at the outset that in all discussions of time measurements of events spatially separated from an observer, such observations are assumed to be corrected for the travel time of light from the events. So, now let us imagine that a number of identically constructed unsynchronized clocks are scattered throughout space. To measure time intervals between events that happen at these spatially separated locations, we have to be able to synchronize them (**see Discussion 5.11**). Einstein explains how to synchronize a remote clock at distance L with a clock at the origin of the same frame as follows: Lets say we retro-reflect a light pulse from the origin off of a mirror at location L so that the pulse leaves the origin at time t_0, reflects from L at time t_1 and arrives back at the origin at time t_2. By the speed-of-light postulate, the outgoing and return trips take the same amount of time: $\Delta t = t_2 - t_1 = t_1 - t_0$ so that $t_1 = \frac{1}{2}(t_0 + t_2)$ is the mid-time between t_0 and t_2 and we can now synchronize the clocks. For example, if we send a light pulse out to L at noon and it returns at 2 o'clock, then we know it must have arrived at L at 1 o'clock and we can, at our leisure, tell a person at L "remember when the light pulse hit your mirror? That was 1 o'clock; you should adjust your clock accordingly."[5] This technique is known as Einstein synchronization (or Poincare–Einstein synchronization).

5.2.2 The Lorentz transformation: covariance among inertial frames

In classical physics, time is absolute so that at any given time, every observer, regardless of his velocity or position, has a common definition of past, present and future events. Within special relativity, however, absolute time is discarded such that different inertial frames can have different definitions of simultaneous events (see Section 5.3.1). We have already seen a clear example of this in the train-car

[5] It can be shown that this method of synchronizing remote clocks is equivalent to initially synchronizing all of the clocks at the origin and then slowly moving the clocks to their ultimate locations. If the clocks are moved rapidly they are not properly synchronized as may be seen in considering the "twin paradox."

thought experiment where the observer aboard the train counted the two light-pulse events as simultaneous while the stationary observer counted one event as occurring prior to the other. More specifically, we will find that the time of an event in a "moving" frame depends on the position of the event in the "stationary" frame. The converse of this, that the spatial coordinate of an event in a moving inertial frame depends on the time of the event in the stationary frame is true even in classical physics: $x' = x - Vt$. This "mixing" of space and time coordinates can be seen to be mathematically similar to the mixing of spatial coordinates resulting from a geometrical rotation but, as we will see, with 3D Euclidean space replaced with "space-time" and the rotation transformation replaced with the "Lorentz boost transformation." Basic relativistic concepts such as time dilation, length contraction and this space-time mixing, while extremely counterintuitive, can be nicely depicted graphically using so-called Minkowski diagrams which reveal general features of Lorentz transformed frames relative to a "stationary" frame (see Section 5.3.1). We now look for a general linear transformation (the Lorentz transformation) that relates coordinates in different inertial frames.[6]

The important new physics is most clearly seen by considering a single spatial dimension and finding the transformation that expresses the space-time coordinates t', x' of the moving K' frame, in terms of t, x of the stationary K frame.[7] First, because of the homogeneity of space-time, the expressions must be linear in both the time and space coordinates. The most general transformation, then, is determined by four parameters $\left[\frac{\partial t'}{\partial t}, \frac{\partial t'}{\partial x}, \frac{\partial x'}{\partial t}, \frac{\partial x'}{\partial x}\right]$ that we indicate as the corresponding partial derivatives:

$$\Delta t'(\Delta t, \Delta x) = \frac{\partial t'}{\partial t}\Delta t + \frac{\partial t'}{\partial x}\Delta x \qquad \Delta x'(\Delta t, \Delta x) = \frac{\partial x'}{\partial t}\Delta t + \frac{\partial x'}{\partial x}\Delta x \tag{5.3}$$

Because of the linearity of the transformation, the four partial derivatives (parameters) depend only on the relative velocity, V, of the two frames and do not depend on the time or position of the coordinates.[8] The coordinate differences (Δ's), are not necessarily small. To determine the four parameters, we will consider four thought experiments utilizing the relativity postulates in reference to Fig. 5.2. These are:

(1) Our first relationship among the coefficients is obtained using the procedure of the previous section to synchronize clocks in the moving frame. In the moving frame, K'–the rest frame of the mirrors–the time difference between the initial launch of the light wave and its arrival at the first mirror, $\Delta t'_1$, may generally be expressed in terms of the time and space differences

[6] Here we assume that space-time as measured in any frame is isotropic and homogenous. This is not so for general relativity in which space-time is affected by the gravity of massive objects.

[7] The approach here largely follows Einstein's, but with modern notation.

[8] To be clear, if the parameter $\frac{\partial t'}{\partial t}$, for example, scales with time as t^S, where $S \neq 0$, then the product $\frac{\partial t'}{\partial t}\Delta t$ would scale as t^{S+1} and so would not be linear.

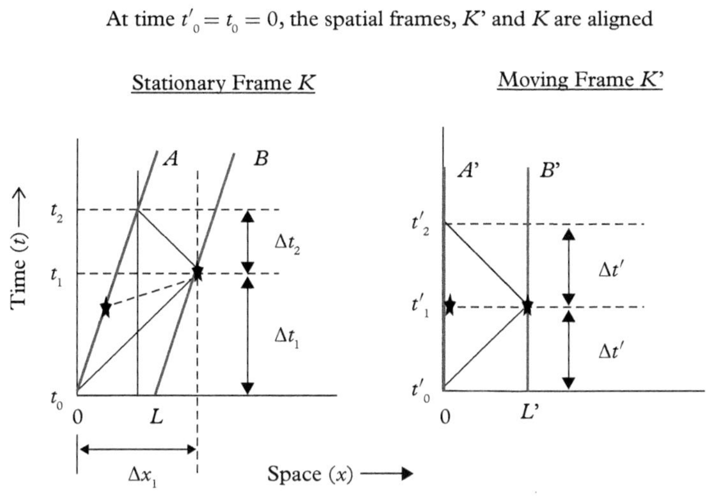

Fig. 5.2 *Relationship of the stationary frame coordinates (K) and the moving frame coodinates (K').*

between the same two events as measured in the stationary frame, K, using the time, Eq. 5.3, $\Delta t_1' = \frac{\partial t'}{\partial t}\Delta t_1 + \frac{\partial t'}{\partial x}\Delta x_1$. A similar expression holds for the light travel time of the return trip, $\Delta t_2'$. In the moving frame both of these time intervals are L'/c, as seen in Fig. 5.2. So, upon subtracting the $\Delta t_1'$ equation from that for $\Delta t_2'$, we have $0 = \frac{\partial t'}{\partial t}(\Delta t_2 - \Delta t_1) + \frac{\partial t'}{\partial x}(\Delta x_2 - \Delta x_1)$. By then working out these stationary frame space and time intervals, a relation between the two t' derivative parameters is obtained (**see Discussion 5.12**):

$$\frac{\partial t'}{\partial x} = -\frac{V}{c^2}\frac{\partial t'}{\partial t} \Rightarrow \Delta t' = \frac{\partial t'}{\partial t}\left[\Delta t - \frac{V}{c^2}\Delta x\right]. \quad (5.4)$$

(2) We can obtain the corresponding relation between the two x' derivative parameters by next considering the right part of Eq. 5.3. Setting $\Delta x' = 0$ corresponds, in the moving frame K', to two events co-located in space but separated in time. In the stationary frame K, these events are separated in both space and time with the deltas related by $\Delta x = V\Delta t$. Inserting this relation into the right ($\Delta x'$) part of Eq. 5.3 for these two events, we immediately obtain a relation between the two x' derivative parameters:

$$\frac{\partial x'}{\partial t} = -V\frac{\partial x'}{\partial x} \Rightarrow \Delta x' = \frac{\partial x'}{\partial x}[\Delta x - V\Delta t] \quad (5.5)$$

(3) To relate $\frac{\partial x'}{\partial x}$ to $\frac{\partial t'}{\partial t}$, consider a light ray along the x-axis started at the origin at $t = t' = 0$. By the second postulate, this ray moves at the speed of light in both frames: $\Delta x = c\Delta t$ and $\Delta x' = c\Delta t'$. Substituting these into the two previous results of Eqs. 5.4 and 5.5, we find that $\frac{\partial x'}{\partial x} = \frac{\partial t'}{\partial t}$ and the results can be summarized in matrix form as

$$\frac{\partial x'}{\partial x} = \frac{\partial t'}{\partial t} \Rightarrow \begin{bmatrix} \Delta t' \\ \Delta x' \end{bmatrix} = \begin{bmatrix} \frac{\partial t'}{\partial t} & -\frac{V}{c^2}\frac{\partial t'}{\partial t} \\ -V\frac{\partial t'}{\partial t} & \frac{\partial t'}{\partial t} \end{bmatrix} \begin{bmatrix} \Delta t \\ \Delta x \end{bmatrix} \tag{5.6}$$

(4) We now have the Lorentz transformation up to a constant depending on the velocity ($\frac{\partial t'}{\partial t} \equiv \frac{\partial t'(V)}{\partial t}$). The inverse transformation is given by taking $V \Rightarrow -V$. Starting in the stationary frame, transforming to the moving frame and then applying the inverse transformation must return the system to the original stationary frame:

$$\begin{bmatrix} \Delta t \\ \Delta x \end{bmatrix} = \begin{bmatrix} \frac{\partial t'(-V)}{\partial t} & +\frac{V}{c^2}\frac{\partial t'(-V)}{\partial t} \\ +V\frac{\partial t'(-V)}{\partial t} & \frac{\partial t'(-V)}{\partial t} \end{bmatrix} \begin{bmatrix} \frac{\partial t'(V)}{\partial t} & -\frac{V}{c^2}\frac{\partial t'(V)}{\partial t} \\ -V\frac{\partial t'(V)}{\partial t} & \frac{\partial t'(V)}{\partial t} \end{bmatrix} \begin{bmatrix} \Delta t \\ \Delta x \end{bmatrix} \tag{5.7}$$

The product of these transformations must be the identity matrix that has a determinant of 1. And since the product of the determinants of two matrices must equal the determinant of the matrix product, we find.[9]

$$\frac{\partial t'}{\partial t} = \left[1 - \frac{V^2}{c^2}\right]^{-1/2} = \gamma \tag{5.8}$$

The sought-for transformation can then be written explicitly as:

$$\begin{bmatrix} \Delta t' \\ \Delta x' \end{bmatrix} = \begin{bmatrix} \gamma & -\gamma\frac{V}{c^2} \\ -\gamma V & \gamma \end{bmatrix} \begin{bmatrix} \Delta t \\ \Delta x \end{bmatrix} \tag{5.9}$$

In what follows, we will write equations in terms of the scaled time variable $c\Delta t$ and express the velocity using $\beta = \frac{V}{c}$. Now both the time and spatial variables have units of length and the Lorentz transformation components are all dimensionless. Concluding: the Lorentz transformations to the moving frame and the Inverse Lorentz transformations back again are found to be:

[9] Since the determinant of the product of the transformation matrices in Eq. 5.7 is equal to the product of the determinants, we have:

$$\frac{\partial t'(V)}{\partial t}\frac{\partial t'(V)}{\partial t}\frac{\partial t'(-V)}{\partial t}\frac{\partial t'(-V)}{\partial t} \times \left[1 - \frac{V^2}{c^2}\right]^2 = 1$$

By symmetry, $\left\|\frac{\partial t'(V)}{\partial t}\right\| = \left\|\frac{\partial t'(-V)}{\partial t}\right\|$. In the limit of small velocities, $\left\|\frac{\partial t'(V)}{\partial t}\right\| \rightarrow 1$. Therefore, $\frac{\partial t'(V)}{\partial t} = \frac{\partial t'(-V)}{\partial t}$ and this equation simplifies to Eq. 5.8.

$$\begin{bmatrix} c\Delta t' \\ \Delta x' \end{bmatrix} = \begin{bmatrix} \frac{\partial(ct')}{\partial(ct)} & \frac{\partial(ct')}{\partial x} \\ \frac{\partial x'}{\partial(ct)} & \frac{\partial x'}{\partial x} \end{bmatrix} \begin{bmatrix} c\Delta t \\ \Delta x \end{bmatrix} = \begin{bmatrix} \gamma & -\gamma\beta \\ -\gamma\beta & \gamma \end{bmatrix} \begin{bmatrix} c\Delta t \\ \Delta x \end{bmatrix} \tag{5.10}$$

$$\begin{bmatrix} c\Delta t \\ \Delta x \end{bmatrix} = \begin{bmatrix} \frac{\partial(ct)}{\partial(ct')} & \frac{\partial(ct)}{\partial x'} \\ \frac{\partial x}{\partial(ct')} & \frac{\partial x}{\partial x'} \end{bmatrix} \begin{bmatrix} c\Delta t' \\ \Delta x' \end{bmatrix} = \begin{bmatrix} \gamma & \gamma\beta \\ \gamma\beta & \gamma \end{bmatrix} \begin{bmatrix} c\Delta t' \\ \Delta x' \end{bmatrix} \tag{5.11}$$

Addition of velocities

In classical mechanics, if you know an object's velocity in one inertial frame, you can determine its velocity in another frame by simply subtracting the relative velocity of the second frame. Things aren't so simple in relativistic space-time. Consider an object moving with an x-velocity, $v' = \frac{dx'}{dt'}$, in the moving frame, K', which itself moves in the $+x$ direction with velocity V relative to stationary frame, K. The particle's x-velocity, $v = \frac{dx}{dt}$, in K is then found by representing the stationary frame differentials, dx and dt, by the inverse Lorentz transform of Eq. 5.11 (with $\Delta x \to dx$, $\Delta t \to dt$):

$$v_{\parallel} = \frac{dx}{dt} = \frac{\frac{\partial x}{\partial t'}dt' + \frac{\partial x}{\partial x'}dx'}{\frac{\partial t}{\partial t'}dt' + \frac{\partial t}{\partial x'}dx'} = \frac{\frac{\partial x}{\partial t'}dt' + \frac{\partial x}{\partial x'}\frac{dx'}{dt'}dt'}{\frac{\partial t}{\partial t'}dt' + \frac{\partial t}{\partial x'}\frac{dx'}{dt'}dt'} \tag{5.12}$$

If we divide out all the dt' and refer again to Eq. 5.11, we identify

$$v_{\parallel} = \frac{\gamma V + \gamma v'}{\gamma + \gamma \frac{Vv'}{c^2}} = \frac{V + v'}{1 + \frac{v'V}{c^2}}$$

where the "$\parallel$" indicates that this is the net speed when the two velocities are parallel and $\gamma = \left(1 - \frac{V^2}{c^2}\right)^{-1/2}$. Generalizing, given two frames K and K' related by a boost, $\vec{V}$, and for an object moving in the K' frame with a velocity, $\vec{v}'$, whose components, $v'_{\parallel}$ and $v'_{\perp}$, are respectively parallel and perpendicular to $\vec{V}$, the components $v_{\parallel}$ and $v_{\perp}$ in the K frame are given by:

$$v_{\parallel} = \frac{v'_{\parallel} + V}{1 + \frac{\vec{V}\cdot\vec{v}'}{c^2}} \qquad v_{\perp} = \frac{v'_{\perp}}{\gamma\left(1 + \frac{\vec{V}\cdot\vec{v}'}{c^2}\right)} \tag{5.13}$$

Note the particular instance for which there is no velocity in the $\vec{V}$ direction within the primed frame ($v'_{\parallel} = 0$) then $\vec{V}\cdot\vec{v}' = 0$ and the equations in Eq. 5.13 reduce to $v_\parallel = V$ and $v_\perp = v'_\perp/\gamma$. This slowing down of the transverse velocity by a factor of $1/\gamma$ is an example of time dilation.

5.3 The invariant interval and the geometry of space-time

The "space-time interval," s_{12} for two events labelled "1" and "2," which, in the rest frame K, are separated a distance Δr_{12} and occur at times differing by Δt_{12}, as measured by properly synchronized clocks at their respective locations, is given by:[10]

$$s_{12}^2 = c^2(\Delta t_{12})^2 - (\Delta r_{12})^2 = c^2(\Delta t_{12})^2 - (\Delta x_{12})^2 - (\Delta y_{12})^2 - (\Delta z_{12})^2 \tag{5.14}$$

Central to special relativity is the result that the space-time interval of two events, so defined, *will be measured the same in any inertial frame.* Essentially, this space-time interval is a measure of the "nearness" of two events in space-time just as the distance Δr_{12} measures the "nearness" of two points in 3-space. In what follows, we will take the invariance of intervals as the starting point for special relativity theory: Lorentz transformations are defined as including rotations in space along with the previously described velocity or "boost" coordinate transformations both of which leave space-time intervals invariant.[11] That this space-time interval is invariant under relativistic boosts may be shown explicitly using the Lorentz transformations of Eq. 5.10 (**see Discussion 5.13**). In contrast to regular space in which distance squared is always positive, in space-time, the quantity s^2 for intervals can, in general, be positive, negative or zero. The intervals are then called "time-like," "space-like," or "light-like," for the three cases, respectively.[12] If the interval, s, is time-like, the quantity

$$\Delta\tau = \frac{s}{c} = \frac{1}{c}\left[c^2(\Delta t)^2 - (\Delta x)^2 - (\Delta y)^2 - (\Delta z)^2\right]^{1/2} \tag{5.15}$$

is real and is called the proper time.[13] The proper time between two events is then immediately seen to be the time Δt measured in a frame where $\Delta x = \Delta y = \Delta z = 0$. In other words, in a frame where the two events occur at the same position, $\Delta\tau = \Delta t$. If, for example, space-time coordinates of a series of events describe the motion of a particle, then the total time elapsed by a clock moving with the particle is the sum of all the little $\Delta\tau$'s between the events–that is, the proper time elapsed. From Eq. 5.15, we can see that the proper time is the shortest

[10] The sign of the invariant interval is arbitrary and taken differently by different authors. The advantage in doing things this way allows for a clear definition of the proper time. Picking the other sign choice shows more clearly the relation between rotations and Lorentz transformations.

[11] In fact, generally, any mathematical "space," whether its regular 3-space, relativistic space-time, or something more exotic, has its own particular mathematical definition of "nearness" and associated "rotation" transformations which leave the "nearness" of two points in that space invariant.

[12] This is the major difference between 3D and 4D space-time: For example, if $r^2 = 0$ in 3D, the two points lie on top of one another. In 4D, if $s^2 = 0$, the two space-time events won't necessarily be coincident but instead can be connected by a light ray.

[13] In keeping with standard notation we use $\Delta\tau$ to denote the proper time here: this τ is not to be confused with the retarded time introduced in Chapter 2.

possible measured time, among all inertial frames, between two time-like events. Thus, time dilation is immediately evident because in any other frame for which the distance between the two events is not zero, the interval is invariant but Δt must be longer and larger than $\Delta\tau$.

5.3.1 Minkowski space-time diagrams

There is a very useful graphical description of special relativity, developed and introduced in 1907 by the mathematician Hermann Minkowski, which is based on the invariant interval. In Euclidean geometry, the collection of all points in space is called a "manifold." Similarly, the collection of all quadruples of coordinates (ct, x, y, z) is the space-time manifold. The "Minkowski" representation of such a manifold is depicted in Fig. 5.3. It visualizes one time and two spatial dimensions in which the time coordinate is scaled by c, as before, and extends vertically.

Events, specified by quadruples of coordinates, can be divided into three types corresponding to whether their interval to the origin is time-like, space-like, or light-like, with the type defining that event's physical relationship with an event at the origin. The coordinates of all events that have light-like intervals to the origin form two "Minkowski light cones" as shown in Fig. 5.3. Generally, events within the cones are time-like and causally connected to the origin and events outside the cones are space-like and causally disconnected from the origin. Space-time is further subdivided by the past/present/future distinction so that (time-like) events occurring at space-time points enclosed within the lower cone are in the past of an event at the origin, in any inertial frame, and can affect that event. And alternatively, any (time-like) events that

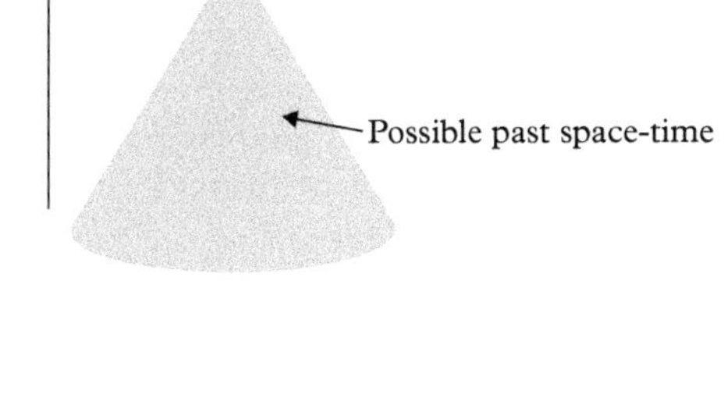

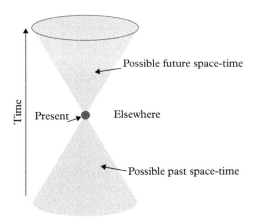

Fig. 5.3 *Minkowski's light cone divides space-time into time-like regions that contain the events causally connected with "Present," and space-like regions for which no causal link is possible.*

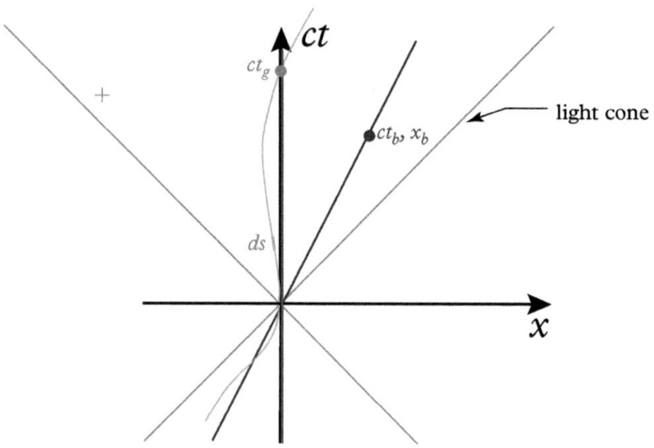

Fig. 5.4 *World-lines for three particles. The red and blue particles are at rest and moving with a constant velocity, respectively. The green particle has a more complicated motion including acceleration.*

occur at space-time points within the upper light cone can be affected by the events occurring at the origin. And again, all events occurring outside the light cones (space-like), regardless of past/present/future, have no causal relationship with the events at the origin. Trajectories of point-objects are just a series of causally connected events and are therefore represented by curves, or so-called "world-lines."

To get at some of the novel physics of special relativity, consider point-like particle motions within a Minkowski diagram restricted to one dimension of space. Figure 5.4 indicates the space-time paths, in the stationary frame K, for several particles. The red particle remains at the origin, the blue particle moves to the right with a constant velocity and the green particle has a more complicated world-line involving some acceleration. Generally, straight world-lines indicate free particles while curved lines represent particles undergoing acceleration. Also, because nothing can exceed the speed of light and, as noted, the time-coordinate is scaled by c, the magnitude of the slope of a particle's trajectory must always be equal to or greater than 1.[14]

An important concept in describing manifolds is the "nearness" of two points, as defined previously, and this is quantified by a metric. In 3D, the metric is the length or distance, the usual square root of the sum of the squares of the coordinate differences for the two points. Differentially, this is written

$$dl^2 = dx^2 + dy^2 + dz^2 \tag{5.16}$$

where it is to be understood in all that follows: $dl^2 \equiv (dl)^2$, and so on. For the Minkowski space-time manifold, the invariant interval is taken as the metric, which in differential form is:

[14] This is equivalent to the statement that causally connected events are within each other's light cones.

$$ds^2 = c^2 dt^2 - dx^2 - dy^2 - dz^2 \tag{5.17}$$

In exactly the same way that the distance between two points is measured to be the same in any two Euclidean frames related by rotation, displacement, or Galilean velocity transformation, the invariant interval in 4D is the same for any two frames related by a Lorentz transformation. Indicated in Fig. 5.4 is the invariant interval, ds, for two differentially close events on the world-line of the green particle. The total interval for general particle motion of finite duration may be found by summing over these differentials. For example, the total interval of the green particle for its motion from the origin to $(ct_g, 0)$ is given by:

$$\begin{aligned} s &= \int_{path} ds = \int_{path} \left[c^2 dt^2 - dx^2\right]^{\frac{1}{2}} \\ &= \int_0^{t_g} \left[c^2 - \left(\frac{d}{dt}x(t)\right)^2\right]^{\frac{1}{2}} dt = c\int_0^{t_g} \frac{dt}{\gamma(t)} \end{aligned} \tag{5.18}$$

where $\frac{dt}{\gamma(t)} = d\tau$ is a differential of proper time. This is consistent with the view, expressed above, that the invariant interval gives the proper time for a moving particle: the total time elapsed for a clock attached to a moving particle as it travels through space is the sum of the proper times for the individual segments of its motion. Finding the proper time of the green particle in Fig. 5.4 thus suggests a general prescription for extending special relativistic kinematics to include accelerations similar to that used in classical physics:

(1) Divide the motion up into small increments during which the velocity relative to an inertial frame is approximately constant.
(2) Use special relativity to describe the kinematics for each increment.
(3) Assemble results from the increments to describe the overall motion.

Free particle: Maximum interval

In Euclidean space, the trajectory taken by a free particle moving from A to B is the straight line between the two points. Mathematically, it can be described as an extremum of the 3D metric for all the possible paths: the shortest distance between two points is a straight line. This is immediately seen by drawing paths between A and B on a diagram since the visualization is in Euclidean space. Referring to Fig. 5.4, the straight line, free particle 4D space-time trajectory of the blue particle in going from the origin to (ct_b, x_b) may also be described as an extremum of a metric between the two points. But in this case, of all the paths, $\{x_{0b}\} \equiv \{x_{0b}(t)\}$, leading from $(0,0)$ to (ct_b, x_b), the straight line of a free particle, $x_{sl}(t) = \frac{x_b}{t_b} t$, is the one that maximizes the total calculated interval:[15]

$$s_{max}[(0,0) \to (ct_b, x_b)] = \int_0^{t_g} \left[c^2 - \left(\frac{d}{dt} x_{sl}(t) \right)^2 \right]^{\frac{1}{2}} dt \tag{5.19}$$

This counter intuitive result, one of many, is reflected in the minus sign in the metric function. That it is not obvious from the picture stems from drawing a non-Euclidean Minkowski space-time diagram in Euclidean space. When we get to dynamics, we will use the important generalization: *the physical trajectory taken by a free particle moving from one point to another is the path between the two points whose space-time interval is a maximum.*

Covariance and Minkowski diagrams

As discussed earlier, in considering covariance of physical quantities under coordinate transformations, Minkowski noted that, classically, spatial rotations are treated very differently than boosts. Covariance under spatial rotations was embedded in geometry whereas covariance under Galilean boosts seemed plausible but was not guaranteed, and indeed, appeared to not hold for Maxwell's equations. In special relativity, however, there can be no doubt about covariance under boosts: it is one of special relativity's postulates. To reveal the formal similarity and make more parallel the treatment of the two types of transformation, boost transformations can be seen, like rotations, to be embedded in the geometry of space-time by defining a "boost parameter" or rapidity, ψ:

$$\frac{V}{c} = \beta = \tanh\psi \tag{5.20}$$

Then, it is readily shown that:

$$\gamma = \cosh\psi \quad \text{and} \quad \gamma\beta = \sinh\psi \tag{5.21}$$

noting that $\psi = \psi(V)$ and that as $V \to 0$, $\psi \to 0$ while if $V \to c$, $\psi \to \infty$. So, with these definitions, the Lorentz transformations of Eq. 5.10 may be written as a pseudo-rotation in space-time:

[15] It is important to keep this concept about an extremum separate from the earlier statements about extremes: namely, that the shortest time duration measured between two events is observed in the inertial frame for which the two events are co-located. We can keep the concepts separate by noting that in the present case we are considering an extremum from among various paths between two events and in the former case we were considering an extremum from among various inertial frames.

$$\begin{bmatrix} ct' \\ x' \end{bmatrix} = \begin{bmatrix} \cosh\psi & -\sinh\psi \\ -\sinh\psi & \cosh\psi \end{bmatrix} \begin{bmatrix} ct \\ x \end{bmatrix} \tag{5.22}$$

and we now have a mathematically compact version of what was earlier stated in words: Lorentz transformations mix time and space coordinates of space-time in a manner similar to the way a 3-space rotation mixes two spatial coordinates.

5.3.2 Physical consequences of special relativity

The violence done by special relativity to our classical intuition is frequently somewhat obscured even within thought problems designed specifically to exploit that violence. For example, although the concept of the relativity of simultaneity, which redefines the meaning of "simultaneous" under the dual constraints of Einstein's postulates, was very well exploited by the train-car thought experiment before, if one were to actually observe a rapidly moving object, the relativistic effects would be mixed with Doppler shifts and differential light-travel delays. The effectiveness of a thought problem is thus measured by how well it isolates relativistic effects from classical effects. Indeed, since much of the novel physics of special relativity is summarized by time dilation and the Lorentz–Fitzgerald contraction, efficient communication of these important new ideas involves well-selected thought experiments which clearly delineate the more intuitive, classical effects. Next, the twin paradox discussion and the rocket problem are concrete examples that isolate and lay bare the conflict between special relativity and classical intuition.

The fast traveling twin

Recall that for two differentially close events separated by a time-like interval, the proper time, the time elapsed for a clock passing through those events, is given by $\frac{ds}{c}$. The previous exercise can then be used to illustrate the relativistic kinematics of the classic twin-paradox. The twin paradox thought experiment is especially valuable because it illustrates in terms immediately humanly accessible, exactly what time dilation means. The problem is as follows: Two twins, John and Jane, live two very different lives. John stays at home on earth. Jane makes a near-light-speed trip on a rocket to a distant star, instantaneously turns around and returns to earth. The world-lines of the two twins are shown in Fig. 5.5. John sees Jane moving at high speed, and thus concludes, because of time dilation, that her clock is running slowly and

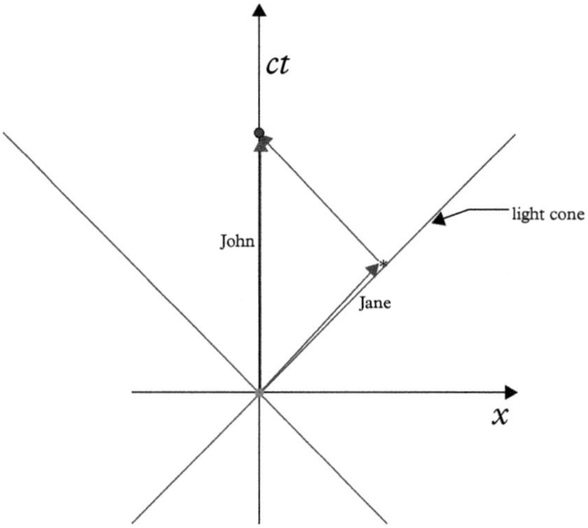

Fig. 5.5 *The twin paradox. John stays at home on Earth. Jane makes a near-light-speed trip in a rocket to a distant star and back. John's interval is longer than Jane's because it is the "free particle" trajectory. Because each twin's interval is simply his/her proper time multiplied by the speed of light, the time measured by a clock co-located with John will be greater than the time measured by a clock co-located with Jane.*

she is aging much less rapidly than himself. This will become especially clear on Jane's return when the twins are standing face-to-face.

The apparent paradox occurs because it would seem that Jane could make a similar argument: John's frame is rapidly moving with respect to hers, thus John should be younger at the end of the trip. The paradox is resolved by noting that only John's rest frame is a constant inertial frame throughout the trip while Jane's rest frame consists of two different inertial frames separated by an extraordinary acceleration at the star. John's conclusion is thus correct: his straight-line, constant velocity ($v = 0$) inertial motion gives the maximum extremum space-time interval between the two events and therefore John will measure the longest time for anyone traveling any path that winds up on earth at the two times indicated. Thus, Jane's excursion leads her to measure a shorter proper time on her clock. On their reuniting, less time has passed for Jane and John is older. For a rocket ship able to travel extremely near to the speed of light, this would appear as time travel to the future: Jane goes to sleep aboard her rocket, and wakes up the next morning coincident with her return; whereupon she finds an Earth that has advanced years into the future of her departure.

The rocket problem

While Einstein's theory and interpretation of the Michelson and Morley result ultimately won out in grand fashion, we feel that for the classically trained mind to understand some of the implications of special relativity, it will be helpful, initially, to start by following Lorentz's thinking. Lorentz recognized that Michelson and Morley's null result could be explained if the interferometer underwent a "contraction."

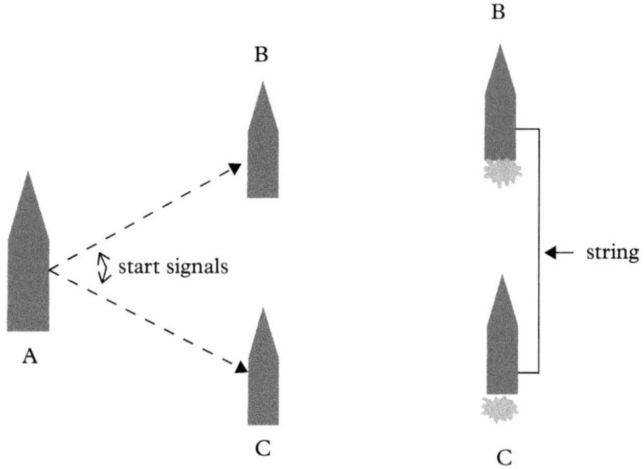

Fig. 5.6 *The rocket problem. Rocket A is initially equidistant from B and C. All rockets are absolutely identical. The problem consists in connecting B and C with a string. After sending a light signal simultaneously to both rockets, rocket A observes that rockets B and C begin accelerating gently, uniformly, and identically. The question: does the hypothetical string connecting rockets B and C break as the rockets accelerate.*

That is, it literally shrank in the direction of the motion relative to the ether, according to $L = L_0\sqrt{1-\frac{V^2}{c^2}} = L_0/\gamma$.[16] Lorentz contraction is sometimes exemplified as a foreshortening of a rocket ship or a rapidly moving car. Exactly what does that mean in a practical sense? To better get at practical consequences of a "Lorentz contraction," we consider the following thought problem. Three identical rockets, A, B, and C, are all initially at rest in frame K. Rocket B, at the front, and rocket C, at the rear, are equal distances from A and are connected tightly by an inelastic string as shown in Fig. 5.6. At time $t = 0$, light signals are received simultaneously by radios within B and C, from rocket A, instructing the two rockets to start their engines. Rocket A observes that rockets B and C then simultaneously begin accelerating gently, uniformly and identically, along y. This acceleration continues until, at time $t = T$ of frame K (Rocket A's frame), Rockets B and C attain the (highly relativistic) velocity V, which, for $t > T$, they maintain–we will call this the "moving" frame K'. The question: will the string break?[17]

The answer to this question can be determined either from frame K of rocket A, or from any of the instantaneous frames of B and C, including their final frame, K', within which they end their acceleration. In frame K, rockets B and C are always separated by a constant displacement as they accelerate. For example, if, at time $t = 0$, C is located at $y_0^C = 0$ and B at $y_0^B = 1000$ and they both undergo the identical acceleration, a, then at any later time, t, C and B will be located at $y^C(t) = \frac{1}{2}at^2$ and $y^B(t) = \frac{1}{2}at^2 + 1000$, respectively, and will still be separated by exactly 1000 m in A's rest frame. There is no more to the kinematics determining their respective positions and separation than that. On the other hand, any moving physical objects, such as

[16] Fitzgerald independently made the same observation; however, importantly, Lorentz produced a theory explaining the origin of the contraction. Working along similar lines, Larmor and Poincare anticipated many of Lorentz's results including time dilation and length contraction.

[17] This thought problem actually goes to the heart of the physical meaning of "foreshortening" in relativity. The reader is forced to confront the difference between an extended object (here the string) versus two points in space time. On the face of it, the observer in A measures the rocket ships B and C to be at a constant distance from each other while simultaneously measuring the string to be shorter.

the rockets B and C, any rulers or "y-oriented" Michelson-Morley interferometers they have on board and, of course, the string, will all be Lorentz contracted along the y–direction in frame K.[18] In particular, at time $t = T$, the length of the string, Δy_s, as measured by A, is given by the Lorentz contraction, $\Delta y_s = L_0\sqrt{1-\frac{V^2}{c^2}}$, which is directly obtained from the Lorentz transformations, Eqs. 5.10, upon setting $\Delta t = 0$.[19] Thus, assuming an ordinary string that is not so incredibly strong that it could significantly change the rockets' motions, *the string must break* (**see Discussion 5.14**).

What do observers on rockets B and C observe? Is their account of the situation consistent with rocket A's account? Consider an instant, $t < T$ in A's frame K, when the rockets have accelerated to some intermediate velocity, $v < V$. So, in this instant of rocket A's time, rocket A perceives B and C to be instantaneously at rest in the intermediate inertial frame k' and still separated by the original length of the string, L_0. That is, in frame K of rocket A, $\Delta y = y^B - y^C = L_0$ and $\Delta t = 0$. If we now choose a space-time origin, for both frames K and k', corresponding to rocket C at this instant, so that $y^C = t^C = 0$ and $y'^C = t'^C = 0$, then we can use the Lorentz transformation to find the k' frame location (y'^B) and time (t'^B) of rocket B corresponding to its location ($y^B = L_0$) and time ($t^B = 0$) in the K frame of rocket A.

$$\begin{bmatrix} \gamma & -\gamma\beta \\ -\gamma\beta & \gamma \end{bmatrix}\begin{bmatrix} 0 \\ L_0 \end{bmatrix} = \begin{bmatrix} -\gamma\beta L_0 \\ \gamma L_0 \end{bmatrix} = \begin{bmatrix} ct'^C \\ y'^C \end{bmatrix}$$

where $\beta = v/c$. So, in the intermediate k' frame, not only are the two rockets further apart ($y'^B - y'^C = \gamma L_0$), but this larger separation occurs between (the rear) rocket C at time $t'^C = 0$, and (the front) rocket B at the earlier time, $t'^B = -\gamma\beta L_0$. Now, this is significant because if we recognize that at these two times the two rockets have the same velocity v relative to rocket A, we must conclude that the velocity of rocket B a little later, at time $t'^B = 0$, will be greater than v. Thus, the situation as perceived by rocket C (or rocket B) at time $t'^C = t'^B = 0$ is that the rocket B to rocket C distance is not only greater than L_0, but the separation is growing and, in fact, has been growing since they started out. That is to say, the string broke almost immediately upon their initial acceleration.

In summary, from the perspective of A, the two points in space-time (rockets B and C) remain separated by L_0, while the physical object (the string) shrinks. From the perspective of the instantaneous rest frame of B and C (and the string), the string remains constant in length at L_0 while the two rockets move apart. The result, that the string

[18] Again, we note the crucial difference between two point-particles bound within an accelerating physical object (such as the string) and two point-particles in space-time constrained to identical acceleration.

[19] The transformation from K to K', with $\Delta y' = L_0$ and $\Delta t = 0$ is,

$$\begin{bmatrix} c\Delta t' \\ L_0 \end{bmatrix} = \begin{bmatrix} \gamma & -\gamma\beta \\ -\gamma\beta & \gamma \end{bmatrix}\begin{bmatrix} 0 \\ \Delta y_s \end{bmatrix} = \begin{bmatrix} -\gamma\beta\Delta y_s \\ \gamma\Delta y_s \end{bmatrix}$$

thus yielding $\Delta t'$ and $\Delta y'$ in terms of Δy_s. The second line gives the Lorentz contraction: $\Delta y_s = L_0/\gamma$.

breaks, is predicted by either reference frame, while the mechanism of that breaking is described very differently.

5.4 Vector space concepts

Before investigating how special relativity enters into electricity and magnetism we need to develop a notation that economically and intuitively includes special relativity: the notation of 4-vectors and 4-tensors.[20] The ideas behind 4-vectors are essentially the same as those of regular 3-vectors. Algebraically, they are both elements of metric vector spaces. The use of 3-vectors stems from the fact that many physical quantities come in sets of three that transform under translations and rotations in the same way as a 3-displacement $(\Delta x, \Delta y, \Delta z)$ and equations written using these 3-vectors are automatically or "manifestly" covariant with respect to translations and rotations.[21] Now, from the above discussion, we can identify the 4-displacement, $(c\Delta t, \Delta x, \Delta y, \Delta z)$, as the appropriate extension to relativistic kinematics and we anticipate that, in exploring the implications of relativity, we will find other 4-quantity sets that transform similarly and form manifestly covariant equations in 4D.

The measure of "nearness" is fundamental to a space and is described quantitatively by the metric of the space. In fact, given only the metric and the manifold itself, all of the geometrical properties of a space can be derived. Thus, it is natural to take the metric as the starting point in the treatment of any given space. For example, the metric for 3-space is the usual dot product of a displacement vector with itself, $dl^2 = dx^2 + dy^2 + dz^2$. Similarly, in space-time, "nearness" is quantified by the invariant interval, $ds^2 = c^2dt^2 - dx^2 - dy^2 - dz^2$, which is likewise represented as a generalized dot product (or "inner" or "scalar" product) of a vector displacement. In general, the symmetry group of transformations is then defined for a given space as "those transformations that leave the metric nearness values unchanged." Displacement vectors are next generalized in that n-space by defining as a vector any n component quantity with transformation properties identical to that of displacements. As an example, consider that under 3-space translations, 3-vector components, by definition, do not change. This statement points to a subtlety: the position coordinates of a point in space do not constitute a 3-vector, as defined, because their values change upon translation. In fact, for most vector discussions where position components of an event are indicated, what is meant is the displacement relative to a fixed point in space that coincides with the coordinate origin.

[20] Technically, although scalars and vectors are 0th and 1st rank tensors, we prefer the former, less formal, terminology while we reserve the term "tensor" for 2^{nd}rank tensors.

[21] The term "covariant" used here in reference to a vector equation means that if the equation is valid in one coordinate frame, then it is valid in any other rotated or translated frame.

Additional key concepts which arise with 4-vectors are the contravector-covector distinction and the related metric tensor. Although these concepts also exist for 3-vectors, they are usually not mentioned as they are numerically trivial. In this section, as an introduction to these central 4-vector concepts, we will expose their existence via block matrix notation for the more familiar 3D case. We will begin to see, for a given metric space–whether it is 3-space, space-time or, indeed, something more exotic–that the metric, the metric tensor and the contravector-covector distinction all essentially encode the properties of the space and are therefore intimately related. As a practical matter. however, it will be important to keep in mind that while these concepts are indispensable in the more sophisticated geometries of General Relativity, for special relativity, or "Minkowski space-time," the point of the formalism, and in particular the subscript-superscript gymnastics in what follows, is mostly just to automatically keep track of the minus signs in complicated 4-vector and 4-tensor expressions.

Before we start, lets consider the various notations to be used in the following sections: We will start off using block matrix notation with the standard conventions of using brackets to indicate vectors and tensors and the rules of row-column multiplication. The vector components will be represented by a capital Roman letter universally subscripted by the indices x, y, z, and, in the case of 4-vectors, ct. As shorthand notation, we may occasionally designate 3-vectors as a capital letter with a top arrow. If we wish to refer to collections of three (or four) quantities without reference to any specific multiplication rules, we will use parentheses. As we know, block matrix notation allows the same vector physical quantity to be written either as a column vector or a row vector. At first glance, for 3-vectors, these appear to be the same components just rearranged; however, we will see shortly that this is a subtle example of the dual representation of a vector quantity and that, in fact, these column and row vector representations have different transformation properties. To emphasize this point and prepare for what follows, we will introduce an additional notation scheme in which we denote column vector quantities with numerical superscripts and row vector quantities with numerical subscripts, labeling them, respectively, "contravariant" and "covariant." For this notation, the associated shorthand representation of a vector will be a capital Roman letter with a subscripted or superscripted index where all 3-vector indices will be capital Roman (e.g., $A^{\mathcal{J}}$) and all 4-vector indices will be lowercase Greek (e.g., A^{μ}). The major change in transitioning to space-time is that the last three components–the spatial components–of the covariant (row) 4-vectors will be opposite in sign to the corresponding contravariant (column) 4-vectors. This is necessary to return correctly signed invariant-interval-type scalar products and represents, in spacetime, the numerical difference between the two vector representations. In practice, we can thus obtain the covariant version of a given contravariant 4-vector by simply adding minus signs to the spatial components of the latter.

5.4.1 Contravariant and covariant vectors

Above, we noted two examples, for 3-space and space-time, that the metric of a space is represented by the scalar product of the associated displacement vector with itself. We will see in this subsection that, in general, the scalar product of vectors always involves two different types of vectors, contravariant and covariant, and any vector quantity can be represented equally by either type. Furthermore, we will see how scalar product invariance under rotation and Lorentz transformations requires that these contravariant and covariant vectors transform in inverse ways. Having established vector transformation properties, we will then explore the transformation of tensors with the basic rule being that a tensor transforms in accord with its component vectors.

Scalar products

The scalar (or inner) product of a pair of numerically indexed 3-vectors, sometimes alternatively called "one-forms," is shown here along with the standard row by column block matrix multiplication convention in which the components are universally subscripted with the x, y and z coordinates:

$$\vec{U}_{\mathcal{J}} \cdot \vec{W}^{\mathcal{J}} = U_{\mathcal{J}} W^{\mathcal{J}} = U_1 W^1 + U_2 W^2 + U_3 W^3 \tag{5.23}$$

$$= \left[U_x, U_y, U_z\right] \begin{bmatrix} W_x \\ W_y \\ W_z \end{bmatrix} = U_x W_x + U_y W_y + U_z W_z \tag{5.24}$$

Here the "$\mathcal{J}$'s" on the first entry just indicate that we are using the row (covariant) version of $\vec{U}$ and the column (contravariant) version of $\vec{W}$. The second entry uses the Einstein convention of summing over repeated indices.

For two 4-vectors, A_α and B^β, the scalar product is again expressed, as it was for 3-vectors in Eqs. 5.23 and 5.24, in numerical index-based or block matrix form:

$$A_\mu B^\mu = A_0 B^0 + A_1 B^1 + A_2 B^2 + A_3 B^3 \tag{5.25}$$

$$= \left[A_{ct}, -A_x, -A_y, -A_z\right] \begin{bmatrix} B_{ct} \\ B_x \\ B_y \\ B_z \end{bmatrix} \tag{5.26}$$

$$= A_{ct}B_{ct} - A_xB_x - A_yB_y - A_zB_z \tag{5.27}$$

where such a product, again, will always involve a (subscripted) covariant vector and a (superscripted) contravariant vector and for which summation, again indicated by repeated indices, now goes from 0 to 3. It is useful at this point to re-iterate the differences between the newly introduced numerical and old standard letter indexing conventions–specifically noting the minus signs and sub/superscript positions. The conventions, either of which can be used redundantly with the block matrix form, are:

$$A^\mu = \left(A^0, A^1, A^2, A^3\right) = \left(A_{ct}, A_xA_y, A_z\right) \qquad \text{and} \tag{5.28}$$

$$A_\mu = (A_0, A_1, A_2, A_3) = \left(A_{ct}, -A_x - A_y, -A_z\right) \tag{5.29}$$

As an important example of a scalar product, consider that of the displacement with its covariant representation. Just as it represents the distance between two points in 3-space, it represents, in spacetime, the invariant interval of two events. It is the space-time metric:

$$\Delta s^2 = \Delta x_\mu \Delta x^\mu = [c\Delta t, -\Delta x, -\Delta y, -\Delta z] \begin{bmatrix} c\Delta t \\ \Delta x \\ \Delta y \\ \Delta z \end{bmatrix} \tag{5.30}$$

$$= c^2\Delta t^2 - \Delta x^2 - \Delta y^2 - \Delta z^2 \tag{5.31}$$

Above, we pointed out that an important difference between 3-space and spacetime vectors is that for 4-vectors the space components of the covariant and contravariant versions differ by a minus sign. Now, with the prototypical form of the "size" of a 4-vector as given by the 4-displacement invariant interval of Eq. 5.31, we can see a second important difference–namely, the physical meaning of the vector magnitudes. For instance, in 3-space, the meaning of a zero displacement magnitude is intuitive; the two positions defining the displacement are identical. For space-time, however, an invariant interval of zero allows the additional possibility that the two events/points are not identical and have a light-like separation–that is, they may lie in the same light-cone. With this in mind, if we define orthogonality in space-time, as in 3-space, as occurring for two vectors having a scalar product of 0, then a 4-vector, such as this light-like displacement, technically can be orthogonal to itself.

Rotations and Lorentz transformations

In the generalization before, we stated that a group of transformations is defined for a given space as "those transformations that leave the metric value unchanged." For Newtonian 3-space, that group is the 10 component Galilean group of translations (including translating the time origin), rotations and Galilean boosts, and for relativistic spacetime it is the 10 component Poincare group of translations, rotations and Lorentz boosts. Moreover, it can be shown that these and only these transformations leave the respective vector scalar products invariant.[22] That being said, for 3-vectors, rotations have the more difficult and relevant transformation properties and so in what follows we will focus on rotations. And because the story for translations and rotations in space-time is nearly identical to that in 3-space, our following discussion of space-time, while drawing from the analogy of spatial rotation, will concentrate on boosts. Like rotations in 3-space, Lorentz boosts are "origin preserving" transformations and can be interpreted as pseudo-rotations in space-time. It is, indeed, through comparison of Lorentz boosts with 3-vector rotations that we gain an understanding of the contravariant/covariant vector distinction.

Scalar product invariance

We know that for 3-vectors the magnitude of a vector quantity is rotation-invariant - as is the scalar product of (and, indeed, the angle between) any pair of 3-vectors. And we know there are similar such invariant constructions for 4-vectors. We will now see how such scalar product invariance under these transformations requires that contravariant and covariant vectors transform in inverse ways. To do this, we examine how the components of the different types of 3D and 4D vectors transform for rotations and Lorentz transformations, respectively. First, we consider two different 3D vectors, $\vec{U}$ and $\vec{W}$, in the unprimed "lab" frame and in a primed frame that is rotated by an angle ϑ about the z-axis. In the lab frame, the components are $\vec{U}_{\mathcal{J}} = (U_1, U_2, U_3) = \left(U_x, U_y, U_z\right)$ and $\vec{W}^{\mathcal{J}} = \left(W^1, W^2, W^3\right) = \left(W_x, W_y, W_z\right)$, while in the rotated frame the same two vectors are described as $\vec{U}'_{\mathcal{J}} = \left(U'_1, U'_2, U'_3\right)$ and $\vec{W}'^{\mathcal{J}} = \left(W'^1, W'^2, W'^3\right)$. For the contravariant, column matrix form of $\vec{W}$, the transformation to the primed frame is obtained through multiplication, from the left, by the rotation matrix, $R(\vartheta)$,

[22] Here, we restrict consideration to continuous transformations. Certain discrete transformations such as reflections will also leave the interval invariant.

$$\vec{W}'^{I} = R^{I}_{\mathcal{J}}(\vartheta) \cdot \vec{W}^{\mathcal{J}} \tag{5.32}$$

$$\begin{bmatrix} W'^{1} \\ W'^{2} \\ W'^{3} \end{bmatrix} = \begin{bmatrix} \cos\vartheta & \sin\vartheta & 0 \\ -\sin\vartheta & \cos\vartheta & 0 \\ 0 & 0 & 1 \end{bmatrix} \begin{bmatrix} W^{1} \\ W^{2} \\ W^{3} \end{bmatrix} \tag{5.33}$$

noting that, in pursuit of clarity, we have, for now, dropped the letter-indexed notation. To transform the row vector version of $\vec{U}$ to the primed frame, block matrix rules require multiplication by a matrix from the right: $\vec{U}'_{I} = \vec{U}_{\mathcal{J}} \cdot M^{\mathcal{J}}_{I}(\vartheta)$. Now, the essence of rotational covariance is the statement, $\vec{U}_{\mathcal{J}} \cdot \vec{W}^{\mathcal{J}} = \left(\vec{U}_{\mathcal{J}} \cdot \vec{W}^{\mathcal{J}}\right)'$. With this constraint of invariance on the product along with the idea, stated earlier in reference to tensors, that a product of vectors transforms in accord with the transformation properties of the individual vectors, we can re-express this statement as $\vec{U}_{\mathcal{J}} \cdot \vec{W}^{\mathcal{J}} = \vec{U}'_{I} \cdot \vec{W}'^{I}$. The appropriate rotation matrix $M(\vartheta)$ can then be found by writing this out as:

$$\vec{U}_{\mathcal{J}} \cdot \vec{W}^{\mathcal{J}} = \vec{U}'_{I} \cdot \vec{W}'^{I} = \vec{U}_{\mathcal{J}} \cdot M^{\mathcal{J}}_{I}(\vartheta) \cdot R^{I}_{\mathcal{J}}(\vartheta) \cdot \vec{W}^{\mathcal{J}} \tag{5.34}$$

For this to hold, $M(\vartheta) \cdot R(\vartheta)$ is necessarily the identity matrix and $M(\vartheta)$ must be the inverse of $R(\vartheta)$, a rotation in the $-\vartheta$ direction, that is, $M(\vartheta) = R(-\vartheta)$. To check this, consider the explicit form of $R(\vartheta)$ given in Eq. 5.33 and recall that rotations are represented by orthogonal matrices for which the inverse of a matrix is its transpose. In summary: column, contravariant vectors transform to a particular rotated frame according to the rotation matrix, $R(\vartheta)$, while covariant, row vectors transform to that same rotated frame by the inverse of that rotation matrix, $R(-\vartheta)$.

It is a similar story for 4-vectors. For two 4-vectors with components A_{μ} and B^{μ} in an unprimed, "lab" frame and A'_{μ} and B'^{μ} in a second frame related to the lab frame by either a rotation or a Lorentz boost, covariance is expressed by invariance of the scalar product, $A_{\mu}B^{\mu} = A'_{\mu}B'^{\mu}$: these so-called "Lorentz" group transformations are the only fixed origin transformations preserving this invariance. As an example, we assume our standard primed and unprimed frames to be related by a boost velocity, V, in the x-direction. The contravariant vector then transforms, by definition, according to the direct Lorentz transformation, $L^{\mu}_{\ \nu}(V)$. The block matrix representation of this transformation is given by the 4D extension of the (2D simplified) Lorentz boost matrix of Eq. 5.10:

$$B'^{\mu} = \frac{\partial x'^{\mu}}{\partial x^{\nu}} B^{\nu} = L^{\mu}_{\ \nu}(V) B^{\nu} \tag{5.35}$$

$$\begin{bmatrix} B'^0 \\ B'^1 \\ B'^2 \\ B'^3 \end{bmatrix} = \begin{bmatrix} \gamma & -\beta\gamma & 0 & 0 \\ -\beta\gamma & \gamma & 0 & 0 \\ 0 & 0 & 1 & 0 \\ 0 & 0 & 0 & 1 \end{bmatrix} \begin{bmatrix} B^0 \\ B^1 \\ B^2 \\ B^3 \end{bmatrix} \tag{5.36}$$

If the boost matrix which Lorentz transforms covariant vectors to the primed frame is labeled as $N^{\alpha}{}_{\beta}$, that is, if $A'_{\mu} = A_{\alpha} N^{\alpha}{}_{\mu}(V)$, then Lorentz invariance of the invariant interval scalar product can be expressed as:

$$A_{\mu} B^{\mu} = A'_{\mu} B'^{\mu} = A_{\alpha} N^{\alpha}{}_{\mu}(V) L^{\mu}{}_{\nu}(V) B^{\nu} \tag{5.37}$$

So, just as in the 3D case, the product of the two transformation matrices must give the identity matrix. In Einstein notation: $N^{\alpha}{}_{\mu}(V) L^{\mu}{}_{\nu}(V) = \delta^{\alpha}_{\nu}$. And therefore the matrix that transforms covariant vectors to the primed frame is again, the inverse of the direct transformation; in this case, the inverse of the Lorentz transformation. For the boost considered, this corresponds to a boost in the opposite direction: $N^{\alpha}{}_{\beta}(V) = L^{\alpha}{}_{\beta}(-V)$. Expressed in terms of numerically-indexed and block matrix notation, covariant 4D vectors are transformed to the primed frame as:[23]

$$A'_{\mu} = A_{\nu} L^{\nu}{}_{\mu}(-V) \tag{5.38}$$

$$[A'_0, A'_1, A'_2, A'_3] = [A_0, A_1, A_2, A_3] \begin{bmatrix} \gamma & \beta\gamma & 0 & 0 \\ \beta\gamma & \gamma & 0 & 0 \\ 0 & 0 & 1 & 0 \\ 0 & 0 & 0 & 1 \end{bmatrix} \tag{5.39}$$

So we have seen, from the requirement of scalar product invariance under rotation and Lorentz transformations, that contravariant and covariant vectors transform in inverse ways. Formally, contravariant vectors may be defined to be objects that transform in the same way as a displacement (for which the quantities are equal to measurable spatial and temporal separations), thus covariant vectors are objects that transform according to the inverse of that transformation.[24] For 4D Minkowski space, as we have seen, this just amounts to a difference of minus signs on the space-like components for the two types of vectors.

Tensors and how they transform

Having compared dot products in 3-space to scalar (or "inner") products of vectors in spacetime, we can now do the same for outer (or "tensor") products of vectors in 3D and 4D. In 3D, a tensor $\overleftrightarrow{Q} = Q^I_{\mathcal{J}}$ can be generated as the outer product of two vectors:

[23] Note: in contrast to the 3D case, Lorentz transformations are not represented by orthogonal matrices and so, here, the inverse transformation is not the same as the transpose.

[24] The prefixes "contra" and "co," specify how the vectors rotate (or transform) relative to the rotation (or transformation) of the coordinate system: "contra-" meaning "against" and "co-" meaning "with." As an example, consider a transformation matrix that represents a coordinate system rotation in one direction (from the system's perspective); then the rotations of the contravariant and covariant vectors (from the coordinate system's perspective), are in the opposite and the same directions, respectively.

$$Q^I_{\mathcal{J}} = \vec{U}^I \otimes \vec{W}_{\mathcal{J}} = \begin{bmatrix} U^1 W_1 & U^1 W_2 & U^1 W_3 \\ U^2 W_1 & U^2 W_2 & U^2 W_3 \\ U^3 W_1 & U^3 W_2 & U^3 W_3 \end{bmatrix} \tag{5.40}$$

$$= \begin{bmatrix} U_x \\ U_y \\ U_z \end{bmatrix} \left[W_x, W_y, W_z\right] = \begin{bmatrix} U_x W_x & U_x W_y & U_x W_z \\ U_y W_x & U_y W_y & U_y W_z \\ U_z W_x & U_z W_y & U_z W_z \end{bmatrix} \tag{5.41}$$

where we note, for the block matrix notation, the required mixture of contravariant and covariant vectors and the required product order. Similarly, in 4D:

$$T^{\mu}{}_{\nu} = A^{\mu} B_{\nu} = \begin{bmatrix} A^0 B_0 & A^0 B_1 & A^0 B_2 & A^0 B_3 \\ A^1 B_0 & A^1 B_1 & A^1 B_2 & A^1 B_3 \\ A^2 B_0 & A^2 B_1 & A^2 B_2 & A^2 B_3 \\ A^3 B_0 & A^3 B_1 & A^3 B_2 & A^3 B_3 \end{bmatrix} \tag{5.42}$$

$$= \begin{bmatrix} A_{ct} \\ A_x \\ A_y \\ A_z \end{bmatrix} \left[B_{ct}, -B_x, -B_y, -B_z\right] \tag{5.43}$$

$$= \begin{bmatrix} A_{ct}B_{ct} & -A_{ct}B_x & -A_{ct}B_y & -A_{ct}B_z \\ A_xB_{ct} & -A_xB_x & -A_xB_y & -A_xB_z \\ A_yB_{ct} & -A_yB_x & -A_yB_y & -A_yB_z \\ A_zB_{ct} & -A_zB_x & -A_zB_y & -A_zB_z \end{bmatrix} \tag{5.44}$$

Thus, just as we did for the 3D and 4D scalar (or inner) products in Eqs. 5.23–5.27, we have here expressed 3D and 4D tensor (or outer) products in both numerically indexed notation (Eqs. 5.40 and 5.42) and letter-indexed block matrix notation (Eqs. 5.41, 5.43, and 5.44). Upon comparing these equations, we can see that while block matrix notation is useful for simultaneously visualizing the sets of physically related quantities that make up vectors and tensors, the alternative numerically indexed notation along with the Einstein convention of summing over repeated indices (inner product multiplication) is considerably more flexible, more compact, and better for doing calculations. To begin with, the order in which the quantities are written in an expression does not matter. But more importantly, flexibility in choosing the contravariant or covariant version of a physical quantity for use in an equation frequently turns out to be convenient. Indeed, block matrix representation of tensors is limited, by row/column multiplication rules, to the mixed variety in which the tensor must be a product of a contravariant vector and a covariant vector, as confirmed in Eqs. 5.41–5.42. In contrast, Einstein notation can represent tensors of all vector type combinations. For example, a fully contravariant version of $T^{\mu}_{\ \nu}$ is simply obtained by taking the outer product of the two contravariant versions of the vectors: $T^{\mu\nu} = A^{\mu}B^{\nu}$.

In the last section, through consideration of the transform invariance of the scalar product, we determined that contravectors and covectors transform inversely. We might turn this around and say that the scalar product adopts the opposing transformation properties of a contravector and a covector combining to result in a quantity whose value is identical in all inertial frames. This observation–that the product adopts the transform properties of its components–naturally extends to tensor products so that transformation of a tensors, upon translation, rotation or boost of the (3D or 4D) coordinate system, can be obtained by working through the different (contravariant and covariant) transformations of the individual vectors in the product. For example, consider the transformation of a fully contravariant version of $Q^{I}_{\ \mathcal{J}}$, the mixed 3-tensor of Eq. 5.40, for a 3-space rotation, $R(\vartheta)$,

$$\begin{aligned} Q'^{I\mathcal{J}} = \vec{U}'^{I} \otimes \vec{W}'^{\mathcal{J}} &= \left(R^{I}_{K}(\vartheta) \cdot \vec{U}^{K}\right) \otimes \left(R^{\mathcal{J}}_{L}(\vartheta) \cdot \vec{W}^{L}\right) \\ &= R^{I\mathcal{J}}_{KL}\left(\vec{U}^{K} \otimes \vec{W}^{L}\right) = R^{I\mathcal{J}}_{KL} Q^{KL} \end{aligned} \tag{5.45}$$

where $R^{I\mathcal{J}}_{KL} \equiv R^{I}_{K}(\vartheta) \otimes R^{\mathcal{J}}_{L}(-\vartheta)$, is a 4^{th}rank, 3D matrix. Similarly, the transformation of a fully contravariant version of $T^{\mu}_{\ \nu}$, the mixed 4-tensor of Eq. 5.42, for a Lorentz boost, $L(V)$,

$$\begin{aligned} T'^{\mu\nu} = A'^{\mu} B'^{\nu} &= \left(L^{\mu}_{\alpha}(V) A^{\alpha}\right)\left(L^{\nu}_{\beta}(V) B^{\beta}\right) \\ &= \Lambda^{\mu\ \nu}_{\ \alpha\ \beta} A^{\alpha} B^{\beta} = \Lambda^{\mu\ \nu}_{\ \alpha\ \beta} T^{\alpha\beta} \end{aligned} \tag{5.46}$$

where $\Lambda^{\mu\ \nu}_{\ \alpha\ \beta} \equiv L^{\mu}_{\alpha}(V) L^{\nu}_{\beta}(V)$, is a 4th rank, 4D matrix. And since, by definition, all 4-tensors of a given subscript-superscript pattern transform identically for a Lorentz transformation, the 4th rank, 4D matrix, $\Lambda^{\mu\ \nu}_{\ \alpha\ \beta}$, is the transformation for any fully contravariant 4-tensor. Equations 5.45 and 5.46 are completely analogous statements in 3D and 4D, respectively. But notice that although both equations use numerically indexed notation, in the 3D equation, Eq. 5.45, the inner and outer product symbols "·" and "⊗" as well as the top arrow have been kept, while in the 4D equation, Eq. 5.46, these last vestiges of 3D formalism have been dropped. The hope and motivation of this gradual notational change is to help provide a smoother, more transparent transition to the 4D notation.

Finally, we can generalize to an mth rank (or order), n-dimensional tensor. Such an object can be viewed as the outer product of m

different n-dimensional vectors, each of which can be either contravariant or covariant. And, again, by definition, all m^{th} rank, n-tensors with the same subscript-superscript pattern transform identically. Thus, the subscript-superscript designation is quite general and can be used with tensors of any rank–with the number of indices matching the tensor rank. Likewise, the scalar product (contraction) rule of opposite indices also extends to include tensors of any order.

5.4.2 The metric tensor

The metric of our 3D space or 4D space-time or, indeed, of any metric space can be represented in terms of the associated metric tensor, $g_{\alpha\beta}$, of that space:

$$ds^2 = g_{\alpha\beta} dx^\alpha dx^\beta \tag{5.47}$$

where we are again using the Einstein convention of summing over repeated indices. The metric tensor of a space can be alternately and equivalently described either as defining the geometry of the space, or as encoding the relationship between the contravariant and covariant vectors of the space.[25] With respect to the first description: while for spaces such as those of General Relativity, the metric tensor varies from point to point resulting in a metric tensor field–akin to the concept of a vector field–extending throughout the spacetime, if a space is uniform, such as those discussed here, then it is wholly described by a single metric tensor. Indeed, the metric tensor is a sort of differentiated version of the metric in the sense that its components are the coefficients within the metric. It is readily seen, for example, through comparison of Eq. 5.47 with Eqs. 5.16 and 5.17, that the metric tensor for 3D space is just the 3×3 identity matrix and the metric tensor for Minkowski spacetime is,

$$g_{\alpha\beta} = \begin{bmatrix} 1 & 0 & 0 & 0 \\ 0 & -1 & 0 & 0 \\ 0 & 0 & -1 & 0 \\ 0 & 0 & 0 & -1 \end{bmatrix} \tag{5.48}$$

So that, in space-time the 4×4 "Minkowski" metric tensor is again diagonal, but now has minus signs on the space-like components, $g_{00} = 1$ and $g_{11} = g_{22} = g_{33} = -1$. In addition to the metric, which is the scalar product of a vector with itself, the scalar product of two different vectors in a vector space may be written using the metric tensor:[26]

$$A_\beta B^\beta = g_{\alpha\beta} A^\alpha B^\beta \tag{5.49}$$

[25] More specifically, it relates the contravariant "components" of a vector to the covariant "projections" of that vector in a particular space. While the distinction of "components" and "projections" is not intuitive in spacetime and completely nonexistent in 3D space, it becomes quite clear in non-orthonormal spaces.

[26] The definition of the scalar product of a metric space is implicit in the metric of that space. More specifically, for an arbitrary metric space, if the magnitude of a displacement vector $\vec{x}$, as specified by the metric of that space in the form of Eq. 5.47, is indicated as $|\vec{x}|$, then the dot product of two different displacement vectors, $\vec{x}$ and $\vec{y}$, is defined completely in terms of the metric-specified displacement vector magnitudes:

$$\vec{x} \cdot \vec{y} = \frac{1}{2}\left[|\vec{x}+\vec{y}|^2 - |x|^2 - |y|^2 \right]$$

While this is completely general, we can consider the simple 3D case in which the metric specifies a displacement magnitude of $|\vec{x}| = \left(x_1^2 + x_2^2 + x_3^2\right)^{1/2}$. Thus

$$\begin{aligned} |\vec{x}+\vec{y}|^2 &= (x_1+y_1)^2 + (x_2+y_2)^2 + (x_3+y_3)^2 \\ &= |x|^2 + |y|^2 + 2\vec{x}\cdot\vec{y} \end{aligned}$$

which rearranges to the previous equation.

where we note, lastly, that the invariance of the quantities, ds^2 and $A_\beta B^\beta$, in Eqs. 5.47 and 5.49, for all inertial frames, requires that the 4×4 metric tensor used, $g_{\alpha\beta}$, is also "Lorentz invariant," which is to say that it is numerically the same in all inertial frames.

In the context of our discussion of Special Relativity, however, it is the second description before, that of $g_{\alpha\beta}$ providing a connection between contravariant and covariant vectors, which is of greater importance and practical value. Previously, we mentioned the convenience of having the flexibility, within the number-indexed convention, of choosing the contravariant or covariant version of a physical quantity for use in an equation and we used the example of obtaining the fully contravariant tensor, $T^{\mu\nu}$, from the mixed tensor, $T^{\mu}_{\ \nu} = A^\mu B_\nu$, in Eq. 5.42. This flexibility is obtained via the Minkowski metric tensor, $g_{\alpha\beta}$, which has the incredibly useful function of lowering indices. From the matrix form of Eq. 5.48, we can see that it takes a contravariant vector and returns its covariant version:

$$A_\alpha = g_{\alpha\beta} A^\beta$$

In fact, the scalar product relations of Eqs. 5.47 and 5.49 can now more clearly be read as the metric tensor's first lowering an index followed by scalar product multiplication. To raise indices, a similar tensor, $g^{\alpha\beta}$, is identified whose numerical entries are identical to those of $g_{\alpha\beta}$. This tensor returns the contravariant version of a vector, given the covariant vector as an input. Consider, in general, the product of the fully contravariant and fully covariant metric tensors of a metric space,

$$g^{\alpha\nu} g_{\nu\beta} = \delta^\alpha_\beta \tag{5.50}$$

This may be read as saying either that the two metric tensors are one another's inverses or that the mixed, contravariant-covariant form of the metric tensor is the identity matrix of the space. Finally, we can note the generally useful rule-of-thumb that in all scalar product based operations (also known as contractions), the number of superscripts minus the number of subscripts is conserved.

5.4.3 Generation of other 4-vectors and 4-tensors

From a set of 4-tensors, additional 4-tensors can be generated by:

(1) Adding two similar 4-tensors together. The resulting 4-tensor adopts the transformation properties of the original tensors.

(2) Taking a scalar product of two 4-tensors: summing over a superscript in one tensor and a subscript in the other. With this and the following, the resulting tensor transformation behavior is readily seen by carrying out operations on two similar tensors created from products of space-time displacements (**see Discussion 5.15**).

(3) Taking the outer product of any two 4-tensors (not necessarily of the same order). For example, from the two 2nd order tensors, $\mathcal{J}^{\alpha\beta} = A^{\alpha}B^{\beta}$ and $K_{\gamma\delta} = C_{\gamma}D_{\delta}$, several different 4th order tensors can be created, including $M^{\alpha\beta}{}_{\gamma\delta} \equiv A^{\alpha}B^{\beta}C_{\gamma}D_{\delta}$ and $N_{\delta}{}^{\beta\alpha}{}_{\gamma} \equiv D_{\delta}B^{\beta}A^{\alpha}C_{\gamma}$.[27]

(4) Contracting an upper and a lower index of a single 4-tensor. This may be considered as a special case of (2).

(5) Multiplication by the Levi–Civita tensor. The components, $\varepsilon^{\alpha\beta\gamma\delta}$, of the Levi–Civita tensor are defined as follows:

$$\varepsilon^{\alpha\beta\gamma\delta} \equiv \begin{cases} +1 \text{ if } (\alpha,\beta,\gamma,\delta) \text{ is an even permutation of } (0,1,2,3) \\ -1 \text{ if } (\alpha,\beta,\gamma,\delta) \text{ is an odd permutation of } (0,1,2,3) \\ 0 \text{ otherwise (one of the indices is repeated)} \end{cases} \tag{5.51}$$

And this can be shown to be a 4-tensor, sometimes called the totally antisymmetric 4-tensor of rank 4. When an antisymmetric 2nd rank 4-tensor, $A_{\gamma\delta}$, is multiplied by the Levi–Civita symbol, the result is called the dual of tensor, $\mathscr{A}^{\alpha\beta}$:

$$\mathscr{A}^{\alpha\beta} = \frac{1}{2}\varepsilon^{\alpha\beta\gamma\delta}A_{\gamma\delta} \tag{5.52}$$

(6) Permuting or raising or lowering indices.[28] It is readily shown that this is a special example of (2) where the scalar product of the metric tensor with a second tensor will raise or lower the corresponding index. If permutation of the indices of a 2nd order tensor yields a tensor which is equivalent to the original, $T^{\alpha\beta} = T^{\beta\alpha}$, then the tensor is said to be symmetric. On the other hand, if $T^{\alpha\beta} = -T^{\beta\alpha}$, the tensor is "antisymmetric." Such a permutation is equivalent, in block notation, to a transpose.

5.5 Some important general 4-vectors

So far, we have only really discussed two related 4-vectors: the displacement and coordinate 4-vectors. We have referred to other

[27] It is important to keep the positions of the subscripts and superscripts straight. Again, the transformation of the resulting outer product tensor is identical to the corresponding outer product created from two tensors that in turn are created from displacement vector outer products. Since the latter–the tensors created from displacement vectors–by definition, transform properly as tensors, the original product vector transforms identically and therefore properly.

[28] Again, any resulting change in the transformation properties of the tensor will correspond to a change in the order or contravariant-covariant nature of the displacements used to create an analogous, identically transforming tensor.

4-vectors and 4-tensors as collections of quantities which have the same transformation properties as their equivalent displacement-based vectors and tensors, but we have not yet been specific. We will now consider a number of important 4-vectors. In this subsection, we consider the more general 4-vectors: the 4-gradient operator, the 4-velocity, the 4-momentum and the 4-force. In the next subsection, we consider 4-vectors specific to electromagnetism: the 4-wavevector, the 4-current density and the Lorenz gauge 4-potential.

5.5.1 The 4-gradient operator

The 4-gradient operator presents an exception to the usual pattern that the "regular" form of a 4-quantity with all plus signs is the contravariant vector. Indeed, we will find below that the set of 4 differential space-time operators

$$\left(\frac{\partial}{\partial\,(ct)},\frac{\partial}{\partial x},\frac{\partial}{\partial y},\frac{\partial}{\partial z}\right)\equiv\left(\frac{\partial}{\partial x^0},\frac{\partial}{\partial x^1},\frac{\partial}{\partial x^2},\frac{\partial}{\partial x^3}\right)$$

do not change under a Lorentz transformation like the displacement set

$$(c\Delta t,\Delta x,\Delta y,\Delta z)$$

Recall, the Lorentz transformation from an un-primed frame to the primed frame can be written in terms of the partial derivatives of the coordinates in the two frames, Eq. 5.10[29]:

$$L^{\alpha}_{\ \beta}\,(V)=\frac{\partial x'^{\alpha}\left(V;x^0,x^1,x^2,x^3\right)}{\partial x^{\beta}} \tag{5.53}$$

Indeed, this equation is little more than the definition of a transformation and would hold true for any linear transformation. Similarly, the inverse transformation is given by

$$\left(L^{-1}\right)^{\beta}_{\ \alpha}(V)=L^{\beta}_{\ \alpha}(-V)=\frac{\partial x^{\beta}\left(V;x'^0,x'^1,x'^2,x'^3\right)}{\partial x'^{\alpha}} \tag{5.54}$$

In 3D, it is useful to identify the gradient vector operator, $\nabla=\left(\frac{\partial}{\partial x},\frac{\partial}{\partial y},\frac{\partial}{\partial z}\right)$. The 4-vector analog is denoted $\partial_\alpha=(\partial_0,\partial_1,\partial_2,\partial_3)=\left(\frac{\partial}{\partial(ct)},\frac{\partial}{\partial x},\frac{\partial}{\partial y},\frac{\partial}{\partial z}\right)$ where we introduce the shorthand notation, ∂_α, for taking the partial derivative with respect to the indexed quantity. The sub-scripted index of this operator indicates that it is covariant. To see

[29] The parentheses here are used just to explicitly indicate the functional dependence of the primed coordinate on the velocity of the moving frame and the unprimed coordinates.

this, consider the action of the 4-gradient operator on a Lorentz scalar quantity in both the un-primed and primed frames:

$$\partial_\alpha \phi = \frac{\partial \phi}{\partial x^\alpha}$$
$$\partial'_\alpha \phi = \frac{\partial \phi}{\partial x'^\alpha} = \frac{\partial x^\mu}{\partial x'^\alpha}\frac{\partial \phi}{\partial x^\mu}$$
$$\Rightarrow \partial'_\alpha = \left(L^{-1}\right)^\mu{}_\alpha \partial_\mu$$

Comparing this with Eq. 5.54, we see the gradient operator with all plus signs transforms according to the inverse transformation and is therefore a covariant vector operator. The contravariant gradient is, therefore,

$$\partial^\alpha = \left(\frac{\partial}{\partial (ct)}, -\frac{\partial}{\partial x}, -\frac{\partial}{\partial y}, -\frac{\partial}{\partial z}\right)$$

As in 3D, the 4-vector gradient operator can be manipulated as any other covariant 4-vector and under Lorentz transformations will transform as expected. In particular, the scalar product of the 4-vector gradient operator and a 4-vector field, or the "4-divergence" of that 4-vector field, can be seen as a time derivative coupled with a regular spatial gradient,

$$\partial_\alpha A^\alpha = \frac{\partial A^0}{c\partial t} + \nabla \cdot \vec{A} = \frac{\partial A^0}{c\partial t} + \frac{\partial A^1}{\partial x} + \frac{\partial A^2}{\partial y} + \frac{\partial A^3}{\partial z} \tag{5.55}$$

which is a Lorentz scalar. The 4-vector divergence operator seems ready-made to write conservation laws. For example, the 4-divergence of the current density 4-vector $\mathcal{J}^\alpha = \left(c\rho, \vec{\mathcal{J}}\right)$, to be formally introduced shortly, is a Lorentz scalar:

$$\partial_\alpha \mathcal{J}^\alpha = \frac{\partial (c\rho)}{c\partial t} + \nabla \cdot \vec{\mathcal{J}} \tag{5.56}$$

So we now see that the charge continuity condition, $\partial_\alpha \mathcal{J}^\alpha = 0$, true in the absence of any current sources or sinks, is valid in all inertial frames–thus establishing the continuity equation as Lorentz invariant. Finally, the 4-vector Laplacian (or D'alambertian), $\partial^\alpha \partial_\alpha$,

$$\partial^\alpha \partial_\alpha \phi = \frac{\partial^2 \phi}{\partial (ct)^2} - \frac{\partial^2 \phi}{\partial x^2} - \frac{\partial^2 \phi}{\partial y^2} - \frac{\partial^2 \phi}{\partial z^2} = \frac{1}{c^2}\frac{\partial^2 \phi}{\partial t^2} - \nabla^2 \phi \tag{5.57}$$

is identified as the scalar wave operator. Note it is identical to the "box" operator, $\Box$.

5.5.2 The 4-vector velocity

By analogy to 3D and given the displacement vector, the next reasonable 4-vector to consider is the 4-velocity. Take, for example, a particle whose motion in space as a function of time is described by the 4-vector, $x^\alpha = (ct, \vec{x}(t))$. Then, apparently the 4-velocity is the time derivative of its coordinates:

$$v^\alpha(t) = \frac{d}{dt}(ct, \vec{x}(t)) = (c, \vec{v}) \tag{5.58}$$

So, do these resulting four components constitute a 4-vector? A necessary condition for four quantities to constitute a 4-vector is that the magnitude of the proposed 4-vector be Lorentz invariant–as is required for 4-vector scalar products in general. Equation 5.8 clearly does not satisfy that requirement, since in the lab frame:

$$|v^\alpha| = \left[(c, \vec{v})(c, -\vec{v})\right]^{1/2} = \left(c^2 - v^2\right)^{1/2} \tag{5.59}$$

whereas in a frame moving with the particle , $\vec{v}' = 0$ and:

$$|v'^\alpha| = [(c, 0)(c, 0)]^{1/2} = c \tag{5.60}$$

This is hardly surprising since in this way of finding the velocity, time assumes a privileged role (i.e., it's referenced to a particular frame–the rest frame) which seems inconsistent with the mixing of time and position that occurs in special relativity. In other words, if a differential displacement vector, dx^α, which we know is Lorentz covariant, is divided by dt, which is not a Lorentz scalar but rather whose magnitude is inertial frame dependent, then it is clear that the resulting magnitude will not be Lorentz invariant and the four quantities will not constitute a 4-vector. A similar 4-vector quantity whose magnitude *is* Lorentz invariant is the derivative of the 4-vector displacement with respect to its proper time, τ. This is called the 4-vector velocity of a particle:[30]

$$U^\alpha = \frac{dx^\alpha}{d\tau} \tag{5.61}$$

which may be written in terms of the time of the reference frame in which it is being observed:

$$U^\alpha = \frac{dx^\alpha}{d\tau} = \frac{dx^\alpha}{dt}\frac{dt}{d\tau} = \frac{dx^\alpha}{dt}\left(\frac{d\tau}{dt}\right)^{-1} = \gamma v^\alpha = \gamma\,(c, \vec{v}) \tag{5.62}$$

[30] Here, we need the *total* differential of the proper time. As before:

$$d\tau = \frac{1}{c}\left[c^2\left(\frac{dt}{dt}\right)^2 - \left(\frac{dx}{dt}\right)^2 - \left(\frac{dy}{dt}\right)^2 - \left(\frac{dz}{dt}\right)^2\right]^{1/2} dt = \frac{1}{c}\left(c^2 - v^2\right)^{1/2} dt = \frac{dt}{\gamma}$$

where we recall that the scalar $d\tau$ is equal to $1/c$ times the magnitude of the differential displacement 4-vector $|dx^\alpha|$ that it is dividing and it is therefore an invariant quantity.

It has an invariant magnitude equal to c,

$$|U^\alpha| = \gamma\left[(c,\vec{v})(c,-\vec{v})\right]^{1/2} = \gamma\left(c^2 - v^2\right)^{1/2} = c$$

In the non-relativistic limit, where $\gamma \sim 1$, the frame-dependent spatial components of the 4-velocity vector return a particle's velocity, $\vec{v}$. The 4-vector acceleration is similarly defined as $a^\alpha = \frac{dU^\alpha}{d\tau}$.

5.5.3 The 4-vector momentum

Momentum and energy appear in classical physics in several places and we expect there to be analogs in special relativity. Relativistic particle momenta would ideally have the following properties:

(1) For small velocities, the limits of relativistic particle momenta and energies are their classical counterparts.

(2) In an isolated system of two or more particles, both total momentum and total energy are conserved. This includes the case of collisions and should hold in all inertial frames.

(3) Given an independent concept of the force acting on a particle, the time derivative of the momentum should equal the force.

(4) When describing a collection of free particles using action principle concepts, the momenta should be conjugate variables to the particle coordinates (see Chapter 7) and the total energy should be the Hamiltonian.

(5) Noether's theorem states that for every symmetry of a system there is a conserved quantity. For a system consisting of a particle or a collection of particles, momentum and energy are identified as the conserved quantities derived from invariance of the system under translations of the spatial and time origins.

Based on the discussion of the relativistic 4-velocity, U^α, it is natural to extend momentum to relativistic momentum by the association of each particle with the 4-vector $p^\alpha = mU^\alpha = \gamma m(c,\vec{v})$ where m is the Lorentz invariant scalar rest mass of the particle. In what follows, we argue that this is, in fact, the correct 4-vector analog with components:

$$p^\alpha = \left(\frac{\mathscr{E}}{c}, \vec{p}\right), \tag{5.63}$$

where, $\vec{p} = \gamma m\vec{v}$, is the relativistic momentum of a particle and $\mathscr{E} = \gamma mc^2$ is its energy.

Indeed, we can immediately demonstrate the first property outlined before: in the limit of small velocities, $v/c \ll 1$, gamma approximates as $\gamma \simeq 1 + v^2/2c^2 \simeq 1$, and $\vec{p}$ and $\mathscr{E}$ reduce to

$$\vec{p} = \gamma m\vec{v} \simeq \left(1 + \frac{1}{2}\frac{v^2}{c^2}\right) m\vec{v} \simeq m\vec{v} \tag{5.64}$$

and

$$\mathscr{E} = \gamma mc^2 \simeq \left(1 + \frac{1}{2}\frac{v^2}{c^2}\right) mc^2 = mc^2 + \frac{1}{2}mv^2 \tag{5.65}$$

So the spatial components clearly return the classical limit for a particle's momentum and the time component of relativistic energy is seen to reduce, in the classical limit, to the kinetic plus rest mass energy of the particle.

Since it is a four vector, the defined quantity of Eq. 5.63 automatially transforms correctly under a Lorentz transformation. In particular, consider two equal mass particles colliding so that the initial and final total 4-momenta, P^α_{ini} and P^α_{fin}, are just the sums of the 4-momenta of the two particles. In the center of mass (COM) frame of Fig. 5.7, $\mathscr{E}_2 = \mathscr{E}_1$ and $\vec{p}_2 = -\vec{p}_1$, so P^α_{ini} and P^α_{fin} are expressed as,

$$P^\alpha_{ini} = \left(\frac{2\mathscr{E}_1}{c}, \vec{p}_1 - \vec{p}_1 = 0\right) \quad and \quad P^\alpha_{fin} = \left(\frac{2\mathscr{E}_1}{c}, -\vec{p}_1 + \vec{p}_1 = 0\right)$$

Thus, in the COM frame, the total 4-momentum through the collision is conserved. This can be expressed by the zero 4-vector difference between the final and inital total 4-momenta: $0^\alpha = P^\alpha_{fin} - P^\alpha_{ini}$. In any other inertial frame, the Lorentz boost transformation will mix the time-like energy and space-like momentum components so that $\mathscr{E}'_2 \neq \mathscr{E}'_1$ and $\vec{p}'_1 \neq -\vec{p}'_2$ and the total 4-momentum in a primed (non-COM)

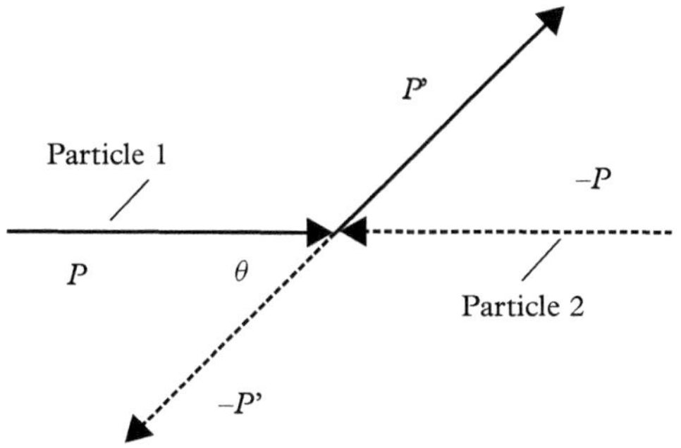

Fig. 5.7 *An elastic collision between two particles in the center of momentum frame.*

frame must be written more generally. For example, the initial total 4-momentum is written

$$P'^{\alpha}_{ini} = \left(\frac{\mathscr{E}'_1}{c} + \frac{\mathscr{E}'_2}{c}, \vec{p}'_1 + \vec{p}'_2 \neq 0\right)$$

with a similar expression for P'^{α}_{fin}, but with different particle energies and momenta. However, because P^{α}_{ini} and P^{α}_{fin} are covariant under Lorentz transformations, their difference, the zero 4-vector (0^{α}), is also Lorentz covariant. Furthermore, it is readily shown that all components of this zero 4-vector remain equal to zero in all frames.[31] That is, not only is $0^{\alpha} = P^{\alpha}_{fin} - P^{\alpha}_{ini}$ Lorentz covariant, its components are Lorentz invariant. Thus, each of the components of $0^{\alpha\prime} = L^{\alpha}{}_{\beta} 0^{\beta}$ will equal zero in any inertial frame - each component of the total 4-momentum during the collision is automatically conserved in all frames. Now, this is consistent with what we know from classical mechanics; namely, that although the individual particle energy and momentum values change upon collision, the total energy and momentum are each individually conserved–and this is true in all inertial frames. Our present relativistic development, however, differs in that it considers relativistic energy and momentum merely to be different components of a single conserved quantity: the 4-momentum. And, in different frames the four components of this energy-momentum "quantity" differ but the magnitude is invariant.

We next examine the time-like component of the 4-momentum of a particle by equating the Lorentz invariant scalar created by the scalar product (covariant times contravariant) of the 4-momentum with itself in the rest frame of the particle to its general form in an arbitrary frame:

$$\frac{\mathscr{E}^2}{c^2} - p^2 = m^2c^2 \tag{5.66}$$

In the rest frame of the particle, this gives Einstein's famous $\mathscr{E} = mc^2$ for the energy related to the particle mass. The rest mass energy has no significant impact on low velocity dynamics and is rarely detectable in everyday experience. The relativistic kinetic energy is the difference between the moving particle's energy and its rest mass energy, $K = \mathscr{E} - mc^2 = (\gamma - 1)\,mc^2$. That this relativistic kinetic energy of a particle in motion converges to that of classical physics may be seen from the low velocity approximation of energy in Eq. 5.65. Alternatively, it may be derived by first using Eq. 5.66 to express the total energy $\mathscr{E}$ in terms of the momentum $\vec{p}$ and then using the binomial approximation

31 Note that in Minkowski space the zero 4-vector (0^{α}) is a special case of zero magnitude vectors in which the two points are colocated (all the components are zero). Contrast this with 3-space in which there is only one zero magnitude vector, the zero 3-vector, $(\vec{0})$.

with $p/mc \ll 1$ to obtain the classical expression for a particle's kinetic energy (**see Discussion 5.16**):

$$K = \left(m^2c^4 + p^2c^2\right)^{1/2} - mc^2 = mc^2\left[1 + \left(\frac{p}{mc}\right)^2\right]^{1/2} - mc^2 \approx \frac{p^2}{2m} \tag{5.67}$$

Here, we have shown that the proposed form of Eq. 5.63 satisfies two of the five properties, outlined previously, expected of relativistic momentum. In the last part of this chapter and in Chapter 7, it will be seen that the definition of a particle's 4-momentum as $p^\alpha = mU^\alpha$ is consistent with properties 3, 4, and 5 in the previous list. Experimentally relevant to property 3–that the force equals the time derivative of the momentum–are Kaufmann's observations (1900–1905) of what he referred to as the increased mass of high energy electrons. These results, which were in better agreement with the relativistic momentum expression, pointed up the need to revise the classical momentum expression. Today, of course, at modern accelerators the force-momentum relationship and the limiting velocity of the speed of light has been experimentally observed and verified to high accuracy.[32] We will see that, for electromagnetism in particular, Eq. 5.63 is indeed the quantity whose time derivative is the Lorentz force on a charged particle.

5.5.4 The 4-vector force

Following the extension to relativistic velocity and then to relativistic momentum, we can similarly extend classical acceleration and forces to relativistic acceleration and 4-forces. Relativistic acceleration, a^α, is thus defined as the derivative of the 4-velocity with respect to the particle's own, or proper, time, τ. And the so-called "Minkowski" 4-force naturally follows as

$$f^\alpha = ma^\alpha = m\frac{dU^\alpha}{d\tau} = \frac{dp^\alpha}{d\tau}$$

where m is the Lorentz invariant scalar rest mass of the particle. Noting that $p^\alpha = \gamma m(c, \vec{v}) = \left(\frac{\mathscr{E}}{c}, \vec{p}\right)$, and the general result that[33]

$$\frac{d\gamma}{d\tau} = \gamma\frac{d\gamma}{dt} = \gamma^4\frac{\vec{a}\cdot\vec{v}}{c^2} \tag{5.68}$$

where $\vec{a}$, $\vec{v}$ and γ are relative to the observer, we can obtain the general equation for the 4-force as

[32] Evenson, K. M. et al. Speed of light from direct frequency and wavelength measurements of the methane-stabilized laser. *Physical Review Letters* 29 (19), 1346–1349 (1972).

[33] The change in gamma over proper time is

$$\frac{d\gamma}{d\tau} = \gamma\frac{d}{dt}\left(1 - \frac{v^2}{c^2}\right)^{-1/2}$$

$$= -\frac{\gamma}{2}\left(1 - \frac{v^2}{c^2}\right)^{-3/2}\left(\frac{-2\vec{v}\cdot\vec{a}}{c^2}\right) = \gamma^4\frac{\vec{a}\cdot\vec{v}}{c^2}.$$

$$f^{\alpha}=\left(\gamma^4 m\frac{\vec{a}\cdot\vec{v}}{c},\ \gamma^4 m\frac{\vec{a}\cdot\vec{v}}{c^2}\vec{v}+\gamma^2 m\vec{a}\right)=\left(\frac{\gamma}{c}\frac{d\mathscr{E}}{dt},\ \gamma\frac{d\vec{p}}{dt}\right) \tag{5.69}$$

which, as the time derivative of the "energy-momentum" 4-vector, would be more appropriately termed a "power-force" 4-vector. As for all 4-vectors, the magnitude of the 4-force is Lorentz invariant. And, as with the 4-momentum, a proper choice of frame allows easy evaluation of the 4-vector magnitude. The frame with a velocity equal to the instantaneous velocity of the accelerating particle is such a choice. At that instant, in that frame, $\vec{v}=0$ and the 4-force magnitude squared evaluates as

$$f^{\alpha}f_{\alpha}=-(ma)^2$$

the square root of which results, interestingly, in an imaginary, "space-like" amplitude equal to the classically defined force. If we now consider the time rate of change of relativistic momentum, we obtain the relativistic force, $\vec{f}=\frac{d\vec{p}}{dt}$,

$$\vec{f}=\gamma^3 m\frac{\vec{a}\cdot\vec{v}}{c^2}\vec{v}+\gamma m\vec{a} \tag{5.70}$$

Thus, we can rewrite Eq. 5.69 in terms of $\vec{f}$ as[34]

$$f^{\alpha}=\left(\frac{\gamma}{c}\vec{f}\cdot\vec{v},\ \gamma\vec{f}\right) \tag{5.71}$$

Comparison of the time components in Eqs. 5.69 and 5.71 confirms that the time rate of change of particle energy is the scalar product of relativistic force and velocity, $\frac{d\mathscr{E}}{dt}=\vec{f}\cdot\vec{v}$. The next point of interest, apparent in Eq. 5.70, is that, unlike for a classical force, the acceleration need not be in the same direction as the force. Indeed, the relativistic force has an additional component in the velocity direction–a direction which, in general, differs from that of the acceleration. This additional term, the $d\gamma/dt$ term in the time derivative of the relativistic momentum, represents the rate of change in momentum due to a change in the relativistic mass, γm (this is in contrast to classical mechanics in which the mass is constant and any change in momentum is due solely to a change in velocity). There are, however, two special cases in which the acceleration is purely parallel to the applied force: constant circular motion and linear acceleration. For circular motion ($\vec{a}\perp\vec{v}$), the first right-hand side term in Eq. 5.70 vanishes, leaving $\vec{f}=\gamma m\vec{a}$. For linear acceleration ($\vec{a}\parallel\vec{v}$), Eq. 5.70 becomes one dimensional resulting in $f=\gamma^3 ma\frac{v^2}{c^2}+\gamma ma=\gamma^3 ma$.[35] So now it is clear why, in general, the

[34] Using Eq. 5.68, the relativistic force, $\vec{f}=\frac{d\vec{p}}{dt}$, can be put in the form,

$$\vec{f}=\gamma^3 m\frac{\vec{a}\cdot\vec{v}}{c^2}\vec{v}+\gamma m\vec{a}$$

Now while it is trivial to see, from Eq. 5.69, that the space part of f^{α} is $\gamma\vec{f}$, the time part is more involved. Working backward from the time component result in Eq. 5.71, we can show it is equivalent to that in Eq. 5.69,

$$\begin{aligned} f^0&=\frac{\gamma}{c}\vec{f}\cdot\vec{v}=\gamma^4 m\frac{\vec{a}\cdot\vec{v}}{c^3}v^2+\gamma^2 m\frac{\vec{a}\cdot\vec{v}}{c}\\ &=\gamma^4 m\frac{\vec{a}\cdot\vec{v}}{c^3}v^2+\gamma^4 m\frac{\vec{a}\cdot\vec{v}}{c^3}\left(\frac{c^2}{\gamma^2}\right)\\ &=\gamma^4 m\frac{\vec{a}\cdot\vec{v}}{c^3}v^2+\gamma^4 m\frac{\vec{a}\cdot\vec{v}}{c^3}\left(c^2-v^2\right)\\ &=\gamma^4 m\frac{\vec{a}\cdot\vec{v}}{c}\end{aligned}$$

[35] More explicitly, decomposing Eq. 5.70 into components parallel and perpendicular to the velocity:

$$\begin{aligned}&\vec{f}_{\parallel}+\vec{f}_{\perp}\\ &=\gamma^3 m\frac{(\vec{a}_{\parallel}\cdot\vec{v})+(\vec{a}_{\perp}\cdot\vec{v})}{c^2}\vec{v}+\gamma m\vec{a}_{\parallel}+\gamma m\vec{a}_{\perp}\end{aligned}$$

we can write this as two separate, orthogonal equations. The parallel force component:

$$\begin{aligned}\vec{f}_{\parallel}&=\gamma^3 m\frac{(\vec{a}_{\parallel}\cdot\vec{v})+(\vec{a}_{\perp}\cdot\vec{v})}{c^2}\vec{v}+\gamma m\vec{a}_{\parallel}\\ &=\gamma^3 m\frac{a_{\parallel}v}{c^2}\vec{v}+\gamma m\vec{a}_{\parallel}=\gamma^3 m\vec{a}_{\parallel}\end{aligned}$$

and the perpendicular force component:

$$\vec{f}_{\perp}=\gamma m\vec{a}_{\perp}$$

acceleration will be in a different direction than the applied force: any applied force can be decomposed into components parallel and perpendicular to the instantaneous velocity of the particle and because the "mass" coefficients are different, the associated accelerations will be proportionally different. For instance, two equal magnitude forces applied parallel and perpendicular to the velocity will result in two different magnitude accelerations in those directions. The two coefficients γm and $\gamma^3 m$, are known, respectively, as the transverse and longitudinal masses. The longitudinal mass is larger because for a given linear acceleration, there is not only a component of force due to a change in velocity but one due to a change in the mass (or energy) of the particle. In Chapter 7, "Relativistic Electrodynamics," we will encounter the specific electromagnetic case of a 4-force: the Lorentz 4-force. In this case, the time and space parts of the 4-force represent, respectively, the rate of the fields imparting relativistic energy and momentum to the charged particles.

5.6 Some important "E&M" 4-vectors

5.6.1 The 4-wavevector

The electric field of a harmonic electromagnetic plane wave in vacuum evolves as:

$$\vec{E}\left(\vec{x},t\right) = \vec{E}_o \cos\left(\omega t - \vec{k}\cdot\vec{x}\right) \tag{5.72}$$

where ω and $\vec{k}$ are the angular frequency and wavevector, respectively, associated with the wave. Now, since the phase of this wave at any given point in space and time, $\phi = \omega t - \vec{k}\cdot\vec{x}$, is known to be a frame independent scalar[36] and since $x^\alpha = (ct, \vec{x})$ is a known 4-vector, we can define the "4-wavevector" for this wave as $k^\alpha = \left(\frac{\omega}{c}, \vec{k}\right)$ so that the phase is recognized as the scalar product: $\phi = k^\alpha x_\alpha$. As a 4-vector, then, this 4-wavevector has a Lorentz invariant (square) magnitude:

$$k^\alpha k_\alpha = \left(\frac{\omega}{c}\right)^2 - k^2 = k^2\left[\frac{1}{c^2}\left(\frac{\omega}{k}\right)^2 - 1\right] = k^2\left(\frac{v_p^2}{c^2} - 1\right) \tag{5.73}$$

where, for the presently considered case, the wave's phase velocity is the speed of light in vacuum, $v_p = \frac{\omega}{k} = c$, and so the magnitude of the 4-wavevector is zero. However, Eq. 5.73 for 4-wavevector magnitude generally applies to any relativistic wave that has no special frame

36 Physically, this means that a particular phase is invariantly associated with every event in space-time. For instance, an event coincident with a maximum or minimum of a wave should remain coincident with that maximum or minimum in all inertial frames.

(i.e., in the absence of a medium) and, in general, the 4-wavevector magnitude of such a relativistic wave need not vanish. Consider, for example, a quantum mechanical free particle probability wave (a "matter" wave) with angular frequency ω and wavevector $\vec{k}$. The DeBroglie equations then relate these wave properties to the total energy and momentum of the particle: $\hbar\vec{k} = \vec{p}$ and $\hbar\omega = \mathscr{E}$. This can be summarized by the resulting relationship between the 4-wavevector and the 4-momentum of the wave/particle:

$$\hbar k^{\alpha} = \left(\frac{\hbar\omega}{c}, \hbar\vec{k}\right) = \left(\frac{\mathscr{E}}{c}, \vec{p}\right) = p^{\alpha} \tag{5.74}$$

Now, we know from Eq. 5.66 that the invariant square magnitude of the 4-momentum yields the energy-momentum relation of a relativistic particle,

$$p^{\alpha} p_{\alpha} = \frac{\mathscr{E}^2}{c^2} - p^2 = m^2 c^2 \tag{5.75}$$

and so, using Eq. 5.74, we immediately obtain the associated invariant square magnitude of the "particle" 4-wavevector,

$$k^{\alpha} k_{\alpha} = \frac{\omega^2}{c^2} - k^2 = \left(\frac{mc}{\hbar}\right)^2 \tag{5.76}$$

where, upon comparison with Eq. 5.73, we can see that in this case, because $k^{\alpha} k_{\alpha} > 0$, the wave's phase velocity exceeds light speed: $v_p = \frac{\omega}{k} > c$. If this seems unphysical, note that it is because we have assumed something unphysical for a massive, localized particle; namely we have assumed a single wavelength probability wave. In fact, a localized particle probability requires a continuum of wavelength components and in such a case the velocity of the localization (the particle) is the group velocity–which is less than the speed of light. As we will see in Chapter 9, the group velocity, v_g, is given by $d\omega/dk$. From Eq. 5.76, this is,

$$v_g = \frac{d\omega}{dk} = \frac{c^2 k}{\omega} = \frac{c^2}{v_p} < c$$

So, it appears that for a particle with $m > 0$, the phase and group velocities will be proportionally above and below light speed, respectively, and will be related by: $v_p v_g = c^2$. Indeed, a localized pulse of electromagnetic energy can be seen as the special case of a massless particle–a photon–in which, as with any particle, Eqs. 5.74 - 5.76 describe the connections between its particle-like and wave-like properties. In the case of light, however, $v_p = v_g = c$.

5.6.2 The 4-current density

Classically, the conservation of charge is expressed by the continuity equation:

$$\frac{\partial \rho}{\partial t} + \nabla \cdot \vec{\mathcal{J}} = 0$$

If we define the 4D current density as $\mathcal{J}^\alpha = (c\rho, \vec{\mathcal{J}})$, then, using 4-vector notation this equation may be more compactly written as $\partial_\alpha \mathcal{J}^\alpha = 0$. Taking $\mathcal{J}^\alpha$ to be a 4-vector ensures that if the continuity equation holds in one frame, it will hold in all frames. That the 4-current density is a 4-vector can be argued as follows: it will be shown in Section 5.7.2 that space-time volume, $\Delta\,(ct)\,\Delta x \Delta y \Delta z = \Delta\,(ct)\,\Delta V$, is a Lorentz invariant scalar. Combining this with the fact that electric charge Q is also a Lorentz scalar, it follows that $\rho = Q/\Delta V$ transforms like time. That is, if $\rho = \rho_0$ is the charge density in its rest frame, then in a moving frame $\rho' = \gamma \rho_0$ and we can rewrite the 4-current density as

$$\mathcal{J}^\alpha = \rho_0\,(\gamma c, \gamma \vec{v}) = \rho_0 U^\alpha$$

and since we know that 4-velocity is a 4-vector this confirms that the above defined 4-current density is also a 4-vector.

5.6.3 The 4-potential (in Lorenz gauge)

In Section 2.3, using the Lorenz gauge condition, we were able to express Maxwell's equations as two inhomogeneous wave equations in terms of the potentials and the sources.

$$\begin{aligned} \varepsilon_o \Box \phi &= -\rho \\ \frac{1}{\mu_o} \Box \vec{A} &= -\vec{\mathcal{J}} \end{aligned} \tag{5.77}$$

These equations, unlike the field equations, neatly decoupled the electric and magnetic parts, thus placing the current density and vector potential on equal footing with the scalar potential and charge density. In the last subsection we saw that the combined source quantities $(c\rho, \vec{\mathcal{J}})$ comprise a 4-vector and so, with respect to Eq. 5.77, it is natural to assume that the potential quantities also represent a 4-vector. In addition, the Lorenz gauge condition, specified as

$$\nabla \cdot \vec{A} + \frac{1}{c^2}\frac{\partial \phi}{\partial t} = 0 \tag{5.78}$$

looks like the 4-divergence of a 4-vector. If we identify the four quantities $A^\mu = \left(\frac{\phi}{c}, \vec{A}\right)$ as a 4-vector, the so-called "4-potential," then the Lorenz gauge equation is the divergence of this 4-potential, which in covariant notation is

$$\partial_\mu A^\mu = 0 \tag{5.79}$$

In these last two subsections, we have begun to see that there is yet another important deep unification that naturally occurs within the framework of special relativity; namely, the unification of electric and magnetic fields as a single entity. Indeed, we first saw that the charge density, ρ, and current density, $\vec{\mathcal{J}}$–the respective sources of electric and magnetic fields–were combined into one entity, the 4-current density, $\mathcal{J}^\alpha$. Likewise, here we have combined the scalar and vector potentials, ϕ and $\vec{A}$, which are respectively associated with the electric and magnetic fields. Our description of this unification will be complete when, in Section 5.9, we unify the E and B fields themselves into a single Lorentz covariant object: the electromagnetic field tensor, $\overleftrightarrow{F}$.

5.7 Other covariant and invariant quantities

5.7.1 The angular momentum 4-tensor

Just as the angular momentum in 3-space is represented by the three independent components (a pseudovector) of the anti-symmetric matrix, $M_{ij} = x_i p_j - x_j p_i$, the angular momentum of a particle in special relativity is given by the 4 tensor:

$$M^{\alpha\beta} = x_\gamma p_\delta - x_\delta p_\gamma \tag{5.80}$$

It is an anti-symmetric tensor whose space-space components are the natural extensions of their classical counterparts and the time-like components–the first column and the top row–have physical meaning but are of no special significance to our discussion. The angular momentum 4-tensor will be discussed in greater detail in Chapter 7.

5.7.2 Space-time volume

In Section 5.4, we extended a one-dimensional integral along an arbitrary worldline in space-time to obtain the Lorentz invariant interval of that path. It might be expected that higher dimensional integrals in space-time are likewise Lorentz invariant. In fact, it can be readily shown that an integrated volume of space-time is also Lorentz invariant. If a particular differential volume of space-time is given as $d^4x = dx^0 dx^1 dx^2 dx^3$, then the primed frame and unprimed frame expressions, d^4x' and d^4x, of this volume are related by the Jacobian of the transformation (**see Discussion 5.17**).

$$d^4x' = dx'^0 dx'^1 dx'^2 dx'^3 = \frac{\partial\left(x'^0, x'^1, x'^2, x'^3\right)}{\partial\left(x^0, x^1, x^2, x^3\right)} dx^0 dx^1 dx^2 dx^3 \tag{5.81}$$

$$= \det\left(L^{\alpha}_{\ \beta}\right) dx^0 dx^1 dx^2 dx^3 = d^4x \tag{5.82}$$

That is, because the determinant of the Lorentz transform (or, equivalently, the Jacobian of the Jacobian Matrix) is unity, the space-time volume element in the primed frame, $dx'^0 dx'^1 dx'^2 dx'^3$, is equal to that in the unprimed frame, $dx^0 dx^1 dx^2 dx^3$–space-time volume is Lorentz invariant. For the case of our standard boost along the x axis, this invariance may be seen as the result of the time dilation of the time differential, dx^0, canceling the Lorentz contraction of the spatial differential, dx^1. Indeed, because directions perpendicular to the relative motion are not contracted, this can be depicted in two dimensions. Figure 5.8 is a Minkowski diagram of the space-time "world-volume"

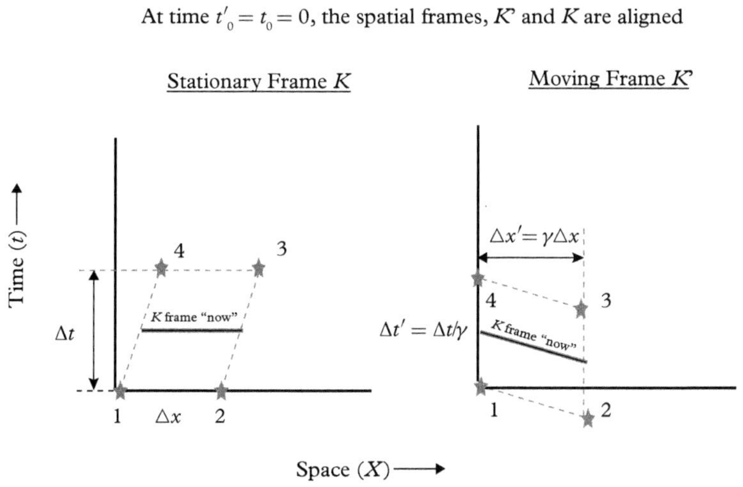

Fig. 5.8 *Space-time volume. The volume (dt)(dx) viewed in two different inertial frames is shown to be invariant.*

traced out by an object of length Δx moving in the $+x$ direction relative to the rest K frame. If we consider the volume enclosed over a duration Δt, we can say this volume is bounded by event 1–4 as designated by the numbered stars. The volume is thus $\Delta x \Delta t$. If we now boost to the K' frame moving with the object, we know the time elapsed $\Delta t'$ from events 1 and 2 to events 4 and 3 is the proper time and thus $\Delta t' = \Delta t/\gamma$. Further, we know the distance measured $\Delta x'$ between events 1 and 4 and events 2 and 3 is $\Delta x' = \gamma \Delta x$. Thus, $\Delta x' \Delta t' = \gamma \Delta x \Delta t/\gamma = \Delta x \Delta t$ and we see volume is Lorentz invariant.

5.7.3 Space-time delta function

The 4-dimensional delta function is defined:

$$\int d^4\tilde{x} \cdot \delta^4\left(\tilde{x}^\alpha - x_0^\alpha\right) \cdot \mathbf{T}\left(\tilde{x}^\alpha\right) = \mathbf{T}\left(x_0^\alpha\right) \tag{5.83}$$

where we are taking $\tilde{x}$ as an integration variable and where $\mathbf{T}(x^\alpha)$ is any single (scalar) or multi-component (vector or tensor) function of the coordinates. For the trivial case of a constant scalar function, $\mathbf{T}(\tilde{x}^\alpha) = 1$,

$$\int d^4\tilde{x} \cdot \delta^4\left(\tilde{x}^\alpha - x_0^\alpha\right) = 1 \tag{5.84}$$

Since, in general, the right-hand side of Eq. 5.83 is a Lorentz scalar and, as shown above, $d^4\tilde{x}$ is a Lorentz scalar, the 4D delta function must also be a Lorentz scalar, vector or tensor.

5.8 Summary of 4-vector results

This table summarizes various 4-vector quantities with the corresponding 3-D analogs:

3-vector: rotations	*4-vector: Lorentz transformations*
$\vec{A} = (A_x, A_y, A_z)$	$a^\alpha = (a_{ct}, a_x, a_y, a_z) = (a^0, a^1, a^2, a^3) = (a^0, \vec{a})$
$\vec{A} \cdot \vec{B} = A_i B_i = A_x B_x + A_y B_y + A_z B_z$	$g_{\alpha\beta} a^\alpha b^\beta = a^\alpha b_\alpha = a_{ct} b_{ct} - a_x b_x - a_y b_y - a_z b_z$
$\nabla = \left(\frac{\partial}{\partial x}, \frac{\partial}{\partial y}, \frac{\partial}{\partial z}\right)$	$\partial_\alpha = (\partial_0, \partial_1, \partial_2, \partial_3) = \left(\frac{\partial}{c\partial t}, \nabla\right)$
$\nabla \cdot \vec{A} = \frac{\partial A_x}{\partial x} + \frac{\partial A_y}{\partial y} + \frac{\partial A_z}{\partial z}$	$\partial_\alpha a^\alpha = \partial_0 a^0 + \partial_1 a^1 + \partial_2 a^2 + \partial_3 a^3 = \frac{1}{c}\frac{\partial a^0}{\partial t} + \nabla \cdot \vec{a}$
$\nabla \cdot \nabla = \nabla^2 = \frac{\partial^2}{\partial x^2} + \frac{\partial^2}{\partial y^2} + \frac{\partial^2}{\partial z^2}$	$\partial^\alpha \partial_\alpha = \partial^0 \partial_0 + \partial^1 \partial_1 + \partial^2 \partial_2 + \partial^3 \partial_3 = \frac{1}{c^2}\frac{\partial^2}{\partial t^2} - \nabla^2$

5.9 Maxwell's equations and special relativity

5.9.1 Manifest covariance of Maxwell's equations

With the notation of 4-vectors and 4-tensors developed in Section 5.4, the special relativity embedded in the Maxwell's equations may be readily identified. As we have seen in Section 2.3, the content of Maxwell's equations may be written as the decoupled differential equations of Lorenz gauge potentials, $\vec{A}$ and ϕ:

$$\nabla^2\phi - \frac{1}{c^2}\frac{\partial^2\phi}{\partial t^2} = -\frac{\rho}{\varepsilon_0} \tag{5.85}$$

$$\nabla^2\vec{A} - \frac{1}{c^2}\frac{\partial^2\vec{A}}{\partial t^2} = -\mu_0\vec{\mathcal{J}} \tag{5.86}$$

with the Lorenz gauge condition

$$\nabla\cdot\vec{A} + \frac{1}{c^2}\frac{\partial\phi}{\partial t} = \partial_\alpha A^\alpha = 0 \tag{5.87}$$

where the Lorenz gauge condition in Eq. 5.87 has been expressed in 4-vector notation as per the discussion leading up to 5.79. In Eqs. 5.85 and 5.86, one can see the ingredients of a single 4-vector equation. Taken together, the left-hand sides look like a 4-vector Laplacian operator, $\partial^\alpha\partial_\alpha = \frac{1}{c^2}\frac{\partial^2}{\partial t^2} - \nabla^2$, operating on the potential 4-vector, A^μ, while the right-hand sides, up to some constants, comprise the current density 4-vector, $\mathcal{J}^\alpha = \left(c\rho, \vec{\mathcal{J}}\right)$. Multiplying Eq. 5.85 by c/c^2, Eqs. 5.85 and 5.86 can be combined into the compact form[37]

$$\partial^\alpha\partial_\alpha A^\beta = \mu_0\mathcal{J}^\beta \tag{5.88}$$

We have expressed all of Maxwell's equations in the two 4-vector expressions of Eqs. 5.87 and 5.88–and they are manifestly Lorentz covariant.

Taking the 4-divergence of Eq. 5.88 and noting that the derivatives all commute:

$$\partial^\alpha\partial_\alpha\,\partial_\beta A^\beta = \mu_0\partial_\beta\mathcal{J}^\beta \tag{5.89}$$

then, by the Lorenz gauge condition of Eq. 5.87, the left-hand side is zero and the charge continuity condition (see Eq. 5.56) is quickly recovered:

$$\partial_\beta\mathcal{J}^\beta = 0 \tag{5.90}$$

[37] Multiplying by c/c^2, Eq. 5.85 is written

$$\nabla^2\left(\frac{\phi}{c}\right) - \frac{1}{c^2}\frac{\partial^2(\phi/c)}{\partial t^2} = -\frac{c\rho}{c^2\varepsilon_0}$$

or $$-\partial^\alpha\partial_\alpha\left(\frac{\phi}{c}\right) = -\mu_o(c\rho)$$

where $-\partial^\alpha\partial_\alpha = \nabla^2 - \frac{1}{c^2}\frac{\partial^2}{\partial t^2}$. In the same way, Eq. 5.86 can be written

$$\nabla^2\vec{A} - \frac{1}{c^2}\frac{\partial^2\vec{A}}{\partial t^2} = -\partial^\alpha\partial_\alpha\left(\vec{A}\right) = -\mu_0\vec{\mathcal{J}}$$

so that with $A^\mu = \left(\frac{\phi}{c}, \vec{A}\right)$ and $\mathcal{J}^\alpha = \left(c\rho, \vec{\mathcal{J}}\right)$, Eqs. 5.85 and 5.86 together can be rewritten:

$$\partial^\alpha\partial_\alpha A^\beta = \mu_0 j^\beta.$$

5.9.2 The electromagnetic field tensor

We have seen that the framework of relativity has unified the scalar/vector potentials and charge/current densities into the 4-potential and 4-current, respectively. We now show that relativity allows a unification of the $\vec{E}$ and $\vec{B}$ fields into a single Lorentz covariant object: the electromagnetic field tensor, $\overleftrightarrow{F} = F^{\alpha\beta}$. With the known transformation properties of this field tensor, we will have the transformation laws for $\vec{E}$ and $\vec{B}$. We start by expressing the electric and magnetic fields in 4-vector, 4-tensor notation by way of the scalar and vector potentials. The idea of this field tensor can be summarized by the statement that, in a limited sense, it represents the 4-space "curl" of the 4-potential. To be clear, in 3D, if $\vec{V}$ is a vector field and a matrix $\overleftrightarrow{T}$ is defined by the nine components $T_{ij} = \frac{\partial V_j}{\partial x_i} - \frac{\partial V_i}{\partial x_j}$, then we find that the diagonal terms all vanish and the other six terms are related: $T_{12} = -T_{21}$, $T_{13} = -T_{31}$, and $T_{23} = -T_{32}$ so that the matrix is antisymmetric and can be represented by just the three terms T_{12}, T_{13}, and T_{23} and it just so happens that these three terms together transform as a 3-vector and we identify them as the curl of $\vec{V}$, a pseudo-vector. That is, $\nabla \times \vec{V} = T_{23}\hat{x} - T_{13}\hat{y} + T_{12}\hat{z}$.

Similarly, in 4-space, if we consider the 4-potential A^α and a fully contravariant 4-tensor $F^{\alpha\beta}$ that is defined by the 16 components $F^{\alpha\beta} = \partial^\alpha A^\beta - \partial^\beta A^\alpha$, then we again find that all the diagonal terms vanish and we have an antisymmetric tensor that can be represented by six independent terms, $F^{01}, F^{02}, F^{03}, F^{12}, F^{13}$, and F^{23}, with well-defined transformation properties. However, this is where the analogue to 3D ends because obviously six terms cannot transform as a 4-vector. In general, the derivative of a potential is a field and so we can guess that the terms of $F^{\alpha\beta}$ will be terms of the fields. To find out, we first consider the magnetic field as the curl of the vector potential, $\vec{B} = \nabla \times \vec{A}$. Recalling that $\left(\frac{\partial}{\partial(ct)}, \frac{\partial}{\partial x}, \frac{\partial}{\partial y}, \frac{\partial}{\partial z}\right) = \left(\frac{\partial}{\partial x^0}, \frac{\partial}{\partial x^1}, \frac{\partial}{\partial x^2}, \frac{\partial}{\partial x^3}\right) = \left(\partial^0, -\partial^1, -\partial^2, -\partial^3\right)$ and $\left(\frac{\phi}{c}, A_x, A_y, A_z\right) = \left(A^0, A^1, A^2, A^3\right)$, the x-component is:

$$B_x = \frac{\partial A^3}{\partial x^2} - \frac{\partial A^2}{\partial x^3} = \partial^3 A^2 - \partial^2 A^3 = F^{32} = -F^{23}$$

We therefore set $F^{32} = +B_x$ and $F^{23} = -B_x$. The lower right part of the field tensor can be filled in using corresponding equations for the components of the magnetic field:

$$F^{\alpha\beta} = \begin{bmatrix} 0 & ? & ? & ? \\ ? & 0 & -B_z & B_y \\ ? & B_z & 0 & -B_x \\ ? & -B_y & B_x & 0 \end{bmatrix} \tag{5.91}$$

We can continue, using the pattern to fill in the remaining entries in the tensor. Recall $\partial^0 A^1 - \partial^1 A^0 = \frac{1}{c}\frac{\partial A_x}{\partial t} + \frac{\partial}{\partial x}\left(\frac{\phi}{c}\right) = -\frac{E_x}{c}$ and therefore $F^{01} = -\frac{E_x}{c}$ and $F^{10} = +\frac{E_x}{c}$. The other tensor components are similarly identified with electric field components giving the full the *electromagnetic field* tensor:

$$F^{\alpha\beta} = \begin{bmatrix} 0 & -\frac{E_x}{c} & -\frac{E_y}{c} & -\frac{E_z}{c} \\ \frac{E_x}{c} & 0 & -B_z & B_y \\ \frac{E_y}{c} & B_z & 0 & -B_x \\ \frac{E_z}{c} & -B_y & B_x & 0 \end{bmatrix} \tag{5.92}$$

Next, we want to use this tensor to recover Maxwell's equations. The divergence of a 4-tensor gives back a 4-vector. Consider the divergence of the EM field tensor: $\partial_\alpha F^{\alpha\beta}$. First, let's look at the $\beta = 0$ component:

$$\partial_\alpha F^{\alpha 0} = \begin{bmatrix} \partial_0 & \partial_1 & \partial_2 & \partial_3 \end{bmatrix} \begin{bmatrix} 0 \\ \frac{E_x}{c} \\ \frac{E_y}{c} \\ \frac{E_z}{c} \end{bmatrix} = \frac{1}{c}\nabla \cdot \vec{E} \tag{5.93}$$

So this is, to within a factor of c, the divergence of the electric field and should call to mind the Gauss's law of Maxwell's equations: $\nabla \cdot \vec{E} = \frac{\rho}{\varepsilon_0} = \mu_0 c^2 \rho$. Now, the charge density has been identified as the time component of the current density 4-vector, so this means:

$$\partial_\alpha F^{\alpha 0} = \mu_0 (c\rho) = \mu_0 \mathcal{J}^0 \tag{5.94}$$

Next, looking at the $\beta = 1$ component,

$$\partial_\alpha F^{\alpha 1} = \begin{bmatrix} \partial_0 & \partial_1 & \partial_2 & \partial_3 \end{bmatrix} \begin{bmatrix} -\frac{E_x}{c} \\ 0 \\ B_z \\ -B_y \end{bmatrix}$$

$$= \left(\nabla \times \vec{B}\right)_x - \frac{1}{c^2}\frac{\partial E_x}{\partial t} = \mu_0 \mathcal{J}^1$$

which is just the x component of Ampere's law part of Maxwell's equations. Thus, these two considerations suggest we write the general statement:

$$\partial_\alpha F^{\alpha\beta} = \mu_0 \mathcal{J}^\beta \tag{5.95}$$

And so from equating the divergence of the field tensor, $\partial_\alpha F^{\alpha\beta}$, with the current density 4-vector we have recovered the two inhomogeneous Maxwell's equations. Moreover, the 4-tensor equations represented by Eq. 5.95 are "manifestly covariant." What about the other two Maxwell's equations? Well, it turns out that there is another way of organizing the field components within an antisymmetric 4-tensor such that the transformation properties of the fields are given. That second antisymmetric configuration is

$$\mathscr{F}^{\alpha\beta} = \begin{bmatrix} 0 & \frac{-B_x}{c} & \frac{-B_y}{c} & \frac{-B_z}{c} \\ \frac{B_x}{c} & 0 & E_z & -E_y \\ \frac{B_y}{c} & -E_z & 0 & E_x \\ \frac{B_z}{c} & E_y & -E_x & 0 \end{bmatrix} \tag{5.96}$$

which is known as the "dual" of the field tensor and can be obtained from the field tensor, $F^{\alpha\beta}$, via the Levi–Civita tensor: $\mathscr{F}^{\alpha\beta} = \frac{1}{2}\varepsilon^{\alpha\beta}_{\ \ \gamma\delta}F^{\gamma\delta}$. The first two homogeneous Maxwell's equations may then be written as the divergence of the dual field tensor: $\partial_\alpha \mathscr{F}^{\alpha\beta} = 0$. We finally note that these expressions of Maxwell's equations in terms of $F^{\alpha\beta}$ and $\mathscr{F}^{\alpha\beta}$ are true regardless of gauge. For example, we can quickly obtain the earlier expression of Eq. 5.88, which assumed the Lorenz gauge, from Eq. 5.95:

$$\partial_\alpha F^{\alpha\beta} = \partial_\alpha \left(\partial^\alpha A^\beta - \partial^\beta A^\alpha\right) = \partial_\alpha \partial^\alpha A^\beta - \partial^\beta \partial_\alpha A^\alpha = \mu_0 \mathcal{J}^\beta$$

where, upon imposing the Lorenz gauge condition $\partial_\alpha A^\alpha = 0$, we retrieve the earlier expression: $\partial_\alpha \partial^\alpha A^\beta = \mu_0 \mathcal{J}^\beta$.

The electric and magnetic fields are now indeed seen to be two parts of a greater whole, the field tensor, $F^{\alpha\beta}$, the components of which transform according to the requirements of special relativity. As discussed earlier in the chapter, Einstein began his paper by observing that a wire that is moving *relative* to a magnetic field induces a current. Before Einstein's relativity, there seemed to be two mechanisms for this current that depended on one's reference frame. Now, with special relativity, the fields can be combined in tensor form and can thus be seen as a single Lorentz covariant object, the field tensor, $F^{\alpha\beta}$. And what was previously considered as two mechanisms for the same physical process is now seen to be a single mechanism as viewed in different inertial frames–analogous to the way any object in 3-space appears different when viewed at different angles. The Lorentz transformation of the field tensor, $F^{\alpha\beta}$, from our standard stationary frame, K, to the frame, K', moving in the $+x$ direction with velocity, $\vec{V} = V_x$, is a straightforward exercise in 4-tensor manipulation given by Eq. 5.46. Indeed, transforming either $F^{\alpha\beta}$ or $\mathscr{F}^{\alpha\beta}$ yields:

$$E'_x = E_x \tag{5.97}$$

$$E'_y = \gamma\left(E_y - V_x B_z\right) = \gamma\left(\vec{E} + \vec{V} \times \vec{B}\right)_y \tag{5.98}$$

$$E'_z = \gamma\left(E_z + V_x B_y\right) = \gamma\left(\vec{E} + \vec{V} \times \vec{B}\right)_z \tag{5.99}$$

$$B'_x = B_x \tag{5.100}$$

$$B'_y = \gamma\left(B_y + \frac{V_x}{c^2}E_z\right) = \gamma\left(\vec{B} - \frac{1}{c^2}\vec{V} \times \vec{E}\right)_y \tag{5.101}$$

$$B'_z = \gamma\left(B_z - \frac{V_x}{c^2}E_y\right) = \gamma\left(\vec{B} - \frac{1}{c^2}\vec{V} \times \vec{E}\right)_z \tag{5.102}$$

where the terms to the far right are obtained when we recognize that because $\vec{V}$ is in the $\hat{x}$ direction alone, we can write the V_iB_j and V_iE_j as components of the cross products, $\vec{V} \times \vec{B}$ and $\vec{V} \times \vec{E}$. Thus, we conclude generally that an electric field observed in one frame is seen as a combination electric/magnetic field in another and vice versa. And so this result provides another example of a general truth provided by special relativity, namely, that any single inertial frame is limited in its perspective to seeing but one aspect of the larger, higher dimensional "thing"–which is, in this case, the electromagnetic tensor. Again, it is exactly analogous to having a limited, single direction view of a complicated 3D object.

5.9.3 Simple field transformation examples

We next consider examples tying these transformations to relativistic kinematics. First, consider an ideal charged capacitor at rest in the unprimed frame and oriented with its symmetry axis, and thus its electric field, along the x direction. If we now perform a boost to a primed frame moving at velocity V in the $+x$ direction and for which the capacitor is moving in the $-x'$ direction, as in Fig. 5.9, we see from the relevant transformation of Eq. 5.97, that the transformed electric field, E'_x, is unchanged. Physically, although the distance between the two plates is Lorentz contracted in the primed frame, the surface charge density on the plates, σ, remains unchanged and thus so does the resulting electric field between the plates, $E = \sigma/\varepsilon_0$. Furthermore, because the capacitor plates are oppositely charged, the currents in the primed frame cancel and the magnetic field $\vec{B}$ is zero.

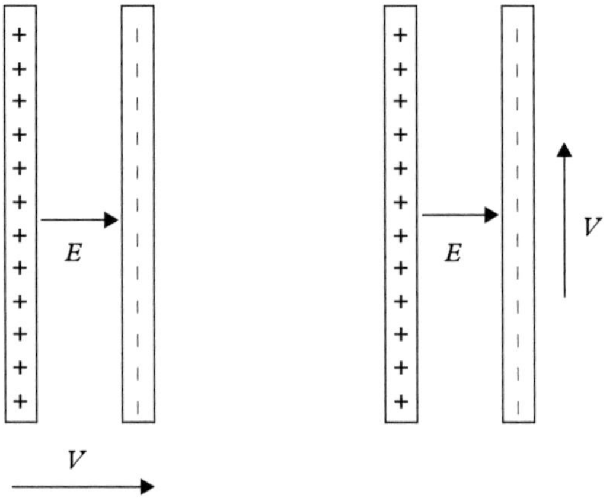

Fig. 5.9 *A capacitor at rest in the unprimed frame whose field is in the direction (left) and perpendicular (right) to the velocity relating the two frames.*

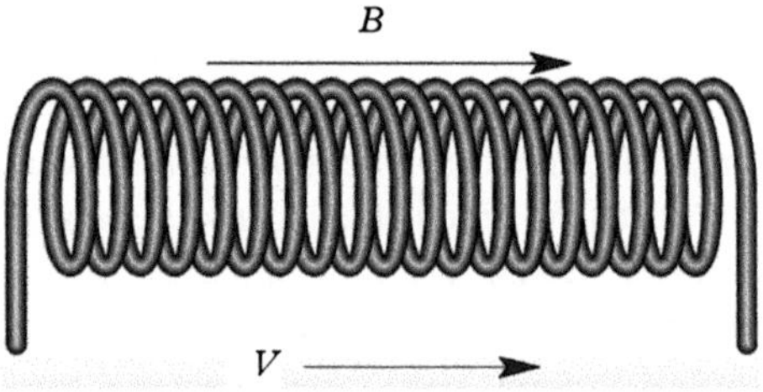

Fig. 5.10 *To relate the transformation of a magnetic field between frames differing by a boost parallel to the field, consider a current carrying coil moving as shown.*

Similarly, we start with a charged capacitor at rest in the unprimed frame but now with the electric field directed perpendicular to the $+x$ direction, in, say, the y direction. If we again perform a $+x$ direction boost to a primed frame moving at velocity V, as in Fig. 5.9, then we see there are two relevant transformation equations, Eqs. 5.98 and 5.102, corresponding, respectively, to the primed frame fields E'_y and B'_z. Thus, we see that a pure electric field, E_y, in the unprimed frame transforms to both an electric field E'_y and a magnetic field B'_z in the primed frame. For this case, physically, the capacitor plates experience a Lorentz contraction that leads to an increase in the surface charge density by γ and a corresponding increase in the electric field strength. In addition, the magnetic field arises from the surface charges that, in the primed frame, are seen as surface currents.

As a final simple example, we illustrate the invariance of the magnetic field component in the direction of a given boost, as given by Eq. 5.100, by considering a boost along the axis of a current carrying ideal coil. For this arrangement, as shown in Fig. 5.10, two effects cancel. First, in the boosted, primed frame, the coil turn density has increased by a factor of γ, which, by itself would lead to a proportional increase in the magnetic field. However, in the boosted frame, there is also a time dilation that can be thought of as causing a slowing down of the flow of charges or a decrease in the coil's current by a factor of γ. Thus, the field is the same in both frames, as indicated in Eq. 5.100.

Relativistically moving charged point particle

Finally, in our discussion of relativistically transformed fields, we look at the important case of the electric field from a relativistic constant velocity charged point particle. Recall this was described before in connection with the string-rocket problem. A point-particle of charge q is at rest at the origin of the K' frame that, as usual, is moving with a velocity V in the $+x$ direction relative to the laboratory or K frame. The two origins are coincident at $t=0$. We wish to find the $\vec{E}$ field measured at the point x, y in the laboratory frame at some time t. In our present treatment, however, we will not consider the magnetic field components, in the laboratory frame, resulting from the transformation of charge density to current density. The Coulomb field of the particle in its rest frame, K', is:

$$E'_x\left(x', y'\right) = \frac{1}{4\pi\varepsilon_0}\frac{q}{\left(x'^2 + y'^2\right)^{3/2}}x' \tag{5.103}$$

$$E'_y\left(x', y'\right) = \frac{1}{4\pi\varepsilon_0}\frac{q}{\left(x'^2 + y'^2\right)^{3/2}}y' \tag{5.104}$$

where we have suppressed a z dimension for simplicity and we note that E' is independent of ct'. These equations thus tell us what the amplitude and direction of the electric field is at a given point and time in the primed frame $\vec{E}'\left(x', y'\right)$. Because the required transformation is not of a simple vector, but rather of a vector field, which is a function of space and time, we must consider the transformation of both the field part $\vec{E}'$ and the coordinate part $\left(x', y'\right)$. So, constructing the associated transformed vector fields in the unprimed frame consists of two steps:

(1) Apply an (inverse) Lorentz transform of the field tensor to the electric field components. This results in a transformation of the amplitude and direction of the electric field to the unprimed frame. However, we still have not properly mapped these transformed electric vector fields from the primed to the unprimed frame.

(2) To properly map the transformed vector fields to the unprimed frame, we express the primed coordinates in terms of the unprimed coordinates: that just means substituting the Lorentz transformations for x' and y'.

Thus, we first apply the inverse Lorentz transforms corresponding to the Lorentz transforms of Eqs. 5.97 and 5.98 to transform vector fields of 5.103 and 5.104 to the laboratory frame. Including the functional dependences explicitly, this gives[38]

$$E_x(ct, x, y) = E'_x\left[x'(ct, x, y), y'(ct, x, y)\right] \tag{5.105}$$

$$E_y(ct, x, y) = \gamma E'_y\left[x'(ct, x, y), y'(ct, x, y)\right] \tag{5.106}$$

We next map these transformed electric vector fields from the primed to the unprimed frame coordinates, first noting that $y' = y$, and then using the Lorentz transformation for x':

$$\begin{bmatrix} ct' \\ x' \end{bmatrix} = \begin{bmatrix} \gamma & -\beta\gamma \\ -\beta\gamma & \gamma \end{bmatrix} \begin{bmatrix} ct \\ x \end{bmatrix} \Longrightarrow x' = \gamma(x - Vt) \tag{5.107}$$

which, upon substitution into Eqs. 5.103 and 5.104, and in turn into Eqs. 5.105 and 5.106, give for the laboratory fields:

$$E_x(ct, x, y) = \gamma \frac{1}{4\pi\varepsilon_0} \frac{q(x - Vt)}{\left[\gamma^2 (x - Vt)^2 + y^2\right]^{3/2}} \tag{5.108}$$

$$E_y(ct, x, y) = \gamma \frac{1}{4\pi\varepsilon_0} \frac{qy}{\left[\gamma^2 (x - Vt)^2 + y^2\right]^{3/2}} \tag{5.109}$$

So, even in the laboratory frame, K, in which the particle is moving relativistically to the right, the electric fields remain everywhere radially directed outward from particle. Directly forward of the particle's motion, where $y = 0$, the field is reduced by γ^{-2}:

$$E_x = \frac{1}{\gamma^2} \frac{1}{4\pi\varepsilon_0} \frac{q}{(x - Vt)^2} \qquad E_y = 0 \tag{5.110}$$

By contrast, for points on the y-axis, perpendicular to the particle's motion, the field is increased by γ. Consider, for example, where $x = Vt$, we have

$$E_x = 0 \quad E_y = \gamma \frac{1}{4\pi\varepsilon_0} \frac{q}{y^2} \tag{5.111}$$

Thus, the field strength in the laboratory frame is concentrated in the plane perpendicular to the particle's motion. For highly relativistic particles at particle accelerators speeds, the lines are compressed into an extremely narrow pancake-like form along the direction of motion.

[38] Using the inverse transforms of Eqs. 5.97 and 5.98 for the field is straightforward, involving only multiplication by γ, because the magnetic field in the primed frame is zero $\vec{B}' = 0$.

We further discuss the consequences and application of this transform in the Chapter 6 discussion of virtual photons.

5.10 The Einstein stress-energy tensor

In Chapter 1, we identified a tensor of electromagnetic field momentum current (flow) densities: the Maxwell stress tensor, $\overleftrightarrow{\sigma}$. We remind the reader that the conventions can vary in every possible way in specifying this stress tensor, and we choose the following: positive microscopic forces act outward relative to the differential volume; the three columns of the stress tensor correspond to the x, y, and z components of the momentum; and the rows give the three orthogonal planes through which the momentum is flowing (i.e., across which the force is acting). See Fig. 5.11.

If there are no constituents of the system beyond those of the fields, conservation of field momentum is expressed by the vector equation

$$\frac{\partial \vec{g}}{\partial t} + \nabla \cdot \overleftrightarrow{\sigma} = 0. \tag{5.112}$$

where $\vec{g}$ is the momentum density vector. Let us now consider the natural relativistic extension of the 3-space Maxwell stress tensor to space-time: the so-called stress-energy 4-tensor, $T^{\alpha\beta}$. While we will discuss this more formally when we examine action-principle relativistic mechanics, here we argue for the general form of $T^{\alpha\beta}$ by analogy with the continuity equation. Considering just one component of Eq. 5.112, we can suggestively rewrite the momentum density derivative so that we have a 4-divergence that involves the x

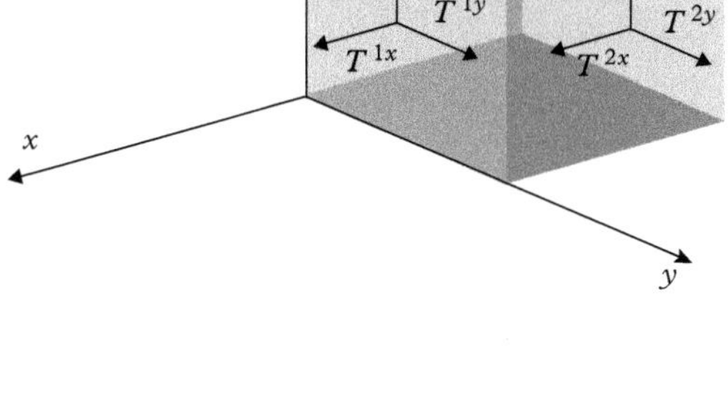

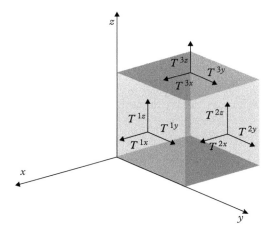

Fig. 5.11 *Our conventions for the Maxwell stress tensor.*

column of the stress tensor and looks very much like the continuity equation:

$$\frac{\partial\left(cg_x\right)}{\partial\left(ct\right)}+\frac{\partial\sigma_{xx}}{\partial x}+\frac{\partial\sigma_{yx}}{\partial y}+\frac{\partial\sigma_{zx}}{\partial z}=0 \tag{5.113}$$

This suggests we extend the stress tensor to 4D by taking cg_x to head the top of the x-column and fill in similarly for the y and z columns. The 4-divergence equation of such a tensor then has the form:

$$\partial_\alpha T^{\alpha\beta}=\left[\partial_0,\partial_1,\partial_2,\partial_3\right]\begin{bmatrix} ? & cg_x & cg_y & cg_z \\ ? & \sigma_{xx} & \sigma_{xy} & \sigma_{xz} \\ ? & \sigma_{yx} & \sigma_{yy} & \sigma_{yz} \\ ? & \sigma_{zx} & \sigma_{zy} & \sigma_{zz} \end{bmatrix}=0 \tag{5.114}$$

Thus, the 4-gradient row vector scalar-multiplies with the space-like columns of the matrix and returns the conservation law of Eq. 5.112. What about the left-most (zeroth) column? Recalling that the zeroth component of the momentum 4-vector is the energy, $p^\alpha=\left(\frac{\mathscr{E}}{c},\vec{p}\right)$ and noting that we may describe the right three space-like columns of $T^{\alpha\beta}$ as "the momentum density and momentum current density components" for the three directions of momentum, it is then plausible to identify the left-most column with the system's energy density and energy-current density components. Although these concepts are not used in continuum mechanics enough to get their own distinctive symbols, for electromagnetic fields the energy current density is the Poynting vector, $\vec{S}$, and the energy density is often denoted by u. So, finally, the microscopic form for momentum and energy conservation is written as:

$$\partial_\alpha T^{\alpha\beta}=\left[\partial_0,\partial_1,\partial_2,\partial_3\right]\begin{bmatrix} u & cg_x & cg_y & cg_z \\ \frac{S_x}{c} & \sigma_{xx} & \sigma_{xy} & \sigma_{xz} \\ \frac{S_y}{c} & \sigma_{yx} & \sigma_{yy} & \sigma_{yz} \\ \frac{S_z}{c} & \sigma_{zx} & \sigma_{zy} & \sigma_{zz} \end{bmatrix}=0 \tag{5.115}$$

Taking the 4-divergence of the first column, then, we obtain Poynting's theorem for the conservation of energy in the absence of charges:

$$\frac{\partial u}{\partial t}+\nabla\cdot\vec{S}=0 \tag{5.116}$$

If we recall the relation, $\vec{S}=c^2\vec{g}$, between energy current density and momentum density, it is clear that $T^{\alpha\beta}$ is symmetric. As we will discuss in Chapter 7, this property will turn out to be important in order to consistently define the angular momentum of the electromagnetic field. Another important property of $T^{\alpha\beta}$ is that it is traceless–this can be shown to be a necessary condition for the photon to have zero mass. Finally, as will be shown in Chapter 7, the Einstein stress-energy tensor may be written in terms of the electromagnetic field tensor:

$$T^{\alpha\beta} = \frac{1}{\mu_0}\left(-g_{\mu\nu}F^{\alpha\nu}F^{\beta\mu} + \frac{1}{4}g^{\alpha\beta}F^{\mu\nu}F_{\mu\nu}\right) \tag{5.117}$$

Exercises

(5.1) Show that a Galilean transformation to a frame moving with velocity v in the x-direction does not leave the form of Eq. 5.2 unchanged.

(5.2) Find the path difference for the Michelson–Morley experiment assuming that the Earth's velocity in its orbit about the sun is the velocity of the apparatus relative to the absolute frame of the ether. Assume the two arms of the apparatus are of equal length $L = 20$ m and are parallel and perpendicular to the earth's velocity $V = 30$ m/s through the ether. Find the size of the fringe shift for 590 nm, yellow light.

(5.3) Argue similarly to that leading to Eq. 5.9 that for a Lorentz boost in the x-direction, $y' = y$, and $z' = z$.

(5.4) Reference Eq. 5.13: find the rest frame (K) components of velocity for a particle moving in the $+y'$ direction with a velocity u'_y as measured in the U' frame.

(5.5) Reference Eq. 5.14: show explicitly that the interval between two events is invariant for the two frames related by a boost that were described before, K and K' .

(5.6) Reference Eq. 5.19: Consider an alternative path going between the two points: $(0,0) \to (ct_1, x_1) \to (ct_b, x_b)$ consisting of two segments joined to a third point not on the physical blue path linking the two points. Show that the sum of the intervals of the two segments of this new path is less than the interval of the physical trajectory. What criteria must be satisfied such that the excursion to (ct_1, x_1) does not violate causality?

(5.7) For a trip to and from a star 25 light-years distant, in a rocket that travels 0.9999 c relative to the Earth, find the time that has passed for the round trip according to (A) John and (B) Jane.

(5.8) Sketch the Minkowski diagrams and find the ages of the twins for the twin paradox problem, where this time Jane's rocket undergoes uniform acceleration from rest to $0.9999c$ as it travels to the 25 light-years distant star. On reaching the star, the rocket instantaneously reverses direction but maintains its speed of $0.9999c$. For the remainder of the trip the rocket executes a uniform deceleration such that the trip home is the opposite of the trip out and they land back on earth with zero velocity. Using Eq. 5.18, solve this problem (A) for a constant acceleration as measured by John and (B) as measured by Jane.

(5.9) Prove the relations of Eq. 5.21.

(5.10) Sketch on a ct, x axis the ct' and x' axis corresponding to the rest frame of the blue particle in Fig. 5.5.

(5.11) Demonstrate, using a Minkowski diagram, that for any two events separated by a time-like interval, there is an inertial frame for which both events occur at the same location AND for any two events separated by a space-like interval, there is an inertial frame in which the two events are simultaneous.

(5.12) Show explicitly that the components of a covariant displacement vector, $x_\alpha = [x^0, -x^1, -x^2, -x^3]$, transform properly from x_α to x'_α for the standard boost in the x direction using the known transformation of the corresponding contravariant vector. (transforming properly means that the covariant and contravariant representations of the displacement in the two frames maintain their defined relationship)

(5.13) The magnitude of the 4-acceleration is a Lorentz scalar. Find it.

(5.14) Find the fields from the 4-potentials of a moving charge and compare with those in the text.

(5.15) Show that for a summed pair of indices in an equation, switching their positions does not affect the result. That is, $A_\mu B^\mu = A^\mu B_\mu$.

(5.16) Write out Eq. 5.34 using Einstein summation notation. (use capital roman letters for summation indices).

(5.17) Show explicitly for the standard unprimed and boosted primed frames that the transformation for the covariant components is given by the inverse Lorentz transformation.

(5.18) Show, using Eq. 5.46 and the definition of $\Lambda^\alpha{}_\gamma{}^\beta{}_\eta$, that the transformed fields are given by Eqs. 5.97–5.102.

(5.19) Show that any massless particle must travel at exactly the speed of light.

(5.20) The Doppler shift is the difference between the observed frequency of a moving light source and the frequency of the source as measured in its own rest frame. Using the fact that the 4-wavevector is a Lorentz 4-vector, find the relativistic Doppler shift of a source moving with velocity, $\vec{V}$.

(5.21) Show that the Levi–Civita symbol is a 4-tensor.

(5.22) Is the identity matrix (1's on the diagonal, 0's off diagonal) a tensor? Is the 2D zero matrix a tensor (all entries, 0)?

(5.23) Show that $\partial_\alpha \mathscr{F}^{\alpha\beta} = 0$ gives the two homogeneous Maxwell equations.

(5.24) Starting with the electrical Coulomb potential of a stationary point charge, use the fact that the 4-potential is a 4-vector to find the potentials in a second frame moving at a constant velocity.

(5.25) Show that the Lorentz transform matrix is not a Lorentz tensor.

5.11 Discussions

Discussion 5.1

In his *Principia* (1687), Newton states clearly his vision of the universe: "Absolute, true and mathematical time, of itself, and from its own nature flows equably without regard to anything external, and by another name is called duration: relative, apparent, and common time, is some sensible and external (whether accurate or unequable) measure of duration by the means of motion, which is commonly used instead of true time; such as an hour, a day, or a year. Absolute space, in its own nature, without regard to anything external, remains always similar and immovable. Relative space is some movable dimension or measure of the absolute spaces; which our senses determine by its position to bodies; and which is vulgarly taken for immovable space; such is the dimension of a subterraneous, an aereal, or a celestial space determined by its position in respect of the earth."

Discussion 5.2

Mathematically, one considers the covariance of classical physics under the ten parameter Galilean group. A general Galilean transformation includes four parameters for translations of the space and time origin, three parameters for describing a general rotation, and three parameters describing boosts along the x-, y-, and z-axis. We will almost always ignore translations of the space-time origin and consider only rotations and boosts, members of the "homogeneous Galilean transformation group." The comparable ten parameter group for relativistic physics is known as the Poincare group or, alternatively, the inhomogeneous Lorentz group.

Discussion 5.3

The motivation for his measurements was to provide a confirmation of Copernicus' claim, made in 1543, that the Earth is in motion around the sun–a proposition for which, even at this late date, there was no direct experimental confirmation. Bradley was carrying out precision measurements of stellar positions with the goal of detecting stellar parallaxes and cementing the heliocentric theory. In fact he did observe a yearly periodic motion of the star Gamma Draconis, see Fig. 5.1. However, the phase of the apparent motion in the sky that he observed is inconsistent with that of a star's parallax. Moreover, all of the other stars in that part of the sky showed identical variations. While Bradley was unsuccessful at measuring stellar parallaxes–these are extremely small and would only be accurately measured over 100 years later–he was able to use his data to directly verify Copernicus' theory that the earth moves around the sun. Additionally, his data measured the speed of light to an accuracy of $\pm 5\%$.

Discussion 5.4

The experimental confirmation of light as an electromagnetic wave came in 1887 when Heinrich Hertz generated UHF radio frequency electromagnetic waves and measured their velocity to be exactly that of light. This experiment together with Heaviside's reformulation of Maxwell's work and publication of Maxwell's equations in their modern form led to its widespread acceptance as providing the definitive description of light.

Discussion 5.5

In the years following Young's double slit experiment, several experiments–notably Arago's prism and experiments of Fizeau–returned a puzzling picture of the ether. Throughout the 1800s, a series of experiments were carried out that studied the motion of light through various dielectrics. The first of these was an experiment in 1810 by Arago in which stellar aberration was measured directly and compared with a measurement in which the light was made to go through a prism. The principle behind this experiment is understood by considering a similar experiment later carried out by George Biddell Airy in 1871–the "water telescope" experiment. In this case, the stellar aberration is compared between a regular telescope, and the same telescope filled with water. If one understands the stellar aberration as arising from the Earth's motion between the arrival of the light at the top of the telescope–the objective lens–and its actual observation at the bottom, then one expects that filling the telescope with water should lead to a greater aberration. Light moves more slowly in water, its travel time increases, and the Earth moves further. In fact, even after filling the telescope with water, Arago's earlier prism experiment turned up no change in angle whatsoever. In 1818, Fresnel reconciled Arago's result by assuming that ether was "partially dragged" by a dielectric: that is, the ether and the waves in it were given a velocity $f v$, where v is the speed of the dielectric (the Earth's speed in these experiments), $f = 1 - 1/n^2$, and n is the dielectric constant.

Discussion 5.6

In fact, the central conclusion drawn from stellar aberration is that it demonstrates the variation in relative motion between the light arriving from a distant star and the Earth in its orbit around the sun. The effect would be absent if the medium carrying the waves moved rigidly with the Earth because the light propagation is assumed to be embedded in the ether. Stokes, an expert in fluid dynamics, advanced a theory that the ether flowed around the Earth as a perfect fluid, irrotationally, and this would allow the rectilinear motion of light required to explain stellar aberration. Lorentz criticized Stokes' theory noting that such flow necessarily required substantial slippage at the interface.

Discussion 5.7

The basic idea behind this work can be seen by considering the observation of light from a "beacon" star that is effectively orbiting a much more massive star, that is, from a binary star system. We imagine that at $t = 0$ the beacon star is moving directly away from the Earth. The emission theory then predicts that for a ray emitted toward the Earth from the beacon star at $t = 0$, the receding motion of the beacon star will subtract from that of the light ray resulting in a slower than normal light speed. Half an orbit later, after traveling behind its massive companion, the beacon star will be moving toward the Earth and now its speed will add to that of the Earthbound light ray, making for fast light. Now if the pair is sufficiently distant, the fast light emitted at the later time will overtake and pass the slow light from $t = 0$, leading to an apparent blinking of the light from the rotating star. An effect that is distinctly not observed in binary star systems. While not all cases will be this readily analyzed, clearly this phenomena would significantly affect observations of orbiting binaries. De Sitter, after making a survey of binary pairs, found no significant discrepancies of this sort showing up in the measurements and concluded that if the speed of light was affected linearly by the velocity of its source, v, and if the measured velocity of the light from the star was given by $c' = c + kv$, then his data put an upper limit on k of only 0.002.

Discussion 5.8

The interferometer operated by splitting a light source into a two perpendicular beams, both of which were reflected and recombined at the original beam-splitter to create an interference pattern. If, for example, one leg lined up with the Earth's velocity through the ether, the travel time for the light in that leg should be slightly longer than the leg perpendicular to the velocity. Rotating the apparatus should produce a shift in the interference pattern of about $1/10^{th}$ of a fringe as the relative orientation between the legs and the Earth-ether velocity change. In the following years, further experiments were carried out designed to detect a relative Ether-earth motion including, notably, the Trouton–Noble experiment and the experiments of Rayleigh and Brace. None of these qualitatively changed the picture described previously.

Discussion 5.9

Lorentz would agree that either of the interpretations of the process is valid–as long as the frame in which it is interpreted is the ether frame. In other words, in his view, both observers couldn't be simultaneously correct. He would claim that the interpretation by an observer in the moving frame, which also predicts the correct force, has been determined by an observer under the influence of false variables. He would not, however, consider the identical predictions by different mechanisms in different frames to be coincidental.

Discussion 5.10

It may be tempting to argue that the second postulate is redundant because the speed of light is a constant predicted by Maxwell's equations and, by the first postulate, the "same laws are valid for all frames." But this argument makes the mistake of assuming the very thing that the 2nd postulate is saying. Namely, that Maxwell's equations are correct physical laws. For clarity, consider an alternative case in which Ritz's "Emission Theory" turned out to be the valid formulation of electrodynamics. Then the 1st postulate of relativity would hold because the theory does not refer to an absolute frame but the second postulate would fail since the speed of light would depend on the motion of the source, as well as on that of the observer. Finally, for an example of a situation in which the first postulate fails but the second holds, consider the non-interacting ether theory: it refers to an absolute frame so the 1st postulate of relativity fails but the speed of light is with respect to the ether and not the source so the second postulate holds.

Discussion 5.11

Einstein lays out the fundamental problem:"If we wish to describe the motion of a material point, we give the values of its coordinates as functions of the time. Now we must bear carefully in mind that a mathematical description of this kind has no physical meaning unless we are quite clear as to what we understand by "time." In many respects the important advance of special relativity is the prescription for defining time for two different locations and following through the implications of that definition. Clearly, it is the central concept in understanding time dilation. And for length contraction, to measure the length of a ruler, for example, it is absolutely essential to know that one is making simultaneous measurements at the ends of the ruler.

Discussion 5.12

As shown in Fig. 5.2, in the stationary frame, the light wave is initially launched from the origin at $t=0$. After time $\Delta t_1 = \Delta x_1/c$, the wave arrives at the position, Δx_1, of the far mirror, reflects, and continues for time $\Delta t_2 = -\Delta x_2/c$ to complete the round trip at the origin of the moving frame. Thus, we see that for the round trip the moving frame origin, moving at speed V, has gone a distance

$$\Delta x_{origin} = V(\Delta t_1 + \Delta t_2) = \frac{V}{c}(\Delta x_1 - \Delta x_2)$$

which can also be expressed directly in terms of the two space differences as

$$\Delta x_{origin} = \Delta x_1 + \Delta x_2$$

noting that $\Delta x_2 < 0$. Equating these relations for Δx_{origin}, we obtain the equation $\Delta x_2 = -\Delta x_1\left(\frac{c-V}{c+V}\right)$. With the equations for Δt_1, Δt_2, and Δx_2 in hand, we can now express

$$0 = \Delta t_2' - \Delta t_1' = \frac{\partial t'}{\partial t}(\Delta t_2 - \Delta t_1) + \frac{\partial t'}{\partial x}(\Delta x_2 - \Delta x_1)$$

completely in terms of Δx_1. That is, after substitution and some manipulation,

$$0 = -\frac{1}{c}\frac{\partial t'}{\partial t}\left(\frac{2V}{c+V}\right)\Delta x_1 + \frac{\partial t'}{\partial x}\left(\frac{2c}{c+V}\right)\Delta x_1$$

and this immediately further simplifies to

$$\frac{\partial t'}{\partial x} = -\frac{V}{c^2}\frac{\partial t'}{\partial t}$$

which is the relation between the t' derivatives.

Discussion 5.13

In what follows, we will almost exclusively consider the "homogeneous" Lorentz group of transformations that includes rotations and boosts, that is, pure rotations or boosts for which the spatial origins coincide at $t = t' = 0$. Translations of the spatial or temporal origins do not change the numeric representations of vectors or tensors.

Discussion 5.14

Alternatively, we can obtain this contracted length, Δy_s, of string measured at time T in A's frame (frame K) by using the Inverse Lorentz transformations, Eqs. 5.11, on corresponding measurements made by observers in the moving frame K'. To be clear: at time $t = T$ in A's frame, the rockets B and C, along with the string, have been accelerated to the velocity $V = \beta c$ and thus are all at rest in the moving frame K'. This means the length of the string, as measured in K', is then simply its proper length ($\Delta y'_s = L_0$). We also know that if we want the length, Δy_s, in A's frame, it is necessary to consider both ends simultaneously (at time T). That is, we must set $\Delta t = 0$. Finally, noting from the relativity of simultaneity that such spatially separated, simultaneous measurements in K do not correspond to simultaneous measurements in K', it must be that $\Delta t' \neq 0$. To summarize: in frame K of rocket A, we have $\Delta y = \Delta y_s$ and $\Delta t = 0$ while in the rest frame K' of the string, $\Delta y' = L_0$ and $\Delta t' \neq 0$. Given this information, then, the length, Δy_s, in the rocket A's frame is obtained from the Inverse Lorentz transformation:

$$\begin{bmatrix} 0 \\ \Delta y_s \end{bmatrix} = \begin{bmatrix} \gamma & \gamma\beta \\ \gamma\beta & \gamma \end{bmatrix}\begin{bmatrix} c\Delta t' \\ L_0 \end{bmatrix} = \begin{bmatrix} \gamma c\Delta t' + \gamma\beta L_0 \\ \beta\gamma c\Delta t' + \gamma L_0 \end{bmatrix} \tag{5.118}$$

Here, from the top line of the vector equation 5.118, we see the relativity of simultaneity: unlike in A's frame, the two "measurements" made on the ends of the string are not simultaneous in the rest frame, K', of the string. Specifically, the "rocket B" end of the string measurement occurs $\Delta t' = -(\beta L_0)/c$ earlier than the "rocket C" end of the string measurement. Substituting $\Delta t' = -(\beta L_0)/c$ into the bottom line of the vector equation 5.118 returns the expected Lorentz contraction for Δy_s. And again the string breaks. Lorentz would argue that the forces binding the molecules of the string would, on moving through the ether, force the molecules closer together. The transformations in his 1904 paper were simply mathematical tricks to describe a moving object for the case in which neither the object nor the frame in which it was being measured were stationary relative to the absolute frame of the ether. But there could be no doubt about his prediction: the string shrinks and therefore breaks.

Discussion 5.15

For example, to see how the scalar product (contraction) of the two mixed 4-tensors,

$$C^{\alpha}{}_{\beta} = A^{\alpha}{}_{\mu} B^{\mu}{}_{\beta} \tag{5.119}$$

transforms, consider how the corresponding displacement 4-tensors, $\Delta x^{\eta} \Delta x_{\nu}$, each created from two displacement 4-vectors, transform:

$$\left(\Delta x'^{\alpha} \Delta x'_{\beta}\right) = \left(L^{\alpha}{}_{\eta}(V)\, \Delta x^{\eta}\right)\left(L^{\nu}{}_{\beta}(-V)\, \Delta x_{\nu}\right) = L^{\alpha}{}_{\eta}(V)\, L_{\beta}{}^{\nu}(V) \left(\Delta x^{\eta} \Delta x_{\nu}\right) = \Lambda^{\alpha}{}_{\eta\beta}{}^{\nu} \left(\Delta x^{\eta} \Delta x_{\nu}\right) \tag{5.120}$$

where we have used the fact that numerically, $L^{\nu}{}_{\beta}(-V) = L_{\beta}{}^{\nu}(V)$ and $\Lambda^{\alpha}{}_{\eta\beta}{}^{\nu} \equiv L^{\alpha}{}_{\eta}(V)\, L_{\beta}{}^{\nu}(V)$ as in Eq. 5.46 but now modified to handle a mixed 4-tensor. Therefore,

$$C'^{\alpha}{}_{\beta} = A'^{\alpha}{}_{\mu} B'^{\mu}{}_{\beta} = \left(\Lambda^{\alpha}{}_{\eta\mu}{}^{\nu} A^{\eta}{}_{\nu}\right)\left(\Lambda^{\mu}{}_{\lambda\beta}{}^{\kappa} B^{\lambda}{}_{\kappa}\right) = L^{\alpha}{}_{\eta} L_{\mu}{}^{\nu} L^{\mu}{}_{\lambda} L_{\beta}{}^{\kappa} A^{\eta}{}_{\nu} B^{\lambda}{}_{\kappa} \tag{5.121}$$

$$= L^{\alpha}{}_{\eta} \delta^{\nu}_{\lambda} L_{\beta}{}^{\kappa} A^{\eta}{}_{\nu} B^{\lambda}{}_{\kappa} = L^{\alpha}{}_{\eta} L_{\beta}{}^{\kappa} A^{\eta}{}_{\nu} B^{\nu}{}_{\kappa} = \Lambda^{\alpha}{}_{\eta\beta}{}^{\kappa} C^{\eta}{}_{\kappa} \tag{5.122}$$

which is the expected transformation for a 2^{nd}order mixed 4-tensor. Creating a similar analogy with the corresponding displacement tensor (or outer) product, a 4th order tensor, proves the general case.

Discussion 5.16

The equivalence of mass and energy implied by the 4-momentum definition can be illustrated by the following kinematically simple process: Atomic hydrogen has an excited 2s state that that lives about $\frac{1}{7}$s and can decay to the ground 1s state via a 2 photon transition. If a 2s hydrogen atom at rest in the lab decays emitting two equal energy photons in exactly opposite directions, then conservation of momentum has the ground state hydrogen atom again at rest. However, conservation of energy, the time-like component of the 4-momentum, demands that $m_{2s}c^2 = m_{1s}c^2 + 2h\nu$: the mass of the excited hydrogen atom is larger than the ground state by the mass equivalent of the transition energy. Much larger differences are seen in nuclear and particle physics.

Discussion 5.17

Generally, the Jacobian matrix $\vec{\mathcal{J}}$ is the $n \times m$ matrix of all first-order partial derivatives of an n–vector-function of m independent variables. In the present case, the vector-function is the set of 4 Lorentz transformations $x'^{\alpha}\left(x^0, x^1, x^2, x^3\right)$ and the Jacobian matrix is, therefore, the $n \times m = 4 \times 4$ Lorentz transformation matrix. The determinant of this matrix is called the "Jacobian" of the matrix and, in the present case, is depicted by

$$\det \vec{\mathcal{J}} = \frac{\partial \left(x'^0, x'^1, x'^2, x'^3\right)}{\partial \left(x^0, x^1, x^2, x^3\right)}$$

In general, the "Jacobian" is used as a multiplication factor connecting the expressions for a volume element as represented in two coordinate systems related by the associated Jacobian matrix coordinate transformation. In the present case, the Jacobian matrix is the Lorentz coordinate transformation and the Jacobian connecting the two volume elements as expressed in the primed and unprimed frames is the determinant of the Lorentz matrix as shown in Eq. 5.81.

6 Radiation from Charges Moving at Relativistic Velocities

- Retarded time and the evaluation of the electric and vector potentials for a charge moving relative to an observer: Lienard–Wiechert potentials
- Derivation of the electric and magnetic fields due to a charge undergoing acceleration while moving at an arbitrary velocity: space and time derivatives accounting for retarded time
- The power radiated from an accelerated charge for accelerations parallel and perpendicular to its velocity
- The spectral content of the power radiated by a moving charge: application to synchrotron radiation
- The special case of a charge moving with constant velocity: electric field lines appearing to come from a charge's current position rather than its retarded position
- Spectral energy density of the fields of a charge passing an observer: approximate calculation of the equivalent number of (virtual) photons
- Calculation of Bremsstrahlung using virtual photons and the method of Weizsacker and Williams

In Chapter 2, we obtained general solutions to the time-dependent Maxwell's Equations in the form of the retarded potentials, V and $\vec{A}$. In Chapter 3, we then obtained from these retarded potentials the Jefimenko equations for the observable fields, $\vec{E}$ and $\vec{B}$, which are essentially time-dependent generalizations of the time-independent Coulomb and Biot–Savart field equations with one important difference being the additional existence of the $1/r_o$ dependent "radiation" terms. In this chapter, we consider the characteristics of the fields, including both the $1/r_o$ and $1/r_o^2$ dependent parts, resulting from an important class of charge distributions: single point charges moving at relativistic speeds along specific trajectories.

Electromagnetic Radiation. Richard Freeman, James King, Gregory Lafyatis,

DOI: 10.1093/oso/9780198726500.001.0001

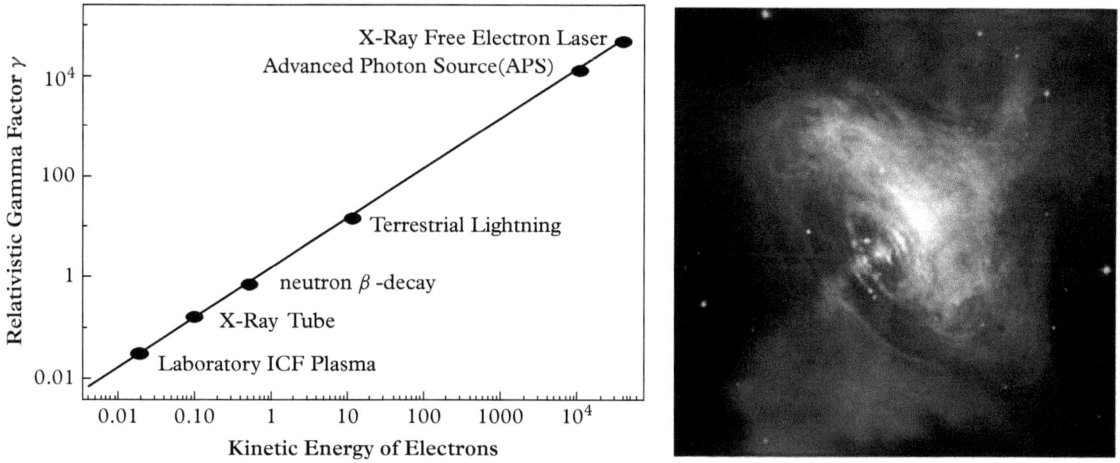

Fig. 6.1 *Left: Various sources of energetic electrons whose radiation is important. Right: The central part of the Crab Nebula. The bright spot near the center is the pulsar that powers the nebula. (Photo Credit NASA-Hubble, Public Domain.)*

In Fig. 6.1 we show typical electron energies and their corresponding "relativity factor" for several important sources of energetic electrons. For Earth-bound sources, the high energy extreme is given by electron storage rings. The Advanced Photon Source, one of many similar devices, has presently the highest energy electrons for such facilities in North America. Insertion devices–wigglers and undulators routinely generate radiation ~ 0.1 nm wavelength for a variety of condensed matter and biophysics purposes. The "fourth generation light source," x-ray free electron laser is the present day state-of-the-art. Astrophysics sources of energetic electrons are similarly far ranging. The inter-cluster medium–that is, the space between clusters of galaxies–has a thermal component of electron energies that is observed via its collisions with ions, so-called Bremsstrahlung radiation. In our celestial neighborhood the Crab Nebula is a remarkable source. Shown in Fig. 6.1 the bright neutron star seen at the center is the remnant of a supernova observed in 1054. Its distance, 6500 ly, means we are seeing radiation generated by electrons based on their positions and motions at their "retarded time"– 6500 years ago. The importance of retarded time is a recurring theme in what follows.

6.1 Lienard–Wiechert potentials

If we seek to compute the electric and magnetic fields from (rapidly) moving point charges, we might attempt to obtain these field equations directly via the Green function method, as was demonstrated in Section 3.4 to obtain the more general Jefimenko field equations.

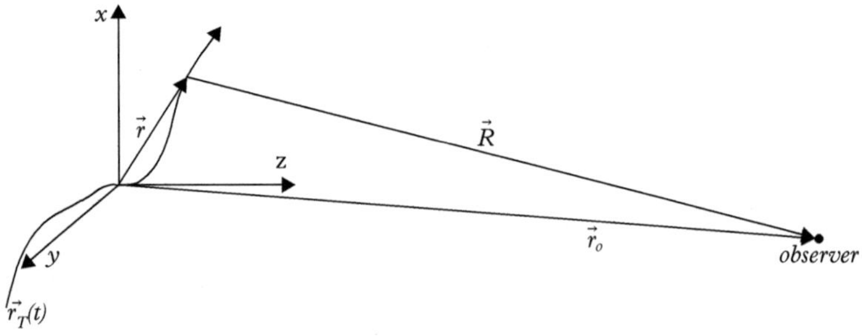

Fig. 6.2 *Definition of the coordinates used in this chapter.*

Instead we take the more instructive route by computing these fields from space and time derivatives of the retarded potentials given in Eqs. 2.19 and 2.20 and are here re-expressed,

$$V(\vec{r}_o, t) = \frac{1}{4\pi\varepsilon_o}\int \frac{\rho(\vec{r}, \tau)}{R}\, d^3r = \frac{1}{4\pi\varepsilon_o}\int \frac{\{\rho\}_{ret}}{R}\, d^3r \tag{6.1}$$

$$\vec{A}(\vec{r}_o, t) = \frac{\mu_o}{4\pi}\int \frac{\vec{\mathcal{J}}(\vec{r}, \tau)}{R}\, d^3r = \frac{\mu_o}{4\pi}\int \frac{\{\vec{\mathcal{J}}\}_{ret}}{R}\, d^3r \tag{6.2}$$

We recall that considerable care is required in the evaluation of integrals of this type because such integration is not merely the sum of all the source contributions in a volume at a single given instant, but rather the sum of all the contributions in a space-time volume which are causally connected to the observation point. This is expressed in Eqs. 6.1 and 6.2 by the symbols $\{\}_{ret}$, around the charge and current densities, which specify that they be evaluated at an earlier "retarded" time, $\tau = t - R/c$, in which the amount of time retardation, $t - \tau$, depends upon the distance, R, of each source volume element from the observation point.

To this point in our discussions, the time dependent charge distributions considered have been confined to macroscopic volumes stationary relative to the observer. Now, in our consideration of microscopic point charges moving at potentially relativistic speeds along arbitrary trajectories, the evaluation of Eqs. 6.1 and 6.2 is further complicated: we first note that a causally constrained integration over such a distribution as a particle trajectory (a worldline) results in just a single source contribution. That is to say, for any given observation point $(\vec{r}_o, t)$, there is no more than one point on the trajectory which is causally connected to it. Additionally, when charge or current distributions - such as point charges - move, relative to the observer, at velocities approaching the speed of light (the speed of causality), then the charge distribution can change in space substantially during the "time of integration." For the case of relativistically moving point charges, this effect results in an overcount/undercount of the total charge of incoming/outgoing particle. The mathematical procedure for correctly evaluating integrals such as Eqs. 6.1 and 6.2 for rapidly moving point charge distributions can be developed in two ways: (1) transforming the integral by explicitly accounting for the dependence of τ in R, and (2) a geometric argument that is more physically motivated.

6.1.1 Derivation by integral transform

Using the coordinate system in Fig. 6.2, we seek to calculate the fields that would be observed at position $\vec{r}_o$ and time t arising from a charged particle moving along a path $\vec{r}_T(t)$ near the origin of coordinates. All space-time quantities are measured in the rest frame of the observer. Most interesting observations discussed in the following will be along or near the z-axis–that is, in the direction of the velocity vector of the particle at its "retarded" position when it "creates" the observed field. The charge distribution of a charged point particle moving along a trajectory $\vec{r}_T(t')$ is represented as $\rho\left(\vec{r}, t'\right) = q\delta[\vec{r} - \vec{r}_T(t')]$. Within the integral of Eq. 6.1, this distribution must be evaluated at the retarded time so it becomes $\rho(\vec{r}, \tau) = q\delta[\vec{r} - \vec{r}_T(\tau)]$. Evaluation of Eq. 6.1 with this distribution will then properly yield the single, causally connected contribution of potential from the trajectory. However, with the distribution in this form, the integration is quite difficult. The integral transform procedure then amounts to the transformation of this delta function distribution to a more integrable form. The moving point charge distribution evaluated at the retarded time can be re-written as

$$\rho(\vec{r}, \tau) = q\int \delta[\vec{r} - \vec{r}_T(t')]\delta(t' - \tau)\, dt'$$

noting that for this form of the retarded density, $\vec{r}_T$ is no longer a function of the retarded time. Insertion of this form into Eq. 6.1

$$V(\vec{r}_o, t) = \frac{q}{4\pi\varepsilon_o}\int \frac{1}{|\vec{r}_o - \vec{r}|}\left(\int \delta[\vec{r} - \vec{r}_T(t')]\delta(t' - \tau)\, dt'\right) d^3r$$

where R has been expressed as $|\vec{r}_o - \vec{r}|$ and we recall that $\tau = t - \frac{1}{c}|\vec{r}_o - \vec{r}|$. Next, we integrate over all space to get,

$$V(\vec{r}_o, t) = \frac{q}{4\pi\varepsilon_o}\int \frac{1}{|\vec{r}_o - \vec{r}_T(t')|}\delta(t' - \tau)\, dt' \qquad (6.3)$$

where now $\tau = t - \left|\vec{r}_o - \vec{r}_T(t')\right|/c = t - R(t')/c$. By using a property of Dirac-delta functions, Eq. 6.3 picks up an additional term in its denominator involving the derivative of $R(t')$:[1]

[1] Starting with Eq. 6.3 we use a property of Dirac-delta functions, that is:

$$\delta(t' - \tau) = \delta\left[t' - (t - R(t')/c)\right] = \frac{\delta(t' - \tau)}{\left[\frac{d}{dt'}(t' - \tau)\right]_{t'=\tau^o}} = \frac{\delta(t' - \tau^o)}{\left[1 - \vec{\beta}(\tau^o)\cdot\hat{R}(\tau^o)\right]}$$

In this equation, our notation means that the result of the transformation of the delta function is a denominator that is not a function of t', but a value related to a particular $\tau = \tau^o$, which is precisely that value of τ associated with the retarded position and velocity giving rise to the fields seen by the observer at t. We will continue to use τ to index the motion of the particle and usually refer to times near τ^o.

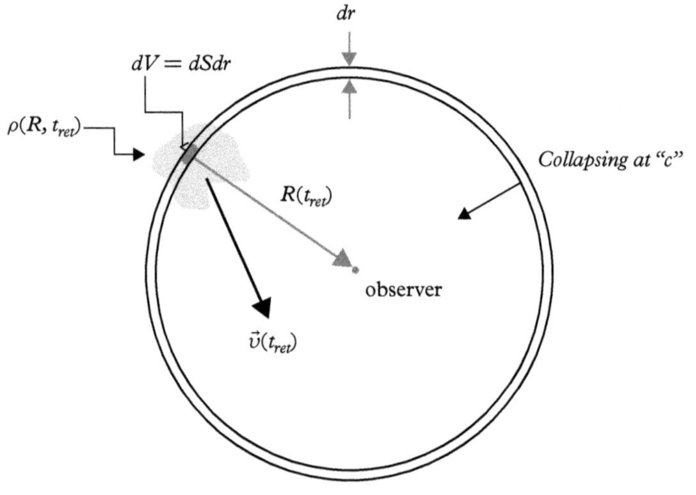

Fig. 6.3 *Information gathering sphere (IGS) for an observation point and a charge distribution moving relative to the observer.*

$$V(\vec{r}_o, t) = \frac{q}{4\pi\varepsilon_o}\left[1 + \frac{1}{c}\frac{dR}{dt'}\right]^{-1}_{t'=\tau^o} \int \frac{1}{R(t')}\delta(t' - \tau^o)\,dt' \tag{6.4}$$

in which τ^o is that singular value of t' that satisfies $t' = t - R(t')/c = \tau$. This immediately evaluates to the Lienard–Wiechert scalar potential:

$$V(\vec{r}_o, t) = \frac{1}{4\pi\varepsilon_o}\left\{\frac{q}{R(1 - \vec{\beta}\cdot\hat{R})}\right\}_{ret} \tag{6.5}$$

And so, as predicted previously, motion of the charged particle toward the observation point $\left(\vec{\beta}\cdot\hat{R} > 0\right)$ enhances the potential there and motion of the charged particle away from the observation point $\left(\vec{\beta}\cdot\hat{R} < 0\right)$ acts to diminish the potential there. The procedures and arguments to obtain $\vec{A}(\vec{r}_o, t)$ follow those for $V(\vec{r}_o, t)$.

6.1.2 Derivation by geometric construction

A physical interpretation of the velocity-dependent term in the denominator of Eq. 6.5 can be obtained by using the "information gathering sphere" (IGS) concept we encountered in Chapter 3. The evaluation of the integral in Eq. 6.1 is visualized as summing the contributions arising from a spherical shell of thickness "*dr*" that collapses onto the observer at the speed-of-information, *c*, and arrives at the observer at *t*. As it collapses, it "gathers information" (sums) from the regions within the shell that are causally connected to the observation.

Figure 6.3 schematically indicates how information about the element of charge density $\rho(\vec{r}, \tau)$ is communicated to the observer via such an IGS[2]. If the charge density $\rho(\vec{r}, \tau)$ is moving at a velocity v ($v \ll c$) relative to the observer there is no issue since the surface of the IGS moves rapidly enough that the result of the integration is essentially the same as for a stationary charge. For $v \sim c$, however, the charge density can move significantly during the passage of the IGS so the result of the integral will depend on the magnitude and direction of the charge's velocity. For a charge density ρ_{ret} moving with velocity $\vec{v}$ at the retarded time and located a distance R from the observation point, the actual amount of charge, dq, swept over by the IGS in a time dt'–during which the IGS moves inward by a radial distance dr–is

$$dq = \rho_{ret} dSdr - \rho_{ret} \left\{ (\vec{v} \cdot \hat{R}) \right\}_{ret} dSdt'$$

where dS is a differential element of area on the IGS. But, since by inspection $dt' = dr/c$ and $dSdr = dV$, we can re-arrange the terms,

$$\rho_{ret} dV = \left\{ \frac{dq}{(1 - \vec{v} \cdot \hat{R}/c)} \right\}_{ret} \tag{6.6}$$

and thus for rapidly moving charges, the "apparent" charge swept out (i.e., collected by the IGS), $\rho_{ret} dV$, is *not equal* to the actual charge swept out, dq, in time dt'. Recalling the general integral form of the retarded scalar potential due to a charge distribution (Eq. 6.1)

$$V(\vec{r}_o, t) = \frac{1}{4\pi \varepsilon_o} \int \frac{\rho_{ret} dV}{R}$$

we obtain the Lienard–Wiechert version by substitution of the "moving distribution" charge element obtained in Eq. 6.6

$$V(\vec{r}_o, t) = \frac{1}{4\pi \varepsilon_o} \int \left\{ \frac{dq}{R(1 - \vec{v} \cdot \hat{R}/c)} \right\}_{ret}$$

For a moving point charge distribution, the denominator is constant over the integration so that we again obtain the Lienard–Wiechert scalar potential,

$$V(\vec{r}_o, t) = \frac{1}{4\pi \varepsilon_o} \left\{ \frac{q}{R(1 - \vec{\beta} \cdot \hat{R})} \right\}_{ret}$$

[2] Here, we follow the discussion by Panofsky and Phillips.

Thus, the prescription for evaluation of the integrals of the type defining V and $\vec{A}$ for a relativistically moving point charge is to substitute

$$q \to \frac{q}{(1-\vec{\beta}\cdot\hat{R})}$$

where $c\vec{\beta} = \vec{v}$ is the speed of the electron.[3] As in the previous section, we note that the procedures and arguments to obtain $\vec{A}(\vec{r}_o, t)$ follow those for $V(\vec{r}_o, t)$. In summary, the scalar and vector potentials resulting from relativistically moving point charges–the Lienard–Wiechert potentials–are:

$$V(r_o, t) = \frac{q}{4\pi\varepsilon_o}\left\{\frac{1}{R(1-\vec{\beta}\cdot\hat{R})}\right\}_{ret} \tag{6.7}$$

$$\vec{A}(r_o, t) = \frac{qc\mu_o}{4\pi}\left\{\frac{\vec{\beta}}{R(1-\vec{\beta}\cdot\hat{R})}\right\}_{ret} \tag{6.8}$$

The placement of the $\{\}_{ret}$ brackets in Eqs. 6.7 and 6.8 is significant. Here the value of $R = |\vec{r}_o - \vec{r}_T(\tau^o)|$ is explicitly the distance from the charge at the retarded time, τ^o, to the observer at time t. As discussed in Section 6.2, this has significant implications for the space and time derivatives used to evaluate the E and B fields from these potentials.

6.2 Radiation fields from a single charge undergoing acceleration

The fields from a (arbitrarily moving) charged particle which produces V and $\vec{A}$ given by Eqs. 6.7 and 6.8 can be obtained by the application of the defining equations for $\vec{E}$ and $\vec{B}$ in terms of V and $\vec{A}$. Recall for example,

$$\vec{E}(\vec{r}_o, t) = \left[-\nabla V(\vec{r}, t) - \frac{\partial}{\partial t}\vec{A}(\vec{r}, t)\right]_{\vec{r}_o, t} \tag{6.9}$$

The reason we are being so careful with our notation is that there are subtleties in using the forms of V and $\vec{A}$ given in Eqs. 6.7 and 6.8. Specifically, finding the gradient operator on $V(\vec{r}_o, t)$ involves evaluating V at two positions near $\vec{r}_o$ at the same time, t, taking their difference and then dividing by the displacement. However, the charged particle locations that are causally connected to the two positions are likely different because of the difference in light-travel-time and the intervening

[3] This expression contains the ubiquitous "forward boost" term $\{\frac{1}{(1-\beta cos\theta)}\}_{ret}$ because $\vec{\beta}\cdot\hat{R} = \beta\, cos\theta$. Here θ is the angle between the particle's velocity and the direction to the observer; for velocities close to c, $\beta \approx 1 - (\frac{1}{2\gamma^2})$ so that for $\gamma >> 1$, the factor $(1-\beta\, cos\theta)^{-1}$ becomes approximately $(1 - cos\theta + \frac{cos\theta}{2\gamma^2})^{-1}$.

motion of the particle. Similarly, the time derivative of $\vec{A}(\vec{r}_o, t)$ involves evaluating it at two differentially different times holding $\vec{r}_o$ constant. But here again the motion of the particle means these correspond to particle locations at different retarded times. We will be careful in what follows to display all the $\vec{r}_o$ and t dependences. To simplify the notation, we write τ while noting that it is a function of r_o and t:

$$\tau = \tau\left[\vec{r}_o, t\right]$$

Specifically, we write the potentials as follows:

$$\begin{aligned} V(\vec{r}_o, t) &= \frac{q}{4\pi\varepsilon_o}\left\{\frac{1}{R(1-\vec{\beta}\cdot\hat{R})}\right\}_{ret} \\ &= \frac{q}{4\pi\varepsilon_o}\left(\frac{1}{R\left[\vec{r}_o,\tau\right] - \vec{\beta}\left[\tau\right]\cdot\vec{R}\left[\vec{r}_o,\tau\right]}\right) \end{aligned} \tag{6.10}$$

$$\begin{aligned} \vec{A}(\vec{r}_o, t) &= \frac{qc\mu_o}{4\pi}\left\{\frac{\vec{\beta}}{R(1-\vec{\beta}\cdot\hat{R})}\right\}_{ret} \\ &= \frac{q}{4\pi\varepsilon_o c}\left(\frac{\vec{\beta}\left[\tau\right]}{R\left[\vec{r}_o,\tau\right] - \vec{\beta}\left[\tau\right]\cdot\vec{R}\left[\vec{r}_o,\tau\right]}\right) \end{aligned} \tag{6.11}$$

We got rid of the "*ret*" subscript by using the retarded time that is causally connected to the measurement time $\vec{r}_o, t$ to evaluate the various quantities. The retarded time, $\tau(\vec{r}_o, t)$, is a function defined implicitly by the equation:

$$\tau = t - \frac{\left|\vec{R}\left[\vec{r}_o,\tau\right]\right|}{c} \tag{6.12}$$

The other symbols are identical to the quantities defined earlier, but with the $\vec{r}_o, t$ dependences displayed. That is,

$$\vec{R}\left[\vec{r}_o,\tau\right] = \vec{r}_o - \vec{r}_T\left[\tau\right] \tag{6.13}$$

and

$$\vec{\beta}\left[\tau\right] = \frac{1}{c}\frac{\partial \vec{r}_T\left[\tau\right]}{\partial\tau} \tag{6.14}$$

We begin by evaluating the vector potential time derivative in Eq. 6.9. By inspection, all of the time dependence in the vector potential enters through the retarded time and thus the time derivative is an application of the chain rule:

$$\frac{\partial \vec{A}(\vec{r}_o, t)}{\partial t} = \frac{q}{4\pi\varepsilon_o c}\frac{\partial}{\partial \tau}\left(\frac{\vec{\beta}[\tau]}{R[\vec{r}_o, \tau] - \vec{\beta}[\tau]\cdot\vec{R}[\vec{r}_o, \tau]}\right)\frac{\partial \tau}{\partial t} \tag{6.15}$$

Pulling this apart:

$$\frac{\partial \vec{\beta}[\tau]}{\partial \tau} = \dot{\vec{\beta}}[\tau] \tag{6.16}$$

This is the acceleration of the particle at the retarded time. From Eq. 6.13

$$\frac{\partial \vec{R}[\vec{r}_o, \tau]}{\partial \tau} = -c\vec{\beta}[\tau] \tag{6.17}$$

We differentiate the quantity

$$R[\vec{r}_o, \tau] = \left|\vec{R}[\vec{r}_o, \tau]\right| = \left(\vec{R}[\vec{r}_o, \tau]\cdot\vec{R}[\vec{r}_o, \tau]\right)^{1/2}$$

as

$$\begin{aligned}\frac{\partial R[\vec{r}_o, \tau]}{\partial \tau} &= \frac{-\vec{r}_o\cdot c\vec{\beta}[\tau] + \vec{r}_T[\vec{r}_o, \tau]\cdot c\vec{\beta}[\tau]}{R[\vec{r}_o, \tau]} \\ &= -c\hat{R}[\vec{r}_o, \tau]\cdot\vec{\beta}[\tau]\end{aligned} \tag{6.18}$$

And finally the required time derivative of the retarded time is found by implicit differentiation of Eq. 6.12:

$$\frac{\partial \tau}{\partial t} = 1 - \frac{1}{c}\frac{\partial (R[\vec{r}_o, \tau]}{\partial \tau}\frac{\partial \tau}{\partial t}$$

whereby using Eq. 6.18 and solving for $\partial\tau(\vec{r}_o, t)/\partial t$ we obtain:

$$\frac{\partial \tau}{\partial t} = \frac{1}{1 - \hat{R}[\vec{r}_o, \tau]\cdot\vec{\beta}[\tau]} \tag{6.19}$$

With these expressions, the time derivative of the vector potential in Eq. 6.11 can be easily found.

Now we move on to find the gradient of the scalar potential in Eq. 6.10 and then collect terms. This calculation proceeds similarly to the previous one except for the fact that some of the various quantities explicitly depend on $\vec{r}_o$ in addition to the implicit dependence on $\vec{r}_o$ through the retarded time $\tau(\vec{r}_o, t)$. We can therefore split this gradient operator into component implicit and explicit gradients with respect to r_o:

$$\nabla V[\vec{r}_o,\tau] = \left[\nabla\tau \frac{\partial}{\partial\tau}\bigg|_{\vec{r}_o\, constant} + \nabla_{r_o}\bigg|_{\tau\, constant} \right] V[\vec{r}_o,\tau] \quad (6.20)$$

$$= \frac{q}{4\pi\varepsilon_o\left(R[\vec{r}_o,\tau] - \vec{\beta}[\tau]\cdot\vec{R}[\vec{r}_o,\tau]\right)^2}$$

$$\times \left[\nabla\tau\frac{\partial}{\partial\tau} + \nabla_{r_o}\right]\left(R[\vec{r}_o,\tau] - \vec{\beta}[\tau]\cdot\vec{R}[\vec{r}_o,\tau]\right)$$

where we have used the chain rule to express the implicit gradient and where the explicit gradient ∇_{r_o} takes the derivative with respect to any explicit $\vec{r}_o$ dependence, ignoring the τ's. This pair of gradients operating on the second term in the parentheses in 6.20 yields:

$$\left[\nabla\tau\frac{\partial}{\partial\tau} + (\nabla_{r_o})\right]\left(\vec{\beta}[\tau]\cdot(\vec{r}_o - \vec{r}_T[\tau])\right)$$

$$= \dot{\vec{\beta}}[\tau]\cdot(\vec{r}_o - \vec{r}_T)[\tau]\nabla - c\beta^2[\tau]\nabla\tau + \vec{\beta}[\tau] \quad (6.21)$$

while acting on the first term yields

$$\left[\nabla\tau\frac{\partial}{\partial\tau} + \nabla_{r_o}\right]R[\vec{r}_o,\tau] = \left[\nabla\tau\frac{\partial}{\partial\tau} + \nabla_{r_o}\right]\left\{\vec{R}[\vec{r}_o,\tau]\cdot\vec{R}[\vec{r}_o,\tau]\right\}^{1/2}$$

$$= -c\hat{R}[\vec{r}_o,\tau]\cdot\vec{\beta}[\tau]\nabla\tau + \hat{R}[\vec{r}_o,\tau] \quad (6.22)$$

where for the first term we have made use of Eq. 6.18. By implicitly differentiating, in this case taking the gradient of Eq. 6.12 similarly to the time derivative taken in Eq. 6.19, we find;

$$\nabla\tau = -\frac{1}{c}\nabla R[\vec{r}_o,\tau]$$

$$= -\frac{1}{c}\left[\nabla\frac{\partial}{\partial\tau} + \nabla_r\right]R[\vec{r}_o,\tau]$$

$$= \frac{-\hat{R}}{c\left(1 - \hat{R}[\vec{r}_o,\tau]\cdot\vec{\beta}[\tau]\right)} \quad (6.23)$$

Equations 6.21, 6.22, and 6.23 allow straightforward evaluation of Eq. 6.9 (which is left as an exercise in the problem set):

$$\vec{E}(\vec{r}_o,t) = \frac{q}{4\pi\varepsilon_o}$$

$$\times\left\{\left(1-\hat{R}\cdot\vec{\beta}\right)^{-3}\left[\frac{(\hat{R}-\vec{\beta})}{\gamma^2R^2} + \frac{[(\hat{R}\cdot\dot{\vec{\beta}})(\hat{R}-\vec{\beta}) - \left(1-\hat{R}\cdot\vec{\beta}\right)\dot{\vec{\beta}}]}{cR}\right]\right\}_{ret} \quad (6.24)$$

An alternate, useful form is[4]

$$\vec{E}(\vec{r}_o,t)=\frac{q}{4\pi\varepsilon_o}\left\{\frac{(\hat{R}-\vec{\beta})}{\left(1-\hat{R}\cdot\vec{\beta}\right)^3\gamma^2R^2}+\frac{\hat{R}\times\left[(\hat{R}-\vec{\beta})\times\dot{\vec{\beta}}\right]}{\left(1-\hat{R}\cdot\vec{\beta}\right)^3cR}\right\}_{ret} \tag{6.25}$$

Next, from the L-W vector potential $\vec{A}$ of Eq. 6.8, we have for the $\vec{B}$ field:

$$\vec{B}=\nabla\times\vec{A}=\frac{q}{4\pi\varepsilon_o c}\left[\nabla\times\left(\frac{\vec{\beta}[\tau]}{R[\vec{r}_o,\tau]-\beta[\tau]\cdot\vec{R}[\vec{r}_o,\tau]}\right)\right] \tag{6.26}$$

We introduce (see problem set) a vector curl transformation analogous to the vector grad transformation of Eq. 6.20:

$$\nabla\times\vec{A}[\vec{r}_o,\tau]=\left(\nabla_{r_o}\times+\nabla\tau\times\frac{\partial}{\partial\tau}\right)\vec{A}[\vec{r}_o,\tau] \tag{6.27}$$

and evaluate $\vec{B}$ using this operator:

$$\vec{B}=-\frac{q}{4\pi\varepsilon_o c}\left(\vec{\beta}\times\nabla_{r_o}\left[\frac{1}{R[\vec{r}_o,\tau]-\vec{\beta}[\tau]\cdot\vec{R}[\vec{r}_o,\tau]}\right]\right.$$
$$\left.+\frac{\vec{R}[\vec{r}_o,\tau]}{c\left(R[\vec{r}_o,\tau]-\vec{\beta}[\tau]\cdot\vec{R}[\vec{r}_o,\tau]\right)}\times\frac{\partial}{\partial\tau}\left[\frac{\vec{\beta}[\tau]}{R[\vec{r}_o,\tau]-\vec{\beta}(\tau)\cdot\vec{R}[\vec{r}_o,\tau]}\right]\right) \tag{6.28}$$

The various derivatives are all readily carried out using the vector identity:

$$\nabla\times\frac{\vec{V}}{g}=\frac{g\nabla\times V-(\nabla g)\times\vec{V}}{g^2}$$

and the results of Eqs. 6.21 and 6.22.

$$\vec{B}(\vec{r}_o,t)=\frac{q}{4\pi\varepsilon_o c}\left\{\frac{\hat{R}}{\left(1-\hat{R}\cdot\vec{\beta}\right)^3}\times\left[\frac{(\hat{R}-\vec{\beta})}{\gamma^2R^2}+\frac{[(\hat{R}\cdot\dot{\vec{\beta}})(\hat{R}-\vec{\beta})-\left(1-\hat{R}\cdot\vec{\beta}\right)\dot{\vec{\beta}}]}{cR}\right]\right\}_{ret}$$
$$=\frac{1}{c}\left\{\hat{R}\right\}_{ret}\times\vec{E}(\vec{r}_o,t) \tag{6.29}$$

[4] obtained by noting $1-\hat{R}\cdot\vec{\beta}=\hat{R}\cdot\left(\hat{R}-\vec{\beta}\right)$ and using the vector identity $\vec{A}\times(\vec{B}\times\vec{C})=(\vec{A}\cdot\vec{C})\vec{B}-(\vec{A}\cdot\vec{B})\vec{C}$.

6.2.1 Moving charge general field characteristics

Summarizing these expressions, we have:

Fields from a point charge moving with arbitrary velocity

$$\vec{E}(\vec{r}_o, t) = \frac{q}{4\pi\varepsilon_o}\left\{\frac{(\hat{R}-\vec{\beta})}{(1-\hat{R}\cdot\vec{\beta})^3\gamma^2 R^2} + \frac{\hat{R}\times[(\hat{R}-\vec{\beta})\times\dot{\vec{\beta}}]}{c(1-\hat{R}\cdot\vec{\beta})^3 R}\right\}_{ret} \tag{6.30}$$

$$\vec{B}(\vec{r}_o, t) = \nabla\times\vec{A}(\vec{r}_o, t) = \frac{1}{c}\left[\hat{R}(\tau^o)\times\vec{E}(\vec{r}_o, t)\right] \tag{6.31}$$

Equations 6.30 and 6.31 reveal a number of important general characteristics of fields from arbitrarily moving charges:

(1) The fields have two components: (a) Velocity Fields: The terms $\sim \frac{1}{R^2}$, which are independent of $\dot{\vec{\beta}}$, the particle's acceleration. (b) Radiation Fields: The terms $\sim \frac{1}{R}$ that are linearly proportional $\dot{\vec{\beta}}$.

(2) All radiation fields appear to be arriving from the *retarded* position of the charge that is causally connected to the observation time and place. The velocity fields, on the other hand, appear to coming from the *current* position of the charge.

(3) The radiation fields at time t are determined by the vector velocity $c\vec{\beta}$ and acceleration $c\dot{\vec{\beta}}$ of the particle and their relative directions to the observation, all evaluated at the earlier time τ. As we shall see, in the highly relativistic limit for the radiation fields, it is the acceleration perpendicular to and velocity parallel to the direction of observation that largely determine the amplitude and polarization of the detected field.

(4) For $\dot{\vec{\beta}} = 0$, the radiation fields disappear and for low velocities the expressions reduce to the approximate radial Coulomb field for $\vec{E}(= \frac{q}{4\pi\varepsilon_o}\frac{\hat{R}}{R^2})$ and concentric rings about the axis of motion for $\vec{B}(= \frac{1}{c}\hat{R}\times\vec{E} \simeq 0)$.

(5) For the radiation component, $\vec{E}$ and $\vec{B}$ are perpendicular to one another and both are perpendicular to the retarded direction of observation, $\hat{R}(\tau)$. Similarly, the radiation component of the $\vec{E}$ field will always be perpendicular to $\hat{R}(\tau)$ and $\vec{B}(\vec{r}_o, t)$. However, the "velocity" component of the $\vec{E}$ field, while always perpendicular to $\vec{B}$, is not perpendicular to $\hat{R}(\tau)$, except in the limit of the extremely relativistic particle velocity.

(6) For $R >> \lambda_{max}$, the radiation zone, when the particle is moving *directly* toward the observer near the speed of light, and experiences an acceleration α, transverse to its velocity, then the observed electric and magnetic fields have a "boosted" value:[5]

$$|\vec{E}_{rad}| = c|\vec{B}_{rad}| \cong \frac{q}{\pi \varepsilon_o c^2} \left\{ \frac{\alpha}{R} \gamma^4 \right\}_{ret}$$

(7) As with its radiation component, the velocity or static component of the B field is perpendicular to both the (retarded) observation direction, $R(\tau)$, and to the observed velocity component of the E field. The velocity component of the E field, however, is never fully perpendicular to $R(\tau)$ but can approach perpendicularity in the limit of relativistic particle velocity.

All of our results so far - the general causal solutions to Maxwell's equations for the potentials, Jefimenko's general equations for the fields and the L-W potentials and associated fields for arbitrarily moving point charges–are correct at relativistic speeds despite having been derived before special relativity. This should not be a surprise since we know that Maxwell's equations are inherently relativistically correct. Before relativity, however, a correct description of electromagnetic phenomena via Maxwell's equations always assumed the observer to be in the rest or "ether" frame. In all such cases the results obtained were correct up to relativistic speeds and effects represented by the factors such as $\frac{d\tau}{dt} \neq 1$ and $1 - \hat{R} \cdot \vec{\beta}$ within the L-W potentials and associated fields were understood to be geometrical effects resulting from motion relative to the ether. Later, with the advent of special relativity and the Lorentz transformations, Maxwell's equations were revealed to provide a correct description of phenomena in all inertial frames, eliminating the need for a special frame of rest.

6.3 Power radiated from an accelerated charge

We now seek to calculate the angular distribution of the power radiated by a charge undergoing acceleration at an arbitrary velocity. For convenience of calculation, we will assume the $r_o >> r$ so that $\vec{R} = \vec{r}_o$. In this section, using the radiation zone approximation, we first construct a relativistically correct general equation. We then consider the low-velocity limit of this equation to obtain the classical angular distribution and total power radiated (Larmor's formula). Finally, we

[5] Because the radiation due to acceleration parallel to the velocity is zero in the exact direction of the velocity, it is radiation from the transverse component of acceleration which dominates.

consider the full relativistic versions of these equations (Lienard's result).[6]

The power per unit solid angle radiated by the charge in direction $\hat{R}$ can be found from the Poynting vector through a surface in that direction, where we note that the E and B fields making up the Poynting vector at observer's current time and position are evaluated using the appropriate retarded time and position of the charge:[7]

$$\begin{aligned}\frac{dP(t)}{d\Omega} &= \hat{R}_{ret}\cdot\vec{S}\left(\vec{R}_{ret},\tau\right)R_{ret}^2\\ &= \hat{R}_{ret}\cdot\frac{1}{\mu_o}\left[\vec{E}\left(\vec{R}_{ret},\tau\right)\times\vec{B}\left(\vec{R}_{ret},\tau\right)\right]R_{ret}^2\\ &= \frac{R_{ret}^2}{Z_0}\left|\vec{E}\left(\vec{R}_{ret},\tau\right)\right|^2\end{aligned}$$

where Z_0 is the characteristic impedance of free space. Here we are using the radiation fields of $\vec{E}$ and $\vec{B}$ from Eqs. 6.30 and 6.31 valid in the far field. Then

$$\frac{dP}{d\Omega} = \frac{q^2\mu_0 c}{16\pi^2}\left\{\left|\frac{\hat{R}\times[(\hat{R}-\vec{\beta})\times\dot{\vec{\beta}}]}{(1-\hat{R}\cdot\vec{\beta})^3}\right|^2\right\}_{ret} \tag{6.32}$$

This, then, gives the relativistically correct angular power distribution from a moving charge, with its instantaneous velocity $\vec{\beta}$ and acceleration $\dot{\vec{\beta}}$ evaluated at the retarded time. The factor $1-\hat{R}\cdot\vec{\beta}$ is raised to the negative sixth power in this expression. And we can see immediately that for $|\beta|\to 1$, highly relativistic particles, the radiation will be strongly "beamed" along the direction of the particle motion.

6.3.1 Low velocities and classical Larmor's formula

Taking the non-relativistic limit ($\beta << 1$) of Eq. 6.32 yields

$$\frac{dP}{d\Omega} = \frac{q^2\mu_0}{16\pi^2 c}\left\{\left|\hat{R}\times(\hat{R}\times\vec{a})\right|^2\right\}_{ret} \tag{6.33}$$

where we have written out $\dot{\vec{\beta}} = \vec{a}/c$. Defining the angle between the direction of the acceleration and the direction from the charge to the observation point as θ (see problem set), we obtain the cylindrically symmetric, dipole-like instantaneous angular distribution of radiation from an accelerating charge moving at non-relativistic speeds.

[6] In what follows we must distinguish between the different types of energy associated with a moving charge. We will therefore designate radiation energy with the symbol $\mathscr{E}$, total energy of the charge with the Roman E and kinetic energy of the charge with the Roman KE (the electric field will continue to be designated with the italic $\vec{E}$).

[7] The notation in Eq. 6.32 is to be read that the value of the power per unit solid angle radiated by the charge as measured by the observer at time t is calculated with fields evaluated at time τ and the charge to observer displacement, $\vec{R}_{ret}$.

Angular distribution of power ($v/c << 1$)

$$\frac{dP}{d\Omega} = \frac{q^2 \mu_o}{16\pi^2 c} a_{ret}^2 \sin^2\theta \tag{6.34}$$

where a_{ret} is the magnitude of the acceleration at the retarded time τ.

The total power radiated is obtained by integrating over 4π steradians:

$$P = \int \frac{dP}{d\Omega'} d\Omega' = \frac{q^2 \mu_o}{16\pi^2 c} a_{ret}^2 \int_0^{2\pi} \int_0^{\pi} \sin^3\theta' d\theta' d\phi' \tag{6.35}$$

In this derivation, we have placed no additional restrictions upon the velocity nor the acceleration, so that Eq. 6.35 is quite general (for $v/c << 1$): The integrals are readily evaluated giving Larmor's Formula for the total radiation emitted by an accelerated charge moving slowly with respect to the speed of light:

Larmor's Formula for Total Radiated Power ($v/c << 1$)

$$P = \frac{q^2 \mu_o}{6\pi c} a_{ret}^2 \tag{6.36}$$

6.3.2 Radiated power for relativistic particles

Generalization of Larmor's formula

We next would like to find the relativistically correct version of Larmor's formula. That is, we need to be able to calculate the amount of radiation energy emission (or energy loss), $\Delta\mathscr{E}$, by an accelerating charge during its period of emission, $\Delta\tau$. One might be tempted to integrate Eq. 6.32 over 4π steradians since it is a relativistically correct equation for the measured angular power distribution coming from the charge. However, this expression is (for a particular direction) the power–energy, $\Delta\mathscr{E}$, per unit time, Δt, per steradian–of the observer. We found in deriving the fields, however, that we needed to account for the particle's motion during the interval. Specifically, the start and end of Δt may correspond to retardation by different amounts and the interval, $\Delta\tau$, over which the particle emitted the energy, $\Delta\mathscr{E}$, may be different from the interval, Δt, over which the observer detected $\Delta\mathscr{E}$. Here, we are interested in the rate at which the particle is losing energy and for this, the angular distribution is:

$$\frac{d^2\mathscr{E}}{d\Omega d\tau} = \frac{d^2\mathscr{E}}{d\Omega dt}\frac{dt}{d\tau}$$

where from Eq. 6.19:

$$\frac{\partial t}{\partial \tau} = (1 - \hat{R} \cdot \vec{\beta})$$

This $1 - \hat{R} \cdot \vec{\beta}$ factor is a scaling associated with Doppler shifts. Recall the discussion leading up to the Lienard potentials. Indeed, it may be understood, for example, as the "bunching up" of energy emitted by a particle in the direction of its motion and the corresponding increase in power measured in that direction.

Relativistically correct angular distribution of power *emitted* by a charged particle

$$\frac{dP}{d\Omega}\left(\frac{\partial t}{\partial \tau}\right) = \frac{dP\left(\hat{R}, \tau\right)}{d\Omega} = \frac{d^2\mathscr{E}}{d\Omega d\tau} = \frac{q^2 \mu_0}{16\pi^2 c}\left[\frac{[\hat{R} \times ((\hat{R} - \vec{\beta}) \times \vec{a})]^2}{(1 - \hat{R} \cdot \vec{\beta})^5}\right] \tag{6.37}$$

Now an expression for the relativistically correct total power $P(\tau)$ emitted by a charged particle can be obtained from Eq. 6.37 by integration over the total solid angle. That integration itself is straightforward but tedious.[8] The result is known as Lienard's generalization of Larmor's result:

Lienard's generalization of Larmor's formula for total radiated power (arbitrary velocity)

$$P(\tau) = \frac{q^2 c \gamma^6 \mu_0}{6\pi}\left[\dot{\vec{\beta}} \cdot \dot{\vec{\beta}} - (\vec{\beta} \times \dot{\vec{\beta}})^2\right] \tag{6.38}$$

Comparison of Eqs. 6.36 and 6.38 shows that in the transition from non-relativistic to relativistic velocities, the total radiated power rises dramatically with increasing particle kinetic energy. Which is to say that, for a given acceleration, it has a very strong dependence on velocity via the γ^6 factor. For example, the ratio of emitted energy for identical acceleration (magnitude, direction and duration) of a 10 MeV kinetic energy particle compared to a 1 MeV kinetic energy particle is:

$$\left(\frac{\mathscr{E}_{10}}{\mathscr{E}_1}\right) \simeq 120{,}000$$

Note also that, in Eq. 6.38, for any acceleration perpendicular to the velocity, the cross product $(\vec{\beta} \times \dot{\vec{\beta}})^2 > 0$ so the power is diminished in

[8] Expand the triple product:

$$\left(\hat{R} \times [(\hat{R} - \vec{\beta}) \times \dot{\vec{\beta}}]\right)^2 = \left((\hat{R} - \vec{\beta})(\hat{R} \cdot \dot{\vec{\beta}}) - \dot{\vec{\beta}}(1 - \hat{R} \cdot \vec{\beta})\right)^2$$

and use $\cos\theta = \hat{R} \cdot \vec{\beta}$; $\cos\phi = \hat{\beta} \cdot \hat{\dot{\beta}}$.

comparison with an equal magnitude of parallel acceleration. Indeed, the distinction between acceleration parallel and perpendicular to the velocity turns out to be an important one in terms of the total power radiated as well as the angular distribution of power. These differences and the previously mentioned strong velocity dependence of the total radiated power will be further explored next.

6.4 Acceleration parallel and perpendicular to velocity

We saw before that radiation generated by accelerating a highly relativistic charge is strongly enhanced or beamed in the direction of the particle's instantaneous motion. As we will discuss in this section, additionally, the *direction* of the acceleration relative to the velocity strongly influences the detailed angular distribution and total emission power. The component of acceleration perpendicular to the velocity will usually be most important. The two relativistic effects of beaming and radiation pattern are results, respectively, of the dependencies in Eq. 6.37 on $(1 - \hat{R} \cdot \vec{\beta})^5$ and $\vec{\beta} \times \vec{a}$.

6.4.1 Angular distribution for acceleration || to velocity

A case of practical interest is one in which the charged particle is undergoing acceleration parallel to its velocity, as in a linear particle accelerator or when the particle is stopped by an abrupt head-on collision. Recall from Fig. 6.2 our z-axis points along the velocity, with a polar angle θ between the direction of the velocity (and acceleration) and the direction to the observer. Noting that $\beta \times \vec{a} = 0$ and $\left|\hat{R} \times \left(\hat{R} \times \vec{a}\right)\right|^2 = a^2 \sin^2\theta$, Eq. 6.37 becomes:

$$\frac{d^2\mathscr{E}}{d\Omega d\tau} = \frac{q^2 c\mu_0}{16\pi^2}\left(\dot{\vec{\beta}} \cdot \dot{\vec{\beta}}\right)\left(\frac{\sin^2\theta}{(1 - \beta\cos\theta)^5}\right) \tag{6.39}$$

This angular distribution of power for the parallel acceleration-velocity case is shown in Fig. 6.4 for $\gamma = 8$. For $\gamma = 1$, or in the instantaneous rest frame of the charge, the angular distribution is the $\sin^2\theta$ that is familiar from dipole radiation. As $\beta \to 1$, the total emission grows rapidly, and its distribution is bent toward the direction of the velocity, although directly along the velocity the power emitted remains zero.

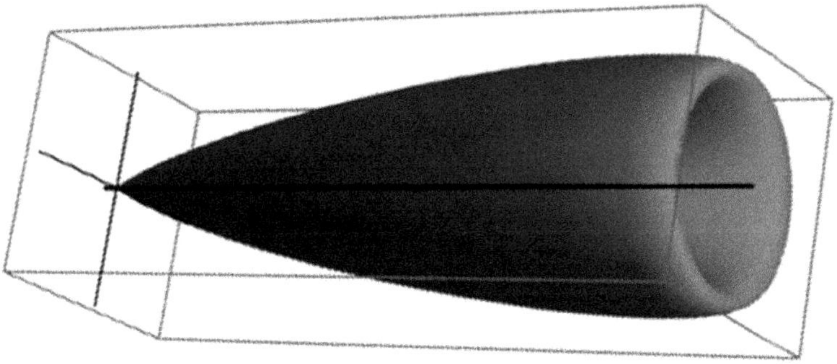

Fig. 6.4 *Angular Distribution of radiation from a relativistic charge moving from left to right with $\gamma = 8$ and undergoing acceleration parallel to its velocity. The power in a direction from the origin is given by the distance of the curve from the origin. The peak power for this velocity is 10^9 greater than the peak power of the charge undergoing the same acceleration but with a γ near 1.*

Looking nearly directly into the oncoming path of a highly relativistic accelerating charge – that is, for $\gamma >> 1$ and $\theta << 1$ – we can approximate Eq. 6.39 as[9]

$$\frac{d^2\mathscr{E}}{d\Omega d\tau} = \frac{2\gamma^{10}q^2\mu_o}{\pi^2 c}a^2\left[\frac{\theta^2}{(1+\gamma^2\theta^2-\frac{1}{2}\theta^2)^5}\right] \tag{6.40}$$

The solid angle Ω subtended by a polar angle θ is:

$$\Omega = 2\pi\,[1-cos(\theta)]$$

which, for small angles, reduces to

$$\Omega \simeq \pi\theta^2 \quad (\theta << 1)$$

so, we can approximate the power emitted within a polar angle θ as:

$$\begin{aligned}\frac{d\mathscr{E}}{d\tau}(into\,\Omega) &\simeq \frac{2\gamma^{10}q^2\mu_o}{\pi c}a^2\left[\frac{\theta^4}{(1+\gamma^2\theta^2-\frac{1}{2}\theta^2)^5}\right]\\ &\overset{\gamma\gg 1}{\approxeq} \frac{2\gamma^6 q^2\mu_o}{\pi c}a^2\left[\frac{(\gamma\theta)^4}{(1+\gamma^2\theta^2)^5}\right]\end{aligned} \tag{6.41}$$

Figure 6.5 shows this power emission distribution as a function of $\gamma\theta$. Note the peak occurs at[10]

$$\theta_{peak} = 1/2\gamma \tag{6.42}$$

while the RMS angle of emission in this relativistic limit is

$$\left\langle\theta^2\right\rangle^{1/2} = \frac{1}{\gamma}$$

[9] Use the small angle expansion $cos\theta \simeq 1-\frac{1}{2}\theta^2$ and write $\beta \approxeq (1-\frac{1}{2}\gamma^{-2})$.

[10] In Eq. 6.39 write $x = cos\theta$; then the maximum of the power per unit solid angle is found by solving for x in the equation

$$\frac{d}{dx}\left(\frac{1-x^2}{(1-\beta x)^5}\right) = 0$$

and using the small angle approximation for $cos\theta$ and $\beta \approxeq (1-\frac{1}{2}\gamma^{-2})$.

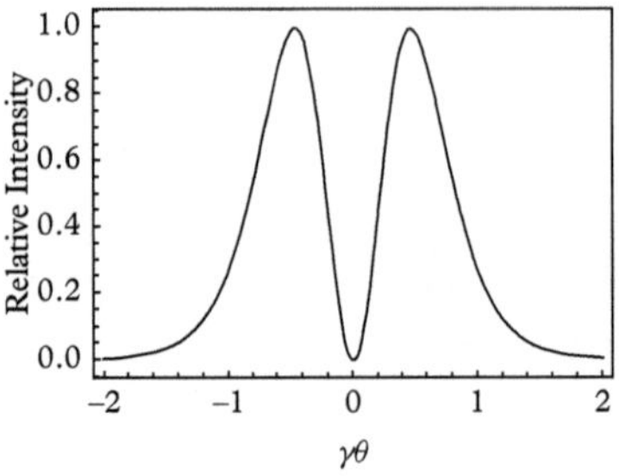

Fig. 6.5 *Total power distribution for acceleration parallel to the velocity as a function of $\gamma\theta$. This drawing is a pseudo-cut through the distribution employing unphysical negative θ to emphasize the conical emission. Compare to the right-hand side of Fig. 6.4.*

This forward "focusing" of the radiation from linear acceleration gives remarkable results for large γ. For a relativistic electron with kinetic energy of 50 MeV, $\gamma \simeq 100$, the peak of the emission when it is accelerated along its velocity is at an angle of only 5 mrad ($\theta_{peak} \sim 0.29°$) relative to its direction of motion.

6.4.2 Angular distribution for acceleration ⊥ to velocity

Referring again to Eq. 6.37, when $\dot{\vec{\beta}}$ is perpendicular to $\vec{\beta}$ all the terms contribute. Expanding the triple cross product yields

$$\frac{d^2\mathscr{E}}{d\Omega d\tau} = \frac{q^2 c\mu_0}{16\pi^2}\left(\frac{1}{(1-\hat{R}\cdot\vec{\beta})^5}\right)\left|(\hat{R}-\vec{\beta})[\hat{R}\cdot\dot{\vec{\beta}}] - \dot{\vec{\beta}}[(\hat{R}-\vec{\beta})\cdot\hat{R}]\right|^2 \tag{6.43}$$

We use the same coordinate system described before. However, with the acceleration perpendicular to the velocity, the cylindrical symmetry we had in the previous case is lost. We choose $\dot{\vec{\beta}}$ to lie along the x-axis with ϕ thus measuring the angle in the x–y plane relative to the x-axis. The direction to the observer is given by $\hat{R} = (\sin\theta\cos\phi)\hat{x} + (\sin\theta\sin\phi)\hat{y} + (\cos\theta)\hat{z}$. The angular distribution of power for this case is:[11]

$$\frac{d^2\mathscr{E}}{d\Omega d\tau} = \frac{q^2 c\mu_0\dot{\beta}^2}{16\pi^2}\left(\frac{1}{(1-\beta\cos\theta)^3}\right)\left[1 - \frac{\sin^2\theta\cos^2\phi}{\gamma^2(1-\beta\cos\theta)^2}\right] \tag{6.44}$$

Comparing the angular distributions of radiation for the two cases of parallel and perpendicular acceleration, Eqs. 6.39 and 6.44, we see that, due to the $(1-\beta\cos\theta)$ terms in the denominators, they share the relativistic beaming in the direction of instantaneous velocity. In sharp contrast to the parallel case, here the distribution peaks in the exact forward direction, $\theta_{peak} = 0$.

[11] Use the directions and angles, as given, to expand the vector relationship in Eq. 6.43

$$\begin{aligned}
&\left|(\hat{R}-\vec{\beta})[\hat{R}\cdot\dot{\vec{\beta}}] - \dot{\vec{\beta}}[(\hat{R}-\vec{\beta})\cdot\hat{R}]\right|^2 \\
&= (1-2\hat{R}\cdot\vec{\beta}+\beta^2)\dot{\beta}^2 \sin^2\theta\cos^2\phi \\
&\quad + (1-\beta\cos\theta)^2\dot{\beta}^2 \\
&\quad - 2\dot{\beta}\sin\theta\cos\phi(1-\beta\cos\theta)\vec{\beta}\cdot(\hat{R}-\vec{\beta}) \\
&= \dot{\beta}^2(1-\beta\cos\theta)^2\left[1+\frac{\sin^2\theta\cos^2\phi}{(1-\beta\cos\theta)^2}\right. \\
&\quad \left.\times(1-2\beta\cos\theta+\beta^2-2+2\beta\cos\theta)\right] \\
&= \dot{\beta}^2(1-\beta\cos)^2\left[1-\frac{\sin^2\theta\cos^2\phi}{\gamma^2(1-\beta\cos\theta)^2}\right]
\end{aligned}$$

As before, we can approximate the highly relativistic ($\gamma >> 1$), small angle ($\theta << 1$) limit of Eq. 6.44:

$$\frac{d^2\mathscr{E}}{d\Omega d\tau} \simeq \frac{\gamma^6 q^2 \mu_o}{2\pi^2 c} a^2 \left(\frac{1}{\left[1+\gamma^2\theta^2 - \frac{1}{2}\theta^2\right]^3} \right) \left[1 - \frac{4\gamma^2\theta^2 \cos^2\phi}{\left[1+\gamma^2\theta^2 - \frac{1}{2}\theta^2\right]^2} \right] \tag{6.45}$$

and Integrating θ^2 over this distribution we find that the RMS angle of emission, a measure of the beam divergence, is identical to the parallel acceleration case: $\langle\theta^2\rangle^{1/2} = \frac{1}{\gamma}$.

6.4.3 Total radiated power for acceleration ∥ and ⊥ to velocity

The Lienard results, Eq. 6.38, for the two cases–writing $a = c\dot{\beta}$ are

$$P_{\parallel} = \frac{q^2 \gamma^6 \mu_o}{6\pi c} a_{\parallel}^2 \tag{6.46}$$

$$P_{\perp} = \frac{q^2 \gamma^4 \mu_o}{6\pi c} a_{\perp}^2 \tag{6.47}$$

Comparison of Eqs. 6.46 and 6.47 shows that for the highly relativistic, $\gamma >> 1$ limit, power for both types of accelerations are strong functions of velocity. For equal magnitudes of acceleration, these equations show that there is more radiated power for the parallel case, specifically $P_{\parallel} = \gamma^2 P_{\perp}$. This comparison is not usually relevant, for in practice we most often should compare the radiation from charges experiencing the same force in either the parallel or perpendicular direction. This changes the result because acceleration produced by equal forces for the two cases will depend differently on the charge's velocity. In other words, for the parallel acceleration case, as the speed of light is approached, further increasing a particle's speed requires increasingly larger forces. For acceleration perpendicular to the velocity, the speed is not changed and, for a given magnitude of acceleration, less force is required to just change the particle's direction. This can be shown by taking the time derivative of the particle's relativistic momentum,[12] $\vec{p} = m\gamma c\vec{\beta}$:

$$\frac{d\vec{p}}{dt} = \vec{F} = mc\left[\vec{\beta}\frac{d\gamma}{dt} + \gamma\frac{d\vec{\beta}}{dt}\right] = mc\left[\gamma^3(\vec{\beta}\cdot\dot{\vec{\beta}})\vec{\beta} + \gamma\dot{\vec{\beta}}\right] \tag{6.48}$$

For a force F_o applied in a direction parallel to the velocity, this becomes[13]:

[12] $\gamma = (1-\beta^2)^{-\frac{1}{2}}$: use the chain rule, noting that $\beta^2 = \vec{\beta}\cdot\vec{\beta}$.

[13] If $\vec{F} \parallel \vec{v}$ then $\dot{\vec{\beta}} \parallel \vec{\beta}$ and Eq. 6.48 becomes

$$\begin{aligned} F_{\parallel} &= mc\left[\gamma^3\beta^2\dot{\beta} + \gamma\dot{\beta}\right] \\ &= mc\gamma\dot{\beta}\left[\gamma^2\beta^2 + 1\right] \\ &= m\gamma c\dot{\beta}\left[\gamma^2\left(1 - \frac{1}{\gamma^2}\right) + 1\right] \\ &= \gamma^3 m a_{\parallel} \end{aligned}$$

$$a_{\|} = \frac{1}{m\gamma^3} F_o$$

On the other hand, for the same force F_o applied in a direction perpendicular:

$$a_{\perp} = \frac{1}{m\gamma} F_o$$

Thus, when Eqs. 6.46 and 6.47 are expressed in terms of force, we have:

Comparison of radiated power for a force F_o applied parallel or perpendicular to the velocity

$$P_{\|} = \frac{q^2 \mu_o}{6\pi m^2 c} F_o^2 \tag{6.49}$$

$$P_{\perp} = \frac{q^2 \mu_o}{6\pi m^2 c} \gamma^2 F_o^2 = \gamma^2 P_{\|} \tag{6.50}$$

For equal magnitude forces, the force perpendicular to the velocity leads to more total radiation power.

Insight into the results of Eqs. 6.49 and 6.50 is gained by considering the time derivative of relativistic momentum or, more specifically, the time derivative of γ.

$$\frac{d\gamma}{dt} = \gamma^3 (\vec{\beta} \cdot \dot{\vec{\beta}})$$

Taking the time derivative of the particle's relativistic kinetic energy, $\mathrm{KE} = (\gamma - 1)mc^2$, and choosing the case of accelerations parallel to the charge's velocity (since perpendicular accelerations do not change the kinetic energy); we then get,

$$\frac{d(\mathrm{KE})}{dt'} = (mc\beta\gamma^3)\, a_{\|} \tag{6.51}$$

Thus for a constant value of the charge's acceleration parallel to its velocity, the rate of change of its kinetic energy increases as γ^3. In practical terms, at higher velocities (higher γ's), much more power must be supplied to achieve the same acceleration as compared to low velocities (in the same way, we note, that much larger forces must be applied, at higher velocities, to maintain the acceleration). Inverting Eq. 6.51 and inserting this expression for $a_{\|}$ into Lienard's result for total radiated power (Eq. 6.38) for the case of the acceleration parallel to the velocity yields:

$$P(t) = \frac{q^2 \mu_o}{6m^2 \pi c^3 \beta^2} \left(\frac{d(\text{KE})}{dt} \right)^2 \tag{6.52}$$

which shows that the radiated power is proportional to the square of the rate of change of the kinetic energy. So, while for higher velocities, much more power has to be supplied and much more force needs to be applied to maintain the same acceleration, in order to maintain a constant power of radiation, one need only supply constant power or apply a constant force at all velocities. Finally, since the time derivative of the particle's relativistic kinetic energy is equal to the time derivative of the particle's total energy, $\frac{d(\text{KE})}{dt} = \frac{d\text{E}}{dt}$ (the rest energy is time independent) and $\frac{1}{c\beta}\frac{d\text{E}}{dt} = \frac{d\text{E}}{dx}$, we can write the power efficiency of a parallel-accelerated charge as

$$\left(P(t) / \frac{d\text{E}}{dt} \right) = \frac{q^2 \mu_o}{6m^2 \pi c^3 \beta^2} \frac{d\text{E}}{dt} = \frac{q^2 \mu_o}{6m^2 \pi c^2 \beta} \frac{d\text{E}}{dx} \tag{6.53}$$

6.5 Spectral distribution of radiation from an accelerated charge

Because of the simple linear relationship of the observed fields to the transverse component of retarded charge acceleration, $E(t) = cB(t) \propto a_{tran}(\tau)$, as expressed in Eqs. 6.30 and 6.31, the simplest spectral distribution of radiation from an accelerating charge is the monochromatic radiation observed from a charge oscillating with completely transverse acceleration and zero average transverse velocity. Any modifications to this simplest case will result in a broadening of the spectral distribution. The most obvious modification is for $a_{tran}(\tau)$, and thus the fields, to have a non-harmonic (non-monochromatic) functional form. If the motion is relativistic, such that Doppler shifting is significant, other modifications to the simplest monochromatic case will affect a broadening of the spectrum. For example, if the oscillation direction is tilted toward the observer, then, alternating with the half cycles, the observed frequency would Doppler shift above and below the original frequency, broadening the spectrum observed. Another example is if there is any relative radial acceleration between the charge and the observer. In such a case, even a simple constant acceleration would impose a frequency chirp onto the observed spectrum. The most interesting and important broadening, however, occurs for highly relativistic motion in which the acceleration is perpendicular to the velocity. As explained in Section 6.3, at relativistic speeds the angular distribution of radiation becomes sharply concentrated in the instantaneous direction of the charged particle's velocity. Thus, as the charge moves along an arbitrary trajectory, this narrow beam sweeps around like a searchlight and an observer in line with a tangent on a curve of this trajectory will see a short pulse of radiation as the beam passes across his location; the more relativistic the velocity and the tighter the arc of the curve, the shorter the pulse.

The shorter a pulse of light in time, the broader its frequency spectrum, roughly by the relation $\Delta\nu\Delta t \simeq 1$. So, the combination of a relativistically narrowed beam and an acceleration perpendicular to the velocity, a curved trajectory, results in a broadened spectrum of the observed radiation.

To provide a more quantitative, yet still general, analysis we will now consider a Fourier transformation of the observed radiation fields resulting from charged particle acceleration. The practical quantity of measurement is generally not the radiation fields themselves but rather the associated energy or intensity of the radiation. In Sections 6.3 and 6.4, we derived and worked with the observable time-domain quantity of energy radiated per unit solid angle per unit time–the angular power distribution. In this section, we would like to obtain a comparable measurable quantity for the frequency domain: the energy radiated per unit solid angle per unit frequency. Thus, we start with the former, the angular power distribution of radiation. In Section 6.3, from the definitions of the radiation fields ($1/R$ part of Eqs. 6.30 and 6.31) and the Poynting vector and noting that the radiation fields are transverse to $\vec{R}$, we found the general result

$$\frac{dP(\vec{r}_o, t)}{d\Omega} = \frac{1}{Z_0}\{R(\vec{r}_o, t)\}^2_{ret}\left|\vec{E}_a(\vec{r}_o, t)\right|^2 \tag{6.54}$$

where $\vec{E}_a$ is the radiation or "acceleration" part of the field that dominates in the radiation zone. We define

$$\vec{\mathbb{E}}(\vec{r}_o, t) = \frac{1}{\sqrt{Z_0}}\{R(\vec{r}_o, t)\}_{ret}\vec{E}_a(\vec{r}_o, t) \tag{6.55}$$

then the angular distribution of power can be expressed as

$$\frac{dP}{d\Omega} = \left|\vec{\mathbb{E}}(\vec{r}_o, t)\right|^2 = \vec{\mathbb{E}}^* \cdot \vec{\mathbb{E}} \tag{6.56}$$

The Fourier transform relationships between $\vec{\mathbb{E}}(\vec{r}_o, t)$ and $\vec{\mathbb{E}}(\vec{r}_o, \omega)$ are then expressed as

$$\vec{\mathbb{E}}(\vec{r}_o, \omega) = \frac{1}{\sqrt{2\pi}}\int dt' e^{i\omega dt'}\vec{\mathbb{E}}(\vec{r}_o, t') \tag{6.57}$$

$$\vec{\mathbb{E}}(\vec{r}_o, t) = \frac{1}{\sqrt{2\pi}} \int d\omega' e^{-i\omega' t} \vec{\mathbb{E}}(\vec{r}_o, \omega') \tag{6.58}$$

The angular power distribution, Eq. 6.56, can now be written in terms of the inverse Fourier transform, Eq. 6.58, and its complex conjugate

$$\frac{dP}{d\Omega} = \frac{1}{2\pi} \int \int d\omega' d\omega'' e^{i(\omega' - \omega'')t} \vec{\mathbb{E}}^* (\omega') \cdot \vec{\mathbb{E}}(\omega'')$$

Integrating this over all time, we obtain the angular distribution of total radiated energy

$$\begin{aligned} \frac{d\mathscr{E}}{d\Omega} &= \int \frac{dP}{d\Omega} dt = \int \int d\omega' d\omega'' \delta(\omega' - \omega'') \vec{\mathbb{E}}^* (\omega') \cdot \vec{\mathbb{E}}(\omega'') \\ &= \int_{-\infty}^{\infty} d\omega' \left| \vec{\mathbb{E}}(\omega') \right|^2 \\ &= \int^{\infty} d\omega' \left[\left| \vec{\mathbb{E}}(\omega') \right|^2 + \left| \vec{\mathbb{E}}(-\omega') \right|^2 \right] \end{aligned} \tag{6.59}$$

where we have used the fact that $\int e^{i(\omega' - \omega'')t} dt = 2\pi \delta(\omega' - \omega'')$ and have equated to an integral over only positive frequency values. $\vec{\mathbb{E}}(\vec{r}_o, t)$ is a real quantity. We can see from Eq. 6.57 that $\vec{\mathbb{E}}(\vec{r}_o, -\omega) = \vec{\mathbb{E}}^*(\vec{r}_o, \omega)$ and re-write the angular distribution of energy as

$$\frac{d\mathscr{E}}{d\Omega} = \int^{\infty} d\omega' 2 \left| \vec{\mathbb{E}}(\omega') \right|^2 \tag{6.60}$$

so that the integrand is identified as the energy radiated per unit solid angle per unit frequency:

$$\frac{d^2\mathscr{E}}{d\Omega d\omega} = \frac{dI}{d\Omega} = 2 \left| \vec{\mathbb{E}}(\omega) \right|^2 \tag{6.61}$$

where $I = I(\vec{r}_o, \omega)$ is here defined as the radiation energy per unit frequency detected by the observer over all time (compare to the power, P). Thus, given the Fourier relationship between the time and frequency domains of the observed fields, we have obtained a general relationship between the angular distributions of power, P, and energy per unit frequency, I, for radiation. That is, referring to Eqs. 6.56 and 6.60,

$$\frac{d\mathscr{E}(\hat{r}_o)}{d\Omega} = \int_{-\infty}^{\infty} dt' \left| \vec{\mathbb{E}}(\vec{r}_o, t') \right|^2 = 2 \int^{\infty} d\omega' \left| \vec{\mathbb{E}}(\omega') \right|^2 \tag{6.62}$$

where we can now see the common sense result that the total energy per solid angle is the same whether summed over frequency or time–an example of Parseval's theorem from Fourier analysis. Up to now, this development has been quite general in that the preceding equations and relations apply not only to single accelerating charges but to arbitrary collections and distributions of such charges. We now specialize the results to the former. Inserting the radiation $(1/R)$ part of the field from a single arbitrarily moving charge, as given from Eq. 6.30, into Eq. 6.55 to get $\vec{\mathbb{E}}(\vec{r}_o, t)$, and then using Eq. 6.57 to represent $\vec{\mathbb{E}}(\vec{r}_o, \omega)$ as the Fourier transform of $\vec{\mathbb{E}}(\vec{r}_o, t)$, we obtain

$$\vec{\mathbb{E}}(\vec{r}_o, \omega) = \left(\frac{q^2}{32\pi^3 c\varepsilon_o}\right)^{1/2} \int dt\, e^{i\omega t} \left\{\frac{\hat{R} \times [(\hat{R} - \vec{\beta}) \times \dot{\vec{\beta}}]}{(1 - \hat{R}\cdot\vec{\beta})^3}\right\}_{ret}$$

Because the terms in the curly brackets are functions evaluated at the retarded time, $\tau = t - R(\tau)/c$, the integration is clearly easier to carry out over τ. Noting, then, that $t = \tau + R(\tau)/c$ and $dt = (1 - \hat{R}\cdot\vec{\beta})d\tau$, integration over τ is represented as

$$\vec{\mathbb{E}}(\vec{r}_o, \omega) = \left(\frac{q^2}{32\pi^3 c\varepsilon_o}\right)^{1/2} \int d\tau' e^{i\omega(\tau' + R/c)} \frac{\hat{R} \times [(\hat{R} - \vec{\beta}) \times \dot{\vec{\beta}}]}{(1 - \hat{R}\cdot\vec{\beta})^2}$$

where there is no longer a need for the notation $\{\}_{ret}$ since the integration is over all time. Furthermore, recall that in the radiation zone we assume $R \gg r$ and we can use the approximation $R \simeq r_o - \hat{r}_o\cdot\vec{r} \simeq r_o - \hat{R}\cdot\vec{r}$ in the exponential

$$\vec{\mathbb{E}}(\vec{r}_o, \omega) = \left(\frac{q^2}{32\pi^3 c\varepsilon_o}\right)^{1/2} e^{i\frac{\omega}{c} r_o} \int dt' e^{i\omega\left(t' - \hat{R}\cdot\vec{r}/c\right)} \frac{\hat{R} \times [(\hat{R} - \vec{\beta}) \times \dot{\vec{\beta}}]}{(1 - \hat{R}\cdot\vec{\beta})^2}$$

where we have dropped the now irrelevant "ret" subscript on the time variable. Inserting this result into Eq. 6.61,

$$\frac{d^2\mathscr{E}}{d\Omega d\omega} = \frac{dI}{d\Omega} = \frac{q^2}{16\pi^3 c\varepsilon_o} \left| \int dt' e^{i\omega\left(t' - \hat{R}\cdot\vec{r}/c\right)} \frac{\hat{R} \times [(\hat{R} - \vec{\beta}) \times \dot{\vec{\beta}}]}{(1 - \hat{R}\cdot\vec{\beta})^2} \right|^2 \tag{6.63}$$

we note that the phase factor has vanished since $e^{i\frac{\omega}{c} r_o} e^{-i\frac{\omega}{c} r_o} = 1$. Thus, we have arrived at the measurable quantity that we have been seeking, namely an expression for the radiation energy per unit solid angle per unit frequency resulting from charged particle acceleration. A further simplified (but less intuitive) expression can be obtained if we recognize that in the radiation zone where, in general, $\hat{R}$ can be approximated as constant in time, the non-exponential part of the integrand in Eq. 6.63 is an exact differential. Specifically (**see Discussion 6.1**),

$$\frac{d}{dt}\left[\frac{\hat{R}\times(\hat{R}\times\vec{\beta})}{1-\hat{R}\cdot\vec{\beta}}\right]=\frac{\hat{R}\times(\hat{R}-\vec{\beta})\times\dot{\vec{\beta}}}{(1-\hat{R}\cdot\vec{\beta})^2} \tag{6.64}$$

Substitution into Eq. 6.63 and a subsequent integration by parts yields the simpler form

$$\frac{d^2\mathscr{E}}{d\Omega d\omega}=\frac{dI}{d\Omega}=\frac{q^2\omega^2}{16\pi^3 c\varepsilon_o}\left|\int dt' e^{i\omega\left(t'-\hat{R}\cdot\vec{r}/c\right)}\left(\hat{R}\times(\hat{R}\times\vec{\beta})\right)\right|^2 \tag{6.65}$$

6.6 Synchrotron radiation

An especially important specialization of acceleration perpendicular to a charge's velocity is the (decaying) circular motion of a charge in a magnetic field. This was first observed in 1947 when Floyd Haber, a staff member in the laboratory of Professor H. C. Pollock, while making adjustments on a "cyclic accelerator"–a synchrotron–removed the opaque metal cover of its chamber and observed a patch of radiation, clearly visible in daylight, that was created by 70 MeV electrons moving circularly in the magnetic field of the accelerator.[14] We next discuss this "synchrotron radiation" in more detail. Figure 6.6 shows the geometry we will use for what follows. The extreme relativisitic regime is especially interesting both for astrophysical sources and man-made terrestrial collimated light sources that can be tuned from the far infrared to extremely hard x-rays. We first find the spectrum of the pulse of radiation of an electron as it sweeps across the observation point. The electron is taken to move in a circle of radius ρ that lies in the x–z plane. Since the electron moves in a circle, we can pick our observation point to lie in the y–z plane without loss of generality. We know that the radiation is concentrated in the forward direction. Figure 6.7 shows an instantaneous "searchlight beam" for a moderately relativisitic electron. The angle of observation, θ, needs to be small to see any significant radiation. We define coordinates such that the electron passes through the origin at time $t=0$. The differential energy spectrum per unit solid angle of the pulse then is found by carrying out the integral in Eq. 6.65 over all time. Here we will just evaluate it over a single pass and later discuss the spectrum

[14] *Am. J. Phys.* 51, 278 (1983).

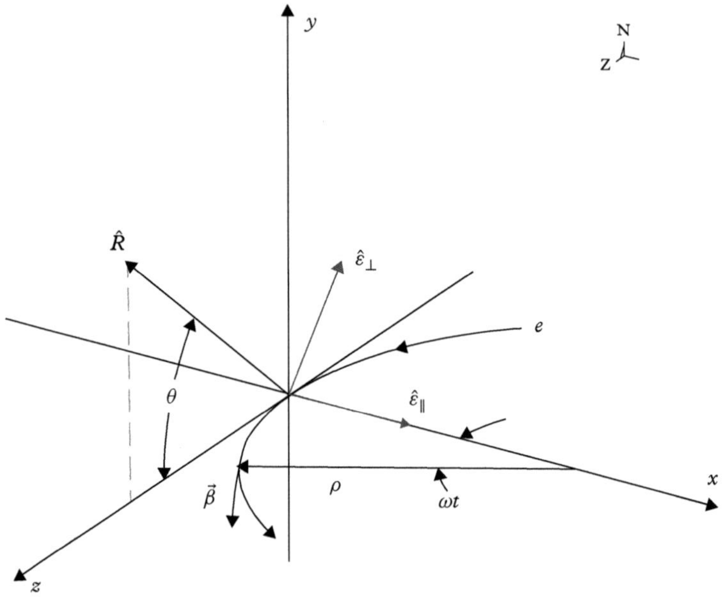

Fig. 6.6 *The geometry used for the synchrotron radiation calculation.*

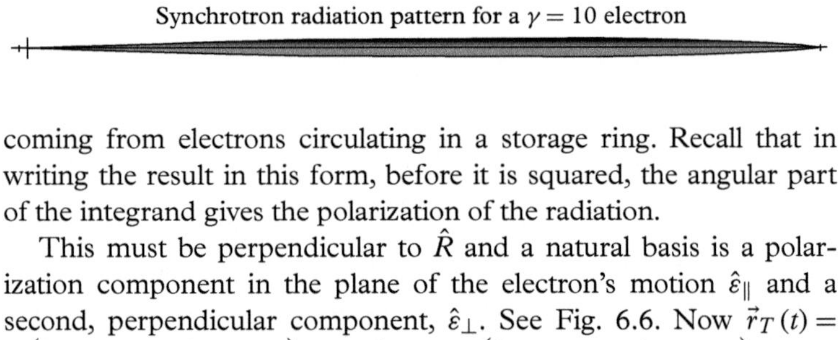

Fig. 6.7 *The "searchlight pattern" of synchrotron radiation from a relativistic electron $\gamma = 10$."The horizontal axis–the direction of beta–and the vertical axis are scaled identically."*

coming from electrons circulating in a storage ring. Recall that in writing the result in this form, before it is squared, the angular part of the integrand gives the polarization of the radiation.

This must be perpendicular to $\hat{R}$ and a natural basis is a polarization component in the plane of the electron's motion $\hat{\varepsilon}_\parallel$ and a second, perpendicular component, $\hat{\varepsilon}_\perp$. See Fig. 6.6. Now $\vec{r}_T(t) = \rho\left(\hat{i}_x \cos\omega_S t + \hat{i}_z \sin\omega_S t\right)$ and $\vec{\beta}(t) = \beta\left(\hat{i}_z \cos\omega_S t + \hat{i}_x \sin\omega_S t\right)$ where $\omega_S = v/\rho$ is the angular frequency of the electron's circular motion. The cross product term in the integrand becomes[15]:

$$\hat{R} \times \left(\hat{R} \times \vec{\beta}\right) = \beta\left[-\hat{\varepsilon}_\parallel \sin\omega_S t + \hat{\varepsilon}_\perp \cos\omega_S t \sin\theta\right] \tag{6.66}$$

For the beaming from a highly relativistic electron, $\omega_S \Delta t$, both the fraction of the arc contributing significantly to the signal and the observation angle, θ, must be small and we expand the exponent in the integrand in Eq. 6.65 accordingly:

15 Use the vector identity $\vec{A} \times (\vec{B} \times \vec{C}) = \vec{B}(\vec{A} \cdot \vec{C}) - \vec{C}(\vec{A} \cdot \vec{B})$ and note the cross product term in Eq. 6.65 represents the vector β component perpendicular to $\hat{R}$.

$$\omega\left(t-\hat{R}\cdot\vec{r}_T/c\right) = \omega\left(t-\rho\frac{\sin\left(\frac{v}{\rho}t\right)\cos\theta}{c}\right)$$

$$\simeq \omega\left(t-\rho\frac{\left[\frac{v}{\rho}t-\frac{1}{6}\left(\frac{v}{\rho}t\right)^3\right]\left(1-\frac{\theta^2}{2}\right)}{c}\right)$$

$$\simeq \frac{\omega}{2}\left[\left(\frac{1}{\gamma^2}+\theta\right)^2 t+\frac{c^2}{3\rho^2}t^3\right]$$

Collecting these results we write:

$$\frac{d^2\mathscr{E}}{d\Omega d\omega} = \frac{q^2\omega^2}{16\pi^3 c\varepsilon_o}\left|-\hat{\varepsilon}_{\|}A_{\|}(\omega)+\hat{\varepsilon}_{\perp}A_{\perp}(\omega)\right|^2 \tag{6.67}$$

where:

$$A_{\|}(\omega) = \frac{v}{\rho}\int_{-\infty}^{\infty} t\,dt\,\exp\left\{\frac{i\omega}{2}\left[\left(\frac{1}{\gamma^2}+\theta^2\right)t+\frac{c^2}{3\rho^2}t^3\right]\right\} \tag{6.68}$$

$$A_{\perp}(\omega) = \theta\int_{-\infty}^{\infty} dt\,\exp\left\{\frac{i\omega}{2}\left[\left(\frac{1}{\gamma^2}+\theta^2\right)t+\frac{c^2}{3\rho^2}t^3\right]\right\}$$

By defining the variables:

$$x = \frac{ct}{\rho\sqrt{\left(\gamma^{-2}+\theta^2\right)}} \text{ and } \xi = \frac{\omega\rho}{3c}\left(\gamma^{-2}+\theta^2\right)^{3/2}$$

the exponents of Eq. 6.68 may be written:

$$\exp\left\{\frac{i\omega}{2}\left[\left(\frac{1}{\gamma^2}+\theta^2\right)t+\frac{c^2}{3\rho^2}t^3\right]\right\} = \exp\left[i\frac{3}{2}\xi\left(x+\frac{1}{3}x^3\right)\right] \tag{6.69}$$

Making use of the Airy integrals (otherwise called the fractional modified Bessel functions):

$$\frac{1}{\sqrt{3}}K_{2/3}(\xi) = \int^{\infty} x\,dx\,\sin\left[\frac{3}{2}\xi\left(x+\frac{1}{3}x^3\right)\right]$$

$$\frac{1}{\sqrt{3}}K_{1/3}(\xi) = \int^{\infty} dx\,\cos\left[\frac{3}{2}\xi\left(x+\frac{1}{3}x^3\right)\right]$$

we have for the differential energy spectrum:

$$\frac{d^2\mathscr{E}}{d\Omega d\omega} = \frac{3q^2}{4\pi^3\varepsilon_0 c}\left(\gamma^{-2}+\theta^2\right)^{-1}\xi^2\left[K_{2/3}^2(\xi)+\theta^2\left(\gamma^{-2}+\theta^2\right)^{-1}K_{1/3}^2(\xi)\right] \tag{6.70}$$

where the first term in the brackets is proportional to the intensity of the parallel polarization and the second to the perpendicular. Integrating this expression over frequency, ω, gives the angular differential emission for the total energy in the pulse:

$$\frac{d\mathscr{E}}{d\Omega} = \int \frac{d^2\mathscr{E}}{d\Omega d\omega} d\omega = \frac{7e^2}{64\pi\varepsilon_0}\left(\gamma^{-2}+\theta^2\right)^{-5/2}\left[1+\frac{5}{7}\left(\gamma^{-2}+\theta^2\right)^{-1}\right] \tag{6.71}$$

Here, again, the first and second terms in the brackets correspond to the energy in the parallel and perpendicular polarizations, respectively. When Eq. 6.71 is integrated over all angles, it is found that there is seven times as much energy radiated in the parallel polarization component as in the perpendicular. Indeed a distinctive feature of synchrotron radiation is its strong polarization. The polarization of radiation from the Crab Nebula has been intensely investigated and is a principal tool in determining electron energies and magnetic fields within the nebula (Fig. 6.8).

We now pull apart the angular and frequency dependence of synchrotron radiation given in Eq. 6.70. In Fig. 6.9 we show the modified fractional Bessel functions, weighted as they appear in Eq. 6.70. Note that they both peak and then fall off rapidly for values exceeding

$$\xi = \frac{\omega\rho}{3c}\left(\gamma^{-2}+\theta^2\right)^{3/2} = \frac{1}{2} \tag{6.72}$$

Generally speaking, we see that as frequencies increase, the value of θ^2 satisfying this relation decreases. That is, qualitatively, high frequencies form a narrow beam about $\theta = 0$ whereas lower frequencies are more spread out. We can define a critical frequency by satisfying Eq. 6.72 in the forward direction:

$$\omega_c = \frac{3c}{2\rho}\gamma^2 = \frac{3}{2}\omega_0\gamma^3 \tag{6.73}$$

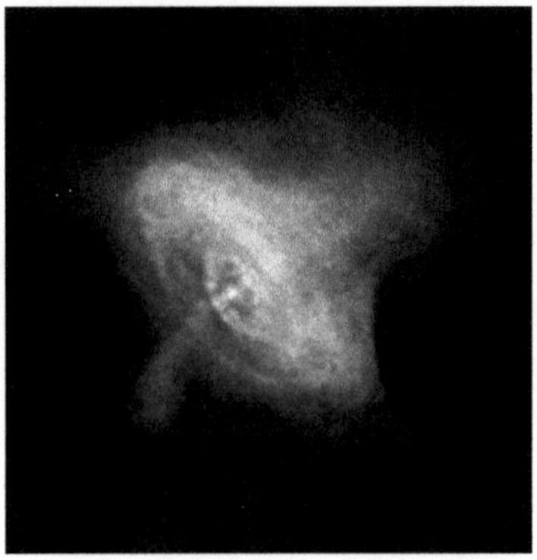

Fig. 6.8 *X-ray emission from the crab nebula, due to synchrotron emission. This is strongly suggested by the image shown taken from the Chandra satellite, of the central part of the crab nebula for photon energies from 0.3 to 3.0 keV. The x-ray radiation is found to be strongly polarization dependent. Credit: NASA Chandra Telescope, Public Domain.*

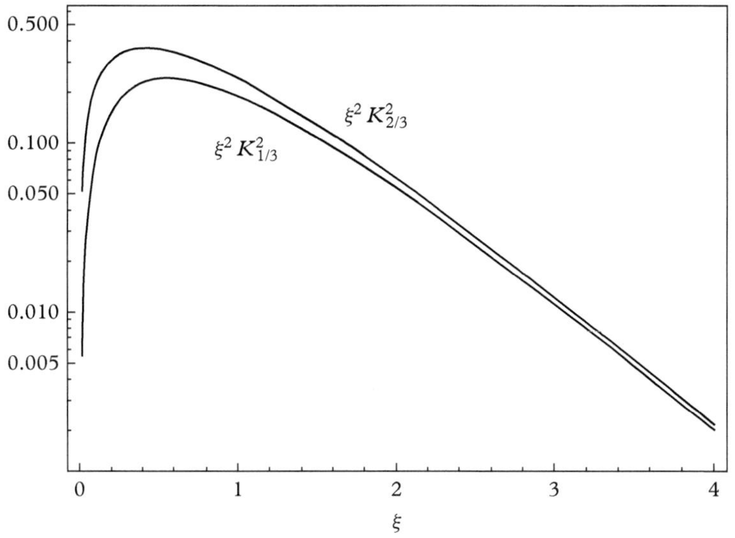

Fig. 6.9 *The squared and weighted modified fractional Bessel functions appearing in the equations for synchrotron radiation patterns.*

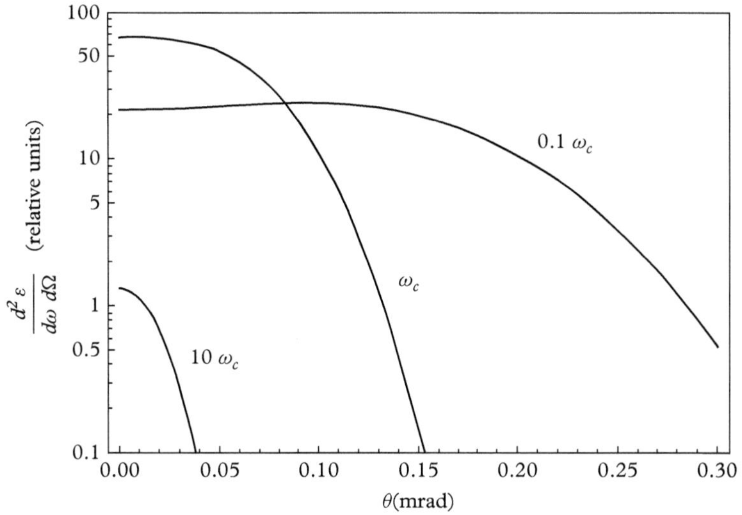

Fig. 6.10 *Synchrotron radiation angular distribution from the Advanced Photon Source for frequencies around the critical frequency. Note that $1/\gamma$ for this energy, 7 GeV, is about 0.073 mrad.*

For frequencies much above this, little radiation is produced at any angle. The Advanced Photon Source is the largest storage ring of its kind in the United States. Its radius is 175 m giving an electron beam circulating frequency equal to $f_0 = \omega_0/(2\pi) = 270$ kHz. When operated at its highest energy, electrons have an energy of 7 GeV, which corresponds to $\gamma = 13,700$. Its critical frequency, $f_c = \omega_c\,(2\pi) = 7.1 \times 10^{17}$ Hz, corresponds to a radiation wavelength of 0.43 nm–a hard x-ray. Figure 6.10 shows the angular distribution of radiation that would be observed by a station on the APS for photon frequencies around ω_c. Early synchrotrons were sources of high energy electrons, and synchrotron radiation was a "parasitic" loss mechanism. However, it quickly found numerous applications in condensed matter and the biological sciences. Subsequently, high energy storage rings were made for the dedicated purpose of generating radiation suited for these applications.

6.7 Fields from a single charge moving with constant velocity

As formally complete as Eqs. 6.30 and 6.31 are, they are of little practical use, because at particle velocities approaching the speed of light, the particle is often far removed from where the observer "sees" it to be. A previous knowledge of the particle's motion (velocity and acceleration) is required to compute the fields at the observation point for a particle located a given distance and direction away from the observer. This issue can be far less significant for the special case of uniform motion of a charge. Here we examine the form of the $\vec{E}$ and $\vec{B}$ fields of a highly relativistic charged particle moving with constant velocity as it passes the observer. Note that these fields are specifically not a case of "free radiation," for the fields associated with the charge in this case have a R^{-2} dependence on the distance from the charge, and remain connected to the charge in the sense that any interactions of these fields with a test charge will affect the otherwise constant velocity. Nevertheless, as we will see, the observer experiences a "wave-like" pulse of electromagnetic fields as the charge passes: Because this circumstance is common we discuss it here in detail.

The task is to express these velocity fields in Eq. 6.30 in coordinates associated solely with the present time, t, rather than in the coordinates of the retarded time $\tau = t - R/c$. This can be done precisely because we know that the velocity will not change, neither in magnitude nor direction. Starting with the expression 6.30 and setting the acceleration $\dot{\vec{\beta}} = 0$

$$\vec{E}_{velocity}(\vec{r}_o, t) = \frac{q}{4\pi\,\varepsilon_o}\left\{\left(\frac{1}{[1-\hat{R}\cdot\vec{\beta}]^3}\right)\left[\frac{(1-\beta^2)(\hat{R}-\vec{\beta})}{R^2}\right]\right\}_{ret} \tag{6.74}$$

As usual, the symbol $\{\}_{ret}$ means that all the quantities are measured at the retarded time in the observer's frame. We can calculate the fields of a uniformly moving charge relative to the observer's frame, in terms of the spatial coordinates and (present) time measured in that frame, by applying the relativistic transformation to the pure Coulomb fields measured in the particle's frame. We have already considered this transformation in Section 5.93. Recalling that case, in which the relative motion is in the x direction only and the particle passes through the origin at $t = t' = 0$, we found:

$$\vec{E}_{velocity}(\vec{r}_o, t) = \frac{q}{4\pi\varepsilon_o} \frac{\gamma}{[(\gamma(x - c\beta t)^2 + y^2 + z^2]^{3/2}} \left[x\hat{x} + y\hat{y} + z\hat{z}\right]$$
$$= \frac{q}{4\pi\varepsilon_o} \frac{1}{\gamma^2} \frac{\vec{R}_o(t)}{R_o^3(t)} \frac{1}{[1 - \beta^2 \sin^2\theta]^{3/2}}$$

here the angle θ is between the particle's velocity and $\vec{R}_o$, the vector from the particle's position *at the present time, t.* to the observer's position. That is, $\vec{R}_o = (x - c\beta t)\hat{x} + y\hat{y} + z\hat{z}$. Together with the corresponding transformation for the B field[16]:

$$\vec{E}_{velocity}\left(\vec{r}_o, t\right) = \frac{q}{4\pi\,\varepsilon_o} \frac{\hat{R}_o(t)}{R_o^2(t)} \frac{1}{\gamma^2} \frac{1}{[1 - \beta^2 sin^2\theta]^{3/2}} \tag{6.75}$$

$$\vec{B}_{velocity}(\vec{r}_o, t) = \frac{q}{4\pi\varepsilon_o c} \frac{1}{\gamma^2} \frac{1}{[1 - \beta^2 sin^2\theta]^{3/2}} \frac{\vec{\beta} \times \hat{R}_o(t)}{R_o^2(t)}$$
$$= \frac{1}{c} [\vec{\beta} \times \vec{E}_{vel}(\vec{r}_o, t)] \tag{6.76}$$

In Eqs. 6.75 and 6.76, all coordinates (R_o, $\hat{R}_o$, θ) are measured in the *present* time in the observer's frame; that is, $\vec{R}_o$ is the vector from the *present* position of the particle, even though the radiation originated from the particle at the retarded time (and position). It is not obvious that the expressions given Eqs. 6.74 and 6.75 are equivalent: However, a simple geometrical argument relating present-time coordinates to retarded-time coordinates shows that they are. Consider the diagram in Fig. 6.11 relating $\vec{R}_{ret}$, the vector distance from the charge to the observer at the retarded time τ $(= t - R_{ret}/c)$ to $\vec{R}_o$, the vector distance from the charge to the observer at the present time t. The position labeled P_{ret} is the position of the charge when it gives off the radiation. The vector distance from P_{ret} to the observer (O) is given by the $\vec{R}_{ret}$. The distance the charge moves during the interval the radiation is traveling from P_{ret} to O is just its velocity $c\beta$ times the time $|\vec{R}_{ret}|/c$; that is, the distance from P_{ret} to P. From the diagram, the distance P_{ret} to Q is clearly $\beta|\vec{R}_{ret}|cos\varphi$, so that the distance Q to O is $QO = |\vec{R}_{ret}|(1 - \beta cos\varphi) = |\vec{R}_{ret}|(1 - \vec{\beta} \cdot \hat{R}_{ret})$; further inspection reveals $|\vec{R}_{ret}|(1 - \vec{\beta} \cdot \hat{R}_{ret}) = \sqrt{[R_o^2 - \beta^2 R_{ret}^2 sin^2\varphi]}$, while the diagram also shows that $|\vec{R}_{ret}|\, sin\varphi = R_o\, sin\theta$. Thus we have

$$|\vec{R}_{ret}|(1 - \vec{\beta} \cdot \hat{R}_{ret}) = R_o(1 - \beta^2 sin^2\theta)^{1/2} \tag{6.77}$$

[16] In the particle's frame, the B field vanishes. In the observer's frame:

$$B_x = B'_{x'} = 0$$
$$B_y = \gamma\,[B'_{y'} + \frac{\beta}{c} E'_{z'}]$$
$$= \frac{\gamma\beta}{c} \frac{q}{4\pi\varepsilon_o} \frac{z}{[(\gamma x - c\beta t)^2 + y^2 + z^2]^{3/2}}$$
$$B_z = \gamma\,[B'_{z'} - \frac{\beta}{c} E'_{y'}]$$
$$= -\frac{\gamma\beta}{c} \frac{q}{4\pi\varepsilon_o} \frac{y}{[(\gamma x - c\beta t)^2 + y^2 + z^2]^{3/2}}$$

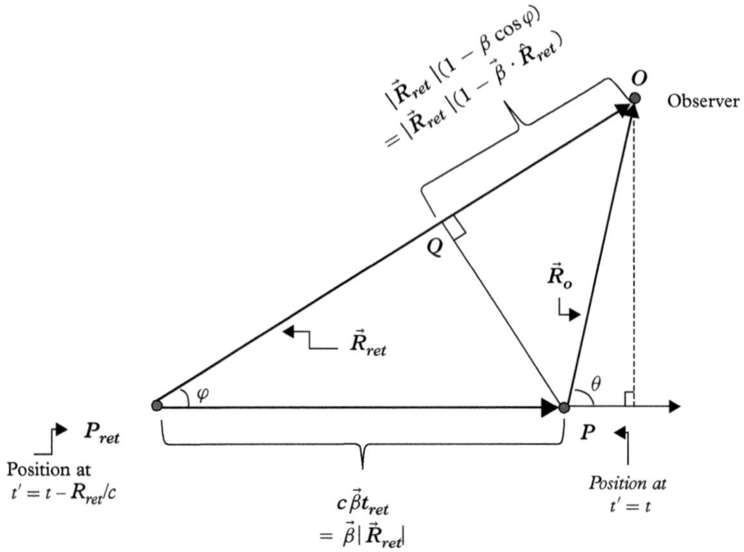

Fig. 6.11 *Geometry diagram relating the retarded position to the present position for a charge moving at a constant (relativistic) velocity.*

and finally, we have again from the diagram, the vector relationship

$$|\vec{R}_{ret}|\left[\hat{R}_{ret}-\vec{\beta}\right]=\vec{R}_{ret}-(R_{ret})\vec{\beta}=\vec{R}$$

$$\text{or}\left\{\hat{R}_o-\vec{\beta}\right\}_{ret}=\frac{\vec{R}_o}{|\vec{R}_{ret}|} \tag{6.78}$$

Substituting the results of Eqs. 6.77 and 6.78 into 6.74 immediately yields 6.75.

An examination of Eqs. 6.75 and 6.76 reveals that the E fields observed when a point charge passes by at nearly the speed of light are reduced by a factor of γ^2 in the direction of motion, and increased by a factor γ for the perpendicular fields (compared to a stationary charge). Figure 6.12 indicates several features of this observed field. Noting that the strength of the E field is indicated by the field line density, it is clear that as the velocity increases, the field strength is substantially less in the horizontal direction and substantially more in the transverse direction. Crucially, the electric field lines always are directed radially from the present position of the charge.

As γ gets to $\gtrsim 10$, the E field appears to become a fan-like structure surrounding the fast electron as the associated B field lies in the fan-like structure, as a circular field falling off transversely as R^{-2}. At any given field position, as the relativistic charge passes, the observed $\vec{E}$ and $\vec{B}$ fields are experienced as those from a plane wave, with E

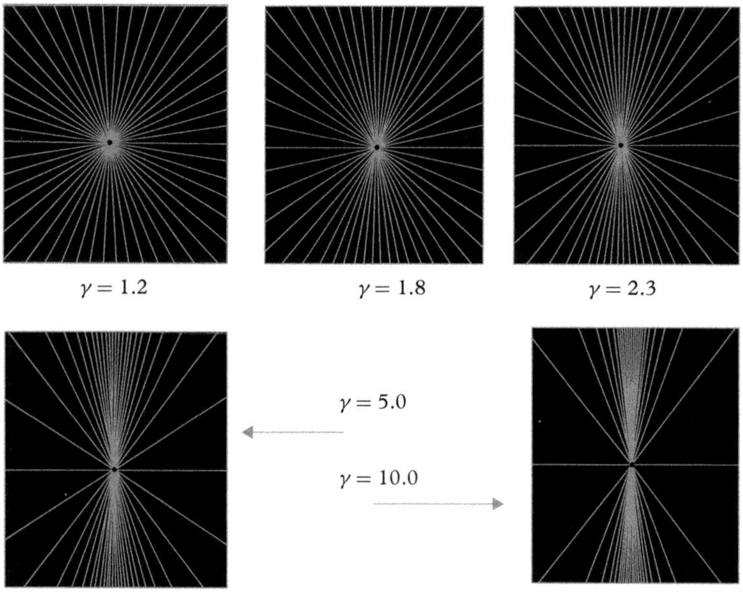

Fig. 6.12 *$\vec{E}$ fields of a charge moving relative to an observer for various values of γ. As shown, the E field lines are always radial from the charge for all values of γ. The relative intensity is judged by the density of lines in a spatial area.*

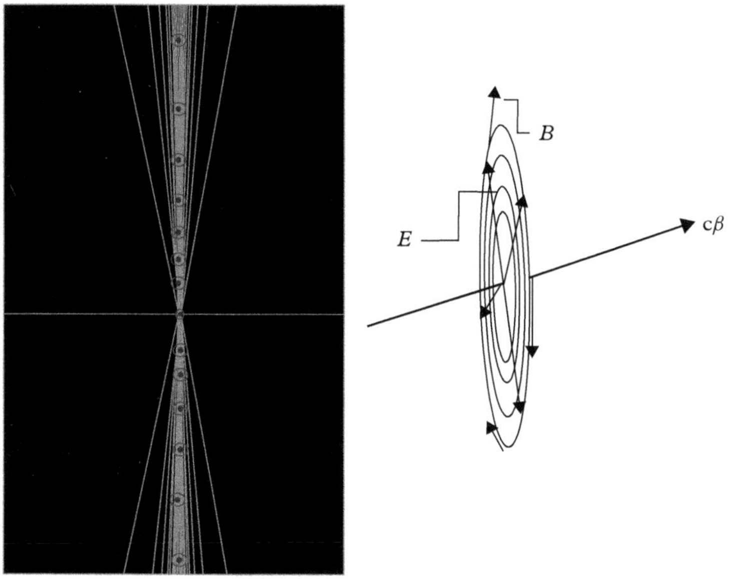

Fig. 6.13 *On the left the electric field lines for positive charge moving with $\beta = 0.999$ ($\gamma = 22$). The associated B fields are indicated by circular dots along the y-axis where the electric field is most intense. On the right, a perspective sketch indicating the circular nature of the B field relative to the E field.*

and cB of the same magnitude, mutually perpendicular, and oriented such that $\hat{E} \times \hat{B}$ is in the direction of $c\vec{\beta}$. This is shown in Fig. 6.13 where the E field lines are plotted for $\gamma = 22$, with the circular B field indicated. Figure 6.14 shows the observed time dependence of the transverse field for a relativistically moving charge q at the position of

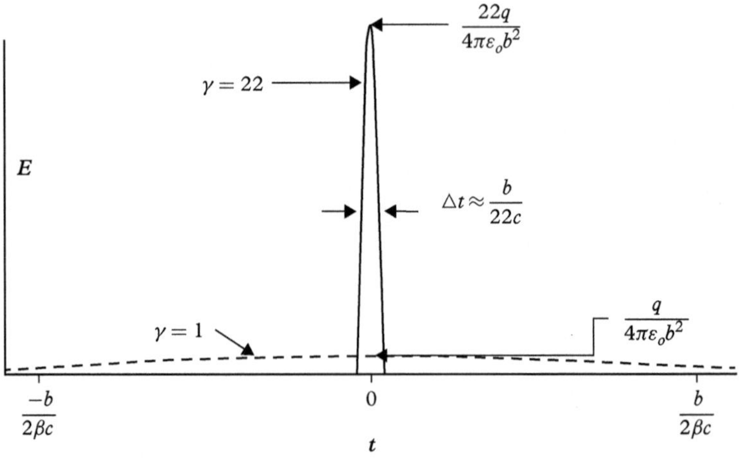

Fig. 6.14 *Transverse fields at a distance b from the closest approach of a slowly moving charge with $\gamma \approx 1$ compared to a rapidly moving charge with $\gamma = 22$.*

its closest approach b. The low velocity limit is sketched as a dashed line; the case for $\gamma = 22$ is also shown. Because the transverse field grows as γ and the longitudinal field decreases as γ^2, the apparent width of the "pulse" will scale as $(\beta\gamma)^{-1}$. Note that while the transverse force on an observer's test charge increases as γ, the duration of this force scales as γ^{-1} (for $\gamma \gg 1$), so that the net transverse momentum exchange between the moving charge and the observer's test charge is approximately independent of the moving charge's kinetic energy. This physical idea of the approximate equivalence of the passage of a relativistic charge and a pulse of radiation was initially recognized by Fermi in 1924.

6.7.1 Parametrization of the fields

We have seen that from the perspective of the observer, the fields appear to originate from the "present" position of the charge, and as the velocity of the charge approaches the speed of light, the fields are seen to be perpendicular to the velocity, and appear to be essentially those of an E&M pulse passing over the observer. Figure 6.15 shows the field orientation of a charge moving at constant velocity passing an observer,[17] in terms of its velocity, and the perpendicular distance of its path from the observer, *b*. Referring to Fig. 6.15 and starting with Eq. 6.75 for the electric fields from a constant velocity charge with a position, at time *t*, given in terms of the observer's coordinates. and choosing the time *t*=0 to correspond to the distance of closest approach, the longitudinal and transverse components of the field are given by

[17] following Panofsky and Phillips.

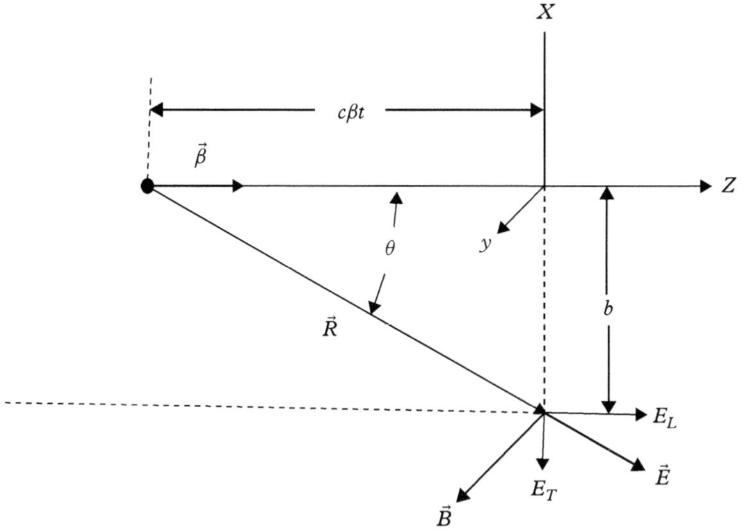

Fig. 6.15 *Diagram of field components from the passage of a point charge with $v \sim c$.*

$$E_{Long} = E_L = \frac{q(-c\beta t)\gamma^{-2}}{4\pi\varepsilon_o}\left[\frac{1}{[b^2+(c\beta t)^2]}\frac{1}{\left[1-\beta^2\left(\frac{b^2}{b^2+(c\beta t)^2}\right)\right]}\right]^{3/2} \tag{6.79}$$

$$E_{Trans} = E_T = \frac{qb\gamma^{-2}}{4\pi\varepsilon_o}\left[\frac{1}{[b^2+(c\beta t)^2]}\frac{1}{\left[1-\beta^2\left(\frac{b^2}{b^2+(c\beta t)^2}\right)\right]}\right]^{3/2} \tag{6.80}$$

Equations 6.80 and 6.79 show that for $\gamma >> 1$, when the charge is further away than approximately $\gamma^{-1}b$, both the longitudinal and transverse fields of the moving charge are much less, by a factor of γ^{-2}, than those of an equivalent stationary charge. The longitudinal field remains small even at distances of closest approach because of the sign reversal, while the transverse field, which can be re-written as

$$E_T(\gamma, b, t) = \frac{qb\gamma^{-2}}{4\pi\varepsilon_o}\frac{1}{[(c\beta t)^2+\gamma^{-2}b^2]^{3/2}}, \tag{6.81}$$

has a peak value at the distance of closest approach of

$$E_T(\gamma, b, t=0) = \gamma\frac{q}{4\pi\varepsilon_o b^2}.$$

From the perspective of an observer, the transverse field of a highly relativistic charge has an abrupt onset and decay with a peak field strength γ times greater than an equivalently placed slowly moving or stationary charge. As indicated in Fig. 6.15, there is an associated B field oriented such that $\frac{\vec{E}\times\vec{B}}{|\vec{E}||\vec{B}|}$ is in the direction of the charge's velocity, with a magnitude $|\vec{B}| = \frac{|\vec{E}|}{c}$. Thus, to order $\sim\gamma^{-1}$, the observer experiences the passage of the charge as an incident transverse E&M wave, one whose peak intensity rises as γ^2.

6.7.2 Spectral energy density of the fields

Our goal at this point is to calculate the energy associated with this equivalent E&M wave, and determine the energy associated with each frequency element $d\omega$. The conversion to virtual photons will be accomplished by invoking the photon discretization rule: $\mathcal{N}(\omega)\hbar\omega d\omega = \mathcal{E}_\omega d\omega$. Because the longitudinal field E_L remains small even at distances of closest approach, we now consider only the contribution of the transverse field, E_T. To find the expression for $\mathcal{E}_\omega$, we calculate the total energy of the pulse by writing it first as $\mathcal{E}=\int^\infty \mathcal{E}_\omega d\omega$ then in terms of the square of the electric field, $\mathcal{E} = \varepsilon_0 \int (E_T)^2 dV$ where $dV = [2\pi b db]\, c\beta\, dt$,

$$\mathcal{E} = \varepsilon_0\, 2\pi c\beta \int_0^\infty \left[\int_{-\infty}^\infty (E_T)^2 dt\right] b\, db. \tag{6.82}$$

Equation 6.82 can be written in terms of the Fourier transform of E, $E_T^F(\gamma, b, \omega)$[18]

$$\mathcal{E} = \varepsilon_0\, 2\pi c\beta \int_0^\infty \left[\int_0^\infty (E_T^F)^2 d\omega\right] b\, db; \tag{6.83}$$

Interchanging the order of integration in Eq. 6.83 and comparing with $\mathcal{E}=\int^\infty \mathcal{E}_\omega d\omega$ yields an expression for the energy density in terms of the Fourier transform of the electric field:

$$\mathcal{E}_\omega = \varepsilon_0 2\pi c\beta \int_0^\infty (E_T^F)^2 b\, db. \tag{6.84}$$

In order to complete the analysis we need to evaluate the Fourier transform of the field:

$$E_T^F(\gamma, b, \omega) = \frac{1}{\sqrt{2\pi}} \int_{-\infty}^\infty E_T(\gamma, b, t) e^{i\omega t} dt \tag{6.85}$$

$$= \frac{qb\gamma^{-2}}{\sqrt{32\pi^3}\varepsilon_0} \int_{-\infty}^\infty \frac{1}{[(c\beta t)^2 + \gamma^{-2}b^2]^{3/2}} e^{i\omega t} dt \tag{6.86}$$

[18] Make use of the following relationship valid for Fourier transforms in general:

$$\int_{-\infty}^{+\infty} f_1(t) f_2(t)\, dt = \frac{1}{2}\int_0^\infty (f_{1\omega}^* f_{2\omega} + f_{1\omega} f_{2\omega}^*) d\omega$$

then,

$$\int_{-\infty}^\infty (E^T)^2\, dt = \int_0^\infty (E_F^T)^2\, d\omega.$$

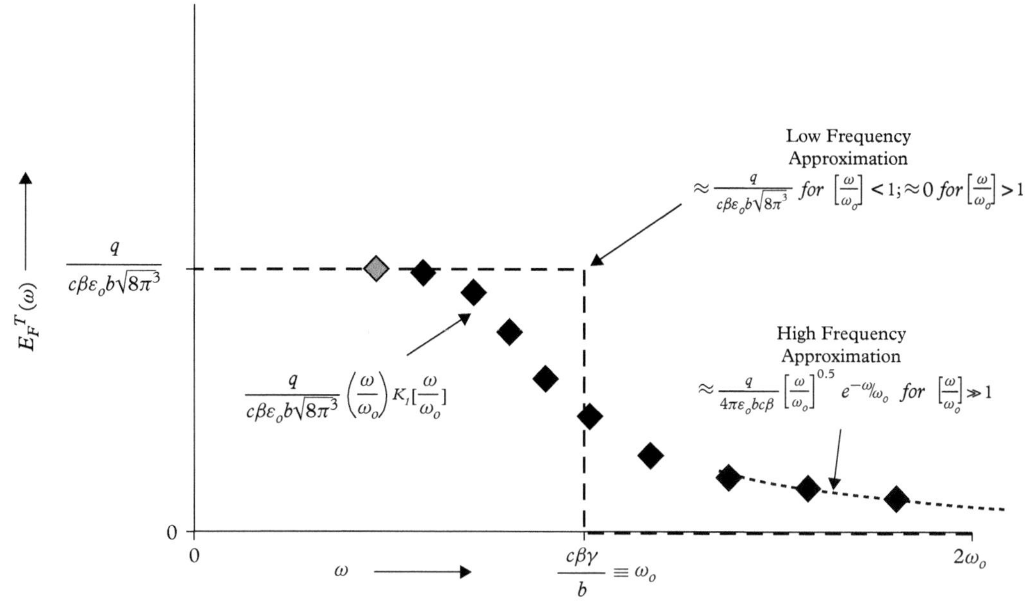

Fig. 6.16 *The frequency spectra of a transverse pulse from a passing charge moving at relativistic velocities (diamonds). The canonical approximation (Low Frequency) has the frequency content constant up to ω_o and zero for ω greater. For frequency content much larger than ω_o, the indicated high frequency approximation is often used.*

where we have used E_T in the form given by Eq. 6.81. Equation 6.86 can be written in terms of the modified Bessel function of the second kind (order 1)[19]

$$E_T^F(\gamma, b, \omega) = \frac{q}{c\beta\varepsilon_o b\sqrt{8\pi^3}}\left\{\frac{\omega b}{\gamma c\beta}K_1\left(\frac{\omega b}{\gamma c\beta}\right)\right\} \tag{6.87}$$

Equation 6.87 shows that by defining $\omega_o \equiv (\gamma c\beta/b)$, the Fourier transform of the transverse field is proportional to $(\omega/\omega_o)K_1(\omega/\omega_o)$; this product function is canonically approximated as $\simeq 1$ for $(\omega/\omega_o) \leq 1$, and ≈ 0 for $(\omega/\omega_o) > 1$, although an examination of Fig. 6.16 suggests that this approximation underestimates the value of the product function from $1 \leq (\omega/\omega_o) \leq 2$. Nevertheless, the canonical approximation serves to give a reasonable expression for the frequency content from the passage of the fast charge. Applying the canonical approximation indicated in Fig. 6.16, we have the approximate values for the Fourier transform of the electric field:

$$\text{For } \omega < \omega_o,\ E_T^F \approx \frac{q}{c\beta\varepsilon_o b\sqrt{8\pi^3}};\text{while for } \omega > \omega_o,\ E_T^F \approx 0 \tag{6.88}$$

[19] The required Fourier transformation is

$$E_T^F(\omega) = \frac{qb\gamma^{-2}}{\sqrt{32\pi^3}\,\varepsilon_o}\int_{-\infty}^{+\infty}\frac{e^{i\omega t'}}{[(c\beta t')^2 + \gamma^{-2}b^2]^{3/2}}dt' = \frac{q}{\sqrt{32\pi^3}\,\varepsilon_o}\frac{1}{bc\beta}\int_{-\infty}^{+\infty}\frac{e^{i(\frac{\omega b}{\gamma c\beta})x}}{(x^2+1)^{3/2}}dx$$

and note that an integral representation of the modified Bessel function of the second kind is:

$$K_1(\omega\tau/\omega_o) = \frac{\tau\omega_o}{2\omega}\int_{-\infty}^{+\infty}\frac{e^{i(\omega/\omega_o)x}}{(x^2+\tau^2)^{3/2}}dx.$$

In our approximation, the frequency components are constant in magnitude up to a cutoff frequency, ω_o, a frequency associated with the observed "transit time" of the pulse across the observer. Using the results of Eq. 6.88 in the expression for $\mathscr{E}_\omega$ (Eq. 6.84), we have the required result of the energy density in the transverse electric field as a function of frequency, velocity, and distance of closest approach:

$$\mathscr{E}_\omega = \varepsilon_o 2\pi c\beta \int_{b_{min}}^{\frac{c\beta\gamma}{\omega}} \left[\frac{q}{c\beta\varepsilon_o b\sqrt{8\pi^3}}\right]^2 b\,db;$$
$$= \frac{q^2}{4\pi^2 c\beta\varepsilon_o} ln\left(\frac{\gamma\beta c}{\omega\, b_{min}}\right) \tag{6.89}$$

6.7.3 Number of photons associated with fields of a passing charge

Having calculated the spectral energy density of the fields of the passing charge, we may now divide by the photon energy, $\hbar\omega$, and describe this result as a density of virtual photons:

$$\mathscr{N}(\omega)d\omega = \frac{q^2}{4\pi^2 c\beta\varepsilon_o}\frac{1}{\hbar\omega}\ln\left(\frac{\gamma\beta c}{\omega\, b_{min}}\right)d\omega \tag{6.90}$$

For the result in Eq. 6.90 to be of use we need to have a value for b_{min}; although since it enters as the log, our result will not be too dependent on its precise value. We have no *a priori* determination for b_{min}, so a reasonable estimate will have to suffice. For an atom within the material, the smallest distance that has physical meaning is approximately the size of an atom; something on the order of the Bohr radius of hydrogen is a reasonable approximation. Using this value and identifying $\alpha = \frac{e^2}{(4\pi\varepsilon_o\hbar c)}$, Eq. 6.90 can be written in a form that clearly shows how many photons/electron are generated:

$$\mathscr{N}(\omega)d\omega = \frac{\alpha}{\beta\pi}\ln\left(\frac{\gamma\beta c}{\omega\, b_{min}}\right)\frac{d\omega}{\omega}, \tag{6.91}$$

This expression depends inversely on frequency: the log term will vary weakly for frequencies that are typically of interest. The expression gives roughly equal numbers of photons in every octave of frequency– the departure from this arising from the log term. Thus for γ of 10, in the octave 500 Hz to 1 kHz, the expression shows that for each electron, 0.059 photons are produced–about 1 photon for every

17 electron collisions. Similarly, looking about 12 orders of magnitude higher in frequency, for an octave that includes the visible spectrum, 700 nm to 350 nm, there are 0.015 photons produced per electron–about 1 for every 60 collisions.

6.8 Bremsstrahlung

One immediate application of our calculation of the fields of the passage of a relativistic charge is an elegant derivation of Bremsstrahlung, the "braking" radiation emitted by a relativistic electron colliding with a stationary ion. What follows is the so-called Weizsacker–Williams method. This method can be broken down into three straightforward steps:

(1) Choose a frame in which the electron is at rest, with the ion approaching at velocity $c\vec{\beta}$ and impact parameter "b." Find the Fourier transform of the electric field pulse seen by the electron to obtain the intensity spectrum–power/unit area/unit frequency–of the approaching ionic field.

(2) The Thomson cross section is the classical result for the scattering of a radiation field by a free electron. (This is discussed in Section 11.2.) Here we use this simple cross section to find the Bremsstrahlung generated by the scattering of the approaching ionic fields from the electron. This calculation is performed in the electron's rest frame.

(3) Transform the scattered fields in step (2) back to the laboratory frame, that is, the rest frame of the ion.

(1) In the frame of the relativistic electron, the "prime" frame, an ion with charge Ze is moving at relativistic velocity toward it. Following the arguments leading to Eq. 6.87, we ignore the small longitudinal field component and find the Fourier transform of the electric field pulse seen by the electron:

$$E_T^F(\gamma, b, \omega) = \frac{Ze\omega'}{\sqrt{8\pi^3}\,\gamma(c\beta)^2\,\varepsilon_o} K_1\left(\frac{\omega' b}{\gamma c\beta}\right) \tag{6.92}$$

The Poynting vector gives the intensity or power/unit area so that

$$|\vec{S}| = \frac{d^2\mathcal{E}}{dAdt} = c\varepsilon_o E^2(t)$$

Using the mathematics surrounding Eq. 6.83, we have:

$$\left[\frac{d^2\mathscr{E}}{dAd\omega}\right]' = 2c\varepsilon_o E^2(\omega') \tag{6.93}$$

Here the factor of two arises because for this expression, we include contributions from $\pm\omega'$ Fourier components. To find the equivalent radiation intensity spectrum incident on the electron corresponding to the ionic field we replace $E(\omega')$ in Eq. 6.93 with $E_T^F(\gamma,b,\omega)$ in Eq. 6.92:

$$\left[\frac{d^2\mathscr{E}}{dAd\omega}\right]'_{inc} = \frac{(Ze)^2(\omega')^2 c}{\gamma^2(c\beta)^4 4\pi^3\varepsilon_o}\left[K_1\left(\frac{\omega' b}{\gamma c\beta}\right)\right]^2 \tag{6.94}$$

The limiting cases of $\omega' << \omega_o$ and $\omega' >> \omega_o$ (see page 221) are obtained by using the limiting forms of the Bessel function[20]

$$\text{For } (\omega' << \omega_o) \quad \left[\frac{d^2\mathscr{E}}{dAd\omega}\right]'_{inc} = \frac{\gamma^2(Ze)^2}{(\gamma^2-1)cb^2 4\pi^3\varepsilon_o} \tag{6.95}$$

$$\text{For } (\omega' >> \omega_o) \quad \left[\frac{d^2\mathscr{E}}{dAd\omega}\right]'_{inc} = \frac{\gamma^2(Ze)^2\omega'}{(\gamma^2-1)^{3/2}c^2 b 8\pi^2\varepsilon_o} exp\left(-\frac{2\omega' b}{\gamma c\beta}\right) \tag{6.96}$$

For both frequency regions, the intensity spectrum of the fields seen by the electron depends inversely upon the impact parameter and yields infinity for an infinitesimal approach to a point charge, a subject we address next.

(2) We now use the classical Thomson scattering cross section to calculate the scattered fields. Thus we limit frequencies seen by the electron to those that satisfy the relation $\hbar\omega' \lesssim m_e c^2$. Then the spectrum of the energy scattered into solid angle $d\Omega$ is given by:

$$\left[\frac{d^2\mathscr{E}}{d\omega d\Omega}\right]'_{scat} = \left[\left(\frac{d\sigma_T}{d\Omega}\right)\right]'\left[\left(\frac{d^2\mathscr{E}}{dAd\omega}\right)\right]'_{inc} \tag{6.97}$$

where σ_T is the Thomson cross section (see Section 11.2)

$$\left[\left(\frac{d\sigma_T}{d\Omega}\right)\right]' = r_e^2\cdot\left(\frac{1+cos^2\theta'}{2}\right) \tag{6.98}$$

[20] for $x << 1, K_1(x) \simeq 1/x$; for $x >> 1$, $K_1(x) \simeq \sqrt{\frac{\pi}{2x}}e^{-x}$.

and the "classical radius of the electron" is:

$$r_e = \frac{e^2}{4\pi\varepsilon_o m_e c^2}$$

Inserting Eqs. 6.94 and 6.98, Eq. 6.97 becomes:

$$\left[\frac{d^2\mathscr{E}}{d\omega d\Omega}\right]'_{scat} = r_e^2 \cdot \left(\frac{1+cos^2\theta'}{2}\right)\left(\frac{(Ze)^2(\omega')^2 c}{\gamma^2(c\beta)^4 4\pi^3\varepsilon_o}\right)\left[K_1\left(\frac{\omega' b}{\gamma c\beta}\right)\right]^2 \tag{6.99}$$

The radiation here is for a single electron collision having an impact parameter b, and averaged over all azimuthal angles. Experimentally, one cannot control "b," and what is usually measured is the scattered energy from a flux of incident electrons on an ion. To calculate the total scattered spectral intensity in such an experiment, we consider an incident flux, ϕ', in electrons per square meter per second and a measurement interval, $\delta t'$. The total number of electrons that pass through the differential area surrounding the impact parameter b is $\delta t' \phi' dA = \delta t' \phi' 2\pi\, bdb$. Multiplying this number by Eq. 6.99 gives their contribution to the energy of frequency ω' scattered in the experiment into angle $d\Omega'$. And the total spectral energy for frequency ω' that is scattered into $d\Omega'$ is found by integrating these results over all b, the cross section of the beam,

$$\begin{aligned}\left[\frac{d^2\mathscr{E}_{tot}}{d\omega d\Omega}\right]'_{scat} &= \delta t' \cdot \phi' \int_{b_{min}}^{\infty} r_e^2 \cdot \left(\frac{1+cos^2\theta'}{2}\right)\left(\frac{(Ze)^2(\omega')^2 c}{\gamma^2(c\beta)^4 4\pi^3\varepsilon_o}\right) \\ &\quad \times \left[K_1\left(\frac{\omega' b}{\gamma c\beta}\right)\right]^2 2\pi\, b\, db \\ &= \delta t' \cdot \phi' \left[\frac{d^2\mathbf{\Sigma}}{d\omega d\Omega}\right]'_{scat}\end{aligned} \tag{6.100}$$

where in the second line we have defined a differential energy cross section $\mathbf{\Sigma}$, spectral energy scattered per unit incident flux per second of the measurement. Note that as b becomes large, the spectral intensity approaches zero exponentially (Eq. 6.96). The minimum value of b, b_{min}, is limited by our approximation that $\hbar\omega' \leq m_e c^2$ and by noting that $b = \gamma c\beta/\omega'$.

(3) To transform the results of Eq. 6.100 back to the laboratory (ion at rest), we make use of the relativistic Doppler, linear angle, and solid angle transformation rules to obtain[21]

[21] The transforms needed here are found in the physics of the relativistic Doppler shift

$$\omega = \gamma\omega'(1+\beta\cos\theta')$$
$$\omega' = \gamma\omega(1-\beta\cos\theta)$$

From these two equations the following relations are immediately obtained:

$$\frac{\omega'}{\omega} = \gamma(1-\beta\cos\theta)$$
$$\cos\theta' = (\cos\theta - \beta)/(1-\beta\cos\theta)$$
$$\frac{d\Omega'}{d\Omega} = \frac{d(\cos\theta')}{d(\cos\theta)} = \gamma^{-2}(1-\beta\cos\theta)^{-2} = \left(\frac{\omega}{\omega'}\right)^2$$

Finally, note that $\frac{d\mathscr{E}}{d\omega}$ is Lorentz invariant because both the numerator and denominator transform in an identical manner.

$$\left[\frac{d^2\Sigma}{d\omega d\Omega}\right]_{lab} = \left\{ r_e^2 \left(\frac{1}{\gamma^2(1-\beta cos\theta)^2} \right) \left(\frac{(1-\beta\cos\theta)^2 + (\cos\theta-\beta)^2}{(1-\beta\cos\theta)^2} \right) \right.$$
$$\left. \times \left(\frac{(Ze)^2 c}{(c\beta)^2 4\pi^2 \varepsilon_0} \right) \int_{\frac{\hbar\gamma\omega(1-\beta cos\theta)}{m_e c^2}}^{\infty} \left[K_1(x) \right]^2 x\,dx \right\} \quad (6.101)$$

where now *all quantities (ω, θ, $d\Omega$) are measured in the ion (laboratory) frame.*

For illustration purposes, we explicitly evaluate Eq. 6.101 using the approximations that for $x \leq 1$, $K_1(x) \backsimeq 1/x$, and that for $x > 1$, $K_1(x) \backsimeq \sqrt{\frac{\pi}{2x}} exp(-x)$:

$$\left[\frac{d^2\Sigma}{d\omega d\Omega}\right]_{lab} = \left\{ r_e^2 \left(\frac{Z^2 e^2}{(\gamma^2-1)4\pi^2 \varepsilon_0 c} \right) \left(\frac{(1-\beta\cos\theta)^2 + (\cos\theta-\beta)^2}{(1-\beta\cos\theta)^4} \right) \right.$$
$$\left. \times \left(ln\left[\frac{m_e c^2}{\hbar\gamma\omega(1-\beta\cos\theta)} \right] + \frac{\pi e^{-2}}{4} \right) \right\} \quad (6.102)$$

This expression is plotted in Fig. 6.17. It shows that the Bremsstrahlung radiation increases rapidly with increasing electron kinetic energy,

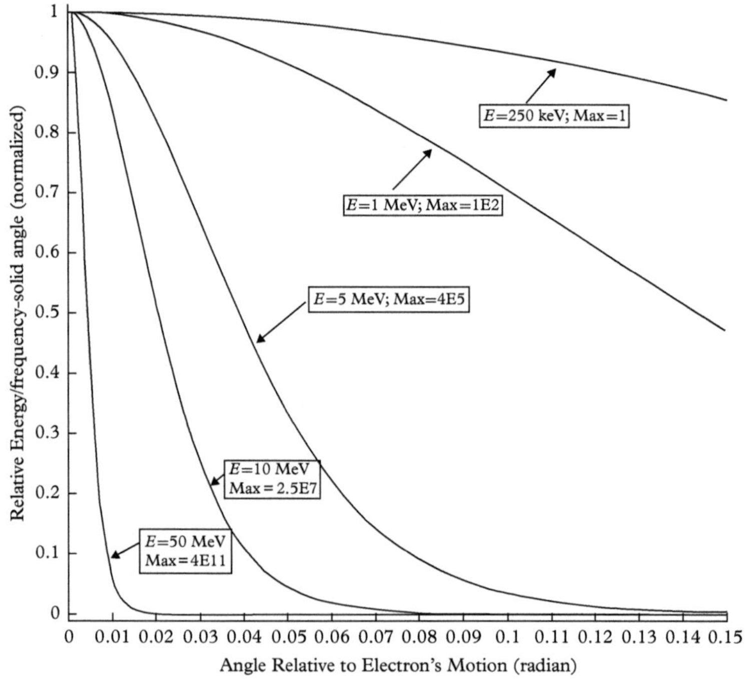

Fig. 6.17 *Angular dependence of the Bremsstrahlung radiation. The total radiated power rises extremely rapidly with increasing electron kinetic energy. (The peak radiated power for 50 MeV electrons is over nine orders of magnitude larger than for electrons with 1 MeV energy.) In addition, the radiated energy becomes tightly confined in a cone centered on the electron's velocity vector.*

and peaks sharply along the direction of the electron's velocity. The approximate expression for $\theta << 1$ and $\gamma >> 1$ is[22]

$$\left[\frac{d^2\Sigma}{d\omega d\Omega}\right]_{lab} = \frac{2\gamma^2 Z^2 e^2}{\pi^2 \varepsilon_o c}\left(\frac{(1+\gamma^4\theta^4)}{(1+\gamma^2\theta^2)^4}\right)\left\{ln\left[\frac{2\gamma m_e c^2}{\hbar\omega(1+\theta^2\gamma^2)}\right] + \frac{\pi e^{-2}}{4}\right\} r_e^2$$

[22] Approximate $cos\theta \approxeq 1 - \frac{\theta^2}{2}$ and $\beta \approxeq 1 - \frac{1}{2\gamma^2}$ and ignore terms $\sim \frac{\theta^2}{\gamma^2}$.

Exercises

(6.1) Proceeding similarly to Section 1.1.1, derive the Lienard–Wiechert expression for the vector potential.

(6.2) Show explicitly that the Lienard–Wiechert potentials satisfy the Lorenz gauge conditions.

(6.3) Supply the missing algebra to formally derive Eq. 6.24.

(6.4) Find the radiation fields and radiated power distribution for a charge carrying out sinusoidal oscillation of amplitude *a*.

(6.5) Find the frequency spectrum from a charge undergoing sinusoidal oscillation for both the case of non-relativistic motion and for a motion having a maximum speed that is extremely relativistic.

(6.6) Show that the expression in Eq. 6.33, $\left|\hat{R}\times\left(\hat{R}\times\vec{a}\right)\right|^2 = a^2 - \left(\vec{a}\cdot\hat{R}\right)^2 = a^2\sin^2\theta$

(6.7) Prove the curl relation of Eq. 6.27.

(6.8) Analytically integrate $\frac{d\mathscr{E}}{d\Omega d\tau} = \frac{2\gamma^{10}q^2\mu_o}{\pi^2 c}a^2\left[\frac{\theta^2}{(1+\gamma^2\theta^2-\frac{1}{2}\theta^2)^5}\right]$ over the solid angle $\Delta\Omega$ subtended by the polar angle $\theta = 1/10$ at $\gamma = 25$ to get $P(t') = (0.166)\frac{q^2\gamma^6\mu_o}{\pi c}a^2$.

(6.9) The Stanford Linear Accelerator (SLAC) is 3.2 km long and can accelerate electrons up to 50 GeV. Assume the electrons experience a constant force during the acceleration. What fraction of the energy used to accelerate an electron is lost to radiation? Discuss in terms of Eq. 6.53.

(6.10) The Chandra x-ray observatory in 2010 found some fraction of the missing mass in the WHIM (warm-hot intergalactic matter) intergalactic plasma. They measured several thermal components with temperatures ~1 million Kelvin. Estimate the Bremsstrahlung spectrum for this plasma. Assume the ions are all hydrogen.

(6.11) Find and graph as a function of frequency: the brightness–power per unit area per unit solid angle per unit frequency–measured 10 meters off the beam-line of the Advanced Photon Source. Assume 100 mA is circulating in the storage ring and do the "single pass" calculation. That is, do not consider the fact that radiation from successive passes past the observation point will add coherently.

(6.12) What fraction of its energy does an electron in the Advanced Photon Source lose via radiation for each orbit?

(6.13) Assume that within the Advanced Photon Source an electron rotates around the ring ten times. Find the spectrum observed and compare with the single pass discussed before.

(6.14) Cosmic ray muons have a typical energy of 6 GeV. As they pass through the atmosphere, which you may assume to be pure nitrogen, Bremsstrahlung radiation will be produced. Predict the Bremsstrahlung spectrum that will be observed.

(6.15) Discuss how the finite size of a nucleus and the electrons in a neutral atom impact the assumptions going into the Bremsstrahlung derivation of Section 1.8.

(6.16) Using the results of Section 1.8, find the total energy radiated by an electron that passes an ion nucleus of charge Ze as a function of impact parameter, b.

(6.17) Derive the expression for the thermal electron Bremsstrahlung of a hydrogen plasma:

$$P(\nu) = 4.163 \cdot 10^{-14} n_e n_i \left[\frac{13.59}{kT}\right]^{1/2} \exp\left(-\frac{h\nu}{kT}\right)\left(\frac{kT}{h\nu}\right)^{1/3}$$

(6.18) For a classical hydrogen atom consisting of an electron orbiting a proton at a radius of a Bohr orbit, find the time it takes for the electron to (classically) radiatively decay to the point that it reaches the proton surface and find the spectrum of the radiation given off.

(6.19) A Bremsstrahlung spectrum is created in neutron beta decay as the electron created experiences the field of the proton decay product. Find the classical spectrum that would be predicted. Discuss any "cutoffs" you make for the calculation.

(6.20) Assume a single electron with kinetic energy of 5 MeV passes through the center of a cell containing singly ionized Li at an equilibrium temperature of 500 degrees Celsius °C (i.e., a vapor pressure consistent with the ideal gas law-you are to make the simplifying assumption that all the Li is singly ionized). Take the longitudinal length of the cell to be 1 cm, and the transverse size of the cell to be 5 cm. You have an exquisitely sensitive diagnostic that can record the kinetic energy of the electron after it passes through the cell. What would you predict the observed energy loss of the electron to be? Comment on the practicality of having to worry about such effects in a realistic experiment.

6.9 Discussions

Discussion 6.1

$$\frac{d}{dt}\left[\frac{\hat{R}\times(\hat{R}\times\vec{\beta})}{1-\hat{R}\cdot\vec{\beta}}\right] = \frac{\hat{R}\times(\hat{R}\times\vec{\beta})}{(1-\hat{R}\cdot\vec{\beta})^2}(\hat{R}\cdot\dot{\vec{\beta}}) + \frac{\hat{R}\times(\hat{R}\times\dot{\vec{\beta}})}{(1-\hat{R}\cdot\vec{\beta})^2}(1-\hat{R}\cdot\vec{\beta})$$

$$= \frac{\hat{R}\times(\hat{R}\times\dot{\vec{\beta}})}{(1-\hat{R}\cdot\vec{\beta})^2} + \frac{\hat{R}\times(\hat{R}\times\vec{\beta})(\hat{R}\cdot\dot{\vec{\beta}}) - \hat{R}\times(\hat{R}\times\dot{\vec{\beta}})(\hat{R}\cdot\vec{\beta})}{(1-\hat{R}\cdot\vec{\beta})^2}$$

The second numerator on the right can be simplified

$$\hat{R}\times(\hat{R}\times\vec{\beta})(\hat{R}\cdot\dot{\vec{\beta}}) - \hat{R}\times(\hat{R}\times\dot{\vec{\beta}})(\hat{R}\cdot\vec{\beta}) = \left[\hat{R}(\hat{R}\cdot\vec{\beta})-\vec{\beta}\right](\hat{R}\cdot\dot{\vec{\beta}}) - \left[\hat{R}(\hat{R}\cdot\dot{\vec{\beta}})-\dot{\vec{\beta}}\right](\hat{R}\cdot\vec{\beta})$$

$$= -\left[\vec{\beta}(\hat{R}\cdot\dot{\vec{\beta}}) - \dot{\vec{\beta}}(\hat{R}\cdot\vec{\beta})\right] = -\hat{R}\times(\vec{\beta}\times\dot{\vec{\beta}})$$

where we have used the vector identity $\vec{A}\times(\vec{B}\times\vec{C}) = \vec{B}(\vec{A}\cdot\vec{C}) - \vec{C}(\vec{A}\cdot\vec{B})$. Replacing the numerator, we get

$$= \frac{d}{dt}\left[\frac{\hat{R}\times(\hat{R}\times\vec{\beta})}{1-\hat{R}\cdot\vec{\beta}}\right] = \frac{\hat{R}\times(\hat{R}\times\dot{\vec{\beta}})}{(1-\hat{R}\cdot\vec{\beta})^2} - \frac{\hat{R}\times(\vec{\beta}\times\dot{\vec{\beta}})}{(1-\hat{R}\cdot\vec{\beta})^2} = \frac{\hat{R}\times(\hat{R}-\vec{\beta})\times\dot{\vec{\beta}})}{(1-\hat{R}\cdot\vec{\beta})^2}$$

Relativistic Electrodynamics

7

- Lagrangian and Hamiltonian mechanics: Concept of action
- The relativistic mechanics of a free particle; free particle canonical 4-momentum and angular momentum 4-tensor
- A charged particle in an external field: Relativistic version of the Lorentz force law
- Description of the electromagnetic field using the action principle: The Lagrange density function and recovery of Maxwell's equations and charge conservation
- The simplest Lagrangian density that can be constructed from a 4-vector field: The Proca Lagrangian and its prediction of a massive photon
- The canonical stress-energy tensor and conservation laws; the angular momentum density of fields; the electromagnetic stress-energy tensor including source terms

7.1 Dynamics using action principles: Lagrangian and Hamiltonian mechanics

We next turn to using action principle formalisms to describe relativistic dynamics. The action-based Lagrangian and Hamiltonian formalisms are elegant and useful ways of reformulating classical, non-relativistic physics under a single unifying principle of minimized action. They show, in a systematic fashion, how various physical concepts are related. For example, for every generalized coordinate, the action formalism automatically defines a corresponding generalized or canonical momentum. Conservation laws are found to result from symmetries in a system's Lagrangian. And the Lagrangian and

Electromagnetic Radiation. Richard Freeman, James King, Gregory Lafyatis,
Oxford University Press (2019). © Richard Freeman, James King, Gregory Lafyatis.
DOI: 10.1093/oso/9780198726500.001.0001

Hamiltonian of a classical, non-relativistic system provide a clear path to the corresponding quantum mechanical description. These formalisms are even more useful for relativistic systems. Special relativity constrains the form of a system's action to the extent that, paraphrasing Einstein, for some simple systems "God didn't have any meaningful choice." In addition, for relativistic systems where native physical intuition is weak to non-existent, one can usually produce a short list of candidate actions that are consistent with special relativity and, by comparisons with observations, whittle this list down, ideally, to a single one. A system's action contains all of the physical information of the system and once it is found, by following the rules of the formalism, the important physics may be identified and studied. Finally, relativistic quantum mechanics and quantum field theory use the action principle descriptions of their classical counterparts.

7.1.1 Concept of action

A basic assumption in dynamics is that any physical system may be completely described by a function called the action, $S(\alpha, \beta, P)$. Here P is a path in configuration space starting at state $\alpha \Leftrightarrow t_\alpha, q(t_\alpha)$ and ending at state $\beta \Leftrightarrow t_\beta, q(t_\beta)$.[1] The actual evolution of the system is described by the path that minimizes S: $\delta S = 0$. This is called Hamilton's principle of least action. Usually the action S is expressed as a time integral of the Lagrangian, L, of the system where, for classical, non-relativistic systems, the Lagrangian is defined as the total kinetic energy, T, minus the total potential energy, U, of the system: $L = T - U$ and the action of a path P is given by:

$$S(\alpha,\beta,P) = \int_{t_\alpha}^{t_\beta} L(q_1(t), q_2(t) \ldots \dot{q}_1(t), \dot{q}_2(t) \ldots, t)\, dt \tag{7.1}$$

where, relative to the path, time t is the independent variable and the generalized coordinates, q_i, and their time derivatives, $\dot{q}_i$, the generalized velocities, are considered to be dependent variables but relative to L, the q_i are the independent variables.[2] The initial and final configurations of the system, α and β are assumed to be known: they are therefore the same for all allowed paths and specified by the generalized coordinates and time values of those configurations, $\alpha \Leftrightarrow q_1(t_\alpha), q_2(t_\alpha), ..., t_\alpha$ and $\beta \Leftrightarrow q_1(t_\beta), q_2(t_\beta), ..., t_\beta$. In what follows, we will use q as an abbreviation for the full set of the generalized coordinates, q_i. So, a specific set of functions, $q(t)$, for the coordinates represents a specific allowed path that begins at α and ends at β. And, for all points in time along a given path, P, the numerical value of the

[1] Note the distinction between configuration space, used here in the Lagrangian formulation, and phase space, to be used in the Hamiltonian formulation: in phase space, $\alpha \Leftrightarrow t_\alpha, q_\alpha$ and $\beta \Leftrightarrow t_\beta, q_\beta$ do not specify single points, as they do in configuration space, but rather a smear of points corresponding to different possible velocities.

[2] The number of generalized coordinates q_i corresponds to the number of degrees of freedom of the system. And, in general, if a system consists of n particles and m constraints, there will be $3n - m$ degrees of freedom.

Lagrangian may be calculated and the time integral, or action, between t_α and t_β in Eq. 7.1 evaluated. There is some ambiguity/flexibility in choosing the Lagrangian of a system: the addition, to a given Lagrangian, of a term that is a total time derivative does nothing to change the path that minimizes S, and thus leaves the dynamics of the system unaffected. This is because a total time derivative, $G(t,q) = \frac{dF(t,q)}{dt}$, upon integration in Eq. 7.1, contributes $F(t_\beta, q(t_\beta)) - F(t_\alpha, q(t_\alpha))$, equally, to all allowed paths. While the formal goal is to find the actual physical path taken by the system, oftener than not, uncovering the important physics does not require carrying that program through to its final end. Indeed, in solving problems, even the a priori knowledge of α and β, the initial and final configurations of the system, is rare and rarely required. Instead we now assume these are known, use the calculus of variations, and consider a small variation of the physical path, $q(t) \to q(t) + \delta q(t)$.[3] The change in the action (Eq. 7.1) for the perturbed path from that of the actual physical path is[4]

$$\delta S = \int_{t_\alpha}^{t_\beta} \sum_i \left(\frac{\partial L}{\partial q_i} \delta q_i + \frac{\partial L}{\partial \dot{q}_i} \delta \dot{q}_i \right) dt \tag{7.2}$$

Integrating the second term on the right by parts:

$$\delta S = \int_{t_\alpha}^{t_\beta} \sum_i \left(\frac{\partial L}{\partial q_i} - \frac{d}{dt} \frac{\partial L}{\partial \dot{q}_i} \right) \delta q_i dt + \sum_i \delta q_i \frac{\partial L}{\partial \dot{q}_i} \Bigg|_{t_\alpha}^{t_\beta} \tag{7.3}$$

Since all the δq_i vanish at the endpoints, the terms in the rightmost summation are all zero. The requirement of Hamilton's principle that $\delta S = 0$ combined with the fact that the δq_i are independent and can each be any arbitrary perturbation from the physical path means that the terms in parentheses, must each vanish identically. Thus we arrive at the Euler–Lagrange equations of motion for the system[5]:

$$\frac{d}{dt} \frac{\partial L}{\partial v_i} - \frac{\partial L}{\partial q_i} = 0. \tag{7.4}$$

where we have written the time-derivatives of the generalized coordinates $\dot{q}_i$ as generalized velocities v_i of the system. Action principles allow wide latitude for selecting different generalized coordinates to describe a system. For example, with a non-relativistic free particle, the Lagrangian is simply the kinetic energy of the system and may either be written using Cartesian coordinates, $L(x,y,z,\dot{x},\dot{y},\dot{z}) = \frac{1}{2} m \left(\dot{x}^2 + \dot{y}^2 + \dot{z}^2\right)$, or alternatively using cylindrical coordinates, $L\left(r,\vartheta,z,\dot{r},\dot{\vartheta},\dot{z}\right) = \frac{1}{2} m \left(\dot{r}^2 + r^2\dot{\vartheta}^2 + \dot{z}^2\right)$. It is important

[3] If the physical path is more explicitly written as

$$q_i(\varepsilon,t) = q_i(0,t) + \varepsilon\eta_i(t)$$

where the $\varepsilon\eta_i(t)$ is a set of perturbing functions of the coordinates parametrized by ε, then, using the shorthand notation,

$$\delta q_i(t) \equiv \frac{\partial q_i}{\partial \varepsilon} d\varepsilon$$

we can write a small variation on that path as

$$\delta q_i(t) = \eta_i(t)\, d\varepsilon$$

so that the slightly perturbed path is expressed as

$$q_i^{pet}(t) = q_i(t) + \eta_i(t)\, d\varepsilon = q_i(t) + \delta q_i(t).$$

[4] Again, using the shorthand notation for a perturbed path,

$$\delta S \equiv \frac{\partial S}{\partial \varepsilon} d\varepsilon$$

we can write

$$\frac{\partial S}{\partial \varepsilon} d\varepsilon = \int_{t_\alpha}^{t_\beta} \sum_i \left(\frac{\partial L}{\partial q_i} \frac{\partial q_i}{\partial \varepsilon} d\varepsilon + \frac{\partial L}{\partial \dot{q}_i} \frac{\partial \dot{q}_i}{\partial \varepsilon} d\varepsilon \right) dt$$

or alternatively

$$\delta S = \int_{t_\alpha}^{t_\beta} \sum_i \left(\frac{\partial L}{\partial q_i} \delta q_i + \frac{\partial L}{\partial \dot{q}_i} \delta \dot{q}_i \right) dt.$$

[5] See Goldstein *Classical Mechanics*, Chapter 2.

to distinguish between the Lagrangian as a function describing relationships among the generalized coordinates of particles in a system–in which these coordinates are considered the independent variables–and the Lagrangian as a physical quantity whose numerical value is to be found for a given configuration of the system–in which time t is considered to be the independent variable.

With each generalized coordinate is associated a generalized or canonical momentum defined by:

$$p_i = \frac{\partial L}{\partial v_i} \tag{7.5}$$

And thus the Euler–Lagrange equations, Eq. 7.4, can be more compactly expressed as[6]

$$\dot{p}_i = \frac{\partial L}{\partial q_i} \tag{7.6}$$

This immediately shows that if the Lagrangian has no explicit dependence on a particular generalized coordinate, $\frac{\partial L}{\partial q_i} = 0$, the corresponding generalized momentum p_i is independent of time–it is a conserved quantity. For example, the y component of canonical momentum for a free particle is the regular kinematic momentum, $p_y = \frac{\partial L}{\partial \dot{y}} = m\dot{y}$, and it is conserved for a free particle, but not for a particle in Earth's gravity (where $L = \frac{1}{2}mv^2 - mgy$). This is an example of Noether's theorem, which states that for every continuous symmetry of a system's action, S, there is a corresponding conserved quantity. A symmetry in the action generally, but not always, corresponds to a symmetry in the Lagrangian. For the case of a free particle, conservation of linear momentum is a consequence of translational symmetry: when written in Cartesian coordinates, the Lagrangian is unchanged upon a translation of the origin of coordinates.

The Hamiltonian formulation is an equivalent description of a system's dynamics obtained by applying a Legendre transformation to the Lagrangian (**see Discussion 7.1**). The Hamiltonian of a system–a function of the generalized coordinates, q, the canonical momenta, p, and time–is defined in phase space as[7]

$$H(q, p, t) = \sum_i p_i \cdot v_i - L(q, v, t) \tag{7.7}$$

where the generalized velocities, v, are to be expressed as functions of q,p, and $t : v_i \equiv v_i(q, p, t)$.[8] The action of a system, Eq. 7.1, may be written in terms of the Hamiltonian:

[6] We note here that while Eq. 7.5 is generally true by definition, Eq. 7.6 is true as a result of the minimization of the action (i.e., in reference to physical systems).

[7] In the Lagrangian formulation, the q are independent variables relative to L and so the natural space to use is configuration space. Similarly, in the Hamiltonian formulation, the q and p are independent variables relative to H and so the natural space to use is phase space.

[8] For a particular system, to obtain the given form of the Hamiltonian function, Eq. 7.7, in terms of the q, p, and t, it is clear that we must first construct the Lagrangian, $L = T - U$, in terms of the q, v, and t for that system. Using Eq. 7.5, we can then obtain the canonical momenta in terms of q, v, and t: $p = p(q, v, t)$. This system of equations can then be inverted to get the generalized velocities expressed in terms of the q, p, and t: $v = v(q, p, t)$. Substitution of these expressions into Eq. 7.7 for the system obtains for us the correct Hamiltonian in terms of the q, p, and t.

$$S = \int_{t_\alpha}^{t_\beta} L(q, v, t)\, dt = \int_{t_\alpha}^{t_\beta} \left(\sum_i p_i \cdot v_i - H(q, p, t) \right) dt \tag{7.8}$$

Invoking Hamilton's principle, for this expression we identify the path that minimizes the integral as the actual, physical path:

$$0 = \delta S = \delta \int_{t_\alpha}^{t_\beta} \left(\sum_i p_i \cdot v_i(q, p, t) - H(q, p, t) \right) dt \tag{7.9}$$

Most commonly, and in all cases considered here, the numerical value of the Hamiltonian H of a system is identified as the system's energy $\mathscr{E}$.[9] For a simple calculation of the numerical value of a system's energy, it does not matter which variables are used in the Hamiltonian, Eq. 7.7. However, in using the Hamiltonian formalism to solve Eq. 7.9 and find the dynamics of a system, it is necessary to express the Hamiltonian function as a mathematical function of the generalized coordinates q and their corresponding canonical momenta p by using Eq. 7.5 to eliminate the velocities from Eq. 7.7. Minimizing the action[10] by carrying out the variation indicated in Eq. 7.9 yields the canonical equations of motion:

$$\frac{dq_i}{dt} = \frac{\partial H}{\partial p_i} \qquad \frac{dp_i}{dt} = -\frac{\partial H}{\partial q_i}. \tag{7.10}$$

The total time derivative of the Hamiltonian is:

$$\frac{dH}{dt} = \sum_i \left(\frac{\partial H}{\partial q_i}\frac{dq_i}{dt} + \frac{\partial H}{\partial p_i}\frac{dp_i}{dt} \right) + \frac{\partial H}{\partial t} \tag{7.11}$$

Substitution of Hamilton's equations into this equation shows that the expression in parentheses is equal to zero. If the time dependence of the Hamiltonian function enters only through the coordinates and momenta–that is, if the Hamiltonian has no explicit time dependence–then its total time derivative is zero, H is constant; energy is conserved. The absence of explicit time dependence in the Hamiltonian of Eq. 7.7 implies the same for the associated Lagrangian,

$$\frac{\partial H}{\partial t} = -\frac{\partial L}{\partial t} \tag{7.12}$$

Here, then, is another example of Noether's theorem: the Lagrangian is invariant with respect to shifts of the temporal origin and the conserved quantity associated with this invariance is the system's energy.

[9] In general, the value of the Hamiltonian is equal to the energy of the system if two conditions hold: (a) the generalized scalar potential is independent of the generalized velocity, $U = U(q, t)$ and (b) the equations connecting the generalized coordinates with the Cartesian coordinates are independent of time, $x = x(q)$. For example, with non-inertial coordinate systems, the second condition does not hold $x = x(q, t)$ and so $H \neq E$. If, in addition to these two conditions, the generalized scalar potential is time independent, $U = U(q)$, then the potential is said to be conservative and the energy of the system is constant in time. Thus, $H = E =$ constant.

[10] In considering slight variations in the path and the associated changes in the action from that of the actual (or physical) path, as we did in the Lagrangian formulation in Eqs. 7.2 and 7.3, we emphasize that it is now the independent variables q and p that are varied in phase space rather than the q varied in configuration space.

7.2 Relativistic mechanics of single point-like particles

We next turn to relativistic mechanics. At the outset, the task is complicated by the following observations:

- An extended rigid body is inconsistent with special relativity. For example, if the near end of a rod is pushed and accelerated, the far end does not simultaneously move since it takes a finite amount of time to communicate that information along the rod. For this reason, we restrict the following discussion to the ideal case of point-like particles. This simplification, however, introduces a second complication when dealing with charged particles:
- Point-like charged particles have infinite potential energy so care must ultimately be taken in treating them. This divergence arises due to charge interaction with itself in a $1/r$ potential. While the topic of self-forces and self-energies is explored at length in Chapter 8, "Field Reactions to Moving Charges," for our present discussion we wish only to convey an awareness of the fact.

In general, for any system's relativistic action we make two requirements. First, as with classical action, if we mathematically break the system's evolution from the state α to the state β into pieces, say from $\alpha \to \gamma$ and $\gamma \to \beta$, we require the total action for the overall motion to be the sum of the actions for the individual pieces: the finite action should be an integral sum of first order differentials, $S = \int dS$. And second, in contrast to the classical case, we require the action to be Lorentz invariant–a Lorentz scalar.

7.2.1 The relativistic mechanics of a free particle

The simplest mechanical system is a point-like free particle of mass m. What quantities can form the basis of the action? The only suitable Lorentz-invariant scalar quantity available is the particle's proper time. The proper time evaluated between an initial, α, and final state β of the free particle system is proportional to the invariant interval between the two states. And thus if this program is to be successful, the action must have the form (**see Discussion 7.2**).

$$S(\alpha, \beta, P) = -\mu s_{\alpha\beta} = -\mu \int_{\alpha}^{\beta} ds \tag{7.13}$$

with μ some constant to be determined and we have included a minus sign for convenience in what follows. We may express $ds = \sqrt{c^2dt^2 - dx^2 - dy^2 - dz^2}$, which clearly varies according to the specific path P between the particle's initial and final states. So, does an extremum of this defined relativistic action correctly describe the motion of a free particle? The answer is yes. To see this, recall when we found, in Chapter 5, that of all the paths between two space-time points such as events α and β, the straight line path of a free particle is the one that maximizes the total calculated interval, $s_{\alpha\beta}$. We were there witnessing a simple example of Hamilton's principle applied to a relativistic free particle. With minimization, the action of Eq. 7.13 thus becomes proportional to the "nearness" of the two events

$$\begin{aligned} S(\alpha, \beta, P) &\Rightarrow S_{min} \\ &= -\mu\Delta s_{\alpha\beta} = -\mu\left(c^2\Delta t^2_{\alpha\beta} - \Delta x^2_{\alpha\beta} - \Delta y^2_{\alpha\beta} - \Delta z^2_{\alpha\beta}\right)^{1/2} \end{aligned} \tag{7.14}$$

It remains to find the constant, μ. Referring to Eq. 7.13 and noting that

$$ds = \sqrt{c^2dt^2 - dx^2 - dy^2 - dz^2} = \frac{c}{\gamma}dt$$

the relativistic Lagrangian to within a constant is found:

$$S = -\int_{\alpha}^{\beta} \frac{\mu c}{\gamma} dt \Longrightarrow L = -\frac{\mu c}{\gamma} \tag{7.15}$$

We compare the low velocity limit of this relativistic Lagrangian with the corresponding classical Lagrangian:

$$L(v \to 0) \simeq -\mu c\left(1 - \frac{1}{2}\frac{v^2}{c^2}\right) \qquad L_{classical} = \frac{1}{2}mv^2 \tag{7.16}$$

The constant term $-\mu c$ is the total time derivative of $-\mu ct$ and so adding it to the Lagrangian has no impact on the dynamics. The choice $\mu = mc$ then makes the low-velocity limit of the relativistic Lagrangian consistent with the classical expression. In conclusion, the relativistic Lagrangian for a free point particle is given by:

$$L = -\frac{mc^2}{\gamma} \tag{7.17}$$

and the corresponding relativistic action is,

$$S = -\int_{\alpha}^{\beta} \frac{mc^2}{\gamma} dt = -\int_{\alpha}^{\beta} mc\,ds \tag{7.18}$$

7.2.2 Free particle canonical 4-momentum

In Chapter 5, we argued for the extension of classical momentum to relativistic momentum through association with the relativistic 4-velocity, U^{α}. Specifically, we claimed that for a particle or system of mass m, the 4-momentum was given by $p^{\alpha} = mU^{\alpha} = m(\gamma c, \gamma\vec{v})$. Now we revisit that discussion using the Lagrangian formulation. The canonical momenta of a system are given by Eq. 7.5. Specifically, for the point-like free particle Lagrangian of Eq. 7.17 written in Cartesian coordinates, the canonical x momentum is:

$$p_x = \frac{\partial L}{\partial v_x} = \gamma m v_x = mU^1 \equiv p^1 \tag{7.19}$$

with similar terms for the y and z momenta. Thus, as claimed, the three canonical linear momenta are scaled to the spatial components of the velocity 4-vector U^{α} by the particle rest mass m, and we can form a momentum 4-vector whose components are relativistically covariant by likewise relating the time component. The 4-momentum is then,

$$p^{\alpha} = mU^{\alpha} = (\gamma mc, \vec{p}) \tag{7.20}$$

We know that the energy of the system is given by the numerical value of the Hamiltonian. As discussed before, in the evaluation of this energy from the Hamiltonian we can use any variables so we do not need to eliminate the velocities in favor of the momenta:

$$\mathscr{E} = H = \sum_i p_i v_i - L = \gamma m v^2 + \frac{mc^2}{\gamma} = \gamma mc^2 \tag{7.21}$$

That is, the relativistic energy of a free particle is given by $\mathscr{E} = \gamma mc^2$. With this result, the 4-momentum, Eq. 7.20, can be expressed as $p^{\alpha} = \left(\frac{\mathscr{E}}{c}, \vec{p}\right)$. That energy is essentially the time-like component of the momentum 4-vector could have been anticipated by noting that the energy and momentum conservation for a system are associated with the invariance of the Lagrangian under temporal and spatial translations, respectively. With every 4-vector comes a Lorentz scalar given by the scalar product of the 4-vector with itself. For relativistic momentum this is:

$$p^{\alpha} p_{\alpha} = \left(\frac{\mathscr{E}}{c}\right)^2 - p^2 = (\gamma mc)^2 - p^2 \tag{7.22}$$

In the rest frame of the particle, $\gamma = 1, p = 0$ and $p^\alpha p_\alpha = m^2c^2$. By definition, a Lorentz scalar has the same value in all frames and so equating the right-hand side of Eq. 7.22 to its rest frame value returns the well-known expression for the relativistic free particle energy in terms of the rest mass and relativistic momentum:

$$\mathscr{E}^2 = m^2c^4 + p^2c^2 \tag{7.23}$$

7.2.3 Free particle angular momentum 4-tensor

Finally, in our relativistic action-based analysis of a point-like free particle, we consider angular momentum. Generally, a conserved quantity resulting from the rotational invariance of a free particle Lagrangian, classical or relativistic, is identified as a component of angular momentum. Classically, it is easy to show that these conserved quantities combine to give $\vec{L}$, an axial vector–whose components are also the three independent components of an antisymmetric tensor, $M^{ij} = x^i p^j - x^j p^i$. By extension, then, we propose that a likely candidate for relativistic angular momentum is the 4-tensor:

$$M^{\alpha\beta} = x^\alpha p^\beta - x^\beta p^\alpha \tag{7.24}$$

Mathematically, this is a fully contravariant, anti-symmetric 4-tensor resulting from the outer product of the coordinate and momentum contravariant 4-vectors.[11] The 4-tensor notation guarantees that the right-hand side construction of Eq. 7.24 is relativistically covariant. Corresponding to the classical case, the relativistic angular momentum vector, $\vec{L}^{rel}$, will be seen to be given by the three independent components within the M^{ij} spatial subtensor: M^{12}, M^{13}, and M^{23}. Note, however, an antisymmetric 4-tensor has six independent components and we will shortly examine the significance of the additional, "spatio-temporal" components, M^{01}, M^{02}, and M^{03}. First we show how the three spatial rotational symmetries of the system, in the $x - y$, $x - z$, and $y - z$ planes, respectively, lead to conservation of relativistic angular momentum, $\vec{L}^{rel} = \vec{r} \times \vec{p}$. For simplicity, we consider a free point particle constrained to move in the $x - y$ plane. By expressing the Lagrangian in the cylindrical coordinate system, (ρ, ϕ, z), we reveal the rotational invariance of this system's Lagrangian to the azimuthal angle ϕ:

$$L = -\frac{mc^2}{\gamma} = -mc^2\left[1 - \frac{\dot{\rho}^2 + \rho^2\dot{\phi}^2}{c^2}\right]^{1/2} \tag{7.25}$$

[11] Technically, this antisymmetric 4-tensor is the outer product of the coordinate and momentum vectors minus the transpose of that outer product–something called the "exterior product."

where $v = (\dot{\rho}^2 + \rho^2\dot{\phi}^2)^{1/2}$ is the particle's velocity in cylindrical coordinates. Because the Lagrangian has no dependence on the generalized coordinate ϕ, the associated conjugate canonical momentum $p_\phi = \frac{\partial L}{\partial \dot{\phi}}$ is a conserved quantity as predicted by the Euler–Lagrange equations (Eq. 7.4). Because this conserved quantity results from a rotational invariance, it is identified as a component of angular momentum–specifically, the z component of $\vec{L}^{rel}$, l_z:

$$p_\phi = l_z = \frac{\partial L}{\partial \dot{\phi}} = -\frac{1}{2}mc^2\left[1 - \frac{v^2}{c^2}\right]^{-1/2}\frac{2v}{c^2}\frac{\partial v}{\partial \dot{\phi}} = \gamma m\rho^2\dot{\phi} \tag{7.26}$$

To compare this with $M^{12} = xp_y - yp_x$ (the component of $M^{\alpha\beta}$ expected to represent l_z), we express it in Cartesian coordinates,

$$\begin{aligned} x &= \rho\cos\phi \quad & \dot{x} &= \dot{\rho}\cos\phi - \rho\dot{\phi}\sin\phi \\ y &= \rho\sin\phi \quad & \dot{y} &= \dot{\rho}\sin\phi + \rho\dot{\phi}\cos\phi \end{aligned} \tag{7.27}$$

Thus:

$$\begin{aligned} x\dot{y} - y\dot{x} &= \rho\cos\phi\left(\dot{\rho}\sin\phi + \rho\dot{\phi}\cos\phi\right) - \rho\sin\phi\left(\dot{\rho}\cos\phi - \rho\dot{\phi}\sin\phi\right) \\ &= \rho^2\dot{\phi} \end{aligned} \tag{7.28}$$

Using this in Eq. 7.26:

$$l_z = \gamma m x\dot{y} - \gamma m y\dot{x} = xp_y - yp_x \tag{7.29}$$

And comparing Eq. 7.24, indeed, $l_z = M^{12}$. Similarly, $l_x = M^{23}$ and $l_y = M^{13}$. So the candidate antisymmetric tensor $M^{\alpha\beta}$ for the angular momentum indeed provides the sought-for conserved quantity resulting from rotational invariance of the Lagrangian.

What about the "spatio-temporal" components? By extension, it seems that they must correspond to "rotational" symmetries in the $t - x$, $t - y$, and $t - z$ planes, respectively. Like the angular momentum components, they do combine to form a vector,

$$\vec{N}^{rel} = \left(ct\vec{p} - \gamma mc\vec{r}\right) = -c\left[\gamma m(\vec{r} - \vec{v}t)\right] \tag{7.30}$$

Focusing on the second form in Eq. 7.30, we can see that conservation of these quantities simply expresses that the particle moves uniformly in a straight line. Apart from the factor c, these quantities have units of mass (γm) times distance ($\vec{r} - \vec{v}t$)–that is, of mass moment. For a collection of moving particles, these mass moments, $\vec{N}_{rel}$, are additive so that the position becomes the center of mass (COM) position,

$\vec{r} \to \vec{r}_{COM}$. In this case, conservation means that however the individual particles move, they do it in such a way that the system's COM moves in a straight line. Furthermore, the three associated symmetries correspond to the invariance of the free particle action, S, to Lorentz transformations that, as pointed out in Chapter 5, are "rotations" in spacetime characterized not by an angle parameter but rather by a boost parameter. Indeed, we know the action is Lorentz invariant because we built it that way!

Finally, at this point it is interesting to note that with the inclusion of the last three conserved quantities of mass moment motion and their association with Lorentz invariance, we now have a complete correspondence between the Poincare Group of ten symmetries within Minkowski spacetime and physical conservation laws: the first four symmetries under space and time translations correspond to conservation of linear momentum and energy. The last six symmetries under Lorentz transformations, the Lorentz sub-group: three rotation and three boost, correspond to conservation of angular momentum and COM motion.

7.2.4 A charged particle in an external electromagnetic field

We next examine a system consisting of a single charged particle interacting with an external electromagnetic field. To the free particle action integral of Eq. 7.18 we need to add an interaction energy term. Basic requirements of this term are that it transform such that the action is a Lorentz scalar and that it has the appropriate units. As before, we require that the action of two or more successive paths is the sum of the action of the individual paths. That is, it should be a first order differential of the path. The simplest putative addition to Eq. 7.18 for our problem would represent the electromagnetic field as a general scalar function, $\varphi(t, \vec{r})$, defined at all positions and times, giving a total action:

$$S = -\int_{\alpha}^{\beta} \left(mc + \frac{q}{c} \varphi(t, \vec{r}) \right) ds \tag{7.31}$$

Here, q is a "charge" that appropriately scales the coupling of the field to the particle. It is left as an exercise to show that adding a general scalar field does not properly describe the influence of an electromagnetic field on a charged particle.[12] The next simplest possibility for an energy interaction term is to represent the electromagnetic field as a 4-vector field, say A_{α}. Again, this field is assumed to be known for all

[12] The scalar $\varphi(t, \vec{r})$ would have to be a Lorentz scalar to satisfy the requirement of Lorentz invariance. For example, each of the components $A^{\alpha}(t, \vec{r})$ of the 4-potential are scalars but they are not Lorentz scalars.

points in space and at all times for the system and it does not include energy due to fields generated by the charged particle itself–the so-called "self-energy" to be discussed in Chapter 8. In order to satisfy the two requirements of Lorentz invariance and proper units, we thus next propose the Lorentz scalar $\frac{q}{c}A^{\alpha}U_{\alpha}$. With this addition, the total action is:

$$S = -\int_{\alpha}^{\beta}\left(mc + \frac{q}{c}A^{\mu}(t,\vec{r})\,U_{\mu}(t,\vec{r})\right)ds \tag{7.32}$$

where, in the following, we will use $A^{\mu} = \left(\frac{\Phi}{c}, \vec{A}\right)$ and $U_{\mu} = (\gamma c, -\gamma\vec{v})$ to separate the time-like and space-like components. At this point we are not identifying q, Φ, and $\vec{A}$ as the electric charge and scalar and vector potentials of classical electromagnetism though, in fact, it will turn out that this is the case. Now we will just go though the formalism using this action and see what it yields. To find the Lagrangian we express the action as an integral over time using $ds = \frac{c}{\gamma}dt$:

$$S = \int_{\alpha}^{\beta}\left(-mc - \frac{q}{c}A^{\mu}U_{\mu}\right)\frac{c}{\gamma}dt = \int_{\alpha}^{\beta}\left(\frac{-mc^2}{\gamma} + q\vec{A}\cdot\vec{v} - q\Phi\right)dt \tag{7.33}$$

The terms in parentheses are identified as the total Lagrangian of the system:

$$L = \frac{-mc^2}{\gamma} + q\vec{A}\cdot\vec{v} - q\Phi \tag{7.34}$$

The canonical momenta are then given by:

$$\vec{P} = \frac{\partial L}{\partial\vec{v}} = \gamma m\vec{v} + q\vec{A} = \vec{p} + q\vec{A} \tag{7.35}$$

We use the convention that the canonical momentum is capitalized and continue to use the lower case for the kinematic momentum, $\vec{p} = \gamma m\vec{v}$.

In quantum mechanics, it is the canonical momentum that is associated with the operator $i\hbar\nabla$.[13] Unmodified from from the classical case, the Euler–Lagrange equations remain the equations of motion for the system.[14] Using Eqs. 7.34 and 7.35 we can combine the Euler–Lagrange equations for the three spatial coordinates into a 3-vector equation:

$$\frac{d}{dt}\left(\vec{p} + q\vec{A}\right) - \nabla\left(\frac{-mc^2}{\gamma} + q\vec{A}\cdot\vec{v} - q\Phi\right) = 0 \tag{7.37}$$

[13] The term $q\vec{A}$ is often referred to as the "potential" momentum in analogy to $q\phi$ being referred to as potential energy.

[14] For example, the x coordinate E-L equation applied to the Lagrangian for the particle in the electromagnertic field, Eq. 7.34, is:

$$\begin{aligned}\frac{d}{dt}\frac{\partial L}{\partial v_x} - \left.\frac{\partial L}{\partial x}\right|_{v_x} &= \frac{d}{dt}(p_x + qA_x) \\ &\quad - \frac{\partial}{\partial x}\left(\frac{-mc^2}{\gamma} + q\vec{A}\cdot\vec{v} - q\Phi\right) \\ &= 0\end{aligned} \tag{7.36}$$

In evaluating the partial derivative with respect to x, we need to keep v_x fixed since the Lagrangian treats it as an independent variable.

Evaluating the field at the position of the particle and using vector identities to find the several gradient terms (**see Discussion 7.3**), we get:

$$\frac{d\vec{p}}{dt} = q\left[-\nabla\Phi - \frac{\partial\vec{A}}{\partial t}\right] + q\vec{v}\times\left(\nabla\times\vec{A}\right). \tag{7.38}$$

Now, we see that if we identify q, Φ, and $\vec{A}$ as the electric charge and scalar and vector potentials of our earlier work, the equation of motion, Eq. 7.38, is a relativistic version of the Lorentz force law:

$$\frac{d\vec{p}}{dt} = q\vec{E} + q\vec{v}\times\vec{B} \tag{7.39}$$

The left-hand side is the derivative of the particle's relativistic kinematic momentum. In the discussion of the expression for the free particle canonical momentum, Eq. 7.19, we found it was the spatial part of a 4-vector whose time-like part was the particle's energy divided by c. We can now rewrite the interacting particle canonical momentum Eq. 7.35 using 4-vector notation:

$$P^\alpha = mU^\alpha + qA^\alpha = p^\alpha + qA^\alpha \tag{7.40}$$

and assume, again, that the time-like component, P^0, is the system's energy $\mathscr{E}$ divided by c. Then the system's energy is

$$\mathscr{E} = \gamma mc^2 + q\phi \tag{7.41}$$

To check this, let's compare this expression to the Hamiltonian function evaluated at a particular time:

$$\begin{aligned} H &= \sum_{i=x,y,z} P_i v_i - L = \gamma mv^2 + q\vec{A}\cdot\vec{v} - \left[\frac{-mc^2}{\gamma} + q\vec{A}\cdot\vec{v} - q\Phi\right] \\ &= \gamma mc^2 + q\phi \end{aligned} \tag{7.42}$$

So, indeed, our assumption is correct and the energy of the system divided by c is the time-like component of the canonical momentum 4-vector.

$$P^0 = \frac{\mathscr{E}}{c} = \frac{H}{c} \tag{7.43}$$

As we noted before, to study the dynamics of a system using the Hamiltonian formalism, the Hamiltonian function must be written in terms of the generalized coordinates and their conjugate momenta and not in terms of the velocity. Thus, the Hamiltonian of Eq. 7.42, with its velocity dependence in the forms of $\vec{v}, v^2$, and $\gamma(v)$, must be transformed to the required dependences. To get an expression free of explicit velocity dependence, consider the particle momentum 4-vector, $p^\alpha = P^\alpha - qA^\alpha = \gamma m(c, \vec{v})$. While this is not a conserved quantity for a charged particle in an external electromagnetic field, it is relativistically covariant and $p^\alpha p_\alpha = m^2c^2$ is a time-independent Lorentz scalar (Eq. 7.22). On the other hand, writing $p^\alpha p_\alpha$ in terms of the canonical momentum P^α and 4-potential A^α:

$$p^\alpha p_\alpha = \left(\frac{H}{c} - q\frac{\phi}{c}\right)^2 - \left(\vec{P} - q\vec{A}\right)^2 = m^2c^2 \tag{7.44}$$

Solving for the H, we get the Hamiltonian expressed in its proper variables:

$$H = \left[m^2c^4 + \left(\vec{P} - q\vec{A}\right)^2 c^2\right]^{1/2} + q\phi \tag{7.45}$$

For small velocities, a Taylor expansion gives[15]

$$H_{classical} = \frac{\left(\vec{P} - q\vec{A}\right)^2}{2m} + q\phi \tag{7.46}$$

Again we have dropped the dynamically insignificant mc^2 constant term in the Taylor expansion. This is the starting point for the low velocity limit, quantum mechanics treatment of the problem. We have already seen from the canonical equations that the time derivative of a particle's momentum gives the Lorentz force law, Eq. 7.39. What about the time derivative of the time-like component, p^0, of the particle's 4-vector momentum? Anticipating that we will identify the particle's mechanical energy as $\mathscr{E}^{particle} = cp^0 = \gamma mc^2$, we have[16]

$$\frac{d\mathscr{E}^{particle}}{dt} = mc^2\frac{d\gamma}{dt} = \vec{v}\cdot\frac{d\vec{p}}{dt} = q\vec{E}\cdot\vec{v} \tag{7.47}$$

Equation 7.39 was used in the final step. This confirms that only the electric field increases a particle's mechanical energy.

We conclude this section by summarizing these results in a manifestly covariant form. Rewriting Eqs. 7.39 and 7.47 as the derivative of the momentum 4-vector with respect to the proper time:

$$\frac{d}{d\tau}\begin{bmatrix} \mathscr{E}^{particle}/c \\ \vec{p}^{particle} \end{bmatrix} = \begin{bmatrix} q\vec{E}\cdot\vec{v}\gamma/c \\ q\left[\vec{E} + \vec{v}\times\vec{B}\right]\gamma \end{bmatrix} \tag{7.48}$$

15 With $\left(\vec{P}-q\vec{A}\right)^2 = p^2 = (\gamma m v)^2$, Eq. 7.45 becomes,

$$\begin{aligned} H &= \left[m^2c^4 + (\gamma m v)^2 c^2\right]^{1/2} + q\phi \\ &= mc^2\left(1 + \frac{(\gamma m v)^2}{m^2c^2}\right)^{1/2} + q\phi \end{aligned}$$

for $v/c \ll 1$, we can approximate this as

$$\begin{aligned} H &\simeq mc^2\left(1 + \frac{(\gamma m v)^2}{2m^2c^2}\right) + q\phi \\ &\simeq mc^2 + \frac{(\gamma m v)^2}{2m} + q\phi \\ &\simeq mc^2 + \frac{\left(\vec{P} - q\vec{A}\right)^2}{2m} + q\phi. \end{aligned}$$

16 The mechanical energy refers to the non-potential energy (the potential energy is stored in the field and potentially available to the particle). For non-relativistic systems this would consist of the particle's kinetic energy alone but for relativistic systems it also includes the rest mass of the particle.

This result can be expressed in terms of the electromagnetic field strength tensor components, $F^{\mu}{}_{\nu}$, obtained in Chapter 5. To see this, first consider the x-component of Eq. 7.38:

$$\frac{dp_x}{\gamma d\tau} = q\left[-\frac{\partial \phi}{\partial x} - \frac{\partial A_x}{\partial t}\right] + q\left[\frac{\partial A_y}{\partial x} - \frac{\partial A_x}{\partial y}\right]v_y - q\left[\frac{\partial A_x}{\partial z} - \frac{\partial A_z}{\partial x}\right]v_z \tag{7.49}$$

where on the left-hand side, we used the relationship between the particle's local and proper time. Rewriting this using 4-vector notation and bringing γ to the right-hand side:

$$\begin{aligned}\frac{dp^1}{d\tau} &= q\left[\partial^1 A_0 - \partial_0 A^1\right]\gamma c + q\left[\partial^1 A_2 - \partial_2 A^1\right]\gamma v_y \\ &\quad + q\left[\partial^1 A_3 - \partial_3 A^1\right]\gamma v_z \end{aligned} \tag{7.50}$$

And so, the Lagrangian formalism applied to a charged particle in an E &M field leads naturally to the field tensor as the three terms in brackets are identified as $F^1{}_0$, $F^1{}_2$, and $F^1{}_3$, respectively. These are multiplied by the zeroth, second and third components of the relativistic 4 velocity U^β and, generalizing, the Lorentz force law in manifestly covariant notation is:

$$\frac{dp^\alpha}{d\tau} = qF^\alpha{}_\beta U^\beta \tag{7.51}$$

which is often referred to as the Lorentz 4-force equation of motion.

7.3 The action principle description of the electromagnetic field

In the previous section, the system described by the Lagrangian and Hamiltonian was a discrete particle and the electromagnetic field was treated as an input to the problem. We now will include the fields themselves in the action functional description of a system. Previously, a discrete number of particle's coordinates and their time derivatives were the dynamical variables entering in the Lagrangian description of the problem, $L(q_i, \dot{q}_i)$. Here, with continuous fields, rather than discrete particles, as the "system," the state of the system is specified by the values of fields, $\varphi^k = \varphi^k(x^\alpha)$. By field, we just mean a quantity defined at all points in space-time and in what follows the "field" of most interest is the 4-vector potential. That is, for electromagnetism $\varphi^k \to A^\mu$. Physical insight can be best gained by mapping the elements of the field theory formalism onto their corresponding discrete variable analogs as is shown in Table 7.1. Now, all the coordinates, (ct, x, y, z), are independent variables; the fields are the dynamical, dependent variables and assume the role formerly played by the particle coordinates. The general goal here is to find the field values at each point in space-time – just as our previous goal was to find the particle coordinate values at each point in time. The "k" superscript index on the field variables identifies individual components of a vector or tensor field and if needed could include an additional indexing of separate fields–with this index distinction being analogous to the difference between coordinate indices for the same particle and those for different particles. A Lagrangian density function, $\mathscr{L}(\varphi^k, \partial_\alpha \varphi^k)$, is constructed from continuous field variables and their derivatives. The Lagrangian, then, is the integral over all space of the Lagrangian density:

$$L = \int \mathscr{L}\left(\varphi^k\left(x^\alpha\right), \partial_\alpha \varphi^k\left(x^\alpha\right), x^\alpha\right) dV \tag{7.52}$$

Thus, if the field Lagrangian represents a quantity, in units of energy, associated with a specified volume of space and which is evolving through time, then the associated Lagrangian density must represent a quantity, in units of energy density, evaluated at every point within that volume, evolving through time. In Eq. 7.52, for completeness and purposes of illustration, we have allowed for explicit dependence of the Lagrangian density on the space-time coordinates but we will frequently consider closed systems for which the Lagrange density will depend on the coordinates only through the fields. This is similar to the particle case where explicit time dependence of the Lagrangian was allowed but only occasionally used. From Eq. 7.52, the action, in turn, is given by:

$$S = \frac{1}{c} \int \int \mathscr{L}\left(\varphi^k\left(x^\alpha\right), \partial_\alpha \varphi^k\left(x^\alpha\right), x^\alpha\right) dV c dt. \tag{7.53}$$

Since both the action and the 4-space differential are Lorentz scalars (see Section 7.5.2), the Lagrangian density $\mathscr{L}$ must also be a Lorentz scalar. While we can arrive at the action via the Lagrangian description, Eq. 7.52, giving special treatment to the time variable is contrary to the spirit of special relativity and taking Eq. 7.53 as the starting point restores the symmetrical treatment of time and space coordinates. This distinction shows up in specifying the integration limits. Taking the Lagrangian route, the Lagrange integral, Eq. 7.52, is specified at initial and final times, t_α and t_β, and for that, the fields themselves must be known at all spatial points at those times and are assumed to go to zero as the spatial coordinates go to infinity. Clearly this is a frame-dependent statement and the time-planes become other surfaces in boosted frames. This is confirmed by the fact that the Lagrangian is

Table 7.1 *Continuous quantities for the Lagrangian density description of the action.*

Quantity	Discrete⟶Continuous
independent variables	$t \longrightarrow [ct, x, y, z] \equiv x^\alpha$
particle/field identifiers/indices	$i \longrightarrow k$
dynamical (dependent) variables	$q_i(t) \longrightarrow \varphi^k(x^\alpha)$
derivatives of dynamical variables	$\dot{q}_i(t) \longrightarrow \partial_\beta \varphi^k(x^\alpha)$
Lagrangian	$L \longrightarrow \mathscr{L}$
generalized momenta	$p = \frac{\partial L}{\partial \dot{q}_i} \longrightarrow \frac{\partial \mathscr{L}}{\partial(\partial_\beta \varphi^k(x^\alpha))}$
equations of motion	$\frac{d}{dt}\left(\frac{\partial L}{\partial \dot{q}_i}\right) = \frac{\partial L}{\partial q_i} \longrightarrow \partial_\beta \left[\frac{\partial \mathscr{L}}{\partial(\partial_\beta \varphi^k)}\right] = \frac{\partial \mathscr{L}}{\partial \varphi^k}$
Hamiltonian	$H = \sum p_i v_i - L \longrightarrow \mathscr{H} = \frac{\partial \mathscr{L}}{\partial(\partial_0 \varphi^k)} \partial^0 \varphi^k - \mathscr{L}$
	generalization: $\mathbb{T}^{\alpha\beta} = \frac{\partial \mathscr{L}}{\partial(\partial_\alpha \varphi^k)} \partial^\beta \varphi^k - g^{\alpha\beta} \mathscr{L}$

not a Lorentz scalar. If, however, we start with the Lagrange density description of the action, we require that the fields be given over a closed 3D hypersurface in space-time, $\mathfrak{R}$, and if this happens to be the two temporal planes and the spatial cylinder at infinity that are used in the Lagrangian description, that is fine. However, the important feature is that the hypersurface $\mathfrak{R}$ encloses the 4D region of space-time to be described. As discussed in the first part of this chapter regarding the Lagrangian treatment of particles, one rarely actually knows a priori the values of $\{t_\alpha, q_\alpha\}$ and $\{t_\beta, q_\beta\}$. Likewise, in the case of the Lagrangian density, the exact hypersurface for a given problem is usually unimportant and one rarely actually knows the values of the fields on that surface. However, in principle they can be specified and are assumed to be known in working through the formalism.

7.3.1 Equations of motion

Now, we derive the equations of motion for the general case of a Lagrangian density made from a set of fields by applying the principle of minimum action to Eq. 7.53. Using the identifications of Table 7.1, the derivation follows closely the calculus of variations derivation of the Euler–Lagrange equations we used for the discrete system.[17]

We consider the system described by:

$$S(\mathfrak{R}, P) = \frac{1}{c} \iint_{\mho} \mathscr{L}\left(\varphi^k(x^\alpha), \partial_\beta \varphi^k(x^\alpha), x^\alpha\right) dV cdt \qquad (7.54)$$

[17] See, for example, chapter 13 in Goldstein, Poole, and Safko *Classical Mechanics*, 3rd edition.

where the values of the fields are given on a space-time surface, $\mathfrak{R}$, and the integral is over the space-time volume $\mathfrak{V}$ enclosed by that surface. Allowed "paths," P, are sets of functions $\{\varphi^k(x^\alpha)\}$ that are defined within $\mathfrak{V}$ and have specified values on the surface $\mathfrak{R}$. Hamilton's principle states that the physical path is the one that minimizes the action. We consider a small variation from the physical path: $\varphi^k(x^\alpha) \to \varphi^k(x^\alpha) + \delta\varphi^k(x^\alpha)$ and $\partial_\beta\varphi^k(x^\alpha) \to \partial_\beta\varphi^k(x^\alpha) + \delta\left[\partial_\beta\varphi^k(x^\alpha)\right]$, subject to the constraint that the variations, $\delta\varphi^k(x^\alpha)$, go to zero on the space-time boundary surface, $\mathfrak{R}$[18]

$$\delta S = \frac{1}{c}\iint_{\mathfrak{V}}\left[\frac{\partial\mathscr{L}}{\partial\varphi^k}\delta\varphi^k + \frac{\partial\mathscr{L}}{\partial\left(\partial_\beta\varphi^k\right)}\delta\left(\partial_\beta\varphi^k\right)\right]dV\,cdt \tag{7.55}$$

(where we note the extensive use of the Einstein summation convention in the notation of Eq. 7.55 and those that follow). The second term may be integrated by parts by noting:

$$\partial_\beta\left[\frac{\partial\mathscr{L}}{\partial\left(\partial_\beta\varphi^k\right)}\delta\varphi^k\right] = \partial_\beta\left[\frac{\partial\mathscr{L}}{\partial\left(\partial_\beta\varphi^k\right)}\right]\delta\varphi^k + \frac{\partial\mathscr{L}}{\partial\left(\partial_\beta\varphi^k\right)}\delta\left(\partial_\beta\varphi^k\right) \tag{7.56}$$

where we used $\partial_\beta\left(\delta\varphi^k\right) = \delta\left(\partial_\beta\varphi^k\right)$. The action variation may be rewritten:

$$\begin{aligned}\delta S = &\frac{1}{c}\iint_{\mathfrak{V}}\left[\frac{\partial\mathscr{L}}{\partial\varphi^k}\delta\varphi^k - \partial_\beta\left[\frac{\partial\mathscr{L}}{\partial\left(\partial_\beta\varphi^k\right)}\right]\delta\varphi^k\right]dV\,cdt\\ &+\frac{1}{c}\iint_{\mathfrak{V}}\partial_\beta\left[\frac{\partial\mathscr{L}}{\partial\left(\partial_\beta\varphi^k\right)}\delta\varphi^k\right]dV\,cdt\end{aligned} \tag{7.57}$$

where the second integral is recognized as a 4-vector divergence. And by the 4D Gauss's theorem this may be expressed as a surface integral over $\mathfrak{R}$. Because the variations for allowed "paths" go to zero at this surface, $\delta\varphi^k(\mathfrak{R}) = 0$, the second integral vanishes. With this, δS can be zero for an arbitrary variation in the field variables $\delta\varphi^k$ only if the following equations of motion are satisfied:

$$\partial_\beta\left[\frac{\partial\mathscr{L}}{\partial\left(\partial_\beta\varphi^k\right)}\right] - \frac{\partial\mathscr{L}}{\partial\varphi^k} = 0 \tag{7.58}$$

Note, that this is exactly the result obtained by starting with the particle Euler-Lagrange equations and making the substitutions in Table 7.1. Similarly to the particle Lagrangian, there is ambiguity/flexibility in the Lagrangian density in that the dynamics of a system are unaffected

[18] If the physical "path" is more explicitly written as

$$\varphi^k(\varepsilon, x^\alpha) = \varphi^k(0, x^\alpha) + \varepsilon\eta\left(x^\alpha\right)$$

where $\varepsilon\eta\left(x^\alpha\right)$ is a perturbing function parametrized by ε. Then, using the shorthand notation,

$$\delta\varphi^k(x^\alpha) \equiv \frac{\partial\varphi^k}{\partial\varepsilon}d\varepsilon$$

we can write a small variation on that path as

$$\delta\varphi^k(x^\alpha) = \eta\left(x^\alpha\right)d\varepsilon$$

so that the slightly perturbed path is expressed as

$$\begin{aligned}\varphi^k(x^\alpha)' &= \varphi^k(x^\alpha) + \eta\left(x^\alpha\right)d\varepsilon\\ &= \varphi^k(x^\alpha) + \delta\varphi^k(x^\alpha).\end{aligned}$$

by adding a term to the Lagrangian density that may be written as 4-divergence of a 4-vector quantity, $\mathcal{G} = \partial_\beta \mathcal{F}^\beta \left(x^\alpha, \varphi^k\right)$ By the 4D Gauss's law, the 4D volume integral over $\mathfrak{V}$ may be converted to a 3D integral of $\mathcal{F}^\beta \left(x^\alpha, \varphi^k\right)$ over the surface $\mathfrak{R}$. Then, because the field values on the space-time surface, $\varphi^k(\mathfrak{R})$, are the same for all "paths," the integral contributes equally to all allowed paths.

7.3.2 Lagrangian density function

Next, we will construct the specific Lagrange density and Lorentz scalar function for the electromagnetic field. Fields for doing this include the 4-vector potential, that is $\varphi^k \Leftrightarrow A^\alpha = \left(\frac{\phi}{c}, \vec{A}\right)$ and its derivatives. An obvious candidate is the Lorentz scalar term $\sim A^\alpha A_\alpha$. It will be seen in the discussion of the Proca Lagrangian that this term corresponds to a photon with a non-zero mass and is inconsistent with observed physics. Scalar terms proportional to the 4-divergence $\partial_\alpha A^\alpha$, as per the discussion before, have no effect on the system dynamics. Next in order of complexity, we consider terms of the form $\partial^\alpha A^\beta \partial_\alpha A_\beta$. To ensure gauge invariance, we use the combinations $F^{\alpha\beta} = \partial^\alpha A^\beta - \partial^\beta A^\alpha$ that define the field tensor and represent the various electric and magnetic field components. These can make a Lorentz scalar of the form $\mathcal{L} \sim F^{\delta\gamma} F_{\delta\gamma} = 2\left(B^2 - E^2/c^2\right)$. We will see that this form does return the observed physics. To obtain the proper units of energy density and to make the physically realized action a minimum, the properly scaled Lagrangian density for the electromagnetic field is:

$$\mathcal{L}^{EM} = -\frac{1}{4\mu_0} F^{\delta\gamma} F_{\delta\gamma} \tag{7.59}$$

The interaction of charged particles and electromagnetic fields represents, ultimately, an exchange of energy between the two. Within the Lagrangian developed above for a particle in a field, Eq. 7.34, this interaction energy manifested itself as the scalar potential or interaction term, $\frac{q}{\gamma} A^\alpha U_\alpha = q\Phi - \vec{A} \cdot q\vec{v}$. Likewise, the field Lagrangians and Lagrangian densities must also include such particle-field interaction energy terms. To study these interaction effects of charges and currents on the fields, we need to understand the coupling of the 4-vector fields A^α to the 4-vector current density, $\mathcal{J}^\alpha = \left(c\rho, \vec{j}\right) = (\rho c, \rho\vec{v})$. The simplest Lagrangian density interaction term that we can construct, which is both Lorentz invariant and has units of energy density, is $\sim A^\beta \mathcal{J}_\beta = \rho\Phi - \vec{A} \cdot \vec{j}$. Again, it is important that these terms be gauge invariant and we will discuss this property below. Finally, a proportionality constant of -1 will turn out to correspond to the physical world, giving for an overall Lagrangian density

$$\mathscr{L} = -\frac{1}{4\mu_0} F^{\delta\gamma} F_{\delta\gamma} - A^{\beta} \mathcal{J}_{\beta} \tag{7.60}$$

In the previous section, where we studied the effect of the fields on charges and currents ultimately resulting in the Lorentz force equation for particle motion, we took the fields as inputs to the problem and solved for the charged particle's motion. Here, we are interested in studying the fields and we will do just the reverse: the current density 4-vector at all space-time points will be considered as a given input to a problem. In either case, when included, these interaction inputs can introduce explicit time dependence to the system. Finally, we note some additional similarities: if we identify the A^{β} terms with the generalized coordinates q and the $F^{\alpha\beta}$ terms with the generalized velocities $\dot{q}$ in the particle Lagrangians, we see in both cases that the free, non-interacting, kinetic terms are quadratic in the velocities and the interaction or potential terms consist of three types of quantities: a charge term, a 4-potential or coordinate term A^{β}, and, noting that $\mathcal{J}^{\beta} = \frac{\rho}{\gamma} U^{\beta}$, a 4-velocity term.

At this point, all that remains is to systematically work through the procedure of the Lagrangian formalism. First, we need to find the equations of motion. Applying Eq. 7.58, with $\varphi^k \to A^{\alpha}$, to our Lagrangian density, Eq. 7.60, we get an equation of motion for each component, A^{α}:

$$-\frac{1}{4\mu_0} \partial_{\beta} \frac{\partial}{\partial \left(\partial_{\beta} A^{\alpha}\right)} \left[F^{\delta\gamma} F_{\delta\gamma}\right] + \frac{\partial}{\partial A^{\alpha}} \left(A^{\beta} \mathcal{J}_{\beta}\right) = 0 \tag{7.61}$$

Evaluating the first term in Eq. 7.61 is an exercise in tensor algebra (**see Discussion 7.4**). We get

$$-\frac{1}{4\mu_0} \partial_{\beta} \frac{\partial}{\partial \left(\partial_{\beta} A^{\alpha}\right)} \left[\left(\partial^{\delta} A^{\gamma} - \partial^{\gamma} A^{\delta}\right)\left(\partial_{\delta} A_{\gamma} - \partial_{\gamma} A_{\delta}\right)\right] = -\frac{1}{\mu_0} \partial^{\beta} F_{\beta\alpha} \tag{7.62}$$

Next, the current density term in Eq. 7.61 is straightforward

$$\frac{\partial}{\partial A^{\alpha}} \left(A^{\beta} \mathcal{J}_{\beta}\right) = \mathcal{J}_{\alpha} \tag{7.63}$$

Combining these, Eq. 7.61 yields:

$$-\frac{1}{\mu_0} \partial^{\beta} F_{\beta\alpha} + \mathcal{J}_{\alpha} = 0 \Longrightarrow \partial^{\beta} F_{\beta\alpha} = \mu_0 \mathcal{J}_{\alpha} \tag{7.64}$$

We have previously obtained this equation in Section 5.9 (Eq. 5.88). There it was shown to be a manifestly relativistically covariant form of the two inhomogeneous Maxwell's equations.

7.3.3 Recovery of Maxwell's equations

We summarize, to this point, our program to formulate electromagnetism with the Hamilton's action principle. We began by constructing a Lagrangian density whose dynamical, dependent variables included just a single 4-vector potential, A^α, and its derivatives, $\partial_\beta A^\alpha$. Specifically, we obtained Eq. 7.60 by combining one of the simplest true Lorentz invariant scalar functions, $F^{\delta\gamma}F_{\delta\gamma}$, with an interaction term that coupled a given current density 4-vector, $\mathcal{J}_\alpha$, to the 4-vector potential, A^α. From this, we found the associated Euler–Lagrange equations of motion (Eq. 7.61) and then recovered the relativistically covariant form of the two inhomogeneous Maxwell equations in terms of the field strength tensor $F_{\beta\alpha}$, Eq. 7.64, with its usual definitions of the electric and magnetic fields. That is, with the proper identification of A^α, it is seen that the Euler–Lagrange equations of motion are the two inhomogenous Maxwell equations:

$$\nabla\cdot\vec{E}=\frac{\rho}{\varepsilon_0} \qquad \nabla\times\vec{B}-\frac{1}{c^2}\frac{\partial\vec{E}}{\partial t}=\mu_0\vec{j}. \tag{7.65}$$

Continuing, the two homogeneous equations may be seen as a result of the antisymmetry of the field tensor as follows. Consider three different indices, α, β, γ chosen from the numbers 0 to 3 used for space-time and the sum formed by cyclically permuting these indices in the expression $\partial^\alpha F^{\beta\gamma}+\partial^\gamma F^{\alpha\beta}+\partial^\beta F^{\gamma\alpha}$. Since $F^{\alpha\beta}\equiv\partial^\alpha A^\beta-\partial^\beta A^\alpha$, it is readily seen that the sum of the permutations is zero. Explicitly writing out the "0, 2, 3" terms, for example, and setting them to zero:

$$\begin{aligned}\partial^0F^{23}+\partial^3F^{02}+\partial^2F^{30}&=-\frac{1}{c}\frac{\partial B_x}{\partial t}+\frac{1}{c}\frac{\partial E_y}{\partial z}-\frac{1}{c}\frac{\partial E_z}{\partial y}\\&=-\frac{1}{c}\frac{\partial B_x}{\partial t}-\frac{1}{c}\left(\nabla\times\vec{E}\right)_x=0\end{aligned} \tag{7.66}$$

This is the x-component of Faraday's equation. From the four possible chosen combinations, four distinct scalar equations result, which correspond to the x, y, and z-components of Faraday's equation and the "1, 2, 3" term is Gauss's law for $\vec{B}$. The two homogeneous Maxwell's equations, then, are implicit in the construction used for the field tensor.

$$\nabla\cdot\vec{B}=0 \qquad \nabla\times\vec{E}+\frac{\partial\vec{B}}{\partial t}=0 \tag{7.67}$$

Finally, charge conservation can be found by taking the 4-vector divergence of Eq. 7.64:

$$\mu_0 \partial^\alpha \mathcal{J}_\alpha = \partial^\alpha \partial^\beta F_{\beta\alpha} = 0 \tag{7.68}$$

where the final equality follows because the operator $\partial^\alpha \partial^\beta$ is symmetric in its indices where $F_{\beta\alpha}$ is antisymmetric. That is, since $\partial^\alpha \partial^\beta = \partial^\beta \partial^\alpha$ and $F_{\beta\alpha} = -F_{\alpha\beta}$, the sum over all $\partial^\alpha \partial^\beta F_{\beta\alpha}$ vanishes. Charge conservation follows in the form (see Eq. 5.90):

$$\partial^\alpha \mathcal{J}_\alpha = \frac{\partial \rho}{\partial t} + \nabla \cdot \vec{\mathcal{J}} = 0 \tag{7.69}$$

We have again witnessed the power and universality of the action formulation. Indeed, the concept of minimum action as defining physics has transcended its initial domain of classical mechanics and discrete particles and entered that of field continuum. We have found that it not only accounts for particle force equations such as the Lorentz 4-force but also the seemingly unrelated field equations such as Maxwell's equations. It is clearly a deeply fundamental principle.

7.3.4 Gauge invariance

Noether's theorem states that any continuous symmetry or invariance of the action of a system leads to a conservation law. Previously, with free particles, we identified conservation of canonical momentum with Lagrangian invariance to the associated generalized coordinate transformation. Here, the dynamical variables are the fields and their gradients so we take up Lagrangian invariance to the gauge of the fields–gauge invariance. We have observed earlier that the dynamics of physical systems do not depend on the gauge we pick for our calculations. How does this show up in our action principle approach? A general gauge transformation is defined by any continuous, differentiable function, Λ, of time and space:

$$\phi' = \phi - \frac{\partial \Lambda}{\partial t} \quad \vec{A}' = \vec{A} + \nabla \Lambda \Longrightarrow A'^\alpha = A^\alpha - \partial^\alpha \Lambda\,(t, \vec{r}) \tag{7.70}$$

Before we consider gauge symmetry in the context of fields and the Lagrangian density, we first consider the effect of a gauge transformation on the action of a charged particle in an electromagnetic field, Eq. 7.32.

$$S' = -\int_a^b \left[mc + \frac{q}{c} \left(A^\alpha - \partial^\alpha \Lambda \right) U_\alpha \right] ds \tag{7.71}$$

$$= -mc \int_a^b ds - q \int_a^b A^\alpha dx_\alpha - q \int_a^b \partial^\alpha \Lambda dx_\alpha \tag{7.72}$$

where we have used $U_\alpha ds = c dx_\alpha$. We can rewrite the action contribution of the additional gauge term:

$$S_\Lambda = \int_a^b \partial^\alpha \Lambda dx_\alpha = \int_a^b \left[\frac{\partial \Lambda}{\partial x}\frac{dx}{dt} + \frac{\partial \Lambda}{\partial y}\frac{dy}{dt} + \frac{\partial \Lambda}{\partial z}\frac{dz}{dt} + \frac{\partial \Lambda}{\partial t} \right] dt$$
$$= \int_a^b \left[\frac{d\Lambda}{dt} \right] dt = \Lambda\left(t_b, \vec{r}_b\right) - \Lambda_a\left(t_a, \vec{r}_a\right) \tag{7.73}$$

Since the gauge term appears in the particle action integral as a total time derivative, it has no effect on the particle dynamics as given by the resulting Lorentz 4-force, Eq. 7.51. Now, getting back to the fields, we ask: How, explicitly, does gauge symmetry enter into the Lagrangian density, Eq. 7.60, which was used in an action principle to find the inhomogeneous field equations? The gauge transformation is a continuous symmetry so we can consider a perturbatively small gauge transformation specified by the gauge function, $\delta\Lambda$. We consider functions that go to 0 on the surface, $\mathfrak{R}$, used to define the space-time volume, $\mathfrak{V}$, of action. Thus, inserting a gauge-transformed Lagrangian density function, Eq. 7.60, into the field action integral of Eq. 7.54, we obtain the gauge-transformed action:

$$S' = -\frac{1}{\mu_0 c} \iint_{\mathfrak{V}} \left[\frac{1}{4} F'^{\delta\gamma} F'_{\delta\gamma} + \mu_0 \left(A^\alpha - \partial^\alpha \delta\Lambda\right) \mathcal{J}_\alpha \right] dV c dt. \tag{7.74}$$

where, for example, $F'_{\alpha\beta} = \partial_\alpha \left(A_\beta - \partial_\beta \delta\Lambda\right) - \partial_\beta \left(A_\alpha - \partial_\alpha \delta\Lambda\right)$. However, since these field tensor entries are components of the gauge invariant electric and magnetic fields they remain unchanged: $F'_{\alpha\beta} = F_{\alpha\beta} = \partial_\alpha A_\beta - \partial_\beta A_\alpha$. What about the coupling term? Noting $\partial^\alpha \left(\delta\Lambda \mathcal{J}_\alpha\right) = \left(\partial^\alpha \delta\Lambda\right) \mathcal{J}_\alpha + \delta\Lambda \left(\partial^\alpha \mathcal{J}_\alpha\right)$, we can integrate the gauge term in Eq. 7.74 by parts:

$$\delta S = \iint_{\mathfrak{V}} \partial^\alpha \left(\delta\Lambda \mathcal{J}_\alpha\right) dV dt - \iint_{\mathfrak{V}} \delta\Lambda \left(\partial^\alpha \mathcal{J}_\alpha\right) dV dt$$
$$= \frac{1}{c} \iint_{\mathfrak{R}} \left(\delta\Lambda \mathcal{J}_\alpha\right) \hat{n}^\alpha d\sigma - \iint_{\mathfrak{V}} \delta\Lambda \left(\partial^\alpha \mathcal{J}_\alpha\right) dV dt \tag{7.75}$$

where, in the first integral, the 4-divergence term was converted to a surface integral using the 4D Gauss's law. Since the perturbed gauge function is taken to be zero on the boundaries, the first term contributes nothing to the perturbed action. The requirement that second term be zero for an arbitrary gauge perturbation $\delta\Lambda$ is satisfied only if:

$$\partial^\alpha \mathcal{J}_\alpha = 0. \tag{7.76}$$

Here, we recognize the 4-vector continuity equation for the conservation of electrical charge. Previously, we saw that charge conservation was intrinsic to Maxwell's equation. This shows why. Charge conservation may be understood as a consequence of the gauge symmetry of the electromagnetic action.

7.3.5 The Proca Lagrangian

Previously, we rejected the simplest Lagrangian density that can be constructed from a 4-vector field, $\mathcal{L}_P = KA^\alpha A_\alpha$, with the claim the it was inconsistent with the observed physical world. This term is called the Proca Lagrangian density and we now examine the consequences of including it in the action for the electromagnetic field system. We consider a total Lagrangian density:

$$\mathcal{L} = -\frac{1}{4\mu_0} F^{\delta\gamma} F_{\delta\gamma} - A^\beta \mathcal{J}_\beta - KA^\alpha A_\alpha \tag{7.77}$$

The vector potential field components and their first derivatives are the independent dynamical variables of the system and the Euler–Lagrange equations are again given by:

$$\partial_\beta \left[\frac{\partial \mathcal{L}}{\partial \left(\partial_\beta A^\alpha \right)} \right] - \frac{\partial \mathcal{L}}{\partial A^\alpha} = 0 \tag{7.78}$$

Since the new term in the Lagrangian density does not depend on the field derivatives, the first term in this equation is identical to that previously evaluated in Eqs. 7.60–7.64, namely, $-\frac{1}{\mu_0}\partial^\beta F_{\beta\alpha}$. Similarly, the current density term contributes as found previously in Eq. 7.63. The Proca term yields $K\frac{\partial}{\partial A^\alpha} A^\gamma A_\gamma = 2KA_\alpha$, giving for the equations of motion:

$$\partial^\beta F_{\beta\alpha} - \mu_0 \mathcal{J}_\alpha + \kappa A_\alpha = 0 \tag{7.79}$$

where $\kappa = -2K\mu_0$. Writing the field tensor $F_{\beta\alpha}$ explicitly in terms of the vector potential:

$$\partial^\beta \partial_\beta A_\alpha - \partial^\beta \partial_\alpha A_\beta - \mu_0 \mathcal{J}_\alpha + \kappa A_\alpha = 0. \tag{7.80}$$

Noting that $\partial^\beta \partial_\alpha A_\beta = \partial_\alpha \partial^\beta A_\beta$ and in the Lorenz gauge, $\partial^\beta A_\beta = 0$, the second term vanishes and the "Proca" equations of motion become:

$$\partial^\beta \partial_\beta A_\alpha + \kappa A_\alpha = \mu_0 \mathcal{J}_\alpha \tag{7.81}$$

First, we study free-space electromagnetic waves by setting the current density 4-vector to zero:

$$\left(\frac{1}{c^2}\frac{\partial^2}{\partial t^2} - \nabla^2\right)A_\alpha + \kappa A_\alpha = 0 \tag{7.82}$$

In the absence of the Proca term ($\kappa = 0$), we recover the regular Lorenz gauge scalar and vector potential wave equations of motion. We can get a Proca dispersion relation from Eq. 7.82 by a plane wave solution, $A_\alpha = a_\alpha \exp\left[i\left(\vec{k}\cdot\vec{r} - \omega t\right)\right]$ for the wave function. The wave equation yields:

$$\frac{\omega^2}{c^2} - k^2 = \kappa \tag{7.83}$$

Recall from quantum mechanics the identifications: $\mathscr{E} = \hbar\omega$ and $p = \hbar k$ for the energy and momentum of a photon. Using these, the dispersion relation may be rewritten:

$$\mathscr{E}^2 = \hbar^2 c^2 \kappa + p^2 c^2 \tag{7.84}$$

Comparing this with the expression for the relativistic energy of a particle, Eq. 7.23, the first term on the right-hand side is identified with the rest mass of the field particle, μ_p, such that

$$\mu_p^2 c^4 = \hbar^2 c^2 \kappa \tag{7.85}$$

and thus, the Proca Lagrangian predicts a massive photon. This is not in harmony with observation. In fact, as of this writing, the most stringent experimental upper limit on the photon mass is that of Luo et al. (PRL 2003): $m_{photon} < 1.2 \times 10^{-54} kg$, about 10^{-24} that of the electron.

We next find Coulomb's Law for the Proca Lagrangian: time independent potential of a point charge at the origin, $\mathcal{J}_\alpha = \left(cq\delta^3(\vec{r}), 0, 0, 0\right)$. Here, the only non-trivial equation-of-motion (Eq. 7.81) is the one for the time-like component, $A_0 = \frac{\Phi}{c}$:

$$-\nabla^2\Phi + \kappa\Phi = \frac{q}{\varepsilon_0}\delta^3(\mathbf{r}) \tag{7.86}$$

Recall $\mu_0 c^2 = \varepsilon_0^{-1}$. The Laplacian of a spherically symmetric potential is:

$$\nabla^2\Phi = \frac{1}{r}\frac{\partial^2}{\partial r^2}(r\Phi) \tag{7.87}$$

and Eq. 7.86 gives:

$$-\frac{\partial^2}{\partial r^2}(r\Phi) + \kappa(r\Phi) = \frac{q}{\varepsilon_0} r\delta(\mathbf{r}) \tag{7.88}$$

We look for a solution of the form $(r\Phi) = ae^{-\lambda r}$:

$$-\lambda^2 ae^{-\lambda r} + \kappa ae^{-\lambda r} = \frac{q}{\varepsilon_0} r\delta(\mathbf{r}) \tag{7.89}$$

So that off origin we have $\lambda^2 = \kappa$. Using Eq. 7.85, in terms of the Proca photon mass, $\lambda = \frac{\mu_p c}{\hbar}$. Thus we can write the potential:

$$\Phi = a\frac{\exp\left(-\frac{\mu_p c}{\hbar} r\right)}{r} \tag{7.90}$$

The constant a is determined by matching at the origin, for Eq. 7.86, the discontinuity of the derivative of Φ on the left-hand side to the δ function on the right-hand side. But we already know the solution to that problem: it is just the Coulomb potential in regular electrostatics. Thus, for the Proca Lagrangian density, the potential of a point charge is

$$\Phi = \frac{1}{4\pi\varepsilon_0}\frac{q}{r}\exp\left(-\frac{\mu_p c}{\hbar} r\right) \tag{7.91}$$

That is, Coulomb's law modified by an exponential fall off at large distances as determined by the photon mass. The range of a force is identified with the distance scale of the exponential fall off of the force. The Proca term changes the infinite range Coulomb force into a finite-range force with scale length $\hbar/(\mu_p c)$.

7.4 The Hamiltonian density and canonical stress-energy tensor

7.4.1 From the Maxwell stress tensor to the 4D stress-energy tensor

Summarizing results from Chapter 5, we have the stress-energy tensor,

$$\mathbb{T}^{\alpha\beta} = \begin{pmatrix} u & cg_x & cg_y & cg_z \\ S_x/c & \sigma_{xx} & \sigma_{xy} & \sigma_{xz} \\ S_y/c & \sigma_{yx} & \sigma_{yy} & \sigma_{yz} \\ S_z/c & \sigma_{zx} & \sigma_{zy} & \sigma_{zz} \end{pmatrix} = \begin{pmatrix} u & c\vec{g} \\ \vec{S}/c & \overleftrightarrow{\sigma} \end{pmatrix}$$

The Maxwell stress tensor is symmetric. Since $\vec{S} = c^2\vec{g}$, the stress-energy tensor is also symmetric. We will find that this is a requirement for a self-consistent definition of angular momentum in special relativity. The divergence of the 0th column returns Poynting's theorem.

$$\partial_\alpha \mathbb{T}^{\alpha 0} = \frac{\partial u}{\partial t} + \nabla \cdot \vec{S} = 0 \tag{7.92}$$

Finally, explicitly

$$\mathbb{T}^{\alpha\beta} =$$

$$\begin{pmatrix} \frac{1}{2}\left(\varepsilon_0 E^2 + \frac{1}{\mu_0}B^2\right) & c\varepsilon_0\left(E_y B_z - E_z B_y\right) & c\varepsilon_0\left(E_z B_x - E_x B_z\right) & c\varepsilon_0\left(E_x B_y - E_y B_x\right) \\ c\varepsilon_0\left(E_y B_z - E_z B_y\right) & \begin{matrix} -\left(\varepsilon_0 E_x^2 + \frac{1}{\mu_0}B_x^2\right) \\ +\frac{1}{2}\left(\varepsilon_0 E^2 + \frac{1}{\mu_0}B^2\right) \end{matrix} & -\varepsilon_0 E_x E_y - \frac{1}{\mu_0}B_x B_y & -\varepsilon_0 E_x E_z - \frac{1}{\mu_0}B_x B_z \\ c\varepsilon_0\left(E_z B_x - E_x B_z\right) & -\varepsilon_0 E_x E_y - \frac{1}{\mu_0}B_x B_y & \begin{matrix} -\left(\varepsilon_0 E_y^2 + \frac{1}{\mu_0}B_y^2\right) \\ +\frac{1}{2}\left(\varepsilon_0 E^2 + \frac{1}{\mu_0}B^2\right) \end{matrix} & -\varepsilon_0 E_y E_z - \frac{1}{\mu_0}B_y B_z \\ c\varepsilon_0\left(E_x B_y - E_y B_x\right) & -\varepsilon_0 E_x E_z - \frac{1}{\mu_0}B_x B_z & -\varepsilon_0 E_y E_z - \frac{1}{\mu_0}B_y B_z & \begin{matrix} -\left(\varepsilon_0 E_z^2 + \frac{1}{\mu_0}B^2\right) \\ +\frac{1}{2}\left(\varepsilon_0 E^2 + \frac{1}{\mu_0}B^2\right) \end{matrix} \end{pmatrix}$$

It may be shown that in relativistic notation this is:

$$\mathbb{T}^{\alpha\beta} = \frac{1}{\mu_0}\left(-g_{\mu\nu}F^{\alpha\nu}F^{\beta\mu} + \frac{1}{4}g^{\alpha\beta}F^{\mu\nu}F_{\mu\nu}\right) \tag{7.93}$$

This will turn out to be the symmetric E & M stress-energy tensor that we obtain next using the Lagrangian formalism. We now resume that thread.

7.4.2 Hamiltonian density: the "00" canonical stress-energy tensor component

We left off the general action principle based description of continuous systems with the Euler–Lagrange equations for fields, Eq. 7.58. We resume that discussion here by defining the Hamiltonian density for continuous systems through comparison with the Hamiltonian expression for discrete systems, Eq. 7.7. Recall that for the continuous case, the fields φ^{k} are the dynamical variables; these are the components of the 4-vector potential, A^κ, in electromagnetism. And all four space-time coordinates are independent variables. That is, $\dot{q} \rightarrow \partial_0\varphi^k$ and the canonical momentum analog for the dynamical variable, φ^{k}, is:

$$\pi^{0}{}_{k} = \frac{\partial \mathscr{L}}{\partial \left(\partial_{0}\varphi^{k}\right)} \tag{7.94}$$

where the "0" superscript specifies that the Lagrangian density is differentiated with respect to the "φ^k–velocity", the field's time derivative. In analogy to the discrete case, the Hamiltonian density function is given by:

$$\mathscr{H} = \sum_{k} \pi^{0}{}_{k}\partial_{0}\varphi^{k} - \delta^{0}_{0}\mathscr{L} \tag{7.95}$$

7.4.3 Canonical stress-energy tensor and conservation laws

In Eq. 7.95 we included the δ^0_0 mutiplying the Lagrangian density to make explicit that the Hamiltonian density is the 00 component of a mixed tensor. The other components of this Lorentz covariant, canonical stress-energy tensor, are found by replacing 00 with other pairs of component labels:

$$\mathbb{T}^{\alpha}_{\text{can }\beta} = \sum_{k} \pi^{\alpha}{}_{k}\partial_{\beta}\varphi^{k} - \delta^{\alpha}{}_{\beta}\mathscr{L} \tag{7.96}$$

The four independent variables ($\alpha = 0, 1, 2, 3$), lead to four different kinds of "generalized velocities" $\partial_\alpha \varphi^k$ and in turn four different kinds of generalized canonical momenta for each field:

$$\pi^{\alpha}{}_{k} = \frac{\partial \mathscr{L}}{\partial \left(\partial_{\alpha}\varphi^{k}\right)} \tag{7.97}$$

We now discuss how this canonical stress-energy tensor is related to the previously discussed stress-energy tensor. Recall the divergence of that tensor was used to express microscopic conservation of energy (Poynting equation) and momentum. Equation 7.58, the Euler–Lagrange equation for the kth field,

$$\partial_{\alpha}\pi^{\alpha}{}_{k} - \frac{\partial \mathscr{L}}{\partial \varphi^{k}} = 0 \tag{7.98}$$

For a closed system the Lagrangian density depends on the fields and their derivatives but has no explicit dependence on the time or position variables:

$$\partial_{\beta}\mathscr{L}\left(\varphi^{k}, \partial_{\alpha}\varphi^{k}\right) = \sum_{k} \frac{\partial \mathscr{L}}{\partial \varphi^{k}}\partial_{\beta}\varphi^{k} + \sum_{k} \pi^{\alpha}{}_{k}\partial_{\beta}\left(\partial_{\alpha}\varphi^{k}\right) \tag{7.99}$$

Using Eq. 7.98 in the first right-hand side term and noting for the second right-hand side term that $\partial_\alpha \partial_\beta = \partial_\beta \partial_\alpha$,

$$\partial_\beta \mathscr{L} = \sum_k \left(\partial_\alpha \pi^\alpha{}_k\right) \partial_\beta \varphi^k + \sum_k \pi^\alpha{}_k \partial_\alpha \left(\partial_\beta \varphi^k\right) = \partial_\alpha \left[\sum_k \pi^\alpha{}_k \left(\partial_\beta \varphi^k\right)\right] \tag{7.100}$$

Noting $\partial_\beta \mathscr{L} = \partial_\alpha \delta^\alpha{}_\beta \mathscr{L}$ and subtracting this from the previous,

$$\partial_\beta \mathscr{L} - \partial_\beta \mathscr{L} = \partial_\alpha \left[\sum_k \pi^\alpha{}_k \left(\partial_\beta \varphi^k\right) - \delta^\alpha{}_\beta \mathscr{L}\right] = 0 \tag{7.101}$$

The term in brackets is the canonical stress-energy tensor of Eq. 7.96. And the equation may be interpreted as microscopic conservation laws written in terms of this tensor:

$$\partial_\alpha \mathbb{T}^\alpha_{\text{can }\beta} = 0 \tag{7.102}$$

7.4.4 Canonical electromagnetic stress-energy tensor

To compare with the previous section's physically-argued stress-energy tensor, we first find $\mathbb{T}^\alpha_{\text{can }\beta}$ components for the electromagnetic field in a source free region. We consider just the field part of the Lagrangian:

$$\mathscr{L}^{\mathscr{E}\mathscr{M}} = -\frac{1}{4\mu_0} F^{\alpha\beta} F_{\alpha\beta} \tag{7.103}$$

The generalized momentum is found to be:

$$\pi^\alpha{}_\kappa = \frac{\partial \mathscr{L}^{\mathscr{E}\mathscr{M}}}{\partial \left(\partial_\alpha A^\kappa\right)} = -\frac{1}{\mu_0} F^\alpha{}_\kappa \tag{7.104}$$

Working with the totally contravariant form of the tensor,

$$\mathbb{T}^{\alpha\beta}_{\text{can}} = \pi^\alpha{}_\kappa \partial^\beta A^\kappa - g^{\alpha\beta} \mathscr{L}^{\mathscr{E}\mathscr{M}} = -\frac{1}{\mu_0} F^\alpha{}_\kappa \partial^\beta A^\kappa - g^{\alpha\beta} \mathscr{L}^{\mathscr{E}\mathscr{M}} \tag{7.105}$$

Note that the unit matrix is the mixed form of the metric tensor, that is, $\delta^\alpha{}_\beta \to g^{\alpha\beta}$. Using the work leading up to Eq. 7.64:

$$\mathbb{T}^{\alpha\beta}_{\text{can}} = -\frac{1}{\mu_0} \left[F^{\alpha\kappa} \partial^\beta A_\kappa - \frac{1}{4} g^{\alpha\beta} F^{\mu\nu} F_{\mu\nu}\right] \tag{7.106}$$

So how does the canonical stress-energy tensor match up with our earlier result, Eq. 7.93? Evaluating some of the terms,

$$\begin{aligned}
\mathbb{T}^{00}_{\text{can}} &= \frac{1}{2}\left(\varepsilon_0 E^2 + \frac{1}{\mu_0}B^2\right) + \left[\varepsilon_0 \nabla \cdot \left(\phi \vec{E}\right)\right] \\
\mathbb{T}^{0i}_{\text{can}} &= \varepsilon_0 c\left(\vec{E} \times \vec{B}\right)_i + \left[\varepsilon_0 c \nabla \cdot \left(A_i \vec{E}\right)\right] \\
\mathbb{T}^{i0}_{\text{can}} &= \varepsilon_0 c\left(\vec{E} \times \vec{B}\right)_i + \left[\varepsilon_0 c\left(\nabla\phi \times \vec{B}\right)_i - \varepsilon_0 E_i \frac{\partial \phi}{c\partial t}\right]
\end{aligned} \tag{7.107}$$

where we have used $\nabla \cdot \vec{E} = 0$ for a charge-free region of space.

7.4.5 Symmetric electromagnetic stress-energy tensor

We see that the canonical stress-energy tensor for the electromagnetic field Lagrangian indeed contains the expected terms of energy and momenta densities, but, in addition, returns unwanted terms bracketed in Eq. 7.107. The additional terms contain the potentials and are not gauge invariant. However, the additions to $\mathbb{T}^{00}_{\text{can}}$ and $\mathbb{T}^{0i}_{\text{can}}$ are divergences and upon integration over all space (to find the total energy and momentum of the fields) these terms can be converted to surface integrals at infinity, by Gauss' law. And assuming that the fields are suitably localized, their contributions vanish: the conservations of the total energy and total momentum are unaffected by the additional terms. But the stress-energy tensor itself has physical significance in terms of the microscopic momentum and energy densities in a system and clearly the canonical tensor not only gets these wrong but has them depend on the gauge used in a calculation. Can this approach be salvaged? In fact, starting with the canonical stress-energy tensor, $\mathbb{T}^{\alpha\beta}_{\text{can}}$, it is readily shown that Eq. 7.102 and the conservations of total energy and momentum are satisfied by any stress-energy tensor of the form:

$$\mathbb{T}^{\alpha\beta} = \mathbb{T}^{\alpha\beta}_{\text{can}} + \partial_\kappa \psi^{\beta\alpha\kappa} \tag{7.108}$$

where the newly introduced tensor is antisymmetric in the final two indices, $\psi^{\beta\alpha\kappa} = -\psi^{\beta\kappa\alpha}$.

To uniquely define the physical stress-energy tensor within the present formalism we require it to be a symmetrical tensor generated from the canonical stress-energy tensor, using a suitable function, $\psi^{\beta\alpha\kappa}$, in Eq. 7.108. The motivation for doing this is that the angular momentum density of a system is found from the stress-energy tensor and microscopic conservation of angular momentum requires the tensor to be symmetric. We will return to this point shortly. Examining Eq. 7.105, with the insight that the potential derivative term looks like it could be part of a field, a symmetrical stress-energy can be created by adding to the canonical tensor the term

$$\frac{1}{\mu_0}\partial_\kappa A^\beta F^{\alpha\kappa} \tag{7.109}$$

But to ensure the new tensor still satisfies the microscopic conservation relation, Eq. 7.102, we must add a term having the proper form. For example:

$$\partial_\kappa \psi^{\beta\alpha\kappa} = \partial_\kappa\left(\frac{1}{\mu_0}A^\beta F^{\alpha\kappa}\right) = \frac{1}{\mu_0}\partial_\kappa A^\beta F^{\alpha\kappa} + \frac{1}{\mu_0}A^\beta \partial_\kappa F^{\alpha\kappa} \tag{7.110}$$

This has the proper form. The first term on the right-hand side is the desired one and referring to the 4-tensor version of Maxwell's equations, Eq. 7.64, for the charge-free and current-free case we are considering, we see that the divergence of the field in the second term vanishes. The resulting symmetric stress-energy tensor for the electromagnetic field is:

$$\mathbb{T}^{\alpha\beta} = -\frac{1}{\mu_0}\left[F^{\alpha\kappa}F^{\beta}_{\ \kappa} - \frac{1}{4}g^{\alpha\beta}F^{\mu\nu}F_{\mu\nu}\right] \tag{7.111}$$

This symmetric tensor is now clearly gauge invariant. In addition, this tensor is traceless. It can be shown, although not here, that for the photon mass to be zero, the trace of the stress-energy tensor must be zero. In fact, this is exactly Eq. 7.93, the 4-tensor found previously by extending Maxwell's stress tensor.

7.4.6 Angular momentum density of fields

In going from the energy-momentum of a particle –a vector– to the stress-energy tensor for fields, we picked up an additional index, the row index of the tensor in Eq. 7.96. The zeroth value of this index (top row of Eq. 7.96) gave the densities of the energy and various momentum directions. The other rows gave the energy and momentum flows in various directions. The second index identified what physical quantity we were considering – 0 for the energy and 1, 2, and 3 for the momentum components. We will set up a similar arrangement for the angular momentum density. Particle angular momentum is described by a second order tensor, rewriting Eq. 7.24:

$$M^{\alpha\beta} = x^{\alpha} p^{\beta} - x^{\beta} p^{\alpha} \tag{7.112}$$

Recall, this is an antisymmetric tensor whose vector components are given by the cyclic permutations. For example,

$$L_z = xp_y - yp_x = x^1 p^2 - x^2 p^1 \tag{7.113}$$

For the angular momentum of fields we will use a 3rd order tensor, the angular momentum density tensor, $\mathbb{M}^{\kappa\alpha\beta}$, whose last two indices indicate the angular momentum component, using the same convention as used before for particles. And similar to the stress-energy or energy-momentum tensor, the first index is 0 for the various component angular momentum densities (divided by c) and $1, 2$, and 3, for angular momentum currents in the x, y, and z. These latter are the same as torque densities. For example, the amount of z- directed angular momentum in a volume, dV, about the point x^{α}, is:

$$\frac{1}{c}\mathbb{M}^{012} dV = \left(x^1 g^2 - x^2 g^1\right) dV = \frac{1}{c}\left(x^1 \mathbb{T}^{02} - x^2 \mathbb{T}^{01}\right) dV \tag{7.114}$$

The covariant generalization of this result gives, for $\kappa \neq 0$ components, the angular momentum–torque density tensor:

$$\mathbb{M}^{\kappa\alpha\beta} = x^{\alpha}\mathbb{T}^{\kappa\beta} - x^{\beta}\mathbb{T}^{\kappa\alpha} \tag{7.115}$$

To see if this is microscopically conserved for the general case, we check whether its 4-divergence is equal to zero:

$$\partial_{\kappa}\mathbb{M}^{\kappa\alpha\beta} = \left(\partial_{\kappa}x^{\alpha}\right)\mathbb{T}^{\kappa\beta} + x^{\alpha}\left(\partial_{\kappa}\mathbb{T}^{\kappa\beta}\right) - \left(\partial_{\kappa}x^{\beta}\right)\mathbb{T}^{\kappa\alpha} - x^{\beta}\left(\partial_{\kappa}\mathbb{T}^{\kappa\alpha}\right) \tag{7.116}$$

The second and fourth terms on the right-hand side are zero by the microscopic conservation of energy and momentum density relation, Eq. 7.102. The other terms are readily evaluated, noting that for any two coordinates, x_u and x_v, $\frac{\partial x^{\mu}}{\partial x^{\nu}}$ is zero unless the top and bottom are the same coordinate, for which the derivative returns 1. The 4-divergence then becomes:

$$\partial_{\kappa}\mathbb{M}^{\kappa\alpha\beta} = \mathbb{T}^{\alpha\beta} - \mathbb{T}^{\beta\alpha} \tag{7.117}$$

That is, microscopic angular momentum conservation holds if and only if the stress-energy tensor is symmetric as found previously.

7.4.7 Electromagnetic stress-energy tensor including source terms

Finally, we consider the stress energy tensor when there are charges and currents in the system. Now the second term in Eq. 7.110 is not zero. The stress-energy tensor's 4-divergence is:

$$\begin{aligned}\partial_\alpha \mathbb{T}^{\alpha\beta} &= -\frac{1}{\mu_0}\left[\left(\partial_\alpha F^{\alpha\kappa}\right)F^{\beta}_{\ \kappa} + F_{\alpha\kappa}\left(\partial^\alpha F^{\beta\kappa}\right) - \frac{1}{2}\left(\partial^\beta F^{\mu\nu}\right)F_{\mu\nu}\right] \\ &= \left[\left(\partial_\alpha F^{\alpha\kappa}\right)F^{\beta}_{\ \kappa} + \left\{\left(\partial^\mu F^{\beta\nu}\right) - \frac{1}{2}\left(\partial^\beta F^{\mu\nu}\right)\right\}F_{\mu\nu}\right] \qquad (7.118)\end{aligned}$$

Using the permutation relation, found in Section 7.3.3, leading to the homogeneous Maxwell's equations and some index rearrangements, the expression in braces is seen to be zero. The remaining term gives:

$$\partial_\alpha \mathbb{T}^{\alpha\beta} = -\mathcal{J}^\kappa F^{\beta}_{\ \kappa} = \mathcal{J}_\kappa F^{\kappa\beta} \qquad (7.119)$$

Writing out the $\beta = 0$ component recovers Poynting's theorem:

$$\frac{\partial u}{\partial t} + \nabla\cdot\vec{S} = -\vec{j}\cdot\vec{E} \qquad (7.120)$$

For $\beta =$ the spatial components, one finds:

$$\frac{\partial\vec{g}}{\partial t} + \nabla\cdot\overleftrightarrow{\sigma} = -\rho\vec{E} - \vec{j}\times\vec{B} \qquad (7.121)$$

where the left-hand side is, as before, the flow of momentum of the fields and the right-hand side may be recognized as the microscopic version of the Lorentz force.

Exercises

(7.1) Show that the Euler–Lagrange (Eq. 7.4) and Newtonian equations of motion are equivalent for the case of a single particle in a conservative system (i.e., $U = U(x_i)$ and $T = T(v_i)$).

(7.2) Using the free particle Lagrangian written in cylindrical coordinates, find the canonical momenta and discuss their conservation.

(7.3) As noted, the value of the Hamiltonian is equal to the energy of the system if the following two conditions hold: (1) the generalized scalar potential is independent of the generalized velocity, $U = U(q,t)$ and (2), the equations connecting the generalized coordinates with the Cartesian coordinates are independent of time, $x = x(q)$. Show that this is true.

$$H = \sum_i \frac{\partial (T-U)}{\partial v_i} v_i - (T-U) =$$
$$2T - (T-U) = T + U = \mathscr{E}$$

(7.4) Show that the absence of explicit time dependence in the Hamiltonian of Eq. 7.7 implies the same for the associated Lagrangian. That is, prove Eq. 7.12.

$$\frac{dH}{dt} = -\frac{\partial L}{\partial t}$$

(7.5) Apply Hamilton's principle to the initially presumed form of the free particle relativistic Lagrangian $L = -\mu c/\gamma$ to show that the path of least action between two arbitrary space-time events (Eq. 7.13) is a straight line. What does comparison of the obtained generalized momentum to the known form of relativistic momentum give for the constant μ?

(7.6) A moving particle is observed in frame K to have a relativistic angular momentum $\vec{L}_{rel} = \vec{r} \times \vec{p}$ with respect to some point. Find the relativistic angular momentum, $\vec{L}'_{rel}$, of that particle relative to that same point as observed from a frame K'moving at velocity $\vec{V}$ with respect to frame K. Hint: the angular momentum 4-tensor, $M^{\alpha\beta} = x^\alpha p^\beta - x^\beta p^\alpha$, is constructed from, and therefore transforms as, the product of two contravariant 4-vectors.

(7.7) Show that the expression of Eq. 7.38 does not change on gauge transformations.

(7.8) Carry out the previous program (from Eq. 7.32 to Eq. 7.39) for the case of a adding a scalar potential field term to the action as in Eq. 7.31. Find the canonical momentum and the equations of motion for the particle and contrast those with the Lorentz force law.

(7.9) Show that the term $\left[m^2c^4 + \left(\vec{P} - q\vec{A}\right)^2 c^2\right]^{1/2}$ in Eq. 7.45 is equal to γmc^2 in Eq. 7.42, thus confirming the equivalence of the two equations for H.

(7.10) Prove the equalities in Eq. 7.47.

(7.11) Fill in the steps to show the equivalence of Eqs. 7.48 and 7.51.

(7.12) Show that the Lorentz invariant Lagrange density function for the electromagnetic field, $\mathscr{L}^{EM} = -\frac{1}{4\mu_0} F^{\delta\gamma} F_{\delta\gamma}$, is equal to the gauge invariant quantity, $-\frac{1}{2\mu_0}\left(B^2 - E^2/c^2\right)$.

(7.13) Evaluate the first term in the interacting field equations of motion of Eq. 7.61.

(7.14) The x-component of Faraday's equation was obtained in Section 7.3.3 by cyclically permuting the numbers "0, 2, 3" as indices in the expression $\partial^\alpha F^{\beta\gamma} + \partial^\gamma F^{\alpha\beta} + \partial^\beta F^{\gamma\alpha} = 0$. Do the same for the three remaining combinations ("0, 1, 3", "0, 1, 2" and "1, 2, 3") to find the y and z components of Faraday's equation and the divergence of $\vec{B}$–that is, the three remaining components of Maxwell's homogeneous equations.

(7.15) Provide an argument that confirms the Hamiltonian density of Eq. 7.95 Lorentz transforms as the "00" component of the symmetric stress-energy tensor, $\mathbb{T}^\alpha_{sym\ \beta}$.

(7.16) Re-express the $\mathbb{T}^{00}$, $\mathbb{T}^{01}$ and $\mathbb{T}^{10}$ components of the canonical stress-energy tensor (Eq. 7.106) in terms of the $\vec{E}$ and $\vec{B}$ fields to confirm Eq. 7.107:

$$\mathbb{T}^{10} = \varepsilon_0 c\left(\vec{E} \times \vec{B}\right)_x + \left[\varepsilon_0 c\left(\nabla\phi \times \vec{B}\right)_x - \varepsilon_0 E_x \frac{\partial\phi}{c\partial t}\right]$$

(7.17) Demonstrate that the alternative stress-energy, so generated, satisfies Eq. 7.102. That is, show that $\partial_\alpha\left(\partial_\kappa \psi^{\alpha\beta\kappa}\right) = 0$.

(7.18) Show that the trace of the symmetric stress-energy tensor of Eq. 7.111 is zero.

(7.19) Using the permutation relation of Section 7.3.3, and some index rearrangements, show that the expression in braces within Eq. 7.118 vanishes.

7.5 Discussions

Discussion 7.1

In general, a Legendre transformation converts a function of a particular set of variables into a related function, in the same units, of a related set of (one or more conjugate) variables. Legendre transformations are most familiar from thermodynamics where they are used to relate the various free energies describing a system. To see, in general, how this works, we consider the differential of the two-variable function, $f(x,y)$

$$df(x,y) = \frac{\partial f}{\partial x}dx + \frac{\partial f}{\partial y}dy = u(x,y)\,dx + w(x,y)\,dy$$

where we have defined two additional variables $u(x,y) = \partial f/\partial x$ and $w(x,y) = \partial f/\partial y$. These related variables, u and w, are the so-called "conjugates" of the original variables, x and y. If we now define a related function in the same two variables

$$g(x,y) = f(x,y) - w(x,y)\,y \tag{7.122}$$

and consider its total differential,

$$dg(x,y) = df - d(wy) = udx + wdy - wdy - ydw = udx - ydw$$

where $dw = \frac{\partial w}{\partial x}dx + \frac{\partial w}{\partial x}dy$; then, with the total differential written in this form, we can identify g as a function of x and w with $u(x,w) = \partial g/\partial x$ and $y(x,w) = -\partial g/\partial w$, and, in particular, note the converse relationships of

$$w = \frac{\partial f}{\partial y} \quad and \quad y = -\frac{\partial g}{\partial w} \tag{7.123}$$

along with the direct relationships,

$$u = \frac{\partial f}{\partial x} \quad and \quad u = \frac{\partial g}{\partial x} \tag{7.124}$$

Thus, if we want to switch to a functional description in terms of a particular variable's conjugate, then we simply subtract the conjugate pair product from the original function. In the present case of converting from the Lagrangian function $L(q,\dot{q})$ to the Hamiltonian function $H(q,p)$, we can make the specific identifications $f(x,y) \Rightarrow L(q,\dot{q})$ and $g(x,w) \Rightarrow -H(q,p)$ with the conjugate pair $yw \Rightarrow \dot{q}p$. Then we can write the Hamiltonian as the related function (Eq. 7.122),

$$H(q,p) = p\dot{q} - L(q,\dot{q})$$

so that the conjugate variable converse relationships (Eq. 7.123) are

$$p = \frac{\partial L}{\partial \dot{q}} \quad and \quad \dot{q} = \frac{\partial H}{\partial p}$$

If we further assume the action has been minimized, then Euler–Lagrange (Eq. 7.6) applies and the direct relationships (Eq. 7.124) are

$$\dot{p} = \frac{\partial L}{\partial q} \text{ and } -\dot{p} = \frac{\partial H}{\partial q}$$

where, in particular, as we will indicate via a slightly different route in the text,

$$\dot{q} = \frac{\partial H}{\partial p} \text{ and } -\dot{p} = \frac{\partial H}{\partial q}$$

are known as "Hamilton's equations" of motion of the system in phase space.

Discussion 7.2

The Lorentz invariant scalar "0" also satisfies both criteria. But it may be seen as a special case of what follows. Alternatively, we can arrive at this form for the action arguing from the form of the Lagrangian as follows. The simplest mechanical system is a point-like free particle of mass m. We know, *a priori*, that the correct physical path, in space and time, of such a particle is a straight line and, indeed, minimizing the classical action for this system yields such a straight path. Applying Hamilton's principle to the relativistic system, we then naturally expect an extremum of the defined relativistic action to also yield a straight path in spacetime. Assuming this to be the case, and with a choice of inertial frames, we can, for simplicity, choose to evaluate the action in the free particle's rest frame. Furthermore, due to the homogeneity of space and time, the associated Lagrangian cannot depend on the spatial coordinates or time of the particle. Thus, with no spatial or temporal dependence and having effectively set $v = 0$, the Lagrangian, in the rest frame at least, has been reduced to some positive constant $L(q, v, t) \Rightarrow \mu c$ with an action given by $S = \int_\alpha^\beta \mu c dt$. If we now require Lorentz invariance of the free particle's relativistic action between the two events, α and β, it must be that the Lagrangian has the general form of $L(v) = \mu c/\gamma(v)$. We can now see that the action is proportional to the fundamental Lorentz invariant quantity associated with the path taken by the particle between the two events–the invariant interval, $s_{\alpha\beta} = \int_\alpha^\beta (c/\gamma(v))\, dt = \int_\alpha^\beta ds$.

Discussion 7.3

The total time derivative of the field is given by:

$$\frac{d\vec{A}}{dt} = \frac{\partial \vec{A}}{\partial x}\frac{dx}{dt} + \frac{\partial \vec{A}}{\partial y}\frac{dy}{dt} + \frac{\partial \vec{A}}{\partial z}\frac{dz}{dt} + \frac{\partial \vec{A}}{\partial t} = (\vec{v} \cdot \nabla)\vec{A} + \frac{\partial \vec{A}}{\partial t} \tag{7.125}$$

For the gradient term: $\nabla \frac{mc^2}{\gamma} = 0$ and we rewrite $\nabla\left(\vec{A}\cdot\vec{v}\right)$ using the vector identity:

$$\nabla\left(\vec{A}\cdot\vec{v}\right) = \left(\vec{A}\cdot\nabla\right)\vec{v} + (\vec{v}\cdot\nabla)\vec{A} + \vec{A}\times(\nabla\times\vec{v}) + \vec{v}\times\left(\nabla\times\vec{A}\right)$$

Since $\vec{v} = \vec{v}(t)$ is not a function of the spatial coordinates, the first and third terms vanish, leaving

$$\nabla\left(\vec{A}\cdot\vec{v}\right) = (\vec{v}\cdot\nabla)\vec{A} + \vec{v}\times\left(\nabla\times\vec{A}\right) \tag{7.126}$$

Substituting the results from Eqs. 7.125 and 7.126 into Eq. 7.37:

$$\frac{d\vec{p}}{dt} + q\left[(\vec{v}\cdot\nabla)\vec{A} + \frac{\partial\vec{A}}{\partial t}\right] - q\left[(\vec{v}\cdot\nabla)\vec{A} + \vec{v}\times\left(\nabla\times\vec{A}\right)\right] + q\nabla\Phi = 0 \tag{7.127}$$

Discussion 7.4

The first term is the more problematic. We note that the upper/lower positions of a summed-over-index (e.g., "β") may be switched:

$$\partial_\beta \frac{\partial}{\partial\left(\partial_\beta A^\alpha\right)}\left[\left(\partial^\delta A^\gamma - \partial^\gamma A^\delta\right)\left(\partial_\delta A_\gamma - \partial_\gamma A_\delta\right)\right] = \partial^\beta \frac{\partial}{\partial\left(\partial^\beta A^\alpha\right)}\left[\left(\partial^\delta A^\gamma - \partial^\gamma A^\delta\right)\left(\partial_\delta A_\gamma - \partial_\gamma A_\delta\right)\right] \tag{7.128}$$

We break this up by letting the derivative with respect to the field gradient successively act on each of the four terms in the brackets:

$$\partial^\beta\left\{\frac{\partial\left(\partial^\delta A^\gamma\right)}{\partial\left(\partial^\beta A^\alpha\right)}\left[\partial_\delta A_\gamma - \partial_\gamma A_\delta\right] + 3 \text{ other terms}\right\} = 4\partial^\beta\left[\partial_\beta A_\alpha - \partial_\alpha A_\beta\right] = 4\partial^\beta F_{\beta\alpha} \tag{7.129}$$

On the left-hand side, since the indices γ and δ are summed over, the derivative, $\frac{\partial\left(\partial^\delta A^\gamma\right)}{\partial\left(\partial^\beta A^\alpha\right)}$, contributes "1" for the pair $\gamma = \alpha, \delta = \beta$ and zero for all other combinations. By reordering indices and exchanging levels of summed indices, it is seen that the remaining three terms in the braces contribute the same as the first.

Discussion 7.5

And so we have obtained the homogeneous 4-divergence form that we sought. As mentioned, such a form is that of a continuity equation and suggests its interpretation as a conservation law. In fact, as a tensor, it represents four continuity equations (specified by the index β). That they are conservation equations can be seen, for a specific value of β, by integrating this divergence relation, at a particular time, over a cubic volume, $\mathcal{V}$, centered at the origin with edges of length $2l$ and sides perpendicular to the three coordinate axes:

$$\frac{d}{dt}\int_{\mathcal{V}} \mathbb{T}^{0}_{\text{can }\beta} dV + \int_{\mathcal{V}} \left(\frac{\partial \mathbb{T}^{x}_{\text{can }\beta}}{\partial x} + \frac{\partial \mathbb{T}^{y}_{\text{can }\beta}}{\partial y} + \frac{\partial \mathbb{T}^{z}_{\text{can }\beta}}{\partial z} \right) dxdydz = 0 \tag{7.130}$$

If $T^{0}_{\text{can }\beta}$ is taken to be the density of some quantity X (i.e., T^{0}_{0}, the Hamiltonian density, is energy per unit volume), then the first term is the rate of change, in X per unit time, of the total amount of X contained in the volume. Next, to better understand the meaning of the second integral term - the spatial part - we consider just the x-derivative term:

$$\int_{-l}^{l}\int_{-l}^{l} dydz \int_{-l}^{l} \frac{\partial \mathbb{T}^{x}_{\text{can }\beta}(ct,x,y,z)}{\partial x} dx = \int_{-l}^{l}\int_{-l}^{l} dydz \left[\mathbb{T}^{x}_{\text{can }\beta}(ct, x=l, y, z) - \mathbb{T}^{x}_{\text{can }\beta}(ct, x=-l, y, z) \right] \tag{7.131}$$

which is to be understood as the outflow of X through the surface at $x = +l$ minus the influx through the surface at $x = -l$. Similarly, the y and z derivatives describe fluxes of X through the corresponding sides of the cube. The second integral, combining these terms, represents the total outflow of X, in X per unit time, from the volume. Thus the first and second integral, in the absence of sources or sinks of X, sum to zero and X is conserved. For a cube whose sides are located well beyond the borders of the physical system, the influxes and outfluxes all vanish leaving just

$$\frac{d}{dt}\int_{\mathcal{V}} \mathbb{T}^{0}_{\text{can }\beta} dV = 0$$

which states that with no flow of X in or out of the volume, the amount of X within the volume is constant in time. For example, with $\beta = 0$, $T^{0}_{\text{can }0}$ is expected to be the Hamiltonian density. The first term, in this case, is the rate of change, in energy per unit time, of the total energy contained in the volume while the second term, in which $\mathbb{T}^{1}_{\text{can }0}$, $\mathbb{T}^{2}_{\text{can }0}$ and $\mathbb{T}^{3}_{\text{can }0}$ are the components of the energy current density–the Poynting vector–represents the total outflow of energy, in energy per unit time, from the volume. Similarly, setting $\beta = 1, 2,$ or 3 in Eq. 7.130 specifies one of the momentum conservation equations in the x, y, or z directions, with the corresponding tensor components giving the momentum density and momentum current densities. Note that the momentum quantities are all scaled by $-c$ in this mixed tensor.

Field Reactions to Moving Charges

8

- Electromagnetic mass and the concept of a self-force on a accelerated charge
- Lorentz's calculation of the self-force: analytical form of the self-force on an accelerating sphere of charge
- Equations of motion for a charge including the self-force: Abraham–Lorentz formula. Sinusoidal motion and constant acceleration, and the issues of runaway and pre-acceleration in the general solutions
- The Landau–Lifshitz approximation and the characteristic time over which self-forces play a significant role
- The problem that the dynamically calculated electromagnetic mass is 4/3 the statically calculated electromagnetic mass, and its reconciliation by the introduction of Poincare stresses
- The infinite electromagnetic mass problem of a point-like charged particle and its possible resolution by considering advanced as well as retarded waves in the potentials

Our experience in electrodynamics, up to now, has essentially involved the solution of two categorically different types of problems:

(1) Given a source distribution of charges and currents, we use Maxwell's equations to find the electric and magnetic fields in space and time. For example, given the world-line of an electron, we can use the methods of Section 6.2 to find the fields at all causally connected points in space-time.

(2) Given the fields in space and time, we use the Lorentz force density equation to determine the time-dependent charge and current density distributions. For example, using the Lorentz force equation we can obtain the world-line of an electron moving through electric and magnetic fields.

Electromagnetic Radiation. Richard Freeman, James King, Gregory Lafyatis,
Oxford University Press (2019).
DOI: 10.1093/oso/9780198726500.001.0001

And as separate problems they have been explored exhaustively. However, it is when we consider a problem that combines these two types that we encounter an important detail that we haven't yet considered, namely that a source distribution determined by an initial specification of fields (via problem type 2), will itself generate fields (via problem type 1) to which it must transfer some of its energy and momentum. Thus, the initial source distribution must also take into account the energy and momentum lost to the fields it generates. As a simple example, the deflection of a fast moving electron passing through a transverse magnetic field is given as a response to the force $\vec{F} = e\vec{v} \times \vec{B}$. However, this does not account for the accelerating electron's loss of energy and momentum to radiation fields and so the actual deflection will be less.

8.1 Electromagnetic field masses

Recalling the discussion in Section 1.9, there are generally two types of fields generated by charge and current sources. First, as specified (in the source frame) by the Coulomb and Biot–Savart laws, there are the "static" $\vec{E}$ and $\vec{B}$ fields. Because the energy density of these fields fall off relatively fast (as $1/R^4$) with distance, they are considered to be "attached" to their source elements. The second type of field, which arises as a result of source acceleration, is the "radiation" field. As we have seen, this type of field is generally specified by the $1/R$ terms in Jefimenko's equations (Eqs. 3.11 and 3.13). Since the energy density of these fields falls off as $1/R^2$, the energy per unit area per unit time is conserved at all radii and the energy, unlike for "static" fields, can escape to infinity. For this reason it is considered to be "detached" from its source elements. Because "radiation" fields are associated with acceleration and "static" fields are associated with constant velocity, the two types of fields are also referred to as "acceleration" and "velocity" fields.

As we have shown in previous chapters, these "velocity" and "acceleration" fields, like matter, carry energy and momentum. As such, given the mass-energy equivalency of special relativity, they also represent an "electromagnetic" form of mass that, like any mass, possesses resistance to changes in its momentum. Thus, any transfer of energy and momentum (i.e., interaction) between charge and current sources and fields must necessarily satisfy Newton's third law of equal and opposite forces. For example, consider a stationary localized distribution of charge. As discussed in Section 1.3, the total energy $\mathscr{E}$ of its electric field, equivalent to the work done to bring the charges in from infinity, is obtained by integrating the field energy density over all space. The associated electromagnetic mass is then given by $M_{em} = \mathscr{E}/c^2$. Also, as it is a field associated with a stationary charge, we expect that it carries no momentum. And it indeed does not since there is no magnetic field and, as known from Section 1.8.3 (Eq. 1.101), the electromagnetic field momentum density is expressed as $\mathscr{P} = \frac{1}{c^2}(\vec{E} \times \vec{H})$. We next apply a force to the charge distribution for a short time and then consider the fields associated with the now moving (with constant velocity) charge distribution. The electric vector field, as we have learned, will remain purely radial relative to the present position of the charge but, more importantly, we now have a magnetic vector field, azimuthally directed about the charge distribution's velocity. This implies that in the process of accelerating the charge distribution, we have transferred momentum ΔP to the "velocity" fields. Likewise, if we were to again integrate the energy density of the fields over all space we would find that the "velocity" field energy has increased. This corresponds, classically, to an increase in the kinetic energy $\Delta\mathscr{E}$ of the electromagnetic mass M_{em}. More to the point, in order to conserve energy and momentum during the acceleration, it must be that the "velocity" field somehow reacts back on the charge distribution with a force, $F_{vel} = -\frac{\Delta P}{\Delta t} = -M_{em}a$. Because we are now referring specifically to the velocity fields that are always localized or "attached" to their sources, we can attribute M_{em} to the source itself. Indeed, when Abraham and Lorentz, in 1903 and 1904, respectively, first considered the problem of field reactions, they assumed that the mass of a charged particle such as an electron was *completely* electromagnetic in nature. We now know that, in fact, the mass is partly mechanical and partly electromagnetic.

In addition to the reaction force due to the velocity fields (which act as an additional attached mass), there is also a reaction force, F_{rad}, due to the "detached" radiation or "acceleration" fields. Again during acceleration, the charge distribution transfers energy and momentum to these radiation fields and so once again there must be a field reaction force that acts back on the charge distribution so that energy and momentum are conserved. Now, however, because the radiation fields detach from the source, they act not like a mass, into which and out of which energy and momentum can be transferred, but rather as a non-conservative damping force.

8.2 Field reaction as a self-force

We have argued that to account for energy and momentum transfer to and from the fields during acceleration, there must be field reaction forces acting back on the charge. To obtain the appropriately modified equations of motion, we need to understand in detail the nature and mechanism of these interactions. If we assume, as Abraham and Lorentz did, that this mechanism is purely electromagnetic and independent of any external source distributions, then we must conclude, as they did, that the reaction forces against which we do work to transfer momentum and energy from the charge to the fields, originate from the charge itself–they are electrical "self-forces." It is natural then to assume the charge is extended and that each little piece of the total charge interacts with (or repels) each other little piece. Consider such a "particle" consisting of a microscopic shell of charge. At rest, this charged sphere feels no net force since the force on each charge element is balanced by the force on an opposite piece of charge. At a constant velocity, its the same–the charged sphere still feels no net force.[1] However, during acceleration this is no longer the case and there is a net force on the sphere. Specifically, at any time t during acceleration, any given pair of charge elements on the sphere will not influence each other with equal and opposite forces and so when all these pair imbalances are added up, there is a net self-force. The mechanism for the "inertia" of the electromagnetic mass, then, is this overall imbalance. Essentially, as we will see, this imbalance arises as a result of the finite propagation speed of the fields between charge pieces (i.e., retardation effects).

8.2.1 Lorentz calculation of the self-force

Let's now calculate the self-force on a spherically symmetric charge distribution of radius a.[2] We start with the equation, determined in Chapter 6 (Eq. 6.30), for the electric field from an arbitrarily moving point charge,

$$d\vec{E}(\vec{r}_o, t) = \frac{dq}{4\pi\varepsilon_o}\left\{\frac{(\hat{R}-\frac{\vec{v}}{c})(1-(\frac{v}{c})^2)}{(1-\hat{R}\cdot\frac{\vec{v}}{c})^3R^2} + \frac{\hat{R}\times[(\hat{R}-\frac{\vec{v}}{c})\times\dot{\vec{v}}]}{c^2(1-\hat{R}\cdot\frac{\vec{v}}{c})^3R}\right\}_{ret} \tag{8.1}$$

where $\{\}_{ret}$ specifies that the $\vec{v}$, $\dot{\vec{v}}$, R, and $\hat{R}$ are to be evaluated at the retarded time, t_{ret}. Thus, we will first be evaluating the electric field $d\vec{E}$ acting on a charge element dq_o at location $\vec{r}_o$ and time t, due to a charge element dq at location $\vec{r}$ and retarded time t_{ret} as shown in Fig. 8.1. A considerable simplification occurs if we impose two further requirements on the motion of the charge distribution:

(1) First, we take the charged sphere's velocity $\vec{v}$ at the evaluation time t to be zero ($\vec{v}(t) = 0$).

[1] The easiest way to see this is by noting that for a constant velocity, we can always transform to the charged sphere's rest frame where the net force is zero. Then, if the net force is zero in one inertial frame, its zero in all inertial frames.

[2] This derivation is due originally to Lorentz and follows the method of Panofsky and Phillips in *Classical Electricity and Magnetism* (pp. 387–389).

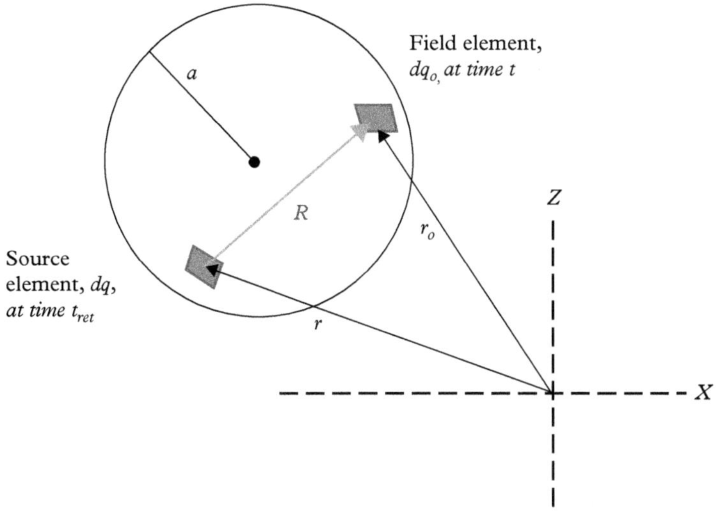

Fig. 8.1 *Spherically symmetric charge distribution for calculating the self-force on an accelerating electron of radius a. Generally, the force on dq_o due to dq is not the same as the force on dq due to dq_o.*

(2) Second, we consider only relatively slowly changing motion. Specifically, given that $\vec{v} = \frac{\Delta \vec{x}}{\Delta t}$, $\dot{\vec{v}} = \frac{\Delta \vec{v}}{\Delta t}$, $\ddot{\vec{v}} = \frac{\Delta \dot{\vec{v}}}{\Delta t}$, and so on, we demand that $a \gg \Delta x$, $v \gg \Delta v$, $\dot{v} \gg \Delta \dot{v}$, and so on. That is, in the time Δt for the field to cross the sphere, the sphere itself will have moved only a small fraction Δx of its radius a, any change in velocity Δv will be small relative to its velocity v, etc. The first requirement is just a restriction to non-relativistic velocities ($\frac{a}{\Delta t} \gg \frac{\Delta x}{\Delta t} \to c \gg v$). Recognizing that $\Delta t = \frac{a}{c}$, these requirements can be written as $a \gg v\frac{a}{c}$, $v \gg \dot{v}\left(\frac{a}{c}\right)$, $\dot{v} \gg \ddot{v}\left(\frac{a}{c}\right)$, and so on, so that the inequalities can be strung together as

$$a \gg v\frac{a}{c} \gg \dot{v}\left(\frac{a}{c}\right)^2 \gg \ddot{v}\left(\frac{a}{c}\right)^3 \gg \dddot{v}\left(\frac{a}{c}\right)^4 \gg \cdots \tag{8.2}$$

which will be useful in terminating the expansions later.

We would like to get the right side of Eq. 8.1, which is in terms of the retarded time, $t_{ret} = t - R/c$, in terms of the evaluation time, t. For the shell velocity and acceleration functions $\vec{v}(t_{ret})$ and $\dot{\vec{v}}(t_{ret})$, we accomplish this with Taylor expansions in powers of R/c,

$$\vec{v}(t_{ret}) = -\dot{\vec{v}}(t)\frac{R}{c} + \ddot{\vec{v}}(t)\frac{1}{2}\left(\frac{R}{c}\right)^2 - \cdots \tag{8.3}$$

$$\dot{\vec{v}}(t_{ret}) = \dot{\vec{v}}(t) - \ddot{\vec{v}}(t)\left(\frac{R}{c}\right) + \cdots \tag{8.4}$$

where we have used the restriction $\vec{v}(t) = 0$. Also, because we are restricting to non-relativistic motion in which $\Delta x \ll a$, we can approximate the $\vec{R}(t_{ret})$ terms in Eq. 8.1, which represent the distance and direction between the retarded position of dq and the evaluation position of dq_o, as constant and equal to the static distance. That is, $\vec{R}(t_{ret}) \simeq \vec{R}(dq, dq_o) = \vec{R}$. With all the terms in Eq. 8.1 now properly expressed in terms of t, it is a straightforward, but tedious, exercise in bookkeeping to obtain $d\vec{E}(\vec{r}_o, t)$, the field on one element due to another. Insofar as the infinite expansions are exact, this solution to $d\vec{E}(\vec{r}_o, t)$ is exact for all time. However, the restriction to slowly changing motion outlined before allows us to accurately calculate this field with just the first few lowest order terms of the expansions. Working through the expansion terms we find that various orders of terms are given by the string of inequalities of Eq. 8.2. If we choose to neglect terms of order $\dddot{v}\left(\frac{a}{c}\right)^4$ or greater, we obtain the result (**see Discussion 8.1**)

$$d\vec{E}(\vec{r}_o, t) \simeq \frac{dq}{4\pi\varepsilon_o R^3}\left[\vec{R} - \frac{2\vec{R}}{c^2}\left(\vec{R}\cdot\dot{\vec{v}}\right) + \frac{R\vec{R}}{2c^3}\left(\vec{R}\cdot\ddot{\vec{v}}\right) + \frac{\ddot{\vec{v}}R^3}{2c^3}\right] \quad (8.5)$$

which, of course, is not our final result but rather just a single contribution from dq to just a single element dq_o. However, because it represents this contribution in terms of motion variables evaluated at the single time t of the evaluation element dq_o, rather than at the various times t_{ret} of the various source elements dq, it can be easily integrated to provide the total electric field contribution at dq_o from all elements.

Next, we consider all possible charge element pairs within the (spherically symmetric) charge distribution in order to determine the average electric field contribution, $\langle d\vec{E}\rangle$, as a function of the distance, R, between the charge elements[3] This is done through term by term and component by component averaging of Eq. 8.5. Starting with the first term, it is easy to see that $\langle\vec{R}\rangle$ vanishes since for every pair of elements separated by R there are two possible cases of dq and dq_o corresponding to equal and opposite $\vec{R}$'s. Next, the fourth term's contribution to $d\vec{E}$ is independent of R and depends, in direction and amplitude, only on the rate of change of acceleration, $\ddot{\vec{v}}$, which is the same for every element, so that $\left\langle\frac{\ddot{\vec{v}}}{2c^3}\right\rangle = \frac{\ddot{\vec{v}}}{2c^3}$. Finally, for the second and third terms of Eq. 8.5, we consider the average of the components. For example, the ith component of the second term can be written as

$$\left[\vec{R}\left(\vec{R}\cdot\dot{\vec{v}}\right)\right]_i = R_i\left(R_i\dot{v}_i + R_j\dot{v}_j + R_k\dot{v}_k\right) \quad (8.6)$$

[3] Note that if the charge distribution were not accelerating and all the force pairs among charge elements were in balance, then all the little field contributions would cancel in pairs and the average electric field contribution over all possible charge element pairs, $\langle d\vec{E}\rangle$, would vanish.

which means we need $\langle R_i^2\rangle$, $\langle R_iR_j\rangle$, and $\langle R_iR_k\rangle$. Now, since R_i, R_j, and R_k are uncorrelated from one possible element pair to the next, it must be that $\langle R_iR_j\rangle$, and $\langle R_iR_k\rangle$ vanish. Also, since $\langle R^2\rangle = \langle R_i^2\rangle + \langle R_j^2\rangle + \langle R_k^2\rangle$, we must have $\langle R_iR_i\rangle = \frac{1}{3}\langle R^2\rangle = \frac{1}{3}R^2$ and so, combining components, $\left\langle \vec{R}\left(\vec{R}\cdot\dot{\vec{v}}\right)\right\rangle = \frac{1}{3}R^2\dot{\vec{v}}$. With these considerations, the average of Eq. 8.5 over all positions, r_o, for a given observer distance, R, is

$$\left\langle d\vec{E}(\vec{r}_o,t,R)\right\rangle \simeq \frac{dq}{4\pi\varepsilon_o}\left[-\frac{2\dot{\vec{v}}}{3c^2R} + \frac{2\ddot{\vec{v}}}{3c^3}\right] \tag{8.7}$$

Given this, we next evaluate the total electric field on one element due to all other elements. In carrying out this integration over dq we note that although this treatment is valid for all spherically symmetric charge distributions, the results will depend on the specific radial distribution. For this calculation, then, we now go back to the simplest case of a charged spherical shell of radius a. Referring to Fig. 8.1, we choose our dq_o along the positive z-axis so that $R = 2a\sin\left(\frac{\theta}{2}\right)$. Defining the surface charge density as $\sigma = q/4\pi a^2$, the integral (over area) can be written

$$\left\langle \vec{E}(\vec{r}_o,t)\right\rangle \simeq \frac{1}{4\pi\varepsilon_o}\int\left[-\frac{\dot{\vec{v}}}{3c^2a\sin\left(\frac{\theta}{2}\right)} + \frac{2\ddot{\vec{v}}}{3c^3}\right]\sigma a^2\sin\theta\, d\theta\, d\phi$$

$$= \frac{q}{8\pi\varepsilon_o}\int\left[-\frac{\dot{\vec{v}}}{3c^2a\sin\left(\frac{\theta}{2}\right)} + \frac{2\ddot{\vec{v}}}{3c^3}\right]\sin\theta\, d\theta \tag{8.8}$$

where we have used the symmetry about the z-axis. Integrating, we get[4]

$$\left\langle \vec{E}(\vec{r}_o,t)\right\rangle \simeq \frac{q}{4\pi\varepsilon_o}\left(-\frac{2}{3}\frac{\dot{\vec{v}}}{ac^2} + \frac{2\ddot{\vec{v}}}{3c^3}\right) \tag{8.9}$$

where, noting that this is the average field value on a charge element in the distribution and not the specific field value of the dq_o we chose for ease of integration of Eq. 8.8, we can immediately obtain the total self-force, to order $\ddot{v}\left(\frac{a}{c}\right)^3$, for an arbitrarily moving shell of charge:

Self-force on an accelerating sphere of charge

$$\vec{F}_{self} = q\left\langle \vec{E}(\vec{r}_o,t)\right\rangle \simeq \frac{q^2}{4\pi\varepsilon_o}\left(-\frac{2}{3}\frac{\dot{\vec{v}}}{ac^2} + \frac{2\ddot{\vec{v}}}{3c^3}\right) \tag{8.10}$$

[4] Starting with Eq. 8.8

$$\left\langle \vec{E}(\vec{r}_o,t)\right\rangle = \frac{q}{8\pi\varepsilon_o}\int\left[-\frac{\dot{\vec{v}}}{3c^2a\sin\left(\frac{\theta}{2}\right)} + \frac{2\ddot{\vec{v}}}{3c^3}\right]\sin\theta\, d\theta$$

Integrating the second term is straightforward,

$$\frac{q}{8\pi\varepsilon_o}\frac{2\ddot{\vec{v}}}{3c^3}\int_0^{\pi}\sin\theta\, d\theta = \frac{q}{4\pi\varepsilon_o}\frac{2}{3}\frac{\ddot{\vec{v}}}{c^3}$$

For the first term, we make the substitution $\theta \to 2\psi$ and use the relation $\sin 2\psi = 2\sin\psi\cos\psi$. We get,

$$\left\langle \vec{E}(\vec{r}_o,t)\right\rangle = \frac{-2q}{8\pi\varepsilon_o}\frac{2\dot{\vec{v}}}{3c^2a}\int_0^{\pi/2}\cos\psi\, d\psi$$

$$= \frac{-q}{4\pi\varepsilon_o}\frac{2}{3}\frac{\dot{\vec{v}}}{ac^2}.$$

where we note that the prefactor of 2/3 in the first term results from our specific choice of radial charge distribution–a shell–whereas the second term is independent of the charge structure. Again, we emphasize that this is actually an approximation to the self-force valid only with the requirement of Eq. 8.2. If we had kept higher order terms we would see that each successive term multiplies by an additional factor of the sphere radius, a,

$$\vec{F}_{self} = \frac{q^2}{4\pi\varepsilon_o}\left(-\frac{2}{3}\frac{\dot{\vec{v}}}{ac^2} + \frac{2\ddot{\vec{v}}}{3c^3} - \alpha\frac{\dddot{\vec{v}}\,a}{c^4} + ...\right) \quad (8.11)$$

Finally, the self-force $\vec{F}_{self}$ can be nicely expressed in compact analytical form as (**see Discussion 8.2**)

Analytical form of self-force on an accelerating sphere of charge

$$\vec{F}_{self} = \frac{q^2}{12\pi\varepsilon_o ca^2}\left[\vec{v}\left(t - \frac{2a}{c}\right) - \vec{v}(t)\right] \quad (8.12)$$

where $2a/c$ is the time it takes light to cross the shell diameter.

8.2.2 Some qualitative arguments for the self-force

We have carried out a fairly involved calculation of the self-force of an extended spherically symmetric charge distribution and we know that it essentially arises due to a force imbalance between charges under acceleration. However, we might wonder why the charge imbalances do not occur for pairs of connected charges moving at a constant velocity. To answer this, and thereby obtain a better understanding, we refer to an earlier discussion in Section 1.93, where we pointed out the oddly coincidental fact that for a constant velocity charge, the associated electric vector field (a velocity field) at all points in space points radially from the *present* position of the charge. That is, the field seems to anticipate the future location of the constant velocity charge. Thus, we can see, from any inertial frame, that force imbalances do not occur for pairs of connected charges moving at a constant velocity because the fields move with the charges and remain equal and opposite (see Fig. 8.2, part a).

However, if the charge pair then accelerates, they will tend to outrun and maneuver relative to the "constant velocity" velocity fields so that the forces on the two charges become unbalanced. As examples,

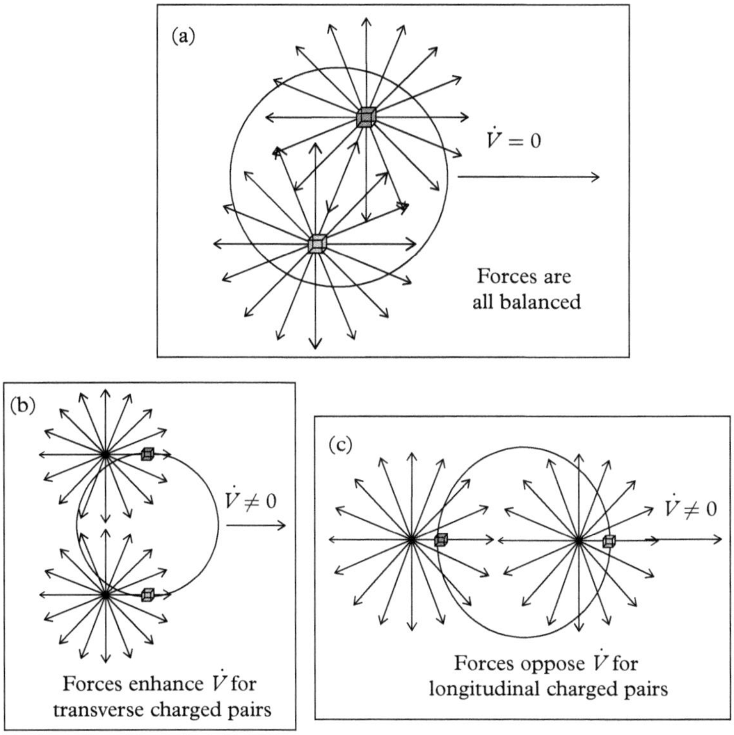

Fig. 8.2 *Charged pairs (a) moving at an uniform velocity will have all forces balanced regardless of the velocity; (b) a charged pair accelerated perpendicular its axis will experience forces that tend to enhance the acceleration; (c) a charged pair accelerated along its axis will experience forces that tend to oppose the acceleration.*

consider the two important simplifying cases of the transversely and longitudinally linearly accelerating charge pairs shown in Fig. 8.2, parts b and c, respectively. For the transverse case, both charges move a bit ahead of the other charge's field and so both feel a slight forward-directed force. This effect contributes to the so-called "runaway" condition, to be discussed. Conversely, in the longitudinal case, we see that while the front charge tends to move a bit ahead of the other's velocity field, the back charge tends to catch up with the field from the front charge. Thus, there is a force imbalance because the front charge feels a diminished forward-directed force while the back charge feels an enhanced rear-directed force. In summary, for charges separated transversely relative to their acceleration, there is a force directed with the acceleration and for longitudinally separated charges, the force imbalance is directed against the acceleration. Inspection of the Abraham–Lorentz force equation (Eq. 8.10) indicates that the former effect is captured by the first ($\dot{\vec{v}}$) term and the latter effect relates to the second ($\ddot{\vec{v}}$) term.

8.3 Abraham–Lorentz formula and the equations of motion

At the beginning of this chapter, we provided a general argument, based on momentum and energy conservation, that there must exist additional forces which act on an accelerating charge as the fields take up energy and momentum. We identified the two forces, F_{vel} and F_{rad}, as being associated with the attached "velocity" and detached "acceleration" or "radiation" fields. The velocity fields exchange energy and momentum conservatively with the charge and therefore act like an additional mass so that F_{vel} is effectively an inertial force. In contrast, the acceleration or "radiation" fields absorb energy and momentum non-conservatively so that F_{rad} is more like a damping force. We now wish to identify these forces within the equation for the self-force (Eq. 8.10).

We start by identifying the inertial force, F_{vel}. Recognizing that this force should have the form $F_{vel} = -M_{em}\dot{\vec{v}}$, where M_{em} is the mass of the velocity field, we see from Eq. 8.10 that it must be represented by the first term and that the electromagnetic mass (for a shell of charge) must therefore be $M_{em} = \frac{q^2}{6\pi\varepsilon_o c^2 a}$. This is confirmed by an integration over all space of the field momentum density, $\vec{g} = \left(\vec{E} \times \vec{H}\right)/c^2$, of a moving extended charge (**see Discussion 8.3**). We will see later, however, that this dynamic calculation (and thus also the first term of the self-force of Eq. 8.10) yields an electromagnetic mass greater by a factor of 4/3 from that calculated by static field energy calculations–the notorious "4/3 problem."

The second term is next tentatively identified as the radiation reaction force, F_{rad}. To confirm this, we again invoke–now more quantitatively–the 1D argument based on conservation of energy. The starting point for this is Larmor's power formula (Eq. 6.36) for total radiated power of a charge q with acceleration $\dot{v}$

$$P = \frac{q^2\mu_o}{6\pi c}\dot{v}^2 \tag{8.13}$$

where we are restricting the motion to linear acceleration. For energy conservation to hold, this power gained by the radiation field must be lost by the charge. In terms of the associated reaction force, call it F_{lrmr}, this power loss is defined by $F_{lrmr}v = -P$,

$$F_{lrmr}v = -\frac{q^2\mu_o}{6\pi c}\dot{v}^2 \tag{8.14}$$

Comparing these results to the rate of work done *on* the charge, P_{on}, by the derived self-force of Eq. 8.10,

$$P_{on} = F_{self}v \simeq (F_{vel} + F_{rad})\,v = \frac{q^2}{4\pi\varepsilon_o}\left(-\frac{2}{3}\frac{\dot{v}v}{ac^2} + \frac{2\ddot{v}v}{3c^3}\right) \tag{8.15}$$

we see that while the first power term ($F_{vel}v$) indeed represents the time rate of change of kinetic energy of the electromagnetic mass ($\frac{dv^2}{dt} = 2\dot{v}v$), the second power term ($F_{rad}v$) does not match the Larmor radiation power loss of Eq. 8.14. That is, $F_{rad}v \neq F_{lrmr}v$. Nonetheless, noting that $\ddot{v}v = \frac{d}{dt}(\dot{v}v) - \dot{v}^2$, we can rewrite Eq. 8.15 as

$$P_{on} = F_{self}v \simeq \frac{q^2}{4\pi\varepsilon_o}\left(-\frac{2}{3}\frac{\dot{v}v}{ac^2} + \frac{2}{3c^3}\frac{d}{dt}(\dot{v}v) - \frac{2\dot{v}^2}{3c^3}\right) \tag{8.16}$$

where the second power loss term of Eq. 8.15 has been expanded to become the last two terms of Eq. 8.16, the second of which represents the Larmor radiation power loss ($F_{lrmr}v$). So, if the first term in Eq. 8.16 is the energy transfer associated with the electromagnetic mass and the third term is that associated with radiation, then what energy transfer does the middle term represent? We know that because its form is similar to the mass term ($\dot{v}v$) in that it can represent power flow to or from the fields, it is likely a conservative power transfer with the velocity fields. In fact, it acts as a higher order (negative) correction to the inertial mass term. Indeed, for periodic motion–just like with the electromagnetic mass term–the middle term averages to zero, or equivalently, integrates to zero over an integral number of periods.[5] Also, because each power loss term is associated with a force, the splitting of the $\ddot{v}v$ "radiation" term in Eq. 8.15 into the last two terms of Eq. 8.16 implies a splitting of the associated self-force,

$$F_{rad} = F_{middle} + F_{lrmr} = \frac{q^2}{4\pi\varepsilon_o v}\left(\frac{2}{3c^3}\frac{d}{dt}(\dot{v}v) - \frac{2\dot{v}^2}{3c^3}\right) \tag{8.17}$$

which, expressed in its contracted form, as in Eq. 8.15, is known as the Abraham–Lorentz formula for the so-called "radiation" reaction force:

Abraham–Lorentz formula

$$F_{rad} = \frac{q^2\ddot{v}}{6\pi\varepsilon_o c^3} \tag{8.18}$$

[5] Thus, having labeled the second terms in Eqs. 8.10 and 8.15, respectively, as the "radiation" reaction force, F_{rad}, and power loss, $F_{rad}v$, we must note that this historic labeling of these as "radiation" terms is somewhat misleading since the terms also include effects due to the exchange of energy with the velocity fields.

8.3.1 The equations of motion

In the previous section, we found Eq. 8.10, with its two terms, as an approximation to the self-force on an accelerating sphere of charge. Accounting for this additional self-force, the general equation of motion is,

$$m\dot{\vec{v}} = \vec{F}_{self} + \vec{F}_{ext}$$
$$m\dot{\vec{v}} \simeq \frac{q^2}{4\pi\varepsilon_o}\left(-\frac{2}{3}\frac{\dot{\vec{v}}}{ac^2} + \frac{2\ddot{\vec{v}}}{3c^3}\right) + \vec{F}_{ext} \tag{8.19}$$

where $\vec{F}_{ext}$ is an externally applied force and m is the non-electromagnetic mass of the charge. Combining the electromagnetic and non-electromagnetic masses, Eq. 8.19 can be written

$$M\dot{\vec{v}} \simeq \frac{q^2\ddot{\vec{v}}}{6\pi\varepsilon_o c^3} + \vec{F}_{ext} \tag{8.20}$$
$$M\dot{\vec{v}} \simeq \vec{F}_{rad} + \vec{F}_{ext} \tag{8.21}$$

where, M, the total measurable mass,

$$M = m + M_{em} = m + \frac{q^2}{6\pi\varepsilon_o c^2 a} \tag{8.22}$$

The derivation of this approximation to the self-force, however, was based on an extended charged particle model that, by definition, is not fundamental. Conversely, a fundamental particle such as an electron, by definition, can have no internal structure and thus must exist as a point. If we wish, then, to apply the theory to describe the self-force of an electron, we must take the limit as the charge radius, a, goes to zero. This, at first, seems to be a blessing since as $a \to 0$, all the higher order terms of the self-force (Eq. 8.11) vanish and the approximation, Eq. 8.10, becomes exact for a point charge. We will find, however, that it is a curse. The general equation of motion for an electron is then,

$$M\dot{\vec{v}} = MT\ddot{\vec{v}} + \vec{F}_{ext}(t) \tag{8.23}$$

where the "approximately equal" sign in Eq. 8.20 has become an equal sign and where $T = q^2/6\pi\varepsilon_o c^3 M$.

Let us now consider some implications of the radiation reaction term $\vec{F}_{rad}$ in Eq. 8.23. Earlier, we argued that because radiation represented a nonconservative loss, the associated reaction force would necessarily be a damping force. In other words, a force that would always oppose the velocity during acceleration (in contrast to an inertial force that always opposes the acceleration). We then defined this nonconservative force as $\vec{F}_{lrmr}$ in relation to the Larmor power loss of Eq. 8.14. If the radiation reaction term, $\vec{F}_{rad}$, in Eq. 8.23 were equivalent to $\vec{F}_{lrmr}$ and thus purely non-conservative, the solution would be simply damped in all cases as with collisional damping, for example. It turns out, however, that this "radiation" term includes another, conservative, mass-like term, $\vec{F}_{middle}$, in addition to $\vec{F}_{lrmr}$, as expressed explicitly in Eq. 8.17. To gain a sense of how these equations of motion, with the special $\vec{F}_{rad}$ "damping" term, are similar to and differ from standard damped solutions, we consider two special cases:

Case 1: Sinusoidal motion

The equation of motion (Eq. 8.23) for a sinusoidally driven ($F_{ext} = A\sin\omega t$) charged particle with a restoring force ($F_{res} = -M\omega_0^2 x$) and a radiation reaction force ($F_{rad} = MT\ddot{v}$) is,

$$M\dot{v} = MT\ddot{v} - M\omega_0^2 x + A\sin\omega t \tag{8.24}$$

Because $x(t)$ is oscillating sinusoidally at the driving frequency, ω, it is generally true that $\ddot{v} = -\omega^2 v$ so we can rewrite Eq. 8.24 as

$$M\dot{v} = -MT\omega^2 v - M\omega_0^2 x + A\sin\omega t \tag{8.25}$$

Thus, for sinusoidal motion at least, the radiation reaction force, $F_{rad} = -MT\omega^2 v$, acts as a damping force since it is at all times in opposition to the velocity and the associated work done on the charge, $F_{rad}v$, is therefore always negative or zero. This is true with or without a driving force since, absent a driving force, the system will still oscillate–only it will oscillate at ω_0 rather than ω. More specifically, this is true for rotational motion. That is, for a charge rotating around some center point, the radiation damping force $\vec{F}_{rad}$ is at all times in exact opposition to the velocity since $\ddot{\vec{v}} = -\omega^2\vec{v}$. How is this to be interpreted in terms of the "velocity" and "radiation" field power loss division within $F_{rad}v$ as expressed by the last two terms of Eq. 8.16? For sinusoidal motion, the third term, the true radiation damping term, varies from zero to $-\alpha\omega^2$, where α is a constant, and the second term, the velocity field term, varies[6] from $-\alpha\omega$ to $\alpha\omega$. So, as expected, the energy transferred to the radiation field is one-way while that transferred to the velocity field in one half cycle is returned in the next half cycle and, overall, energy is lost to the radiation field confirming that F_{rad} acts as a damping force for the case of sinusoidal motion. For circular motion, the velocity field term vanishes and the true radiation damping term is constant in time.[7] Thus, for circular motion the

[6] For sinusoidal motion: From Eq. 8.16 we take the work done on the charge by the F_{rad} part of the self-force:

$$P_{rad} = F_{rad}v = \alpha\frac{d}{dt}(\dot{v}v) - \alpha\dot{v}^2$$

where

$$\alpha = \frac{q^2}{4\pi\varepsilon_o}\frac{2}{3c^3}$$

Now if $v = \sin\omega t$, then $\dot{v} = \omega\cos\omega t$ and we discard the constant α,

$$\begin{aligned} P_{rad} &= \frac{d}{dt}(\omega\sin\omega t\cos\omega t) - \omega^2\cos^2\omega t \\ &= \omega^2\left(1 - 2\sin^2\omega t\right) - \omega^2\cos^2\omega t \end{aligned}$$

so over one cycle, the second term, the true radiation damping term, which represents work done by the radiation field, varies from zero to $-\omega^2$, and the second term, the velocity field term, varies from $-\omega$ to ω. We note also that the total work done by F_{rad} on the charge at any given time is $P_{rad} = -\omega^2[sin(\omega)]^2$, which is always negative or zero.

[7] For circular motion, the only work done on the charge is by the $\vec{F}_{rad}$ part of the self-force:

$$P_{rad} = \vec{F}_{rad}\cdot\vec{v} = \alpha\frac{d}{dt}\left(\dot{\vec{v}}\cdot\vec{v}\right) - \alpha\dot{v}^2$$

where $\vec{v} = r\omega\left(-\sin\omega t\hat{x} + \cos\omega t\hat{y}\right)$ and $\dot{\vec{v}} = -r\omega^2\left(\cos\omega t\hat{x} + \sin\omega t\hat{y}\right)$ so the first term vanishes and the second term is the constant $-\alpha r^2\omega^4 = -\alpha\dot{v}^2$ that, with $\alpha = \frac{q^2}{4\pi\varepsilon_o}\frac{2}{3c^3}$, is the Larmor Power formula (Eq. 8.13).

power supplied is at all times equal to the power radiated while for sinusoidal motion the two are equal only when considered over an integral number of cycles.

Case 2: Constant acceleration–radiation without "radiation reaction"

An uncharged particle acted upon by a force of constant direction and amplitude will have an acceleration of constant direction and amplitude. Is this true for an electron and its additional self-forces? The general solution of Eq. 8.23 for the case of a constant applied force, $\vec{F}_{ext}$, is easily obtained as[8]

$$\dot{\vec{v}}(t) = \vec{A}\exp\left(\frac{t}{T}\right) + \frac{\vec{F}_{ext}}{M} \qquad (8.26)$$

where, as usual, we use $T = q^2/6\pi\varepsilon_o c^3 M$. Thus, in the general case, where $\vec{A} \neq 0$, the answer is–oddly–no, since the acceleration, starting at its initial value of $\vec{A}+\vec{F}_{ext}/M$, exponentially increases or decreases in time depending on the sign of $\vec{A}$. This general result is clearly unphysical as it blatantly violates the conservation laws. The trivial-yet physical-case in which $\vec{A} = 0$, represents a constant acceleration solution, $\dot{\vec{v}} = \vec{F}_{ext}/M$, that is easily seen to satisfy Eq. 8.23. However, with a charged particle accelerating at a constant rate, by Larmor's power formula (Eq. 8.13), it must be radiating and if it's radiating there must be an associated reaction force. And thus, even in the $\vec{A} = 0$ case, we seem to have a dilemma because $\vec{F}_{rad}$, the radiation reaction force, vanishes with constant acceleration. What is going on? To resolve this apparent violation of the conservation laws we need to recall that the power loss $\vec{F}_{rad} \cdot \vec{v}$ is actually two terms (the 2nd and 3rd terms in Eq. 8.16) where one of the terms is the nonconservative loss due to radiation (Larmor power formula) and the other is conservative and represents a correction to the energy exchange with the velocity fields. This means that there could be a coincidence of nonzero acceleration (and thus radiation) and a zero net radiation reaction power loss (i.e., $\vec{F}_{rad} \cdot \vec{v} = 0$) if the two power terms had equal and opposite values. The corresponding forces, $\vec{F}_{middle}$ and $\vec{F}_{lrmr}$ in Eq. 8.17, would then also be necessarily equal and opposite. This is indeed the case for a charge under constant acceleration: apparently, while $\vec{F}_{lrmr}$ is doing negative work on the charge to account for the radiation, $\vec{F}_{middle}$ is doing an equal amount of positive work with the effective result being a continuous transfer of energy from the velocity fields to the radiation fields.

[8] We can rewrite Eq. 8.23

$$\frac{d\dot{\vec{v}}}{dt} = \frac{1}{T}\left(\dot{\vec{v}} - \frac{\vec{F}_{ext}}{M}\right)$$

where $T = q^2/6\pi\varepsilon_o c^3 M$. Rearranging and making the substitution $\vec{\xi} = \dot{\vec{v}} - \vec{F}_{ext}/M$,

$$\frac{d\vec{\xi}}{\vec{\xi}} = \frac{1}{T}dt$$

Solving for $\vec{\xi}$ and replacing the original variable we get

$$\dot{\vec{v}} = \vec{A}e^{\frac{t}{T}} + \frac{\vec{F}_{ext}}{M}.$$

General solution: Runaway condition or pre-acceleration

In general, the acceleration of a charged particle is neither sinusoidal nor constant. And, unlike for true damping forces such as the collisional damping force, the radiation reaction force, $\vec{F}_{rad}$, does not, in general, oppose the velocity. This leads to surprising and unphysical conditions. We have already discussed this odd behavior before in the case of a constant applied force: for solutions with $\vec{A} > 0$, the amplitude of the acceleration increases exponentially with time. This is an example of a "runaway" condition. In fact, as indicated in Eq. 8.26, this condition is independent of the external force so that even with $\vec{F}_{ext} = 0$, there will be a runaway condition unless the coefficient, $\vec{A}$, is set to zero. A runaway condition exists when the $\ddot{v}v$ term in Eq. 8.15 is greater than zero–corresponding to positive work done on the charge by $\vec{F}_{rad}$. This, in turn, means that $\vec{F}_{middle}$, the conservative part of $\vec{F}_{rad}$, is completely responsible for this condition since $\vec{F}_{lrmr}$ only does negative work on the charge. Thus, to obtain a physically consistent solution it seems that $\vec{A}$ must be generally set to zero. The consequence of this, however, is even more unbelievable: pre-acceleration or the violation of causality.

For a more definite description of the runaway condition and its alternative, pre-acceleration, we now find the solution to the general equation of motion for an arbitrary time-dependent external force term. Equation 8.23 can be rewritten in the form of a linear first order D.E.,

$$\frac{d\dot{\vec{v}}}{dt} - \frac{1}{T}\dot{\vec{v}} = -\frac{\vec{F}_{ext}(t)}{MT} \tag{8.27}$$

for which, using standard methods, we get the solution in terms of a definite integral[9]

$$\dot{\vec{v}}(t_2)\exp\left(-\frac{t_2}{T}\right) - \dot{\vec{v}}(t_1)\exp\left(-\frac{t_1}{T}\right) = -\frac{1}{MT}\int_{t_1}^{t_2} \exp\left(-\frac{t'}{T}\right)\vec{F}_{ext}(t')\,dt' \tag{8.28}$$

Next, multiplying by $\exp\left(\frac{t_2}{T}\right)$, rearranging and letting $t_2 \to t$, we obtain the form

$$\dot{\vec{v}}(t) = \dot{\vec{v}}(t_1)\exp\left(\frac{t-t_1}{T}\right) - \frac{1}{MT}\int_{t_1}^{t} \exp\left(-\frac{(t'-t)}{T}\right)\vec{F}_{ext}(t')\,dt' \tag{8.29}$$

[9] The D.E. integrating factor, μ, for Eq. 8.27 is given by

$$\mu(t) = \exp\left(-\int \frac{1}{T}d5t\right) = \exp\left(-\frac{t}{T}\right)$$

which is multiplied with Eq. 8.27 to yield,

$$\frac{d\dot{\vec{v}}}{dt}\exp\left(-\frac{t}{T}\right) - \frac{1}{T}\exp\left(-\frac{t}{T}\right)\dot{\vec{v}} = -\frac{\exp\left(-\frac{t}{T}\right)}{MT}\vec{F}_{ext}(t)$$

but the left-hand side is a total differential so we can write

$$\frac{d}{dt}\left[\dot{\vec{v}}\exp\left(-\frac{t}{T}\right)\right] = -\frac{\exp\left(-\frac{t}{T}\right)}{MT}\vec{F}_{ext}(t)$$

and the general definite integral is

$$\dot{\vec{v}}(t_2)\exp\left(-\frac{t_2}{T}\right) - \dot{\vec{v}}(t_1)\exp\left(-\frac{t_1}{T}\right) = -\frac{1}{MT}\int_{t_1}^{t_2}\exp\left(-\frac{t'}{T}\right)\vec{F}_{ext}(t')\,dt'.$$

and so we see in this general solution the same exponential acceleration growth in time that is, again, independent of $\vec{F}_{ext}$. And just as we set the coefficient $\vec{A}$ to zero, as in the previous case of constant F_{ext}, to avoid this runaway condition, we now, equivalently, set $t_1 \to \infty$ to get

$$\dot{\vec{v}}(t) = \frac{1}{MT}\int_t^{\infty} \exp\left(-\frac{(t'-t)}{T}\right)\vec{F}_{ext}(t')\,dt' \tag{8.30}$$

where the limits of integration have been reversed. So this is a generalization of case 2–the constant acceleration case–and it does indeed reduce to the constant acceleration solution, $\dot{\vec{v}} = \vec{F}_{ext}/M$, for a constant $\vec{F}_{ext}$. However, now we see an even more counterintuitive result.[10] Evidently, the present acceleration, $\dot{\vec{v}}$, at time t depends on what the (now time-dependent) external force does at future times, t'. Because of the exponential weighting, this effect is valid only within a time interval on the order of $T \simeq 10^{-23}s$. This clearly acausal condition is known as "pre-acceleration." Thus it seems, classically at least, we are stuck with one or the other of these un-physical conditions of runaway or pre-acceleration–a violation of either energy conservation or causality. Indeed, this problem persists even in the relativistic formulation for which the starting point is Lienard's generalization of the Larmor power formula.

It is interesting to note that were it not for the requirement of a point charge of zero radius, these failures could all be avoided. It can be shown[11] that for solutions to the general equations of motion using the exact analytical self-force form of Eq. 8.12 (i.e., where all the terms are kept in the Lorentz calculation), and with a charge radius larger than the classical radius ($a > a_{class}$), there is no problem with runaway or pre-acceleration.

8.3.2 Landau–Lifshitz approximation

An alternative approach that overcomes the difficulties of the Abraham–Lorentz formula was introduced by Landau and Lifshitz as an approximation to the Abraham–Lorentz formula[12]. Essentially, the Landau–Lifshitz equation approximates the radiation reaction, $\vec{F}_{rad}$ (Eq. 8.20), as proportional to the time rate of change of the applied force, $\dot{\vec{F}}_{ext}$:

$$M\dot{\vec{v}} \approx T\dot{\vec{F}}_{ext} + \vec{F}_{ext} \tag{8.31}$$

More specifically, referring to Eq. 8.23, the time rate of change of acceleration, $\ddot{\vec{v}}$, and thus the radiation reaction, $\vec{F}_{rad}$, is approximated

[10] For a constant external force, Eq. 8.30 can be expressed

$$\dot{\vec{v}}(t) = \frac{\vec{F}_{ext}}{MT}\int_t^{\infty} \exp\left(-\frac{(t'-t)}{T}\right)dt' = \frac{\vec{F}_{ext}}{MT}\exp\left(\frac{t}{T}\right)\int_t^{\infty}\exp\left(-\frac{t'}{T}\right)dt'$$

And the simple integral evaluates as

$$\int_t^{\infty}\exp\left(-\frac{t'}{T}\right)dt' = -T\left[0-\exp\left(-\frac{t}{T}\right)\right] = T\exp\left(-\frac{t}{T}\right)$$

so,

$$\dot{\vec{v}}(t) = \frac{\vec{F}_{ext}}{M}.$$

as before.

[11] Griffiths, D. J. Proctor, T. C., and Schroeter, D. F. Abraham–Lorentz versus Landau–Lifshitz, *American Journal of Physics* 78, 391 (2010).

[12] Landau, L. D. and Lifshitz, E. M. *The Classical Theory of Fields*. Pergamon, Oxford (1971), Sec. 75.

to be due solely to the time rate of change of the applied force, $\dot{\vec{F}}_{ext}$, ignoring that small additional part of $\ddot{\vec{v}}$ due to the time rate of change of the radiation reaction itself, $\dot{\vec{F}}_{rad} \simeq \dddot{\vec{v}}$. Thus, the Landau–Lifshitz equation prevents the radiation reaction from feeding back on itself and generating a runaway condition. So the validity of the Landau–Lifshitz approximation depends on the requirement that $\left|\dot{\vec{F}}_{rad}\right| / \left|\dot{\vec{F}}_{ext}\right| \ll 1$. Consider the 1D example of a sinusoidal external force, $F_{ext} = F_o \sin \omega t$. The time rate of change of this gives the external force contribution, $\ddot{v}_{ext}$, to $\ddot{v}$, and thus to F_{rad}:

$$\dot{F}_{ext} = \omega F_o \cos \omega t = M\ddot{v}_{ext} \tag{8.32}$$

If we assume, for the moment, that the Landau–Lifshitz approximation is valid to first order, and that this contribution dominates over the first order radiation reaction self-contribution, $\dot{F}_{rad} = M\ddot{v}_{rad}$, in the representation of F_{rad}, then we can express the second order self-contribution in terms of $\dddot{v}_{ext}$ as $\dot{F}_{rad} = MT\dddot{v}_{ext}$ or, referring to Eq. 8.32,

$$\dot{F}_{rad} = -T\omega^2 F_o \sin \omega t \tag{8.33}$$

Now, applying the condition for validity to the second term,

$$\left|\dot{\vec{F}}_{rad}\right| / \left|\dot{\vec{F}}_{ext}\right| = \frac{T\omega^2 F_o}{\omega F_o} = \frac{2\pi T}{T_o} \ll 1 \tag{8.34}$$

where T_o is the period of oscillation of the system. Equation 8.34 then suggests that the regime of validity for the Landau–Lifshitz approximation of the radiation reaction for an oscillating charge is for oscillations whose period is much longer than the characteristic time, T. In other words, It's a good approximation for motion that does not change significantly in times on the order of T. With all the references to the characteristic time, it seems rather fundamental and so requires some elaboration.

8.3.3 Characteristic time

Within our discussion we have been representing the constant $q^2/6\pi\varepsilon_o c^3 M$ with the symbol T. This is not merely a matter of convenience since T, in fact, represents a characteristic time over which the effects of the self-force play a significant role and for which, if $t \gg T$, these effects are negligible. For most cases encountered in electromagnetism, these effects are negligible and this explains why we have been able to ignore self-forces up to now. Indeed, for an electron, $T = 6.26 \times 10^{-24}$ seconds. We can estimate the timescale of significant motion necessary for non-negligible self-forces (and thus T) through a comparison of radiation energy lost to system energy. In fact, we have already done this for sinusoidal motion in a previous section when we found that as $T_o \to T$, self-forces begin to have a significant effect. We can do something similar for constant acceleration of a charge from rest. In this case, the comparison is of the energy radiated to the kinetic energy of the charge. At early times, the radiation energy is much greater than the kinetic energy but at some point, say T, the kinetic and radiated energies are equal. Subsequently, the radiation loss becomes fractionally less and less significant. This can be quantified since for constant acceleration, Larmor's power formula, Eq. 8.13, gives us the energy radiated in time t,

$$\mathscr{E}_{lrmr} = \frac{q^2}{6\pi\varepsilon_o c^3}\dot{v}^2 t \tag{8.35}$$

and since the charge started from rest, the kinetic energy after time t is

$$\mathscr{E}_{kin} = \frac{1}{2}M(\dot{v}t)^2 \tag{8.36}$$

With the radiation reaction self-force significant when $\mathscr{E}_{lrmr} \gg \mathscr{E}_{kin}$, Eqs. 8.35 and 8.36 yield the associated timescale

$$t \ll 2\frac{q^2}{6\pi\varepsilon_o c^3 M} \sim T \tag{8.37}$$

so just as for sinusoidal motion where radiation reaction is significant for $T_0 \ll T$, here we find that for constant linear acceleration from rest the radiation reaction is significant for early times, $t \ll T$.

8.4 The 4/3 problem, instability, and relativity

We come now to three seemingly unrelated issues with the presently obtained Abraham–Lorentz model: the so-called "4/3 problem," the instability of the charge distribution, and the lack of relativistic covariance of the model. We will in this section see that these are indeed intimately related issues that are simultaneously resolvable through the inclusion of non-electromagnetic forces or "Poincare stresses."

The first is a very considerable issue that has, up to now, only been hinted at. In Section 8.3 we identified the acceleration coefficient in the first term of the self-force of an extended shell of charge q with radius a (Eq. 8.10) as the electromagnetic mass:

$$M_{em} = \frac{q^2}{6\pi\varepsilon_o ac^2} \tag{8.38}$$

A further confirmation came through comparison of this coefficient with the electromagnetic mass obtained by field momentum density integration of a non-relativistically moving charge. The problem arises, however, when we calculate the electromagnetic mass of the charge in a co-moving frame. Consider the example of a stationary spherical shell of charge q with radius a: first, we consider the simple calculation of the electromagnetic *energy* of this charge distribution by integration of the associated electromagnetic energy density over all space where for $r > a$, $\vec{E} = \frac{q}{4\pi\varepsilon_o r^2}\hat{r}$. This is easily found to be $\mathscr{E}_{em}^{stat} = q^2/8\pi\varepsilon_o a$. Comparison with Eq. 8.38 then yields the energy-mass relationship: $\mathscr{E}_{em}^{stat} = \frac{3}{4}M_{em}c^2$. While on the one hand, this result is historically relevant because it pointed to the general c^2 relationship between mass and energy, indicating again the inherent consistency of classical electrodynamics with special relativity. On the other hand, relativity demands the energy-mass relationship: $\mathscr{E}_{em}^{stat} = M_{em}c^2$. In other words, the statically calculated electromagnetic energy, according to relativity, predicts the electromagnetic mass to be

$$M_{em}^{stat} = \frac{\mathscr{E}_{em}^{stat}}{c^2} = \frac{q^2}{8\pi\varepsilon_o c^2 a} \tag{8.39}$$

so we see that the dynamically calculated mass of Eq. 8.38 is larger than the statically calculated mass of Eq. 8.39 by a factor of 4/3.[13]

$$\frac{4}{3}M_{em}^{stat} = M_{em} \tag{8.40}$$

This "4/3 problem" was, in the formative days of special relativity, a source of much confusion.

In addition, there is the issue of stability. Because the Abraham–Lorentz (A-L) model is purely electromagnetic, a stationary charge distribution, for example, with only electric forces acting between all the pieces, is highly unstable and will quickly explode from mutual repulsion. It was Poincare who, in 1905, resolved this issue by pointing out that there must exist non-electromagnetic attractive forces acting to hold the distribution together. Such binding forces, or "Poincare

[13] Remember that although, relative to the observer, M_{em} corresponds to a moving charge and M_{em}^{stat} is stationary, the velocities considered so far are non-relativistic so that $M_{em} > M_{em}^{stat}$ has nothing to do with a relativistic mass increase.

stresses," whatever their origin, then not only provide stability but also modify the work required to assemble the pieces of charge from infinity. In other words, the fields of these binding forces represent a non-electromagnetic correction to the energy/mass of the charge distribution. The total "field" mass is therefore the electromagnetic field mass plus the "stress" field mass.[14] And, indeed, we expect that just like the electric forces between the pieces of charge, under acceleration these stresses between pieces of charge also become unbalanced. That is to say, associated with this "stress" field component of mass are components of self-force and momentum. We will see that the addition of Poincare stresses and the accompanying modifications to the total energy, momentum, and self-force of the Abraham–Lorentz charge distribution can effectively resolve the 4/3 problem. We say "effectively" because with this modification the static and dynamic calculations of the *total* masses (respectively, from the total energy and total momentum or total self-force) will agree but the disagreement of the static and dynamic calculations of the individual mass contributions (i.e., electromagnetic or stress alone) will persist. Thus, despite Lorentz's hope that the mass of an electron be purely electromagnetic in nature, the need for inclusion of Poincare stresses showed that at least part of the mass was of a different origin.

Finally, there is the important issue of relativistic covariance. Up to now, our discussion of field reactions in this chapter has been non-relativistic so the fact that the model is not Lorentz covariant has not been appreciated. However, the addition of the stress component solves this issue as well and makes the theory relativistically correct. To understand this, we must first be clear about what is required for relativistic covariance of this model. Since the A-L model is that of a particle, albeit with an attached charge, we require that its total energy/mass and total momentum transform like those of an uncharged point particle–that is, like the quantities of a 4-vector. Within the A-L model, the charge distribution's energy/mass and momentum are purely electromagnetic and, as shown in Chapters 5 and 7, given such a configuration of charges and currents and their resulting fields, the electromagnetic energy/mass and momentum densities (and current densities) are mathematically specified at each point in spacetime by the 16 elements of the symmetric electromagnetic stress-energy tensor $T^{\alpha\beta}$ that, as in Eq. 5.117, may be written in terms of the electromagnetic field tensor elements, $F^{\alpha\kappa}$,

$$T^{\alpha\beta} = -\frac{1}{\mu_0}\left[F^{\alpha\kappa}F^{\beta}{}_{\kappa} - \frac{1}{4}g^{\alpha\beta}F^{\mu\nu}F_{\mu\nu}\right] \tag{8.41}$$

[14] With a proton, for example, which is not a point charge and has internal structure, the Poincare stresses are represented by the strong (gluon mediated) force and so part of the proton mass is due to the strong force.

or less abstractly, in its matrix form,

$$T^{\alpha\beta} = \begin{pmatrix} u & cg_x & cg_y & cg_z \\ S_x/c & \sigma_{xx} & \sigma_{xy} & \sigma_{xz} \\ S_y/c & \sigma_{yx} & \sigma_{yy} & \sigma_{yz} \\ S_z/c & \sigma_{zx} & \sigma_{zy} & \sigma_{zz} \end{pmatrix} \tag{8.42}$$

where $u =$ energy density, $\vec{g}$ =momentum density vector, $\vec{S}$ =energy current density vector, and $\overleftrightarrow{\sigma}$ = Maxwell stress tensor (of momentum current densities). Thus, for the A-L model to be relativistically covariant we require that the four quantities $G^\beta = \int T^{0\beta} d^3x$ transform as a 4-vector,

$$\begin{aligned}\left[G^0, G^1, G^2, G^3\right] &= \left[\int T^{00} d^3x, \int T^{01} d^3x, \int T^{02} d^3x, \int T^{03} d^3x\right] \\ &= \left[\mathscr{E}, c\vec{P}\right] \end{aligned} \tag{8.43}$$

Now, it can be shown[15] that these volume integrals of the $T^{0\beta}$ components transform as a 4-vector only if the 4-divergence of the rank 2 tensor, $T^{\alpha\beta}$, vanishes everywhere ($\partial_\alpha T^{\alpha\beta} = 0$) and the tensor components are non-zero only within a finite volume of space. And since, as we found in Section 7.4.7, the 4-divergence for the symmetric stress-energy tensor $T^{\alpha\beta}$ representing electromagnetic fields in the presence of charge and current densities, $\mathcal{J}_\kappa = \left[c\rho, c\vec{v}\right]$, is not zero,

$$\partial_\alpha T^{\alpha\beta} = (\partial_0, \partial_1, \partial_2, \partial_3) \begin{pmatrix} u & cg_x & cg_y & cg_z \\ S_x/c & \sigma_{xx} & \sigma_{xy} & \sigma_{xz} \\ S_y/c & \sigma_{yx} & \sigma_{yy} & \sigma_{yz} \\ S_z/c & \sigma_{zx} & \sigma_{zy} & \sigma_{zz} \end{pmatrix} = \mathcal{J}_\kappa F^{\kappa\beta} \tag{8.44}$$

The quantities $\int T^{0\beta} d^3x$ for the A-L charged particle model will not transform as a 4-vector. This is not to say that the energy and momentum continuity equations represented by Eq. 8.44,

$$\partial_\alpha T^{\alpha 0} = \mathcal{J}_\kappa F^{\kappa 0} \Rightarrow \frac{\partial u}{\partial t} + \nabla \cdot \vec{S} = -\vec{\mathcal{J}} \cdot \vec{E} \tag{8.45}$$

$$\partial_\alpha T^{\alpha(1,2,3)} = \mathcal{J}_\kappa F^{\kappa(1,2,3)} \Rightarrow \frac{\partial \vec{g}}{\partial t} + \nabla \cdot \overleftrightarrow{\sigma} = -\rho\vec{E} - \vec{\mathcal{J}} \times \vec{B} \tag{8.46}$$

are not Lorentz covariant. Indeed, $\mathcal{J}_\kappa F^{\kappa\beta} = f^\beta = [\vec{\mathcal{J}} \cdot \vec{E}, \rho\vec{E} + \vec{\mathcal{J}} \times \vec{B}]$ is the relativistic Lorentz force density 4-vector and Eq. 8.44 is correspondingly, Lorentz covariant. Rather, the field energy and

15 Panofsky and Phillips 2nd edn, pages 309–310.

field momentum of the A-L model do not transform as they should and as they do for an uncharged point particle (i.e., as a 4-vector). If we think about it, it must be this way since the A-L model of a charged particle assumes the energy/mass and momentum is purely electromagnetic in nature and will remain that way for all time (i.e., it was incorrectly assumed to be stable). This requires that there never be any transfer to "mechanical" energy or momentum which would correspond to a nonzero Lorentz force density $\mathcal{J}_\kappa F^{\kappa\beta}$ in Eq. 8.44. In fact, the A-L model is not stable without the binding Poincare stresses included and so there will necessarily be work done on and momentum transferred to the charges as they fly apart so that $\mathcal{J}_\kappa F^{\kappa\beta}$ is necessarily nonzero. Thus, the addition of Poincare stress, which is in the mathematical form of an additional non-electromagnetic stress-energy tensor, $T_P^{\alpha\beta}$, to the electromagnetic stress-energy tensor will stabilize the system so that the electric forces on the charges are everywhere balanced by an opposing non-electric force[16] and the energy and momentum, now completely contained by the electromagnetic and stress fields, is never converted to mechanical energy and momentum. This additional Poincare stress component, then, effectively cancels out the Lorentz force density term, $\mathcal{J}_\kappa F^{\kappa\beta}$, of Eq. 8.44 and the 4-divergence of the resulting stress-energy tensor, $S^{\alpha\beta} = T^{\alpha\beta} + T_P^{\alpha\beta}$, vanishes ($\partial_\alpha S^{\alpha\beta} = 0$). This condition of zero 4-divergence, or equivalently, the vanishing - in the rest frame - of the volume integrated components of the 3×3 stress subtensor[17]

$$\int S^{ij} d^3x = 0 \quad \text{for } i,j = 1,2,3$$

then, means the four quantities of total energy and momentum transform as a 4-vector and the stabilized charge distribution is relativistically covariant.

Now consider, in contrast, the example of a spatially finite pulse of pure electromagnetic energy. In this source-free case, the right side of Eq. 8.44 vanishes, $\partial_\alpha T^{\alpha\beta} = 0$, and so the quantities $\int T^{0\beta} d^3x$ transform as a 4-vector as expressed in Eq. 8.43. This means that the total energy and momentum of the pulse (i.e., within the finite volume) transforms as an uncharged point particle–a photon of electromagnetic energy is covariant. Furthermore, it can be shown that this "particle" has zero rest mass by noting that the Lorentz scalar created by the scalar product of the 4-momentum of this "particle" with itself vanishes.

$$G^\beta G_\beta = \mathscr{E}^2 - c^2 P^2 = 0 \tag{8.47}$$

[16] The imbalance of forces referred to here must not be confused with the imbalance of forces, referred to earlier, which give rise to the self-forces. In the present case, we are referring to the sum of forces on a charge element at a point whereas in the earlier case we were referring to the sum of (electric only) forces over the whole charge distribution of the extended electron model.

[17] The S^{ij} represent flow from field momentum density to mechanical momentum density so any nonzero $\int S^{ij} d^3x$ would indicate an unbalanced field stress and thus an unstable charge distribution (such as an exploding shell of charge).

All this combined with the light speed propagation of the pulse is consistent with the properties of a photon.

So we have seen that the addition of a non-electromagnetic "Poincare" stress component to the purely electromagnetic A-L model simultaneously resolves the issues of stability and covariance. We also argued that it resolves the inconsistency between the static and dynamic electromagnetic mass calculations by providing additional components of "stress" mass to both calculations. More precisely, with the "Poincare stress" stress-energy tensor of the extended charged particle added to the pure electromagnetic stress-energy tensor, the static and dynamic electromagnetic-only mass calculations still disagree but when these energy-based and momentum-based (or self-energy-based) calculations include the "Poincare stress" components the resulting total masses are equal and that total mass is covariant. To see an example of this, we summarize the results of a model, due to Schwinger[18] and reviewed in detail within Jackson, implementing a specific Poincare energy-stress tensor. Schwinger first defines the electromagnetic 4-potential of this stabilized charged particle as

$$A^{\alpha} = \frac{e}{c} f(z) \left[U^0, U^1, U^2, U^3 \right] \tag{8.48}$$

where U^{α} is the 4-velocity and $f(z)$ is an arbitrary function of the invariant $z = -x^{\beta} x_{\beta} + \frac{1}{c^2} \left(U^{\beta} x_{\beta} \right)^2$ with $f(z) \rightarrow 1/r$ as $r \rightarrow \infty$. From this he obtains, in turn, specific functions for the field strength tensor elements, $F^{\alpha\beta}$, and the current density 4-vector, $\mathcal{J}_{\alpha}$. He then uses these within Eq. 8.44 to find an expression for the 4-divergence of the associated electromagnetic symmetric stress-energy tensor. To cancel the resulting non-zero Lorentz force density 4-vector, he chooses a Poincare stress tensor

$$T_P^{\alpha\beta} = g^{\alpha\beta} t(z) \tag{8.49}$$

where $t'(z) = -\frac{e^2}{2\pi} \left[3 \left(f' \right)^2 + 2 z f' f'' \right]$, such that the 4-divergence of the total stress-energy tensor vanishes,

$$\partial_{\alpha} \left(T^{\alpha\beta} + T_P^{\alpha\beta} \right) = \partial_{\alpha} S^{\alpha\beta} = 0 \tag{8.50}$$

Thus, the quantities $\int S^{0\beta} d^3x$ transform as a 4-vector and this stabilized charged particle is covariant. The energy-based mass calculations for the electromagnetic and Poincare stress components result in the equations,

[18] Schwinger, J. *Found. Phys.* 13, 373 (1983).

$$M_{em}^{energy} = M_{em}^{stat}\left(\frac{4}{3}\gamma - \frac{1}{3\gamma}\right) \quad M_{p}^{energy} = M_{em}^{stat}\left(\frac{1}{3\gamma}\right) \tag{8.51}$$

while the momentum-based mass calculations for the electromagnetic and Poincare stress components result in the equations,

$$M_{em}^{mom} = M_{em}^{stat}\left(\frac{4}{3}\gamma\right) \quad M_{p}^{mom} = 0 \tag{8.52}$$

where γ is the relativistic factor and M_{em}^{stat} is obtained from the rest frame volume integral of the electromagnetic energy density,

$$\mathscr{E}_{em}^{stat} = M_{em}^{stat}c^2 = \int T^{00}d^3x \tag{8.53}$$

which leads to Eq. 8.39. We first note that for non-relativistic velocities ($\gamma \simeq 1$), the Eqs. 8.51 and 8.52 reduce to

$$M_{em}^{energy} \simeq M_{em}^{stat} \quad M_{p}^{energy} = \frac{1}{3}M_{em}^{stat} \tag{8.54}$$

and

$$M_{em}^{mom} = \frac{4}{3}M_{em}^{stat} \quad M_{p}^{mom} = 0 \tag{8.55}$$

so that the energy-based and momentum-based total nonrelativistic masses M_{tot}^{nonrel} are now in agreement

$$M_{tot}^{nonrel} = M_{em}^{energy} + M_{p}^{energy} = M_{em}^{mom} + M_{p}^{mom} = \frac{4}{3}M_{em}^{stat} \tag{8.56}$$

However, as cautioned earlier, the disagreement between the energy-based and momentum-based electromagnetic masses is still not resolved: $M_{em}^{mom} = \frac{4}{3}M_{em}^{energy}$. Thus, we conclude that the total electromagnetic field of a charge does not follow the same laws relating energy and momentum as does an uncharged point particle. Finally, in considering the total relativistic mass, the sum of terms in Eq. 8.51,

$$\begin{aligned} M_{tot}^{rel} &= M_{em}^{energy} + M_{p}^{energy} \\ &= M_{em}^{stat}\left(\frac{4}{3}\gamma - \frac{1}{3\gamma}\right) + M_{em}^{stat}\left(\frac{1}{3\gamma}\right) = \frac{4}{3}\gamma M_{em}^{stat} \end{aligned} \tag{8.57}$$

and we see that it transforms as a mass should.

Next, we consider something that is far more problematic than the preceding issues. Indeed, as mentioned at the start of this chapter, it is "a curse" that shakes the whole foundation of classical electromagnetism and extends through to QED where it is solved but in a somewhat unsatisfactory way.

8.5 Infinite mass of the Abraham–Lorentz model

Previously, we commented that the unphysical runaway and pre-acceleration conditions of the A-L model could be avoided if the model assumed a charge radius larger than the classical radius ($a > a_{class}$) while keeping all the terms in the Lorentz calculation of the self-force (the series expansion of Eq. 8.12). In the last section, we further found that inclusion of binding forces correct the model to make it stable, covariant, and mass-consistent. Thus, if we modify the Lorentz calculation of the self-force to include these "Poincare stresses," and keep all the terms while assuming $a > a_{class}$, it seems we could have a very robust theory of a stable, extended charged particle and its interactions with the fields. The problem is that, as before, we want to use this (now modified) A-L model to describe a zero radius fundamental particle[19] so we must contend with the conditions of runaway and pre-acceleration. However, there is a far more serious problem that plagues all zero-radius charged particle models considered so far: the problem of divergent mass. If we take the radius a to zero in the model, all the higher order terms of the self-force vanish, and the second term, the radiation reaction self-force, is unaffected but the first term, the field mass term, scales as $1/a$ and so diverges to infinity.

Many attempts have been made to overcome this problem with most concepts involving a modification to either Maxwell's equations or the Lorentz force law. However, we will here briefly discuss just two: a concept, suggested by Paul Dirac, which includes the effect of "advanced" waves, and a closely related but more radical concept, the so-called "absorber theory," proposed by John Wheeler and Richard Feynman.

Recalling that Maxwell's field equations of electrodynamics are symmetric in time and thus predict not only "retarded wave" solutions but also the so-called "advanced wave" solutions,[20] Dirac submitted the following idea: *Both the retarded and the advanced wave solutions should be used in the Lorentz calculation of the self-force* (Eq. 8.11).

To obtain the advanced component of the self-force, $\vec{F}^{adv}_{self}$, we replace the retarded time, $t_{ret} = t - R/c$, with the advanced time, $t_{adv} = t + R/c$, within the calculations starting at Eq. 8.1. In particular,

[19] As we have pointed out, a fundamental particle, by definition, can have no internal structure and thus must exist as a point ($a = 0$).

[20] Advanced wave solutions to Maxwell's equations are spherical waves which would appear to an observer as a regular expanding spherical wave but played backwards in time. They are generally rejected based on the lack of experimental evidence. In the same way that we never see a spilt glass of milk reassemble itself and unspill, we, for example, never see a spherical wave converging on a charge and shaking it. That is, advanced waves are physically possible but are generally considered to be highly unlikely.

the Taylor expansions of the charge shell velocity and acceleration functions $\vec{v}(t_{ret})$ and $\dot{\vec{v}}(t_{ret})$, in powers of R/c, become,

$$\vec{v}(t_{adv}) = \dot{\vec{v}}\frac{R}{c} + \ddot{\vec{v}}\frac{1}{2}\left(\frac{R}{c}\right)^2 - \cdots \tag{8.58}$$

$$\dot{\vec{v}}(t_{adv}) = \dot{\vec{v}} + \ddot{\vec{v}}\left(\frac{R}{c}\right) + \cdots \tag{8.59}$$

with the final result

$$\vec{F}^{adv}_{self} \simeq \frac{q^2}{4\pi\varepsilon_o}\left(-\frac{2}{3}\frac{\dot{\vec{v}}}{ac^2} - \frac{2\ddot{\vec{v}}}{3c^3} - \alpha\frac{\dddot{\vec{v}}a}{c^4} - \ldots\right) \tag{8.60}$$

whereas the retarded component (Eq. 8.11) is

$$\vec{F}^{ret}_{self} \simeq \frac{q^2}{4\pi\varepsilon_o}\left(-\frac{2}{3}\frac{\dot{\vec{v}}}{ac^2} + \frac{2\ddot{\vec{v}}}{3c^3} - \alpha\frac{\dddot{\vec{v}}a}{c^4} + \ldots\right) \tag{8.61}$$

where we see that the signs of all the even time derivatives of $v(t)$ have been switched. Dirac then specified the total self-force of an extended charged particle to be

$$\vec{F}_{self} = \frac{1}{2}\left(\vec{F}^{ret}_{self} - \vec{F}^{adv}_{self}\right) = \frac{q^2}{4\pi\varepsilon_o}\left(\frac{2\ddot{\vec{v}}}{3c^3} + \text{even derivatives}\right) \tag{8.62}$$

And since all the higher order (even) time derivatives are in higher powers of a, they also vanish in the limit of a point particle,

$$\vec{F}_{self} = \frac{q^2}{4\pi\varepsilon_o}\left(\frac{2\ddot{\vec{v}}}{3c^3}\right) \tag{8.63}$$

Thus with this characteristically brilliant stroke, Dirac used the (seemingly acausal) effect of the advanced component self-force to effectively cancel the offensively divergent inertial force but yet keep the absolutely necessary radiation reaction force. The electromagnetic mass thus vanishes in this scheme. One problem with the Dirac formulation is that, other than eliminating the divergent mass problem, there is no reason for the choice of Eq. 8.62.

A second, related scheme which also includes the effect of advanced waves is known as the Feynman–Wheeler absorber theory. In this scheme, unlike in Dirac's, the electron *simply does not interact with itself.* While it is clear that this alone resolves the infinite mass problem associated with the first self-force term, it also completely eliminates all other self-force terms including the radiation reaction, $\vec{F}_{rad}$, which apparently is essential to the process of radiation. This means $\vec{F}_{rad}$ must then be provided by other charges–indeed, according to the theory, all other causally connected charges in the universe! At first sight, it would seem that the time delay for any effects from the surrounding charges would not allow for an $\vec{F}_{rad}$ that acts at the charge simultaneously with its acceleration. However, if advanced waves are allowed this is precisely what, according to the theory, occurs: As schematically shown in Fig. 8.3, a charge undergoes a pulse of acceleration at time t and radiation is emitted. At time $t' = t + R/c$, this pulse of radiation interacts with and accelerates another charge a distance R away that then radiates both a retarded wave forward in time (not shown) and an advanced wave backward in time. The advanced wave then interacts with the first charge at time $t = t' - R/c$ during its acceleration pulse thus effectively providing part of the radiation reaction self-force. In addition to this, during the pulse of acceleration

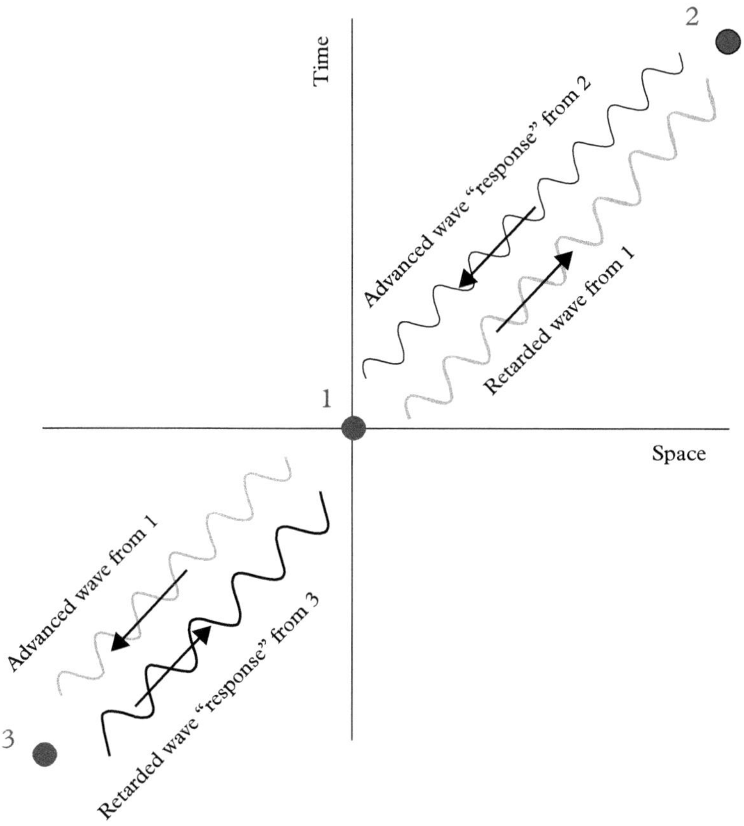

Fig. 8.3 *Feynman–Wheeler concept of the interaction of all charges causally, and anti-causally, connected to the accelerated charge (1). Charge (2) receives a retarded wave from charge (1) and responds with an advanced wave. Charge (3) receives an advanced wave from charge (1) and responds with a retarded wave.*

the first charge also radiates an advanced wave into its past where it will again interact with all charges in the universe (as indicated in the left quadrant of Fig. 8.3). And again, they will then radiate both retarded and advanced waves but this time it will be the retarded waves from the other charges that react back on the first charge during its initial acceleration. The accelerating charge interacts in this way with all the charges causally (and anti-causally) connected in the universe so that their combined response provides the total radiation reaction force which reacts on the charge at the exact instant, t, that it begins to accelerate. So, there are ultimately two components of self-force from the rest of the universe: one from the all the advanced wave responses (from the future) and one from the all the retarded wave responses (from the past). In terms of an equation, we have

$$\vec{F}_{rad} = \frac{1}{2}\left(\vec{F}^{ret}_{othercharges} + \vec{F}^{adv}_{othercharges}\right) \tag{8.64}$$

in contrast to Dirac's result of Eq. 8.63.

Exercises

(8.1) (A) Integrate the energy density of the field of a stationary shell of charge q of radius a over all space to obtain the field energy, $\mathscr{E}$, and electromagnetic mass, M_{em}. (B) Do the same for a solid sphere of charge q and radius a. (C) Assume the charge is now moving at a non-relativistic velocity, v. Again, integrate the energy density of the resulting fields. What can the increased field energy be attributed to?

(8.2) (A) Write the general equation of motion (EOM) for an accelerating shell of charge, Eq. 8.19, in terms of the analytical form of $\vec{F}_{self}$ as given by Eq. 8.12. This form is known as the Sommerfeld–Page equation of motion. (B) If we set the external force to zero and assume, as Lorentz did in 1904, that the mass is purely electromagnetic, what form does this EOM take? Describe the sort of motion, apart from the trivial one of constant velocity, this equation yields? Does the motion diminish with time? What does this say about the radiation fields?

(8.3) In Section 8.3.1, we considered the equation of motion for a sinusoidally driven charged particle with a restoring force. Extend this 1D problem to 2D by considering circular motion. Specifically, for a charge, q, in constant velocity circular motion with radius R: (A) Find $\vec{F}_{rad}$ (as defined in Eq. 8.18) in terms of the charge velocity, $\vec{v}$. What is the direction of $\vec{v}$ relative to the reaction force, $\vec{F}_{rad}$? (B) Referring to Eq. 8.17, show that, for this 2D problem, the velocity term ($\vec{F}_{middle}$) vanishes and the radiation term ($\vec{F}_{lrmr}$) is constant in time. What does this mean physically? (C) Show that the power supplied to maintain a constant velocity (i.e., counteract the radiation reaction) is equal to the power radiated according to the Larmor formula of Eq. 8.13.

(8.4) (A) In the text, we concluded that for a charge undergoing constant acceleration, the particle radiates despite a vanishing $\vec{F}_{rad}$ and that this can be explained by a continuous transfer of energy from the velocity fields to the radiation fields. Consider, now, a charge in a constant gravitational field, g, and evaluate the terms in Eq. 8.17 to show that $\vec{F}_{rad} = 0$. What is the power radiated? (B) By the principle of equivalence (POE), all objects in a gravitational field, charged or uncharged, follow the same accelerating trajectory. Thus our conclusion applies, right? However, consider an observer in free fall with the charge. Does he/she observe this radiation? Indeed, according to General Relativity, this observer and the falling charge are in a proper inertial frame while the Earthbound observers are not. In fact, it may be further argued that a charge at rest upon the Earth's surface should be radiating since, according to the POE, it's in an accelerating frame. Refer to the discussion in this reference and comment: The radiation of a uniformly accelerated charge is beyond the horizon: A simple derivation *American Journal of Physics* 74, 154 (2006).

(8.5) In Section 8.3.1, we obtained the Abraham–Lorentz corrected form (Eq. 8.23) of Newton's force equation for an electron:

$$M\dot{\vec{v}} = MT\ddot{\vec{v}} + \vec{F}_{ext}(t)$$

(A) Solve this equation for $\dot{\vec{v}}(t)$ (as a first order D.E.) for the following constant, but temporary, external force,

$$\vec{F}_{ext}(t) = \begin{cases} 0 & t < 0 \\ F_0 & 0 \leq t < T \\ 0 & T \leq t \end{cases}$$

in the three temporal regions. The solutions should be continuous at $t = 0$ and $t = T$. (B) In this solution, identify the effects of pre-acceleration (in the region $t < 0$) and runaway (in the region $T \leq t$). What should the remaining constant be set to in order to eliminate pre-acceleration? Runaway? Is there a value for this constant that will eliminate both effects?

(8.6) Carry out Problem 8.4 for the Landau–Lifshitz approximation (Eq. 8.32) of the Abraham–Lorentz force equation for an electron:

$$M\dot{\vec{v}} = T\dot{\vec{F}}_{ext}(t) + \vec{F}_{ext}(t)$$

Compare this solution to that of problem 5 and discuss the differences.

(8.7) What is the characteristic time $T = q^2/6\pi\varepsilon_o c^3 M$? In Section 8.3.3, we demonstrated, by a comparison of radiation energy lost to system energy, that it was the time over which the effects of the self-force are significant. If we assume that the characteristic time, T, is the approximate timescale for transit of an electric field to cross the extended charge (i.e., $T \simeq a/c$) then it is on this timescale that the imbalance of self-forces during acceleration, as derived in Section 8.2.1, is revealed. Show that this is indeed the case by defining the classical radius of an electron, r_e, as the radius beyond which the electric field energy equals the electron mass energy, and then comparing this r_e to cT.

(8.8) The relativistic generalization of the Abraham–Lorentz force was found by Paul Dirac, in 1938, to be:

$$F_\nu^{rad} = \frac{q^2\gamma^2}{6\pi\varepsilon_o c^3}\left[\frac{d^2 p_\nu}{dt^2} - \frac{p_\nu}{m^2c^2}\left(\frac{dp_\mu}{dt}\frac{dp_\mu}{dt}\right)\right]$$

where $p_\nu = (\gamma mc, \gamma m\vec{v})$. Show the validity of this force by using the time average equation

$$\frac{1}{\Delta t}\int_0^t P_{rad}dt = \frac{1}{\Delta t}\int_0^t \vec{F}\cdot\vec{v}dt$$

and Lienard's relativistic generalization of Larmor's power formula (Eq. 6.38) at an instant for which $\vec{v} = 0$,

$$P_{rad} = \frac{q^2\gamma^6}{6\pi\varepsilon_o c^3}\left(\frac{d\vec{v}}{dt}\cdot\frac{d\vec{v}}{dt}\right)$$

8.6 Discussions

Discussion 8.1

With $\vec{R}$ assumed constant in Eq. 8.1, we can extract the factor $1/R^3$ from the brackets,

$$d\vec{E}(\vec{r}_o,t)=\frac{dq}{4\pi\varepsilon_0 R^3}\left\{\frac{(\vec{R}-\frac{\vec{v}}{c}R)(1-(\frac{v}{c})^2)}{(1-\hat{R}\cdot\frac{\vec{v}}{c})^3}+\frac{\vec{R}\times[(\vec{R}-\frac{\vec{v}}{c}R)\times\dot{\vec{v}}]}{c^2(1-\hat{R}\cdot\frac{\vec{v}}{c})^3}\right\}_{ret} \tag{8.65}$$

Using the expansions of Eqs. 8.3 and 8.4,

$$\vec{v}(t_{ret})=-\dot{\vec{v}}\frac{R}{c}+\ddot{\vec{v}}\frac{1}{2}\left(\frac{R}{c}\right)^2-\cdots \tag{8.66}$$

$$\dot{\vec{v}}(t_{ret})=\dot{\vec{v}}-\ddot{\vec{v}}\left(\frac{R}{c}\right)+\cdots \tag{8.67}$$

we start by approximating the common term $\left\{(1-\hat{R}\cdot\frac{\vec{v}}{c})^{-3}\right\}_{ret}\simeq\left\{1+\frac{3}{c}\hat{R}\cdot\vec{v}\right\}_{ret}$ and expressing it in terms of the $\vec{v}(t_{ret})$ expansion,

$$\left\{1+\frac{3}{c}\hat{R}\cdot\vec{v}\right\}_{ret}\simeq 1-\frac{3}{c^2}\left(\vec{R}\cdot\dot{\vec{v}}\right)+\frac{3R}{2c^3}\left(\vec{R}\cdot\ddot{\vec{v}}\right)-\cdots \tag{8.68}$$

Next, multiplying out the numerators of the two terms in brackets, we get four terms for each

$$Left\ Numerator=LN=\left\{\vec{R}-\frac{R}{c}\vec{v}-\vec{R}\left(\frac{v}{c}\right)^2+\frac{R}{c}\left(\frac{v}{c}\right)^2\vec{v}\right\}_{ret} \tag{8.69}$$

and

$$Right\ Numerator=RN=\left\{\vec{R}\left(\vec{R}\cdot\dot{\vec{v}}\right)-\frac{R}{c}\left(\vec{R}\cdot\dot{\vec{v}}\right)\vec{v}-R^2\dot{\vec{v}}+\frac{R}{c}\left(\vec{R}\cdot\vec{v}\right)\dot{\vec{v}}\right\}_{ret} \tag{8.70}$$

Expressing the four terms of the left numerator, Eq. 8.69, in terms of the expansions of 8.66 and 8.67, we get

$$\begin{aligned}
LN1&=\vec{R}\\
LN2&=-\frac{R}{c}\left[-\dot{\vec{v}}\frac{R}{c}+\ddot{\vec{v}}\frac{1}{2}\left(\frac{R}{c}\right)^2-\cdots\right]\\
LN3&=-\frac{\vec{R}}{c^2}\left[-\dot{\vec{v}}\frac{R}{c}+\ddot{\vec{v}}\frac{1}{2}\left(\frac{R}{c}\right)^2-\cdots\right]^2\\
LN4&=\frac{R}{c^3}\left[-\dot{\vec{v}}\frac{R}{c}+\ddot{\vec{v}}\frac{1}{2}\left(\frac{R}{c}\right)^2-\cdots\right]^2\left[-\dot{\vec{v}}\frac{R}{c}+\ddot{\vec{v}}\frac{1}{2}\left(\frac{R}{c}\right)^2-\cdots\right]
\end{aligned} \tag{8.71}$$

Likewise expanding the four terms of the right numerator, Eq. 8.70,

$$
\begin{aligned}
RN1 &= \vec{R}\left[\left(\vec{R}\cdot\dot{\vec{v}}\right) - \frac{R}{c}\left(\vec{R}\cdot\ddot{\vec{v}}\right) + \cdots\right] \\
RN2 &= -\frac{R}{c}\left[\left(\vec{R}\cdot\dot{\vec{v}}\right) - \frac{R}{c}\left(\vec{R}\cdot\ddot{\vec{v}}\right) + \cdots\right]\left[-\dot{\vec{v}}\frac{R}{c} + \ddot{\vec{v}}\frac{1}{2}\left(\frac{R}{c}\right)^2 - \cdots\right] \\
RN3 &= -R^2\left[\dot{\vec{v}} - \ddot{\vec{v}}\left(\frac{R}{c}\right) + \cdots\right] \\
RN4 &= \frac{R}{c}\left[-\left(\vec{R}\cdot\dot{\vec{v}}\right)\frac{R}{c} + \left(\vec{R}\cdot\ddot{\vec{v}}\right)\frac{1}{2}\left(\frac{R}{c}\right)^2 - \cdots\right]\left[\dot{\vec{v}} - \ddot{\vec{v}}\left(\frac{R}{c}\right) + \cdots\right]
\end{aligned}
\tag{8.72}
$$

So that the bracketed part of Eq. 8.65 is given by the expansion of the common denominator, Eq. 8.68, multiplied by the expansions in 8.71 and 8.72,

$$
\{\}_{ret} = \left[1 - \frac{3}{c^2}\left(\vec{R}\cdot\dot{\vec{v}}\right) + \frac{3R}{2c^3}\left(\vec{R}\cdot\ddot{\vec{v}}\right) - \cdots\right]\left[\left(\sum LN\right) + \frac{1}{c^2}\left(\sum RN\right)\right]
$$

Inspecting these expansion products (especially noting the additional $1/c^2$multiplying the RN) and anticipating the neglecting of terms of order $\dddot{v}\left(\frac{a}{c}\right)^4$ or greater, we see that only $LN1, LN2, RN1$, and $RN3$ contribute terms. Thus we can represent Eq. 8.65

$$
d\vec{E}(\vec{r}_o, t) = \frac{dq}{4\pi\varepsilon_o R^3}\left[1 - \frac{3}{c^2}\left(\vec{R}\cdot\dot{\vec{v}}\right) + \frac{3R}{2c^3}\left(\vec{R}\cdot\ddot{\vec{v}}\right) - \cdots\right]\left[(LN1 + LN2) + \frac{1}{c^2}(RN1 + RN3)\right]
$$

Finally, multiplying the expansions and neglecting terms of order $\dddot{v}\left(\frac{a}{c}\right)^4$ or greater, we get

$$
\begin{aligned}
d\vec{E}(\vec{r}_o, t) = \frac{dq}{4\pi\varepsilon_o R^3}&\left[\vec{R} - \frac{3\vec{R}}{c^2}\left(\vec{R}\cdot\dot{\vec{v}}\right) + \frac{3R\vec{R}}{2c^3}\left(\vec{R}\cdot\ddot{\vec{v}}\right) + \frac{R^2}{c^2}\dot{\vec{v}}\right. \\
&\left. - \frac{R^3}{2c^3}\ddot{\vec{v}} + \frac{\vec{R}}{c^2}\left(\vec{R}\cdot\dot{\vec{v}}\right) - \frac{R\vec{R}}{c^3}\left(\vec{R}\cdot\ddot{\vec{v}}\right) - \frac{R^2}{c^2}\dot{\vec{v}} + \frac{R^3}{c^3}\ddot{\vec{v}}\right]
\end{aligned}
$$

and upon simplifying

$$
d\vec{E}(\vec{r}_o, t) = \frac{dq}{4\pi\varepsilon_o R^3}\left[\vec{R} - \frac{2\vec{R}}{c^2}\left(\vec{R}\cdot\dot{\vec{v}}\right) + \frac{R\vec{R}}{2c^3}\left(\vec{R}\cdot\ddot{\vec{v}}\right) + \frac{R^3}{2c^3}\ddot{\vec{v}}\right]
$$

Discussion 8.2

From Griffiths, D. J., Proctor, T. C., and Schroeter, D. F., "Abraham–Lorentz versus Landau–Lifshitz, *American Journal of Physics* 78, 391 (2010), it is claimed that

$$\vec{F}_{self} = \frac{q^2}{12\pi\varepsilon_o ca^2}\left[\vec{v}\left(t - \frac{2a}{c}\right) - \vec{v}(t)\right] \tag{8.73}$$

is the exact analytical form of the self-force. We can show this to be true to within the approximation of Eq. 8.10. We first expand $\vec{v}\left(t - \frac{2a}{c}\right)$ around t, the evaluation time,

$$\vec{v}\left(t - \frac{2a}{c}\right) = \vec{v}(t) - \frac{2a}{c}\dot{\vec{v}}(t) + \frac{1}{2}\left(\frac{2a}{c}\right)^2 \ddot{\vec{v}}(t) - \ldots$$

Then, plugging this into Eq. 8.73, we obtain

$$\vec{F}_{self} = \frac{q^2}{12\pi\varepsilon_o ca^2}\left[-\frac{2a}{c}\dot{\vec{v}}(t) + \frac{1}{2}\left(\frac{2a}{c}\right)^2 \ddot{\vec{v}}(t) - \ldots\right]$$

whereupon rearranging and truncating,

$$\vec{F}_{self} = \frac{q^2}{4\pi\varepsilon_o}\left[-\frac{2}{3ac^2}\dot{\vec{v}}(t) + \frac{2}{3c^3}\ddot{\vec{v}}(t)\right]$$

we obtain Eq. 8.10.

Discussion 8.3

The electric vector field of a spherical shell of charge q and diameter a moving at a constant, non-relativistic velocity, $\vec{v} = v\hat{z}$, and instantaneously at the origin, is essentially the same as the field of a stationary charge q at the origin,

$$\vec{E} = \frac{q}{4\pi\varepsilon_o r^2}\hat{r} = \frac{q}{4\pi\varepsilon_o r^3}(x\hat{x} + y\hat{y} + z\hat{z})$$

The magnetic field is

$$\vec{H} = \frac{\vec{B}}{\mu_o} = \frac{q\vec{v}\times\hat{r}}{4\pi r^2} = \frac{qv\hat{z}\times(x\hat{x}+y\hat{y}+z\hat{z})}{4\pi r^3} = \frac{qv\varepsilon_o}{4\pi\varepsilon_o r^3}(-y\hat{x} + x\hat{y})$$

and the field momentum density is

$$\begin{aligned}\vec{g} &= \frac{\left(\vec{E}\times\vec{H}\right)}{c^2} = \frac{E^2}{r^2}\frac{v\varepsilon_o}{c^2}\left[(x\hat{x}+y\hat{y}+z\hat{z})\times(-y\hat{x}+x\hat{y})\right] \\ &= \frac{E^2}{r^2}\frac{v\varepsilon_o}{c^2}\left[-zx\hat{x} - zy\hat{y} + \left(x^2+y^2\right)\hat{z}\right]\end{aligned}$$

The volume integral of $\vec{g}$ over all space carried out in spherical coordinates is then set up as

$$\vec{G} = \int \vec{g} dV = \frac{v\varepsilon_o}{c^2}\left(\frac{q}{4\pi\varepsilon_o}\right)^2 \int \frac{1}{r^2}\left[-\cos\theta\sin\theta\left(\cos\phi\hat{x} + \sin\phi\hat{y}\right) + \left(\sin^2\theta\right)\hat{z}\right]\sin\theta d\theta d\phi dr$$

Upon integration over θ, the $\hat{x}$ and $\hat{y}$ terms vanish. This is consistent with the cylindrical symmetry about the direction of motion. Thus we are left with

$$\vec{G} = \frac{2\pi v\varepsilon_o}{c^2}\left(\frac{q}{4\pi\varepsilon_o}\right)^2 \hat{z}\int \frac{1}{r^2}\sin^3\theta d\theta dr$$

but it is easily shown that $\int \sin^3\theta d\theta = \frac{1}{3}\cos^3\theta - \cos\theta$ evaluated from 0 to π is 4/3 and $\int \frac{1}{r^2}dr = -\frac{1}{r}$ evaluated from a to ∞ is $\frac{1}{a}$ so that the total electromagnetic momentum is

$$\vec{G} = \frac{2\pi v\varepsilon_o}{c^2}\left(\frac{q}{4\pi\varepsilon_o}\right)^2 \frac{4}{3a}\hat{z} = \frac{q^2}{6\pi\varepsilon_o c^2 a} v\hat{z}$$

and completely in the $\hat{z}$ direction. The factor proportional to v is thus identified as the electromagnetic mass, $M_{em} = \frac{q^2}{6\pi\varepsilon_o c^2 a}$.

Part IV

Radiation in Materials

Properties of Electromagnetic Radiation in Materials

9

- Electromagnetic radiation in the presence of materials: macroscopic vectors $\vec{D}$ and $\vec{H}$
- Electric and magnetic fields in materials with real and imaginary responses: propagation versus diffusion
- Description of a pulse of radiation and its propagation in dispersive materials: phase versus group velocity
- Reflection and transmission of light at an interface between materials with differing indices of refraction: Fresnel equations
- Response of materials to applied electric and magnetic fields in the time domain, and its correlation to the frequency response of the material polarization
- Kramers–Kronig equations: fundamental relationship between the real and imaginary parts of any material's polarizability
- Practical measurement techniques for extracting the real and imaginary components of a material's index of refraction

In Chapter 1, within our review of Maxwell's equations, we briefly discussed the main concepts associated with the response of matter to static electric and magnetic fields, including polarization and magnetization, bound charges and currents, and electric and magnetic susceptibilities, concluding with a derivation of the time-dependent macroscopic Maxwell's equations within matter. In the subsequent chapters of Parts I, II and III, these concepts played no role since we restricted our discussions of the radiation fields to free space, or vacuum, within which there is no material to polarize or magnetize and the electric and magnetic susceptibilities, $\tilde{\chi}_e$ and $\tilde{\chi}_m$, vanish. In this chapter and the next, we revisit and extend those material concepts

Electromagnetic Radiation. Richard Freeman, James King, Gregory Lafyatis,
Oxford University Press (2019).
DOI: 10.1093/oso/9780198726500.001.0001

discussed in Chapter 1 and expand our equations describing the transmission of electromagnetic energy to encompass the effects of a non-vacuum medium on the radiation.

9.1 Polarization, magnetization, and current density

The propagation of electromagnetic radiation through a material which responds to the electric and magnetic fields of the wave is complicated by the fact that the fields within the wave are, in turn, partially determined by the altered state of the matter. This problem is handled by a systematic analysis of the time-dependent response of the material to a passing wave.

In Chapter 1, we found Maxwell's equations in a material system:

$$\nabla \times \vec{E}(\vec{r},t) = -\frac{\partial \vec{B}(\vec{r},t)}{\partial t} \tag{9.1}$$
$$\nabla \times \vec{H}(\vec{r},t) = \vec{\mathcal{J}}_f(\vec{r},t) + \frac{\partial \vec{D}(\vec{r},t)}{\partial t}$$
$$\nabla \cdot \vec{D}(\vec{r},t) = \rho_f(\vec{r},t)$$
$$\nabla \cdot \vec{B}(\vec{r},t) = 0$$

where $\vec{E}(\vec{r},t)$ and $\vec{H}(\vec{r},t)$ are the electric and magnetic fields, $\vec{D}(\vec{r},t)$ and $\vec{B}(\vec{r},t)$ are the electric displacement and magnetic induction, and $\rho_f(\vec{r},t)$ and $\vec{\mathcal{J}}_f(\vec{r},t)$ are the free charge and free current density distributions in the system under study. These quantities are understood to be local averages over the atomic scale charges and fields making up the system. In Chapter 1, we introduced the macroscopic field concepts of polarization $\vec{P}(\vec{r},t)$ and magnetization $\vec{M}(\vec{r},t)$ (both, hereafter, generically referred to as "polarization"). The macroscopic fields $\vec{D}$ and $\vec{H}$ are defined by:

$$\vec{D}(\vec{r},t) = \varepsilon_o \vec{E}(\vec{r},t) + \vec{P}(\vec{r},t) \tag{9.2}$$
$$\vec{H}(\vec{r},t) = \frac{1}{\mu_o}\vec{B}(\vec{r},t) - \vec{M}(\vec{r},t) \tag{9.3}$$

Upon substituting these into the curl equations of Eq. 9.1 and performing the standard manipulations and substitutions, we obtain the general equations[1]

$$\nabla \times \nabla \times \vec{E} + \frac{1}{c^2}\frac{\partial^2 \vec{E}}{\partial t^2} = -\mu_0 \frac{\partial}{\partial t}\left[\vec{\mathcal{J}}_f + \frac{\partial \vec{P}}{\partial t} + \nabla \times \vec{M}\right] \tag{9.4}$$
$$\nabla \times \nabla \times \vec{H} + \frac{1}{c^2}\frac{\partial^2 \vec{H}}{\partial t^2} = \nabla \times \mathcal{J}_f + \nabla \times \frac{\partial P}{\partial t} - \frac{1}{c^2}\frac{\partial^2 M}{\partial t^2}$$

[1] To obtain the first equation, take the curl of the first Maxwell's equation of Eq. 9.1 and substitute $\vec{B} = \mu_o\left(\vec{H} + \vec{M}\right)$ from Eq. 9.3,

$$\nabla \times \nabla \times \vec{E} = -\mu_o \frac{\partial}{\partial t}\left[\nabla \times \left(\vec{H} + \vec{M}\right)\right]$$

Substitute the second Maxwell's equation of Eq. 9.1,

$$\nabla \times \nabla \times \vec{E} = -\mu_o \frac{\partial}{\partial t}\left[\vec{\mathcal{J}}_f + \frac{\partial \vec{D}}{\partial t} + \nabla \times \vec{M}\right]$$

Finally substitute the definition of $\vec{D}$ (Eq. 9.2).

Similarly, to obtain the second equation take the curl of the second Maxwell's equation of Eq. 9.1 and use Eq. 9.2,

$$\nabla \times \nabla \times \vec{H} = \nabla \times \mathcal{J}_f + \frac{\partial}{\partial t}\left[\varepsilon_o\left(\nabla \times \vec{E}\right) + \nabla \times \vec{P}\right]$$

Next, substitute the first Maxwell's equation of Eq. 9.1,

$$\nabla \times \nabla \times \vec{H} = \nabla \times \mathcal{J}_f - \varepsilon_o \frac{\partial^2 \vec{B}}{\partial t^2} + \nabla \times \frac{\partial \vec{P}}{\partial t}$$

Finally substitute $\vec{B} = \mu_o\left(\vec{H} + \vec{M}\right)$ from Eq. 9.3.

As in Chapter 1, we define the term in brackets to be the "total material current density":

$$\vec{\mathcal{J}}_t = \vec{\mathcal{J}}_f + \frac{\partial \vec{P}}{\partial t} + \nabla \times \vec{M} \tag{9.5}$$

where $\frac{\partial \vec{P}}{\partial t}$, the "polarization current density" and $\nabla \times \vec{M}$, the "magnetization current density" can be interpreted as arising from the "bound" charges in the material. We now specialize to the case of plane wave solutions in materials that are linear, isotropic and homogeneous. For such media, it is useful to consider harmonic (single frequency) solutions to the equations:

$$\begin{aligned} \vec{E}(r,t) &= \vec{E}(r)\exp(-i\omega t) \\ \vec{P}(r,t) &= \vec{P}(r)\exp(-i\omega t) \\ \vec{\mathcal{J}}_f(r,t) &= \vec{\mathcal{J}}_f(r)\exp(-i\omega t) \\ &\vdots \end{aligned} \tag{9.6}$$

Substituting these approximations into Maxwell's equations (Eq. 9.1) gives:

$$\begin{aligned} \nabla \times \vec{E}(\vec{r}) &= i\omega \vec{B}(\vec{r}) \\ \nabla \times \vec{H}(\vec{r}) &= \vec{\mathcal{J}}_f(\vec{r}) - i\omega \vec{D}(\vec{r}) \\ \nabla \cdot \vec{D}(\vec{r}) &= \rho_f(\vec{r}) \\ \nabla \cdot \vec{B}(\vec{r}) &= 0 \end{aligned} \tag{9.7}$$

where the time dependent exponentials have been divided out. By linear, we mean that the polarization $\vec{P}$ and magnetization $\vec{M}$ are linearly related to the electric and magnetic fields by the generally complex electric and magnetic susceptibilities, $\tilde{\chi}_e(\omega)$ and $\tilde{\chi}_m(\omega)$,

$$\begin{aligned} \vec{P} &= \varepsilon_0 \tilde{\chi}_e \vec{E} \\ \vec{M} &= \tilde{\chi}_m \vec{H} \end{aligned} \tag{9.8}$$

With these relations, the expressions of Eqs. 9.2 and 9.3 can then be written as the constitutive equations,

$$\begin{aligned} \vec{D} &= \tilde{\varepsilon}(\omega)\vec{E} \\ \vec{B} &= \tilde{\mu}(\omega)\vec{H} \end{aligned} \tag{9.9}$$

where $\tilde{\varepsilon}(\omega) = \varepsilon_0(1 + \tilde{\chi}_e)$ and $\tilde{\mu}(\omega) = \mu_0(1 + \tilde{\chi}_m)$ are the material's permittivity and permeability, respectively. Similarly, the current density is linearly related to the electric field by the complex conductivity, $\tilde{\sigma}(\omega)$,

$$\vec{\mathcal{J}}_f = \tilde{\sigma}(\omega)\vec{E} \tag{9.10}$$

which is known as Ohm's law.

9.2 A practical convention for material response

As introduced and discussed in Chapter 1, the "bound" and "free" types of material response to electric fields are associated with the permittivity $\tilde{\varepsilon}$ and the conductivity $\tilde{\sigma}$, respectively. Although this is the traditional and intuitive notation, in this section we will describe more general notation for material response which makes no reference to bound or free components. Thus, in the following, to distinguish the original notation from the new notation, we now subscript the original notation with "f" for free and "b" for bound. That is, for the original notation, $\tilde{\sigma} \rightarrow \tilde{\sigma}_f$ and $\tilde{\varepsilon} \rightarrow \tilde{\varepsilon}_b$ so that, for example, Eqs. 9.9 and 9.10 are now written as $\vec{D} = \tilde{\varepsilon}_b\vec{E}$ and $\vec{\mathcal{J}}_f = \tilde{\sigma}_f\vec{E}$. The full material response to an electric field can then be associated with an unsubscripted total permittivity,

$$\tilde{\varepsilon} = \tilde{\varepsilon}_b + i\frac{\tilde{\sigma}_f}{\omega} = (\varepsilon_{b1} + i\varepsilon_{b2}) - \frac{1}{\omega}\left(\sigma_{f2} - i\sigma_{f1}\right) \tag{9.11}$$

In general then, as defined, the quantities $\tilde{\sigma}_f$ and $\tilde{\varepsilon}_b$ do not individually represent the full response of the material and their real and imaginary components do not, therefore, necessarily constitute a proper pair of optical constants. So, while this division into free ($\tilde{\sigma}_f$) and bound ($\tilde{\varepsilon}_b$) contributions is useful from a theoretical standpoint, it does not lend itself well to experimental measurement. For example, given a measurement of the absorption (and thickness) of an otherwise unknown material, we can easily work out the value of the extinction coefficient n_2, which is a proper optical constant. But without knowing the type or state of the material, we cannot determine how much of the absorption is due to bound electrons and how much is due to free electrons.

A second, more useful, convention, which we now present, represents the same total permittivity, $\tilde{\varepsilon}$, by generically dividing this material response into real and imaginary parts, with no attention paid to designating bound and free contributions, thus reducing the number of independent quantities from four to two. The total permittivity then represents the total material response of Eq. 9.11 simply as

$$\tilde{\varepsilon} = \varepsilon_1 + i\varepsilon_2 \tag{9.12}$$

where ε_1, for example, includes the real terms in Eq. 9.11 and ε_2 includes the imaginary terms. With this convention then, ε_1 and ε_2 are a proper pair of optical constants for any material. Conversely, we can represent–with equal generality–the same total material response in terms of a generalized conductivity,

$$\tilde{\varepsilon} = i\frac{\tilde{\sigma}}{\omega} = -\frac{\sigma_2}{\omega} + i\frac{\sigma_1}{\omega} \tag{9.13}$$

where, in this representation, σ_1 and σ_2 are also a proper pair of optical constants. Comparing Eqs. 9.12 and 9.13, we identify,

$$\varepsilon_1 = -\frac{\sigma_2}{\omega} \quad \text{and} \quad \varepsilon_2 = \frac{\sigma_1}{\omega} \tag{9.14}$$

from which Eq. 9.11, the total permittivity, can be rewritten generally, and usefully, in yet a third way

$$\tilde{\varepsilon} = \varepsilon_1 + i\frac{\sigma_1}{\omega} \tag{9.15}$$

so that ε_1 and σ_1 are another proper pair of optical constants. In summary, we accept this new convention with the understanding that from now on the symbols ε_1, ε_2, σ_1, and σ_2 are all proper optical constants each of which may include both bound and free contributions; and the subscripts f or b will be attached whenever referring to the original convention of purely "free" electron conductivity and purely "bound" electron permittivity. Which of the alternate forms of the total material response (Eqs. 9.12, 9.13, and 9.15) is used depends on the particular problem and material.

9.3 E&M propagation within simple media

Most materials of interest are non-magnetic at optical frequencies ($\tilde{\chi}_m \simeq 0 \rightarrow \tilde{\mu} \simeq \mu_o$) and we will assume such in what follows. If we further assume the material has zero total charge density ($\rho = 0$), then by taking the curl of the first equation of Eq. 9.7 and substituting the second for the right-hand side, we find[2]

$$\nabla^2 \vec{E}(\vec{r}) + \omega^2 \mu_0 \varepsilon_0 \left[\frac{\tilde{\varepsilon}_b}{\varepsilon_0} + i \frac{\tilde{\sigma}_f}{\varepsilon_0 \omega} \right] \vec{E}(\vec{r}) = 0 \tag{9.16}$$

where, according to Eq. 9.11, the quantity in square brackets is associated with the total response of the material.

A similar result is found for the magnetic field. For a pure dielectric, the quantity $\tilde{\varepsilon}_b/\varepsilon_0$ is called the dielectric constant. Mathematically, Maxwell's equations for harmonic fields in a material return the Helmholtz equation of free space with the substitution:

$$\varepsilon_0 \to \varepsilon_0 \left[\frac{\tilde{\varepsilon}_b}{\varepsilon_0} + i \frac{\tilde{\sigma}_f}{\varepsilon_0 \omega} \right] \equiv \tilde{\varepsilon}(\omega) = \varepsilon_1(\omega) + i\varepsilon_2(\omega) \tag{9.17}$$

So we see that the response of an isotropic, homogeneous, linear material to an electric field–which is entirely described by two real numbers, the real and imaginary parts of the general complex permittivity, $\tilde{\varepsilon}(\omega)$–plays an essential role in the form of electromagnetic propagation within the material. Further, if the material is non-magnetic, then the electric field response is the only field response of the material. Helmholtz equations of the form of Eq. 9.16 for the single-frequency electric and magnetic fields may then be conveniently written in terms of the general complex permittivity as,

$$\nabla^2 \vec{E}(\vec{r}) + k_0^2 \frac{\tilde{\varepsilon}(\omega)}{\varepsilon_0} \vec{E}(\vec{r}) = 0 \tag{9.18}$$

$$\nabla^2 \vec{H}(\vec{r}) + k_0^2 \frac{\tilde{\varepsilon}(\omega)}{\varepsilon_0} \vec{H}(\vec{r}) = 0 \tag{9.19}$$

where $k_0 = \omega\sqrt{\mu_0 \varepsilon_0} = \omega/c$ is the magnitude of the free space wavenumber.

As discussed in Section 9.2, the information contained in the real and imaginary parts of the general complex permittivity may be expressed by various pairs of proper optical constants. Generally, for a wave propagating in a material, the essential elements of physics to be conveyed are the dispersion–casually speaking, the variation of the wavelength of the wave–and the losses or spatial decay of the wave as it propagates. The frequency dependent real permittivity and real conductivity are usually defined by Eq. 9.15. And, as discussed in the previous section, ε_1 and σ_1 may each include terms from both bound electrons and free electron conduction currents. Alternatively, we can

[2] Take the curl of the first Maxwell equation of Eq. 9.7 and substitute $\vec{B} = \mu\vec{H}$ from Eq. 9.8,

$$\nabla \times \nabla \times \vec{E} = i\omega\tilde{\mu}\left[\nabla \times \vec{H}\right]$$

Next, expressing $\nabla \times \nabla \times \vec{E} = \nabla\left(\nabla \cdot \vec{E}\right) - \nabla^2 \vec{E}$ and substituting the second Maxwell equation of Eq. 9.7, we get

$$\nabla\left(\nabla \cdot \vec{E}\right) - \nabla^2 \vec{E} = i\omega\tilde{\mu}\left[\vec{\mathcal{J}}_f - i\omega\vec{D}\right]$$

Finally, using $\vec{D} = \tilde{\varepsilon}_b \vec{E}$, $\vec{\mathcal{J}}_f = \tilde{\sigma}_f \vec{E}$, $\nabla \cdot \vec{E} = \rho = 0$, and $\tilde{\mu} \simeq \mu_o$ results in

$$\nabla^2 \vec{E} + \omega^2 \mu_0 \left[\tilde{\varepsilon}_b + i\frac{\tilde{\sigma}_f}{\omega}\right] \vec{E} = 0$$

which can be written as in Eq. 9.16.

express the material response in terms of the generalized complex conductivity, $\tilde{\sigma}(\omega)$. For example, we can rewrite the single-frequency Ampere–Maxwell contribution to Eq. 9.7 as,

$$\nabla \times \vec{H}(\vec{r}) = \vec{J}_f(\vec{r}) - i\omega\vec{D}(\vec{r}) = \left(\tilde{\sigma}_f - i\omega\tilde{\varepsilon}_b\right)\vec{E}(\vec{r}) \equiv \tilde{\sigma}(\omega)\vec{E}(\vec{r}) \quad (9.20)$$

where, as defined in Eqs. 9.13 and 9.14, the relation of the complex generalized conductivity to the complex generalized permittivity is $\tilde{\sigma}(\omega) = -i\omega\tilde{\varepsilon}(\omega)$.

The electric field plane wave solution of Eq. 9.18 for a wave traveling in the $+z$ direction is:

$$\vec{E}(z) = \vec{E}_0 \exp\left[i\tilde{k}z\right] \quad (9.21)$$

where the complex wave number is given by,

$$\tilde{k}(\omega) = k_1 + ik_2 = k_0\sqrt{\frac{\tilde{\varepsilon}(\omega)}{\varepsilon_0}} \quad (9.22)$$

and the square root of the dielectric constant is the complex index of refraction:

$$\tilde{n}(\omega) \equiv n + i\kappa = \sqrt{\frac{\tilde{\varepsilon}(\omega)}{\varepsilon_0}} \quad (9.23)$$

which for a lossless dielectric returns the familiar real index of refraction, $n(\omega)$. The imaginary part of the index of refraction or the "extinction coefficient," $\kappa(\omega)$, arises due to absorption losses and scattering in the material.

Although we have previously defined a number of pairs of "proper" optical constants, these two quantities are frequently referred to as the optical constants of the material. They represent properties of a material that are most directly related to measurements. Squaring Eq. 9.23, the dielectric coefficient is expressed in terms of these optical constants,

$$\frac{\varepsilon_1(\omega)}{\varepsilon_0} = n^2(\omega) - \kappa^2(\omega)$$
$$\frac{\varepsilon_2(\omega)}{\varepsilon_0} = 2n(\omega)\kappa(\omega)$$

We can thus rewrite the Helmoltz equations (Eqs. 9.18 and 9.19) for a harmonic wave propagating in a material (labelled as "A"):

$$\nabla^2 \vec{E}(\vec{r}) + k_0^2\, \tilde{n}_A^2(\omega)\, \vec{E}(\vec{r}) = 0 \tag{9.24}$$
$$\nabla^2 \vec{H}(\vec{r}) + k_0^2\, \tilde{n}_A^2(\omega)\, \vec{H}(\vec{r}) = 0 \tag{9.25}$$

By comparison with the vacuum equations, plane wave solutions in material are thus described by a complex wave vector, $\vec{k}_A = \vec{k}_0 \tilde{n}_A$, that specifies the direction of the wave's propagation, and for a lossy (or scattering) material, the characteristic scalelength of decay. The wave vector magnitude in material "A" satisfies $k_A = k_0\,|\tilde{n}_A| = (\omega/c)\,|\tilde{n}_A|$. A second complex vector constant, $\vec{E}_0$, specifies the polarization of the wave and its magnitude. Generally,

$$\vec{E}(\vec{r}) = \vec{E}_0 \exp\left(i\vec{k}_A \cdot \vec{r}\right)$$
$$\vec{B}(\vec{r}) = \sqrt{\mu_A \varepsilon_A}\left(\hat{k}_A \times \vec{E}_0\right) \exp\left(i\vec{k}_A \cdot \vec{r}\right)$$

where $\hat{k}_A \equiv \vec{k}_A / k_A$. From Eq. 9.21, a wave propagating in a lossy (or scattering) material exponentially decays over a characteristic distance

$$\delta = \frac{1}{k_2(\omega)} = \frac{c}{\omega\kappa(\omega)} \tag{9.26}$$

which is known as the "skin depth." The intensity of the wave $\sim |E^2|$ and its value drops by a factor of $1/e$ over a distance $\delta/2 = 1/\alpha_{abs}$ where $\alpha_{abs}(\omega)$ is the material's "absorption coefficient." The associated time-dependent harmonic solutions, Eqs. 9.6, can now be written,

$$\vec{E}(\vec{r},t) = \vec{E}_0 \exp\left(i\vec{k}_A \cdot \vec{r} - i\omega t\right) \tag{9.27}$$
$$\vec{B}(\vec{r},t) = \sqrt{\mu_A \varepsilon_A}\left(\hat{k}_A \times \vec{E}_0\right) \exp\left(i\vec{k}_A \cdot \vec{r} - i\omega t\right)$$

so that we can identify the phase velocity of plane waves traveling through material as

$$v_p(\omega) = \frac{\omega}{k_1(\omega)} = \frac{c}{n(\omega)} \tag{9.28}$$

9.4 Frequency dependence

9.4.1 $\omega \to \infty$

Much of the business of applied physics is to measure, understand, and control the electromagnetic properties of materials as a function of frequency. This is the science of spectroscopy and is a central activity of the remainder of this text. Here, we briefly discuss two limits: $\omega \to \infty$ and $\omega \to 0$. The absolute $\omega \to \infty$ limit is simply understood as follows. The complex generalized permittivity of Eq. 9.11 may be written:

$$\tilde{\varepsilon}(\omega) = \varepsilon_0 (1 + \tilde{\chi}_e) \tag{9.29}$$

where the electric susceptibility describes the general response of the material to an incident electric field. For a sufficiently large ω, the field changes so rapidly that the material medium is unable to "keep up." In that case, $\tilde{\chi}_e \to 0$, $\tilde{\varepsilon}(\omega) \to \varepsilon_0$ and in turn $n(\omega) \to 1$ and $\kappa(\omega) \to 0$. What constitutes a "sufficiently large" frequency? This will be discussed at more length in the next chapter; roughly, the frequency must be much larger than many, but not all, atomic transitions in the material. For example, to get above all transitions for gold, including core excitations, would require photons in the gamma ray part of the spectrum. However, Fig. 9.1 shows that the high frequency is reached for radiation of wavelength ~ 10 nm–a fairly soft x-ray.

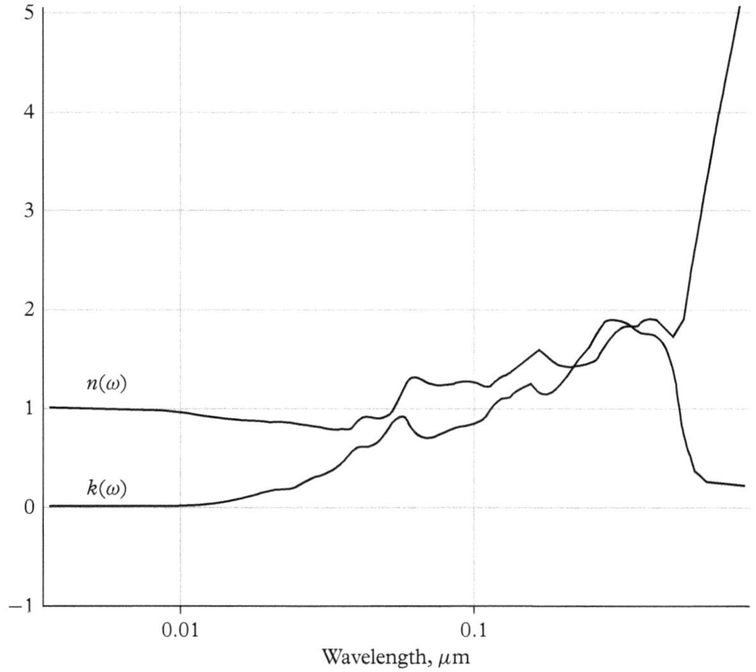

Fig. 9.1 *The optical constants for gold from the near IR to x-rays.*

9.4.2 $\omega \to 0$

The low frequency limit ($\omega \to 0$) may be studied in the familar setting of linear lumped circuit components as shown in Fig. 9.2. The physical makeup is a slab of material with cross section area A and thickness d between two parallel conductors. Depending upon the material, this could be either a resistor or a capacitor, or some combination of both. If the material is a conductor, then on applying a voltage, current flows even at DC. The resistance of such a device is given by:

$$R = \frac{d}{A\sigma_{f1}} \tag{9.30}$$

where $\tilde{\sigma}_f \simeq \sigma_{f1}$ is here taken to be constant and real–the low frequency conductivity of the material (see Eq. 9.11). On the other hand, if the material is a perfect dielectric, the device behaves as a capacitor with

$$C = \frac{\varepsilon_{b1} A}{d} \tag{9.31}$$

where $\tilde{\varepsilon}_b \simeq \varepsilon_{b1}$ is here considered to be constant and real for the limited range of low frequencies we consider. For real materials, upon imposing a voltage V across the plates, the current, I, flowing between the plates is due to both "free" electron conduction and to displacement currents, with the latter due to the polarization of the material. We can sum these using the lumped circuit impedance, Z, for the two elements in parallel[3]

[3] At low frequencies, since $\tilde{\sigma}_f \simeq \sigma_{f1}$ and $\tilde{\varepsilon}_b \simeq \varepsilon_{b1}$, the relation of Eq. 9.15 becomes,

$$\tilde{\varepsilon} = \varepsilon_1 + i\frac{\sigma_1}{\omega} \Rightarrow \varepsilon_{b1} + i\frac{\sigma_{f1}}{\omega}$$

and the associated relation,

$$\tilde{\sigma} = \sigma_1 + i\omega\varepsilon_1 \Rightarrow \sigma_{f1} + i\omega\varepsilon_{b1}$$

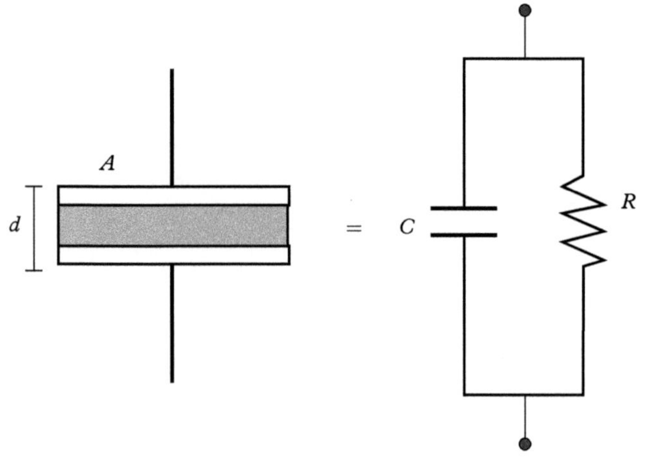

Fig. 9.2 *A generic lumped circuit element. On the right is a linear electric model for the physical capacitor on the left.*

$$I = V\frac{1}{Z} = V\left(\frac{1}{R} + i\omega C\right) = V\frac{A}{d}\left(\sigma_{f1} + i\omega\varepsilon_{b1}\right) = V\frac{A}{d}\tilde{\sigma} \qquad (9.32)$$

At low frequencies, then, using a complex conductivity is analogous to the electronics treatment of a capacitor as being "a frequency dependent resistor."

An alternative but less intuitive analogy could be made between the complex permittivity and conduction within a capacitor. The material distinction between a conductor and a dielectric is simply that for an imposed electric field at $\omega = 0$, a finite current flows in a conductor and no current flows for a perfect dielectric. Real materials form a continuum between the two limits. For non-zero frequencies, we can ask whether a given real material is "more a dielectric" or "more a conductor." This can be phrased in the lumped circuit system by comparing the ratio of the current flowing through the equivalent capacitor to that flowing through the equivalent resistor:

$$\left|\frac{I_C}{I_R}\right| = \frac{\omega\varepsilon_{b1}}{\sigma_{f1}} = \omega\tau \qquad (9.33)$$

where τ is the characteristic response time[4] of the material and at low frequencies is the same as τ_{RC}, the lumped circuit RC time constant. Thus, at low frequencies, if $\omega\tau > 1$, then the material is "more a dielectric;" and if $\omega\tau < 1$, it is "more a conductor."

9.4.3 Plane waves versus diffusion

The form of the propagation of electromagnetic fields within simple media ultimately depends on how fast the material can respond to the changing fields. Again, for ω sufficiently large, the field changes so rapidly that the material medium is unable to "keep up." To quantify this, we need to define and compare the respective characteristic times, T and τ, for the fields and the material. For the fields, it is $T = 2\pi/\omega$– the period of the E&M wave. For the material, it was shown in Section 9.4.2 that for sufficiently low frequencies where $\tilde{\varepsilon}_b \simeq \varepsilon_{b1}$ and $\tilde{\sigma}_f \simeq \sigma_{f1}$, the characteristic response time of the material is given by $\tau = \varepsilon_{b1}/\sigma_{f1}$. Because we expect different forms of field propagation for when the material responds very fast ($\tau < T$) and very slow ($\tau > T$) relative to the fields, we expect the form of the governing differential equation (Eq. 9.16) to likewise change. Indeed, we can express Eq. 9.16 in terms of the characteristic time τ and thus identify the limiting forms of the differential equation and its solutions. At sufficiently low frequencies, then, Eq. 9.16 can be re-expressed as,

[4] Here, we use τ as the symbol for the characteristic time to follow the convention concerning time responses in materials. It is not to be confused with the use of τ to represent the retarded time in potentials as introduced in Chapter 2.

$$\nabla^2 \vec{E} + \omega^2 \mu_0 \varepsilon_{b1} \left[1 + i \frac{\sigma_{f1}}{\varepsilon_{b1} \omega} \right] \vec{E} = 0 \tag{9.34}$$

where $\sigma_{f1}/\varepsilon_{b1}\omega = T/2\pi\tau$ and we immediately see a dependence of the form of the equation on the term $\sigma_{f1}/\varepsilon_{b1}\omega$. Specifically, we can define two limits: $\sigma_{f1}/\varepsilon_{b1}\omega \ll 1$ and $\sigma_{f1}/\varepsilon_{b1}\omega \gg 1$. In the first limit, the second term in the bracket becomes negligible and Eq. 9.34 approximates a wave equation; and in the second limit, the first term on the right becomes negligible and Eq. 9.34 approximates a diffusion equation. Note also that these limits can be characterized by the domination of one of the two source terms for $\vec{H}$ in the macroscopic Maxwell–Ampere equation (Eq. 9.20). For low frequencies, where $\tilde{\varepsilon}_b \simeq \varepsilon_{b1}$ and $\tilde{\sigma}_f \simeq \sigma_{f1}$, Eq. 9.20 can be written,

$$\nabla \times \vec{H} = \vec{\mathcal{J}}_f - i\omega \vec{D} = \left(\sigma_{f1} - i\omega\varepsilon_{b1}\right) \vec{E}$$

So that in the wave propagation limit, $\sigma_{f1}/\varepsilon_{b1}\omega \ll 1$, the conduction current $\vec{\mathcal{J}}_f$ is negligible compared to the displacement current $\vec{\mathcal{J}}_d = -i\omega\vec{D}(\vec{r})$ and in the diffusion limit in which, for example, we have a relatively low frequency electric field within a relatively good conducting medium, we expect that conduction currents will play the dominant role compared to the displacement current. Indeed, we see that the ratio of these two currents, for a given harmonic component ω, is

$$\left| \frac{\vec{\mathcal{J}}_f}{\vec{\mathcal{J}}_d} \right| = \frac{\sigma_{f1}}{\omega\varepsilon_{b1}} = \frac{1}{2\pi} \left(\frac{T}{\tau} \right) \tag{9.35}$$

These two limits of diffusion and wave propagation also correspond to the strong and weak absorption limits, respectively. In the $\vec{\mathcal{J}}_f$-dominated diffusion limit, the average electromagnetic power absorbed by the material is maximized because the total current density $\vec{\mathcal{J}} \simeq \vec{\mathcal{J}}_f$ is in phase with the driving field

$$\text{Avg power absorbed } = \left\langle \vec{\mathcal{J}}_f \cdot \vec{E} \right\rangle = \sigma_{f1} E_0^2 \left\langle \cos^2\left(\vec{k} \cdot \vec{r}\right) \right\rangle = \frac{1}{2} \sigma_{f1} E_0^2$$

while in the $\vec{\mathcal{J}}_d$-dominated wave propagation limit, on average, virtually no electromagnetic power is absorbed by the material because the total current density $\vec{\mathcal{J}} \simeq \vec{\mathcal{J}}_d$ is 90° out of phase with the driving field,

$$\text{Avg power absorbed } = \left\langle \vec{\mathcal{J}}_d \cdot \vec{E} \right\rangle = \omega\varepsilon_{b1} E_0^2 \left\langle \cos\left(\vec{k} \cdot \vec{r}\right) \sin\left(\vec{k} \cdot \vec{r}\right) \right\rangle \simeq 0$$

both of which are consistent with the fast/slow material response argument since fast responding materials will tend to keep up with the phase of the driving field while slow responding materials will tend to fall out of phase.

Finally, we can generalize these diffusion/plane wave (or strong/weak absorption) limits for all frequencies by noting that these limits are generally defined by the ratio of the imaginary and real parts of the material response. With the adoption of the notation of Eq. 9.15, in which we represent the generalized permittivity as

$$\tilde{\varepsilon} = \varepsilon_1 + i\sigma_1/\omega$$

and where, by Eq. 9.11, it is seen that $\varepsilon_1 = \varepsilon_{b1} - \sigma_{f2}/\omega$ and $\sigma_1/\omega = \varepsilon_{b2} + \sigma_{f1}/\omega$, we can rewrite Eq. 9.16 as

$$\nabla^2\vec{E}(\vec{r}) + \omega^2\mu_0\varepsilon_0\left[\frac{\varepsilon_1}{\varepsilon_0} + i\frac{\sigma_1}{\varepsilon_0\omega}\right]\vec{E}(\vec{r}) = 0$$

and thus generalize Eq. 9.34 to,

$$\nabla^2\vec{E} + \omega^2\mu_0\varepsilon_1\left[1 + i\frac{\sigma_1}{\varepsilon_1\omega}\right]\vec{E} = 0$$

so that all of this applies generally and the diffusion/plane wave (or strong/weak absorption) limits correspond, respectively, to $\sigma_1/\varepsilon_1\omega \gg 1$ and $\sigma_1/\varepsilon_1\omega \ll 1$.

In summary, at low frequencies the ratio in Eq. 9.35 is small for a weakly conducting medium, or a situation in which the period of the E&M wave is short enough. Under these conditions the second term in Eq. 9.34 is negligible and an E&M wave will propagate (perhaps with some loss). When the ratio is large, the first term in Eq. 9.34 is negligible and an E&M wave will quickly damp away, and the only transport of an electric field or magnetic field is by diffusion. Magnetic field diffusion in metals is a well-established experimental phenomena; Electric field diffusion is more difficult to observe because the electric fields move charges within the conducting medium in precisely the manner necessary to cancel themselves out. Nevertheless, Eq. 9.35 predicts the phenomenon, which we will discuss in greater detail in Chapter 10, of "ultra-violet transparency," which occurs in most metals: Because the conductivity of most common metals is limited, for a sufficiently high frequency the ratio in Eq. 9.35 can become much less than 1, and can thus be within the wave propagation limit, so that the E&M wave can penetrate well inside of the medium.

9.4.4 Transient response in a conductor

To further explore the idea of the characteristic time of a material and separate it conceptually from that of the fields, we consider an example that makes no explicit reference to the fields. Recall the continuity equation

$$\nabla \cdot \vec{\mathcal{J}}_f + \frac{\partial \rho_f}{\partial t} = 0 \tag{9.36}$$

that is valid in all cases for which there are no charge sources or sinks. In special cases for which there is no charge accumulation, we would have $\nabla \cdot \vec{\mathcal{J}}_f = \nabla \cdot \vec{\mathcal{J}}_{stat} = 0$, which represents a stationary condition of current flow. In general, if $\vec{\mathcal{J}}_f$ is briefly perturbed from its stationary condition, for a characteristic transient period of time, τ, we will have $\nabla \cdot \vec{\mathcal{J}}_f \neq 0$ and different regions will variously collect and deplete free charge, as shown mathematically in Eq. 9.36, to compensate for this. Eventually, after multiple characteristic times, the current will settle to the stationary condition.

We can represent the continuity equation as a first order non-homogeneous ODE. Starting with the continuity Eq. 9.36 and using $\nabla \cdot \vec{D} = \rho_f$ yields,

$$\nabla \cdot \left(\vec{\mathcal{J}}_f + \frac{\partial \vec{D}}{\partial t} \right) = \nabla \cdot \vec{\mathcal{J}}_{stat} = 0$$

where we have defined $\vec{\mathcal{J}}_{stat} = \vec{\mathcal{J}}_f + \frac{\partial \vec{D}}{\partial t}$. Next, noting that $\vec{D} = (\varepsilon_{b1}/\sigma_{f1})\vec{\mathcal{J}}_f$, we write,

$$\nabla \cdot \left(\vec{\mathcal{J}}_f + \frac{\varepsilon_{b1}}{\sigma_{f1}} \frac{\partial \vec{\mathcal{J}}_f}{\partial t} \right) = \nabla \cdot \vec{\mathcal{J}}_{stat} = 0$$

and we have the non-homogeneous ODE,

$$\vec{\mathcal{J}}_f(t) + \frac{\varepsilon_{b1}}{\sigma_{f1}} \frac{\partial \vec{\mathcal{J}}_f}{\partial t} = \vec{\mathcal{J}}_{stat} \tag{9.37}$$

where, by definition, $\vec{\mathcal{J}}_{stat}$ is constant in time and has zero divergence. A particular solution to this ODE is $\vec{\mathcal{J}}_f(t) = \vec{\mathcal{J}}_{stat}$ since $\frac{\partial \vec{\mathcal{J}}_{stat}}{\partial t} = 0$. Combining this with the solution to the associated homogeneous equation, $\vec{\mathcal{J}}_o exp\left(-\frac{t}{\varepsilon/\sigma}\right)$, yields the solution:

$$\vec{\mathcal{J}}_f(t) = \vec{\mathcal{J}}_o exp\left(-\frac{t}{\varepsilon_{b1}/\sigma_{f1}}\right) + \vec{\mathcal{J}}_{stat} \tag{9.38}$$

which means a perturbation of amplitude $\vec{\mathcal{J}}_o$ occurring at time $t = 0$ causes a transient non-stationary condition in $\vec{\mathcal{J}}_f$ with a characteristic relaxation time given by $\tau = \varepsilon_{b1}/\sigma_{f1}$. Specifically, if there is initially no current in the medium (trivial stationary condition in which $\vec{\mathcal{J}}_{stat} = 0$) and a free current is excited, the characteristic time for it to decay is given by τ.

9.4.5 Temporal wave-packet

Previous to now, we have considered single frequency E-field plane waves within matter. We now consider general plane waves that are a superposition of monochromatic waves of different frequencies. In the Drude model discussion in Chapter 10, we will find frequency dependent forms of the complex refractive index $\tilde{n}(\omega)$ and wavenumber $\tilde{k}(\omega)$ within matter. Combining this with the definition of the phase velocity of a plane wave solution, it is clear that each frequency component of a general plane wave has a phase velocity given by Eq. 9.28,

$$v_p(\omega) = \frac{\omega}{k(\omega)} = \frac{c}{n(\omega)} \tag{9.39}$$

where both k and n are the real parts of their associated complex functions. If one is measuring the motion of a wave composed of several frequencies, the apparent motion of this collection is described not by the phase velocity, but rather by the group velocity,

$$v_g(\omega) = \frac{\mathrm{d}\omega}{\mathrm{d}k} \tag{9.40}$$

which, indeed, is the speed of energy transfer at frequency ω. This is a distinction without a difference when considering wave propagation in a vacuum or any medium with a frequency-independent refractive index. However, it is a well established experimental observation that there are important differences between the phase and group velocities for wave propagation in virtually all dispersive non-vacuum environments. To explore this difference we first invert the relation $k(\omega) = \frac{\omega}{c}n(\omega)$ and write $\omega(k) = ck/n(k)$. A temporal "wave-packet" or pulse, with a "carrier envelope" having a central wavenumber k_o and associated central frequency $\omega_o = \omega(k_o)$ can then be formed as an integral over k of our plane wave solutions to the Helmholtz equation (take $\vec{k}$ along the x-axis for simplicity) (**see Discussion 9.1**)

$$E(x,t) = \int_{-\infty}^{\infty} \tilde{E}(k)e^{-i(kx-\omega(k)t)}\mathrm{d}k. \tag{9.41}$$

where the E-field has been reduced to a scalar. Thus we have expressed the wave-packet occupying a length Δx at some time t_o as a coherent addition of plane waves with a spread in the wave vector's Δk.

The Fourier complementarity between Δx and Δk is generally prescribed as $\Delta k \Delta x \geq 1/2$. We could have just as well specified the pulse as being composed of a spread in frequencies $\Delta\omega$ and existing for the duration Δt about time t_o. In this case, the Fourier result is stated, equivalently, as $\Delta\omega \Delta t \geq 1/2$. If we restrict our discussion to the physically reasonable requirement that the spatial spread of the wave-packet obeys

$$\lambda \ll \Delta x \tag{9.42}$$

then we limit $\tilde{E}(k)$ to a narrow band of frequencies and k-vectors (such that $\Delta\omega/\omega$ and $\Delta k/k$ are $\ll 1)^5$. In general, $\omega(k)$ is not a linear function of k. However, within the restriction of Eq. 9.42, $\Delta\omega(k) = \omega(k) - \omega_o$ can be approximated as

$$\Delta\omega(k) \simeq \Delta\omega' = \omega'(k) - \omega_o = \left.\frac{d\omega}{dk}\right|_{k0} \Delta k \tag{9.43}$$

and we can rewrite Eq. 9.41 as

$$E(x,t) = e^{-i(k_o x - \omega_o t)} \int_{-\infty}^{\infty} \tilde{E}(k) e^{-i[(k-k_o)x - (\omega'(k)-\omega_o)\,t]} dk \tag{9.44}$$

$$= e^{-i(k_o x - \omega_o t)} \int_{-\infty}^{\infty} \tilde{E}(k) e^{-i[(\Delta k)x - (\Delta\omega')\,t]} dk \tag{9.45}$$

$$= e^{-i(k_o x - \omega_o t)} \int_{-\infty}^{\infty} \tilde{E}(k) e^{-i(\Delta k)[x - (\Delta\omega'/\Delta k)\,t]} dk \tag{9.46}$$

where k_o is the central wave number and $\omega_o = \omega(k_o)$ is its associated frequency and, as before, $k - k_o = \Delta k$ and $\omega'(k) - \omega_o = \Delta\omega'$. Now letting

$$\mathscr{E}\left[x - \left(\left.\frac{d\omega}{dk}\right|_{k0}\right)t\right] = \int_{-\infty}^{\infty} \tilde{E}(k) e^{-i(\Delta k)[x - (\Delta\omega'/\Delta k)\,t]} dk \tag{9.47}$$

where we note that this function depends on position and time only through the combination $x - (\Delta\omega'/\Delta k)\,t$ and from Eq. 9.43, $\Delta\omega'/\Delta k = \frac{d\omega}{dk}|_{k_o}$ is a constant independent of the integration variable, k. Substituting this into Eq. 9.44, we have

$$E(x,t) = \mathcal{E}\left[x - \left(\left.\frac{d\omega}{dk}\right|_{k0}\right)t\right] e^{-i(k_o x - \omega_o t)} \tag{9.48}$$

[5] If $\lambda \ll \Delta x$ and if $\Delta k \Delta x \simeq 1/2$, then $\Delta k \lambda \ll 1/2$ or $\Delta k/k \ll 1/4\pi$.

If $\mathcal{E}(x,t)$ is a slowly varying function of x and t, then Eq. 9.48 represents a "wave-packet" or envelope of a finite train of oscillations that propagates at velocity $\left.\frac{d\omega}{dk}\right|_{k_0}$ through the material with (nearly) constant amplitude and width.

9.4.6 Group velocity versus phase velocity

Now we can specify a rather unambiguous definition of group velocity (see Fig. 9.3)

> *The group velocity is the velocity that an observer must move in the material such that $\mathcal{E}(x,t)$ appears to be roughly constant in time.*

which is to say, if we write $\mathcal{E}(x,t) = \mathcal{E}\left[x - \left(\left.\frac{d\omega}{dk}\right|_{k_0}\right)t\right] = \mathcal{E}(\chi)$,

$$\frac{d\mathcal{E}}{dt} = \frac{\partial \mathcal{E}}{\partial \chi}\frac{\partial \chi}{\partial x}\frac{dx}{dt} + \frac{\partial \mathcal{E}}{\partial \chi}\frac{\partial \chi}{\partial t} = 0 \qquad (9.49)$$

yielding, from Eq. 9.47[6]

$$\frac{\partial \mathcal{E}}{\partial \chi}\frac{dx}{dt} - \frac{\partial \mathcal{E}}{\partial \chi}\left(\left.\frac{d\omega}{dk}\right|_{k_0}\right) = 0$$

or

$$v_g \equiv \frac{dx}{dt} = \left.\frac{d\omega}{dk}\right|_{k_0} = \frac{\Delta\omega'}{\Delta k} \qquad (9.50)$$

where we have used the fact that $\chi = x - \left.\frac{d\omega}{dk}\right|_{k_0}$ is independent of the integration variable k in Eq. 9.47 so that its derivatives $\left(\frac{\partial \chi}{\partial x} \text{ and } \frac{\partial \chi}{\partial t}\right)$ can be brought outside the integral.

[6] In general, x and t are independent variables within $\mathcal{E}(x,t) = \mathcal{E}\left[x - \left(\left.\frac{d\omega}{dk}\right|_{k_0}\right)t\right] = \mathcal{E}(\chi)$. However, setting the constraint

$$d\mathcal{E} = \frac{\partial \mathcal{E}}{\partial x}dx + \frac{\partial \mathcal{E}}{\partial t}dt = 0$$

or

$$d\mathcal{E} = \frac{\partial \mathcal{E}}{\partial \chi}\frac{\partial \chi}{\partial x}dx + \frac{\partial \mathcal{E}}{\partial \chi}\frac{\partial \chi}{\partial t}dt = 0$$

forces a relationship between x and t, namely,

$$\frac{\partial \mathcal{E}}{\partial \chi}dx - \frac{\partial \mathcal{E}}{\partial \chi}\left(\left.\frac{d\omega}{dk}\right|_{k_0}\right)dt = 0$$

$$\text{or} \quad \frac{dx}{dt} = \left.\frac{d\omega}{dk}\right|_{k_0}$$

so to satisfy the constraint that $\mathcal{E}(x,t)$ is constant in time, we must move at velocity $dx/dt = v_g = \left.\frac{d\omega}{dk}\right|_{k_0}$.

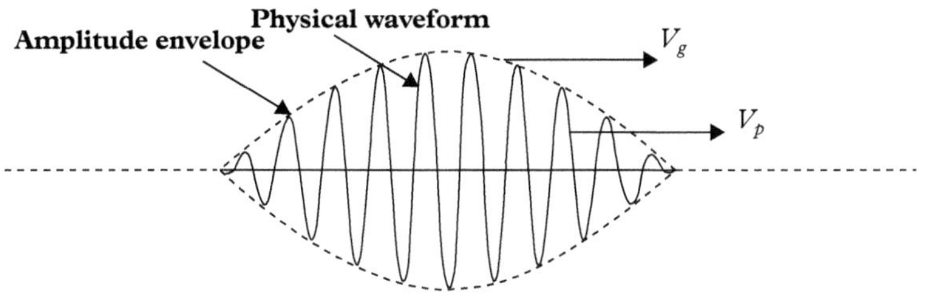

Fig. 9.3 *When the phase velocity V_p is not equal to the group velocity V_g, the high frequency oscillations will move relative to the envelope, which in turn moves at the group velocity through the medium.*

Our restriction that the envelope $\mathcal{E}(x,t)$ of Eq. 9.47 is a slowly changing function of position is necessary and more significant than may be obvious at first approach: To define a group velocity and simultaneously identify this with the rate of energy transfer through the material, the spatial form of the envelope must not vary at least for the time $\Delta x/c$. We thus have two distinctly different velocity concepts. In a material that has a defined relationship of $\omega(k)$ versus k, at $\omega=\omega_0$ and $k=k_0$,

$$\text{phase velocity} = v_p = \frac{\omega_0}{k_0},$$

$$\text{group velocity} = v_g = \left.\frac{d\omega}{dk}\right|_{k_0} \tag{9.51}$$

An expression for the group velocity in terms of the index of refraction is obtained by noting that $k(\omega)=\frac{\omega n(\omega)}{c}$, so

$$(v_g)^{-1} = \left.\frac{dk}{d\omega}\right|_{\omega_0} = \frac{1}{c}\left[n(\omega)+\omega\frac{dn(\omega)}{d\omega}\right]_{\omega_0} \tag{9.52}$$

$$v_g = \frac{c}{\left[n(\omega)+\omega\frac{dn(\omega)}{d\omega}\right]_{\omega_0}} \tag{9.53}$$

$$= \frac{v_p}{\left[1+\frac{\omega}{n(\omega)}\frac{dn(\omega)}{d\omega}\right]_{\omega_0}} \tag{9.54}$$

These equations make it clear that the group velocity and the phase velocity will differ in any medium for which the index of refraction is not independent of the frequency. For most materials, at most frequencies, $\frac{dn(\omega)}{d\omega}>0$ (so-called normally dispersive), with the result that the group velocity is less than or equal to the phase velocity, and certainly less than c. Nevertheless, Eq. 9.53 does admit the possibility that $v_g>c$ when $\frac{dn(\omega)}{d\omega}<0$ (so-called anomalous dispersion that occurs in the vicinity of a resonance). However, under these special circumstances, the index has a large (and dominant) imaginary term giving rise to strong absorption, resulting in just the kind of change in the envelope function our formalism cannot describe. Careful analysis of these situations reveals that the flow of energy never exceeds c (**see Discussion 9.2**).

9.4.7 Pulse broadening

One of the characteristics of a dispersive medium is that a pulse of radiation will change shape as it propagates. The reason is clear from the previous discussion: A radiation pulse necessarily has a spectrum of frequency components with a width inversely proportional to its temporal width as given by simple Fourier analysis. For a temporal pulse that is formed at t_o such that it spans a distance Δx, there is a spread of wave numbers, Δk, about the central wave number, k_o, whose magnitude satisfies:

$$\Delta k \gtrsim \frac{1}{\Delta x} \tag{9.55}$$

From Eq. 9.40, $v_g = \dfrac{d\omega(k)}{dk}$, and the rate of change of this velocity with k is $\dfrac{dv_g}{dk} = \dfrac{d^2\omega(k)}{dk^2} = v_g'$. Using Eq. 9.55, we find the range of effective group velocities around frequency $\omega_o = \omega(k_o)$ is determined by

$$\Delta v_g \simeq v_g'(\omega_o)\,\Delta k \gtrsim \frac{v_g'(\omega_o)}{\Delta x} \tag{9.56}$$

Clearly in a non-dispersive medium such as free space, or in a dispersive material at a specific frequency where the second derivative of the frequency with respect to k is zero, there will be no spread of the velocities around v_g (Fig. 9.4). However, in general, for a pulse

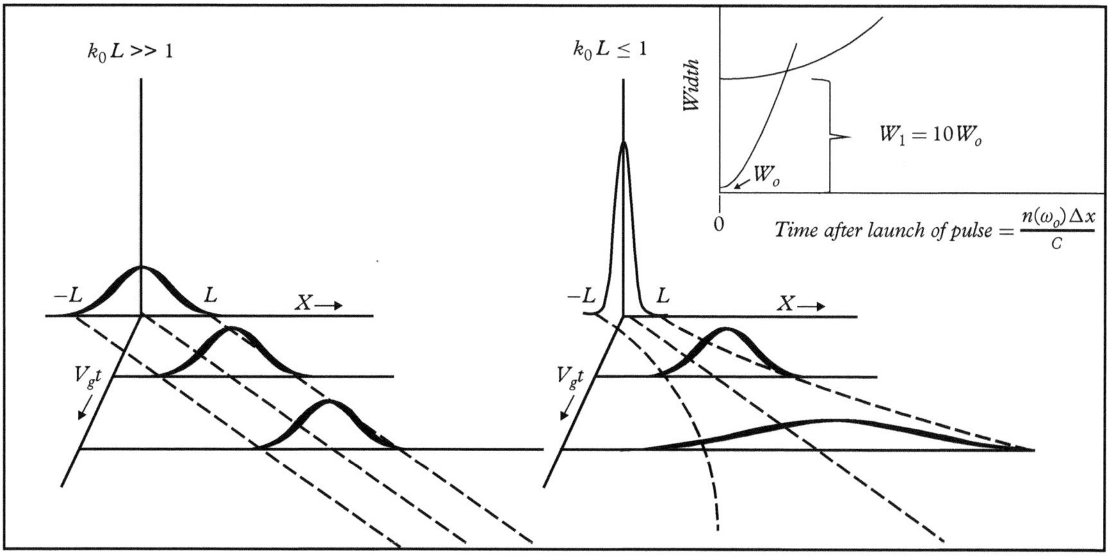

Fig. 9.4 *Pulse broadening in a dispersive medium. The narrower the original pulse is in space at the launch, the faster it broadens.*

of radiation passing through a dispersive material, the spread S in the initial width Δx after a time t is

$$S(\omega_o) = \Delta v_g t \gtrsim \frac{v_g'(\omega_o)}{\Delta x} t \tag{9.57}$$

Thus, as represented in Fig. 9.4, the narrower in length the original shape of the pulse is, the faster the pulse will spread with time (due to a broader spectrum and thus broader range of wave speeds). Maintaining extremely short pulses in propagation through materials that have a nonlinear dependence of the frequency on the k vector is difficult.

9.5 Plane waves at interfaces

9.5.1 Boundaries

In general, when electromagnetic radiation passes across a sharp boundary from one material to another, each with its own permittivity and permeability, $\tilde{\varepsilon}$ and $\tilde{\mu}$, and thus different indices of refraction $\tilde{n}(\omega)$, part of the incident radiation transmits, part of it reflects and we might well expect that there will be some absorptive loss in transmission. In this discussion, we consider a derivation of the reflection/transmission properties of an interface resulting in equations in which the refractive indexes (and thus the wavevectors) of the two media are taken as complex. In particular, we will maintain our prior assumption of a complex permittivity, $\tilde{\varepsilon}$, but now we also allow for the possibility of a magnetic material with complex permeability, $\tilde{\mu}$.

We begin with the electromagnetic plane wave solutions to the Helmholtz equations (Eqs. 9.18 and 9.19). As covered in Section 1.6, any arbitrary polarization of such a plane wave can be decomposed into two orthogonal linear polarizations. Because of this, and since the reflection/transmission from a surface is highly dependent on the direction of polarization, we will separately consider two special cases in which the $\vec{E}$ field is either in the plane of incidence ("P" polarization) or perpendicular to it ("S" polarization).

For each $\vec{E}$ field polarization, the corresponding $\vec{B}$ field is given by $\frac{1}{\omega}(\vec{k} \times \vec{E})$, where the wavevector $\vec{k}$ is potentially complex. Figure 9.5 shows the orientation and notation of the E fields and interface for this discussion: In general, an "S" or "P" polarized plane wave, incident from the upper left and moving along $\vec{k}$, encounters the horizontal interface at the origin where it partially reflects along $\vec{k}''$

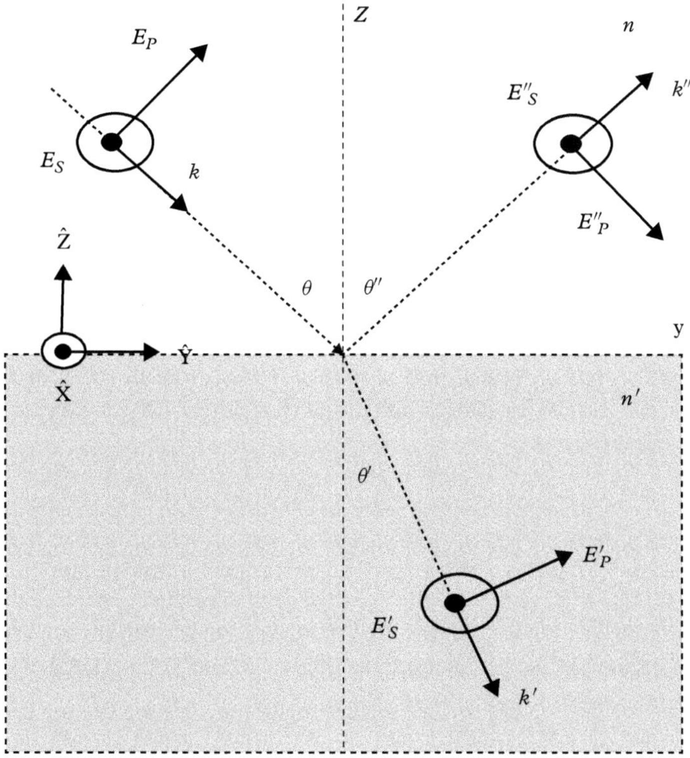

Fig. 9.5 *Reflection and transmission of fields at a boundary. There are two polarizations shown: "P," in which the incident E field is in the plane of incidence, and "S," in which the incident E field is perpendicular to this plane (it is in the plane of the interface) and directed out of the figure. It is important to note our initial choices of the positive directions of the various transmitted E′ and reflected E″ vectors in the figure because they ultimately determine the signs in the Fresnel Equations. The associated B fields (not shown) are related to the E fields, in amplitude and direction, by* $\vec{B} = \frac{1}{\omega}(\vec{k} \times \vec{E})$.

and partially transmits along $\vec{k}'$. The associated $\vec{B}$ fields are not shown. The directions $\hat{x}$, $\hat{y}$, and $\hat{z}$ define the mutually orthogonal coordinate basis set, with $\hat{x}$ perpendicular to the plane of incidence and directed out of the page, $\hat{y}$ in the plane of the interface and in the direction of propagation, and $\hat{z}$ perpendicular to the interface and directed upward, as shown. For material above the surface, the permittivity and permeability are given, respectively, by $\tilde{\varepsilon}$ and $\tilde{\mu}$ while for the material below the surface these quantities are given as $\tilde{\varepsilon}'$ and $\tilde{\mu}'$. In the figure, the indices of refraction are indicated above and below the surface as $\tilde{n}$ and $\tilde{n}'$, respectively.

The reflection/transmission equations we seek are obtained when the fields ($\vec{D}$, $\vec{E}$, $\vec{B}$, and $\vec{H}$) of the three plane waves satisfy specific conditions, derived from Maxwell's equations, everywhere and at all times on the interface between the two materials. These polarization-independent "boundary conditions" are given as (for no free charges or currents):

Table 9.1

(1) Continuity of $\vec{D}$ normal to interface	$\left(\tilde{\varepsilon}(\vec{E}+\vec{E}'')-\tilde{\varepsilon}'\vec{E}'\right)\cdot\hat{z}=0$
(2) Continuity of $\vec{B}$ normal to interface	$\left(\vec{B}+\vec{B}''-\vec{B}'\right)\cdot\hat{z}=0$
(3) Continuity of $\vec{E}$ parallel to interface	$\left(\vec{E}+\vec{E}''-\vec{E}'\right)\times\hat{z}=0$
(4) Continuity of $\vec{H}$ parallel to interface	$\left(\frac{1}{\tilde{\mu}}(\vec{B}+\vec{B}'')-\frac{1}{\tilde{\mu}'}\vec{B}'\right)\times\hat{z}=0$

where $\tilde{\mu}''=\tilde{\mu}$, and $\tilde{\varepsilon}''=\tilde{\varepsilon}$ since the incident and reflected waves are in the same material. Common to these boundary conditions is the implied condition that along the interface surface the phases of the incident, reflected and transmitted waves (all in the form of Eq. 9.27) must be equal at all points and at all times.[7] This is expressed mathematically as

$$\vec{k}\cdot y\hat{y}-\omega t=\vec{k}'\cdot y\hat{y}-\omega' t=\vec{k}''\cdot y\hat{y}-\omega'' t \tag{9.58}$$

Equation 9.58 yields a number of important relationships that do not depend upon the details of the wave. First, by considering the point of the origin ($y=0$), we confirm that the frequencies of all three waves are the same: $\omega''=\omega'=\omega$. Next, setting $t=0$, and noting that $\vec{k}\cdot y\hat{y}=(k_r+ik_i)\sin\theta$ and so on, we obtain

$$\tilde{k}\sin\theta=\tilde{k}'\sin\theta'=\tilde{k}''\sin\theta'' \tag{9.59}$$

which, because $\tilde{k}=\tilde{k}''$, shows the equality of the incident and reflected angles,

Law of reflections

$$\theta=\theta'' \tag{9.60}$$

And noting the complex relation: $\tilde{k}=\tilde{n}\omega/c$, Eq. 9.59 also gives a complex version of Snell's law,

$$\frac{\tilde{n}\omega}{c}\sin\theta=\frac{\tilde{n}'\omega}{c}\sin\theta' \tag{9.61}$$

within which the equality of the real components gives the more familiar version,

Snell's law of refraction

$$n\sin\theta=n'\sin\theta' \tag{9.62}$$

[7] For example, the surface $(x,y,0)$ boundary condition stated in the text as

$$\left(\vec{E}+\vec{E}''-\vec{E}'\right)\times\hat{z}=0$$

is more explicitly written as

$$\Big[\vec{E}_0\exp\left(i\vec{k}\cdot\vec{r}-i\omega t\right)+\vec{E}_0''\exp\left(i\vec{k}''\cdot\vec{r}-i\omega'' t\right)-\vec{E}_0'\exp\left(i\vec{k}'\cdot\vec{r}-i\omega' t\right)\Big]\times\hat{z}=0$$

and for this to hold at all surface positions and times, it follows that

$$\vec{k}\cdot(x\hat{x}+y\hat{y})-\omega t=\vec{k}''\cdot(x\hat{x}+y\hat{y})-\omega'' t=\vec{k}'\cdot(x\hat{x}+y\hat{y})-\omega' t$$

and, in particular, because $\vec{k}$ has no component in the $\hat{x}$ direction, this is written

$$\vec{k}\cdot y\hat{y}-\omega t=\vec{k}''\cdot y\hat{y}-\omega'' t=\vec{k}'\cdot y\hat{y}-\omega' t.$$

9.5.2 Fresnel transmission and reflection amplitude coefficients

Using the boundary continuity relations in Table 9.1 along with the chosen field reference directions of Fig. 9.5, we are now in a position to calculate the complex reflected and transmitted fields at a sharp interface resulting from an S or P polarized plane wave incident to that interface at angle θ. The various complex electric waves in the figure all have the general form

$$\vec{E} = \vec{E}_0 e^{i\phi} \exp\left(i\vec{k}\cdot\vec{r} - i\omega t\right)$$

where we will identify the associated complex amplitude in the following as $\tilde{E}_0 = \left|\vec{E}_0\right| e^{i\phi}$.

(1) "***S*-Polarization**:" $\vec{E}_s$ is perpendicular to the plane of incidence and parallel to the interface, so we make use of the continuity of the parallel components of $\vec{E}$ and $\vec{H}$ (third and fourth rows of Table 9.1):

(a) The third boundary condition in $\vec{E}$ directly yields

$$\tilde{E}_{0s} + \tilde{E}''_{0s} - \tilde{E}'_{0s} = 0 \tag{9.63}$$

(b) For the fourth boundary condition in $\vec{H} = \frac{1}{\tilde{\mu}}\vec{B}$, employ the general relation $\vec{B} = \frac{\tilde{k}}{\omega}(\hat{k}\times\vec{E})$ to yield (**see Discussion 9.3**)

$$\frac{\tilde{n}}{\tilde{\mu}}\left(-\tilde{E}_{0s} + \tilde{E}''_{0s}\right)\cos\theta + \frac{\tilde{n}'}{\tilde{\mu}'}\tilde{E}'_{0s}\cos\theta' = 0 \tag{9.64}$$

(c) Solve Eqs. 9.63 and 9.64 and use the complex version of Snell's law, Eq. 9.61, to obtain the relative transmitted and reflected complex amplitudes, $\tilde{E}'_{0s}/\tilde{E}_{0s}$ and $\tilde{E}''_{0s}/\tilde{E}_{0s}$, completely in terms of the incident angle θ (**see Discussion 9.4**):

$$\frac{\tilde{E}'_{0s}}{\tilde{E}_{0s}} = \frac{2\tilde{n}\cos\theta}{\tilde{n}\cos\theta + (\tilde{\mu}/\tilde{\mu}')\sqrt{\tilde{n}'^2 - \tilde{n}^2\sin^2\theta}} \tag{9.65}$$

$$\frac{\tilde{E}''_{0s}}{\tilde{E}_{0s}} = \frac{\tilde{n}\cos\theta - (\tilde{\mu}/\tilde{\mu}')\sqrt{\tilde{n}'^2 - \tilde{n}^2\sin^2\theta}}{\tilde{n}\cos\theta + (\tilde{\mu}/\tilde{\mu}')\sqrt{\tilde{n}'^2 - \tilde{n}^2\sin^2\theta}} \tag{9.66}$$

(2) "***P*-Polarization**:" $\vec{E}_p$ is in the plane of incidence so we can again use the continuity of the parallel component of $\vec{E}$ and the parallel component of $\vec{H}$ (third and fourth rows of Table 9.1):

(a) The third boundary condition in $\vec{E}$ yields

$$\tilde{E}_{0p}\cos\theta + \tilde{E}''_{0p}\cos\theta - \tilde{E}'_{0p}\cos\theta' = 0 \tag{9.67}$$

(b) Again, for the fourth boundary condition in $\vec{H} = \frac{1}{\tilde{\mu}}\vec{B}$, we employ the general relation $\vec{B} = \frac{\tilde{k}}{\omega}(\hat{k}\times\vec{E})$. Referring to Fig. 9.5 and noting the directions of $\vec{E}_p$, $\vec{E}'_p$ and $\vec{E}''_p$ at the boundary, we find (**see Discussion 9.5**)

$$-\frac{\tilde{n}}{\tilde{\mu}}\left(\tilde{E}_{0p} - \tilde{E}''_{0p}\right) + \frac{\tilde{n}'}{\tilde{\mu}'}\tilde{E}'_{0p} = 0 \tag{9.68}$$

(c) Solve Eqs. 9.67 and 9.68, and again use the complex version of Snell's law, Eq. 9.61, to obtain the relative transmitted and reflected complex amplitudes, $\tilde{E}'_{0p}/\tilde{E}_{0p}$ and $\tilde{E}''_{0p}/\tilde{E}_{0p}$, completely in terms of the incident angle θ (**see Discussion 9.6**).[8]

$$\frac{\tilde{E}'_{0p}}{\tilde{E}_{0p}} = \frac{2\tilde{n}\tilde{n}'\cos\theta}{(\tilde{\mu}/\tilde{\mu}')\,\tilde{n}'^2\cos\theta + \tilde{n}\sqrt{\tilde{n}'^2 - \tilde{n}^2\sin^2\theta}} \tag{9.69}$$

$$\frac{\tilde{E}''_{0p}}{\tilde{E}_{0p}} = \frac{-(\tilde{\mu}/\tilde{\mu}')\,\tilde{n}'^2\cos\theta + \tilde{n}\sqrt{\tilde{n}'^2 - \tilde{n}^2\sin^2\theta}}{(\tilde{\mu}/\tilde{\mu}')\,\tilde{n}'^2\cos\theta + \tilde{n}\sqrt{\tilde{n}'^2 - \tilde{n}^2\sin^2\theta}} \tag{9.70}$$

(3) **Four Angles of Importance** $(\theta = 0^o, \theta = 90^o, \theta = \theta_B, \theta = \theta_C)$ Referring to Fig. 9.6, we can identify four special angles:

(a) At an angle of incidence of $\theta = 0$ ("normal incidence"), S and P polarizations are the same so that with $\tilde{\mu} \simeq \tilde{\mu}' \simeq \mu_0$, both (1) and (2) simplify to

$$\left(\frac{\tilde{E}'_0}{\tilde{E}_0}\right)_{norm} = \frac{2\tilde{n}}{\tilde{n}+\tilde{n}'} \quad \text{and} \quad \left(\frac{\tilde{E}''_0}{\tilde{E}_0}\right)_{norm} = \frac{\tilde{n}-\tilde{n}'}{\tilde{n}+\tilde{n}'} \tag{9.71}$$

(b) At $\theta = 90°$, the interface is 100% reflective for both S- and P-polarizations. This is evident in the forms of Eqs. 9.65 and 9.66 for S-polarization and Eqs. 9.69 and 9.70 for P-polarization.

(c) For the general case of P polarized light, the reflection coefficient reaches a minima at the so-called "Brewster's

[8] Our expressions are correct for the directions of the field vectors as given in our figure 9.5. If it is found that such solutions in other texts differ in sign then these differences are likely due to differences in the reference directions initially chosen for the various field vectors. For example, for P polarization light, our reflection coefficients appear to be the negative of the expressions given in Jackson. Jackson's equations are correct for the field reference directions defined in his drawings (Jackson, J.D., *Classical Electrodynamics*, 3rd edition, Wiley, NY., 1999. Fig. 7.6 (b)).

Angle." In particular, for P-polarized light incident to a boundary in which indices are essentially real ($\tilde{n}' = n'$ and $\tilde{n} = n$), and the materials are non-magnetic ($\tilde{\mu} = \tilde{\mu}' = \mu_0$), the reflection coefficient completely vanishes at this angle. In such a case, the numerator of Eq. 9.70 vanishes:

$$-n'^2 \cos\theta + n\sqrt{n'^2 - n^2 \sin^2\theta} = 0 \tag{9.72}$$

and the solution for this "Brewster's Angle" is obtained as

$$\theta_B = \tan^{-1}\left(\frac{n'}{n}\right)$$

This special case of zero P-polarized light reflection is depicted in Fig. 9.6 where θ_B is indicated. We also note from this figure (and from the numerator of Eq. 9.70) that for angles less than Brewster's angle, the incident E field undergoes a 180° phase change upon reflection. Contrast this with the S-polarization field, which undergoes a 180° degree phase change, upon reflection, for all incident angles (see the numerator of Eq. 9.66);[9] and to the transmitted light that, for both S and P cases, remains in phase with the incident light. When light having both S and P components strikes such an interface at the Brewster's angle, both components have a transmitted fraction, while only the S component has a reflected component. This property can be exploited to "polarize" a beam of light (with loss); propagating P-polarized through interfaces with no loss is also a frequent application of this phenomenon.

(d) For the special case of real n and n', when $n > n'$ as on the right-hand side of Fig. 9.6, there is an incident angle ($< 90°$) for which the refracted angle equals 90° (parallel to the surface). At this point, the reflection coefficient is 1, and the transmission coefficient is 0, for both P and S polarizations. This angle is called the Critical Angle:

$$\theta_C = \sin^{-1}\left(\frac{n'}{n}\right)$$

Beyond the critical angle the reflection coefficient remains 1, but the phase of the reflected beams depends strongly upon the polarization and angle of incidence ($\theta_C < \theta_i < 90$).

[9] To avoid confusion, one must keep in mind the initially defined reference field directions in Fig. 9.5. It is these reference directions which dictate the signs of the amplitude coefficients shown in Eqs. 9.65, 9.66, 9.69, and 9.70 and that are depicted in Fig. 9.6.

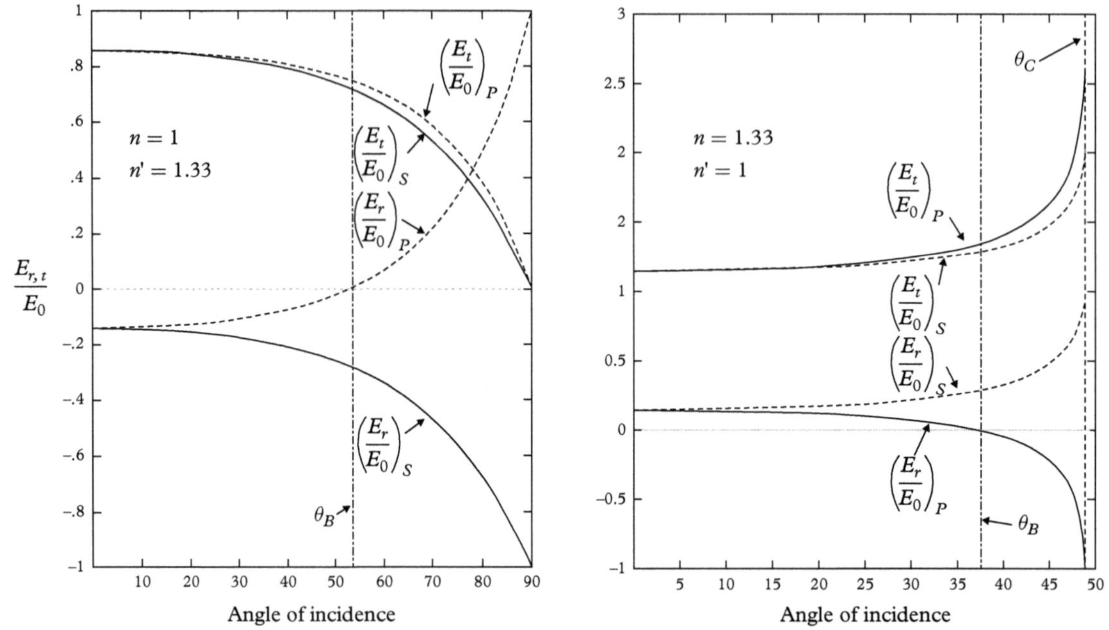

Fig. 9.6 *The ratio of the S and P reflected fields (subscript r) and transmitted fields (subscript t) to the incident field as a function of the angle. On the left the incident index is $n = 1$, with a final index of $n' = 1.33$, while on the right the indexes are interchanged. The angle for which the reflected P-polarized light changes signs is "Brewster's Angle" (θ_B). For the case on the right of $n > n'$, the angle beyond which the solutions to the Fresnel equations are no longer real is marked as the "critical angle" (θ_C). We discuss such cases in Section 9.5.3.*

The transmission and reflection Eqs. of 9.65, 9.66, 9.69, and 9.70 are known as the Fresnel amplitude coefficients. They are fully complex equations and are, of course, equally valid in the purely real context such as depicted in Fig. 9.6. However, as fully complex equations the resulting angles can be complex, making a straightforward geometric interpretation impossible in such cases. In practice, most applications use real indices of refraction, especially on the incidental side, which is frequently considered to be a vacuum.

9.5.3 Total internal reflection

Total internal reflection (TIR) is the name given to the phenomenon described in (d) of Section 9.5.2 when the indices are essentially real ($\tilde{n}' = n'$ and $\tilde{n} = n$) and $n > n'$. For angles greater than the critical angle, the incident radiation is completely reflected back into the initial medium. TIR is depicted in Fig. 9.7.

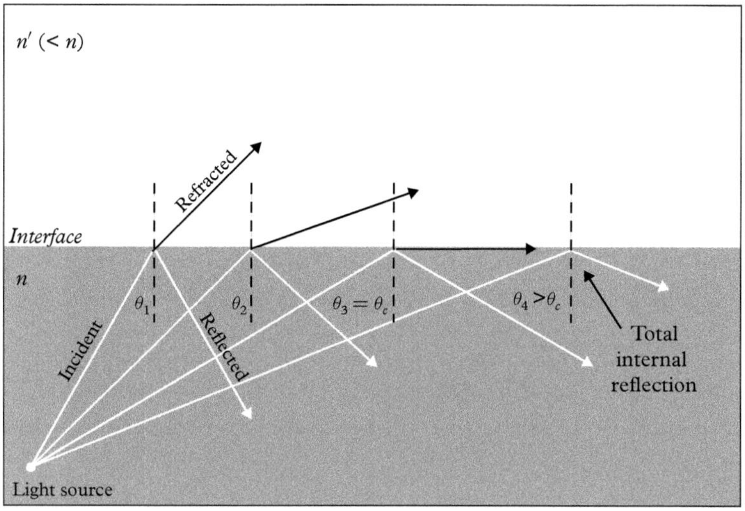

Fig. 9.7 *For the case of purely real indices of refraction in which $n > n'$, the light is refracted at an angle larger than the incident angle. Consequently, there is an incident angle ($< 90°$) for which the angle of refraction equals 90° (parallel to the surface). For this angle (θ_3 in the figure) the reflection coefficient is 1, and the transmission coefficient is 0, for both P and S polarizations. This angle is called the Critical Angle, θ_C. At angles larger than the Critical Angle, the reflection coefficient remains 1–that is, the light is totally internally reflected.*

We can examine the phase change this reflected wave suffers by formally extending the Fresnel equations beyond the critical angle and generating a reflection ratio that is complex. In both reflectivity Eqs. 9.66 and 9.70, for $\theta > \theta_C$ we have

$$\left(\frac{\tilde{E}''_0}{\tilde{E}_0}\right)_{S/P} = \pm\left(\frac{A^{S/P} - iB^{S/P}}{A^{S/P} + iB^{S/P}}\right) = \pm\exp(-2i\tan^{-1}(B^{S/P}/A^{S/P}) \tag{9.73}$$

where for S polarization,

$$A^S = n\cos\theta \quad \text{and} \quad B^S = \sqrt{n^2\sin^2\theta - n'^2}$$

and for P polarization,

$$A^p = n'^2\cos\theta \quad \text{and} \quad B^P = n\sqrt{n^2\sin^2\theta - n'^2}$$

in which we have used the standard approximation that $\tilde{\mu}/\tilde{\mu}' \simeq 1$ and for $n > n'$, the A's and B's are both real. From the exponential representation of the complex number in Eq. 9.73, we have the expected result that for either S or P polarized light incident at $\theta > \theta_C$, the reflectivity coefficient has a magnitude of 1.

Also, the reflected fields have an angle dependent phase shift relative to the input, and further, this phase shift differs for the two polarizations. As shown in Fig. 9.8, for the case of an initial index of $n = 1.6$ and a final index of $n' = 1$, the difference in phase shifts upon

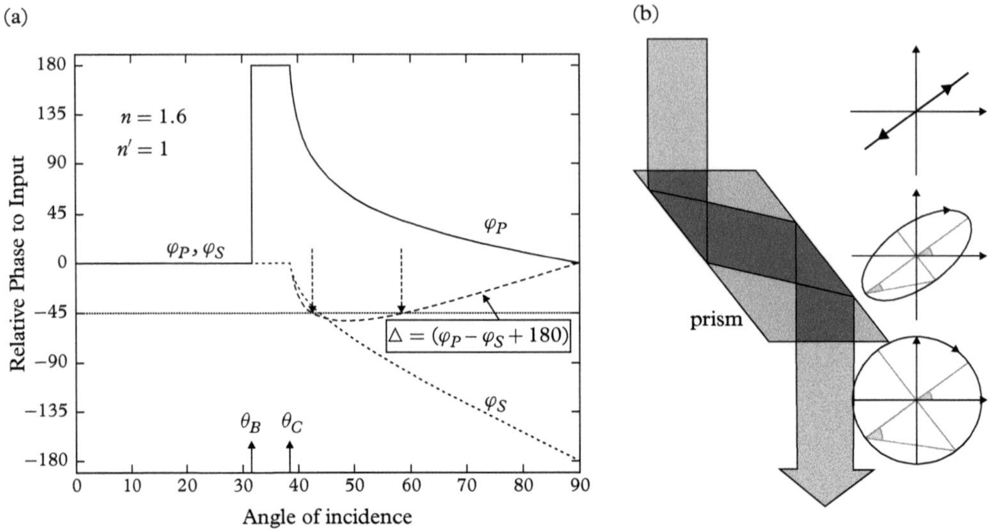

Fig. 9.8 *Part A shows the phase of reflected light relative to the incident wave as a function of angle for* $n = 1.6$ *and* $n' = 1$*. The difference between the phase shifts for the two polarizations,* Δ*, is plotted as the dashed line. Brewster's angle is indicated by* θ_B *and the Critical Angle by* θ_C*. The two incident angles for which* $|\Delta| = 45°$ *(modulo* 180*) are marked with dotted arrows at* 42.5 *and* 58.6°*. In part B, within a Fresnel Rhomb, a light beam linearly polarized at* 45° *with respect to the plane of incidence reflected off the rhomb at one of these two angles will be converted, upon two reflections, into a beam of circularly polarized light.*

reflection are 45° at incident angles of 42.5° and 58.6°. This effect permits the construction of a glass piece with two internal reflections (at either value of incidence) to generate a total phase difference of 90° between the S and P components. Such a device is shown pictorially in part B of Fig. 9.8. It is known as a Fresnel Rhomb because it is a phenomenon first recognized by Fresnel.

The transmitted wave

When θ reaches the critical angle, the transmitted wave runs along the interface between the two media. Because the reflection coefficient goes to 1 and remains at 1 for all $\theta > \theta_C$, there is no transmitted energy across the interface. Following in the spirit of this discussion we can investigate some general properties of this wave: the $\vec{k}'$ vector of the refracted wave in the second material has a component parallel to the interface given by

$$k'_y = k' \sin\theta' = \frac{\omega}{c} n \sin\theta$$

from Snell's law in the form of Eq. 9.59. The perpendicular component of the $\vec{k}'$ vector is given by $k'_z = -k' \cos\theta'$ (in the negative z direction). Using $k' = n'\omega/c$ and making formal use of Snell's law for $n > n'$ and $\theta > \theta_C$, we can express the perpendicular component as the imaginary quantity,

$$k'_z = -k' \cos\theta' = -\frac{i\omega}{c} n' \sqrt{\left(\frac{n}{n'}\right)^2 \sin^2\theta - 1} = -\frac{i\omega}{c} n \sqrt{\sin^2\theta - \left(\frac{n'}{n}\right)^2}$$

and write the z component of the transmitted wave in this region as

$$E'_z = E' \exp\left(i\vec{k}' \cdot \vec{r}\right) = E' \exp\frac{\omega}{c} n \left[(i \sin\theta)\, y + \left(\sin^2\theta - \left(\frac{n'}{n}\right)^2 \right)^{1/2} z \right] \tag{9.74}$$

So, referring to Fig. 9.5, the transmitted wave is a traveling plane wave along the interface in the $\hat{y}$ direction and an exponentially decaying wave, or evanescent wave, in the direction, $-\hat{z}$, normal to the interface. In order for the total field to satisfy $\nabla \cdot \vec{E}' = 0$, we must add the necessary $\hat{y}$ component of the field (**see Discussion 9.7**) to Eq. 9.74.

$$E'_y = -\frac{1}{i \sin\theta} \left(\sin^2\theta - \left(\frac{n'}{n}\right)^2 \right)^{1/2} E' \exp\left(i\vec{k}' \cdot \vec{r}\right) \tag{9.75}$$

Then writing $\mathscr{K} = \sqrt{\sin^2\theta - (n'/n)^2}$, we have an expression for the total $\vec{E}$ field that satisfies Maxwell's equations:

$$\vec{E}'(\vec{r}, t) = \left(\hat{z} + \frac{\mathscr{K}\, e^{i\pi/2}}{\sin\theta} \hat{y} \right) E' \exp\left(i\vec{k}' \cdot \vec{r}\right) \tag{9.76}$$

Equation 9.76 describes a traveling wave with unique characteristics:

(1) It is confined to the interface for $\theta > \theta_C$, exponentially decaying into the second medium with a distance, $\delta \sim 1/\mathscr{K}$. This confined region varies from a fraction of the wavelength for θ near 90 degrees, to an infinite extent for $\theta \simeq \theta_C$.

(2) It represents a plane wave that travels in the $\hat{y}$ direction, but is not transverse, rather it has a component in the direction of travel ($\hat{y}$) whose magnitude ($\propto \mathscr{K}$) is near zero at $\theta \simeq \theta_C$, but quickly rising to be of comparable magnitude to the $\hat{z}$ component for larger angles of incidence.

(3) The phase fronts of the wave are not constant in magnitude, but rather drop exponentially as a function of distance into the second material.

(4) The $\hat{y}$ component is 90° out of phase with the $\hat{z}$ component, resulting in an elliptically polarized wave in the $\hat{y}$–$\hat{z}$ plane, not in the usual $\hat{x}$ - $\hat{z}$ plane (perpendicular to the direction of travel).

(5) As mentioned previously, since the reflection coefficients for both polarizations have magnitude of unity, we expect there to be no energy flow across the surface, an expectation that is confirmed by noting the time-averaged normal component of the Poynting vector across the interface has a value of zero.

9.5.4 Fresnel transmission and reflection intensity coefficients

For any interface, the intensity of the light incident to that interface is given by the component of the incident radiation's Poynting vector normal to the interface. That is

$$I = \frac{1}{2Z_r}|\tilde{E}_0|^2 \cos\theta \tag{9.77}$$

$$I'' = \frac{1}{2Z_r''}|\tilde{E}_0''|^2 \cos\theta'' \tag{9.78}$$

$$I' = \frac{1}{2Z_r'}|\tilde{E}_0'|^2 \cos\theta' \tag{9.79}$$

where Z_r is the real part of $\tilde{Z} = \sqrt{\tilde{\mu}/\tilde{\varepsilon}}$, the so-called "wave impedance." The transmission, T, and the reflection, R, coefficients for intensity are then defined in terms of these intensities as

$$R \equiv \frac{I''}{I} \quad \text{and} \quad T \equiv \frac{I'}{I}$$

where $R + T = 1$. In Fig. 9.9, the transmission T and reflection R coefficients for P- and S-polarized light are plotted as a function of angle for the special, yet common, case in which both $\tilde{n}$ and $\tilde{n}'$ are purely real. Noting that $\theta'' = \theta$, and $Z_r'' = Z_r$, the reflection coefficients are easily expressed from Eqs. 9.78 and 9.77 as,

$$R(\theta) = \left|\frac{\tilde{E}_0''}{\tilde{E}_0}\right|^2 \tag{9.80}$$

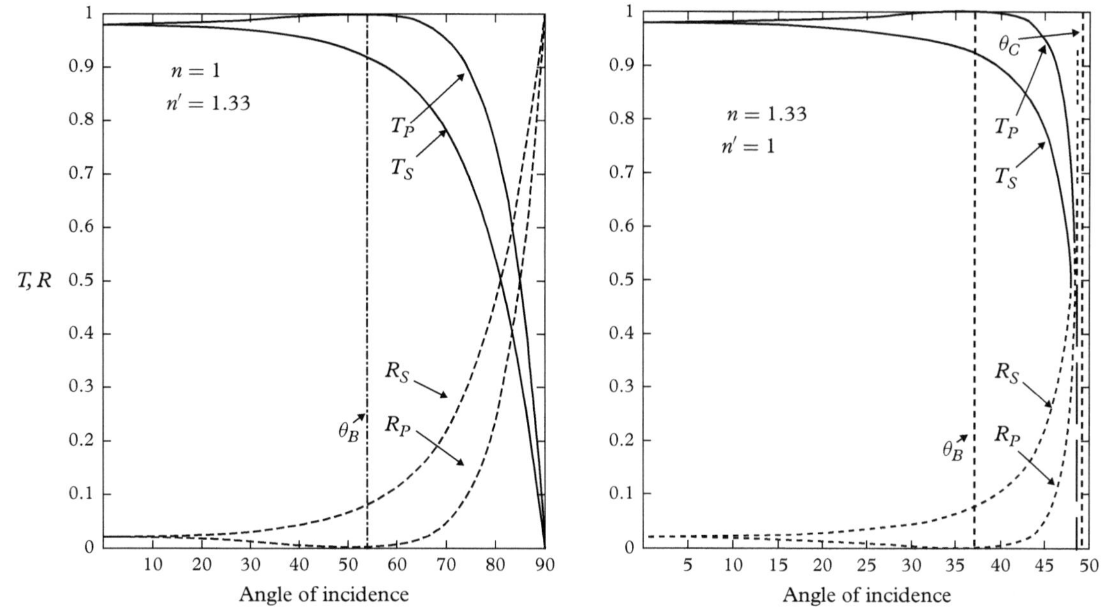

Fig. 9.9 *The transmission T and reflection R coefficients for P and S-polarized light as a function of angle. On the left the incident index is $n = 1$, with a final index of $n' = 1.33$, while on the right the indices are interchanged. Brewster's angle is marked at θ_B and the critical angle, for $n_1 > n_2$, is marked at θ_C.*

And if we restrict our attention to non-magnetic materials in which $\tilde{\mu}'' = \tilde{\mu}' \simeq \mu_0$ (so that $Z'_r/Z_r \simeq n/n'$) and use Snell's law to express $n' \cos\theta' = \sqrt{n'^2 - n^2 \sin^2\theta}$, then the transmission coefficients are given by Eqs. 9.79 and 9.77 as,

$$T(\theta) = \left|\frac{\tilde{E}'_0}{\tilde{E}_0}\right|^2 \left(\frac{\sqrt{n'^2 - n^2 \sin^2\theta}}{n \cos\theta}\right) \qquad (9.81)$$

The transmission and reflection Eqs. 9.80 and 9.81 are known as the Fresnel (intensity) coefficients. They apply to either S- or P-polarized light and thus, more generally, to light of any polarization.

9.5.5 Fresnel transmission and reflection: vacuum/material interface

An important and common situation occurs when a plane wave of radiation is incident from near vacuum (with $\tilde{n} \simeq 1$) onto a sharp non-magnetic ($\tilde{\mu} = \tilde{\mu}' = \mu_0$) surface. For example, consider laser light incident from vacuum at angle θ onto a material with a sufficiently sharp density profile. The Fresnel transmission and reflection intensity coefficients, Eqs. 9.81 and 9.80, can then be used with Eqs. 9.65 and 9.66 to obtain $T(\theta)$ and $R(\theta)$. For S-polarized light, these are:

$$T_S(\theta) = \left| \frac{2\cos\theta}{\cos\theta + \sqrt{\tilde{n}'^2 - \sin^2\theta}} \right|^2 \left(\frac{\sqrt{n'^2 - \sin^2\theta}}{\cos\theta} \right) \tag{9.82}$$

$$R_S(\theta) = \left| \frac{\cos\theta - \sqrt{\tilde{n}'^2 - \sin^2\theta}}{\cos\theta + \sqrt{\tilde{n}'^2 - \sin^2\theta}} \right|^2 \tag{9.83}$$

where $\tilde{n}'$ is the complex index of refraction of the material and the index of vacuum $\tilde{n}$ has been set to unity. Likewise, $T(\theta)$ and $R(\theta)$ for P-polarized incident light is obtained by using Eqs. 9.69 and 9.70 within Eqs. 9.81 and 9.80,

$$T_P(\theta) = \left| \frac{2\tilde{n}'\cos\theta}{\tilde{n}'^2\cos\theta + \sqrt{\tilde{n}'^2 - \sin^2\theta}} \right|^2 \left(\frac{\sqrt{n'^2 - \sin^2\theta}}{\cos\theta} \right) \tag{9.84}$$

$$R_P(\theta) = \left| \frac{-\tilde{n}'^2\cos\theta + \sqrt{\tilde{n}'^2 - \sin^2\theta}}{\tilde{n}'^2\cos\theta + \sqrt{\tilde{n}'^2 - \sin^2\theta}} \right|^2 \tag{9.85}$$

For normal incidence ($\theta = 0$), these S and P equations merge and are polarization independent:

$$T_\perp = \frac{4n'}{|1+\tilde{n}'|^2} = \frac{4n'}{n'^2 + \kappa'^2 + 1 + 2n'} \tag{9.86}$$

$$R_\perp = \left| \frac{1-\tilde{n}'}{1+\tilde{n}'} \right|^2 = \frac{n'^2 + \kappa'^2 + 1 - 2n'}{n'^2 + \kappa'^2 + 1 + 2n'} \tag{9.87}$$

for which it is easy to show that $R_\perp + T_\perp = 1$. For the case in which $\tilde{n}'$ is also purely real, Eqs. 9.82–9.87 are represented in the transmission T and reflection R coefficients plots of Fig. 9.9.

In general, however, $\tilde{n}'$ is complex. In the normal incidence equations previously, we have expressed $T_\perp$ and $R_\perp$ in terms of the real and imaginary components of $\tilde{n}'$. This is useful because it is then straightforward to map $T_\perp(n', \kappa')$ and $R_\perp(n', \kappa')$. The same can be done for the general, angle-dependent equations $T_S(\theta)$, $R_S(\theta)$, $T_P(\theta)$, and $R_P(\theta)$ through considerably more involved treatment (**see Discussion 9.8**). For example, the simplest case is the S and P polarized reflectivity of Eqs. 9.83 and 9.85 in which the treatment expresses $R_S(\theta)$ and $R_P(\theta)$ as

$$R_S(\theta) = \frac{\cos^2\theta - 2p\cos\theta + p^2 + q^2}{\cos^2\theta + 2p\cos\theta + p^2 + q^2} \tag{9.88}$$

$$R_P(\theta) = \frac{\left[\left(p^2 - q^2 + \sin^2\theta\right)\cos\theta - p\right]^2 + (2pq\cos\theta - q)^2}{\left[\left(p^2 - q^2 + \sin^2\theta\right)\cos\theta + p\right]^2 + (2pq\cos\theta + q)^2} \tag{9.89}$$

where the two quantities p and q are given in terms of n' and κ' as,

$$p = \frac{1}{\sqrt{2}}\left\{\left(n'^2 - \kappa'^2 - \sin^2\theta\right) + \sqrt{\left(n'^2 - \kappa'^2 - \sin^2\theta\right)^2 + (2n'\kappa')^2}\right\}^{1/2}$$

$$q = \frac{1}{\sqrt{2}}\left\{-\left(n'^2 - \kappa'^2 - \sin^2\theta\right) + \sqrt{\left(n'^2 - \kappa'^2 - \sin^2\theta\right)^2 + (2n'\kappa')^2}\right\}^{1/2}$$

Thus, given a material's complex index of refraction and the incident angle, Eqs. 9.88 and 9.89 will provide the reflectivities for S- and P-polarized light.

9.6 Some practical applications

9.6.1 The two-surface problem

The physical ideas discussed in Section 9.5 are readily generalized to reflections from two or more parallel surfaces. The solution to this problem is useful in several applications. We begin by generalizing the previous geometry to two surfaces as shown in Fig. 9.10. Here, the wave incident on the first surface through material "A" passes through a thin film of material "B" with thickness Δx and passes through a second surface through which it exits into material "C." In addition to the incident, reflected, and transmitted fields of before, we identify two internal fields in material "B," which are labelled with the subscripts "$I1$" and "$R1$," since geometrically they appear as incident and reflected waves on the second, "B-C" interface. We apply boundary conditions similar to those applied in the single surface problem where, after using the law of reflection and Snell's law, we were left with three unknown complex scalars, $\tilde{E}_0$, $\tilde{E}'_0$, and $\tilde{E}''_0$, whose ratios were found by applying boundary conditions 3 and 4 from Table 9.1. Again using the law of reflection and Snell's law, we are now left with two additional unknowns, $\tilde{E}_{I1}$ and $\tilde{E}_{R1}$, the complex amplitudes for the electric fields internal to the thin film. To find ratios of these five fields, we need four boundary conditions. Examining the figure, we see that

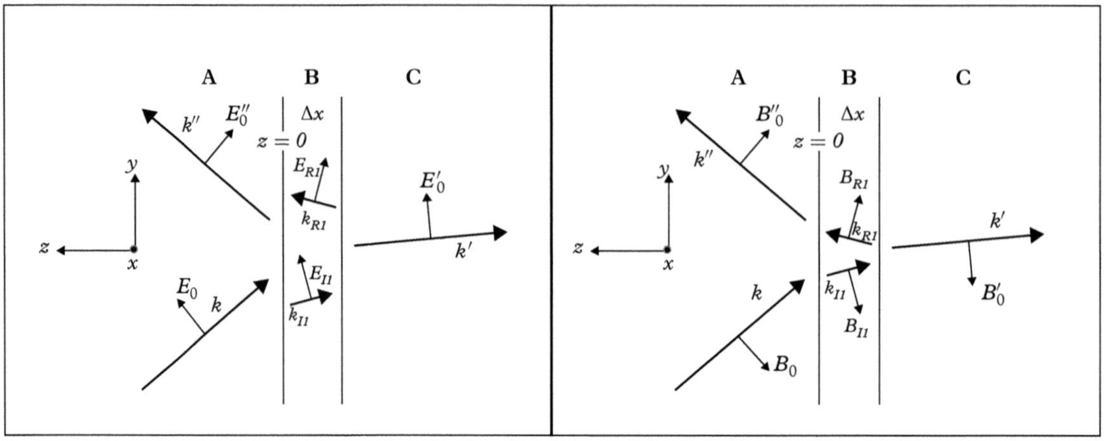

Fig. 9.10 *The vectors used for the two surface problem: To the left is the P-polarized case and on the right is the S-polarized case. Note that the conventions for the reference directions of the fields are the natural extensions of those used in Fig. 9.5.*

the right, "B-C" surface geometry is identical to that of the single surface problem of Fig. 9.5. So, choosing to solve the S-polarization case, we immediately get two equations from the corresponding single surface problem by making the replacements $\tilde{E}_{0s} \to \tilde{E}_{I1}$ and $\tilde{E}''_{0s} \to \tilde{E}_{R1}$ within Eqs. 9.63 and 9.64,

$$\tilde{E}_{I1} + \tilde{E}_{R1} - \tilde{E}'_{0s} = 0$$

$$\frac{\tilde{n}_B}{\tilde{\mu}_B}\left(-\tilde{E}_{I1} + \tilde{E}_{R1}\right)\cos\theta_{I1} + \frac{\tilde{n}_C}{\tilde{\mu}_C}\tilde{E}'_{0s}\cos\theta' = 0$$

An additional two independent equations are found by directly applying to the first surface the boundary conditions 3 and 4 from Table 9.1. This is very similar to the single surface problem with the substitution $\tilde{E}'_{0s} \to \tilde{E}_{I1}$. That is, the wave transmitted by the first surface is the wave incident to the second surface. In addition, however, a fourth wave, $\tilde{E}_{R1}$, must be included in these boundary conditions. At the first surface then,

$$\tilde{E}_{0s} + \tilde{E}''_{0s} - \tilde{E}_{I1} - \tilde{E}_{R1} = 0$$

$$\frac{\tilde{n}_A}{\tilde{\mu}_A}\left(-\tilde{E}_{0s} + \tilde{E}''_{0s}\right)\cos\theta + \frac{\tilde{n}_B}{\tilde{\mu}_B}\left(\tilde{E}_{I1} - \tilde{E}_{R1}\right)\cos\theta_{I1} = 0$$

By eliminating variables, we could write the closed-form expression for the reflection and transmission coefficients for the two surface problem. However, little insight is provided by viewing these and solving systems of linear equations. We will therefore proceed more qualitatively in what follows. The possibility of interference between the waves interacting with the two surfaces leads to several physical realizations of the two surface problem. First, we consider the symmetric example of a thin sheet of glass in air in which materials "A" and "C" (Fig. 9.10) are the same. Such a configuration can serve as a wavelength selective "etalon" for use in a laser cavity. The basic idea is illustrated in Fig. 9.11a, in which the S-wave reflectivity of a $50\,\mu$m thick sheet of BK7 glass is shown for light incident at a range of near-normal angles. Plotted in black is 589 nm radiation and light with a 10% longer wavelength, 650 nm, is shown in gray. Note that although normal incidence reflection loss at a typical air-glass surface is $\sim 4\%$, the two surface reflectivity varies between 0 and 16% as the fields interfere. Consider the 589 nm light incident at an angle of 0.1 radians: the reflectivity vanishes and so essentially all of the radiation at 589 nm is transmitted through the etalon. If this etalon were inserted in a laser cavity and tilted at that angle, this wavelength

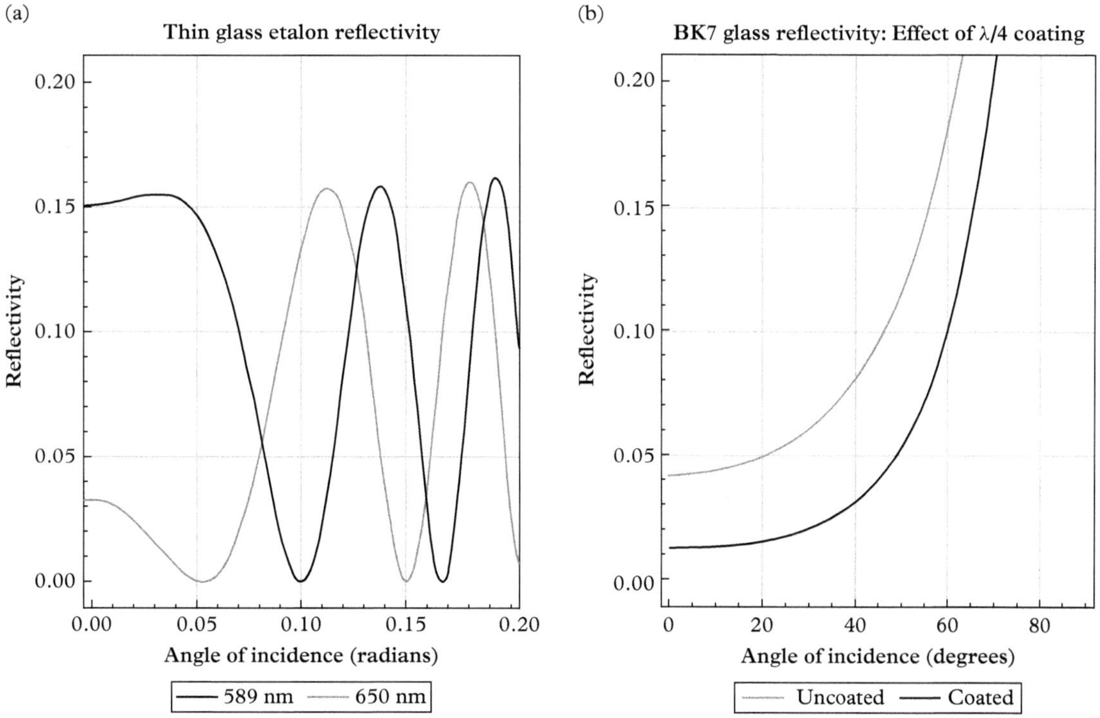

Fig. 9.11 *Two applications of the thin film solution. (a) Thin etalon and (b) anti-reflective (AR) coating.*

would be strongly selected for lasing since nearby wavelengths, such as 650 nm light, would suffer repeated etalon-reflection losses as the laser light circulated in the laser cavity. Tilting the etalon to 0.15 radians would allow for lasing at 650 nm.

While reflection losses of 4% per surface may seem small, precision imaging optics are frequently made up of many lenses. A simple zoom lens has at least four elements and high end camera lenses may have two or three times this number. Good anti-reflective (AR) coatings are essential for the lenses in most optical systems to both reduce the loss from the signal of interest as well as to reduce the amount of light "bouncing around" in a complex lens due to undesired surface reflections. The simplest and still widely used AR coating is a quarter-of-a-wavelength of magnesium fluoride, whose relatively low index of refraction of $n = 1.378$ makes it useful for this application. In Fig. 9.11b we show the S-wave reflectivity of a BK7 surface first without and then with a $\lambda/4$ coating of magnesium fluoride. The destructive interference reduces the reflection losses from 4% down to just over 1%. In a multi-lens system this could result in a significant improvement in performance. State-of-the-art coatings use many layers and for these, the previous two-surface problem is readily extended to many surfaces. Mathematically, each surface provides additional reflected and transmitted waves–two more complex scalars–and adding the corresponding two boundary conditions at the new surface makes the problem solvable.

9.6.2 Lossy dielectrics and metals

These results for reflectivity and transmission coefficients do not rely on the index of refraction or, equivalently, the permittivity to be real. Gold is occasionally used in optical elements. Gold films may be used as attenuators for light. At a wavelength of 633 nm, gold has a complex index of refraction of $\tilde{n} = 0.495 + 3.29i$. In Fig. 9.12 we show the transmission and reflection of a free-standing gold film in air. These are both around 40% for normal incidence and the fact that they do not sum to one is due to the new feature a complex index of refraction introduces: absorption losses. For this film about 20% is absorbed at normal incidence.

By varying the thickness of the coating, a specific amount of attenuation can be obtained. At this wavelength, for example, a gold thickness of 70 nm gives 1% transmission. For mechanical stability, metal coatings of these thicknesses are only found on glass or some other material substrate and so for accurate attenuation estimates this is a three surface problem.

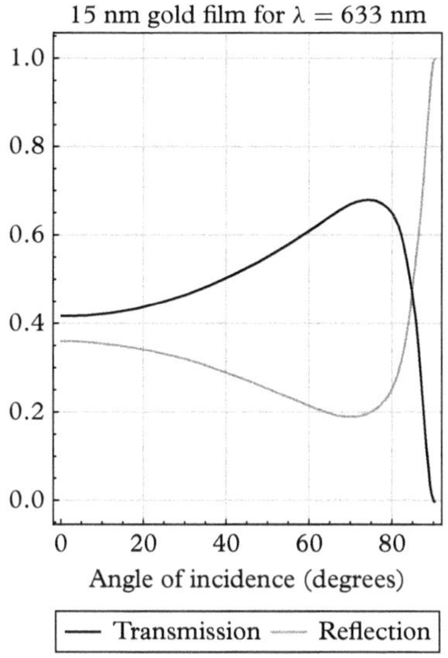

Fig. 9.12 *The transmission and reflection of a free-standing 15 nm gold film.*

9.7 Frequency and time domain polarization response to the fields

As explained earlier, for the case of a monochromatic E&M wave passing through a material, the polarizations $\vec{P}$ and $\vec{M}$ will oscillate at the same frequency as the $\vec{E}$ and $\vec{B}$ fields of the wave. And the susceptibilities, $\tilde{\chi}_e(\omega)$ and $\tilde{\chi}_m(\omega)$, are the complex frequency dependent functions which relate, at frequency ω, the magnitudes and phases of the polarizations to those of the driving fields. For example, the single-frequency components of an electric polarization and an electric field are mathematically related by $\vec{P}(\vec{x},\omega)\,e^{-i\omega t} = \varepsilon_o\tilde{\chi}_e(\omega)\,\vec{E}(\vec{x},\omega)\,e^{-i\omega t}$. If we divide out the time dependence, we can then express the frequency domain polarization response to the fields as

$$\begin{aligned}\vec{P}(\vec{x},\omega) &= \varepsilon_o\tilde{\chi}_e(\omega)\,\vec{E}(\vec{x},\omega)\\ \vec{M}(\vec{x},\omega) &= \tilde{\chi}_m(\omega)\vec{H}(\vec{x},\omega)\end{aligned}$$

In the case of a general time-dependent field, such as $\vec{E}(\vec{x},t)$, which is a superposition of components with various frequencies and amplitudes, the general polarization response, $\vec{P}(\vec{x},t)$, is an integration of component polarization responses, $\vec{P}(\vec{x},\omega)$, over frequency,

$$\begin{aligned}\vec{P}(\vec{x},t) &= \frac{1}{\sqrt{2\pi}}\int_{-\infty}^{\infty}\vec{P}(\vec{x},\omega)\,e^{-i\omega t}\mathrm{d}\omega\\ &= \frac{1}{\sqrt{2\pi}}\int_{-\infty}^{\infty}\varepsilon_o\tilde{\chi}_e(\omega)\vec{E}(\vec{x},\omega)e^{-i\omega t}\mathrm{d}\omega \qquad (9.90)\end{aligned}$$

which relates, essentially by Inverse Fourier transform (IFT), the time dependence of the polarization to the various frequency components $\vec{E}(\vec{x},\omega)$, each weighted by $\tilde{\chi}_e(\omega)$, of a general $\vec{E}(\vec{x},t)$. The question is: Can we represent a more direct relationship between $\vec{P}(\vec{x},t)$ and $\vec{E}(\vec{x},t)$? To do this we must account for the fact that the full response of a material to the application of a field at some point in space and time is not only not instantaneous, it endures beyond the time of application–that is, the material rings. In particular, the response of the material to a sharp impulse of the field (a Dirac delta) is a sinusoidal, exponentially diminishing function in time that we label $g_{e/m}(t-t')$, where t' is the time of the impulse and t the time of the observed response. So now, in general, the $\vec{E}$ and $\vec{B}$ fields can be seen as a continuous time series of impulses occurring at point $\vec{x}$ and the polarization in the material at $\vec{x}$ at a time t can be seen to depend, through $g_{e/m}(t-t')$, upon the history, t', of the fields at that point. This temporally non-local response can be related mathematically to the frequency dependence of the susceptibility, $\tilde{\chi}_{e/m}$. Roughly speaking, if $\tilde{\chi}_{e/m}$ were independent of frequency, the response of the material would be completely local temporally and $\vec{P}(\vec{x},t)$ would exactly follow $\vec{E}(\vec{x},t)$ in time (again, apart from a constant phase factor). In particular, the response to a field impulse at point $\vec{x}$ and time t would be an impulse of polarization at the same point and time - corresponding to an infinitely short impulse response function, $g_{e/m}(t-t')$[10]. In complete contrast, consider the case of a susceptibility with a extremely sharp and narrow peak around a particular frequency, ω_o. In this case, all the frequency components of an arbitrary $\vec{E}(\vec{x},t)$ except for those in the region of ω_o would be greatly attenuated and the material would ring at those frequencies near ω_o for a relatively long time corresponding to a temporally long impulse response function, $g_{e/m}(t-t')$.[11]

We can address this physical reality for the case of polarization and obtain the exact form of $g_e(t-t')$ in our formalism, thereby finding a more direct relationship between $\vec{P}(\vec{x},t)$ and $\vec{E}(\vec{x},t')$. The Fourier transform (FT) of $\vec{E}(\vec{x},t)$ is $\vec{E}(\vec{x},\omega)$ so Eq. 9.90 can be written

$$\vec{P}(\vec{x},t)=\frac{1}{\sqrt{2\pi}}\int_{-\infty}^{\infty}\varepsilon_o\tilde{\chi}_e(\omega)\left[\frac{1}{\sqrt{2\pi}}\int_{-\infty}^{\infty}\vec{E}(\vec{x},t')e^{i\omega t'}dt'\right]e^{-i\omega t}d\omega \tag{9.91}$$

exchanging orders of integration

$$\vec{P}(\vec{x},t)=\frac{\varepsilon_o}{\sqrt{2\pi}}\int_{-\infty}^{\infty}dt'\vec{E}(\vec{x},t')\left[\frac{1}{\sqrt{2\pi}}\int_{-\infty}^{\infty}\tilde{\chi}_e(\omega)e^{-i\omega(t-t')}d\omega\right] \tag{9.92}$$

where the term in brackets is identified as $\tilde{\chi}_e(t-t')$, the IFT of $\tilde{\chi}_e(\omega)$ in terms of time delay $T=t-t'$. So we have

[10] Mathematically, if $\chi_{e/m}$ is independent of frequency, Eq. 9.90 can be written

$$\vec{P}(\vec{x},t)=\frac{\varepsilon_o\chi_e}{\sqrt{2\pi}}\int_{-\infty}^{\infty}\vec{E}(\vec{x},\omega)e^{-i\omega t}d\omega = \varepsilon_o\chi_e\vec{E}(\vec{x},t)$$

and $\vec{P}(\vec{x},t)$ exactly follows $\vec{E}(\vec{x},t)$ in time.

[11] If we take the extreme case in which $\chi_{e/m}$ is infinitely sharp and narrow as in a Dirac delta, Eq. 9.90 is written

$$\vec{P}(\vec{x},t)=\frac{1}{\sqrt{2\pi}}\int_{-\infty}^{\infty}\varepsilon_o\chi_e\delta(\omega_o-\omega)\vec{E}(\vec{x},\omega)e^{-i\omega t}d\omega = \frac{\varepsilon_o\chi_e}{\sqrt{2\pi}}\vec{E}(\vec{x},\omega_o)e^{-i\omega_o t}$$

so that $\vec{P}(\vec{x},t)$ is proportional to $\varepsilon_o\chi_e\vec{E}(\vec{x},\omega_o)$ and oscillates at frequency ω_o for all times.

$$\vec{P}(\vec{x},t) = \frac{\varepsilon_o}{\sqrt{2\pi}} \int_{-\infty}^{\infty} \tilde{\chi}_e(t-t')\vec{E}(\vec{x},t')dt' \tag{9.93}$$

which is the more direct relationship between $\vec{P}(\vec{x},t)$ and $\vec{E}(\vec{x},t')$ that we sought. And further, we can see that $\tilde{\chi}_e(t-t') \equiv g_e(t-t')$. That is, the IFT of the electric susceptibility of a material is the impulse response (or susceptibility kernel) of that material.

$$g_e(t-t') = \frac{1}{\sqrt{2\pi}} \int_{-\infty}^{\infty} \tilde{\chi}_e(\omega)e^{-i\omega(t-t')}d\omega \tag{9.94}$$

and by a similar line of reasoning,

$$\vec{M}(\vec{x},t) = \frac{1}{\sqrt{2\pi}} \int_{-\infty}^{\infty} \tilde{\chi}_m(t-t')\vec{H}(\vec{x},t')dt' \tag{9.95}$$

where

$$\tilde{\chi}_m(t-t') = g_m(t-t') = \frac{1}{\sqrt{2\pi}} \int_{-\infty}^{\infty} \tilde{\chi}_m(\omega)e^{-i\omega(t-t')}d\omega \tag{9.96}$$

We have now, therefore, the time domain polarization response to the field–a generalization of the static relationships between the material's polarization and the applied field in the integral forms,

$$\vec{P}(\vec{x},t) = \frac{\varepsilon_o}{\sqrt{2\pi}} \int_{-\infty}^{\infty} g_e(t-t')\vec{E}(\vec{x},t')dt' \tag{9.97}$$

$$\vec{M}(\vec{x},t) = \frac{1}{\sqrt{2\pi}} \int_{-\infty}^{\infty} g_m(t-t')\vec{H}(\vec{x},t')dt' \tag{9.98}$$

where, again, $g(t-t')$ describes the response of the material to a delta-function impulse at time t' (i.e., $g(t-t')$ is essentially a temporal Green function) and the polarization in the material at time t depends upon the time history of the electric and magnetic fields of the wave in the medium at times t' before time t; clearly the polarization at time t can have no dependence upon fields occurring after t. So causality provides us the important result that for any physically realistic response, $g_{e/m}(t-t') = 0$ for $t' > t$. This physical reasoning can be readily confirmed mathematically by performing the contour integrations in Eqs 9.94 or 9.96. Using Eqs. 9.2 and 9.3, we have the time dependent forms for $\vec{D}$ and $\vec{H}$ as well

$$\vec{D}(\vec{x},t) = \varepsilon_0\vec{E}(\vec{x},t) + \vec{P}(\vec{x},t) \tag{9.99}$$

$$\vec{H}(\vec{x},t) = \frac{1}{\mu_0}\vec{B}(\vec{x},t) - \vec{M}(\vec{x},t) \tag{9.100}$$

Thus, these two relationships can now be interpreted as the response of the material to the applied field expressed in the time domain: When a material is subjected to an applied $\vec{E}$ or $\vec{B}$ field, at time t, the response is due in part to the value of the applied field at time t, but also a residual response due to the history of the fields in the material at (in principle all) times before t, modulated by the relaxation or "memory" of the material.

9.7.1 Example

Problem: Given the simple susceptibility $\tilde{\chi}$ for a medium with a single transition resonance with damping constant γ and resonant frequency ω_o,

$$\tilde{\chi}(\omega) = \frac{\omega_p^2}{\omega_o^2 - \omega^2 - i\gamma\omega}$$

show that

$$g(t - t') = g(T) = \sqrt{2\pi}\,\omega_p^2 \exp\left(-\frac{\gamma T}{2}\right) \frac{\sin(\alpha_o T)}{\alpha_o} \theta(T)$$

where $\alpha_o = \sqrt{\omega_o^2 - \gamma^2/4}$ and $\theta(T)$, the Heaviside step function, is defined such that $\theta(T < 0) = 0$ and $\theta(T > 0) = 1$.

Solution: First write the impulse response function as the inverse Fourier transform of $\tilde{\chi}(\omega)$

$$g(T) = \frac{\omega_p^2}{\sqrt{2\pi}} \int_{-\infty}^{\infty} \frac{e^{-i\omega T}}{\omega_o^2 - \omega^2 - i\gamma\omega} d\omega$$

The integration is most easily accomplished via infinite semicircle contour integrals in the complex plane. In order for the exponential in the integrand to vanish rather than blow up, we need to close the contour in the upper half plane (UHP) for $T < 0$ and in the lower half plane (LHP) for $T > 0$. This is readily seen from $|exp(-i\omega T)| = |exp(-i(x + iy)T)| = |exp(-ixT + yT)| = exp(yT)$. To find the singularities, solve $\omega_o^2 - \omega^2 - i\gamma\omega = 0$ to get $\omega = -\frac{i\gamma}{2} \pm \sqrt{\omega_o^2 - \gamma^2/4} = -\frac{i\gamma}{2} \pm \alpha_o$. Thus, the singularities are found to be in the lower half plane and so for $T < 0$, closing in the UHP encloses no singularities and the contour integral is zero. Since the infinite semicircle part of the integral is also zero, we conclude $g(T < 0) = 0$. For $T > 0$, the contour in the LHP encloses both singularities so we have, from Cauchy's residue theorem,

$$\oint \frac{e^{-i\omega T}}{\left[\omega-(-\frac{i\gamma}{2}+\alpha_o)\right]\left[\omega-(-\frac{i\gamma}{2}-\alpha_o)\right]}d\omega$$
$$=\int_{-\infty}^{\infty} \frac{e^{-i\omega T}}{\left[\omega-(-\frac{i\gamma}{2}+\alpha_o)\right]\left[\omega-(-\frac{i\gamma}{2}-\alpha_o)\right]}d\omega+$$
$$+\int_{arc} \frac{e^{-i\omega T}}{\left[\omega-(-\frac{i\gamma}{2}+\alpha_o)\right]\left[\omega-(-\frac{i\gamma}{2}-\alpha_o)\right]}d\omega$$
$$=2\pi i\sum Res \tag{9.101}$$

where, again, the infinite semicircle part of the integral is zero, and we get

$$\frac{\omega_p^2}{\sqrt{2\pi}}\int_{-\infty}^{\infty} \frac{e^{-i\omega T}}{\left[\omega-(-\frac{i\gamma}{2}+\alpha_o)\right]\left[\omega-(-\frac{i\gamma}{2}-\alpha_o)\right]}d\omega$$
$$=\sqrt{2\pi}\,\omega_p^2 exp\left(-\frac{\gamma T}{2}\right)\frac{\sin(\alpha_o T)}{\alpha_o}$$

We combine this result for $T > 0$ with that for $T < 0$

$$g(T)=\sqrt{2\pi}\,\omega_p^2\exp\left(-\frac{\gamma T}{2}\right)\frac{\sin(\alpha_o T)}{\alpha_o}\theta(T)$$

where we make use of the Heaviside function to clamp the function to zero for $T < 0$.

9.8 Kramers–Kronig relationships

All of the characteristics of the material that give rise to an electric or magnetic response are contained in the various forms of the $\tilde{\chi}_{e/m}(\omega)$ as a complex function of the complex variable, ω. For example, the response of an atom to radiation is critically dependent on the frequency of the radiation because of the atom's sharply defined energy states: The polarization response of the atom is peaked near quantum allowed transitions between these states. All of this frequency structure in the polarization response is captured in the atom-specific form of $\tilde{\chi}_{e/m}(\omega)$. Before we consider example forms and discuss their consequences, in this section we derive a universal relationship between the real and imaginary parts of $\tilde{\chi}$ which will prove invaluable in understanding how electromagnetic energy is transported through

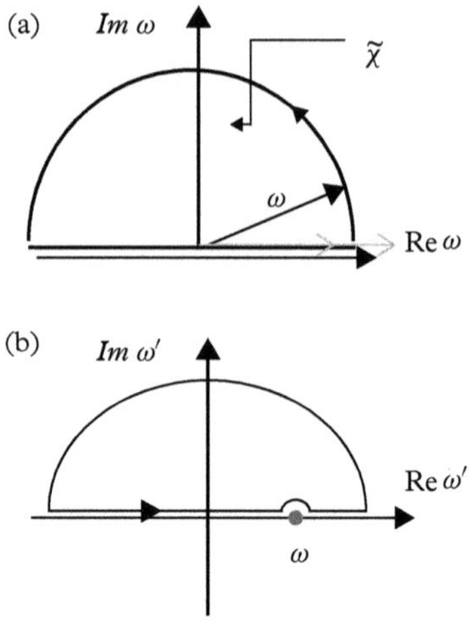

Fig. 9.13 *Contour Integration in the complex ω plane for (a) $\oint \chi_{e/m}(\omega)e^{-i\omega T}d\omega$ with $(T<0)$ and (b) $\oint \frac{\chi(\omega)}{\omega-\omega_o}d\omega$, where $\chi_{e/m}$ is any physically realizable material susceptibility. In (a) the UHP contour integral must be zero to satisfy causality. The contour integral in (b) is easily evaluated because of the results in (a).*

materials. In order to develop this relationship, we need first to show that any physically realizable $\tilde{\chi}$ has a mathematical representation which is necessarily analytic everywhere in the upper half complex ω plane. To accomplish this, consider the following contour integral

$$\oint \tilde{\chi}_{e/m}(\omega)e^{-i\omega T}d\omega = \int_{-\infty}^{\infty} \tilde{\chi}_{e/m}(\omega)\,e^{-i\omega T}d\omega + \int_{arc} \tilde{\chi}_{e/m}(\omega)\,e^{-i\omega T}d\omega \tag{9.102}$$

In this equation, which is typically used to solve for the functional form of $g_{e/m}(T)$ in Eq. 9.94, evaluation of the contour integral for $T=t-t'<0$ will always first require that it is closed in the UHP where, for large complex ω, the exponential term $exp(-i\omega T) = exp[-i(\omega_x + i\omega_y)T] = exp(-i\omega_x T)exp(\omega_y T)$ in the second (arc) term on the right vanishes–rather than explodes–as $\omega \to \infty$ (see Fig. 9.13a). Furthermore, because the first term on the right is proportional to $g_{e/m}$ (Eq. 9.94) and because, as noted previously, any physically realistic, causally constrained response function, $g_{e/m}(T)$, must vanish for $T<0$, this term must also vanish. Thus, for any physically realizable $\tilde{\chi}$,

$$\oint \tilde{\chi}_{e/m}(\omega)\,e^{-i\omega T}d\omega = 0 \quad (\textit{in the U.H.P.}) \tag{9.103}$$

and we have the important intermediate result that causality requires that all physically realizable $\tilde{\chi}(\omega)$ have a mathematical representation with no poles in the upper half of the complex ω plane. With this result, we are now in the position to consider the contour integral relevant to our derivation, namely,

$$\oint \frac{\tilde{\chi}(\omega)}{\omega - \omega_o} d\omega \qquad (9.104)$$

in which the integrand has a pole at $\omega = \omega_o$ along the real axis but no poles anywhere in the upper half plane. The general physical properties of $\tilde{\chi}(\omega)$ suggest that as the frequency increases, there will be a point at which the physical system will fail to respond fully, due to its inertia. For again, polarization in material is the result of movement of masses in response to an applied time varying field. Thus we expect that as $\omega \to \infty$, $\tilde{\chi} \to 0$ (consider the simple $\tilde{\chi}_e(\omega)$ in the example in Section 9.7.1) and assuming $\tilde{\chi}$ falls to zero rapidly enough with ω, we may readily evaluate the integral in Eq. 9.104 for a contour taken along the real axis from $-\infty$ to $+\infty$ and then completed at extremely large ω in the upper half plane (see Fig. 9.13b). Such an integral is evaluated as (**see Discussion 9.9**)

$$\oint \frac{\tilde{\chi}(\omega)}{\omega - \omega_o} d\omega = \mathrm{P}\left(\int_{-\infty}^{\infty} \frac{\tilde{\chi}(\omega)}{\omega - \omega_o} d\omega\right) - i\pi\,\tilde{\chi}(\omega_o) = 0,$$

so that

$$\tilde{\chi}(\omega_o) = \frac{1}{i\pi}\mathrm{P}\left(\int_{-\infty}^{\infty} \frac{\tilde{\chi}(\omega)}{\omega - \omega_o} d\omega\right) \qquad (9.105)$$

Resolving into real and imaginary components,

$$\tilde{\chi}(\omega_o) = \chi_1(\omega_o) + i\chi_2(\omega_o) = \frac{1}{i\pi}\mathrm{P}\left(\int_{-\infty}^{\infty} \frac{\chi_1(\omega) + i\chi_2(\omega)}{\omega - \omega_o} d\omega\right)$$

and separating so that

$$\chi_1(\omega_o) = \frac{1}{\pi}P\left(\int_{-\infty}^{\infty} \frac{\chi_2(\omega)}{\omega - \omega_o} d\omega\right) \qquad (9.106)$$

$$\chi_2(\omega_o) = -\frac{1}{\pi}P\left(\int_{-\infty}^{\infty} \frac{\chi_1(\omega)}{\omega - \omega_o} d\omega\right) \qquad (9.107)$$

These two equations–which quite generally apply to response functions in many different physical systems and are used to calculate the real part from the imaginary part (or vice versa) of such response functions–in this case provide a powerful tool for the analysis of the polarization response of materials irradiated by electromagnetic waves. Taking advantage of the symmetry properties[12] of $\tilde{\chi}(\omega)$ for real ω, we can rewrite Eqs. 9.107 and 9.106 in their canonical form[13]

[12] If, in general, the impulse response function $g(T)$ is assumed real and ω is assumed to be complex and we know from Eq. 9.94 that

$$\begin{aligned}
\tilde{\chi}(\omega) &= \int_0^{\infty} g(T)e^{i\omega T}dT \\
\tilde{\chi}(-\omega) &= \int_0^{\infty} g(T)exp(-i\omega T)dT \\
&= \int_0^{\infty} g(T)exp(\omega_y T - i\omega_x T)dT \\
&= \left[\int_0^{\infty} g(T)exp(\omega_y T + i\omega_x T)dT\right]^* \\
&= \left[\int_0^{\infty} g(T)exp(i\omega^* T)dT\right]^* \\
&= \tilde{\chi}^*(\omega^*)
\end{aligned}$$

and if ω is pure real then $\tilde{\chi}(-\omega) = \tilde{\chi}^*(\omega)$ so that the symmetry properties are

$$\chi_1(-\omega) = \chi_1(\omega) \; and \; \chi_2(-\omega) = -\chi_2(\omega).$$

[13] Rewriting Eq. 9.106

$$\begin{aligned}
\chi_1(\omega_o) = \frac{1}{\pi}P\left(\int_{-\infty}^{0} \frac{\chi_2(\omega')}{\omega' - \omega_o} d\omega'\right) \\
+ \frac{1}{\pi}P\left(\int_0^{\infty} \frac{\chi_2(\omega')}{\omega' - \omega_o} d\omega'\right)
\end{aligned}$$

setting $+\omega'$ to $-\omega'$ in the first integral and using $\chi_2(-\omega') = -\chi_2(\omega')$,

$$\begin{aligned}
\chi_1(\omega_o) &= \frac{1}{\pi}P\left(\int_0^{\infty} \frac{\chi_2(\omega')}{\omega' + \omega_o} d\omega'\right) \\
&\quad + \frac{1}{\pi}P\left(\int_0^{\infty} \frac{\chi_2(\omega')}{\omega' - \omega_o} d\omega'\right) \\
&= \frac{2}{\pi}P\left(\int_0^{\infty} \frac{\omega'\chi_2(\omega')}{\omega'^2 - \omega_o^2} d\omega'\right)
\end{aligned}$$

In the same way, we can show

$$\chi_2(\omega_o) = -\frac{2\omega_o}{\pi}\mathrm{P}\left(\int_0^{\infty} \frac{\chi_1(\omega')}{\omega'^2 - \omega_o^2} d\omega'\right)$$

Kramers–Kronig relationships for the polarizabilities

$$\chi_1(\omega_o) = \frac{2}{\pi}\mathrm{P}\left(\int_0^{\infty} \omega' \frac{\chi_2(\omega')}{\omega'^2 - \omega_o^2} d\omega'\right) \tag{9.108}$$

$$\chi_2(\omega_o) = -\frac{2\omega_o}{\pi}\mathrm{P}\left(\int_0^{\infty} \frac{\chi_1(\omega')}{\omega'^2 - \omega_o^2} d\omega'\right) \tag{9.109}$$

9.9 Measuring the response of matter to fields

Within the equations originally published by Maxwell in 1865 there was no presupposition of free or bound charges and so the ideas of polarization and magnetization–that is, the effects of fields within materials–were not part of the original set of Maxwell's equations. In their complex form, the original "microscopic" Maxwell equations merely show us the various mathematical forms of the field sources: $\vec{J}_f$, ρ_f, dE/dt, and dB/dt, with no particular reference to material mechanisms. Only later, with the understanding of the constituents of matter, did the macroscopic versions of Maxwell's equations come along. The sources were then modified to include material effects of polarization $\vec{P}$ and magnetization $\vec{M}$. In particular, we have seen (Section 9.3, Eq. 9.27) that comparison of an electromagnetic wave in vacuum with a wave in material yields $k_0 \rightarrow \tilde{k} = k_0\tilde{n} = (\omega/c)\,\tilde{n}$. Thus, for a wave traveling in material in the x direction

$$\vec{E}(\vec{r},t) = \vec{E}_0 \exp\left(i\frac{\omega n}{c}x - i\omega t\right)\exp\left(-\frac{\omega\kappa}{c}x\right)$$

which therefore directly relates measurable properties of the wave such as the phase velocity ($v_p = \omega/\frac{\omega}{c}n = c/n$) and attenuation length ($\delta = c/\omega\kappa$) to the material properties. Similarly, we have found that the reflectivity, R, of a material, another measurable quantity, is given in terms of the complex index of refraction. For example, Eq. 9.87 expresses reflectivity at normal incidence from a sharp vacuum-material interface as:

$$R_{\perp} = \left|\frac{1-\tilde{n}}{1+\tilde{n}}\right|^2 = \frac{n^2+\kappa^2+1-2n}{n^2+\kappa^2+1+2n} \tag{9.110}$$

If we further assume that the response of the material is dependent upon the frequency, ω, of the electromagnetic wave so that, for example, $\tilde{\chi}_e \rightarrow \tilde{\chi}_e(\omega)$, $\tilde{\varepsilon} \rightarrow \tilde{\varepsilon}(\omega)$ and $\tilde{n} \rightarrow \tilde{n}(\omega)$, then a full spectral measurement can yield the material response as a function of frequency.

Phenomenologically, then, with such a measurement we would have a complete frequency mapping of the response of the material to the field–in the forms of polarization, magnetization, and current density–to the response of the field to the material in the forms of reflectivity, transmission, absorption and phase shift. But what is it exactly that we learn about the material? Without a physical model, the best we can do is obtain the functional forms of the real and imaginary parts of the response. For example, if we look at the Maxwell–Ampere equation,

$$\nabla \times \vec{B} = \mu\tilde{\sigma}_f\vec{E} + i\omega\mu\tilde{\varepsilon}_b\vec{E}$$

it is clear that the two material responses to the sinusoidally driving E field are 90° out of phase and presumably correspond to different physical processes (**see Discussion 9.10**). However, they are both linear responses, each with real and imaginary components and are otherwise indistinguishable without a model. A full continuum measurement of $\tilde{n}(\omega) = n(\omega) + i\kappa(\omega)$ will not differentiate between the two types of response seen in the Maxwell–Ampere equation. In Chapter 10, we will consider such measurements in light of some simple classical models that discern free and bound responses, but next we consider single-frequency and broadband measurements in the absence of a material model. For these cases, the general notation–introduced in Section 9.2–for representing material response in the absence of a model will be used.

9.9.1 Measuring the optical constants of a material

As discussed in Section 9.2, the response of a material to light (i.e., its optical properties) can be quantified by just two components which are generally represented as a complex frequency-dependent function. And, as we saw, this response can be represented either as the complex index of refraction, $\tilde{n}(\omega) = n + i\kappa$, or as any of the related generalized forms of Eqs. 9.12, 9.13, and 9.15. Furthermore, when light encounters matter it can be reflected, transmitted and/or absorbed depending on the values of the real and imaginary components of the complex dielectric function or, equivalently, the complex index of refraction of the material. And, because of the frequency dependence of the material response, the light can be dispersed, or differentially phase-shifted, within the material, resulting in an apparent frequency-dependent light speed. Various measurements of reflectivity, transmission, absorption, and phase shifting of the light can thus be made to obtain the real and imaginary components of the index of refraction and dielectric function.

9.9.2 Single frequency measurements

In general, when monochromatic light is incident upon a material interface, it is partially reflected and partially transmitted. Some or all of the light which is transmitted is absorbed within the material. And both the reflected part and transmitted part suffer additional phase shifts. Measurements of these various quantities of reflectivity, transmission, absorption, and phase shifting yield information about the complex optical response of the material at a single frequency, ω_l. Because the optical response consists of a real and an imaginary part, simultaneous pairs of measurements are required. In general, the complex index of refraction $\tilde{n}(\omega_l)$ is first obtained and then the complex dielectric function $\tilde{\varepsilon_r}(\omega_l) = \tilde{\varepsilon}/\varepsilon_o \simeq \tilde{n}^2$. This 'response at ω_l' can then alternatively be expressed in terms of the generalized conductivity, $\tilde{\sigma}$ (see Exercises at the end of this chapter), where again, "generalized" means there is no distinction between bound and free contributions. The particular technique may then be repeated at multiple frequencies to obtain the rough functional form of the material response in frequency space. We now outline common techniques of this sort.

Reflection-transmission measurements of thin targets

As a first example of single frequency measurement of optical constants, we consider simultaneous measurements of the reflected and transmitted fractions, $R_\perp$ and $T_\perp$, of the intensity of a laser pulse normally incident from vacuum onto a thin wafer of material of unknown complex refractive index $\tilde{n} = n + i\kappa$. In this case, the energy balance equation $R_\perp + T_\perp + A = 1$ is used where A is the fraction of incident laser energy absorbed within the material. If we assume, for simplicity, that the light reflects only off the front surface and thus makes only one pass through the wafer, then the reflected intensity is simply the Fresnel normal incidence result:

$$R_\perp(n,\kappa) = \left|\frac{1-\tilde{n}}{1+\tilde{n}}\right|^2 = \frac{n^2+\kappa^2+1-2n}{n^2+\kappa^2+1+2n} \tag{9.111}$$

with the unreflected fraction of intensity passing through the wafer where it is partially absorbed, A, and exiting with the fraction $T_\perp$. That is, $1 - R_\perp = A + T_\perp$. The fraction of power absorption, A, depends on the imaginary part of the wavevector, $k_2 = (\omega/c)\kappa$. With intensity proportional to the square of the field, the absorption coefficient is then given by $\alpha = 2k_2 = 2(\omega/c)\kappa$ and the transmitted intensity can be expressed as

$$T_{\perp}(n,\kappa) = (1 - R_{\perp})e^{-\alpha l} = (1 - R_{\perp})\exp\left(-2\frac{\omega\kappa}{c}l\right) \quad (9.112)$$

where l is the wafer thickness. Thus, with the measured values of $R_{\perp}$ and $T_{\perp}$, the two Eqs. 9.111 and 9.112 can be solved to obtain the complex index of refraction of the unknown material.

This method requires a wafer of specific and highly uniform thickness. If the wafer is too thick, the transmitted intensity vanishes and if the thickness is less than or equal to one absorption length, $l \lesssim \alpha^{-1}$, then multiple reflections within the target must be taken into account.[14]

Measuring transmittance and phase shift of a thin target

From Eq. 9.71, the complex relative amplitude for normal transmission of monochromatic light across a vacuum-material $(v - m)$ interface is given by

$$\tilde{T}_{v-m} = \left(\frac{\tilde{E}_0'}{\tilde{E}_0}\right)_{norm} = \left(\frac{\left|\vec{E}_0'\right| e^{i\Delta}}{\left|\vec{E}_0\right|}\right)_{norm} = \frac{2}{1+\tilde{n}} \quad (9.113)$$

where $\Delta = \phi' - \phi$ is the phase shift introduced by the transmission. For a thin wafer of thickness l, this complex amplitude is further modified by passage through the material so that at the rear surface of the wafer it has become,

$$\tilde{T}_B = \tilde{T}_{v-m}\exp\left(i\frac{\omega}{c}\tilde{n}l\right) = \frac{2}{1+\tilde{n}}\exp\left(i\frac{\omega}{c}\tilde{n}l\right) \quad (9.114)$$

and finally, upon exiting the rear side material-vacuum interface– and again assuming for simplicity just a single passage of light– we get the complex relative amplitude for normal transmission of monochromatic light through a thin wafer of unknown material index,

$$\tilde{T} = \tilde{T}_B\tilde{T}_{m-v} = \frac{2}{1+\tilde{n}}\exp\left(i\frac{\omega}{c}\tilde{n}l\right)\frac{2\tilde{n}}{\tilde{n}+1} \quad (9.115)$$

which is

$$\tilde{T} = Te^{i\delta_t} = \frac{\left|\vec{E}_{0t}\right| e^{i\delta_t}}{\left|\vec{E}_0\right|} = \frac{4(n+i\kappa)}{(n+i\kappa+1)^2}\exp\left(-\frac{\omega}{c}\kappa l\right)\exp\left(i\frac{\omega}{c}nl\right) \quad (9.116)$$

where $T = \left|\vec{E}_{0t}\right| / \left|\vec{E}_0\right|$ is the ratio of wafer-transmitted amplitude to incident amplitude and $\delta_t = \phi_t - \phi$ is the total phase shift introduced

[14] Forsman, A. *et al.*, Interaction of femtosecond laser pulses with ultrathin foils, *Phys. Rev. E*, 58, R1248 (1998).

Widmann, K. *et al.*, Single-state measurement of electrical conductivity of warm dense gold, *Phys. Rev. Lett.* 92, 125002 (2004).

Ping, Y. *et al.*, Broadband dielectric function of non-equilibrium warm dense gold, *Phys. Rev. Lett.* 96, 255003 (2006).

by passage through the wafer. Noting that the fraction of intensity transmitted through the wafer is,

$$T_\perp = \tilde{T}\tilde{T}^* = T^2 \tag{9.117}$$

we see that a measurement of $T_\perp$ gives the amplitude ratio, $T = \sqrt{T_\perp}$. Combining this with a simultaneous measurement of the phase shift, δ_t, then yields $\tilde{T}$ and so solves Eq. 9.116 for the unknown optical constants, n and κ. Measurement of phase shift is experimentally difficult for thin targets as it requires the control of lengths to within a small fraction of the wavelength.

Reflection measurements of thick targets

In many instances, the target is too thick for a transmission measurement and other techniques must be used to obtain a pair of optical constants. Often this involves a pair of off-normal reflectivity measurements so we must consider the more general Fresnel equations for reflection of S- and P-polarized light. In Section 9.5, Eqs. 9.88 and 9.89 expressed the angle-dependent reflection intensities, $R_S(\theta)$ and $R_P(\theta)$, in terms of the real and imaginary components of $\tilde{n}'$ so that given the incident angle and the complex index, the S and P reflectivities are obtained. For the analysis of experimentally measured reflectivities of S- and P-polarized light, this process is inverted: For a given angle, θ_o, the P-polarized reflectivity is measured and then projected as a reflectivity contour onto a P-polarized reflectivity map, $R_P(n', \kappa', \theta_o)$, given by Eq. 9.89. This is repeated for either a different angle and the same polarization or for the same angle and S polarization (using Eq. 9.88) in order to generate a second reflectivity contour. The point where the two contours graphically intersect in the space of (n', κ') determines the material's complex index of refraction. Figure 9.14 shows this process for reflectivities obtained with S- and P-polarized light incident at 80° onto the surface of room temperature aluminum ($\tilde{n} = 0.884 + i6.16$).

Measuring reflectance and phase shift of a thick (or thin) target

As with measuring the phase shift on transmission through a thin wafer, the relevant equation is the complex field amplitude rather than the intensity. The complex relative amplitude for normal reflection from a Fresnel vacuum-material interface is given by Eq. 9.71 as,

$$\tilde{\rho} = re^{i\delta_r} = \left(\frac{\tilde{E}_0''}{\tilde{E}_0}\right)_{norm} = \left(\frac{\left|\vec{E}_0''\right| e^{i\delta_r}}{\left|\vec{E}_0\right|}\right)_{norm} = \frac{1-\tilde{n}}{1+\tilde{n}} \tag{9.118}$$

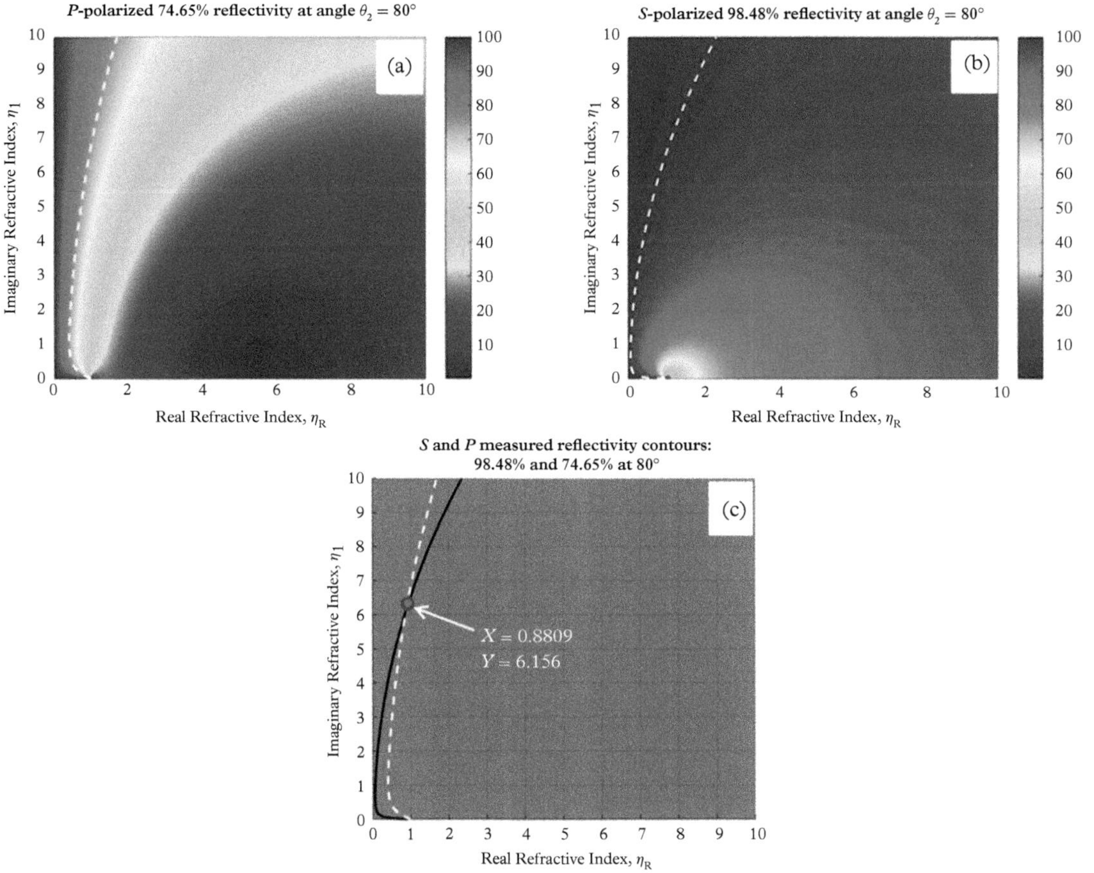

Fig. 9.14 *The top two maps are of P and S-polarized reflectivity in the domain of n and κ for an incidence angle of 80º as determined from Eqs. 9.89 and 9.88. The color shading shows the reflectivity at all combinations of real and imaginary components of the index of refraction. The dashed lines trace out the locus of pairs of real and imaginary index values that would give the stated reflectivity at the indicated incident angle. However, there is only one pair of real and imaginary values that will produce both the P and S reflectivity indicated. In an experiment, the P and S reflectivities would be measured at a given angle and then the associated contours would be projected onto the two maps through Eqs. 9.89 and 9.88, and their intersection would reveal the real and imaginary values of the index of the material responsible for the two observed reflectivity values. In this test case the "measured" reflectivities (75 and 98%) were obtained by a separate calculation for a material index of 0.884 + i6.16. The intersection of the two contours are shown overlaid in the bottom map where they intersect at exactly* $\tilde{n} = 0.881 + i6.16$*, as expected.*

where $r = \left(\tilde{E}_0''/\tilde{E}_0\right)_{norm}$ is the ratio of reflected to incident amplitudes and $\delta_r = \phi'' - \phi$ is the phase shift introduced by the reflection. Equation 9.118 can be inverted to yield

$$\tilde{n} = \frac{1 - re^{i\delta_r}}{1 + re^{i\delta_r}} \tag{9.119}$$

Noting that the fraction of normally reflected intensity, a measurable quantity, is,

$$R_\perp = \tilde{\rho}\tilde{\rho}^* = r^2 \tag{9.120}$$

we see that a measurement of $R_\perp$ will give the amplitude ratio, $r = \sqrt{R_\perp}$. We can then write Eq. 9.119 in terms of this normal reflectance. Specifically, the real and imaginary components of the complex index of refraction can be expressed as,

$$n = \frac{1 - R_\perp}{1 + R_\perp + 2\sqrt{R_\perp}\cos\delta_r} \tag{9.121}$$

$$\kappa = \frac{-2\sqrt{R_\perp}\sin\delta_r}{1 + R_\perp + 2\sqrt{R_\perp}\cos\delta_r} \tag{9.122}$$

So, combining the reflectance measurement with a simultaneous measurement of the phase shift δ_r yields the complex index of refraction of the material. And, although not as difficult as in the transmission configuration, measurement of the phase shift upon reflection still requires sensitivity of lengths small compared to the wavelength.

Ellipsometry

Ellipsometry is the most widely used technique to study the dielectric properties of materials or, more frequently, to characterize a sample. Ellipsometry considers the modification of the polarization of a laser pulse upon reflection from a sharp interface. A primary application is characterizing thin films. If a linearly polarized pulse is incident to the surface with a combined S- and P-polarization, reflection will modify the amplitudes and phases of the S and P components differently. This will, in general, result in a reflected pulse that is elliptically polarized. Analysis by comparison of incident and reflected polarizations then yields the complex index of refraction, $\tilde{n}$.

There are several ways of carrying out measurements but a frequently used method is "RAE," rotating analyzer ellipsometry. The technique is shown schematically on the left-hand side of Fig. 9.15. Polarized light is reflected off a sample and reflectance is measured as a function of polarization. The incident light is often linearly polarized

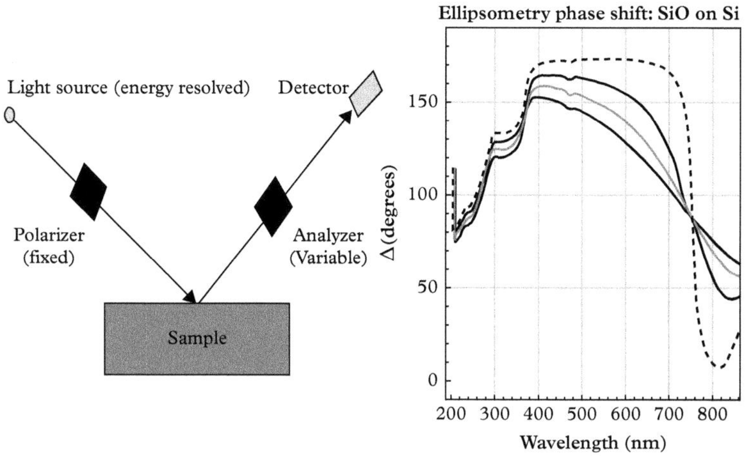

Fig. 9.15 *Left: A schematic of the "RAE" or rotating analyzer ellipsometry technique. Right: The relative phase angle Δ for several very thin layers of* SiO_2 *on* Si. *The dashed curve is for uncoated* Si *and, looking at the right side axis, the curves are for 1, 2 and 3 nm coatings.*

at 45° to the plane of incidence with a polarization half-way between the S- and P-polarizations. If we note that in this case the incident S and P amplitudes are related to the total incident complex amplitude as $\tilde{E}_0 = \sqrt{2}\tilde{E}_{0s} = \sqrt{2}\tilde{E}_{0p} = \sqrt{2}\tilde{E}_{sp}$, then the total reflected amplitude vector normalized to the total incident amplitude is described by:

$$\frac{\vec{E}''_0}{\tilde{E}_0} = \frac{\alpha}{\sqrt{2}}\left[\frac{\tilde{E}''_{0s}}{\tilde{E}_{sp}}\hat{\varepsilon}_s + \frac{\tilde{E}''_{0p}}{\tilde{E}_{sp}}\hat{\varepsilon}_p\right] \tag{9.123}$$

where α is an overall scaling constant that includes factors such as the divergence of the beam–real beams are not infinitely wide plane waves. We recognize the terms in the brackets as the effective Fresnel coefficients, Eqs. 9.66 and 9.70, for this surface. In general, $\hat{\varepsilon}_s$ and $\hat{\varepsilon}_p$ are unit vectors for S and P polarizations. Referring to the coordinate system of Fig. 9.5, the former is along $\hat{x}$ and the latter is in the plane of incidence, orthogonal to the beam's direction. We can write

$$\frac{\tilde{E}''_{0p}}{\tilde{E}''_{0s}} = \tan\psi \exp[i\Delta] \tag{9.124}$$

where Δ represents the phase difference between $\tilde{E}''_{0p}$ and $\tilde{E}''_{0s}$ and $\tan\psi$ is the ratio of their moduli.[15] With this, Eq. 9.123 becomes:

$$\frac{\vec{E}''_0}{\tilde{E}_0} = \frac{\alpha\tilde{E}''_{0s}}{\sqrt{2}}\left[\frac{1}{\tilde{E}_{sp}}\hat{\varepsilon}_s + \frac{\tan\psi \exp[i\Delta]}{\tilde{E}_{sp}}\hat{\varepsilon}_p\right] \tag{9.125}$$

The analyzer rotates its polarization in the plane perpendicular to the beam's propagation (that is, in the plane containing the two

[15] If the phase difference of the S and P reflections were zero, $\Delta = 0$, then the reflected light would remain linearly polarized and then ψ would be the angle of that polarization relative to $\hat{\varepsilon}_s = \hat{x}$.

polarization components). An analyzer polarization making an angle ϕ with respect to $\hat{x} = \hat{\varepsilon}_s$ may be written:

$$\hat{\varepsilon}_{AN}(\phi) = \hat{\varepsilon}_s \cos\phi + \hat{\varepsilon}_p \sin\phi \qquad (9.126)$$

The intensity of the beam transmitted through this polarizer is then proportional to:

$$\left| \hat{\varepsilon}_{AN}(\phi) \cdot \frac{\vec{E}_0''}{\tilde{E}_0} \right|^2 = \left| \frac{\alpha \tilde{E}_{0s}''}{\sqrt{2}\tilde{E}_{sp}} \left[\cos\phi + \tan\psi \exp(i\Delta) \sin\phi \right] \right|^2 \qquad (9.127)$$

$$= \beta \Big[\cos^2\phi + \tan^2\psi \sin^2\phi + 2\tan\psi \cos\Delta \cos\phi \sin\phi \Big] \qquad (9.128)$$

where β is some other overall scaling constant. Thus, by measuring the reflected polarized signal as a function of ϕ, the constants $\tan\psi$ and Δ can be determined. Analysis of such information for a spectral scan will yield the complex index of refraction over the scanned range of wavelengths. Note that:

- Only the relative reflectivity as a function of angle need be measured. It is not necessary to determine β.
- The "constants" $\tan\psi$ and Δ may be functions of the incidence angle in the apparatus and–most importantly–of wavelength.
- Having measured $\tan\psi$ and θ, these two parameters can be immediately inverted to give the dielectric function of the "pseudo-material" they correspond to.

$$\tilde{\varepsilon} = \sin^2\theta \left[1 + \tan^2\theta \left(\frac{1 - \tan\psi \exp[i\Delta]}{1 + \tan\psi \exp[i\Delta]} \right)^2 \right] \qquad (9.129)$$

Usually, measurements are made as a function of wavelength and occasionally incidence angle. The results are then modeled and fit to the data using whatever is known about the sample under study to return whatever parameters are desired to be determined. An especially important system for study is the formation or growth of a silicon dioxide film, SiO_2, on a fresh silicon surface. In Fig. 9.15 we show how exquisitely sensitive the ellipsometry measurements–particularly the measurement of Δ–are to surface conditions. For this idealized problem, using known values for the silicon and SiO_2, the curves show the predicted values for different values of the thin oxide film. Thicknesses of 0, 1, 2, and 3 nm are readily distinguished. Note, 1 nm corresponds to roughly three monolayers of SiO_2.

9.9.3 Spectral measurements

While measurements of the optical constants at discrete frequencies can be useful in filling in experimental gaps in data and in determining certain frequency independent quantities,[16] it is far more desirable to obtain, in a single shot, or at least a single set of identical shots, a broadband measurement of the material response. With broadband measurements, response features–such as resonance widths–are quickly identified and models can be rapidly confirmed or disputed. With a discrete-frequency measurement we know simply how the material responds at that frequency whereas with a broadband measurement of the response, because of the Fourier transform relationship of the time and frequency domain response functions, we can predict the time domain response of the material to an arbitrarily varying field.

Though time consuming, discrete measurements such as those previously outlined can be built up to give a weakly resolved picture of the frequency response of the material. However, measurements of broadband continuum reflection or transmission and phase shifting provides all the information in a single shot. Each of the single frequency measurement techniques in the previous section can be generalized to measurements over a broad continuous range of frequencies. Any of these generalized techniques can then yield the functional form of the complex index of refraction over that range of frequencies. From $\tilde{\varepsilon} = \tilde{n}^2$, the response curve structure (real and imaginary) specific to the probed material can then be found. Indeed, the imaginary component of the response curve is the characteristic absorption spectrum of the material. In addition, as we found in Section 9.8, there is a close mathematical relationship between the functional forms of the real and imaginary parts of a response function. Thus, for a sufficiently broad spectral measurement of one optical constant, say the absorption coefficient $\kappa(\omega)$, this relationship will provide the functional form of the conjugate optical constant, in this case $n(\omega)$.

Kramers–Kronig analysis of reflectance–phase measurements

As an example of how the results of a single frequency measurement technique generalized to broadband can be treated using the Kramers–Kronig relations, we next consider the reflectance–phase shift measurements discussed previously. In Section 9.8 we derived the Kramers–Kronig relations between the real and imaginary

[16] For example, given a measurement-based determination of optical constants such as $\tilde{n} = n + i\kappa$, an assumption of the Drude model can provide an estimation of the electron density n_e or collision frequency, γ, of the material.

components of the frequency domain polarization response function, $\tilde{\chi}_e$ (in the time domain, the response function is generally real). In fact, this is a very general relationship resulting from the necessary causality (cause always precedes effect) of any physical response to a stimulus and it can be extended to any pair of the frequency dependent optical constants. Many of these associated relationships can be obtained from the relations for $\tilde{\chi}_e$, Eqs. 9.108 and 9.109, by simple substitution. For example, with the bound dielectric constant represented as $\tilde{\varepsilon}_b = 1 + \tilde{\chi}_e$, the relations between $\varepsilon_{b1}(\omega)$ and $\varepsilon_{b2}(\omega)$ are

$$\varepsilon_{b1}(\omega_o) = 1 + \frac{2}{\pi}\mathrm{P}\left(\int_0^\infty \omega' \frac{\varepsilon_{b2}(\omega')}{\omega'^2 - \omega_o^2}\mathrm{d}\omega'\right) \quad \text{and}$$

$$\varepsilon_{b2}(\omega_o) = -\frac{2\omega_o}{\pi}\mathrm{P}\left(\int_0^\infty \frac{\varepsilon_{b1}(\omega') - 1}{\omega'^2 - \omega_o^2}\mathrm{d}\omega'\right) \tag{9.130}$$

and similarly, with the *generalized* conductivity as $\tilde{\sigma} = -i\varepsilon_o\omega\tilde{\varepsilon}$[17]

$$\sigma_1(\omega_o) = \frac{2\omega_o^2}{\pi}\mathrm{P}\left(\int_0^\infty \frac{\sigma_2(\omega')}{\omega'\left(\omega'^2 - \omega_o^2\right)}\mathrm{d}\omega'\right) \quad \text{and}$$

$$\sigma_2(\omega_o) = -\varepsilon_o\omega_o - \frac{2\omega_o}{\pi}\mathrm{P}\left(\int_0^\infty \frac{\sigma_1(\omega')}{\omega'^2 - \omega_o^2}\mathrm{d}\omega'\right) \tag{9.131}$$

The method, described previously, for the single frequency measurements of normal reflectance and phase shift of a thick (or thin) target can be generalized to broadband continuum measurements. By Eq. 9.120, a broadband measurement of $R_\perp(\omega)$ will yield the amplitude ratio, $r(\omega) = \sqrt{R_\perp}$ over a continuum of frequencies. A simultaneous broadband measurement of the phase shift, $\delta_r(\omega)$ determines $n(\omega)$ and $\kappa(\omega)$, via the inverted Eq. 9.119, over the range of frequencies measured. However, if we could apply Kramers–Kronig, this method could be reduced to just a single broadband measurement.

Consider the natural log of the complex reflection amplitude, $\tilde{\rho}$, in Eq. 9.118,

$$\ln\tilde{\rho} = \ln\left(re^{i\delta_r}\right) = \ln r + i\delta_r \tag{9.132}$$

If we note that normal reflectance, $R_\perp$, and the amplitude for normal reflection $\tilde{\rho}$, are both causal, we can write basic Kramers–Kronig relations for the real and imaginary parts, $\ln r(\omega)$ and $\delta_r(\omega)$, of Eq. 9.132,

[17] Note that these relations for the generalized conductivity differ from those for the conductivity as defined in Eq. 9.10. For the latter, more familiar definition, $\tilde{\sigma}_f$ represents only the "free" contributions to the material response. The Kramers–Kronig relations for this conductivity are obtained by the same method as for the case of the electric susceptibility, $\tilde{\chi}_e$. They are therefore,

$$\sigma_1(\omega_o) = \frac{2}{\pi}\mathrm{P}\left(\int_0^\infty \omega' \frac{\sigma_2(\omega')}{\omega'^2 - \omega_o^2}\mathrm{d}\omega'\right)$$

and

$$\sigma_2(\omega_o) = -\frac{2\omega_o}{\pi}\mathrm{P}\left(\int_0^\infty \frac{\sigma_1(\omega')}{\omega'^2 - \omega_o^2}\mathrm{d}\omega'\right)$$

$$\ln r(\omega_o) = \frac{1}{\pi}\mathrm{P}\int_{-\infty}^{\infty} \frac{\delta_r(\omega')}{\omega' - \omega_o} d\omega' \quad \text{and}$$
$$\delta_r(\omega_o) = -\frac{1}{\pi}\mathrm{P}\int_{-\infty}^{\infty} \frac{\ln r(\omega')}{\omega' - \omega_o} d\omega' \tag{9.133}$$

It can be shown that $\tilde{\rho}(\omega)$ is Hermitian, $\tilde{\rho}(-\omega)=\tilde{\rho}^*(\omega)$, so that $\ln r(\omega)$ is even and $\delta_r(\omega)$ is odd[18]. As we did in Section 9.2 for the susceptibility relations, we can use these symmetry properties to eliminate the negative frequencies and put the Kramers–Kronig relations of Eq. 9.133 in usable form,

$$\ln r(\omega_o) = \frac{2}{\pi}\mathrm{P}\left(\int_0^{\infty} \omega' \frac{\delta_r(\omega')}{\omega'^2 - \omega_o^2} d\omega'\right) \quad \text{and}$$
$$\delta_r(\omega_o) = -\frac{2\omega_o}{\pi}\mathrm{P}\left(\int_0^{\infty} \frac{\ln r(\omega')}{\omega'^2 - \omega_o^2} d\omega'\right) \tag{9.134}$$

Thus, referring to Eq. 9.119 and the experimental determination of the complex index of refraction $\tilde{n}(\omega) = n + i\kappa$, we see that because the real and imaginary parts of the functions $\ln r(\omega)$ and $\delta_r(\omega)$ are related by Kramers–Kronig, it is not necessary to make a broadband measurement of both. And, in this case, because it is a simpler measurement to make than the phase shift, the normal reflectivity is measured across a *sufficiently* broad continuum of frequencies. The second Kramers–Kronig relation of Eq. 9.134 then provides $\delta_r(\omega)$ thus solving for the complex index of refraction in Eq. 9.119. And just as with the reflectivity-phase measurements, the single-frequency transmission-phase measurement technique can be extended to broadband. The Kramers–Kronig relations are in that case used to relate the real and imaginary parts of the natural log of Eq. 9.116: $\ln \tilde{\tau} = \ln\left(te^{i\delta_t}\right) = \ln t(\omega) + i\delta_t(\omega)$.

Experimental limitations

There is one practical limitation to the use of the Kramers–Kronig relations with these measurements: the Kramers–Kronig integrals range from DC to infinity while any acquired broadband data is necessarily finite with the lowest measured frequency much greater than zero. Ideally, then, the data should be extrapolated from the lowest measured frequency to zero and from the highest measured frequency to infinity. It is seen from any of the Kramers–Kronig relations, however, that the greatest contributions to the integrals come from behavior near $\omega' = \omega_o$ so that the further from the evaluated frequency, ω_o, the less accuracy required in the extrapolation. The

[18] If we take as given that $\tilde{n}(\omega)$ is Hermitian, $\tilde{n}(-\omega)=\tilde{n}^*(\omega)$, then Eq. 9.118 can be written

$$\tilde{\rho}(\omega) = re^{i\delta_r} = \frac{1-\tilde{n}}{1+\tilde{n}} = \frac{(1-n) - i\kappa}{(1+n) + i\kappa} =$$
$$\frac{\left[(1-n)(1+n) - \kappa^2\right] - i\left[(1-n)\kappa + (1+n)\kappa\right]}{(1+n)^2 + \kappa^2}$$

and using the Hermiticity of $\tilde{n}(\omega)$, we can write,

$$\tilde{\rho}(-\omega) = \frac{1-\tilde{n}^*}{1+\tilde{n}^*} = \frac{(1-n) + i\kappa}{(1+n) - i\kappa} =$$
$$\frac{\left[(1-n)(1+n) - \kappa^2\right] + i\left[(1-n)\kappa + (1+n)\kappa\right]}{(1+n)^2 + \kappa^2}$$

so $\tilde{\rho}(-\omega) = \tilde{\rho}^*(\omega)$ and $\tilde{\rho}$ is indeed Hermitian.

extrapolations are generally based on the most appropriate models for the expected measurements in the associated frequency range. We will discuss the simple classical models of Drude and Lorentz in the following chapter.

Finally, we reiterate that without a model and some knowledge of the type of material (conductor, insulator, etc.), these single-frequency and spectral measurements cannot yield any further information about the material. Specifically, as noted previously, these measurements will not discriminate between bound and free contributions to the material response. We, therefore, require a model for bound and free electrons which will attach specific physical meanings to the complex quantities of susceptibility, conductivity, dielectric constant, and index of refraction.

Exercises

(9.1) There is corollary to a rule, the quotient rule, which states that if contraction of a 4×4 entity produces a known 4-vector, then that entity is a 4-tensor. Using the known bound charge and current relations from Chapter 1,

$$\nabla \cdot \vec{P} = \rho_b \text{ and } \nabla \times \vec{M} = \vec{J}_b$$

and the fact that $\vec{J}^\alpha = \left(c\rho, \vec{J}\right)$ is a known 4-vector, show that

$$M^{\alpha\beta} = \begin{pmatrix} 0 & -cP_x & -cP_y & -cP_z \\ cP_x & 0 & M_z & -M_y \\ cP_y & -M_z & 0 & M_x \\ cP_z & M_y & -M_x & 0 \end{pmatrix}$$

is a 4-tensor. Thus, the 3-space polarization and magnetization vectors can be combined into a relativistic 4-tensor.

(9.2) Derive expressions for the real and imaginary parts of the refractive index in terms of: (A) the real and imaginary parts of the generalized dielectric constant. (B) the real and imaginary parts of the generalized conductivity.

(9.3) Consider the following three different electron responses $g\left(t - t'\right) = g(T)$ to a field within a material: (A) electrons in material are free and undamped (collisionless): $g(T) = \chi_o \Theta(T)$, where $\Theta(T)$ is the Heaviside step function. (B) electrons in material are free and damped: $g(T) = \chi_o \Theta(T) \exp(-\gamma T)$, where γ is a damping constant. (C) electrons in material are bound and damped: $g(T) = \chi_o \Theta(T) \exp(-\gamma T) \exp(i\omega_o T)$, where ω_o is the resonant frequency. Noting that the susceptibility, $\chi(\omega)$, is the Fourier transform of the response function, $g(T)$, find $\chi(\omega)$ for these three material responses.

(9.4) Show that for the response of a material to a plane wave, the imaginary components of the permittivity ε_{b2} and conductivity σ_{f2} lead, respectively, to energy absorption and no energy absorption.

(9.5) It is generally true, according to the result of Eq. (Eq. after 10.23) that to have a nonzero $\varepsilon_2(\omega)/\varepsilon_0$ (and thus material absorption of field energy) both n and κ (or, equivalently, both k_1 and k_2) must be nonzero. Explain why, for a plane wave passing into a material, this makes physical sense, in terms of $\tilde{k}=\tilde{n}\omega/c$, by invoking a conservation of energy argument. That is, consider three cases: (a) $k_1 \neq 0, k_2 = 0$, (b) $k_1 = 0, k_2 \neq 0$ and (c) $k_1 \neq 0, k_2 \neq 0$ and explain why only case (c), in which both k_1 and k_2 are nonzero, involves material absorption of field energy.

(9.6) As we will discuss in Chapter 10, the dispersion relation for plasmas and conductors at sufficiently high frequencies is given by

$$k^2 = \omega^2 \mu\varepsilon \left(1 - \frac{\omega_p^2}{\omega^2}\right)$$

where ω_p is known as the plasma frequency of the medium. (a) Find the group and phase velocities, v_g and v_p, in terms of k, for this dispersion relation. Show that $v_g v_p = (\mu\varepsilon)^{-1}$. (b) Find the group and phase velocities, in terms of ω, for this dispersion relation, and plot $v_g(\omega)$ and $v_p(\omega)$ for $\omega > \omega_p$.

(9.7) In the text, we derived the Fresnel transmission and reflection amplitude coefficients for both S and P polarized light using the third and fourth of the four Maxwell boundary continuity conditions given in Section 10.5.1. In that case, the other two BC's were redundant and, in general, only two of the four boundary conditions is required to obtain these equations. In a similar way, obtain these four Fresnel equations ($\tilde{E}'_{0s}/\tilde{E}_{0s}$, $\tilde{E}''_{0s}/\tilde{E}_{0s}$, $\tilde{E}'_{0p}/\tilde{E}_{0p}$ and $\tilde{E}''_{0p}/\tilde{E}_{0p}$) using a different pair–choose one for $\vec{E}$ and one for $\vec{B}$ (or $\vec{H}$) - of the four boundary continuity conditions.

(9.8) Referring to Fig. 9.16, and using Snell's law and the definition of critical angle, find the range of incident angles $\Delta\theta_1$ for which rays within the fiber will be totally internally reflected.

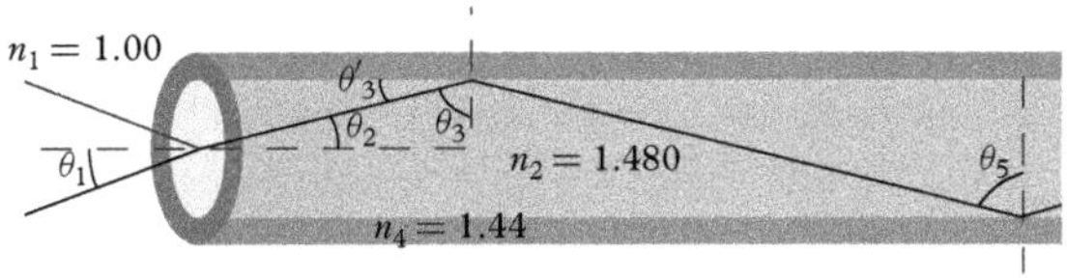

Fig. 9.16

(9.9) For the transmitted wave part of a total internal reflection, use the Poynting vector to calculate the energy flow in the $\hat{y}$ direction and show that there is no energy flow in the $-\hat{z}$ direction.

(9.10) Given the general susceptibility function

$$\chi(\omega) = \frac{S\omega_0^2}{\omega_0^2 - \omega^2 - i\omega\gamma}$$

show that the real and imaginary parts of this response function are related by the Kramers–Kronig relations (Eqs. 9.108 and 9.109).

(9.11) Show that we can obtain Eq. 9.16 directly from the general electric field of Eq. 9.4 by substitution of the harmonic solutions and with the same assumptions of a material which is non-magnetic ($\mu \simeq \mu_o$) and has zero total charge density ($\rho = 0$).

(9.12) Given Eq. 9.13 with $\tilde{\varepsilon}/\varepsilon_o = \tilde{n}^2$, express σ_1 and σ_2 in terms of the real and imaginary components, n and κ, of the complex refractive index. Invert this to express n and κ in terms of the real and imaginary components, σ_1 and σ_2, of the generalized AC conductivity.

9.10 Discussions

Discussion 9.1

Because a general plane wave can be expressed as an integration of component monochromatic traveling waves, each in the form of Eq. 9.27, its analysis requires the tools of Fourier and with this we are given the choice to work with either the time-frequency or the space-wavenumber Fourier space pairs. Since a snapshot in time of a spatially extended traveling waveform and its decomposition into spatially extended traveling harmonic waves is more easily envisaged, we choose the latter. Neglecting, for the moment, the effect of dispersion within Eq. 9.41 so that $\omega(k) = kc/n$, we can claim that although the different wavenumber components are generally out of phase with each other and thus largely cancel, there will, at any given time t_o, be some point x_o in space for which all the phases are aligned and add constructively to form a peak. And because all the components are traveling together–without dispersion–at speed $v_p = \omega/k = c/n$, this peak also moves at v_p. So we can imagine this "wave-packet," which occupies a length Δx about some position x_o, as a coherent addition of plane waves with a spread in wavenumbers Δk. Furthermore, for a wider spread in wavenumbers, the aligned phases will decohere at a shorter distance from x_o resulting in a spatially narrower pulse. Indeed, a general result from Fourier analysis relates the wavenumber spread and pulse width as $(\Delta k)(\Delta x) \geq 1/2$. In Quantum Mechanics, this result manifests as the famous uncertainty principle where $(\hbar\Delta k)(\Delta x) = (\Delta p)(\Delta x) \geq \hbar/2$.

Discussion 9.2

The initial assumption, which enabled the definition of group velocity, was that n (and thus k) changed relatively slowly with ω. In an anomalous dispersion region this assumption and thus the validity of the group velocity is lost. In a material that has dispersion, the definition of the velocity of energy, or the signal velocity, must be carefully considered. When the wave enters the absorptive medium, part of the energy of the wave is moved to and stored within the medium. This arises from the work the fields do on the polarization charges in the material. This energy remains in the material, and does not propagate. The energy velocity is the speed with which the energy not absorbed moves through the material. As shown by Diener (Ann. Phys. 7, 639, 1998), the energy velocity is given by:

$$v_E = \frac{2n}{n^2+1}c < c$$

while the group velocity is

$$v_g = \frac{c}{n + \omega\frac{dn}{d\omega}}$$

The group velocity does indeed represent the apparent speed of the (distorting) wave packet through the absorbing material, but the speed of the transmitted energy, or signal velocity, is always less than c.

Discussion 9.3

Insert $\vec{B}=\frac{\tilde{k}}{\omega}(\hat{k}\times\vec{E})$ into the fourth boundary condition and note that the complex wavenumbers for the incident and reflected waves are equal, $\tilde{k}=\tilde{k}''$,

$$\left(\frac{\tilde{k}}{\omega\tilde{\mu}}\left[(\hat{k}\times\vec{E}_s)+(\hat{k}''\times\vec{E}''_s)\right]-\frac{\tilde{k}'}{\omega\tilde{\mu}'}(\hat{k}'\times\vec{E}'_s)\right)\times\hat{z}=0$$

but $\tilde{k}/\omega\tilde{\mu}=\tilde{n}/c\tilde{\mu}$, so

$$\left(\frac{\tilde{n}}{\tilde{\mu}}\left[(\hat{k}\times\vec{E}_s)+(\hat{k}''\times\vec{E}''_s)\right]-\frac{\tilde{n}'}{\tilde{\mu}'}(\hat{k}'\times\vec{E}'_s)\right)\times\hat{z}=0$$

Use the vector identity $(\hat{k}\times\vec{E})\times\hat{z}=(\hat{k}\cdot\hat{z})\vec{E}-(\vec{E}\cdot\hat{z})\hat{k}$ and Fig. 9.5, to see, for example, that $\vec{E}_s\cdot\hat{z}=0$ and $\hat{k}\cdot\hat{z}=-\cos\theta$, so that we can write

$$\frac{\tilde{n}}{\tilde{\mu}}\left(-\tilde{E}_{0s}+\tilde{E}''_{0s}\right)\cos\theta+\frac{\tilde{n}'}{\tilde{\mu}'}\tilde{E}'_{0s}\cos\theta'=0$$

Discussion 9.4

Solving Eqs. 9.63 and 9.64, we obtain

$$\frac{\tilde{E}'_{0s}}{\tilde{E}_{0s}}=\frac{2\tilde{n}\cos\theta}{\tilde{n}\cos\theta+(\tilde{\mu}/\tilde{\mu}')\tilde{n}'\cos\theta'} \tag{9.135}$$

$$\frac{\tilde{E}''_{0s}}{\tilde{E}_{0s}}=\frac{\tilde{n}\cos\theta-(\tilde{\mu}/\tilde{\mu}')\tilde{n}'\cos\theta'}{\tilde{n}\cos\theta+(\tilde{\mu}/\tilde{\mu}')\tilde{n}'\cos\theta'} \tag{9.136}$$

and using the complex version of Snell's law, Eq. 9.61, to express $\tilde{n}'\cos\theta'=\sqrt{\tilde{n}'^2-\tilde{n}^2\sin^2\theta}$, we obtain an alternate form, completely in terms of the incident angle θ,

$$\frac{\tilde{E}'_{0s}}{\tilde{E}_{0s}}=\frac{2\tilde{n}\cos\theta}{\tilde{n}\cos\theta+(\tilde{\mu}/\tilde{\mu}')\sqrt{\tilde{n}'^2-\tilde{n}^2\sin^2\theta}} \tag{9.137}$$

$$\frac{\tilde{E}''_{0s}}{\tilde{E}_{0s}}=\frac{\tilde{n}\cos\theta-(\tilde{\mu}/\tilde{\mu}')\sqrt{\tilde{n}'^2-\tilde{n}^2\sin^2\theta}}{\tilde{n}\cos\theta+(\tilde{\mu}/\tilde{\mu}')\sqrt{\tilde{n}'^2-\tilde{n}^2\sin^2\theta}} \tag{9.138}$$

Discussion 9.5

Insert $\vec{B} = \frac{\tilde{k}}{\omega}(\hat{k} \times \vec{E})$ into the fourth boundary condition and note that the complex wavenumbers for the incident and reflected waves are equal, $\tilde{k} = \tilde{k}''$,

$$\left(\frac{\tilde{k}}{\omega\tilde{\mu}} \left[(\hat{k} \times \vec{E}_p) + (\hat{k}'' \times \vec{E}''_p) \right] - \frac{\tilde{k}'}{\omega\tilde{\mu}'} (\hat{k}' \times \vec{E}'_p) \right) \times \hat{z} = 0$$

but $\tilde{k}/\omega\tilde{\mu} = \tilde{n}/c\tilde{\mu}$, so

$$\left(\frac{\tilde{n}}{\tilde{\mu}} \left[(\hat{k} \times \vec{E}_p) + (\hat{k}'' \times \vec{E}''_p) \right] - \frac{\tilde{n}'}{\tilde{\mu}'} (\hat{k}' \times \vec{E}'_p) \right) \times \hat{z} = 0$$

Noting the directions of $\vec{E}_p$, $\vec{E}'_p$, and $\vec{E}''_p$ in Fig. 9.5, this can be rewritten as,

$$\left(\frac{\tilde{n}}{\tilde{\mu}} \left(\tilde{E}_{0p}\hat{x} - \tilde{E}''_{0p}\hat{x} \right) - \frac{\tilde{n}'}{\tilde{\mu}'} \tilde{E}'_{0p}\hat{x} \right) \times \hat{z} = 0$$

and the boundary condition for $\vec{H} = \frac{1}{\tilde{\mu}}\vec{B}$ parallel to the surface in terms of E fields of the "P" polarized situation is obtained from the cross product with $\hat{z}$,

$$-\frac{\tilde{n}}{\tilde{\mu}} \left(\tilde{E}_{0p} - \tilde{E}''_{0p} \right) + \frac{\tilde{n}'}{\tilde{\mu}'} \tilde{E}'_{0p} = 0$$

Discussion 9.6

Solving Eqs. 9.67 and 9.68, we get

$$\frac{\tilde{E}'_{0p}}{\tilde{E}_{0p}} = \frac{2\tilde{n}\tilde{n}' \cos\theta}{(\tilde{\mu}/\tilde{\mu}')\,\tilde{n}'^2 \cos\theta + \tilde{n}\tilde{n}' \cos\theta'}, \tag{9.139}$$

$$\frac{\tilde{E}''_{0p}}{\tilde{E}_{0p}} = \frac{-(\tilde{\mu}/\tilde{\mu}')\,n'^2 \cos\theta + \tilde{n}\tilde{n}' \cos\theta'}{(\tilde{\mu}/\tilde{\mu}')\,\tilde{n}'^2 \cos\theta + \tilde{n}\tilde{n}' \cos\theta'} \tag{9.140}$$

Again, expressing $\tilde{n}' \cos\theta' = \sqrt{\tilde{n}'^2 - \tilde{n}^2 \sin^2\theta}$, we obtain an alternate form, completely in terms of the incident angle, θ

$$\frac{\tilde{E}'_{0p}}{\tilde{E}_{0p}} = \frac{2\tilde{n}\tilde{n}' \cos\theta}{(\tilde{\mu}/\tilde{\mu}')\,\tilde{n}'^2 \cos\theta + \tilde{n}\sqrt{\tilde{n}'^2 - \tilde{n}^2 \sin^2\theta}}, \tag{9.141}$$

$$\frac{\tilde{E}''_{0p}}{\tilde{E}_{0p}} = \frac{-(\tilde{\mu}/\tilde{\mu}')\,\tilde{n}'^2 \cos\theta + \tilde{n}\sqrt{\tilde{n}'^2 - \tilde{n}^2 \sin^2\theta}}{(\tilde{\mu}/\tilde{\mu}')\,\tilde{n}'^2 \cos\theta + \tilde{n}\sqrt{\tilde{n}'^2 - \tilde{n}^2 \sin^2\theta}} \tag{9.142}$$

Discussion 9.7

Noting that since there are no x or y components in Eq. 9.74, $\partial E'_x/\partial x = \partial E'_y/\partial y = 0$, and

$$\begin{aligned}\nabla\cdot\vec{E}' &= \frac{\partial E'_z}{\partial z}\\ &= \frac{\omega}{c}n\left(\sin^2\theta - \left(\frac{n'}{n}\right)^2\right)^{1/2} E'\exp\left(i\vec{k}'\cdot\vec{r}\right)\neq 0\end{aligned}$$

and so the divergence is not zero. There must therefore be a y component for which $\partial E'_y/\partial y$ combines with $\partial E'_z/\partial z$ to give zero divergence. If we add the $\hat{y}$ term,

$$E'_y = -\frac{1}{i\sin\theta}\left(\sin^2\theta - \left(\frac{n'}{n}\right)^2\right)^{1/2} E'\exp\left(i\vec{k}'\cdot\vec{r}\right)$$

then

$$\frac{\partial E'_y}{\partial y} = -\frac{\omega}{c}n\left(\sin^2\theta - \left(\frac{n'}{n}\right)^2\right)^{1/2} E'\exp\left(i\vec{k}'\cdot\vec{r}\right)$$

so that

$$\nabla\cdot\vec{E}' = \frac{\partial E'_x}{\partial x} + \frac{\partial E'_y}{\partial y} + \frac{\partial E'_z}{\partial z} = 0$$

Discussion 9.8

If we let

$$\sqrt{\tilde{n}'^2 - \sin^2\theta} = p + iq \tag{9.143}$$

Then we can write the general Fresnel Eqs. 9.82–9.85 as,

$$\begin{aligned}T_S(\theta) &= \left|\frac{2\cos\theta}{\cos\theta + p + iq}\right|^2\left(\frac{n'^2 - \sin^2\theta}{\cos^2\theta}\right)^{1/2}\\ R_S(\theta) &= \left|\frac{(\cos\theta - p) - iq}{(\cos\theta + p) + iq}\right|^2 = \frac{\cos^2\theta - 2p\cos\theta + p^2 + q^2}{\cos^2\theta + 2p\cos\theta + p^2 + q^2}\\ T_P(\theta) &= \left|\frac{2\cos\theta\sqrt{(p+iq)^2 + \sin^2\theta}}{\left[(p+iq)^2 + \sin^2\theta\right]\cos\theta + p + iq}\right|^2\left(\frac{n'^2 - \sin^2\theta}{\cos^2\theta}\right)^{1/2}\end{aligned} \tag{9.144}$$

$$R_P(\theta) = \left| \frac{\left[(p+iq)^2 + \sin^2\theta\right]\cos\theta - p - iq}{\left[(p+iq)^2 + \sin^2\theta\right]\cos\theta + p + iq} \right|^2 = \frac{\left[\left(p^2 - q^2 + \sin^2\theta\right)\cos\theta - p\right]^2 + (2pq\cos\theta - q)^2}{\left[\left(p^2 - q^2 + \sin^2\theta\right)\cos\theta + p\right]^2 + (2pq\cos\theta + q)^2}$$

We can find p and q in terms of n' and κ' by first squaring both sides of Eq. 9.143,

$$n'^2 - \kappa'^2 + 2in'\kappa' - \sin^2\theta = p^2 - q^2 + 2ipq \tag{9.145}$$

Then equating the real and imaginary parts, we have

$$n'^2 - \kappa'^2 - \sin^2\theta = p^2 - q^2 \tag{9.146}$$

$$n'\kappa' = pq \tag{9.147}$$

Solving Eq. 9.147 for p and substituting into Eq. 9.147, we eliminate q and obtain a quadratic equation of the form $a\left(p^2\right)^2 + b\left(p^2\right) + c = 0$. The solution is found to be

$$p^2 = \frac{1}{2}\left\{\left(n'^2 - \kappa'^2 - \sin^2\theta\right) + \left[\left(n'^2 - \kappa'^2 - \sin^2\theta\right)^2 + \left(2n'\kappa'\right)^2\right]^{1/2}\right\} \tag{9.148}$$

and then from Eq. 9.146,

$$q^2 = \frac{1}{2}\left\{-\left(n'^2 - \kappa'^2 - \sin^2\theta\right) + \left[\left(n'^2 - \kappa'^2 - \sin^2\theta\right)^2 + \left(2n'\kappa'\right)^2\right]^{1/2}\right\} \tag{9.149}$$

where we have chosen the "+" signs for the following reason: since it is always true that $n'^2 - \kappa'^2 - \sin^2\theta < \sqrt{\left(n'^2 - \kappa'^2 - \sin^2\theta\right)^2 + (2n'\kappa')^2}$, and, by definition p and q must be real, the quantities in brackets must be positive.

Discussion 9.9

This evaluation is handled by breaking the integral into three parts: The first, a ccw integration along an arc in the UHP vanishes as $\omega \to 0$. The second, an integration along the real axis from $-\infty$ to $\omega_o - \varepsilon$ and from $\omega_o + \varepsilon$ to $+\infty$ with $\varepsilon \to 0$. This is the principle part of the integration designated with a "P". Finally, a clockwise integration around a semicircle at the real axis pole, ω_o, completes the integration. This part contributes 1/2 of the residue or $-i\pi\tilde{\chi}(\omega_o)$, with the minus sign originating from traversing the half circle of radius ε in a clockwise manner. It is of some interest to show this last part more explicitly: considering just the clockwise integration around the pole

$$\int_{\pi}^{0} \frac{\tilde{\chi}(\omega)}{\omega-\omega_o} \mathrm{d}\omega = -\int_{0}^{\pi} \frac{\tilde{\chi}(\omega)}{\omega-\omega_o} \mathrm{d}\omega$$

now the contour can be represented as an infinitesimal semicircle of radius ε about ω_o. That is, $\omega = \omega_o + \varepsilon e^{i\theta}$ where $d\omega = i\varepsilon e^{i\theta} d\theta$. Substituting these into the integral,

$$-\int_{0}^{\pi} \frac{\tilde{\chi}\left(\omega_o + \varepsilon e^{i\theta}\right)}{\varepsilon e^{i\theta}} i\varepsilon e^{i\theta} d\theta = -i \int_{0}^{\pi} \tilde{\chi}\left(\omega_o\right) d\theta = -i\pi \tilde{\chi}\left(\omega_o\right)$$

Discussion 9.10

Indeed, we know that for DC fields, free ($\tilde{\sigma}_f$) and bound ($\tilde{\varepsilon}_b$) responses are easily discerned since free charges will move as long as the field acts–resulting in a DC current–while bound charges will stretch from their positions until a restoring force stops them–resulting in a DC polarization. For AC fields, however, this distinction can be lost. Indeed, the free charges no longer move arbitrary distances but, like the bound charges, oscillate about a point and so their response can be interpreted as more AC polarization. Conversely, the motion of the bound charges is a current and so their response could be interpreted as just more AC current. At low frequencies, the distinction is somewhat preserved in that the two responses maintain different phase relations with the driving field. However, at higher frequencies even this distinction is blurred.

10 Models of Electromagnetic Response of Materials

- Physical models that relate the generalized complex permittivity and complex conductivity to the electronic structure of matter
- Classical models of Drude and Lorentz; expressions for the complex index of refraction in a nominal insulator
- Reflection and transmission properties of interfaces in insulators in terms of the complex index of refraction; properties of electromagnetic propagation in the frequency vicinity of an atomic resonance
- Model of a metal represented in a Drude model as a plasma
- Measurement techniques applicable to the determination of the parameters of a Lorentz/Drude representation of matter

10.1 Classical models of Drude and Lorentz

In the previous chapter, we introduced conventions for representing the complex response of a material to electromagnetic radiation in terms of a generalized complex permittivity, $\tilde{\varepsilon}$, or a generalized complex conductivity, $\tilde{\sigma}$. These representations made no reference to "bound" or "free" contributions but rather divided the material response generically into real and imaginary components. The advantage of this convention, as we saw, is that it is the real and imaginary components that are obtained directly through experimental measurements. Furthermore, because the imaginary component of the response function to the field, bound or free, is always in phase with the field, and the real component is always 90° out of phase, the ratio of these components characterizes the degree to which the material

Electromagnetic Radiation. Richard Freeman, James King, Gregory Lafyatis,
Oxford University Press (2019). © Richard Freeman, James King, Gregory Lafyatis.
DOI: 10.1093/oso/9780198726500.001.0001

absorbs the electromagnetic energy and, by extension, whether the transport of electromagnetic energy is dominantly wavelike or diffusive. This model-independent convention, however, provides little insight into what is going on in the material. Indeed, the first step toward a physical understanding is to assume the atomic model of electrons orbiting a nucleus of protons and neutrons. This allows for two basic states of electrons within a material: electrons that are localized or "bound" to the nuclei and those that are itinerant or "free". For such a model, the best representation of the material response is the original representation, expressed as the first equality in Eq. 9.11 that divides the response into free $\tilde{\sigma}_f$ and bound $\tilde{\varepsilon}_b$ contributions. For example, with DC fields, the "free current" source response to the field, $\vec{\mathcal{J}}_f = \tilde{\sigma}_f \vec{E}$, is clearly associated with free electrons while the polarization and magnetization responses are seen to arise due to the atomically bound electrons. If we focus on the two types of material response to the electric field, $\vec{P} = \varepsilon_o \tilde{\chi}_e \vec{E}$ and $\vec{\mathcal{J}}_f = \tilde{\sigma}_f \vec{E}$, and we take the fields as monochromatic, the macroscopic Maxwell–Ampere equation can be written as,

$$\begin{aligned}\nabla \times \vec{H} &= -i\omega\varepsilon_o (1 + \tilde{\chi}_e) \vec{E} + \tilde{\sigma}_f \vec{E} \\ &= -i\omega \left(\tilde{\varepsilon}_b + i\frac{\tilde{\sigma}_f}{\omega} \right) \vec{E}\end{aligned} \tag{10.1}$$

where the complex quantities of bound permittivity $\tilde{\varepsilon}_b$ and free conductivity $\tilde{\sigma}_f$ together represent the total response of the material to the electric field and the quantity in parentheses is identified as the generalized permittivity,

$$\tilde{\varepsilon} = \tilde{\varepsilon}_b + i\frac{\tilde{\sigma}_f}{\omega} \tag{10.2}$$

We have seen that full spectral measurements reveal the total complex functional response of the material thus distinguishing the real and imaginary components of the response. Next, with the atomic model and association of its bound and free electronic states with the two types of electric field response within Maxwell's equations, we will see that comparison of specific model predictions with measurements can be used to discern the "free" contributions of the conductivity $\tilde{\sigma}_f$ and the "bound" contributions of the permittivity, $\tilde{\varepsilon}_b$. This will later be discussed in Section 10.4. And so it is with this description that materials can be characterized variously as conductors/plasmas, semiconductors and insulators.

We continue in this section by examining the classical models of Drude and Lorentz and their representation of the general material response function or dielectric function, $\tilde{\varepsilon}(\omega)/\varepsilon_o$. In Section 10.2, we will discuss the response of an example Lorentz insulator (in the dense limit of solids or liquids) in various frequency regions and limits and consider the associated electromagnetic transmission, absorption and reflection. In Section 10.3, we repeat this analysis for a Drude metal. In Section 10.4, we will consider methods for extracting free and bound contributions, and thus material parameters, from broadband measurements of complex response functions.

10.1.1 The Drude model of free electrons

The simplest model for the free electron contribution to material response is the classical Drude conductivity model. For many materials, such as metals, plasmas and narrow-gap or degenerate semiconductors, there is a non-zero conductivity, $\tilde{\sigma}_f$. In these materials, according to the model, some electrons are not bound to the atoms or molecules and so rather than rotating or stretching bonds to produce polarization, the response of these free electrons to an electric field is current flow within the material.

The canonical Drude conductivity derivation is based on the classical equations of motion of an electron in a harmonically varying electric field. It assumes the electrons in the medium are free from, yet suffer collisions with, the immobile ionic cores. The velocity of an electron is given by the solution to

$$m_e \frac{\mathrm{d}\vec{v}(t)}{\mathrm{d}t} + m_e \gamma \vec{v}(t) = -e\vec{E}(t), \tag{10.3}$$

where γ, the electron-ion collision frequency, serves as a damping constant and $\vec{E}(t) = \vec{E}_o e^{-i\omega t}$ is the sinusoidally driving field. Expressing the time derivatives and rearranging, we get

$$\vec{v}_o = \frac{-e\vec{E}_o}{m_e(\gamma - i\omega)} \tag{10.4}$$

where the sinusoidal time dependence has been divided out leaving a simple, frequency-dependent relation between the amplitudes of the driving field and the electron velocity. In a medium of n_e bound electrons per unit volume, the current density $\vec{\mathcal{J}}_f$ is related to the velocity by $\vec{\mathcal{J}}_f(t) = -n_e e\vec{v}(t)$ and so with Eq. 10.4,

$$\vec{\mathcal{J}}_f(\omega) = -n_e e\vec{v}_o \equiv \frac{n_e e^2}{m_e(\gamma - i\omega)} \vec{E}_o(\omega) = \tilde{\sigma}_f(\omega)\vec{E}(\omega) \tag{10.5}$$

so that from Ohm's law the "AC" Drude conductivity is now identified as,

$$\tilde{\sigma}_f(\omega) = \frac{n_e e^2}{m_e(\gamma - i\omega)} \tag{10.6}$$

and the free electron response due to the electric field is given as a current density, $\vec{\mathcal{J}}_f$. If we set $\omega = 0$ (corresponding to a constant DC field), we obtain the "DC" Drude conductivity,

$$\sigma_o = \frac{n_e e^2}{m_e \gamma} = \frac{n_e e^2 \tau}{m_e} \tag{10.7}$$

where $\tau = 1/\gamma$ is the mean time between collisions or "relaxation time." In terms of this frequency-independent "DC" conductivity, Eq. 10.6 can be more compactly expressed as,

$$\tilde{\sigma}_f(\omega) = \frac{\sigma_o}{(1 - i\omega\tau)} \tag{10.8}$$

Furthermore, this may be separated into real and imaginary parts by multiplying top and bottom by $1 + i\omega\tau$, so that

$$\sigma_{f1} = \frac{\sigma_o}{1 + (\omega\tau)^2} \quad \text{and} \quad \sigma_{f2} = \frac{\omega\tau\sigma_o}{1 + (\omega\tau)^2} \tag{10.9}$$

The example of copper

The room temperature (300 K) "DC" conductivity of copper (Cu) in SI units is measured to be $\sigma_o \sim 6 \times 10^7 \Omega^{-1} m^{-1}$. With a cold Cu electron number density of $n_e \sim 8.5 \times 10^{28}\ m^{-3}$ and with $e = 1.6 \times 10^{-19} C$ and $m_e = 9.1 \times 10^{-31} kg$, Eq. 10.7 can be solved to obtain a relaxation time of $\tau = 2.5 \times 10^{-14} s$ and collision frequency of $\gamma = 1/\tau = 4 \times 10^{13} Hz$. Now, looking at the real and imaginary components of Eq. 10.9, it is clear that the transition of the AC conductivity from mostly real to mostly imaginary occurs at $\omega\tau = 1$ (or when $\omega = \gamma$). Indeed, as indicated in Fig. 10.1, the behavior of the conductivity can be divided into two regions $\omega < \gamma$ and $\omega > \gamma$ with two associated limits: the low frequency limit ($\omega \ll \gamma$) where the free current density $\vec{\mathcal{J}}_f$ is in phase with the driving field, resulting in strong absorption; and the high frequency limit ($\omega \gg \gamma$) where the free current density is 90° out of phase with the driving field and the absorption is thus weak. The characterization of a "low" or "high" frequency, therefore, refers to the field frequency relative to the collision frequency, $\gamma = 1/\tau$. For visible light upon copper, $\omega\tau \simeq 70 \to 110$, so it is nearly in the high frequency, weak absorbing limit.

Figure 10.1 shows the real and imaginary parts of the Drude conductivity of T = 300 K copper plotted against frequency, ω. At

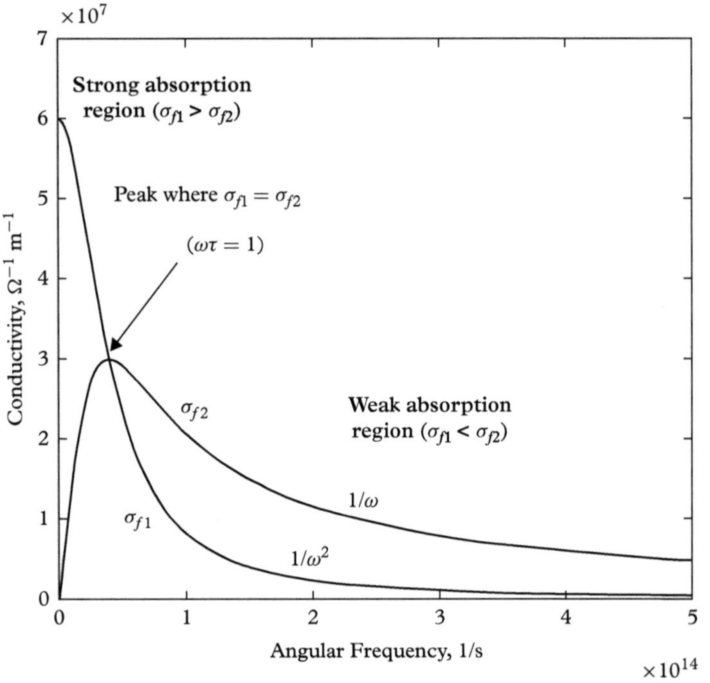

Fig. 10.1 *Real and imaginary Drude conductivity components plotted against angular frequency, ω, for room temperature Cu. Note the peak in σ_{f1} at $\omega = 0$ and the peak in σ_{f2} at $\omega = 1/\tau$.*

$\omega = 0$ and in the low frequency limit region, the real part σ_{f1} completely dominates and remains constant for many orders of magnitude. As ω increases beyond this region, the imaginary part, σ_{f2}, increases linearly from zero while σ_{f1} begins to fall off. At the frequency $\omega = 1/\tau$, the two components are equal, $\sigma_{f1} = \sigma_{f2}$, with the imaginary part having reached a maximum value. Beyond this, both components fall off for increasing ω. With both σ_{f1} and σ_{f2} components reaching their maximum fall-off of $1/\omega^2$ and $1/\omega$, respectively, in the high frequency limit, the conductivity becomes essentially imaginary.

10.1.2 The lorentz model of bound electrons

The simplest model for atomically bound electrons within an insulator or semiconductor is the classical Lorentz model. Earlier, in our Chapter 1 discussion of the response of matter to electric fields, we obtained a general steady-state expression (Eq. 1.18) for the polarization $\vec{P}$ in terms of the *microscopic polarizing* field $\vec{E}_p$ within all linearly responding materials,

$$\vec{P} = \varepsilon_o N \alpha \vec{E}_p \tag{10.10}$$

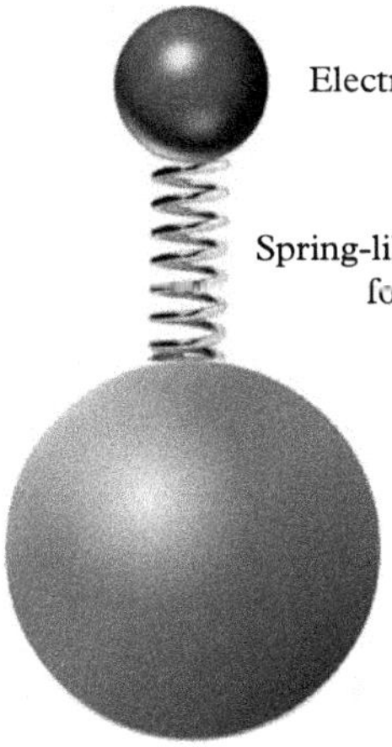

Fig. 10.2 *Lorentz model in which the electrons in the medium are harmonically bound to immobile ionic cores.*

where α is the atomic (or molecular) DC polarizability and N is the number per volume of polarizing elements. For time-varying fields, the canonical microscopic polarizability is based on the Lorentz model in which the electrons in the medium are harmonically bound to immobile ionic cores as in Fig. 10.2. If we assume a single binding frequency, ω_o, and a single damping constant, γ, between all the electrons and their ionic cores, then the displacement of any bound electron is given by the solution to

$$\frac{\mathrm{d}^2\vec{r}(t)}{\mathrm{d}t^2}+\gamma\frac{\mathrm{d}\vec{r}(t)}{\mathrm{d}t}+\omega_o^2\vec{r}(t)=-\frac{e}{m_e}\vec{E}_p(t) \tag{10.11}$$

where $\vec{E}_p(t)=\vec{E}_p e^{-i\omega t}$ is a harmonic polarizing field, assumed for now to be independent of position. Expressing the time derivatives and rearranging, we get

$$\vec{r}=\frac{-e\vec{E}_p}{m_e\left(\omega_o^2-\omega^2-i\gamma\omega\right)} \tag{10.12}$$

where the sinusoidal time dependence has been divided out leaving a simple, frequency-dependent relation between the amplitudes of the driving field and the electron displacement. For a medium in which there are $N=n_e$ bound electrons per unit volume, the electric polarization $\vec{P}$ is given by noting that $\vec{P}=-n_e e\vec{r}$,

$$\vec{P}(\omega)=\frac{n_e e^2\vec{E}_p}{m_e\left(\omega_o^2-\omega^2-i\gamma\omega\right)}=\varepsilon_o n_e\tilde{\alpha}\vec{E}_p \tag{10.13}$$

According to this model the polarization of a linear medium by a harmonic field follows it exactly in time, apart from a constant frequency-dependent phase factor. From Eq. 10.13, we define the complex, frequency-dependent "AC" polarizability as

$$\tilde{\alpha}(\omega) = \frac{\omega_p^2/n_e}{\left(\omega_0^2 - \omega^2 - i\gamma\omega\right)} \tag{10.14}$$

where we here introduce $\omega_p^2 \equiv n_e e^2/\varepsilon_o m_e$ as the "plasma frequency."[1] The "DC" polarizability is obtained by setting $\omega = 0$,

$$\alpha = \frac{\omega_p^2/n_e}{\omega_0^2} \tag{10.15}$$

A related and equally general expression is the more commonly known relation between the polarization $\vec{P}$ and the *macroscopic applied* field amplitude, $\vec{E}$,

$$\vec{P} = \varepsilon_o \tilde{\chi}_e \vec{E} \tag{10.16}$$

where $\tilde{\chi}_e$ is the complex electric susceptibility. And just as discussed in Chapter 1 for DC fields, the AC polarizing field ($\vec{E}_p$ of Eq. 10.13) felt by the atoms and molecules in matter is not generally of the same amplitude as the macroscopic applied field amplitude ($\vec{E}$ of Eq. 10.16) within a material. Specifically, we have

$$\vec{E}_p = \vec{E} + \vec{P}/3\varepsilon_o \tag{10.17}$$

which is due to the specific accounting, by $\vec{E}_p$, of fields from other nearby dipoles. When dealing with insulators or semiconductors, it is thus important to keep in mind this difference between the applied field $\vec{E}$ and the local polarizing field $\vec{E}_p$. We can obtain the macroscopic AC susceptibility $\tilde{\chi}_e$ in terms of the microscopic AC polarizabilty $\tilde{\alpha}$ for dense materials. Substituting Eq. 10.17 into Eq. 10.13 and comparing to Eq. 10.16, we identify

$$\tilde{\chi}_e = \frac{n_e\tilde{\alpha}}{\left(1 - \frac{1}{3}n_e\tilde{\alpha}\right)} \tag{10.18}$$

And, inserting into this equation the frequency-dependent "AC" polarizability of Eq. 10.14, we obtain,

$$\tilde{\chi}_e(\omega) = \frac{\frac{\omega_p^2}{\left(\omega_0^2 - \omega^2 - i\gamma\omega\right)}}{1 - \frac{\omega_p^2}{3\left(\omega_0^2 - \omega^2 - i\gamma\omega\right)}} \tag{10.19}$$

[1] The units of ω_p^2 are indeed inverse seconds squared. First, by noting the unit relations of Coulomb's law

$$\left[F = ma = \frac{q^2}{4\pi\varepsilon_o r^2}\right] \Rightarrow \frac{kg\cdot m}{s^2} = \frac{(coul)^2}{m^2}[1/\varepsilon_o]$$

the units of $1/\varepsilon_o$ are found to be

$$[1/\varepsilon_o] = \frac{kg\cdot m^3}{(coul)^2 s^2}$$

The units of ω_p^2 are then immediately found to be that of a frequency squared,

$$\left[\omega_p^2 = \frac{n_e e^2}{\varepsilon_o m_e}\right] = [1/\varepsilon_o]\frac{(coul)^2}{kg\cdot m^3} = s^{-2}$$

Note the close relationship the plasma frequency has with the DC conductivity: $\varepsilon_o\omega_p^2 = \gamma\sigma_o$.

which, with a little algebra, can be written as

$$\tilde{\chi}_e(\omega) = \frac{\omega_p^2}{\left(\omega_o^2 - \omega_p^2/3\right) - \omega^2 - i\gamma\omega} = \frac{\omega_p^2}{\omega_{os}^2 - \omega^2 - i\gamma\omega} \tag{10.20}$$

where labelling $\omega_{os}^2 \equiv \omega_o^2 - \omega_p^2/3$ identifies a correction to the binding frequency for solids. The susceptibility of Eq. 10.20 and the polarizability of Eq. 10.14 are thus identical in form–both Lorentzian–with ω_{os}^2 in $\tilde{\chi}_e(\omega)$ replacing ω_o^2 in $\tilde{\alpha}(\omega)$. We will see shortly that these terms represent resonant response frequencies within gases and solids, respectively, where within solids ω_{os}^2 has been reduced from the bare resonant frequency, ω_o^2, due to nearby field corrections. We can separate Eq. 10.20 into its real and imaginary components by multiplying top and bottom by $\omega_{os}^2 - \omega^2 + i\gamma\omega$ so that

$$\chi_{e1}(\omega) = \frac{\omega_p^2\left(\omega_{os}^2 - \omega^2\right)}{\left(\omega_{os}^2 - \omega^2\right)^2 + (\gamma\omega)^2} \tag{10.21}$$

$$\chi_{e2}(\omega) = \frac{\gamma\omega\omega_p^2}{\left(\omega_{os}^2 - \omega^2\right)^2 + (\gamma\omega)^2} \tag{10.22}$$

The example of an insulator

In this example, we consider a typical insulator such as crystalline salt or liquid water. A typical binding energy for valence electrons in insulators is ~6 eV. This corresponds to a binding frequency of $\omega_{os} \simeq 9.1 \times 10^{15}\,Hz$. If we use this value in Eqs. 10.21 and 10.22, along with a plasma frequency of $\omega_p \simeq 9.8 \times 10^{15}\,Hz$–corresponding to a typical bound electron number density of $n_e \simeq 3.0 \times 10^{28}\,m^{-3}$–and a reasonable collision frequency of $\gamma \simeq 6 \times 10^{13}\,Hz$, we obtain Fig. 10.3, the real and imaginary components of the Lorentz susceptibility for a single-resonance insulator. Near ω_{os}, the imaginary part has the shape of a resonance while the real part follows something like the derivative of a resonance. As with conductivity, the behavior can be divided into regions of strong and weak absorption as determined by the relative absolute amplitudes of the real and imaginary components and the resulting phase relationship with the driving field. And because the collision frequency, γ, determines the width of the resonant/anomalous dispersion region, we can define the strong and weak absorption regions as

$$\textit{Region of Strong Absorption} \Longrightarrow \omega_{os} - \gamma < \omega < \omega_{os} + \gamma$$

$$\textit{Regions of Weak Absorption} \Longrightarrow \omega < \omega_{os} - \gamma \;\; \textit{and} \;\; \omega > \omega_{os} + \gamma$$

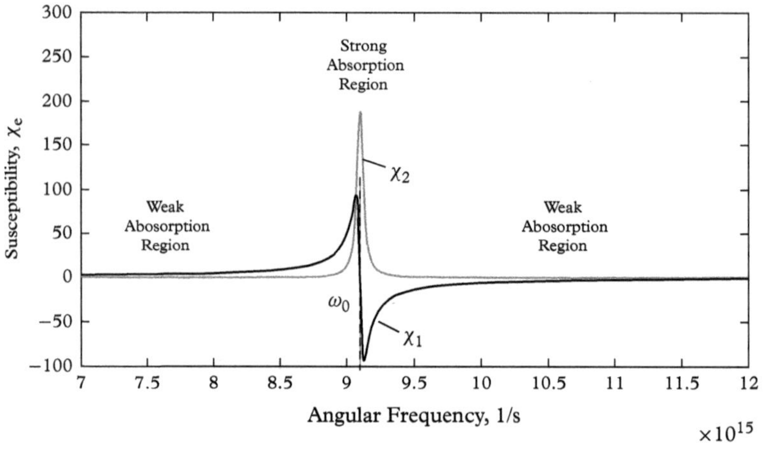

Fig. 10.3 *Real and imaginary Lorentz susceptibility components plotted against angular frequency, ω, for an ideal single-resonance insulator. Note that strong absorption corresponds to the region about the resonant peak in* χ_2 *at* $\omega = \omega_{os}$. *Compare to the conductivity curve of Fig. 10.1 that effectively has the same resonant structure, but with* ω_{os} *shifted to zero.*

That is, at frequencies far from ω_{os}, the imaginary component, χ_2, essentially vanishes while χ_1 remains non-zero – staying positive to the left and negative to the right of the resonance. We finally note that the conductivity plot of Fig. 10.1 (or, more accurately, a plot of $i\sigma(\omega)/\varepsilon_o\omega$) can be interpreted as a particular case of the susceptibility plot of Fig. 10.3 for $\omega_{os} = 0$ (free electrons). Indeed, setting ω_{os} to zero and excluding negative frequencies ($\omega < \omega_{os} = 0$), we obtain the strong and weak absorption regions of a conductor, as discussed previously.

10.1.3 The combined model: Lorentz–Drude

The Lorentz results for polarizability and susceptibility (Eqs. 10.14 and 10.20) obtained in Section 10.1.2 are too simple in that they assume just a single binding frequency, ω_o (or ω_{os}), and a single damping constant, γ, between the electrons and the ionic cores. In reality, each ionic core binds to its surrounding electrons with multiple binding frequencies and damping constants. A more realistic expression of the susceptibility is therefore expressed as a sum

$$\tilde{\chi}_e(\omega) = \frac{n_i e^2}{\varepsilon_o m_e} \sum_j z_j (\omega_j^2 - \omega^2 - i\gamma_j \omega)^{-1} \tag{10.23}$$

where n_i is the number density of ionic cores and z_j is the number of electrons per atom with binding frequency, ω_j, and damping constant, γ_j, so that $\sum z_j = Z$ is the total number of electrons per atom. Now, if we note that the conductivity of Eq. 10.6, with a prefactor $i/\omega\varepsilon_o$, can be written as

$$\frac{i\tilde{\sigma}_f(\omega)}{\omega\varepsilon_o} = \frac{n_i e^2}{\varepsilon_o m_e}\frac{z_0}{\left(-\omega^2 - i\omega\gamma\right)} \tag{10.24}$$

then we see that conductivity can be included in Eq. 10.23 as a $j=0$ susceptibility term in which the zeroth binding frequency (binding energy) vanishes ($\omega_{j=0} = 0$) and z_0 is the number of free electrons per atom. Thus, the $\tilde{\sigma}_f(\omega)$ term can be viewed as a term of $\tilde{\chi}_e(\omega)$ with zero binding energy. This is the classical Lorentz–Drude result.[2]

10.1.4 Lorentz and Drude model response functions

In Chapter 9, we explored the connections between measurable quantities - such as absorption, reflectance, transmittance and phase shifting - and the two optical constants, n and κ, of the complex refractive index $\tilde{n}$ through the Fresnel equations and plane wave solutions to Maxwell's equations. As we saw, $\tilde{n}$ is in turn related to the complex dielectric function (or response function) of the material by $\tilde{n}^2 = \tilde{\varepsilon}/\varepsilon_o = \tilde{\varepsilon}_r$. Now, having obtained the classical frequency-dependent Drude and Lorentz models describing, respectively, the free and bound electron responses in terms of material properties, we can look in detail at response functions for specific types of materials. In Sections 10.2 and 10.3, we thus consider:

(1) Materials such as insulators and semiconductors whose response to fields is dominated by the bound electrons and are described by the Lorentz model.
(2) Materials such as metals and plasmas whose response to fields is dominated by free electrons and are described by the Drude model.

In particular, in the next two sections we will carry out a descriptive analysis of the interaction of these materials with incident electromagnetic waves and will identify in these materials a number of characteristic frequency regions defined by a dominance of either transmittance, reflectance or absorption.

The core response

In all of the following cases of conductors, insulators and semiconductors, there are electrons which are so tightly bound to their nuclei that their response to the field, though non-zero, is effectively

[2] Note that in the Drude model there is no need for the distinction between a "local" field, such as the polarization field, and the applied (average) field. This is because the free electrons, unlike the bound electrons, are not localized and so the average field they "feel" will always be the applied field, $\vec{E}$.

frequency independent. This response, often referred to as the "core" contribution, χ_{ec}, is combined with the vacuum for a purely real "core" dielectric constant,

$$\frac{\varepsilon_c}{\varepsilon_o} = \varepsilon_{rc} = 1 + \chi_{ec} \tag{10.25}$$

Classically, as long as the resonant frequencies (or binding energies) for these core electrons are far larger than the field frequencies encountered, $\omega_j \gg \omega$, then their associated Lorentzian components within the Lorentz–Drude susceptibility of Eq. 10.23 are seen to be essentially independent of frequency. In terms of band theory, the core electrons exist in neither the conduction or the valence bands and thus do not contribute to either the free current or transition current responses. In the limit of the high frequencies, for which the Lorentz and Drude model terms (Eqs. 10.23 and 10.24) vanish and the frequency independent core response remains (i.e., when $\omega \gg \omega_j \gg \gamma_j$ for the "responding" electrons yet $\omega \ll \omega_j$ is still true for the "core" electrons), we have $\tilde{n}^2 \simeq \varepsilon_{rc}$. For this reason, the core contribution is often labeled as $\varepsilon_{r\infty}$.

10.2 Lorentz insulators

For Lorentz insulators and cold, wide-band semiconductors, the permittivity of Eq. 10.2 has no "free" electron contribution. In terms of the constant "core" contribution and the frequency-dependent single-resonance bound contribution of Eq. 10.20, it can be written,

$$\tilde{\varepsilon}(\omega) \simeq \tilde{\varepsilon}_b = \varepsilon_o \left(1 + \chi_{ec} + \frac{\omega_p^2}{\omega_{os}^2 - \omega^2 - i\gamma\omega}\right) \tag{10.26}$$

And the associated dielectric (or response) function, $\tilde{\varepsilon}/\varepsilon_o = \tilde{\varepsilon}_r$, is then written as,

$$\tilde{\varepsilon}_r(\omega) = \varepsilon_{rc} + \frac{\omega_p^2}{\omega_{os}^2 - \omega^2 - i\gamma\omega} \tag{10.27}$$

which is just the Lorentz susceptibility of Eq. 10.20 plus the core response of Eq. 10.25. It then follows from Eqs. 10.21 and 10.22 that the real and imaginary parts are,

$$\varepsilon_{r1}(\omega) = \varepsilon_{rc} + \frac{\omega_p^2\left(\omega_{os}^2 - \omega^2\right)}{\left(\omega_{os}^2 - \omega^2\right)^2 + (\gamma\omega)^2} \tag{10.28}$$

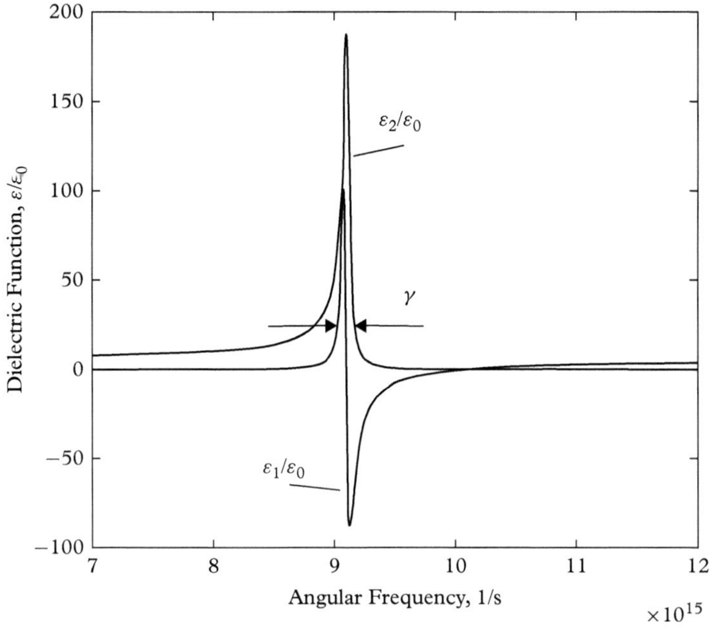

Fig. 10.4 *Real and imaginary components of the Lorentz dielectric function plotted against angular frequency, ω.*

$$\varepsilon_{r2}(\omega) = \frac{\gamma\omega\omega_p^2}{\left(\omega_{os}^2 - \omega^2\right)^2 + (\gamma\omega)^2} \tag{10.29}$$

Again using our insulator example of Section 10.1 in which we chose a binding energy of ~6 eV, a plasma frequency of $\omega_p \simeq 9.8 \times 10^{15} Hz$ and a collision frequency of $\gamma \simeq 6 \times 10^{13} Hz$, the real and imaginary parts, Eqs. 10.28 and 10.29, are plotted–with ε_{rc} set to the typical value of 5 – in Fig. 10.4.

In such a simplified plot of material response, ω_{os} is the location of the resonance, the collision frequency (or damping rate), γ, determines the width of this resonance and the plasma frequency, ω_p, roughly marks the high frequency location where ε_{r1} changes sign. With these parameters, we can define four regions of interest:

(1) **The low frequency region here defined as $\omega < \omega_{os} - \gamma$:** In this region, ε_{r2} is very small, vanishing at $\omega = 0$ and increasing linearly with ω while ε_{r1} asymptotically approaches the positive value $\varepsilon_{r1} = \varepsilon_{rc} + \omega_p^2/\omega_{os}^2$ as $\omega \to 0$. Because the response is essentially real in this region, there is very little absorption by the material. That is, the polarization $\vec{P}$ is in phase with the field or, equivalently, the bound current density $\vec{J}_b$ is 90^oout of phase with the field.

(2) **The region in the vicinity of the resonance, here defined as $\omega_{os} - \gamma < \omega < \omega_{os} + \gamma$:** In this region, ε_{r2} is very large, reaching its peak at $\omega = \omega_{os}$ while ε_{r1} abruptly reverses its normally increasing trend and plummets through zero at $\omega = \omega_{os}$, reaching a minimum at $\omega = \omega_{os} + \gamma/2$ (for this reason, it is called the region of anomalous dispersion). Thus, in this region about ω_{os}, the response is largely imaginary with the bound current density $\vec{\mathcal{J}}_b$ mostly in phase with the field and so the material is strongly absorbing.

(3) **The high frequency region defined as $\omega_{os} + \gamma < \omega < \omega(\varepsilon_{r1} = 0)$:** As in region 1, ε_{r2} is very small, again vanishing for $\omega \gg \omega_{os}$. And again, the response is essentially real with very little absorption by the material. However, the distinguishing feature of this right-hand-side region of the resonance is that ε_{r1} is negative. Thus, while the bound current density $\vec{\mathcal{J}}_b$ is still 90^oout of phase with the field, the polarization $\vec{P}$ is now exactly 180° out of phase.

(4) **The very high frequency region defined as $\omega > \omega(\varepsilon_{r1} = 0)$:** In this region, ε_{r2} is essentially zero, and so there is still little or no absorption, but ε_{r1}, while small, has become positive. The phase relationships are thus the same as in region 1. This transition can be seen mathematically by the domination of the positive-valued core response, ε_{rc} , over the ever-diminishing negative-valued susceptibility at frequencies above $\omega(\varepsilon_{r1} = 0)$. In this region, we can generally assume that $\omega \gg \omega_{os} \gtrsim \gamma$ so that ε_{r1}, as expressed in Eq. 10.28, can be approximated by

$$\varepsilon_{r1}(\omega) \simeq \varepsilon_{rc} - \frac{\omega_p^2}{\omega^2} \tag{10.30}$$

and at this transition frequency, also known as the "screened plasma," the frequency, $\omega(\varepsilon_{r1} = 0) = \omega_{sp}$, is approximately,

$$\omega_{sp} \simeq \frac{\omega_p}{\sqrt{\varepsilon_{rc}}} \tag{10.31}$$

Equation 10.30 also shows that the response of the material goes to the "core" response in the limit of very high frequencies: $\varepsilon_{r1}(\infty) \to \varepsilon_{rc}$.

So far, using field-matter phase considerations, the Lorentz model has shown us but one facet of material interaction: absorption. We have said nothing yet about transmission and reflection. For this, we need the Lorentz model expression for the complex index of refraction. As discussed in the previous chapter, the square root of the dielectric constant is the complex index of refraction:

$$\tilde{n}(\omega) \equiv n + i\kappa = \sqrt{\tilde{\varepsilon}_r(\omega)} \tag{10.32}$$

Squaring Eq. 10.32, the dielectric coefficient can be expressed in terms of the optical constants,

$$\varepsilon_{r1}(\omega) = n^2(\omega) - \kappa^2(\omega) \tag{10.33}$$

$$\varepsilon_{r2}(\omega) = 2n(\omega)\kappa(\omega) \tag{10.34}$$

And these equations can be inverted[3] to express n and κ in terms of ε_{r1} and ε_{r2}:

$$n(\omega) = \sqrt{\frac{1}{2}}\left(\sqrt{\varepsilon_{r1}(\omega)^2 + \varepsilon_{r2}(\omega)^2} + \varepsilon_{r1}(\omega)\right)^{1/2} \tag{10.35}$$

$$\kappa(\omega) = \sqrt{\frac{1}{2}}\left(\sqrt{\varepsilon_{r1}(\omega)^2 + \varepsilon_{r2}(\omega)^2} - \varepsilon_{r1}(\omega)\right)^{1/2} \tag{10.36}$$

The Lorentz model is then connected to the observables of wave absorption/propagation and phase shifts through an electric field solution, given in Section 9.3, which can take the form

$$\vec{E}(x,t) = \vec{E}_o exp(-\frac{\kappa(\omega)\omega}{c}x) exp(i\frac{n(\omega)\omega}{c}x - i\omega t) \tag{10.37}$$

in which κ and n represent the dissipative and traveling components, respectively, of the field solution. Similarly, a connection to the observables of wave transmittance and reflectance was given, in Section 9.5.5, by the Fresnel equations. For the case of a wave normally incident to a vacuum-material interface of index $\tilde{n} = n + i\kappa$, we obtained the polarization independent equations,

$$T(\omega)_\perp = \frac{4n(\omega)}{|1+\tilde{n}(\omega)|^2} = \frac{4n(\omega)}{n(\omega)^2 + \kappa(\omega)^2 + 1 + 2n(\omega)} \tag{10.38}$$

and

$$R(\omega)_\perp = \left|\frac{1-\tilde{n}(\omega)}{1+\tilde{n}(\omega)}\right|^2 = \frac{n(\omega)^2 + \kappa(\omega)^2 + 1 - 2n(\omega)}{n(\omega)^2 + \kappa(\omega)^2 + 1 + 2n(\omega)} \tag{10.39}$$

Continuing with our example insulator, in Fig. 10.5 we show frequency plots of the refractive index components, $n(\omega)$ and $\kappa(\omega)$, Eqs. 10.35 and 10.36, and the reflectance $R(\omega)$, Eq. 10.39, beneath that of the associated dielectric function of Fig. 10.4. Since absorption only takes place around the resonant frequency, electromagnetic energy must be conserved elsewhere such that the interaction is either

[3] First, from Eq. 10.34, κ is expressed in terms n and plugged into Eq. 10.33. The result is then multiplied by n^2 and rearranged into quadratic form in n^2 ,

$$\left(n^2\right)^2 - \varepsilon_{r1} n^2 - \frac{1}{4}(\varepsilon_{r2})^2 = 0$$

and this is solved using the quadratic equation (where, e.g., $n^2 \to x$). The result is

$$n^2 = \frac{1}{2}\left\{\varepsilon_{r1} \pm \sqrt{(\varepsilon_{r1})^2 + (\varepsilon_{r2})^2}\right\}$$

Next, using Eq. 10.33,

$$\kappa^2 = \frac{1}{2}\left\{-\varepsilon_{r1} \pm \sqrt{(\varepsilon_{r1})^2 + (\varepsilon_{r2})^2}\right\}$$

Taking the square roots,

$$n = \sqrt{\frac{1}{2}}\left\{\varepsilon_{r1} + \sqrt{(\varepsilon_{r1})^2 + (\varepsilon_{r2})^2}\right\}^{1/2}$$

$$\kappa = \sqrt{\frac{1}{2}}\left\{-\varepsilon_{r1} + \sqrt{(\varepsilon_{r1})^2 + (\varepsilon_{r2})^2}\right\}^{1/2}$$

where the "+" signs were chosen for the following reason: since it is always true that $\varepsilon_{r1} < \sqrt{(\varepsilon_{r1})^2 + (\varepsilon_{r2})^2}$, and, by definition n and κ must be real, the quantities in braces must be positive.

dominantly transmissive, or dominantly reflective. Regions 1–4 have been relabeled as T, A, R, and T to indicate the dominant type of response where "T" stands for transmissive, "A" stands for absorptive, and "R" stands for reflective.

Region T (low frequency): Dominantly transmissive

In the low frequency region of Fig. 10.5, extending from DC to $\omega_{os} - \gamma$, we see that both the dielectric constant and the complex index of refraction are essentially real: $\varepsilon_{r1} \gg \varepsilon_{r2}$ and $n \gg \kappa$. Thus, the form of electromagnetic propagation in the material is wavelike with an absorption length, $1/\alpha = c/2\omega\kappa = \lambda_o/4\pi\kappa$, many times longer than the vacuum wavelength, λ_o. Indeed, in the limit of $\kappa \to 0$, the insulator becomes perfectly transparent. Also, in this region, because $n(\omega)$ increases with frequency, broadband radiation will experience normal, positive dispersion. The reflectivity is largely determined by the dominant real component of n so that Eq. 10.39 can be approximated in this low frequency range as

$$R(\omega)_\perp \simeq \left(\frac{1-n}{1+n}\right)^2 = 1 - \frac{4n}{(n+1)^2} \tag{10.40}$$

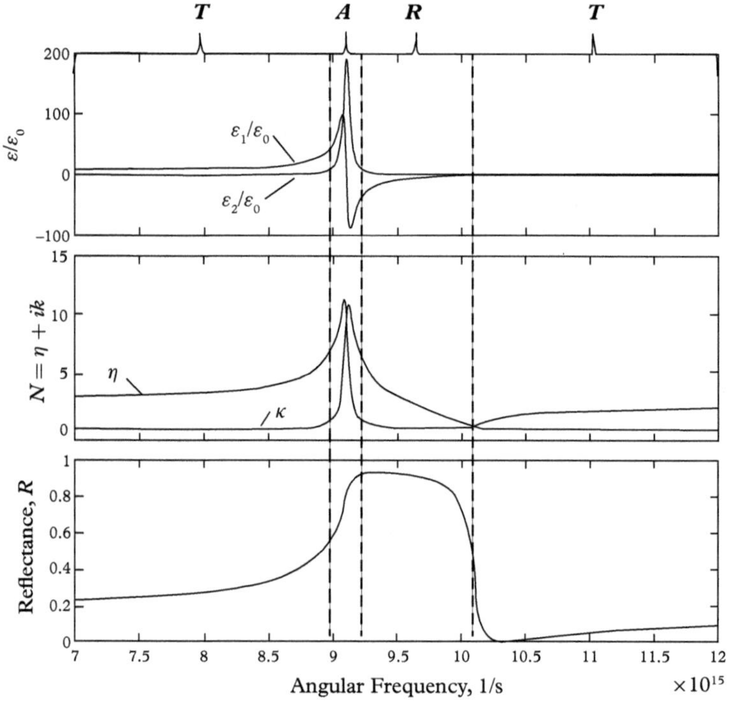

Fig. 10.5 *Frequency plots of the refractive index components, $n(\omega)$ and $\kappa(\omega)$, and the reflectance $R(\omega)$ beneath that of the associated dielectric function. The plots are divided into four regions dominated by transmissive (T), absorptive (A), reflective (R), and transmissive (T) behavior.*

For the example insulator used, Fig. 10.5 shows $n \simeq 3$ that yields a reflectivity of 25%. Because $n(\omega)$ increases with frequency, the reflectivity will increase, according to Eq. 10.40, as ω approaches the resonance region.

Region A: Dominantly absorptive

In the vicinity of resonance where $\omega_{os} - \gamma < \omega < \omega_{os} + \gamma$, according to Fig. 10.5, $\varepsilon_{r2} = \varepsilon_2/\varepsilon_o$ nears its peak and $\varepsilon_{r1} = \varepsilon_1/\varepsilon_o$ passes through zero. If we represent this by setting $\varepsilon_{r1} = 0$ in Eqs. 10.35 and 10.36, we find that $n = \kappa = \sqrt{\varepsilon_{r2}/2}$. This condition of maximum absorption is shown graphically in the middle plot of Fig. 10.5 and is consistent with Eq. 10.33 that shows that ε_{r1} vanishes when $n = \kappa$. Equal values of n and κ result, according to Eq. 10.37, in a form of field propagation which has both damped and traveling components. This, too, is physically consistent since a complete description of absorption requires both a traveling part to continuously transport the energy in and a damped part to represent attenuation due to absorption.

What about the reflectivity? We can rewrite Eq. 10.39 as

$$R(\omega)_\perp = 1 - \frac{4n}{(n+1)^2 + \kappa^2} \tag{10.41}$$

If we further note that near the resonance $n \simeq \kappa \simeq \sqrt{\varepsilon_{r2}/2} \gg 1$, then Eq. 10.41 simplifies to

$$R(\omega)_\perp \simeq 1 - \frac{2}{n(\omega)} \simeq 1 - 2\sqrt{\frac{2}{\varepsilon_{r2}}} \tag{10.42}$$

and so as ε_{r2} rises to its peak at resonance, the reflectivity rises to near unity. In this region, we can therefore associate an increase in absorption with an increase in reflectivity. This trend is indicated in the bottom plot of Fig. 10.5. Note finally that the abrupt decrease in $\varepsilon_{r1}(\omega)$ near resonance translates, according to Eqs. 10.35 and 10.36, to a sharp fall-off of $n(\omega)$ and a sharp rise of $\kappa(\omega)$. This behavior of $n(\omega)$ is referred to as "anomalous dispersion."

Region R: Dominantly reflective

In the region $\omega_{os} + \gamma < \omega < \omega_{sp}$, as in the low frequency region, the amplitude of ε_{r2} is very small and there is little or no absorption. However, the non-negligible ε_{r1} is now negative. This leads, according to Eqs. 10.35 and 10.36, to a role reversal of n and κ: Now, along with $|\varepsilon_{r1}| \gg \varepsilon_{r2}$, we have $\kappa \gg n \simeq 0$ so that according to Eq. 10.37, the field propagation is strongly damped with essentially no traveling component.[4] With $\varepsilon_{r1} < 0$ and $|\varepsilon_{r1}| \gg \varepsilon_{r2}$, Eq. 10.36 approximates to

$$\kappa(\omega) \simeq \sqrt{|\varepsilon_{r1}|} \tag{10.43}$$

If we now consider a mid-region frequency in which $\gamma \ll \omega_{os} \ll \omega \ll \omega_p$, then ε_{r1}, as given by Eq. 10.28, is approximated as $\varepsilon_{r1}(\omega) \simeq -\omega_p^2/\omega^2$ and,

$$\kappa(\omega) \simeq \frac{\omega_p}{\omega} \tag{10.44}$$

so that in the mid-region, $\kappa \gg 1$, with an absorption length, $1/\alpha = c/2\omega\kappa = \lambda_o/4\pi\kappa$, many times shorter than the vacuum wavelength, λ_o. In this region, if there is negligible absorption and the material is strongly opaque, then where does the field energy go? It would seem that it must be reflected. Indeed, a quick, rough estimate of the reflectivity in this region is obtained by setting $n = 0$ in Eq. 10.39 to yield

$$R(\omega)_\perp \simeq \frac{\kappa(\omega)^2 + 1}{\kappa(\omega)^2 + 1} = 1 \tag{10.45}$$

Referring to Fig. 10.5, we can see that this is maybe too ideal. A better estimate involves again considering the mid-region in which $\gamma \ll \omega_{os} \ll \omega \ll \omega_p$ to approximate ε_{r2} from Eq. 10.29 as

$$\varepsilon_{r2}(\omega) \simeq \frac{\gamma\omega_p^2}{\omega^3} \tag{10.46}$$

Furthermore, with $|\varepsilon_{r1}| \gg \varepsilon_{r2}$, Eq. 10.35 can be approximated as[5]

$$n(\omega) \simeq \frac{\varepsilon_{r2}(\omega)}{2\sqrt{|\varepsilon_{r1}(\omega)|}} = \frac{\gamma\omega_p}{2\omega^2} \tag{10.47}$$

And with $\kappa \gg n$, Eq. 10.39, written in the form of Eq. 10.41 and using Eqs. 10.44 and 10.47, is approximated as

$$R(\omega)_\perp = 1 - \frac{4n}{\kappa^2} = 1 - \frac{2\gamma}{\omega_p}$$

which is nearly unity since it is generally the case that $\gamma \ll \omega_p$.

Region T (high frequency): Dominantly transmissive

With $\varepsilon_{r2}(\omega)$ negligible and $\varepsilon_{r1}(\omega)$ approximated by Eq. 10.30, a frequency–the "screened plasma" frequency, ω_{sp}, of Eq. 10.31–is

[4] It is interesting to note that we have here a case in which $n < 1$. Indeed, in this region $n \ll 1$ and $\kappa \simeq \sqrt{-\varepsilon_{r1}} > 0$.

[5] Eq. 10.35 can be rewritten

$$n(\omega) = \left(\frac{|\varepsilon_{r1}|}{2}\sqrt{1+\left(\frac{\varepsilon_{r2}}{\varepsilon_{r1}}\right)^2} + \frac{\varepsilon_{r1}}{2}\right)^{1/2}$$

And with $|\varepsilon_{r1}| \gg \varepsilon_{r2}$, this can be approximated as

$$n(\omega) \simeq \left(\frac{|\varepsilon_{r1}|}{2}\left(1+\frac{1}{2}\left(\frac{\varepsilon_{r2}}{\varepsilon_{r1}}\right)^2\right) + \frac{\varepsilon_{r1}}{2}\right)^{1/2} = \frac{\varepsilon_{r2}}{2\sqrt{|\varepsilon_{r1}|}}$$

reached where ε_{r1} becomes positive. As in the low frequency transmissive region, Eqs. 10.36 and 10.35 then yield $\kappa(\omega) \simeq 0$ and $n(\omega) \simeq \sqrt{\varepsilon_{r1}(\omega)}$. Thus, as the frequency passes through ω_{sp}, the material abruptly goes from $\kappa \gg n \simeq 0$ and highly reflective to $n \gg \kappa \simeq 0$ and highly transmissive. This is known as the "plasma edge" and is shown clearly in the middle and bottom plots of Fig. 10.5. For the example insulator, $\omega_{sp} \simeq 1.01 \times 10^{16}$ Hz, which is well into the ultraviolet range. At some point, as ω increases further, $\varepsilon_{r1}(\omega)$ of Eq. 10.30 and $n(\omega) \simeq \sqrt{\varepsilon_{r1}(\omega)}$ both reach unity and the normally incident reflectivity vanishes completely. In Fig. 10.5, that point is $\omega \simeq 1.03 \times 10^{16}$Hz. Finally, in the limit of $\omega \to \infty$, $n \to \sqrt{\varepsilon_{rc}} = \sqrt{5}$.

10.2.1 Multiple binding frequencies

A more practical Lorentz model of an insulator (or wide-band semiconductor) includes multiple binding frequencies, ω_j, and collision frequencies, γ_j, as discussed previously in Section 10.1.3. Using Eq. 10.23 in place of the single-resonance Lorentz susceptibility of Eq. 10.27, we obtain an expression for a dielectric function that includes multiple resonances:

$$\tilde{\varepsilon}_r(\omega) = 1 + \chi_{ec} + \tilde{\chi}_e(\omega) = \varepsilon_{rc} + \frac{n_i e^2}{\varepsilon_o m_e} \sum_j z_j (\omega_j^2 - \omega^2 - i\gamma_j \omega)^{-1} \tag{10.48}$$

and noting that electrons with different binding energies have different number densities, $n_{ej} = z_j n_i$, and, therefore, different plasma frequencies, ω_{pj}, we can write Eq. 10.48 to reflect this

$$\tilde{\varepsilon}_r(\omega) = 1 + \chi_{ec} + \tilde{\chi}_e(\omega) = \varepsilon_{rc} + \sum_j \frac{\omega_{pj}^2}{\omega_j^2 - \omega^2 - i\gamma_j \omega} \tag{10.49}$$

where for insulators, $j = 1, 2, 3...$ and runs from the least tightly bound electrons to the most tightly bound electrons. And again, the tightest bound electrons, the "core" electrons, are represented by the constant ε_{rc}. If plotted against frequency, the real and imaginary components of Eq. 10.49 have the form shown in Fig. 10.6. In this case there are four different binding frequencies and j runs from 1 to 4. Here, the resonances are separated enough in frequency that $\varepsilon_{r2}(\omega)$ has a peak and $\varepsilon_{r1}(\omega)$ has a derivative-like form centered on each resonance. Note that while $\varepsilon_{r1}(\omega)$ is almost always increasing (normal dispersion), its overall tendency is to decrease (as indicated by the dotted line in the figure). Indeed, as the frequency passes each resonance, the derivative-like form will step down a bit.

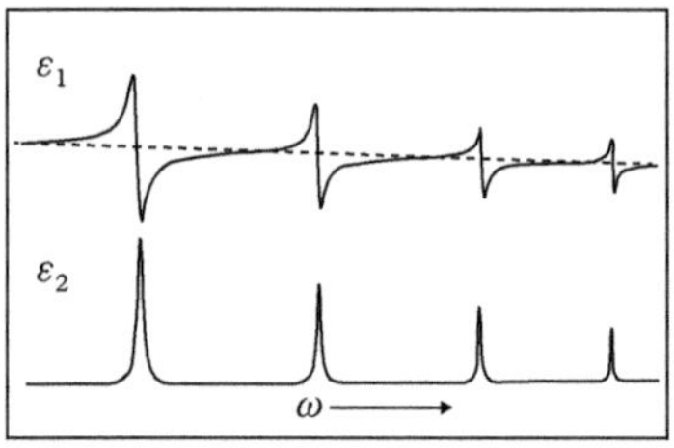

Fig. 10.6 *Real and imaginary components of the dielectric function plotted against frequency, ω, for an insulator with four binding frequencies. Note that as the frequency passes each resonance, the derivative-like form of ε_{r1} steps down a bit.*

10.3 Drude metals and plasmas

For metals, the greatest contribution to the material response, by far, is that of the free electrons with negligible contribution from the frequency-dependent bound contribution of Eq. 10.20. The complex permittivity, Eq. 10.2, is therefore well approximated by just two terms: The frequency-independent, purely real core response term, $\varepsilon_c = \varepsilon_o(1 + \chi_{ec})$, of Eq. 10.25 and the Drude-based conductivity term of Eq. 10.6,

$$\tilde{\varepsilon}(\omega) \simeq \varepsilon_c + i\frac{\tilde{\sigma}_f}{\omega} = \varepsilon_o\left(1 + \chi_{ec} + i\frac{n_e e^2}{m_e\varepsilon_o\omega(\gamma - i\omega)}\right) \tag{10.50}$$

which, if we recall the definition, from Section 10.2, of the plasma frequency

$$\omega_p^2 = \frac{n_e e^2}{m_e \varepsilon_o} \tag{10.51}$$

then we can write the related dielectric function, $\tilde{\varepsilon}/\varepsilon_o = \tilde{\varepsilon}_r$, as

$$\tilde{\varepsilon}_r(\omega) = \varepsilon_{rc} - \frac{\omega_p^2}{\omega^2 + i\gamma\omega} \tag{10.52}$$

and multiplication, top and bottom, by $\omega^2 - i\gamma\omega$ yields the general real and imaginary parts of the complex Drude dielectric constant as,

$$\varepsilon_{r1}(\omega) = \varepsilon_{rc} - \frac{\omega_p^2}{\omega^2 + \gamma^2} \quad \text{and} \quad \varepsilon_{r2}(\omega) = (\gamma/\omega)\frac{\omega_p^2}{\omega^2 + \gamma^2} \tag{10.53}$$

which, we note, are equivalent to Eqs. 10.28 and 10.29 for a zero-frequency electron resonance, $\omega_{os} = 0$, otherwise known as a "free" electron. Note also the minus sign for the real part of the Drude contribution.

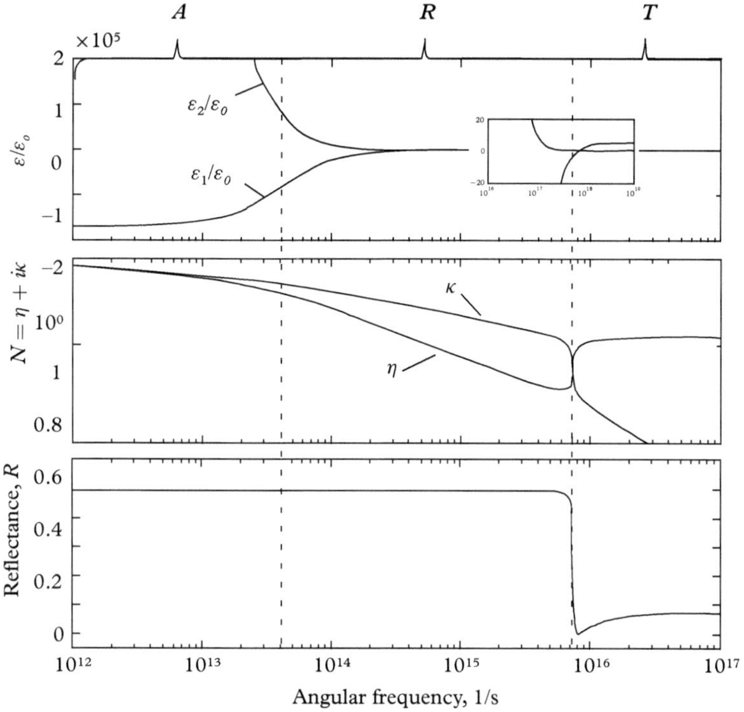

Fig. 10.7 *Frequency plots of the refractive index components, $n(\omega)$ and $\kappa(\omega)$ and the reflectance $R(\omega)$ beneath that of the associated dielectric function for room temperature copper. The plots are divided into three regions dominated by absorptive (A), reflective (R) and transmissive (T) behavior.*

Continuing with our example of copper: given the values, quoted in Section 10.1, of the copper electron number density, n_e, the electron charge, e, and the electron mass, m_e, along with the permittivity of free space, $\varepsilon_o = 8.85 \times 10^{-12}\ coul^2 s^2/kg \cdot m^3$, we find from Eq. 10.51 that the normal density, room temperature Cu plasma frequency is $\omega_p = 1.6 \times 10^{16} s^{-1}$. We can compare this with a typical visible light frequency of $\omega_{light} = 4 \times 10^{15} s^{-1}$ and with the collision frequency found earlier to be $\gamma = 4 \times 10^{13} s^{-1}$. For copper, the frequency-independent contribution, χ_{ec}, from the core electrons to the dielectric function is about four so, from Eq. 10.25, $\varepsilon_{rc} \simeq 5$. Inserted into the equations of 10.53, these numbers result in the real and imaginary Drude response function components for room temperature Cu shown in the top plot of Fig. 10.7. For the Lorentz model of insulator response, we defined four characteristic regions associated with the dominance of absorption, transmission, and reflection. We can now do the same for the Drude model of conductor response. We will find that the dielectric response function for a conductor is the same as that of an insulator but with the resonant region–that is, the "absorption"–centered at the origin, $\omega_{os} = 0$. In Fig. 10.7, we define three regions of interest labeled as A, R, and T:

Region A: Dominantly absorptive

The low frequency region of $\omega < \gamma$ can be compared to the right side of the resonance region of the Lorentz insulator in Fig. 10.5. In the limit of $\omega \to 0$, the core term in Eq. 10.53, $\varepsilon_{rc} \simeq 5$, is negligible compared to the large negative Drude term so that $\varepsilon_{r1} \simeq -\omega_p^2/\gamma^2 = -1.6 \times 10^5$. Meanwhile, $\varepsilon_{r2} \simeq \omega_p^2/\gamma\omega$ diverges to positive infinity at $\omega = 0$ so that at very low frequencies the dielectric function is large and effectively imaginary. Here, the free current density $\vec{\mathcal{J}}_f$ is mostly in phase with the field so the material is strongly absorbing–just as it is near the peak of a resonance. As ω increases from 0, both ε_{r1} and ε_{r2} decrease in amplitude, with ε_{r2} falling more quickly as $1/\omega$.

With $\varepsilon_{r1} \ll \varepsilon_{r2}$ for very low frequencies in which $\omega \ll \gamma$, Eqs. 10.35, 10.36, and 10.53 indicate that

$$n(\omega) = \kappa(\omega) \simeq \sqrt{\frac{\varepsilon_{r2}(\omega)}{2}} = \sqrt{\frac{\omega_p^2}{2\gamma\omega}} \gg 1 \tag{10.54}$$

So that once again high absorption is accompanied by large values of n and κ that are equal. Thus, for a very low frequency field, propagation within and normal reflectance from a metal are of the same forms as those for a field in an insulator with a frequency near a resonance: First, the equation for field propagation, Eq. 10.37, has both damped and traveling components. In particular, the amplitude of the field decays exponentially, with a skin depth into the metal,

$$\delta = \frac{2}{\alpha} = \frac{c}{\omega\kappa} = \sqrt{\frac{2\gamma c^2}{\omega_p^2\omega}}$$

which grows to infinity as $\omega \to 0$. The reflectance, likewise, has the same simplified form as Eq. 10.42 and so for very low frequencies, we use Eq. 10.54 to specify the Drude reflectivity from a metal as,

$$R(\omega)_\perp \simeq 1 - \frac{2}{n(\omega)} \simeq 1 - \sqrt{\frac{8\gamma\omega}{\omega_p^2}}$$

so that in the limit of $\omega \to 0$, a metal such as copper is a perfect reflector.

Region R: Dominantly reflective

The high frequency region defined as $\gamma < \omega < \omega(\varepsilon_{r1} = 0)$ can likewise be compared to that of the Lorentz insulator. In the limit of $\omega \gg \gamma$ within this region, the Eqs. of 10.53 show that both components tend to zero with ε_{r1} rising to zero as $1/\omega^2$ and ε_{r2} falling to zero faster, as $1/\omega^3$. Thus, at these high frequencies the imaginary part quickly becomes negligible compared to the real part. The free current density $\vec{\mathcal{J}}_f$ is then nearly 90° out of phase with the field and very little absorption by the material occurs. Also, because ε_{r1} is negative, the polarization $\vec{P}$ is now exactly 180° out of phase.

As in the dominantly reflective region of a Lorentz insulator, $\kappa \gg n \simeq 0$ so that the field propagation is strongly damped with very little traveling component. And again considering the middle of this reflective region in which $\gamma \ll \omega \ll \omega_p$, we again approximate

$$\kappa(\omega) \simeq \sqrt{|\varepsilon_{r1}|} = \frac{\omega_p}{\omega}$$

So, in this strongly damped (but weakly absorbing) region the skin depth is now given simply by

$$\delta = \frac{2}{\alpha} = \frac{c}{\omega\kappa} \simeq \frac{c}{\omega_p}$$

And with $\omega_p = 1.6 \times 10^{16} s^{-1}$ within our example of copper, the associated skin depth is $\delta \simeq 19$ nm. Comparing this with the wavelength, λ_o, of a field in the range of $\gamma \ll \omega \ll \omega_p$–approximately the visible range–we see that $\delta \ll \lambda_o$. As we did for the reflective region of a Lorentz insulator, we here obtain a rough estimate of the reflectivity by setting $n = 0$ in Eq. 10.39 to yield $R(\omega)_\perp \simeq 1$. And the more accurate estimate again involves using the approximations $\kappa(\omega) \simeq \omega_p/\omega$ and $n(\omega) \simeq \gamma\omega_p/2\omega^2$ within the $\kappa \gg n$ approximation of Eq. 10.41. We again obtain,

$$R(\omega)_\perp = 1 - \frac{4n}{\kappa^2} = 1 - \frac{2\gamma}{\omega_p}$$

which is nearly unity since it is generally the case that $\gamma \ll \omega_p$.

Region T (high frequency): Dominantly transmissive

The very high frequency region defined as $\omega > \omega(\varepsilon_{r1} = 0)$ is again comparable to that for the Lorentz insulator. As in region 2, the imaginary part is negligible compared to the real part and there is little to no absorption, but ε_{r1} has become positive. Mathematically, for $\omega \gtrsim \omega_p \gg \gamma$, the dielectric function takes the form of Eq. 10.30,

$$\tilde{\varepsilon}_r(\omega) \simeq \varepsilon_{r1} \simeq \varepsilon_{rc} - \frac{\omega_p^2}{\omega^2} \tag{10.55}$$

and so here again the sign change occurs at the "screened plasma" frequency, $\omega(\varepsilon_{r1}=0)=\omega_{sp}\simeq\omega_p/\sqrt{\varepsilon_{rc}}$. And, again, as with insulators, in the limit of very high frequencies, $\varepsilon_{r1}(\infty)\rightarrow\varepsilon_{rc}$. For the simplified case of a plasma where there is no core contribution to the response, we take $\varepsilon_{rc}=1$ (see Eq. 10.25), and the screened plasma frequency is the plasma frequency, $\omega_{sp}=\omega_p$.

Just as in a Lorentz insulator, as the frequency passes through ω_{sp} and enters into this region, the material abruptly goes from $\kappa\gg n\simeq 0$ and highly reflective to $n\gg\kappa\simeq 0$ and highly transmissive. In metals, this transition at the "plasma edge" is known as "ultraviolet transparency." The transition features at this point are indicated clearly in all three plots of Fig. 10.7.

10.4 Measuring the Lorentz–drude response of matter to fields

In Section 9.9, we found that with the field measurements of reflectivity, transmission and phase shifting, we could determine the response of the material to the field–in the forms of the real and imaginary components of $\tilde{n}=n+i\kappa$. And we saw further that, admitting the frequency dependence of the response, full broadband measurements of these quantities could yield a complete frequency mapping of the material response to the field. We related this complex index, $\tilde{n}(\omega)$, to the macroscopic Maxwell's equations by expression in terms of a generalized conductivity or generalized permittivity but that was as far as we could go without a physical model. Now, with the Drude conductor and Lorentz insulator models of Eqs. 10.6 and 10.20, we have a means of discerning free and bound responses and we can take our measurements of reflectivity, transmission, and phase shifting one step further to determine the specific electron density, n_e, collision frequency, γ, and binding frequencies (or resonances), ω_{os}, of the material.

10.4.1 Single frequency measurements

Because the optical response consists of a real and an imaginary part, simultaneous pairs of measurements are required. In Chapter 9, we outlined a number of such pairs of measurements including:

(1) Reflectance plus transmittance measurements at normal incidence from thin targets.

(2) Transmission coefficient plus phase-shift at normal incidence from thin targets.

(3) Reflection coefficient plus phase-shift at normal incidence from thick targets
(4) Double reflectivity measurements from thick targets–in which the two measurements differ by either polarization or incident angle.

The particular technique may then be repeated at multiple frequencies to obtain the rough functional form of the material response. To implement the Drude and/or Lorentz models, we carry out the measurements and obtain $\tilde{n} = n + i\kappa$ at a particular frequency as before. Using Eqs. 10.33 and 10.34, we can then determine the real and imaginary components of the dielectric constant, $\tilde{\varepsilon}_r(\omega) = \varepsilon_{r1} + i\varepsilon_{r2}$. The Lorentz and Drude expressions for $\tilde{\varepsilon}_r(\omega)$, Eqs. 10.28, 10.29 and 10.53, along with the definition of the plasma frequency, Eq. 10.51, are then used to extract information about n_e, γ, ω_{os}, and ε_{rc} of the material. Carefully observe, however, that because these models generally introduce more than two material parameters, two single-frequency measurements will not generally suffice. For example, with insulators we may need to additionally identify the locations of nearby resonances, ω_{os}, and the value of ε_{rc} if it differs significantly from 1. Or, it may sometimes be safe to assume a limit: For example, in either the strong or weak absorption limit for conductors, $\omega/\gamma \ll 1$ or $\omega/\gamma \gg 1$, Eq. 10.52 simplifies and the number of required measurements is effectively reduced by one. We will next, however, consider as an example a special case in which there are only two material parameters, n_e and γ, and so only two measurements are required.

10.4.2 Dual polarization Fresnel reflectivity measurement

We now consider the fourth method in which, for example, the target is too thick for a transmission measurement and a phase measurement proves too difficult. This method involves off-normal reflectivity measurements and so we consider the more general Fresnel equations Eqs. 9.88 and 9.89. To simplify the set-up to a single probe beam as shown in Fig. 10.8, we choose to consider reflection measurements of S- and P-polarized light at a single angle of incidence θ rather than measuring a single polarization at separate angles. Eqs. 9.88 and 9.89 then provide reflection intensities, $R_S(\theta)$ and $R_P(\theta)$, in terms of the real and imaginary components of the complex index, $\tilde{n}$, so that given the measured S and P reflectivities at the incident angle, n and κ are determined either by inverting the equations or by the graphical method of crossed reflectivity contours described in Chapter 9.

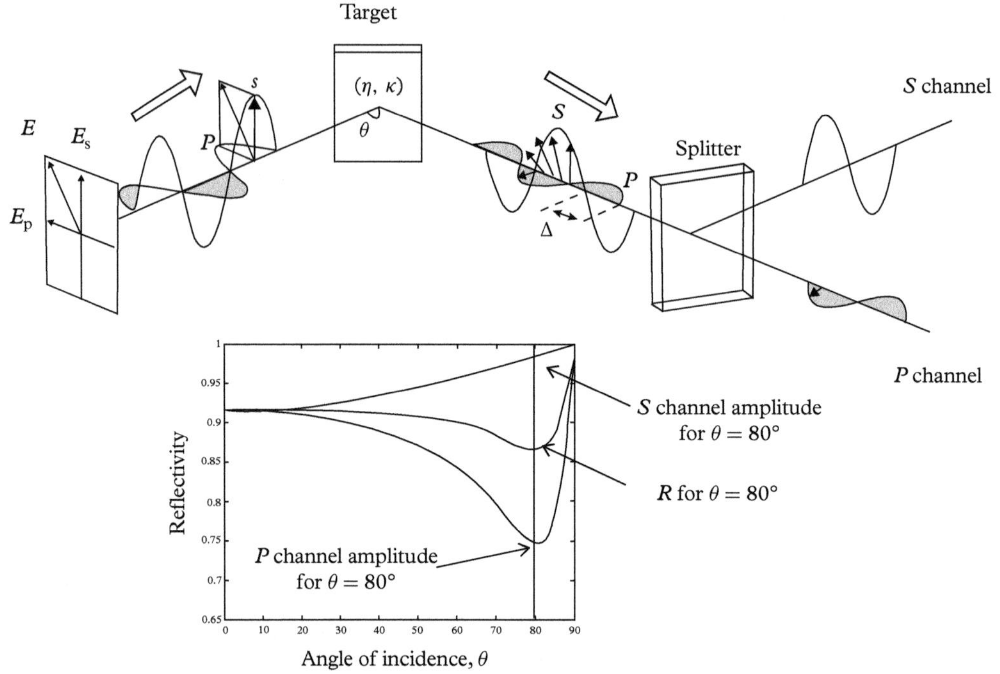

Fig. 10.8 *Diagram of the dual polarization Fresnel reflectivity measurement set-up.*

If the target surface is heated with, say, a burst of x-rays to a high enough temperature, say 20 eV, fast enough and the probe beam arrives at this surface before any significant expansion starts, then the laser light is effectively interacting with an isochorically heated near-solid-density Fresnel surface plasma whose complex index is well approximated by,

$$\tilde{n}(\omega) = \sqrt{\tilde{\varepsilon}_r(\omega)} \simeq \left(1 - \frac{\omega_p^2}{\omega^2 + i\gamma\omega}\right)^{1/2} \tag{10.56}$$

where $\omega_p = n_e e^2/\varepsilon_o m_e$. In comparison to Eq. 10.52 for conductors, $\varepsilon_{rc} = 1$ because a plasma of free ions and electrons provides essentially zero bound response. Thus, in such a case there are only two material parameters, n_e and γ, and so only two reflectivity measurements are required.

In the absence of any other models, the Drude model combined with the measured reflectivities would then inform the experimenter of the electron density n_e and collision frequency γ. However, as pointed out before, these classical models are essentially nineteenth century models which include no quantum or relativistic corrections–that is, Drude and Lorentz are far from the last word in material response. In fact, with such corrections things quickly get very complicated and potential solutions quickly move from analytical to numerical. And with many alternate means of accurately measuring and modeling material parameters, the goal appropriately changes to determining the most accurate models of material response.

To this end, we could start with a simulation according to well-established physics that includes initial conditions of the experiment such as target and x-ray source parameters and their relative configurations. For example, in the P-polarized case, known equations of state (EOS) and x-ray energy deposition processes will proceed within a simulation to provide the electron density and material temperature of the target at the time of probe reflection. Also, a collision frequency, γ (or collision time, τ) model is chosen. In general, γ is a function of the density and temperature of the material. It has been found[6] that at "low" temperatures, γ rises rapidly with temperature and plateaus in the range of 10–100 eV for most all materials. In this "cold" regime, γ has a strong inverse dependence on the density. A more sophisticated model for γ at near solid density and $T \simeq 20\ eV$ is provided by Lee and More.[7] At this point, assuming accurate x-ray deposition/EOS modeling of n_e and T and accurate collisional modeling of $\gamma\ (n_e, T)$, each of the electrical conductivity models, including the Drude model, can be tested by comparison of their predicted values of reflectivity against the measured reflectivities. Alternatively–rather than a full simulation–actual measurements of the electron density and temperature may be performed at the time of probe reflection. This is possible in the simplified geometry of a sharp interface because the reflection is from a single surface with well a defined density and temperature.[8]

10.4.3 Broadband measurements and response models

We have found that discrete frequency measurements can determine the optical constants. And using models, we can further determine certain frequency independent material parameters. We now explore broadband measurements. Broadband measurements have the advantage that response features such as the locations, amplitudes and widths of resonances are readily identified and models can be quickly evaluated. Broadband measurements of continuum reflection, transmission and phase shifting provide a large amount of information in a single shot. Each of the single frequency measurement techniques in the previous section can be generalized to broadband measurements.

[6] Milchberg, H. M., Freeman, R. R., Davey, S. C., and More, R. M., *Phys. Rev. Lett.* 61, 2364 (1988).

[7] Lee, Y. T. and More, R. M., An electron conductivity model for dense plasmas. *Physics of Fluids* (1958–1988), 27(5) (1984).

[8] In practice, however, the Fresnel approximation of a sharp interface is difficult to achieve and plasma expansion inevitably occurs upon measurement. This, then, necessitates additional numerical or analytical modeling of hydrodynamic expansion to provide a temperature/density profile at the time of reflection.

Any of these generalized techniques can then yield the functional form of the complex index of refraction, $\tilde{n}(\omega)$, over the measured range of frequencies. Then, from $\tilde{\varepsilon}_r = \tilde{n}^2$, the response curve structure (real and imaginary) specific to the probed material can be found. However, simply extending these techniques, which involve pairs of measurements, to broadband continuum measurements will provide redundant information. Indeed, in this section we will show that a broadband determination of the functional form of either the real or imaginary component of $\tilde{\varepsilon}_r$ and use of the Kramers–Kronig relationship will yield the full complex form. Next, application of specific response models such as those of Drude and Lorentz, will determine not only the frequency-independent parameters such as n_e and γ, but also the locations ω_o, amplitudes A_o and widths $\Delta\omega$ of resonances.

As an example, we start with a broadband extension to the first measurement technique outlined previously: reflectance plus transmittance measurements at normal incidence from thin targets. If we consider the very high frequency range response of an insulator where $\omega \gg \omega_{os} \gtrsim \gamma$, then a considerable simplification occurs. According to the Lorentz model of Eqs. 10.28 and 10.29, in this range we can generally assume that $\varepsilon_{r2} \ll 1$ and ε_{r1} is approximated by Eq. 10.30. In fact, as more and more resonances are exceeded, as in the case of x-ray radiation, the core response, ε_{rc}, approaches unity and Eq. 10.30 goes to

$$\varepsilon_{r1}(\omega) \simeq 1 - \frac{\omega_p^2}{\omega^2} \tag{10.57}$$

where $\omega_p^2/\omega^2 \ll 1$ and most–but not all–resonances have been exceeded. If we now express $\tilde{\varepsilon}_r$ as

$$\tilde{\varepsilon}_r(\omega) = \tilde{n}^2 = 1 + \tilde{\chi}_e(\omega)$$

we see that $\tilde{\chi}_e \ll 1$ and using the binomial approximation we get that

$$\tilde{n}(\omega) = n + \kappa \simeq 1 + \frac{1}{2}\tilde{\chi}_e(\omega)$$

Recall from the previous chapter that if we ignore contributions from multiple internal reflections, the transmitted intensity can be approximated as,

$$T_\perp(\omega) = (1 - R_\perp(\omega)) \exp\left(-2\frac{\omega\kappa(\omega)}{c}l\right) \tag{10.58}$$

where l is the wafer thickness and the absorption coefficient is given by $\alpha = 2k_2 = 2(\omega/c)\kappa$. With $\tilde{n}(\omega) \simeq 1$, the Fresnel normal incidence result of Eq. 9.87 yields $R_\perp \simeq 0$ and Eq. 10.58 simplifies to

$$T_\perp(\omega) = \exp\left(-2\frac{\omega\kappa(\omega)}{c}l\right) \tag{10.59}$$

which is to say that at x-ray frequencies there is essentially no reflection and so all energy that is not transmitted has necessarily been absorbed. Thus, with a broadband measurement of the transmittance in this range, we directly obtain $\kappa(\omega)$.

With some knowledge of the type of material–conductor, insulator, semiconductor, and so on–we can assume an appropriate response model. Such models include Drude for the dominantly free responses of metals, Lorentzians for the dominantly bound responses of insulators, often a sum of several Lorentzians corresponding to multiple resonances and sometimes a combination Drude-Lorentz response.[9] In this case, we have assumed that we have an insulator exposed to x-rays with a frequency above all but one resonance. We will thus use a single complex Lorentzian as our model. That is, from our previous simplification and Eq. 10.27, near the last resonance the complex index is

$$\tilde{n}(\omega) \simeq 1 + \frac{1}{2}\tilde{\chi}_e(\omega) = 1 + \frac{1}{2}\left(\frac{\omega_p^2}{\omega_{os}^2 - \omega^2 - i\gamma\omega}\right) \tag{10.60}$$

so that,

$$n(\omega) \simeq 1 + \frac{1}{2}\left[\frac{\omega_p^2\left(\omega_{os}^2 - \omega^2\right)}{\left(\omega_{os}^2 - \omega^2\right)^2 + (\gamma\omega)^2}\right] \tag{10.61}$$

and

$$\kappa(\omega) \simeq \frac{1}{2}\left[\frac{\gamma\omega\omega_p^2}{\left(\omega_{os}^2 - \omega^2\right)^2 + (\gamma\omega)^2}\right] \tag{10.62}$$

It is important now to point out that the Kramers–Kronig relations, such as those shown in Section 9.9.3, that exist for optical constant pairs such as $\sigma_1(\omega)$ and $\sigma_2(\omega)$ and $\varepsilon_{r1}(\omega)$ and $\varepsilon_{r2}(\omega)$ ultimately derive from the Kramers–Kronig relation for the complex susceptibility $\tilde{\chi}_e(\omega)$ and therefore the two terms must be formed to represent the real and imaginary components of $\tilde{\chi}_e(\omega)$. In this simplified case the optical constant pair of n and κ are linearly related to $\tilde{\chi}_e(\omega)$ so that the associated Kramers–Kronig relations are:

[9] As noted before, many other, more sophisticated model functions have been developed that account for quantum and relativistic effects but these models are beyond the scope of this book and would not add to our present discussion.

$$n(\omega_o) = 1 + \frac{1}{\pi}\mathrm{P}\left(\int_0^\infty \omega' \frac{\kappa(\omega')}{\omega'^2 - \omega_o^2} \mathrm{d}\omega'\right)$$

$$\kappa(\omega_o) = -\frac{\omega_o}{\pi}\mathrm{P}\left(\int_0^\infty \frac{n(\omega') - 1}{\omega'^2 - \omega_o^2} \mathrm{d}\omega'\right) \qquad (10.63)$$

Next, inserting $\kappa(\omega)$, derived from our broadband measurement of the transmittance (Eq. 10.59) in the range of the last resonance, ω_{os}, into $n(\omega_o)$ within Eq. 10.63, we carry out the integral to obtain the real part of the index. Now that we can refer to a specific model, it is interesting to look more closely at this Kramers–Kronig integral. Referring to the plots of $n(\omega)$ and $\kappa(\omega)$ in Fig. 10.5, and to the Eqs. of 10.63, we note that when the integral for $n(\omega_o)$ is evaluated at points just below resonance, $\omega_o < \omega_{os}$, it is positive. This is because for such points, the Lorentzian, $\kappa(\omega')$, of Eq. 10.62 is just rising to its peak and because as $(\omega'^2 - \omega_o^2)^{-1}$ integrates through ω_o, it first swings negative and then positive, it samples a greater value on the high side than on the low side. Similarly, for points just above resonance $\kappa(\omega')$ is falling from its peak and the integral for $n(\omega_o)$ evaluates to be negative. This results in the familiar derivative-like form of the real part of the complex index of refraction in the vicinity of a resonance. Finally, as discussed in the previous chapter, because broadband measurements are necessarily finite while the Kramers–Kronig integrals range from zero to infinity, it is often necessary in practice to extrapolate the data from the lowest measured frequency to zero and from the highest measured frequency to infinity. It is seen from the Kramers–Kronig relations, however, that the greatest contributions to the integrals come from behavior near $\omega' = \omega_o$ so that the further we get from the evaluated frequency, ω_o, the less accuracy we are required to have in the extrapolation. And, as we have seen, the extrapolations are generally based on the most appropriate models for the expected measurements in the associated frequency range.

Exercises

(10.1) A conductor may be seen as a gas of electrons interpenetrating a lattice of positive ions such that there is overall charge neutrality. If the electrons are all momentarily displaced from equilibrium, an electric field restoring force will act to bring the electrons back to their equilibrium positions, thus producing simple harmonic motion at a frequency known as the electron plasma frequency, ω_p. In this problem, we derive ω_p. For simplicity, consider a metal slab of thickness T and area A, as in Fig. 10.9

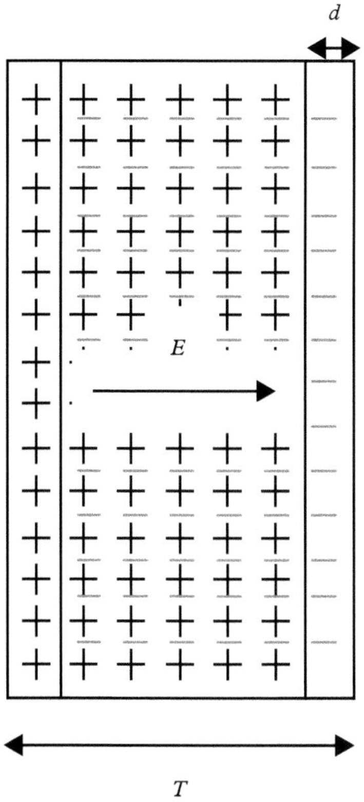

Fig. 10.9

(a) Given that n_e is the number density of conduction electrons in the slab and that e is the electron charge magnitude, use Gauss's law to show that a small displacement $d \ll T$ of the electron gas (as shown) will result in an electric field, within the slab, of magnitude:

$$E = \frac{n_e e}{\varepsilon_o} d$$

(b) Next, given that the force on the electrons is $-eE$, and m_e is the electron mass, solve the equation of motion for the electrons and identify the oscillation frequency, ω_o of this harmonic solution as

$$\omega_p = \sqrt{\frac{n_e e^2}{\varepsilon_o m_e}}$$

This oscillation is a normal mode of the electrons in the metal slab and ω_p is known as the plasma frequency.

(c) Given that n_e in copper is $8.4 \times 10^{28}\, m^{-3}$, calculate the plasma frequency of copper. Where does this frequency sit relative to typical frequencies of visible light?

(10.2) Using the Drude and Lorentz models for conductivity and susceptibility:

(a) Show that setting the collisional absorption factor, $\gamma = 1/\tau$, to zero ($\omega\tau \to \infty$ and $\omega\gamma \to 0$) leads to a conductivity $\tilde{\sigma}_f$ that is purely imaginary and a susceptibility $\tilde{\chi}_e$ that is purely real.

(b) Show that a purely imaginary conductivity σ_{f2} and a purely real susceptibility χ_{e1} leads to zero absorption of the field energy by the material.

(c) Repeat (a) and (b) for the case of an exceedingly large collisional absorption factor ($\omega\tau \to 0$ and $\omega\gamma \to \infty$) to show that the converse is true.

(10.3) It was noted that in the Drude model there is no need for the distinction between a "local" field, such as the polarization field, E_p and the applied (average) field, E, because the free electrons, unlike the bound electrons, are not localized and so the average field they "feel" will always be the applied field. Confirm that this is the case by considering how far a lightwave-driven electron travels during one cycle of the lightwave. Assume the maximum velocity of the electrons corresponds to an energy of ~10 eV. Is this distance large enough that the electrons, on average, feel the macroscopic average field, E?

(10.4) If we express the more general, multi-resonance susceptibility of Eq. 10.23 in terms of "bound" and "free" electron susceptibilities, $\tilde{\chi}_e = \tilde{\chi}_{eb} + \tilde{\chi}_{ef}$, then the associated general definition of the displacement is,

$$\vec{D} = \varepsilon_o\left[1 + \tilde{\chi}_e(\omega)\right]\vec{E}$$
$$= \varepsilon_o\left[1 + \tilde{\chi}_{eb}(\omega) + \tilde{\chi}_{ef}(\omega)\right]\vec{E}$$

From the relation between free current density and conductivity, $\vec{\mathcal{J}}_f = \tilde{\sigma}(\omega)\vec{E}$, obtain the relationship between the conductivity, $\tilde{\sigma}_f(\omega)$, and the "free electron" susceptibility, $\tilde{\chi}_{ef}(\omega)$.

(10.5) While the dielectric response in Lorentz insulators is due solely to the response of bound electrons, the model-independent notation introduced in Chapter 9 defined a generalized conductivity, $\tilde{\sigma} = \sigma_1 + i\sigma_2$, in which σ_1 and σ_2 each encompassed both free and bound contributions. Given the dielectric response function of a (single-resonance) Lorentz insulator (Eq. 10.27), derive a frequency dependent expression for $\tilde{\sigma} = \sigma_1 + i\sigma_2$.

(10.6) For dense insulating materials, the dielectric response function is expressed in terms of the microscopic polarizability, α, as

$$\tilde{\varepsilon}_r = 1 + \tilde{\chi}_e = 1 + \frac{n_e\alpha}{1 - n_e\alpha/3}$$

(a) Solve this for α in terms of $\tilde{\varepsilon}_r$ to obtain the Clausius–Mossotti equation

$$\frac{\tilde{\varepsilon}_r - 1}{\tilde{\varepsilon}_r - 2} = \frac{n_e\alpha}{3}$$

(b) Under what condition does this Clausius–Mossotti relationship reduce to the relationship between thc dielectric constant and the electric susceptibility given for a gas, $\tilde{\varepsilon}_r = 1 + n_e\alpha$?

(10.7) The Lorentz susceptibility for solids, as given by Eq. 10.20, has resonant frequencies, ω_{os}, which are reduced from the bare resonant frequencies, ω_o, due to nearby field corrections. Obtain Eq. 10.20 directly by solving Eq. 10.11 with the substitution $\vec{E}_p = \vec{E} + \vec{P}/3\varepsilon_o$, given by Eq. 10.17.

(10.8) A useful alternative for the expression of the general refractive index $\tilde{n}$ is the polar form

$$\tilde{n} = n + i\kappa = \|\tilde{n}\|\left(\cos\psi + i\sin\psi\right)$$
$$= \|\tilde{n}\|\, exp\left(i\psi\right)$$

(a) Show that, assuming the combined model susceptibility, $\tilde{\chi}_e$, of Section 10.1.3 and the general relation, $\tilde{n}^2 = 1 + \tilde{\chi}_e$, the absolute value, $\|\tilde{n}\|$, and phase, ψ, can be expressed as:

$$\|\tilde{n}\| = \left[(1 + \chi_{e1})^2 + (\chi_{e2})^2\right]^{1/4} \quad \tan\psi$$
$$= (\chi_{e2})^{-1}\left[\sqrt{(1+\chi_{e1})^2 + (\chi_{e2})^2} - (1+\chi_{e1})\right]$$

(b) Next, show that these quantities can be expressed in terms of the conductivity $\tilde{\sigma}_f$ and permittivity $\tilde{\varepsilon}_b$, of Eqs. 10.6 and 10.26, as

$$\|\tilde{n}\| = \left[\left(\frac{\varepsilon_{b1}}{\varepsilon_o} - \frac{\sigma_{f2}}{\varepsilon_o\omega}\right)^2 + \left(\frac{\varepsilon_{b2}}{\varepsilon_o} + \frac{\sigma_{f1}}{\varepsilon_o\omega}\right)^2\right]^{1/4}$$

$$\tan\psi = \left(\frac{\varepsilon_{b2}}{\varepsilon_o} + \frac{\sigma_{f1}}{\varepsilon_o\omega}\right)^{-1} \times$$
$$\left[\sqrt{\left(\frac{\varepsilon_{b1}}{\varepsilon_o} - \frac{\sigma_{f2}}{\varepsilon_o\omega}\right)^2 + \left(\frac{\varepsilon_{b2}}{\varepsilon_o} + \frac{\sigma_{f1}}{\varepsilon_o\omega}\right)^2} - \left(\frac{\varepsilon_{b1}}{\varepsilon_o} - \frac{\sigma_{f2}}{\varepsilon_o\omega}\right)\right]$$

(10.9) (a) Show that the region of anomalous dispersion, $\omega_{os} - \gamma/2 < \omega < \omega_{os} + \gamma/2$, can be determined by setting the derivative of Eq. 10.28 with respect to frequency to zero and solving for ω.

(b) Show that the exact solution to the "screened plasma frequency" is given by:

$$\omega_{sp}^2 = \omega_{os}^2 + \frac{1}{2}\left(\frac{\omega_p^2}{\varepsilon_{cr}} - \gamma^2\right)$$
$$\left(1 + \sqrt{1 - \left(\frac{2\omega_{os}\gamma}{\omega_p^2/\varepsilon_{cr} - \gamma^2}\right)^2}\right)$$

(10.10) (a) Obtain Lorentz model expressions for the following parameters of an E-M wave passing through a Lorentz insulator in the first dominantly transmissive frequency range (region T): (i) the relative phase shift between the E and B fields; (ii) the absorption coefficient; (iii) the Poynting vector.

(b) Do the same for the dominantly absorptive, reflective and transmissive frequency ranges and compare and discuss the results from all four regions.

(10.11) Consider the case of a plasma in which there is no core contribution to the response (i.e., we take $\frac{\varepsilon_c}{\varepsilon_o} = 1$). Determine the group velocity, $v_g = \frac{d\omega}{dk}$, as a function of ω, in the region around the plasma frequency, $\omega = \omega_p$, where the response function is given by:

$$\tilde{n}^2(\omega) \simeq 1 - \frac{\omega_p^2}{\omega^2}$$

(a) What is v_g for ω just below ω_p? What does an imaginary group velocity correspond to?

(b) What is v_g for ω at ω_p? What does this say about the propagation of energy into the plasma at this frequency?

(c) What is v_g for ω just above ω_p? What value does it asymptotically approach as $\omega \to \infty$?

(10.12) (a) Obtain Drude model expressions for the following parameters of an E-M wave passing through a Drude conductor in the dominantly absorptive frequency range (region A): (i) the relative phase shift between the E and B fields; (ii) the absorption coefficient; (iii) the Poynting vector.

(b) Do the same for the dominantly reflective and dominantly transmissive frequency ranges and compare and discuss the results from all three regions.

(10.13) For matter in the plasma state, $\varepsilon_c^d \to 1$ and the dielectric response function is given by

$$\tilde{n}^2 = \tilde{\varepsilon}^d = 1 + i\frac{\sigma_f(\omega)}{\varepsilon_0\omega}$$
$$= 1 + \frac{i}{\varepsilon_0\omega}\frac{n_e e^2}{m_e(\gamma_0 - i\omega)}$$

(a) In Chapter 9, we found that the ratio $\varepsilon_2^d/\varepsilon_1^d$ provided a model-independent measure of the relative absorption occurring in the material. That is, $\varepsilon_2^d/\varepsilon_1^d \gg 1$ and $\varepsilon_2^d/\varepsilon_1^d \ll 1$ correspond, respectively, to the strong and weak absorption limits. Show that, within the Drude model, such a measure can similarly be provided by the quantity $\gamma_0/\omega = 1/\omega\tau$ where $\tau = 1/\gamma_0$ is the time between electron-ion collisions.

(b) Show that in the strong absorption limit ($\omega\tau \ll 1$), Eq. 9.13 can be expressed as

$$\tilde{n}^2 = 1 + i\frac{\sigma_0}{\varepsilon_0\omega}$$

and in the weak absorption limit ($\omega\tau \gg 1$), Eq. 9.13 can be expressed as

$$\tilde{n}^2 = 1 - \frac{\omega_p^2}{\omega^2}$$

where $\sigma_0 = n_e e^2/m_e\gamma_0$ is the purely real "DC" conductivity and $\omega_p^2 = n_e e^2/\varepsilon_0 m_e$ is the plasma frequency.

(10.14) Consider Eq. 9.13 with the strong absorption approximation $\omega\tau \ll 1$. Show that for a plasma in the strong absorption limit, the normal reflectivity of Eq. 9.87 simplifies to

$$R_\perp = \frac{n-1}{n+1}$$

Note that only one reflectivity measurement is required under these limiting conditions to determine the "DC" conductivity, σ_0.

(10.15) As was done for the top K-K integral of Eq. 11.64, $n(\omega_o)$, in the paragraph following Eq. 11.64, provide a qualitative description of how the bottom K-K integral leads to a resonant form of $\kappa(\omega_o)$.

11 Scattering of Electromagnetic Radiation in Materials

- Scattering problem definition; previously derived multipole expansions using the language of scattering theory. Applications of these results including resonant scattering and plasmon resonances
- Formal scalar scattering theory. The integral scattering equation is derived and used to find the Born expansion and to prove the optical theorem
- Partial wave analysis for the scalar scattering problem with a discussion connecting quantum (wave theory) and classical views
- Vector spherical harmonics and the extension of partial wave analysis to scattering of the vector fields of electromagnetic waves
- Mie theory. Scattering from a homogeneous sphere: solution with several applications including glory scattering and whispering gallery mode resonances

11.1 Scattering

The interaction of an electromagnetic wave with a material target and its later evolution is alternatively described as scattering or diffraction. The former usually refers to an interaction with a compact material target and oftener than not the primary focus of a problem is to learn something of the target. Indeed, most of our knowledge of the world around us comes from observing light scattering off surrounding objects. "Diffraction," on the other hand, generally focuses on the uniquely wave-like features of the radiation itself that follows its interaction with a more extended target such as a slit

Electromagnetic Radiation. Richard Freeman, James King, Gregory Lafyatis,
Oxford University Press (2019).
DOI: 10.1093/oso/9780198726500.001.0001

or a lens. Despite the essential identity of the two processes at a fundamental level, two different sets of tools have been developed to describe the two phenomena. In this chapter we cover scattering theory.

Historically, almost every aspect of scattering theory is in some way associated with John William Strutt known otherwise as Lord (third Baron) Rayleigh. While in optics he is best known for the scattering of electromagnetic radiation from small particles–Rayleigh scattering–he worked on almost every topic covered in this chapter including the optical theorem, the so-called Born approximation, and the partial wave expansion. His contributions went far beyond his explanation as to why the sky is blue–a subject he turned to at least three times in his career. Indeed much of what we would now recognize as scattering theory is found in his *Theory of Sound* and elastic-solid investigations.[1]

Here, we will take a catholic approach to scattering, often going beyond electromagnetic problems. As we saw in the sections on relativity, the vector or polarization variables in electromagnetic waves frequently complicate problems to the point of obfuscating essential underlying physics. In the following, we usually start a topic by looking at the scalar-field analog and only later include the polarization degrees of freedom. By this way of viewing the electromagnetic problem in a simplified context, we also automatically learn something of the scattering of sound waves. Moreover, as we will see, the quantum mechanical Schrodinger wave function of a structureless particle is a classical scalar field, the free particle wave function is a solution of the Helmholtz equation and, so long as we only consider time-independent problems, quantum potential scattering is the scalar theory analog of Maxwell's equations. That is, we get non-relativistic quantum mechanics scattering for free. Some important fine print: the Schrodinger equation and Maxwell's (wave) equations do have different time dependences or, in Fourier space, dispersion relations. And including spin in non-relativistic quantum mechanics is importantly different from adding polarization to electromagnetic problems.

Figure 11.1 shows schematically, the basic idea behind solving scattering problems. A monochromatic plane wave is incident on a scatterer and the (time independent) scattered flux is sought. The cross section describes the amount of power taken out of the incident beam as an equivalent cross sectional area of the beam. The total scattering cross section, σ_{scat}, is defined by:

$$P_{scat} = \sigma_{scat} I_{inc} \tag{11.1}$$

where I_{inc} is the intensity of the incident plane wave and P_{scat} is the total power scattered from the beam by the scatterer. Similarly, if the

[1] A review of Rayleigh's contributions to scattering theory is found in: Twersky, V., *Rayleigh Scattering, App. Opt.*, 3, 1150 (1964).

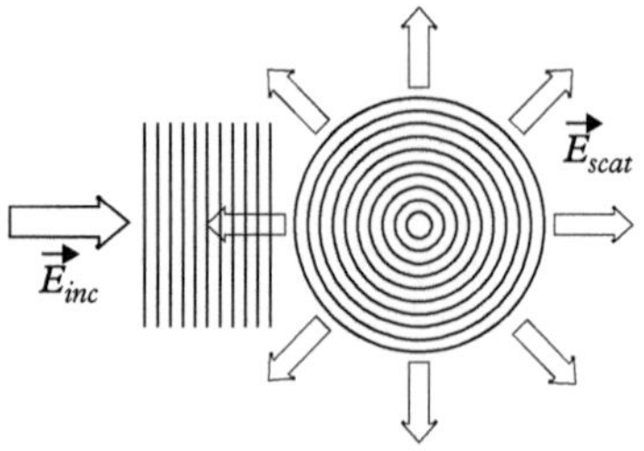

Fig. 11.1 *Schematic of the scattering wave function. An incident plane wave interacts with the target and produces a scattered outgoing wave.*

target can absorb energy, the extinction cross section is defined by replacing the left-hand side of Eq. 11.1 with P_{ext}, the total power removed from the incident beam by both scattering and absorption. The differential cross section is defined by the power scattered into the solid angle, $d\Omega$. A detector located at a distance r from the scattering center and oriented with its sensitive area, $dA = r^2 d\Omega$, perpendicular to the scattered beam will detect power

$$I_{scat}(\theta, \phi, r)\, r^2 d\Omega = d\sigma_{scat} I_{inc} \Leftrightarrow \frac{d\sigma}{d\Omega} = \frac{I_{scat}(\theta, \phi, r)}{I_{inc}} r^2.$$

noting that on the right we have dropped the "scattering" subscripts so that hereafter it will be assumed, unless otherwise specified, that σ refers to the total scattering cross section. To the extent that the material parameters of the environment and the scatterer are independent of frequency, scattering problems are characterized by the size parameter $\beta = 2\pi a_0/\lambda = a_0 k$. Here, a_0 is a characteristic length describing the scatterer. For the spheres that we study in most detail, a_0 is always taken to be the radius of the sphere and λ is the wavelength of the incident electromagnetic radiation. Two problems having the same size parameter will have solutions for the various physical quantities of interest that are trivially scaled versions of one another. There are two limits, $\beta \gg 1$ and $\beta \ll 1$, for which the full blown scattering theory formalism is unnecessary. For $\beta \gg 1$, the short wave limit, geometrical optics can be used. We will study geometric optics scattering toward the end of this chapter to give context to some of the formal scattering theory calculations. In the next section of this chapter, the long wavelength limit is studied, $\beta \ll 1$. The scattering for these systems are simply multipole fields. By far, the most important is the electric dipole field. Since the scattered multipole fields have already been found in Chapter 4, in a very real sense the hard work for these systems is already done.

In Section 11.3, we start formal scattering theory. The basic idea is to find time independent solutions for the problem that asymptotically look like the expected solutions of the Helmholtz equation: an incident plane wave and an outgoing spherical wave that is modulated in angle to give the angle-dependent scattered flux. Again, referring to

Fig. 11.1, we will always use the geometry of a $+z$-directed plane wave incident on a target at the coordinate origin. The total wave function may be written:

$$\vec{E}_{tot} = \vec{E}_{inc} + \vec{E}_{scat} = \hat{\varepsilon} e^{ikz} + \vec{f}(\hat{\varepsilon},\theta,\phi)\frac{e^{ikr}}{r} \tag{11.2}$$

We are assuming an incident wave of unit field strength: $\hat{\varepsilon}$ is a polarization vector with unit amplitude and $\vec{f}\left(\hat{\varepsilon},\theta,\phi\right)$ is the scattering amplitude. The scalar version of this equation has $\hat{\varepsilon} \to 1$ and the scattering amplitude is a scalar quantity:

$$\varphi(\vec{r}) = e^{ikz} + f(\theta,\phi)\frac{e^{ikr}}{r} \tag{11.3}$$

In the asymptotic region, the field of the scattered wave locally appears to be an outward directed plane wave and so for both the incident and the asymptotic scattered wave, the accompanying magnetic fields are readily determined from the electric field and the direction of propagation using Eq. 1.79.

The intensity of any plane wave is proportional to the absolute square of its amplitude. For an electromagnetic plane wave traveling in a uniform medium the time-averaged intensity is given by (**see Discussion 11.1**)

$$I = \left\langle \vec{S} \right\rangle = \frac{1}{2}\left| Re\vec{S}_c \right| = \frac{1}{2}\left| \mathrm{Re}\left[\vec{E} \times \vec{H}^* \right] \right| = \frac{1}{2Z}\left| \vec{E} \right|^2 \tag{11.4}$$

where $\vec{S}_c = \vec{E} \times \vec{H}^*$ is the complex form of the Poynting vector and $Z = \sqrt{\mu/\varepsilon}$ is the characteristic impedance of the medium. Thus, in general, for a scattering wave function with an incident wave normalized as in Eq. 11.2:

$$\frac{d\sigma}{d\Omega} = \frac{I_{scat}(\theta,\phi,R)}{I_{inc}} r^2 = \frac{\left| \vec{f}\left(\hat{\varepsilon},\theta,\phi\right) \right|^2}{\left| \hat{\varepsilon} \right|^2} = \left| \vec{f}\left(\hat{\varepsilon},\theta,\phi\right) \right|^2 \tag{11.5}$$

The total scattering cross section is then just the square of the scattering amplitude integrated over 4π steradians. To find the cross section for scattering into a specific final polarization, $\hat{\varepsilon}'$, the component of the scattered field having that polarization is found by taking the dot product of $\hat{\varepsilon}'^*$ with the field and that component is used to calculate the intensity of the outgoing wave of that polarization. In Eq. 11.5 this means replacing the right-hand side with $\left| \hat{\varepsilon}'^* \cdot \vec{f}\left(\hat{\varepsilon},\theta,\phi\right) \right|^2$. Section 11.3 also discusses two general results that derive from the mathematical form of the Helmholtz equation: the Born approximation and the optical theorem. The partial wave expansion is the starting point of many real-life scattering calculations, whether in nuclear or atomic physics or in the optics of water droplets that we will examine in some detail. In Section 11.4, we begin with the scalar field version of the partial wave expansion. The generalization of the partial wave expansion to include polarization requires a few mathematical side trips. We are fortunate to have one widely applicable, exactly soluble scattering system: the "Mie" problem is the scattering of an electromagnetism wave by a uniform dielectric sphere of any size. The partial wave expansion developed by Gustav Mie, among other things, realizes an exact solution for water droplets. It can serve as a bridge from the dipole limit to the geometrical optics limit. The present day capability of rapidly generating Mie-theory solutions with thousands of partial waves using just a laptop computer makes Mie theory an invaluable "numerical laboratory" for studying scattering problems. In Section 11.5, we look at some representative examples.

11.2 Scattering by dielectric small particles

The important case of a scatterer whose extent is much smaller than the wavelength of the radiation is straightforward given the multipole fields. The scattering is broken up into two steps: the generation of oscillating charge and current multipoles by the incident radiation field and the subsequent scattering creations of these moments, the multipole radiation fields. Since we already have the multipole fields, the additional work in the scattering problem is to describe the generation of the moments by the incident wave. The most important case is scattering via electric dipole radiation. The time averaged radiation from a system generating an harmonically oscillating electric dipole in vacuum is given by Eqs. 4.56 and 4.57:

$$\left\langle \frac{dP}{d\Omega} \right\rangle_{ed} = \omega^4 \frac{|\vec{p}|^2}{32\pi^2 \varepsilon_0 c^3} \sin^2 \vartheta$$

$$\langle P \rangle_{ed} = \omega^4 \frac{|\vec{p}|^2}{12\pi \varepsilon_0 c^3} \tag{11.6}$$

where $\vec{p}e^{-i\omega t}$ describes the oscillation of the electric dipole and ϑ is the observation angle of the scattering referenced to the direction of $\vec{p}$ (not the spherical coordinate, θ, which is referenced to the z-axis). Thus, using the relation of Eq. 11.1, we can give general expressions for the differential and total cross sections for dipole scattering:

$$\frac{d\sigma}{d\Omega} = \frac{\omega^4}{16\pi^2\varepsilon_0^2 c^4}\frac{|\vec{p}|^2}{E_0^2}\sin^2\vartheta \tag{11.7}$$

$$\sigma = \frac{\omega^4}{6\pi\varepsilon_0^2 c^4}\frac{|\vec{p}|^2}{E_0^2} \tag{11.8}$$

where we have used that the incident intensity is $I_{inc} = \frac{1}{2}\varepsilon_0 c E_0^2$. In the top part of Fig. 11.2 we show a 3D polar plot for the electric dipole scattering of a circularly polarized field. The meaning of the plot is that a surface is generated such that for each direction, the distance from the origin to the surface is proportional to the differential cross section in that direction. In much of what follows, we will either consider a circularly polarized incident field on an isotropically polarizable target or a scalar field scattering, both of which give axially (ϕ-referenced about the direction of the incident field) symmetric scattering from a central potential. In contrast, scattering represented by Eq. 11.7 is symmetric about the dipole axis. Thus, an important–albeit after the fact–obvious consequence of a linearly polarized incident field is to break the axial symmetry of the problem. Indeed, as seen in Eq. 4.40, the field created by an oscillating electric dipole in the far-field, radiation zone is:

$$\vec{E}(\vec{r},t) = k^2\frac{1}{4\pi\varepsilon_0}\frac{e^{ikr}}{r}\left[\left(\hat{r}\times\vec{p}(t)\right)\times\hat{r}\right] \tag{11.9}$$

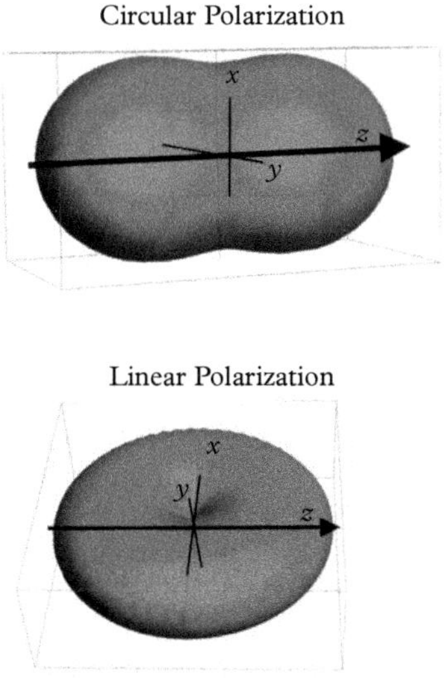

Fig. 11.2 *Differential cross section for electric dipole scattering. The plots are surfaces that are generated such that for each direction, the distance from the origin to the surface is proportional to the differential cross section in that direction. Top: The cross section for incident σ^+ polarization. Bottom: The cross section for linearly y-polarized light. In both cases the incident light propagates in the +z direction: roughly from left to right in the figure.*

A y-linearly polarized field, for example, will produce the classic dipole pattern about the y-axis, which will give the corresponding (clearly not axially symmetric) differential cross section shown at the bottom of Fig. 11.2.

11.2.1 Scattering by a free electron: Thomson scattering

As a first example of long wavelength (or dipole) limit scattering, we consider Thomson scattering. Thomson scattering refers to the elastic scattering of electromagnetic radiation by a free electron. Classically, the oscillation of an electron in a harmonic field is described by Newton's second law. For an $x-$polarized wave, $E_x(t) = E_0 \exp[-i\omega t]$, an oscillating dipole is created, $p_x(t) = -ex(t) = -ex_0 \exp[-i\omega t]$, where:

$$m\ddot{x} = -m\omega^2 x = -eE_x \Rightarrow p_{x0} = -ex_0 = \frac{-e^2 E_0}{m\omega^2} \tag{11.10}$$

On substituting this into Eq. 11.6, Larmor's formula, Eq. 6.36, is returned for the total dipole scattered power and Eq. 11.8, the total cross section for dipole scattering yields:

$$\sigma_{Th} = \frac{e^4}{6\pi\varepsilon_0^2 m^2 c^4} \tag{11.11}$$

This result for the Thomson cross section is frequently expressed as:

$$\sigma_{Th} = \frac{8\pi}{3} r_e^2 \tag{11.12}$$

where,

$$r_e = \frac{1}{4\pi\varepsilon_0} \frac{e^2}{mc^2} = 2.818 \times 10^{-15}\text{m} \tag{11.13}$$

is the "classical radius of the electron." This is roughly the size the electron would be if its mass was entirely due to its electrostatic energy. We note that for Thomson scattering the long wavelength (dipole) limit is always satisfied since r_e is much shorter than all but the very shortest gamma wavelengths. The Thomson scattering cross section is independent of the wavelength of the radiation and quantitatively:

$$\sigma_{Th} = 6.652 \times 10^{-29}\text{m}^2. \tag{11.14}$$

Similarly the differential cross section is found to be:

$$\frac{d\sigma_{Th}}{d\Omega} = r_e^2 \sin^2 \vartheta \tag{11.15}$$

again, with the caveats regarding axial symmetry. Important corrections to the Thomson scattering cross section include:

- The Compton wavelength shift for short wavelength scattering. The observation of Compton scattering is compelling evidence of the quantum nature of the electromagnetic field. The most profound result is that the frequency of the scattered radiation is changed according to the Compton shift in wavelength of the scattered radiation:

$$\lambda' = \lambda + \lambda_C (1 - \cos\theta) \tag{11.16}$$

  Here, λ is the wavelength of the incident radiation, λ' is the wavelength of the scattered radiation, θ is the angle of the scattered radiation and $\lambda_C = h/(mc) = 2.43 \times 10^{-12}$m is the Compton wavelength of the electron. This is a kinematic effect resulting from the non-zero momentum carried by the photon.
- At high intensities, nowadays achievable by focused high power lasers, the equation of motion of the electron, Eq. 11.10 must be corrected to include special relativity. Specifically, the magnetic field of the laser becomes important for high velocity motion and relativistic mass effects must be included. The scattering problem is no longer linear in the fields and harmonics of the incident field are generated in the scattered radiation.

11.2.2 Scattering by a harmonically bound electron

An electron-on-a-spring is a classical model for an atomic or molecular resonance transition. The one dimensional equation of motion of a driven, damped electron-on-a-spring is given by Newton's second law:

$$m\ddot{x} = -kx - \alpha\dot{x} - eE_0 e^{-i\omega t} \tag{11.17}$$

The forces on the right-hand side of the equation include a harmonic restoring force, characterized by a spring constant, k, a viscous-type damping force with coefficient α, and the driving force of the electric field in the incident electromagnetic wave. We rewrite this:

$$\ddot{x} = -\omega_0^2 x - \gamma \dot{x} - \frac{eE_0}{m} e^{-i\omega t} \tag{11.18}$$

Now, $\omega_0 = \sqrt{k/m}$ is the natural, resonant frequency of the transition being modeled, and $\gamma = \alpha/m$ is a damping term. It is assumed that γ is much smaller than both the driving and natural frequencies of the system: $\gamma \ll \omega, \omega_0$. Assuming the transients in the system have all died out, a harmonic solution

$$x = x_0 e^{-i\omega t} \tag{11.19}$$

returns the algebraic expression for the complex amplitude:

$$x_0 = \frac{-eE_0}{m} \frac{1}{\left(\omega_0^2 - \omega^2 - i\gamma\omega\right)} \tag{11.20}$$

And for the amplitude of the oscillating dipole:

$$|\vec{p}| = |p_{x0}| = |-ex_0| = \left| \frac{e^2 E_0}{m} \frac{1}{\left(\omega_0^2 - \omega^2 - i\gamma\omega\right)} \right| \tag{11.21}$$

We examine the various limits for this expression. For high frequencies of the incident radiation,[2] the rapidly driven oscillation changes the sign of the displacement so rapidly that the restoring force of the spring has little net effect:

$$|\vec{p}| = |-ex_0| = \left| \frac{e^2 E_0}{m\omega^2} \right| \tag{11.22}$$

This limit is the free-electron Thomson scattering result of Eq. 11.10. In the low-frequency, long-wavelength limit,

$$|\vec{p}| = |-ex_0| = \left| \frac{e^2 E_0}{m\omega_0^2} \right| \tag{11.23}$$

The total scattering cross section (Eq. 11.8) in this limit may be written in terms of the wavelength of the radiation and:

$$\sigma_{Ra} = \frac{\omega^4}{\omega_0^4} \sigma_{Th} = \frac{\lambda_0^4}{\lambda^4} \sigma_{Th} \tag{11.24}$$

where $\lambda_0 = 2\pi c/\omega_0$ is the wavelength corresponding to a molecular transition being modeled by the spring system. This is Rayleigh scattering. Short wavelength radiation is more strongly scattered than long

[2] Such "high frequencies" are necessarily still in the long wavelength (dipole) limit since a_o is assumed very small.

wavelength light. For the nitrogen and oxygen molecules making up most of the atmosphere, the lowest energy electronic transitions have frequencies in the deep ultraviolet: for nitrogen, the Lyman–Birge–Hopfield bands have wavelengths 140–170 nm and the molecular oxygen Schumann–Runge bands extend from 176 to 193 nm. Hence, the long wavelength limit is the appropriate one for describing the scattering of sunlight in the atmosphere. Using Eq. 11.24 to model the electronic molecular transitions, the scattering of blue light, 480 nm, is about four times that of red light, 680 nm. This is why the sky appears blue. For an optical path in the z-direction through a number density n of independent scatterers, the attenuation of light is given by:

$$I(z) = I_0 e^{-n\sigma z} \equiv I_0 e^{-z/\Lambda} \tag{11.25}$$

where we have identified the optical attenuation length $\Lambda = (n\sigma)^{-1}$. If in Eq. 11.24 we take $\lambda_0 = 150$ nm as typical of molecular nitrogen and oxygen resonance transitions and use for n the atmospheric number density at STP, we find attenuation lengths of $\Lambda_{680} = 240$ km and $\Lambda_{480} = 60$ km for red light and blue light, respectively. The decrease in pressure/number density with distance above the surface of the Earth is characterized by the atmospheric scale height. This is the height at which the pressure ($\approx$number density) is down by a factor of e. For the Earth, this value is about 8.5 km. On comparing this to the attenuation lengths, we conclude that at noon when the sun is directly overhead, for all visible (and near UV) wavelengths, sunlight intensity is attenuated very little by atmospheric scattering. This is when sunburns can occur. On the other hand, when the sun is low in the sky, at sunrise or sunset, its extended path through the atmosphere is long enough to significantly decrease the intensity of the shorter wavelength light making it to an observer and the sun appears red. Rayleigh observed that, similarly, for a complex sound in the forest, trees will scatter the second harmonic 16 times stronger than the fundamental making the back-reflected sound more brilliant and the sound transmitted through the forest duller than the original.[3]

11.2.3 Scattering near resonance

For near resonance scattering, $\omega \approx \omega_0$, we write the detuning of the driving field as $\delta\omega = \omega - \omega_0$. Eq. 11.21 may then be approximated

$$|\vec{p}| \cong \left| \frac{e^2 E_0}{2\omega_0 m} \frac{1}{(\delta\omega - i\gamma/2)} \right|. \tag{11.26}$$

[3] *The Theory of Sound, Vol II,* John William Strutt, Baron Rayleigh, MacMillan & Co, London, 1896.

giving:

$$|\vec{p}|^2 = \left(\frac{e^2 E_0}{2\omega_0 m}\right)^2 \frac{1}{\delta\omega^2 + (\gamma/2)^2} \tag{11.27}$$

and from Eq. 11.8 for the total scattering cross section:

$$\sigma = \frac{\omega_0^2 e^4}{24\pi\varepsilon_0^2 m^2 c^4} \frac{1}{\delta\omega^2 + (\gamma/2)^2}. \tag{11.28}$$

This expression gives a general expression for the elastic scattering cross section for any damping mechanism or mechanisms that are characterized collectively by the damping coefficient, γ. The lineshape described by the cross section is a "Lorentzian." We know that at the very least the emission of radiation itself is a damping mechanism. But additional loss mechanisms such as collisions with other particles in the system may contribute to the damping. For the case in which the only damping is the radiation emission, we may use Eq. 11.6, the radiation power emitted from an harmonically oscillating electric dipole, to find γ (**see Discussion 11.2**). The damping coefficient for this case is then:

$$\gamma = \frac{\omega_0^2 e^2}{6\pi\varepsilon_0 m c^3} \tag{11.29}$$

Using this in Eq. 11.28:

$$\sigma = 6\pi \left(\frac{\lambda_0}{2\pi}\right)^2 \times \frac{(\gamma/2)^2}{\delta\omega^2 + (\gamma/2)^2} \tag{11.30}$$

On resonance, this cross section is $6\pi(\lambda_0/2\pi)^2$. For an atomic system, spontaneous emission is the quantum analog of the classical dipole radiation used here.

11.2.4 Plasmon resonance

For small radii spheres in the long wavelength (dipole) regime, $a_0 \ll \lambda/2\pi$, at time t, the instantaneous electric dipole, $\vec{p}(t)$, created within a dielectric sphere by a wave's electric field, $\vec{E}(t)$, is well-approximated by the static and uniform polarization of the sphere in a constant electric field of the same strength. This is given by the well-known expression[4] for the dipole moment induced by an electric field, E_0, within a dielectric sphere with permittivity ε_2 embedded in a medium with permittivity ε_1:

[4] Jackson, J. D., *Classical Electrodynamics*, 3rd edition, Wiley, NY., 1999, Eq. 4.56.

$$p_0 = 4\pi \left(\frac{\varepsilon_2 - \varepsilon_1}{\varepsilon_2 + 2\varepsilon_1} \right) a_0^3 E_0 \tag{11.31}$$

The direction of the dipole moment is that of the electric field's polarization. Using this result in Eq. 11.8 predicts the total cross section for a light wave scattered from a small dielectric sphere to be[5]

$$\sigma = \frac{8}{3} \left(\frac{2\pi a_0}{\lambda_1} \right)^4 \left(\frac{\varepsilon_2 - \varepsilon_1}{\varepsilon_2 + 2\varepsilon_1} \right)^2 \pi a_0^2 \tag{11.32}$$

An especially interesting case is for real conductors whose complex permittivity frequently has a negative real part. This is most familiar in the Drude expression for the complex permittivity of a metal as derived in Section 10.3 "Drude metals and plasmas,"

$$\tilde{\varepsilon} = \varepsilon_\infty - \frac{\omega_p^2}{\omega(\omega + i\gamma)} \varepsilon_0 \tag{11.33}$$

Here, ε_∞ is the conductor's permittivity at infinite frequency (i.e., the core response to the field as described in Section 10.1), ω_p is the plasma frequency of the conductor, γ is its damping constant, and ω is the electromagnetic radiation frequency. For γ small, and it usually is, Eq. 11.32 predicts a resonance in the scattering cross section of a small conducting sphere at a frequency[6]

$$\omega = \sqrt{\frac{\varepsilon_0}{2\varepsilon_1 + \varepsilon_\infty}} \omega_p \tag{11.34}$$

where ε_1 is the permittivity of the medium in which the small conducting sphere is embedded. This resonance is known as a surface plasmon. Referring to Fig. 11.3, the origin of the resonance is easily understood as arising from the restoring force set up by the polarization of the sphere.

For silver spheres in water, $\omega_p = 1.4 \cdot 10^{16} s^{-1}$, $\gamma = 3.2 \cdot 10^{13} s^{-1}$, and the permittivity of water is $1.77\varepsilon_0$. In the standard Drude model, ε_∞ is taken equal to ε_0, giving a surface plasmon resonance for light of wavelength 290 nm. The Drude model for silver is somewhat unreliable in this region of the spectrum due to interband transitions but may be modified by taking as a constant offset $\varepsilon_\infty \sim 5\varepsilon_0$ to empirically account for those transitions.[7] Doing so predicts a resonance for silver spheres in water at about 390 nm and this is reasonably consistent with observations. Alternatively, the bottom part of Fig. 11.3 shows a calculation based on direct measurements of the dielectric constant.

[5] Replacing $\varepsilon_0 \to \varepsilon_1$ and $c^4 \to (c/n_1)^4$ so Eq. 11.8 for dipole scattering within a medium with permittivity ε_1, permeability μ_1 and index n_1 can be written:

$$\sigma = \frac{1}{6\pi\varepsilon_1^2} \left(\frac{\omega n}{c} \right)^4 \frac{|\vec{p}|^2}{E_0^2}$$

Inserting Eq. 11.31 into this

$$\begin{aligned} \sigma &= \frac{1}{6\pi\varepsilon_1^2} \left(\frac{\omega n}{c} \right)^4 16\pi^2 \left(\frac{\varepsilon_2 - \varepsilon_1}{\varepsilon_2 + 2\varepsilon_1} \right)^2 a_0^6 \\ &= \frac{8}{3\varepsilon_1^2} \left(\frac{2\pi a_0}{\lambda_1} \right)^4 \left(\frac{\varepsilon_2 - \varepsilon_1}{\varepsilon_2 + 2\varepsilon_1} \right)^2 \pi a_0^2 \end{aligned}$$

[6] Equation 11.34 is obtained with ε_2 within Eq. 11.32 given by Eq. 11.33. The denominator of Eq. 11.32 is then set to zero and solved for ω.

[7] Cai, W. and V. Shalaev, V., *Optical Metamaterials*, Springer, New York (2010).

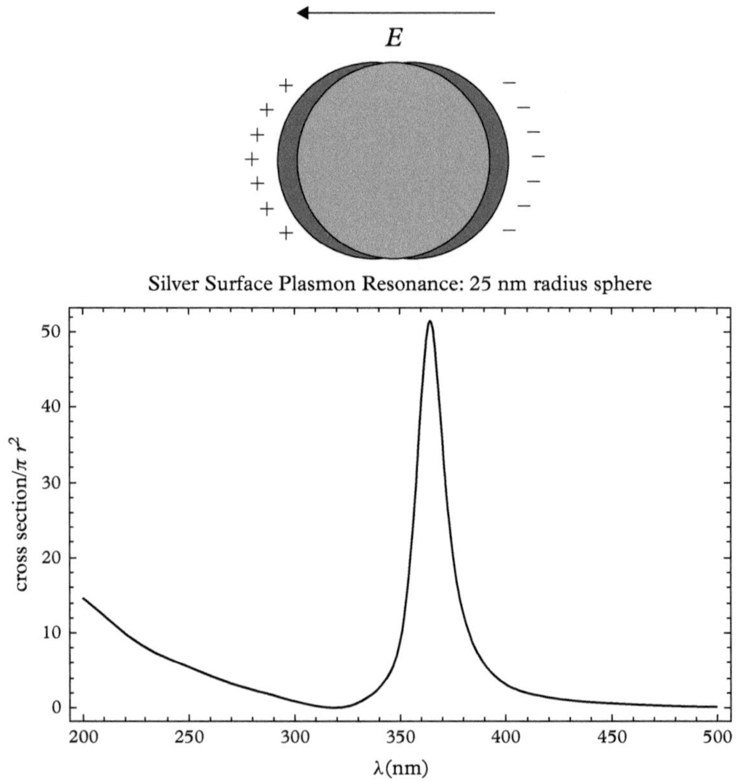

Fig. 11.3 *Physics behind the surface plasmon resonance of a nanosphere. The electric field from an incident wave instantaneously drives a displacement of electrons to one side of the sphere as shown in the top figure. The attraction of the charges on opposite sides of the sphere is a restoring force. The surface plasmon is the resonance of that system. The bottom figure is the plasmon cross section, normalized to the geometric value, for scattering off silver spheres in water, showing a plasmon resonance.*

Note that the scattering cross section is over an order of magnitude larger than the sphere's geometric cross section.

While the overall scaling of the cross section depends on the size of the sphere, the relative wavelength dependence of the resonance does not as long as the sphere is small compared to the wavelength. In determining the wavelength of the resonance, what is important is the shape of the scatterers. The frequency of the resonance can be engineered by modifying either the shape of the nanoparticles or their composition. Conducting rods and spheroids exhibit well defined resonances at frequencies different from their spherical counterparts. Again, simple pictorial-based models serve to predict the resonant frequencies.

The best known application of surface plasmon resonances is the Lycurgus cup in Fig. 11.4, "perhaps made in Alexandria" about 300 AD. The story goes that King Lycurgus tried to kill Ambrosia, a follower of Dionysus. And, as shown on the cup, having been turned

Fig. 11.4 *The Lycurgus cup: When lit from the front, the glass appears green; when lit from the top, the transmitted light is red. Its red color results from a surface plasmon resonance in light scattering off silver-gold alloy nanospheres embedded in the glass of the cup. [Credit: Johnbod (own work, Wikimedia commons) British Museum]*

into a vine, she entwines the enraged king and ultimately kills him. The king turns a red color on transmitted light and this is attributed to the surface plasmon resonance of ~70 nm particles embedded in the glass. The nanoparticles consist of an alloy of silver with a very small amount of gold.

11.3 Integral equations, the Born approximation and optical theorem

A scattering target may be described as a local variation in the permittivity and/or permeability of an otherwise homogeneous material. We assume harmonic fields $\sim \exp(-i\omega t)$ so Maxwell's equations, allowing for spatial variations in the permittivity, $\varepsilon(\vec{r})$, and magnetic permiability, $\mu(\vec{r})$, are:

$$\nabla \cdot \vec{D} = 0 \tag{11.35}$$

$$\nabla \times \vec{E} - i\omega \vec{B} = 0 \tag{11.36}$$

$$\nabla \cdot \vec{B} = 0 \tag{11.37}$$

$$\nabla \times \vec{H} + i\omega \vec{D} = 0 \tag{11.38}$$

where:

$$\vec{D} = \varepsilon(\vec{r})\vec{E} \qquad \vec{B} = \mu(\vec{r})\vec{H} \tag{11.39}$$

Here, we assume no bare charges or free currents (we will later include currents via a complex permittivity). Carrying out the usual operations for obtaining the Helmholtz equation for the electric field from these Maxwell's equations, but this time allowing for variations in the material parameters, yields[8]

$$\nabla^2\vec{E} + n^2(\vec{r})k_0^2\vec{E} = -\nabla\left[\vec{E}\cdot\nabla(\ln\varepsilon_r)\right] - i\omega\left(\nabla\mu\times\vec{H}\right) \tag{11.40}$$

where $k_0 \equiv \omega/c$ and we have identified the index of refraction, $n = c\sqrt{\varepsilon\mu}$ and the dielectric constant, $\varepsilon_r = \varepsilon/\varepsilon_o$, which, for this problem, are functions of position. Note that only the right-hand side of this equation couples the different Cartesian coordinates. At optical frequencies, most materials are non-magnetic, $\mu = \mu_0$, and the second term on the right-hand side is negligible. The effects of the other right-hand side term's mixing of polarizations is generally significant at material boundaries or for small particles in the long wave limit, discussed in the previous section. On the other hand, if the dielectric constant does not vary rapidly over distances on the order of a wavelength, then the problem reduces to solving three separate equations for scalar wave functions, one for each Cartesian coordinate. Scalar field scattering theory should be adequate. If we identify $\varphi(\vec{r})$ as one field component, the equation it satisfies is:

$$\nabla^2\varphi + n^2(\vec{r})k_0^2\varphi = 0 \tag{11.41}$$

If we consider the scattering target to be a variation in the dielectric constant localized at the origin and embedded in an otherwise homogeneous environment, we can pull out the spatially varying part:

$$n^2(\vec{r}) = n_1^2 + \delta n^2(\vec{r}) \tag{11.42}$$

where n_1 is the index of the embedding medium. Eq. 11.41 may then be written:

$$-\nabla^2\varphi + U(\vec{r})\varphi = k_1^2\varphi \tag{11.43}$$

Here, $k_1 = n_1 k_0$ is the magnitude of the wavevector in the embedding medium and we have identified $U(\vec{r}) = -\delta n^2(\vec{r})k_0^2$ as a potential energy term. This is exactly the time-independent Schrodinger

[8] Taking the curl of the second Maxwell's Eq. 11.36:

$$\nabla\times\nabla\times\vec{E} = i\omega\left[\nabla\times\left(\mu\vec{H}\right)\right]$$
$$= i\omega\left(\mu\nabla\times\vec{H} + \nabla\mu\times\vec{H}\right)$$
$$\left[\nabla\left(\nabla\cdot\vec{E}\right) - \nabla^2\vec{E}\right]$$
$$= i\omega\left[\mu\left(-i\omega\vec{D}\right) + \nabla\mu\times\vec{H}\right]$$

Using the first Maxwell's Eq. 11.35 and inserting the constitutive relation, Eq. 11.39, we get $\nabla\cdot\left(\varepsilon\vec{E}\right) = 0 \Rightarrow \nabla\cdot\vec{E} = -\vec{E}\cdot\nabla\ln\varepsilon_r$. Rearrange the terms in this and use the constitutive relations.

equation for a scalar particle[9] with $\frac{\hbar^2}{2m} \to 1$ and $E \to k_1^2$. For a dielectric sphere embedded in vacuum, $U(\vec{r}) \to -\delta n^2(\vec{r})\,k_0^2$ acts as an attractive potential.[10]

11.3.1 Scalar theory

The Born approximation

The Green function method for solving the wave equation may be used to formulate a general approach to solving scattering problems: the Born approximation. In what follows, we will drop the subscript "1" in the notation used before to identify the scattering medium with the understanding that the wave-vector and related quantities refer to the medium surrounding the scattering. Recall that the normalized outgoing spherical wave:

$$G_F\left(\vec{R}\right) = \frac{e^{ikR}}{4\pi R} \tag{11.44}$$

is the Green function for the inhomogeneous, or driven, Helmholtz equation:

$$\left(\nabla^2 + k^2\right)\psi(\vec{r}) = -F(\vec{r}) \tag{11.45}$$

with the outgoing wave boundary condition. Specifically, this means a general solution to this equation is given by:

$$\psi(\vec{r}) = \psi_0(\vec{r}) + \int dV' F\left(\vec{r'}\right) \frac{\exp\left(ik\left|\vec{r}-\vec{r'}\right|\right)}{4\pi\left|\vec{r}-\vec{r'}\right|} \tag{11.46}$$

where $\psi_0(\vec{r})$ is a solution to the homogeneous Helmholtz equation. If we apply this to Eq. 11.43, we have:

$$\varphi(\vec{r}) = e^{ikz} - \int dV' U\left(\vec{r'}\right)\varphi\left(\vec{r'}\right) \frac{\exp\left(ik\left|\vec{r}-\vec{r'}\right|\right)}{4\pi\left|\vec{r}-\vec{r'}\right|} \tag{11.47}$$

where $\varphi_0(\vec{r}) = \exp(ikz)$. The integral here has contributions only from the region localized about the origin for which the dielectric function varies, $\delta n^2(\vec{r})$, from that of the embedding material or vacuum. If we take $\varphi_0 = \exp(ikz)$ to be the incident plane wave for the scattering problem, then $\varphi(\vec{r})$ is indeed the solution with the proper scattering boundary conditions. To evaluate the integral in this expression for

[9] As the wave passes into the target, n^2 increases by δn^2, $U = -\delta n^2(\vec{r})k_0^2$ thus decreases, and the wavevector in the target (from $-\nabla^2\varphi$) increases, resulting in a shorter wavelength within the (higher index) target - as expected.

[10] Note that electromagnetism and quantum texts usually differ by the sign of the potential and in turn differ on signs for the Green function definitions used in these sorts of problems.

$r \gg r'$, we make the approximation $\left|\vec{r}-\vec{r'}\right| \simeq r - r' \cos\vartheta$ and the exponential function can then be written,

$$\exp\left(ik\left|\vec{r}-\vec{r'}\right|\right) \simeq \exp(ikr)\exp\left[-ikr' \cos\vartheta\left(\hat{i}_r, \hat{i}_{r'}\right)\right] \tag{11.48}$$

where $\hat{i}_r$ and $\hat{i}_{r'}$ are the directions of observation and of the volume element in the integrand and ϑ is the angle between them. Asymptotically, then, Eq. 11.47 is an integral expression for the scattering wave function:

$$\varphi(\vec{r}) = e^{ikz} - \frac{e^{ikr}}{4\pi r}\left\{\int dV' U\left(\vec{r'}\right)\varphi\left(\vec{r'}\right)\exp\left[-ikr' \cos\vartheta\left(\hat{i}_r, \hat{i}_{r'}\right)\right]\right\} \tag{11.49}$$

where, upon comparison with Eq. 11.3, we see that the term in braces is the scattering amplitude. If the scattering is weak, in the sense that in the volume of the target the total wave function required in the integral is dominated by the incident wave function, then we can approximate the former by the latter:

$$\varphi_B(\vec{r}) = e^{ikz} - \frac{e^{ikr}}{r}\left\{\frac{1}{4\pi}\int dV' U\left(\vec{r'}\right)e^{ikz'}\exp\left[-ikr' \cos\vartheta\left(\hat{i}_r, \hat{i}_{r'}\right)\right]\right\} \tag{11.50}$$

This is the Born approximation or the "first order" Born approximation. Comparing this expression with Eq. 11.3, the expression in braces is the scattering amplitude, $f(\theta,\phi)$, in the direction of $\hat{i}_r$, and its magnitude squared is the differential cross section in that direction. It is possible to iterate this process by using $\varphi_B(\vec{r})$, obtained here, as an improved estimate for the wave function in Eq. 11.47, thereby producing the "second-order" Born approximation. Successive iterations of this process are possible but not necessarily useful. Unless the series converges rapidly, the results are unreliable.

The optical theorem

Starting from Eq. 11.43, we may derive a useful result that relates the total extinction cross section to the imaginary part of the forward scattering amplitude, $f(\theta = 0)$. Of course, this is a scalar for a scalar field. We begin by multiplying Eq. 11.43 by φ^*, multiplying the conjugate of Eq. 11.43 by φ, and subtracting the latter from the former:

$$\varphi^* \nabla^2 \varphi - \varphi \nabla^2 \varphi^* = 2i|\varphi|^2 \,\mathrm{Im} U \tag{11.51}$$

On integrating this equation over the volume of a very large sphere of radius R and applying the divergence theorem to the left-hand side, we have:[11]

$$\oint_R \mathrm{Im}\left[(\varphi^* \nabla\varphi)\cdot\hat{r}\right] dS = \int_R \left|\varphi^2\right| \mathrm{Im}U\, dV \tag{11.52}$$

where $\hat{r}$ is a unit vector in the radial direction and is thus perpendicular at each point on the large sphere's surface. An imaginary part of the scatterer's effective potential energy corresponds to absorption. It can be shown that the integral on the right-hand side corresponds to the total power absorbed by the scatterer. However, here we assume that the potential is real, the scattering is lossless, and there is no absorption. We take φ to be a solution to Eq. 11.43 satisfying scattering boundary conditions:

$$\varphi = e^{ikz} + \frac{e^{ikr}}{r} f(\theta,\phi) \tag{11.53}$$

The radial unit vector in the brackets on the left-hand side of Eq. 11.52 means we need consider only the radial derivatives of the wave function and thus we can write the integral over the sphere's surface (**see Discussion 11.3**)

$$\begin{aligned} &R^2 \oint_R \mathrm{Im}\left[(\varphi^*\nabla\varphi)\cdot\hat{r}\right] d\Omega \\ &= R^2 \mathrm{Im} \oint_R \left\{ ik\cos\theta + \left(ik\frac{e^{ikR-ikR\cos\theta}}{R} - \frac{e^{ikR-ikR\cos\theta}}{R^2} \right) f(\theta,\phi) \right. \\ &\left. + \left(ik\cos\theta \frac{e^{ikR\cos\theta-ikR}}{R} \right) f^*(\theta,\phi) + \left(\frac{ik}{R^2} - \frac{1}{R^3} \right) \left| f^2(\theta,\phi) \right| \right\} \sin\theta\, d\theta\, d\phi \end{aligned} \tag{11.54}$$

The angular integral over the first term in the braces on the right-hand side is zero. Furthermore, in the limit $R \to \infty$, $R \gg \lambda = 2\pi/k$ and Eq. 11.54 therefore simplifies to

$$\begin{aligned} &R^2 \oint_R \mathrm{Im}\left[(\varphi^*\nabla\varphi)\cdot\hat{r}\right] d\Omega \\ &\simeq R^2 \mathrm{Im} \oint_R ik \left\{ \left(\frac{e^{ikR-ikR\cos\theta}}{R} \right) f + \left(\cos\theta \frac{e^{ikR\cos\theta-ikR}}{R} \right) f^* + \frac{\left|f^2\right|}{R^2} \right\} \sin\theta\, d\theta\, d\phi \end{aligned} \tag{11.55}$$

[11] To prepare to apply the divergence theorem, we can rewrite the left-hand side of Eq. 11.51 as

$$\nabla\cdot(\varphi^*\nabla\varphi) - \nabla\cdot(\varphi\nabla\varphi^*)$$

and because for a complex number, $z - z^* = 2iIm(z)$, we have

$$2iIm\left[\nabla\cdot(\varphi^*\nabla\varphi)\right]$$

and so Eq. 11.51 can be written

$$Im\left[\nabla\cdot(\varphi^*\nabla\varphi)\right] = |\varphi|^2\,\mathrm{Im}U.$$

Now consider the contribution from the R^{-1} terms in Eq. 11.55:

$$I_1 = \mathrm{Im}\left[ikRe^{ikR}\int_0^{2\pi} d\phi \int_0^{\pi} \left(f(\theta,\phi)\sin\theta e^{-ikR\cos\theta}\right) d\theta\right]$$

$$I_2 = \mathrm{Im}\left[ikRe^{-ikR}\int_0^{2\pi} d\phi \int_0^{\pi} \left(f^*(\theta,\phi)\cos\theta\sin\theta e^{ikR\cos\theta}\right) d\theta\right]$$

For $R \to \infty$, these are similar to integrals evaluated using the stationary phase method and we use a variation of that method here. Note, there are increasingly rapid oscillations of the exponential in the integrand as $R \to \infty$ and these cause cancellation for all contributions of the integrand everywhere except where the exponent is stationary–that is, where has a vanishing first derivative. Specifically, there are significant contributions in the vicinities of $\theta = 0$ and $\theta = \pi$. We first consider the $\theta = 0$ contribution to I_1. In the spirit of the stationary phase method we carry out an expansion of the integrand for small θ (**see Discussion 11.4**)

$$\begin{aligned} I_1(\theta = 0) &\simeq \mathrm{Im}\left[ikRf(0,\phi)\int_0^{2\pi} d\phi \int_0^{\pi} \exp\left(\frac{ikR}{2}\theta^2\right)\theta d\theta\right] \\ &= \mathrm{Im}\left[2\pi ikRf(0,\phi)\frac{1}{2}\int_0^{\pi^2} \exp\left(\frac{ikR}{2}x\right)dx\right] \\ &= \mathrm{Im}\left[2\pi ikRf(0,\phi)\frac{1}{ikR}\left(e^{ikR\pi^2/2} - 1\right)\right] \end{aligned} \tag{11.56}$$

where we have assumed that $f(\theta,\phi)$ is symmetric about ϕ and have made the substitution $x = \theta^2$. If we recognize that the wavevector has an infinitesimally small imaginary part corresponding to the small assumed loss, then the exponential term vanishes for $R \to \infty$[12]

$$I_1(\theta = 0) \simeq \mathrm{Im}\left[-2\pi ikRf(0,\phi)\frac{1}{ikR}\right] = -2\pi \mathrm{Im}f(0) \tag{11.57}$$

By a similar argument, the contribution to I_2 about $\theta = 0$ is identical[13]

$$I_1(\theta = 0) + I_2(\theta = 0) = -2\pi \mathrm{Im}\left[f(0) - f^*(0)\right] = -4\pi \mathrm{Im}f(0) \tag{11.58}$$

The two contributions about the second stationary point, $I_1(\theta = \pi)$ and $I_2(\theta = \pi)$, combine to cancel out.[14] The contribution of the third term, I_3, in the brackets of Eq. 11.55 is:

[12] If we let $k \to k + i\varepsilon$ within the term in parentheses of Eq. 11.56,

$$\exp\left(\frac{ikR}{2}\pi^2\right) - 1 \to$$

$$\exp\left(\frac{ikR}{2}\pi^2\right)\exp\left(-\frac{\varepsilon R}{2}\pi^2\right) - 1$$

and then we take the limit as $R \to \infty$,

$$\exp\left(\frac{ikR}{2}\pi^2\right)\exp\left(-\frac{\varepsilon R}{2}\pi^2\right) - 1 \to -1$$

so that even an infinitesimally small imaginary component to the wavevector will cause the whole exponential term to vanish.

[13] The integrand of the second term in Eq. 11.55 is $\cos\theta$ multiplied by the complex conjugate of the integrand of I_1 and so when evaluated about $\theta = 0$, $\cos\theta \simeq 1$ and the contribution of I_2 is the same as that of I_1.

[14] This is most easily seen by changing variables to $\theta' = \theta - \pi$, turning the calculation about $\theta = \pi$ into a calculation about $\theta' = 0$ which is evaluated similar to the previous. The difference is that the $\cos\theta'$ term acquires a minus sign so that the I_1 and I_2 contributions cancel at $\theta = \pi$.

$$I_3 = k \oint_R d\Omega \left| f^2(\theta, \phi) \right| = k \oint_R d\Omega \frac{d\sigma}{d\Omega} = k\sigma_{scat} \quad (11.59)$$

where we have used the relation of Eq. 11.5. Collecting these results into Eq. 11.52, that is, $I_1 + I_2 + I_3 = 0$, yields the optical theorem:

$$\sigma_{scat} = \frac{4\pi}{k} \mathrm{Im} f(0) \quad (11.60)$$

Although not shown here, for a lossy scatterer, the cross section for absorption is given by the right-hand side of Eq. 11.52. More generally, it is the extinction cross section that is given by the optical theorem

$$\sigma_{ext} = \sigma_{scat} + \sigma_{abs} = \frac{4\pi}{k} \mathrm{Im} f(0) \quad (11.61)$$

In addition to being fundamentally important by giving a physically measurable meaning to the complex scattering amplitude, at least in the forward direction, the optical theorem is a convenient way to find total cross sections. This circumvents having to calculate differential cross sections for all angles and integrating.

11.3.2 Vector theory

Vector integral equation

We start with the source-free Maxwell's equation in matter:

$$\begin{aligned} \nabla \cdot \vec{D} &= 0 \\ \nabla \times \vec{E} &= -\frac{\partial \vec{B}}{\partial t} \\ \nabla \cdot \vec{B} &= 0 \\ \nabla \times \vec{H} &= \frac{\partial \vec{D}}{\partial t} \end{aligned}$$

with the constitutive relations $\vec{D} = \varepsilon(\vec{r})\vec{E}$ and $\vec{B} = \mu(\vec{r})\vec{H}$. This time, it is convenient to work with the electric displacement field instead of the electric field. The permittivity and permeability are assumed constant over most space, having the values, ε_1 and μ_1 characteristic of the medium in which the system is embedded. In the region near the origin, they may vary with position, $\varepsilon(\vec{r}) = \varepsilon_1 + \delta\varepsilon(\vec{r})$ and $\mu(\vec{r}) = \mu_1 + \delta\mu(\vec{r})$, where the fractional variations of the two terms drives the scattering. The scattering problem requires finding the solutions of the inhomogeneous Helmholtz equation with the asymptotic form:

$$\vec{D}_{tot} = \vec{D}_{inc} + \vec{D}_{scat} = \hat{\varepsilon} e^{ikz} + \vec{f}(\hat{\varepsilon}, \theta, \phi) \frac{e^{ikr}}{r} \tag{11.62}$$

Following the treatment at the beginning of this section for the electric field, wave equations for the fields may be derived that, on assuming a time variation of $\sim e^{-i\omega t}$, lead to the Helmholtz equation for the displacement field

$$\left(\nabla^2 + k^2\right) \vec{D} = -\nabla \times \nabla \times \left[\delta\varepsilon\left(\vec{r}\right) \cdot \vec{E}\right] - i\varepsilon_1 \omega \nabla \times \left[\delta\mu\left(\vec{r}\right) \cdot \vec{H}\right] \tag{11.63}$$

where $k = \omega\sqrt{\varepsilon_1 \mu_1}$. Now this may be read as a separate Helmholtz equation for each component in the vector equation. If we knew the solution for the fields, we could treat the right-hand side of this equation as a driving function. That is, a formal solution of the equation may obtained by using an Eq. 11.46 for each component:

$$\begin{aligned} \vec{D}\left(\vec{r}\right) = {} & \vec{D}_0\left(\vec{r}\right) \\ & + \int dV' \frac{\exp\left(ik\left|\vec{r} - \vec{r}'\right|\right)}{4\pi\left|\vec{r} - \vec{r}'\right|} \left\{\nabla' \times \nabla' \times \left[\delta\varepsilon\left(\vec{r}'\right) \cdot \vec{E}\left(\vec{r}'\right)\right] \right. \\ & \left. + i\varepsilon_1 \omega \nabla' \times \left[\delta\mu\left(\vec{r}'\right) \cdot \vec{H}\left(\vec{r}'\right)\right]\right\} \end{aligned} \tag{11.64}$$

where $\vec{D}_0$ is a solution of the homogeneous equation.

The Born approximation

The asymptotic solution for the displacement field for the scattering of an electromagnetic wave is:

$$\vec{D}\left(\vec{r}\right) = \vec{D}_{inc}\left(\vec{r}\right) + \vec{D}_{sc}\left(\vec{r}\right) = \vec{D}_{inc}\left(\vec{r}\right) + \vec{f}_{sc} \frac{e^{ikr}}{r} \tag{11.65}$$

If $\vec{D}_0\left(\vec{r}\right)$ in Eq. 11.64 is taken to be an incident plane wave with polarization unit vector, $\hat{\varepsilon}$, then $\vec{D}_0\left(\vec{r}\right) = \vec{D}_{inc}\left(\vec{r}\right) = \hat{\varepsilon} \exp(ikz)$ and the calculated displacement field is the scattering solution for $r \to \infty$. Expanding the exponential using the approximation $\left|\vec{r} - \vec{r}'\right| \simeq r - \hat{r} \cdot \vec{r}'$, as in the scalar case of Eq. 11.48, the scattering amplitude in the direction specified by the unit vector $\hat{r}(\theta, \phi)$ is identified as:

$$\begin{aligned} \vec{f}_{sc}(\theta, \phi) = {} & \frac{1}{4\pi} \int dV' \exp\left(-ik\hat{r} \cdot \vec{r}'\right) \left\{\nabla' \times \nabla' \times \left[\delta\varepsilon\left(\vec{r}'\right) \cdot \vec{E}\left(\vec{r}'\right)\right] \right. \\ & \left. + i\varepsilon_1 \omega \nabla' \times \left[\delta\mu\left(\vec{r}'\right) \cdot \vec{H}\left(\vec{r}'\right)\right]\right\} \end{aligned} \tag{11.66}$$

Integrating this by parts (**see Discussion 11.5**)

$$\vec{f}_{sc}\left(\hat{r} \leftrightarrow \theta, \phi\right) = \frac{1}{4\pi} \int dV' \exp\left(-ik\hat{r} \cdot \vec{r'}\right) \left\{k^2 \hat{r'} \times \left[\delta\varepsilon\left(\vec{r'}\right) \cdot \vec{E}\left(\vec{r'}\right)\right] \times \hat{r}' \right.$$
$$\left. - \varepsilon_1 \omega k \hat{r'} \times \left[\delta\mu\left(\vec{r'}\right) \cdot \vec{H}\left(\vec{r'}\right)\right]\right\} \qquad (11.67)$$

The first order Born approximation is found by approximating the fields, actual fields in the problem, $\vec{E}\left(\vec{r'}\right)$ and $\vec{H}\left(\vec{r'}\right)$, by the values of the incident fields,

$$\vec{E}\left(\vec{r'}\right) \simeq \frac{\hat{\varepsilon}}{\varepsilon\left(\vec{r'}\right)} \exp\left(ikz'\right) \qquad \vec{H}\left(\vec{r'}\right) \simeq -\frac{i}{\mu\left(\vec{r'}\right)\omega} \nabla' \times \vec{E}\left(\vec{r'}\right) \qquad (11.68)$$

As in the scalar case, higher order Born approximations may be obtained by iterating Eq. 11.64.

Mathematical machinery for vector scattering

In generalizing the scalar results to real 3D electromagnetic field scattering, it is useful to introduce a two component object for very generally describing the two polarization degrees of freedom of an electromagnetic wave. The object is sort of in-between a scalar and a 3D vector and so far as notation, it is similar to that often used in quantum mechanics to keep track of the spin components of a wave function. We will use the Fraktur font for these.[15] As an example, consider expressing the electric field of a wave moving in the $+z$ direction at a particular point in space. Using a linear polarization basis of $\hat{x}$ and $\hat{y}$, we would specify this field:

$$\vec{\mathfrak{E}}_{lin} = \begin{bmatrix} E_x \\ E_y \end{bmatrix} \qquad (11.69)$$

To recover the Cartesian coordinate vector form of this field we multiply it by a 3×2 matrix, $\overleftrightarrow{I}_{lin}$, that has the form:

$$\overleftrightarrow{I}_{lin} = \begin{bmatrix} 1 & 0 \\ 0 & 1 \\ 0 & 0 \end{bmatrix} \Leftrightarrow \vec{E} = \overleftrightarrow{I}_{lin} \cdot \vec{\mathfrak{E}}_{lin} \qquad (11.70)$$

Alternatively, we could write the same vector in terms of a circular polarization basis, $\hat{e}_{\perp}^{\pm}\left(k\hat{z}\right) = \frac{1}{\sqrt{2}}(\hat{x} \pm i\hat{y})$, for the given propagation

[15] Here we are following the approach and notation of Newton, R., *Scattering Theory of Waves and Particles*, McGraw-Hill Inc., New York (1966).

vector, $k\hat{z}$, by setting $E^+ = \frac{1}{\sqrt{2}}\left(E_x - iE_y\right)$ and $E^- = \frac{1}{\sqrt{2}}\left(E_x + iE_y\right)$ and now express the field,

$$\vec{E} = \hat{e}_\perp^+ \left(k\hat{z}\right) E^+ + \hat{e}_\perp^- \left(k\hat{z}\right) E^- = \overleftrightarrow{I}_{circ}\left(k\right) \cdot \vec{\mathfrak{E}}_{circ} \tag{11.71}$$

where here,

$$\vec{\mathfrak{E}}_{circ} = \begin{bmatrix} E^+ \\ E^- \end{bmatrix} \tag{11.72}$$

Now $\overleftrightarrow{I}_{circ}$ is a 3×2 matrix made up of the unit vector components of the circular polarization basis. For example, in Cartesian coordinates:

$$\overleftrightarrow{I}_{circ}\left(\vec{k}\right) = \begin{bmatrix} e_x^+ & e_x^- \\ e_y^+ & e_y^- \\ e_z^+ & e_z^- \end{bmatrix} \tag{11.73}$$

And in this example, for $\vec{k}$ in the z direction:

$$\overleftrightarrow{I}_{circ}\left(k\hat{z}\right) = \frac{1}{\sqrt{2}}\begin{bmatrix} 1 & 1 \\ i & -i \\ 0 & 0 \end{bmatrix} \tag{11.74}$$

Further, we could specify an x-polarized field of magnitude E_0 in this circular basis by having equal components of $\hat{e}_\perp^+$ and $\hat{e}_\perp^-$, each of magnitude $E_0/\sqrt{2}$:

$$\vec{\mathfrak{E}}_{circ} = \frac{1}{\sqrt{2}}\begin{bmatrix} E_0 \\ E_0 \end{bmatrix} = \begin{bmatrix} 1/\sqrt{2} \\ 1/\sqrt{2} \end{bmatrix} E_0 = E_0 \hat{\mathfrak{E}}_{circ}\left(x\right) \tag{11.75}$$

where here, the notation $\hat{\mathfrak{E}}_{circ}\left(x\right)$ represents a unit vector in the x direction. Normally, we omit the "circ" subscript in which case $\hat{\mathfrak{E}}\left(x\right)$ specifies a unit vector in the x direction in whatever polarization basis we have agreed upon. On the one hand, in the present case one could not use this basis to represent a vector in the z direction. On the other, if the wavevector itself is in the z direction, the notation automatically includes only allowed polarizations. For the case of an incident wave with wavevector $\vec{k}$, scattering into a final direction $\vec{k}'$, it is frequently convenient to use the plane defined by $\vec{k} \times \vec{k}'$ and take as basis vectors the linear polarizations for the $\parallel$ (= in-plane, "p") and $\perp$ (= out-of-plane, "s") directions. Note that since the incident and scattered wavevectors are different, the basis vectors for the $\vec{k}$-directed incident wave will be different than those for the scattered wave in the $\vec{k}'$

direction. For the scattering problem, we can write the scattering wave function in the asymptotic region away from the scatterer, Eq. 11.2, using the polarization objects. For a particular incident polarization, for example linearly polarized in the x direction, we write:

$$\begin{aligned}\vec{E}_{tot}(\vec{r}) &= \vec{E}_{inc}(\vec{r}) + \vec{E}_{scat}(\vec{r})\\ &= \overleftrightarrow{I}_{inc}\left(\vec{k}\right)\cdot\hat{\mathfrak{E}}_{inc}(x)\,e^{i\vec{k}\cdot\vec{r}} + \overleftrightarrow{I}_{scat}\left(\vec{k}'\right)\cdot\vec{\mathfrak{E}}_{scat}\left(x,\vec{k}'\right)\frac{e^{ikr}}{r}\end{aligned} \quad (11.76)$$

where, referring back to Eq. 11.2, we see that $\hat{\varepsilon} = \hat{x} \Rightarrow \overleftrightarrow{I}_{inc}\left(\vec{k}\right)\cdot\hat{\mathfrak{E}}_{inc}(x)$ and $\vec{f}(\hat{\varepsilon},\theta,\phi) \Rightarrow \overleftrightarrow{I}_{scat}\left(\vec{k}'\right)\cdot\vec{\mathfrak{E}}_{scat}\left(x,\vec{k}'\right)$. Here, $\vec{\mathfrak{E}}_{scat}(x,\vec{k}')$ has the dimensions of electric field multiplied by length. It consists of the two scattering amplitudes for scattering from the x-polarized incident beam in the $\vec{k}$ direction into the two polarizations of whatever basis is chosen for the scattered beam in the $\vec{k}'$ direction. We can generalize this result by defining the *scattering amplitude matrix*, $\overleftrightarrow{\mathscr{F}}(\vec{k},\vec{k}')$, a 2×2 matrix that describes how the two incident field polarization components of an incident field propagating in the direction of $\vec{k}$ are mapped onto the two polarization components for the field scattered in the direction of $\vec{k}'$. The scattering amplitude is defined by the relation:

$$\vec{\mathfrak{E}}_{scat}\left(\vec{k},\vec{k}'\right) = \overleftrightarrow{\mathscr{F}}\left(\vec{k},\vec{k}'\right)\cdot\hat{\mathfrak{E}}_{inc}\left(\vec{k}\right) \quad (11.77)$$

for an incident polarization. $\overleftrightarrow{\mathscr{F}}\left(\vec{k},\vec{k}'\right)$ fully describes the scattering. The individual matrix elements depend on the polarization basis chosen for the incident and scattered fields.

The optical theorem revisited with vector fields

Here we extend the *optical theorem* to vector fields by using the previous formalism in examining the scattering amplitude in the forward direction. We consider an experiment in which an x-polarized beam propagating in the z-direction is incident on a scatterer located at the origin of coordinates. *Very* far downstream at the location z, the intensity of the beam is measured on a screen that is set up perpendicular to the incident wave. The total field in this forward scattered direction is given by:

$$\begin{aligned}\vec{E}_{tot}(\vec{r}) &= \vec{E}_{inc}(\vec{r}) + \vec{E}_{scat}(\vec{r})\\ &= \left[\overleftrightarrow{I}_{inc}\left(\vec{k}\right) + \overleftrightarrow{I}_{scat}\left(\vec{k}'\right)\cdot\overleftrightarrow{\mathscr{F}}(0)\frac{e^{ik(r-z)}}{r}\right]\cdot\hat{\mathfrak{E}}_{inc}\left(\vec{k}\right)e^{ikz}\end{aligned} \quad (11.78)$$

where $\mathscr{F}(0)$ is the scattering amplitude in the forward direction. The intensity of the beam at a point on the screen is given by:

$$|S| = \frac{\left|\vec{E}_{tot}\right|^2}{2Z} = \frac{1}{2Z}\left(\vec{E}^*_{tot}\cdot\vec{E}_{tot}\right) = \frac{1}{2Z}\vec{E}^\dagger_{tot}\cdot\vec{E}_{tot} \tag{11.79}$$

where in the final equality we have used matrix notation in which the Hermitian conjugate takes the electric field, to this point expressed as column vector, to a complex coefficient row vector. For scattering in the forward direction, we can use the same polarization basis for both the incident and the scattered waves, $\overleftrightarrow{I}_{inc}=\overleftrightarrow{I}_{scat}$. Substituting in the total field[16]

$$|S| = \frac{1}{2Z}\vec{\mathfrak{E}}^\dagger_{inc}\left(\vec{k}\right)\cdot\left[1+\overleftrightarrow{\mathscr{F}}^\dagger(0)\frac{e^{-ik(r-z)}}{r}\right]\cdot\overleftrightarrow{I}^\dagger_{inc}\left(\vec{k}\right)\cdot\overleftrightarrow{I}_{inc}\left(\vec{k}\right)\cdot\left[1+\overleftrightarrow{\mathscr{F}}(0)\frac{e^{ik(r-z)}}{r}\right]\cdot\vec{\mathfrak{E}}_{inc}(k)\approx\frac{1}{2Z}\left\{\vec{\mathfrak{E}}^\dagger_{inc}\left(\vec{k}\right)\cdot\vec{\mathfrak{E}}_{inc}\left(\vec{k}\right)+\frac{2}{r}\mathrm{Re}\left[e^{+ik(r-z)}\vec{\mathfrak{E}}^\dagger_{inc}\left(\vec{k}\right)\cdot\overleftrightarrow{\mathscr{F}}(0)\cdot\vec{\mathfrak{E}}_{inc}\left(\vec{k}\right)\right]\right\} \tag{11.80}$$

Far downstream, $r\to z$ in the denominator of the second term. The exponent of the second term must be treated more carefully. It is responsible for creating a diffraction pattern downstream. There, we approximate:

$$r - z = \left(x^2+y^2+z^2\right)^{1/2} - z \simeq \frac{\left(x^2+y^2\right)}{2z} \tag{11.81}$$

The total power on the screen is found by integrating Eq. 11.79 over the surface of the screen. If the screen has area, a:

$$P = \frac{1}{2Z}\left\{\vec{\mathfrak{E}}^\dagger_{inc}\left(\vec{k}\right)\cdot\vec{\mathfrak{E}}_{inc}\left(\vec{k}\right)a + \frac{2}{z}\mathrm{Re}\left[\vec{\mathfrak{E}}^\dagger_{inc}\left(\vec{k}\right)\cdot\overleftrightarrow{\mathscr{F}}(0)\cdot\vec{\mathfrak{E}}_{inc}\left(\vec{k}\right)\int_{screen}dxdy\exp\left(ik\frac{x^2+y^2}{2z}\right)\right]\right\} \tag{11.82}$$

If the screen is large enough to receive the forward scattering but not so large that the scattering amplitude varies,[17] the integral can be evaluated by taking the limits, $x,y\to\pm\infty$. Converting the integral to cylindrical coordinates:

$$\int_{-\infty}^{+\infty}dy\int_{-\infty}^{+\infty}dx\exp\left(ik\frac{x^2+y^2}{2z}\right) = 2\pi\int_0^{+\infty}\rho d\rho\exp\left(ik\frac{\rho^2}{2z}\right) = \frac{2\pi iz}{k} \tag{11.83}$$

[16] Substituting Eq. 11.80 into Eq. 11.81 produces an equation of the form $|S|\sim\vec{\mathfrak{E}}^\dagger_{inc}\left[1+\overleftrightarrow{\mathscr{F}}^\dagger\frac{e^{-ik(r-z)}}{r}\right]\cdot\left[1+\overleftrightarrow{\mathscr{F}}\frac{e^{ik(r-z)}}{r}\right]\vec{\mathfrak{E}}_{inc}$. We note that the term $\overleftrightarrow{I}^\dagger_{inc}\left(\vec{k}\right)\cdot\overleftrightarrow{I}_{inc}\left(\vec{k}\right)$ cancels since the transpose of an orthogonal matrix is its inverse and the $\overleftrightarrow{\mathscr{F}}^\dagger\cdot\overleftrightarrow{\mathscr{F}}$ term becomes vanishingly small at infinity and is discarded.

[17] If the size of the screen $\sim d$, then we are assuming $d\ll z$.

giving for the total power on the screen:

$$P = \frac{1}{2Z}\left\{\vec{\mathfrak{E}}^{\dagger}_{inc}\left(\vec{k}\right)\cdot\vec{\mathfrak{E}}_{inc}\left(\vec{k}\right)a + \frac{4\pi}{k}\mathrm{Re}\left[i\vec{\mathfrak{E}}^{\dagger}_{inc}\left(\vec{k}\right)\cdot\overleftrightarrow{\mathcal{F}}(0)\cdot\vec{\mathfrak{E}}_{inc}\left(\vec{k}\right)\right]\right\} \tag{11.84}$$

We evaluate the second term for the case being considered: an x-polarized incident beam. Taking x and y linear polarizations for a polarization basis, the matrix multiplication in the braces gives:

$$\begin{aligned}&\begin{bmatrix} E_x^* & E_y^* \end{bmatrix}\cdot\begin{bmatrix} f_{xx} & f_{xy} \\ f_{yx} & f_{yy}\end{bmatrix}\cdot\begin{bmatrix} E_x \\ E_y \end{bmatrix}\\ &= \begin{bmatrix} E_{inc}^* & 0 \end{bmatrix}\cdot\begin{bmatrix} f_{xx} & f_{xy} \\ f_{yx} & f_{yy}\end{bmatrix}\cdot\begin{bmatrix} E_{inc} \\ 0 \end{bmatrix} = f_{xx}\left|E_{inc}^2\right|\end{aligned} \tag{11.85}$$

With this, Eq. 11.84 gives the total power incident on the screen,

$$P = \frac{\left|\vec{E}_{inc}\right|^2}{2Z}a - \frac{4\pi}{k}Im\left[f_{xx}(0)\right]\frac{\left|\vec{E}_{inc}\right|^2}{2Z} \tag{11.86}$$

The first term on the right-hand side is just the power in the incident beam that would hit the screen in the absence of the scatterer. Clearly, the second term is the *total* power removed from the beam. This must necessarily include both scattering *and* absorption by the target. We identify the "total" or "extinction" cross section as the area corresponding the power taken from the beam, which on dividing by the intensity of the incident beam gives the total scattering cross section:

$$\sigma_{tot} = \frac{4\pi}{k}\mathrm{Im}\left[f_{xx}(0)\right] \tag{11.87}$$

This proof is general in that, for circular or some other unusual polarization, we can always take the basis for the scattering matrix to be that polarization and its orthogonal polarization and so the argument goes through unmodified.

11.4 Partial wave analysis

A source of mathematical complexity in scattering theory is the mixing of the different natural coordinates of the two terms in the scattering wave function. The incident wave is most conveniently described using either Cartesian or cylindrical coordinates whereas spherical coordinates are the natural choice for the scattered wave. These are reconciled in the partial wave formalism that begins by expanding the incident plane wave in terms of spherical harmonics solutions of the Helmholtz equation. While, most generally, the technique is applicable to a wide range of systems, we restrict our consideration here to spherically symmetric targets. We will sketch the basics of the approach by working through the scalar field version of the theory and then extend it to electromagnetic vector fields.

11.4.1 Scalar theory

We recall Eq. 11.43, the wave equation governing the scalar scattering problem:

$$-\nabla^2\psi + U(r)\psi = k_1^2\psi \tag{11.88}$$

Previously, $U(r)$ derived from the variation in the dielectric constant of the material through which a light wave was passing. Here, we will keep it general with the restrictions that it is spherically symmetric and non-zero only for a confined region about the origin. We look for a solution of this equation that satisfies the scattering boundary condition: far from the origin the wave function is the sum of an incident plane wave and an outgoing spherical wave:

$$\psi(\vec{r}) = e^{ikz} + f(\theta)\frac{e^{ikr}}{r} \tag{11.89}$$

where $f(\theta)$ is the scattering amplitude: for a scalar field scattering from a spherically symmetric scatterer, the wave function can have no ϕ dependence.

Basis functions for the scalar expansion of a scattering wave function in the asymptotic region

We next identify a suitable basis for expanding the scattering wave function in the asymptotic region. Here, assuming $U \to 0$ sufficiently rapidly as $r \to \infty$, the wave equation in that region is a homogeneous Helmholtz equation and so we begin by finding solutions to the scalar Helmholtz equation in spherical coordinates.[18] First, we consider solutions of the form:

$$\psi = R(r)\,\Theta(\theta)\,\Phi(\phi) \tag{11.90}$$

The Helmholtz equation in spherical coordinates is well known and may be readily shown to be separable for this wave function: three ordinary differential equations are produced with separation variables l^2 and m^2:

[18] This is identical to the procedure used to solve central potential problems in quantum mechanics.

$$\frac{d^2\Phi}{d\phi^2}+m^2\Phi=0 \tag{11.91}$$

$$\frac{1}{\sin\theta}\frac{d}{d\theta}\left(\sin\theta\frac{d\Theta}{d\theta}\right)+\left(l^2-\frac{m^2}{\sin^2\theta}\right)\Theta=0 \tag{11.92}$$

$$r^2\frac{d^2R}{dr^2}+2r\frac{dR}{dr}+\left(k^2r^2-l^2\right)R=0 \tag{11.93}$$

The solutions for the angular variables are the spherical harmonics, $Y_{lm}(\theta,\phi)=\Theta_{lm}(\theta)\,\Phi_m(\phi)$, where $\Theta_{lm}(\theta)$, the solutions to Eq. 11.92, are the associated Legendre Polynomials, $\Theta_{lm}(\theta)=P_l^m(cos\theta)$. The solutions of the radial equations are spherical Bessel and Hankel functions in the variable $\rho=kr$ (see Appendix A). For these, the spherical Bessel functions of the first and second kind are a complete set. Importantly, the large argument asymptotic forms of these are:

$$j_l\sim\frac{\sin(\rho-l\pi/2)}{\rho}\qquad \eta_l\sim-\frac{\cos(\rho-l\pi/2)}{\rho} \tag{11.94}$$

Similarly, the spherical Hankel functions form a second complete set:

$$h_l^{\binom{1}{2}}=j_l(\rho)\pm i\eta_l(\rho)\qquad h_l^{\binom{1}{2}}\rightarrow(\mp i)^{l+1}\frac{e^{\pm i\rho}}{\rho}\quad \text{for}\ \rho\rightarrow\infty \tag{11.95}$$

Summarizing, in the asymptotic region, we can expand the scattering wave function using basis functions of the form:

$$\psi_{lm}^{(a)}=z_l^{(a)}(\rho)\,Y_{lm}(\theta,\phi) \tag{11.96}$$

where we introduce the notation $z_l^{(a)}(\rho)=R(r)$ for the radial solution to indicate some type of spherical Bessel or Hankel function, with $a=1,2,3,4$ used for j_l, η_l, $h_l^{(1)}$, $h_l^{(2)}$, respectively. Central to the partial wave approach is the expression of the incident plane wave in terms of these basis functions. We next turn to that.

Scalar spherical harmonic expansion

We know that a general solution of the Helmholtz equation in spherical coordinates is

$$\psi_0(r,\theta,\phi)=\sum_{l,m}\left[A_{lm}j_l(kr)+B_{lm}\eta_l(kr)\right]Y_{lm}(\theta,\phi) \tag{11.97}$$

We now consider the expansion for $\psi_0(r,\theta,\phi)=exp(ikz)=exp(ikr\cos\theta)$, the incident plane wave of our scattering problem. We immediately observe two simplifying characteristics of the plane wave to be expanded: first, the wave has cylindrical symmetry about its direction of travel, z, so that we need only consider $Y_{l0}(\theta)$ terms $(m=0)$ in the expansion; and second, because $exp(ikz)$ is finite at the origin and the $\eta_l(kr)$'s go to infinity there, the B_{lm}'s all vanish. We can then express Eq. 11.97 as

$$\exp(ikr\cos\theta) = \sum_{l=0}^{\infty} A_l j_l(kr)\, Y_{l0}(\theta) \tag{11.98}$$

We take advantage of the orthogonality of the $Y_{l0}(\theta)$ for different l's and attempt to obtain the (radially dependent) coefficients $A_l j_l(kr)$. Combining[19] Eq. 11.98 with the orthonormality of the $Y_{l0}(\theta)$, we arrive at expressions for the coefficients, $A_l j_l(kr)$

$$A_l j_l(kr) = \int \exp(ikr\cos\theta)\, Y_{l0}^{*}(\theta)\, d\Omega \tag{11.99}$$

which can be evaluated to find $A_l = i^l\sqrt{4\pi(2l+1)}$ (**see Discussion 11.6**). Equation 11.98 becomes

$$\exp(ikz) = \sum_{l=0}^{\infty} i^l\sqrt{4\pi(2l+1)}\, j_l(kr)\, Y_{l0}(\theta) \tag{11.100}$$

which is the Cartesian harmonic plane wave solution as an expansion of a product of scalar spherical harmonics and spherical Bessel functions. Equation 11.100 is known as the Rayleigh equation. The spherical harmonic addition theorem can be used to extend this result for expanding plane waves traveling in any direction (**see Discussion 11.7**). The partial wave index, l in quantum mechanics, is the quantum number of angular momentum associated with a partial wave. The partial wave wavefunction is an eigenstate of angular momentum with $L = \hbar\sqrt{l(l+1)}$. It serves a similar role in electromagnetism. The angular momentum of a photon with wave number k, whose trajectory has a transverse (and thus minimum) distance from the origin of r, has angular momentum magnitude $L = pr = \hbar kr$. In the top of Fig. 11.5 a spherical Bessel function of the variety that shows up in the plane wave expansion, $j_{l=100}$, is plotted as a function of $\rho = kr$. We observe that for values of ρ somewhat less than l, the radial part of the partial wave is nearly zero. This informs the convergence of the partial wave expansion for a potential with a finite range, R_0. Partial waves in the plane wave expansion significantly larger than $l \simeq \rho_0 = kR_0$ do not even know that the potential is there. Adding $\hbar$'s allows phrasing this as a quantum mechanical statement about the maximum angular momentum a photon can have and still interact with the potential, $L_{max} \equiv \hbar l_{max} = \hbar kR_0$. The connection between l and the transverse

[19] Substitute Eq. 11.98 into the rhs of Eq. 11.99 and use the orthogonality of the $Y_{l0}(\theta)$: $\int Y_{l'0}(\theta)\, Y_{l0}^{*}(\theta)\, d\Omega = \delta_{l'l}$

$$\int \left[\sum_{l'=0}^{\infty} A_{l'} j_{l'}(kr)\, Y_{l'0}(\theta)\right] Y_{l0}^{*}(\theta)\, d\Omega$$
$$= \sum_{l'=0}^{\infty} A_{l'} j_{l'} \left[\int Y_{l'0}(\theta)\, Y_{l0}^{*}(\theta)\, d\Omega\right]$$
$$= \sum_{l'=0}^{\infty} A_{l'} j_{l'}(kr)\, \delta_{l'l} = A_l j_l(kr).$$

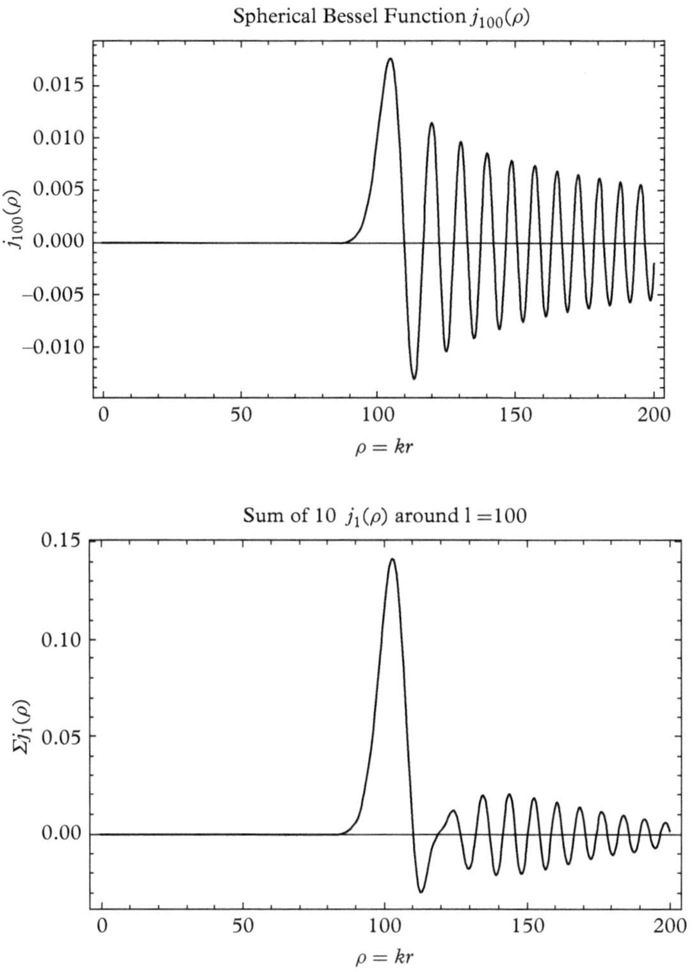

Fig. 11.5 *Top: The spherical Bessel function, $j_l(\rho)$, is essentially 0 for $\rho < l$: this gives a cutoff for the partial wave expansion. Bottom: An angular momentum wave packet created by ten partial waves about $l = 100$. Waves about a value l correspond classically to the impact parameter, b, where $kb = l$.*

distance of a photon from the origin can be further illustrated by adding together several partial waves with values around l as shown in the bottom of Fig. 11.5. Here a wave-packet is formed by the previously seen hard cutoff for ρ small, and interference among the waves for ρ large. Partial waves $\approx l$ correspond to an impact parameter $r = l/k$ in the classical scattering picture. This impact-parameter-radius is usually called "b" in classical scattering theory.

Solving the scattering problem

In the asymptotic region, the scattered wave function is, by itself, also a solution of the (homogeneous) Helmholtz equation. The Hankel functions of the first type are the clear choice[20] for expanding the scattered wave function: they have the desired outgoing spherical wave

[20] Recall $h_l^{\binom{1}{2}} = j_l(\rho) \pm i n_l(\rho) \sim (-i)^{l+1} \frac{e^{\pm i\rho}}{\rho}$. Since the scattered wave is only evaluated for $r \to \infty$, the functions blowing up at the origin are inconsequential.

asymptotic dependence. The full scattering solution in the asymptotic limit will have the form:

$$
\begin{aligned}
\psi(r \to \infty) &= e^{ikz} + f(\theta)\frac{e^{ikr}}{r} \\
&= \sum_{l=1}^{\infty}\left[\sqrt{4\pi(2l+1)}\, i^l j_l(\rho)\, Y_{l0}(\theta)\right] \\
&\quad + \sum_{l=1}^{\infty}\left[a_l\sqrt{4\pi(2l+1)}\, i^l h_l^{(1)}(\rho)\, Y_{l0}(\theta)\right]
\end{aligned} \tag{11.101}
$$

which is the desired asymptotic form. Here, the constant factors after the a_l on the scattered part of the partial wave are included to simplify the algebra in what follows. The first term in brackets produces the incident wave. The a_l expansion coefficients may be determined by matching boundary conditions in a systematic fashion as we describe next. In adding to the Helmholtz equation a spherically symmetric scattering target at the origin, $U(r)$ in Eq. 11.88, the resulting PDE is again separable. The angular equations are unchanged: only the radial equations, Eq. 11.93, are altered by the addition of a potential term:

$$
r^2\frac{d^2R_l}{dr^2} + 2r\frac{dR_l}{dr} + \left(k^2r^2 - l^2 - r^2U(r)\right)R_l = 0 \tag{11.102}
$$

and therefore each spherical harmonic term in the wave function may be treated separately. For a given spherical harmonic, a solution to this radial equation may be found that behaves suitably at the coordinate origin. “Behaves suitably” for most problems simply means the solution is finite at the origin. Once this outward radial solution passes into the asymptotic region–where the potential goes to zero–it is matched to the asymptotic one given in Eq. 11.101

$$
\begin{aligned}
R_l(r_{match}) &\Longleftrightarrow \sqrt{4\pi(2l+1)}\, i^l j_l(kr_{match}) \\
&+ a_l\sqrt{4\pi(2l+1)}\, i^l h_l^{(1)}(kr_{match})
\end{aligned} \tag{11.103}
$$

The needed degrees of freedom for continuity of the wave function and its first derivative at the matching radius are a_l and the normalization of R_l. In carrying out the matching, it is useful to rewrite the asymptotic forms of Bessel functions, explicitly showing their decomposition into incoming and outgoing waves. For large $\rho = kr$:

$$
j_l(\rho) = \frac{1}{\rho}\frac{1}{2i}\left[(-i)^l e^{i\rho} - i^l e^{-i\rho}\right] \tag{11.104}
$$

Using this, along with Eq. 11.95, Eq. 11.101 becomes:

$$\psi(r \to \infty) = \sum_{1}^{\infty} \sqrt{4\pi(2l+1)}\, i^l Y_{l0}(\theta) \frac{1}{\rho}\frac{1}{2i}\left\{(-i)^l (1+2a_l) e^{i\rho} - i^l e^{-i\rho}\right\} \tag{11.105}$$

Examining the terms in the braces, the first corresponds to an outgoing wave and the second to an incoming wave. If there is no absorption in the problem (or spontaneous generation), the coefficients on these terms must have equal magnitude. In particular, $(1+2a_l)$ must have a magnitude of 1. We accordingly define the partial wave "phase shift":

$$e^{2i\delta_l} = 1 + 2a_l \Leftrightarrow a_l = \frac{1}{2}\left(e^{2i\delta_l} - 1\right) = ie^{i\delta_l}\sin\delta_l \tag{11.106}$$

and therefore the scattering amplitude is:[21]

$$f(\theta) = \frac{1}{k}\sum \sqrt{4\pi(2l+1)} e^{i\delta_l} \sin\delta_l Y_{l0}(\theta) \tag{11.107}$$

And the procedure of matching the boundary conditions is now seen to be equivalent to finding the phase shifts, δ_l. The orthogonality of the spherical harmonics allows us to write the total cross section:

$$\sigma = \int d\Omega \frac{d\sigma}{d\Omega} = \int d\Omega |f(\theta)|^2 = \frac{4\pi}{k^2}\sum (2l+1)\sin^2\delta_l \tag{11.108}$$

For most purposes, the partial wave expansion can be cut off a little beyond the value, $L_{\max}$, discussed before.

11.4.2 Vector partial wave analysis

We next extend the partial wave technique to vector fields. Especially important is the ability of the technique to handle problems involving targets whose size is on the order of the wavelength of the radiation. In such problems, the scattering targets are too large to be handled by the multipole techniques discussed before and too small to be treated by geometrical optics. The novel feature of the vector theory in contrast to the scalar theory discussed before is polarization. Some of the bookkeeping choices to keep track of polarizations are subtle and we begin by outlining those. We seek scattering wave function solutions having the asymptotic form:

$$\vec{E}_{tot} = \vec{E}_{inc} + \vec{E}_{scat} = \hat{\varepsilon} e^{ikz} + \vec{f}(\hat{\varepsilon}, \theta, \phi)\frac{e^{ikr}}{r} \tag{11.109}$$

[21] Consider the scattering part of Eq. 11.101 and use the asymptotic form of the Hankel function from Eq. 11.95 and the relation in Eq. 11.106. Next, identify $f(\theta)$ with terms multiplying $\frac{e^{ikr}}{r}$.

Here, $\vec{f}(\hat{\varepsilon},\theta,\phi)$ is the vector scattering amplitude: the vector part of this specifies the polarization of the scattered wave and a natural polarization basis for this wave is the spherical coordinate system unit vectors, $\hat{\theta},\hat{\phi}$, orthogonal to the direction of propagation of the scattered wave. A complete analysis of the scattering problem includes finding solutions for all incident polarizations, $\hat{\varepsilon}$. The linearity of the problem means that it is necessary and sufficient to find solutions for incident waves of two orthogonal polarizations. The spherical wave expansion of an incident plane wave is most straightforward for circular polarizations and so we will describe the incident plane waves using $\pm$ circular polarizations. In the formal notation developed before, the complete solution to the problem is the scattering amplitude or transition matrix, a 2D polarization matrix object that is based on the polarization basis vectors for the incident and scattered waves. These matrices have the form,

$$\overleftrightarrow{\mathcal{F}}(\theta,\phi)=\begin{bmatrix} f_{+,\theta}(\theta,\phi) & f_{-,\theta}(\theta,\phi) \\ f_{+,\phi}(\theta,\phi) & f_{-,\phi}(\theta,\phi) \end{bmatrix}. \tag{11.110}$$

where, for example, $f_{+,\theta}(\theta,\phi)$ is the amplitude for wave scattering in the θ,ϕ direction, with linear polarization in the associated $\hat{\theta}$ direction, due to an incident wave of pure positive circular polarization. As previously, we assume the incident field is a plane wave traveling in the $+z$ direction. The scattered field for any particular incident polarization is described by:

$$\vec{\mathfrak{E}}_{scat}\left(\vec{k}'\right)=\overleftrightarrow{\mathcal{F}}_{kk'}(\theta,\phi)\cdot\hat{\mathfrak{E}}_{inc}\left(\vec{k}\right) \tag{11.111}$$

where $\vec{k}=k\hat{z}$ and $\hat{\mathfrak{E}}_{inc}\left(\vec{k}\right)$ is the two component object giving the incident field in terms of the circular polarization basis, $\hat{e}_{\perp}^{\pm}$, and $\vec{\mathfrak{E}}_{scat}\left(\vec{k}'\right)$ describes the scattered field in the $\hat{\theta},\hat{\phi}$ linear polarization basis for direction $\vec{k}'$.

The discussion of the vector partial wave analysis parallels that of the scalar section. We begin by identifying suitable basis functions for expanding the scattering vector wave function. Spherical coordinate eigensolutions of the vector Helmholtz equation, analogous to the $\psi_{lm}^{(a)}$ in the scalar problem, form a basis for expanding any general vector function solution of the electric and magnetic fields within electromagnetic waves. These eigensolutions are labeled $\vec{M}_{lm}^{(a)}$ and $\vec{N}_{lm}^{(a)}$. Note that, in addition to solving the vector Helmholtz equation, for these functions to represent the incident or scattered electric or magnetic fields in regions absent of charge they must also have zero divergence. Expansions for an incident plane wave along the z-axis are found for both circular polarizations. Finally, the scattering problems for both polarizations are solved term by term by matching boundary conditions at a surface where the "scattering potential" falls to zero. This allows construction of the transition matrix, $\overleftrightarrow{\mathscr{F}}(\theta,\phi)$. We illustrate this procedure by solving the "Mie" problem of scattering from a uniform dielectric sphere.

Stratton's $\vec{M}$ and $\vec{N}$ solutions to the vector Helmholtz equation

We need to find vector functions to expand electric and magnetic vector wave functions. Since these fields are divergence-free, our expansion basis should also be diverence-free. They will be based on the vector spherical harmonics discussed in Appendix A. For a spherically symmetric potential, one could identify divergence free combinations of the vector spherical harmonics. Instead we use a general approach for finding suitable functions originally outlined by Stratton.[22]

Away from the scatterer, the scattering wave functions are solutions to the homogeneous Helmholtz equations in a material with permittivity and permeability, ε_1, μ_1, respectively, and corresponding index of refraction, $n_1 = c\sqrt{\varepsilon_1\mu_1}$. The region is free of charges and currents and therefore the fields are simply related:

$$\begin{aligned}\nabla\times\vec{E} &= i\omega\mu_1\vec{H}\\ \nabla\times\vec{H} &= -i\omega\varepsilon_1\vec{E}\end{aligned} \tag{11.112}$$

And in the usual fashion, the fields are readily shown to be solutions of the Helmholtz equation:

$$\left(\nabla^2+k^2\right)\vec{E}=\left(\nabla^2+k^2\right)\vec{H}=0 \tag{11.113}$$

with $k = n_1\omega/c$. Given any solution to the scalar Helmholtz equation, ψ, and any spatially constant vector, $\vec{C}$, we can define:

$$\vec{M}=\nabla\times\vec{C}\psi \tag{11.114}$$

where "$\vec{C}$" in this context is referred to as the "pilot vector" and ψ is the "generating function." Using the vector identity,

$$\nabla\times\left(\nabla\times\vec{M}\right)=\nabla\left(\nabla\cdot\vec{M}\right)-\nabla^2\vec{M} \tag{11.115}$$

[22] Stratton, J. A., *Electromagnetic Theory*, McGraw-Hill, New York (1941).

and noting that the divergence of any curl vanishes, $\nabla \cdot \vec{M} = 0$, we find that $\vec{M}$ is a solution to the vector Helmholtz equation (**see Discussion 11.8**). It could therefore, by itself, represent the electric or magnetic field of an electromagnetic wave. We next define,

$$\vec{N} = \frac{1}{k}\nabla \times \vec{M} \tag{11.116}$$

which is a second solution to the vector Helmholtz equation.[23] Again, this has zero divergence, $\nabla \cdot \vec{N} = 0$, and also could by itself represent an electromagnetic wave. In addition, using the result of Eq. 11.179 in Discussion 11.8 with Eq. 11.116, it is seen that:

$$\vec{M} = \frac{1}{k}\nabla \times \vec{N} \tag{11.117}$$

The similarity between the symmetry of these expressions for $\vec{M}$ and $\vec{N}$ and the cross product expressions, Eq. 11.112, relating the electric and magnetic fields make these useful functions for expanding electromagnetic waves. Specifically, if a general electromagnetic wave that, on expanding, has an electric field with the form,

$$\vec{E} = \ldots + A\vec{M} \ldots + B\vec{N} \tag{11.118}$$

then by Eqs. 11.112, 11.116, and 11.117, the magnetic field will have a corresponding form of:

$$\vec{H} = \ldots - \frac{i}{Z_1}A\vec{N} \ldots - \frac{i}{Z_1}B\vec{M} \tag{11.119}$$

where $Z_1 = \sqrt{\mu_1/\varepsilon_1}$ is the wave impedance of the material.

$\vec{M}$ and $\vec{N}$ functions for spherical coordinates

The scattered waves are outgoing spherical waves and therefore wave solutions in spherical polar coordinates are the most suitable basis for solving the problem. Here, it is useful to take $\vec{r}$ as the pilot vector and define:

$$\vec{M} = \nabla \times (\vec{r}\psi) \tag{11.120}$$
$$\vec{N} = \frac{1}{k}\nabla \times \vec{M}$$

Although, here, the pilot vector is not constant, it is readily shown (**see Discussion 11.9**) that these $\vec{M}$ and $\vec{N}$ vector functions are a suitable extension of the Stratton approach for spherical coordinates: they are solutions to the vector Helmholtz equations and also have

[23] Using the fact that $\nabla \times (\nabla \times \vec{N}) = -\nabla^2 \vec{N}$, and, from Eq. 11.116, that $\nabla \times \vec{N} = k\vec{M}$, we see that $\nabla \times (\nabla \times \vec{N})$ also equals $k^2 \vec{N}$.

the desirable cross product symmetry of Eqs. 11.116 and 11.117. For generating functions, we take the previously found solutions to the scalar wave equations in polar coordinates, $\sim \psi_{lm}^{(a)} = z_l^{(a)}(kr)\, Y_{lm}(\theta,\phi)$ (see Eq. 11.96), and define:

$$\vec{M}_{ml}^{(a)} = -\frac{1}{\sqrt{l(l+1)}} \nabla \times \left(\vec{r}\psi_{lm}^{(a)}\right) \tag{11.121}$$

Including the prefactor makes $\vec{M}_{lm}^{(a)}$ simply the vector spherical harmonic, $\vec{\Phi}_{lm}$, with a radial spherical Bessel function tacked on, $\vec{M}_{ml}^{(a)} = z_l^{(a)}\vec{\Phi}_{lm}$ (**see Appendix and Discussion 11.10**). Similarly, we define the functions

$$\vec{N}_{lm}^{(a)} = \frac{1}{k} \nabla \times \vec{M}_{lm}^{(a)} \tag{11.122}$$

The orthogonality of all the $\vec{M}$'s and $\vec{N}$'s follows immediately from curl relations for the vector harmonics and the orthogonality of the different types of vector spherical harmonics. From Eq. 11.117, we have the important symmetrical relation:

$$\vec{M}_{lm}^{(a)} = \frac{1}{k} \nabla \times \vec{N}_{lm}^{(a)} \tag{11.123}$$

Written explicitly (**see Discussion 11.11**)

$$\vec{M}_{lm}^{(a)} = \frac{z_l^{(a)}(\rho)}{\sqrt{l(l+1)}} \left[-\frac{im}{\sin\theta} Y_{lm}(\theta,\phi)\hat{\theta} + \frac{\partial Y_{lm}(\theta,\phi)}{\partial\theta}\hat{\phi} \right] \tag{11.124}$$

where $\rho = kr$, and

$$\begin{aligned} \vec{N}_{lm}^{(a)} = &-\frac{l\sqrt{(l+1)}}{\rho} z_l^{(a)} Y_{lm}\hat{r} \\ &-\frac{1}{\rho\sqrt{l(l+1)}} \frac{\partial}{\partial\rho}\left(\rho z_l^{(a)}\right)\left(\frac{\partial Y_{lm}}{\partial\theta}\hat{\theta} + \frac{im}{\sin\theta} Y_{lm}\hat{\phi}\right) \end{aligned} \tag{11.125}$$

Note that the $\vec{N}$ functions have radial components where the $\vec{M}$ functions do not. Consequently, for outward propagating waves, the $\vec{M}$ solutions are "transverse" solutions. An electromagnetic field mode whose electric field is given by $\vec{M}_{lm}^{(a)}$ is called a "transverse electric" or TE mode and magnetic dipole fields are an important example. In contrast, a mode whose electric field is given by $\vec{N}_{lm}^{(a)}$ will have a magnetic field that is transverse. An electric dipole field represents such a mode and is called "transverse magnetic" or TM. The radial fields decrease with distance by a power of $\frac{1}{r}$ more rapidly than the transverse components and are negligible in the asymptotic region of the scattering problem. Sometimes $\vec{N}_{lm}^{(a)}$ are referred to as "electric multipole" solutions and the $\vec{M}_{lm}^{(a)}$ are referred to as "magnetic multipole" solutions. Complete sets of functions for expanding any electromagnetic wave in free space may be constructed using the radial functions, $j_l(kr)$ and $\eta_l(kr)$. That is, the functions $\vec{M}_{lm}^{(1)}$, $\vec{N}_{lm}^{(1)}$, $\vec{M}_{lm}^{(2)}$, and $\vec{N}_{lm}^{(2)}$ for all l, m each constitute a complete set of orthogonal functions for expanding arbitrary electric and magnetic fields that satisfy the Helmholtz equation for a specific value of k. An alternative set may be constructed using the Hankel functions. For example, the first variety of Hankel functions, $h_l^{(1)}(\rho) = z_l^{(3)}(\rho)$, describe outgoing waves and the associated $\vec{N}_{l\pm1}^{(3)}$ and $\vec{M}_{l\pm1}^{(3)}$ describe the fields from oscillating electric and magnetic multipole sources. Finally, a third set of solutions to the vector Helmholtz equation, that are longitudinal with zero-curl, are defined by $\vec{L}_{ml}^{(a)} = \nabla\left(\vec{r}\psi_{lm}^{(a)}\right)$. These are appropriate for studying sound waves.

Vector spherical wave expansions of vector plane waves

In the general case of a vector plane wave field, $\vec{V}$, which varies in space in a manner consistent with the Maxwell's equations,[24] we need to resort to the more powerful method of vector spherical harmonic expansion by using the divergence free solutions to the vector Helmholtz equation–the $\vec{M}$ and $\vec{N}$ vector functions–as an expansion basis. Proceeding in a manner analogous to the scalar spherical harmonic expansion of a plane wave to find the electric field, we identify a general solution to the free-space Maxwell's equations for harmonic waves in terms of an expansion of the electric field:

$$\vec{E}(r,\theta,\phi) = \sum_{lm}\left[A_{lm}\vec{M}_{lm}^{(a)} + B_{lm}\vec{N}_{lm}^{(a)}\right] \tag{11.126}$$

where the $A_{lm}\vec{M}_{lm}^{(a)}$ $\left(B_{lm}\vec{N}_{lm}^{(a)}\right)$ term is the contribution to the electric field due to the lm^{th} magnetic (electric) multipole. We will work with waves having circular polarization. Doing so facilitates the math both because the magnetic and electric fields are simply related and the functional form of the circularly polarized fields allows using angular momentum results familiar from quantum mechanics. The two circular polarization unit vectors may be written in terms of unit vectors in the x and y directions:

$$\hat{e}_{\perp}^{\pm} = \frac{1}{\sqrt{2}}(\hat{x} \pm i\hat{y}) \tag{11.127}$$

[24] Maxwell's equations require that the vector field contain no components in the direction of wave propagation and that the electric and magnetic fields be everywhere mutually perpendicular.

where the upper "+" and lower "−" refer to right and left circular polarization, respectively. In what follows, we will continue this convention where the top row of an index is the σ^+ result and the bottom row is to be used for σ^- polarization. For a wave incident along the z-axis, the curl relations may be used to show that the electric fields and magnetic fields are related[25]

$$\vec{E} = E_o \hat{e}_\perp^\pm \exp[ikz] \tag{11.128}$$

$$\vec{H} = \frac{\mp i}{Z} E_o \hat{e}_\perp^\pm \exp[ikz] = \frac{\mp i}{Z} \vec{E} \tag{11.129}$$

where the 90° phase shift of the complex electric field due to the i accounts for the constant orthogonality of $\vec{H}$ and $\vec{E}$ as they spiral. We observe that, as with the scalar expansion, since $\exp(ikz)$ is finite at the origin the j_l's must be used for the radial parts of the expansion functions: all other spherical Bessel functions are infinite at the origin. Recalling that the j_l spherical Bessel functions correspond to the (1) superscripts on the $\vec{M}$ and $\vec{N}$ functions, Eq. 11.126 can be written as

$$E_o \hat{e}_\perp^\pm \exp(ikz) = \sum_{lm} \left[A_{lm}^\pm \vec{M}_{lm}^{(1)} + B_{lm}^\pm \vec{N}_{lm}^{(1)} \right] \tag{11.130}$$

Unlike for the scalar plane wave, a circularly polarized wave is not cylindrically symmetric. Its field vectors trace out helices along its direction of travel. However, the coefficients for the $\vec{M}$'s may be found by taking the scalar product of $\vec{\Phi}_{lm}^*$ and Eq. 11.130. We exploit the inherent orthogonalities existing among all the right-hand side terms of Eq. 11.130 to obtain (**see Discussion 11.12**):

$$\int \vec{\Phi}_{lm}^* \cdot \left[E_o \hat{e}_\perp^\pm \exp(ikr\cos\theta) \right] d\Omega = A_{lm}^\pm j_l(kr) \tag{11.131}$$

Using the angular momentum operator familiar from quantum mechanics, $\widehat{L} = \frac{1}{i}(\vec{r} \times \nabla)$, and writing it in terms of raising and lowering operator combinations in Cartesian coordinates, we can express Eq. 11.131 as (**see Discussion 11.13**)

$$\frac{-i}{\sqrt{2l(l+1)}} \int \left(\widehat{L}_\mp Y_{lm} \right)^* E_o \exp(ikr\cos\theta) d\Omega = A_{lm}^\pm j_l(kr) \tag{11.132}$$

The raising and lowering operators act on the spherical harmonics to effectively "raise" or "lower" them by one unit of m with the introduction of an l, m dependent coefficient. That is,

$$\widehat{L}_\mp Y_{lm} = \sqrt{l(l+1) - m(m \mp 1)}\, Y_{l,m\mp1} \tag{11.133}$$

25
$$\vec{H} = \frac{1}{i\omega\mu} \nabla \times \left(E_0 \hat{e}_\perp^\pm \exp[ikz] \right) = \frac{E_0}{ikZ} \left\{ \frac{1}{\sqrt{2}} (\hat{y} \mp i\hat{x}) \frac{\partial}{\partial z} (\exp[ikz]) \right\} = \frac{E_0}{Z} (\mp i \hat{e}_\perp^\pm) \exp[ikz]$$
which leads to Eq. 11.129.

So applying Eq. 11.133 to Eq. 11.132 yields

$$\frac{-iE_o\sqrt{l(l+1)-m(m\mp 1)}}{\sqrt{2l(l+1)}}\int \exp(ikr\cos\theta)\,Y^*_{l,m\mp 1}d\Omega = A^{\pm}_{lm}j_l(kr) \tag{11.134}$$

and so we have arrived at an integral that promises to extract the $Y_{l,m\mp 1}$ component of $\exp(ikr\cos\theta)$ and thus solve the equation for the $A^{\pm}_{lm}$, the + and − coefficients of the contributions to the electric field by the $(l,m)^{th}$ magnetic multipole. This is accomplished by making use of the scalar spherical harmonic expansion of a plane wave found in the previous section. Substituting this expansion, Eq. 11.100, into Eq. 11.134, we obtain

$$\frac{-iE_o\sqrt{l(l+1)-m(m\mp 1)}}{\sqrt{2}\sqrt{l(l+1)}}$$
$$\int\left(\sum_{l'=0}^{\infty} i^{l'}\sqrt{4\pi\,(2l'+1)}j_{l'}(kr)\,Y_{l',0}\right)Y^*_{l,m\mp 1}d\Omega = A^{\pm}_{lm}j_l(kr) \tag{11.135}$$

which, due to the orthogonality of the Y_{lm}'s, first tells us that for $m\neq\pm 1$, the coefficient $A^{\pm}_{lm}j_l(kr)$ vanishes. In other words, the scalar spherical harmonic, $Y^*_{l,m\mp 1}$, associated with the projection of the right/left circularly polarized radiation onto the lm^{th} magnetic multipole contribution to the electric field (essentially $\vec{\Phi}^*_{lm}\cdot\hat{e}^{\pm}_{\perp}$) is further constrained by the fact that a scalar plane wave is composed of only Y_{l0} (i.e., its cylindrically symmetric), thus limiting $Y^*_{l,m\mp 1}$ and $A^{\pm}_{lm}$ to those for which $m=\pm 1$. Carrying out the integral in Eq. 11.135 with the assumption that $m=\pm 1$, only the $l'=l$ term survives while $\sqrt{l(l+1)-m(m\mp 1)}/\sqrt{l(l+1)}=1$,

$$A^{\pm}_{l,\pm 1}=\frac{E_o}{\sqrt{2}}i^{l-1}\sqrt{4\pi\,(2l+1)} \tag{11.136}$$

And so we obtain the coefficients, $A^{\pm}_{lm}$, for the $\vec{M}$ vector functions in the expansion (Eq. 11.130) of a circularly polarized electromagnetic plane wave in which these coefficients represent the amplitudes of the various *magnetic* multipole contributions to the electric field. To obtain the corresponding amplitudes, $B^{\pm}_{lm}$, of the various electric multipole contributions, the $\vec{N}$ vector functions, we exploit the particularly simple relation between electric and magnetic fields for circular polarizations, Eq. 11.129, and the symmetry between the electric and magnetic field expansion coefficients for corresponding $\vec{N}$ and $\vec{M}$ functions displayed in Eqs. 11.118 and 11.119. Comparing the circularly polarized plane wave expansion of Eq. 11.130 to the general electric field expansion of Eq. 11.118, we see that the $A's \leftrightarrow A^{\pm}_{lm}$ and the $B's \leftrightarrow B^{\pm}_{lm}$. With these identifications and the specific expansion for the electric field, we substitute 11.119 for the left-hand side of Eq. 11.129 and in the right-hand side we substitute Eq. 11.118 for the electric field. The magnetic field expansion is then,

$$\vec{H}^{\pm} = \sum_{lm}\left[\left(-\frac{i}{Z}\right)B^{\pm}_{lm}\vec{M}^{(1)}_{lm} + \left(-\frac{i}{Z}\right)A^{\pm}_{lm}\vec{N}^{(1)}_{lm}\right]$$

$$= \frac{\mp i}{Z}\sum_{lm}\left[A^{\pm}_{lm}\vec{M}^{(1)}_{lm} + B^{\pm}_{lm}\vec{N}^{(1)}_{lm}\right] \tag{11.137}$$

The orthogonality of the $\vec{M}$'s and $\vec{N}$'s requires the coefficients multiplying a particular function be the same on both sides of the equation. In particular, for the $\vec{M}$ functions:

$$-\frac{i}{Z}B^{\pm}_{lm} = \frac{\mp i}{Z}A^{\pm}_{lm} \tag{11.138}$$

$$B^{\pm}_{l,\pm 1} = \pm A^{\pm}_{l,\pm 1} = \pm\frac{E_o}{\sqrt{2}}i^{l-1}\sqrt{4\pi\,(2l+1)} \tag{11.139}$$

with the final result for the electric field expansion of Eq. 11.130,

$$\vec{E}^{\pm}_{inc} = E_o\hat{e}^{\pm}_{\perp}\exp(ikz) = \frac{E_o}{\sqrt{2}}\sum_l i^{l-1}\sqrt{4\pi\,(2l+1)}\left[\vec{M}^{(1)}_{l,\pm 1} \pm \vec{N}^{(1)}_{l,\pm 1}\right] \tag{11.140}$$

The magnetic counterpart expansion is then directly found from Eq. 11.129 to be

$$\vec{H}^{\pm}_{inc} = -\frac{i}{Z}\frac{E_o}{\sqrt{2}}\sum_l i^{l-1}\sqrt{4\pi\,(2l+1)}\left[\vec{N}^{(1)}_{l,\pm 1} \pm \vec{M}^{(1)}_{l,\pm 1}\right] \tag{11.141}$$

Combined, these make up the vector spherical harmonics multipole expansions of the electric and magnetic parts of free-space circularly polarized plane waves.

Expansion of incident and scattered vector wave functions in vector spherical harmonics

To study scattering of a plane wave off an arbitrary target, we seek solutions of the scattering problem with the asymptotic form given in Eq. 11.109. The partial wave method expands this solution using functions based on the vector spherical harmonics. For the incident wave, a plane wave propagating in the z-direction, we use Eqs. 11.140

and 11.141, with $E_0 = 1$, for $\vec{E}_{inc}$ and $\vec{H}_{inc}$, respectively. Linear combinations of the circularly polarized solutions may be used to make an incident wave of any polarization.

The scattered field has the form of an outgoing spherical wave and can be expanded using the complete set obtained from the Hankel functions. Only the $h_l^{(1)}$ functions will contribute, since they correspond to outgoing waves. The most general expansion for the electric field of the scattered wave may be conveniently written:

$$\vec{E}^{\pm}_{scat} = \vec{f}_{\pm}(\theta,\phi)\frac{e^{ikr}}{r} = \frac{E_0}{\sqrt{2}}\sum_{lm}^{\infty} i^{l-1}\sqrt{4\pi(2l+1)}\left(a^{\pm}_{lm}\vec{M}^{(3)}_{lm} \pm b^{\pm}_{lm}\vec{N}^{(3)}_{lm}\right) \tag{11.142}$$

where the general solution coefficients of Eq. 11.126 are here written as $A_{lm} = \frac{E_0}{\sqrt{2}}i^{l-1}\sqrt{4\pi(2l+1)}a^{\pm}_{lm}$ and $B_{lm} = \pm\frac{E_0}{\sqrt{2}}i^{l-1}\sqrt{4\pi(2l+1)}b^{\pm}_{lm}$ with the various prefactors included to simplify the algebra that follows. Recall, the superscript (3) indicates that the outgoing spherical Hankel functions are to be used with the vector spherical harmonics in the $\vec{M}$ and $\vec{N}$ functions. We use $\pm$ to allow the simultaneous solution to the scattering of both positively and negatively circularly polarized incident waves. The magnetic field may be determined using Eq. 11.119:

$$\vec{H}^{\pm}_{scat} = -\frac{i}{Z}\frac{E_0}{\sqrt{2}}\sum_{lm}^{\infty} i^{l-1}\sqrt{4\pi(2l+1)}\left(a^{\pm}_{lm}\vec{N}^{(3)}_{lm} \pm b^{\pm}_{lm}\vec{M}^{(3)}_{lm}\right) \tag{11.143}$$

11.4.3 Solution of scattering from a homogeneous sphere: Mie scattering

Solving the scattering problem as in the scalar case means smoothly joining the asymptotic scattering solution to a solution in the region of the scatterer and extracting the scattering amplitude from the former. We do this for the important special case of a plane wave traveling in a medium with permittivity ε_I and permeability μ_I scattering off a homogeneous sphere of radius $r = a$ with permittivity ε_{II} and permeability μ_{II}. We include consideration of complex ε_{II}. The index of refraction of the sphere is $n_{II} = \sqrt{\varepsilon_{II}\mu_{II}}c$, the impedance of the sphere's material, $Z_{II} = \sqrt{\mu_{II}/\varepsilon_{II}}$, and wavevectors have magnitude $k_{II} = n_{II}k_0 = n_{II}\omega/c$. We naturally locate the origin of coordinates at the center of the sphere.

The field in the sphere must be constructed from functions that are finite at the origin: the basis functions within the sphere must use the spherical Bessel functions $j_l(k_{II}r)$ in their radial dependence. That is, the functions $\vec{M}_{lm}^{(1)}(k_{II}r)$ and $\vec{N}_{lm}^{(1)}(k_{II}r)$ form a complete, orthogonal basis for allowed waves within the sphere where "$k_{II}r$" is a reminder to use the material parameters of the sphere. The electric field for an arbitrary wave within the sphere is expanded as:

$$\vec{E}_{sphere}^{\pm} = \frac{E_0}{\sqrt{2}} \sum_{lm}^{\infty} i^{l-1} \sqrt{4\pi(2l+1)} \left(c_{lm}^{\pm} \vec{M}_{lm}^{(1)}(k_{II}r) \pm d_{lm}^{\pm} \vec{N}_{lm}^{(1)}(k_{II}r) \right) \tag{11.144}$$

and the magnetic field is found, as before, using Eq. 11.119:

$$\vec{H}_{sphere}^{\pm} = -\frac{i}{Z_{II}} \frac{E_0}{\sqrt{2}} \sum_{lm}^{\infty} i^{l-1} \sqrt{4\pi(2l+1)} \left(c_{lm}^{\pm} \vec{N}_{lm}^{(1)}(k_{II}r) \pm d_{lm}^{\pm} \vec{M}_{lm}^{(1)}(k_{II}r) \right) \tag{11.145}$$

To find the scattered wave, we impose the boundary conditions of the fields across the surface of the dielectric: the transverse $\vec{E}$ and $\vec{H}$ fields are continuous. Outside the sphere the wave function consists of the incident $\vec{E}_{inc}$ and $\vec{H}_{inc}$ (Eqs. 11.140 and 11.141) plus the scattered waves $\vec{E}_{scat}$ and $\vec{H}_{scat}$ (Eqs. 11.142 and 11.143) and we match these to the solution inside the sphere (Eqs. 11.144 and 11.145) to determine the various expansion coefficients. In applying the boundary conditions, the problem is similar to finding the Fresnel coefficients that describe reflection and transmission at the planar boundary between two dielectrics. Here the incident, scattered and internal waves correspond, respectively, to the incident, reflected, and transmitted waves in the Fresnel solutions. Indeed in the limit of a large radius sphere, these are the same problems.

For the homogeneous sphere, these boundary conditions are especially easy to apply. The coefficients, a_{lm}, b_{lm}, c_{lm}, and d_{lm} may be found by separately solving for the continuity of the electric and magnetic fields across the boundary of the dielectric sphere for the $\vec{M}_{lm}$ and $\vec{N}_{lm}$ functions. See Eqs. 11.124 and 11.125. This gives four inhomogeneous equations for each l. For the $m = \pm 1$ components these are (**see Discussion 11.14**)

$$c_{l,\pm 1}^{\pm} j_l(k_{II}r) - a_{l,\pm 1}^{\pm} h_l^{(1)}(k_I r) = j_l(k_I r) \tag{11.146}$$

$$\frac{Z_I}{Z_{II}} d_{l,\pm 1}^{\pm} j_l(k_{II}r) - b_{l,\pm 1}^{\pm} h_l^{(1)}(k_I r) = j_l(k_I r), \tag{11.147}$$

$$\frac{n_I}{n_{II}} d_{l,\pm 1}^{\pm} \frac{\partial}{\partial(k_{II}r)} \left[k_{II}r \cdot j_l(k_{II}r)\right] - b_{l,\pm 1}^{\pm} \frac{\partial}{\partial(k_I r)} \left[k_I r \cdot h_l^{(1)}(k_I r)\right] = \frac{\partial}{\partial(k_I r)} \left[k_I r \cdot j_l(k_I r)\right] \tag{11.148}$$

$$\frac{\mu_I}{\mu_{II}} c_{l,\pm 1}^{\pm} \frac{\partial}{\partial (k_{II} r)} \left[k_{II} r \cdot j_l (k_2 r) \right] - a_{l,\pm 1}^{\pm} \frac{\partial}{\partial (k_I r)} \left[k_I r \cdot h_l^{(1)} (k_I r) \right] = \frac{\partial}{\partial (k_I r)} \left[k_I r \cdot j_l (k_I r) \right], \tag{11.149}$$

These equations are to be evaluated for $r = a$. The incident field components have been written on the right-hand side of these equations: they appear as inhomogeneous "driving" terms. Since there are no $m \neq \pm 1$ terms in the incident field expansion, there is no scattering into $m \neq \pm 1$ components. In going forward, for notation, we will drop the $\pm$ superscript on the coefficients since the $m = \pm 1$ subscript on a coefficient is adequate for identifying the relevant polarization. The $\vec{M}$ and $\vec{N}$ terms in the electric field are analogous to the S- and P-polarizations in the Fresnel coefficients solutions. Finally, these equations are solved for the coefficients of the scattered wave:

$$a_{l,\pm 1} = \left. \frac{j_l (k_{II} r) \frac{\partial}{\partial (k_I r)} \left[k_I r \cdot j_l (k_I r) \right] - \frac{\mu_I}{\mu_{II}} j_l (k_I r) \frac{\partial}{\partial (k_{II} r)} \left[k_{II} r \cdot j_l (k_{II} r) \right]}{\frac{\mu_I}{\mu_{II}} h_l^{(1)} (k_I r) \frac{\partial}{\partial (k_{II} r)} \left[k_{II} r \cdot j_l (k_{II} r) \right] - j_l (k_{II} r) \frac{\partial}{\partial (k_I r)} \left[k_I r \cdot h_l^{(1)} (k_I r) \right]} \right|_{r=a}$$

$$b_{l,\pm 1} = \left. \frac{\frac{n_I}{n_{II}} j_l (k_I r) \frac{\partial}{\partial (k_{II} r)} \left[k_{II} r \cdot j_l (k_{II} r) \right] - \frac{Z_I}{Z_{II}} j_l (k_{II} r) \frac{\partial}{\partial (k_I r)} \left[k_I r \cdot j_l (k_I r) \right]}{\frac{Z_I}{Z_{II}} j_l (k_{II} r) \frac{\partial}{\partial (k_I r)} \left[k_I r \cdot h_l^{(1)} (k_I r) \right] - \frac{n_I}{n_{II}} h_l^{(1)} (k_I r) \frac{\partial}{\partial (k_{II} r)} \left[k_{II} r \cdot j_l (k_{II} r) \right]} \right|_{r=a}$$

Note that by defining the internal and scattered wave functions with the $\pm$ signs between the $\vec{M}$ and $\vec{N}$ terms in the partial wave expansion using the convention identical to the plane wave expansion, the coefficients, a_l, b_l, c_l, and d_l are independent of the polarization sense, $\pm$, and so we drop that dependence going forward. The problem has been set up in a circularly polarized basis for the incident wave and a scattered wave that in any given direction is specified by the linear polarization basis unit vectors, $\hat{\theta}$, $\hat{\phi}$ for that direction. As always, we take the incident wave to be propagating in the $+z$ direction. On taking the $r \to \infty$ limit for the Hankel functions in Eq. 11.142 (indicated by a (3) superscript), the scattering amplitude matrix may be written explicitly (**see Discussion 11.15**)

$$\overleftrightarrow{\mathscr{F}} (\theta, \phi) = \begin{bmatrix} f_{+,\theta} (\theta, \phi) & f_{-,\theta} (\theta, \phi) \\ f_{+,\phi} (\theta, \phi) & f_{-,\phi} (\theta, \phi) \end{bmatrix}$$

$$= \frac{\sqrt{2\pi}}{k_I} \sum_l \sqrt{\frac{2l+1}{l(l+1)}} \begin{bmatrix} i \left(a_l \frac{Y_{l,1}}{\sin\theta} - b_l \frac{\partial Y_{l,1}}{\partial \theta} \right) & -i \left(a_l \frac{Y_{l,-1}}{\sin\theta} - b_l \frac{\partial Y_{l,-1}}{\partial \theta} \right) \\ -\left(a_l \frac{\partial Y_{l,1}}{\partial \theta} - b_l \frac{Y_{l,1}}{\sin\theta} \right) & -\left(a_l \frac{\partial Y_{1,-1}}{\partial \theta} - b_l \frac{\partial Y_{1,-1}}{\sin\theta} \right) \end{bmatrix} \tag{11.150}$$

The two component scattering amplitude in the $\hat{\theta},\hat{\phi}$ basis, then, is found by multiplying this matrix with the unit-normalized polarization vector of the incident field, $\hat{\mathfrak{E}}_{inc}\left(\vec{k}\right)$

$$\vec{\mathfrak{E}}_{scat}\left(\hat{\varepsilon}_{in},\theta,\phi\right)=\left[\begin{array}{c}\mathfrak{E}_{scat,\theta}\left(\hat{\varepsilon}_{in},\theta,\phi\right)\\ \mathfrak{E}_{scat,\phi}\left(\hat{\varepsilon}_{in},\theta,\phi\right)\end{array}\right]=\overleftrightarrow{\mathcal{F}}\left(\theta,\phi\right)\left[\begin{array}{c}\varepsilon_{in}^{+}\\ \varepsilon_{in}^{-}\end{array}\right] \tag{11.151}$$

where $\hat{\boldsymbol{\varepsilon}}_{in}$ is a possibly complex unit vector for the incident field in the circular polarization basis. $\vec{\mathfrak{E}}_{scat}\left(\theta,\phi\right)$ can be used to find the total scattering in the θ,ϕ direction:

$$\frac{d\sigma\left(\hat{\varepsilon}_{in},\theta,\phi\right)}{d\Omega}=\left|\vec{\mathfrak{E}}_{scat}\left(\hat{\varepsilon}_{in},\theta,\phi\right)\right|^{2} \tag{11.152}$$

Alternatively, the differential cross section for scattering into a given output polarization, $\hat{\boldsymbol{\varepsilon}}_{out}$, can be found by projecting out that polarization component:

$$\frac{d\sigma\left(\hat{\varepsilon}_{in},\theta,\phi,\hat{\boldsymbol{\varepsilon}}_{out}\right)}{d\Omega}=\left|\vec{\mathfrak{E}}_{scat}\left(\hat{\varepsilon}_{in},\theta,\phi\right)\cdot\hat{\varepsilon}_{out}^{*}\right|^{2} \tag{11.153}$$

Here, $\hat{\boldsymbol{\varepsilon}}_{out}$ is the two component unit vector for the desired output polarization written in $\hat{\theta},\hat{\phi}$ basis. Finally, the three component vector valued scattering amplitude may be found by multiplying $\vec{\mathfrak{E}}_{scat}\left(\hat{\varepsilon}_{in},\theta,\phi\right)$ by the 3×2 matrix whose two columns are the Cartesian or other coordinate system unit vectors for the polarization basis:

$$\vec{f}\left(\theta,\phi,\hat{\varepsilon}_{in}\right)=\overleftrightarrow{I}\left(\hat{\theta},\hat{\phi}\rightarrow x,y,z\right)\cdot\vec{\mathfrak{E}}_{scat} \tag{11.154}$$

where

$$\overleftrightarrow{I}\left(\hat{\theta},\hat{\phi}\rightarrow x,y,z\right)=\left[\begin{array}{cc}\cos\theta\cos\phi & -\sin\phi\\ \cos\theta\sin\phi & \cos\phi\\ -\sin\theta & 0\end{array}\right] \tag{11.155}$$

expresses the unit vectors $\hat{\theta},\hat{\phi}$ in terms of Cartesian coordinates. The conversion necessarily depends on the direction, $\vec{k}'$, (whose direction relative to the origin is θ, ϕ) since the polarization basis is necessarily perpendicular to the wavevector of the observed scattering (see Eqs. 11.72–11.74). The scattered electric field for a given problem, then, is found by multiplying this vector scattering amplitude by the (scalar) amplitude of the incident field, E_0. Total cross sections may be found by integrating the corresponding differential cross section over all solid angles. Working directly from Eq. 11.142, the angular orthogonality af the $\vec{M}$ and $\vec{N}$ functions leaves only contributions from diagonal terms. These fall off as the expected $1/r^2$ and, on multiplying by r^2, the total cross section is seen to be given by:

$$\sigma_{tot} = \frac{2\pi}{k_1^2} \sum (2l+1) \left(|a_l|^2 + |b_l|^2 \right) \tag{11.156}$$

In examining the scattering amplitude matrix, Eq. 11.150, there are two types of angle θ dependent terms that are frequently referred to as π and τ functions. These are:

$$\pi_l(\theta,\phi) = \frac{Y_{l,\pm 1}}{\sin\theta} \quad \text{and} \quad \tau_l(\theta,\phi) = \frac{\partial Y_{l,\pm 1}}{\partial \theta} \tag{11.157}$$

In Fig. 11.6, we show polar plots for the lowest few of these in the plane $\phi = 0$. As always in what follows, the incident beam travels from left to right. In these plots, for a given θ, the distance of the curve from the origin gives the value of the function in that direction. Black codes are for positive values and gray are for negative values. For the smallest scatterers, the expected dipole patterns are found in the lowest functions. π_1 is a circular symmetric function that describes the angular dependence of a field generated by a dipole at the origin oscillating perpendicular to the plane of the paper. For higher partial waves, the forward scattering is positive for all functions: these add constructively. Generally, for non-forward directions, partial waves contribute with differing signs and overall tend to cancel. This is especially true of backward scattering where two successive partial waves are observed to have similar shapes but opposite signs. Taken

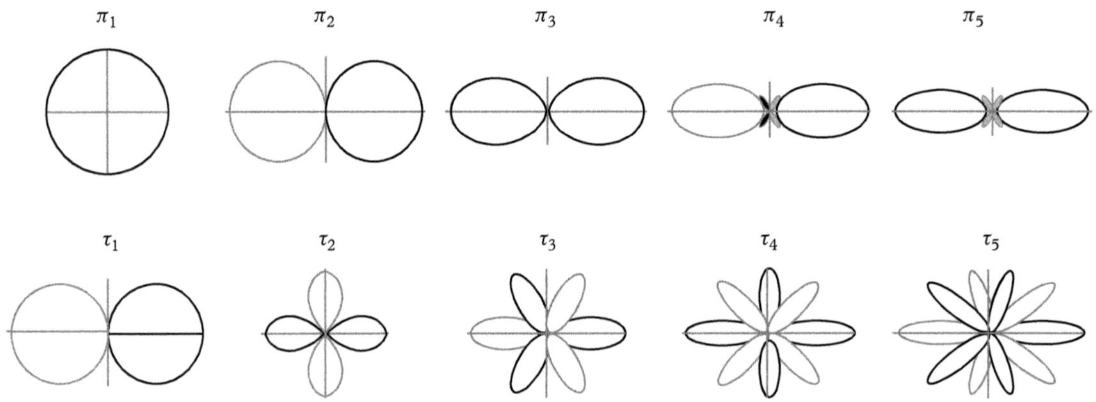

Fig. 11.6 *Polar plots of the lowest order angular dependent functions for the partial wave expansion.*

together, the dominant scattering of a large object is the diffraction peak around the forward direction.

11.5 Some results

It is nice to know that the computer understands the problem. But I would like to understand it too.

(*attributed to Eugene Wigner*)[26]

The Mie solution is invaluable as an exact solution of Maxwell's equations for scattering electromagnetic waves by a homogeneous sphere. But, it brings with it little in the way of physical insight. Next, we discuss a few specific solutions.

11.5.1 The long wavelength limit

For long wave scattering from a small dielectric particle of radius a_0, $\beta = 2\pi a_0/\lambda \ll 1$, the Mie solution returns exactly the same result as seen in the previous discussion. When a dielectric sphere is excited by a $\pm$ circularly polarized wave, only b_1 in the Mie expansion of the electric field, Eq. 11.142, is non-zero, giving $\vec{N}^{(3)}_{1,\pm 1}$ for the electric field and correspondingly giving $\vec{M}^{(3)}_{1,\pm 1}$ for the magnetic field. Scattering of linearly polarized waves is found by superposition.

We can examine, and compare with Mie results, the case of a small, perfectly conducting sphere embedded in a medium with permittivity ε_1 in the long wavelength limit. We include conductivity in the problem using a complex permittivity:

$$\tilde{\varepsilon}_2 = \varepsilon + i\frac{\sigma_c}{\omega} \simeq i\frac{\sigma_c}{\omega} \tag{11.158}$$

where $|\tilde{\varepsilon}_2| \gg |\varepsilon_1|$ and we have here denoted the electrical conductivity of the sphere with σ_c. With this approximation, Eq. 11.32 predicts a total scattering cross section for a small, perfectly conducting sphere to be:

$$\sigma \simeq \frac{8}{3}\left(\frac{2\pi a_0}{\lambda_1}\right)^4 \pi a_0^2 = \frac{8\pi}{3k_1^2}\beta^6 \tag{11.159}$$

Yet the Mie calculation gives a 25% larger result and, in fact, the differential cross section is weighted in the backward direction. See Fig. 11.7. To understand this incomplete result, we note that for the perfect conductor calculation, in addition to b_1 being non-zero

[26] Nussenzveig, H. M., *Diffraction Effects in Semiclassical Scattering*. Cambridge University Press (1992).

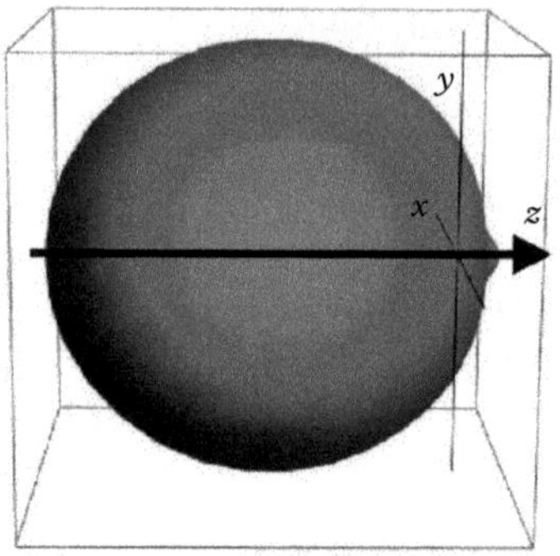

Fig. 11.7 *Mie theory calculation of the differential cross section for light scattering off a small, perfectly conducting sphere. The incident wave travels from left to right along the dark z-axis. The scattering is weighted in the backward direction.*

in Eq. 11.142, so too is a_1. The fields associated with this latter coefficient - the electric field of $\vec{M}^{(3)}_{1,\pm1}$ and the magnetic field of $\vec{N}^{(3)}_{1,\pm1}$– are exactly the reverse of those originating from the electric dipole and so we see that, indeed, a magnetic dipole field is also a large component in the solution. The short story is that the magnetic field of the incident wave creates circulating shielding surface currents–a perfect conductor is also a perfect diamagnet. Using this fact, the magnetic dipole generated by the sphere is $\vec{m} = -2\pi a_0^3 \vec{H}$. Noting also Eq. 11.31 for the electric dipole, $\vec{p}$, generated by the sphere, the total scattered field at position $\vec{r}$, including both the electric and magnetic dipoles, is found from Eqs. 4.10 and 4.19 for fields in the radiation zone:

$$\begin{aligned}\vec{E}_{scat} &= -\frac{k^2}{4\pi\varepsilon_1}\frac{e^{ikr}}{r}\left[(\vec{p}\times\hat{r})\times\hat{r} + \frac{1}{c}\hat{r}\times\vec{m}\right] \\ &= k^2 a_0^3 E_1 \frac{e^{ikr}}{r}\left[(\hat{r}\times\hat{\varepsilon}_{in})\times\hat{r} - \frac{1}{2}(\hat{z}\times\hat{\varepsilon}_{in})\times\hat{r}\right] \qquad (11.160)\end{aligned}$$

where E_1 and $\hat{\boldsymbol{\varepsilon}}_{in}$ are the magnitude and polarization of the incident field within the embedding material. Here, the first and second terms are for the electric and magnetic dipoles, respectively. The differential cross section is proportional to $\left|\vec{E}_{scat}\right|^2$. The first term in the brackets is just that from the electric dipole and it can be seen that the two dipoles add constructively when $\vec{r}$, the observation point, is in the $-\hat{z}$ direction and destructively when it's in the $+\hat{z}$ direction. Integrating $\left|\vec{E}_{scat}\right|^2$ over all space, the contribution from squaring the first term in the brackets is the electric dipole result, Eq. 11.159, and the contribution from the second term adds an additional one-fourth of the electric dipole value from squaring the one-half factor. The cross terms are odd in z and integrate to 0. Thus, the total cross section for the perfectly conducting sphere is:

$$\sigma = \frac{10\pi}{3k_1^2}\beta^6 \tag{11.161}$$

and so we note again that the long wavelength treatment of scattering is in agreement with the quoted Mie results.

11.5.2 Scattering off dielectric spheres: water droplets

Classical scattering

Geometric optics scattering frequently provides interpretations of Mie calculations. Figure 11.8 outlines how the differential cross section is determined from the classical scattering parameters, the impact parameter, b, and the deflection angle, θ. The figure is drawn for a general potential as might be used for potential scattering of particles. For light, it would represent the geometrical optics description of scattering from a centrally symmetric gradient index sample. For the actual uniform spheres that we will consider in detail, deflection happens only at the sphere's surface. Referring to the figure, the differential cross section may be written

$$\frac{d\sigma}{d\Omega} = \frac{b}{\sin\theta}\left|\frac{d\theta}{db}\right|^{-1} \tag{11.162}$$

This expression assumes symmetry about the axial (ϕ) axis. From the discussion on the spherical harmonic expansion of the incident plane wave, Section 11.4, the classical scattering for an impact parameter b corresponds, in the wave picture, to the scattered partial waves, $j_l(kr)$, for which $l \approx kb$.

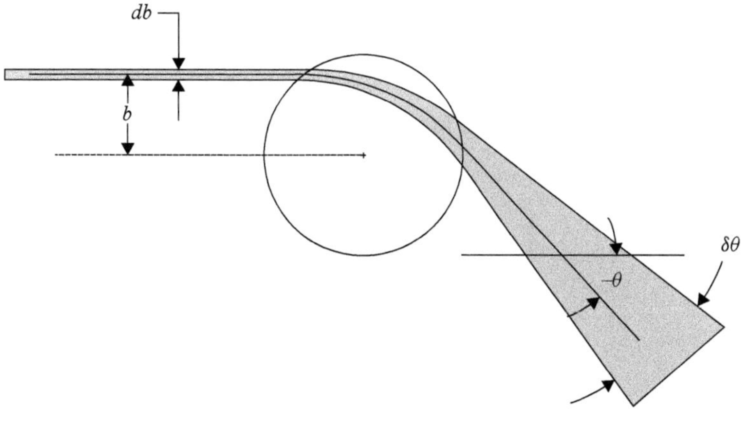

Fig. 11.8 *Classical differential cross section for scattering from a central potential. Incident particles/radiation with a range of impact parameters, db, centered about b are scattered into angles, $\delta\theta$, about θ. The cross section, $d\sigma = 2\pi b\, db$, for scattering into solid angle, $d\Omega = 2\pi \sin\theta\delta\theta$, gives the expression for the classical differential cross section: $\frac{d\sigma}{d\Omega} = \frac{b}{\sin\theta}\frac{db}{d\theta}$.*

Among the most important and interesting features in scattering are resonances, conditions for which $d\sigma/d\Omega$ is enhanced. The geometrical optics pictures for the three best known of these are shown in Fig. 11.9. The language used for light scattering from water droplets carries over into other scattering problems: rainbow and glory scattering are seen in both atomic and nuclear physics. In atomic physics, autoionization is analogous to the classical orbiting resonances and predissociation is an orbiting analog in molecular physics. Note that in contrast to the previous figure, Fig. 11.9 shows the reflected deflection of a beam from a uniform sphere only when it hits the sphere's surface. Thus, in these pictures, while it is to be understood that at each interface there is both a reflected and a transmitted beam, the only rays shown will be those relevant to the immediate discussion. The top drawing

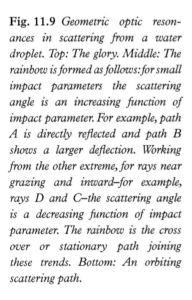

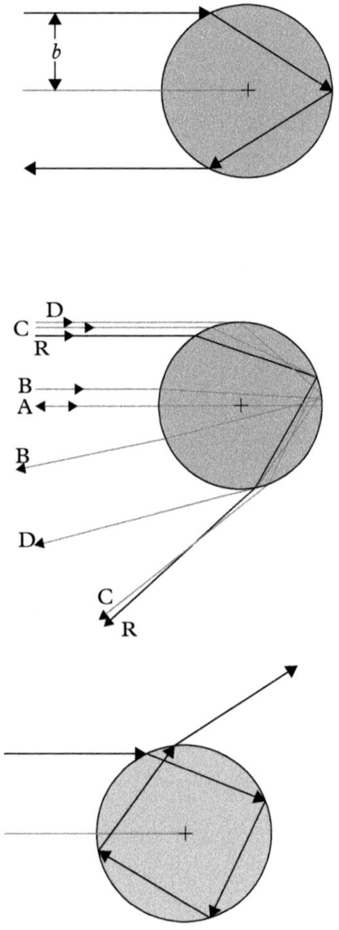

Fig. 11.9 *Geometric optic resonances in scattering from a water droplet. Top: The glory. Middle: The rainbow is formed as follows: for small impact parameters the scattering angle is an increasing function of impact parameter. For example, path A is directly reflected and path B shows a larger deflection. Working from the other extreme, for rays near grazing and inward–for example, rays D and C–the scattering angle is a decreasing function of impact parameter. The rainbow is the cross over or stationary path joining these trends. Bottom: An orbiting scattering path.*

in Fig. 11.9 is glory scattering. This occurs when rays with an impact parameter greater than zero scatter either directly forward ($\theta = 0^o$) or, as shown in the figure, directly backward ($\theta = 180^o$). In Eq. 11.162, the $\sin\theta$ in the denominator causes the classical differential cross section to become infinite in the backward (or forward) direction. In the wave picture, diffraction will soften the scattering and keep the differential cross section finite. The most famous resonance, the rainbow (middle drawing), may be understood as being created by a change in direction–either positive to negative or vice versa in the dependence of the deflection angle on impact parameter. Mathematically, $d\theta/db$ passes through zero for that situation and according to Eq. 11.162 the classical cross section becomes infinite. Again, diffraction keeps the corresponding wave physics cross-section finite. To see the origin of the formation of the primary rainbow of water droplets we consider the various one-internal-reflection ray paths shown in the middle drawing of Fig. 11.9. Starting at the ray labeled A whose deflection angle is 180° degrees and then considering ray B and so on, we see that the deflection angle is an increasing function of the impact parameter. On the other hand, considering near grazing rays, D and C, it is seen that the scattering angle is a decreasing function of impact parameter. For some impact parameter the function turns around, $d\theta/db = 0$, and when that happens the scattering cross section peaks and, through refractive dispersion, a rainbow is created. The bottom drawing in Fig. 11.9 shows an orbiting resonance. A sufficiently strong resonance may be thought of, classically, as orbiting several times. In this way, energy is stored within the sphere and the fields are enhanced. To some extent, orbiting occurs for any incident ray since every ray is partially reflected every time it hits a boundary. In most cases, however, a large fraction of the light is transmitted and so generating a strong resonance requires special circumstances. Generally, near-grazing rays are best for making a resonance since reflection coefficients usually increase with angle of incidence. Classically, it is not possible for rays in the incident beam to access a path having total internal reflections since rays that get in will necessarily be able to get out. As drawn in the figure, the path just misses closing on itself. However, one possibility is that a ray's path leads to retracing. In such a case, the droplet acts as a miniature resonator with constructive interference reducing the reflection losses.

Total cross sections, diffraction, and coronas

The first and most studied application of Mie theory is for light scattering by water droplets in air. In the optical region of the electromagnetic spectrum, water is essentially non-magnetic and characterized by an index of refraction of $n = 1.33$. Shown in Fig. 11.10 are total cross

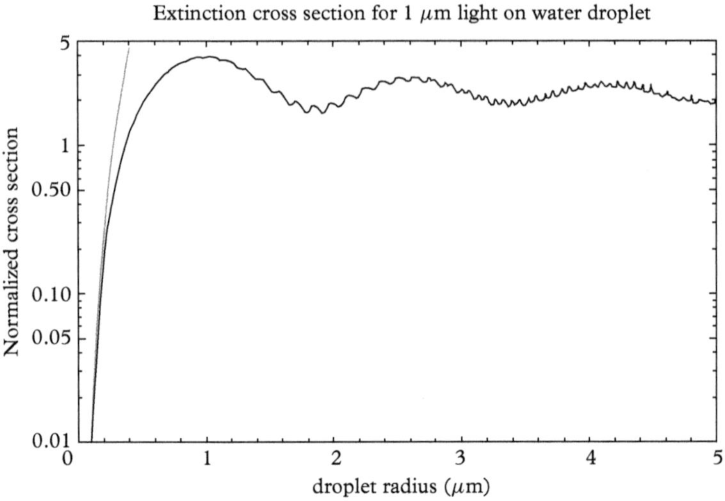

Fig. 11.10 *Total cross section for scattering of 1 μm light off various sized water droplets. The total cross section is scaled at each point by dividing by the corresponding geometric cross section, πr^2, of the water droplet. The black curve shows results of Mie theory calculations using 100 partial waves. The gray curve gives the dipole result for a small, electrically polarizable sphere.*

section Mie theory results in which we consider scattering of 1 μm light off various sized water droplets in the range of 1–5 μm so that $\beta \approx 1$. To the extent that dispersion may be neglected, the size dependence of a problem is specified entirely by the scaling due to β. And so, with this caveat these results may be equally understood as the wavelength dependence for scattering from a single sized sphere. The calculated cross sections in Fig. 11.10 are normalized by dividing by the corresponding geometrical cross section at each point. Although these calculations were carried out for σ^+ (circular polarized) radiation, the total cross sections are, in fact, independent of the polarization state of the incident radiation. Also, shown in gray on the graph are the electric dipole results from the long wavelength cross section Eq. 11.32. There is good agreement with the exact Mie results for radii less than about one tenth of the radiation wavelength, $\beta \lesssim 1$. For larger radii, the long wavelength approximation begins to fail. The overestimation, by the small sphere expression, of the cross section for larger spheres ($> 0.2\mu$m) can be explained: at some instant of time, the field varies significantly or even changes sign within the dielectric sphere, resulting in a smaller polarization than went into the long wavelength approximation and this, in turn, results in a smaller cross section than predicted.

For larger radii spheres three qualitative features of the Mie results are evident from the graph:

(1) The value of the total scattering cross section approaches twice the geometric cross section.

(2) Large scale "shape" resonances are observed.

(3) Tight wiggles further modulate the cross section.

That the asymptotic value of the cross section is nearly twice the geometric cross section seems inconsistent with the understanding that geometrical optics supplies the short wavelength limit for electromagnetic waves. This discrepancy is referred to as the extinction paradox. However, the important physics is that geometric optics absolutely does not include is diffraction. It is easy to see that cross sections larger than the geometric cross section occur all the time by considering light incident on a large, totally absorbing disk. Geometric optics correctly predicts the short wavelength absorption cross section to be the πR^2 area of the disk. However, additional light is scattered from the incident beam by diffraction at the boundary of the disk. The amount of light scattered into the diffraction pattern may be estimated using Babinet's principle which states that a target and its complement generate the same diffraction pattern. In the case of an absorbing disk, the complementary target is a screen with a hole in it the same size as the disk. For a hole that is large compared with the wavelength of the light, when illuminated by a plane wave, the light in the downstream diffraction pattern is simply the same as that passing through the πR^2 area of the hole. Thus the total extinction cross section for the disk is $2\pi R^2$: half from absorption and half from diffraction. Similarly, for a water droplet, in addition to the light intercepted and redirected by the droplet, additional light is bent by diffraction. And for a spherical droplet large relative to the light wavelength, diffraction adds about πR^2 to the geometrical cross section. In atmospheric optics, this diffracted light creates a corona as seen in Fig. 11.11. The large ripples in Fig. 11.10 appear to oscillate with a period of about 1.5 μm. These may be qualitatively understood as due to the downstream interference between the incident beam passing through the sphere and that passing by the sphere but not interacting. The difference in optical path length for two such beams across the diameter of the sphere is related to the phase shift between the two paths:

$$\frac{\delta\phi}{2\pi} = \frac{2a_0}{\lambda_{water}} - \frac{2a_0}{\lambda_0} = \frac{2a_0}{\lambda_0}(n_{water} - 1) \approx 0.66\frac{a_0}{\lambda_0} \tag{11.163}$$

This predicts resonance structure between these two paths, that is, $\delta\phi = 2\pi$ as a function of sphere radii for $\delta a_0 \approx 1.5\lambda_0 = 1.5\,\mu$m — exactly the periodicity seen in the graph. Finally, the small scale wiggles can be seen as arising from the vanishing of the denominators for individual a_l and b_l coefficients in the partial wave expansion. We

Fig. 11.11 *Optical corona. These rings around the moon are the diffraction off water droplets in the atmosphere. In this case, the well defined colors in the bands indicate that the droplets scattering the light are all nearly identical in size. Credit: W. Salzmann (Own Work via Wikimedia Commons).*

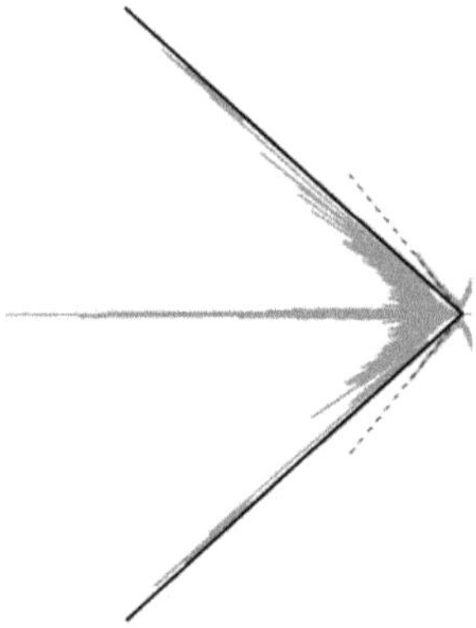

Fig. 11.12 *Polar plot for the backward scattering from a water droplet. The water droplet is located at the convergence near the right hand edge of the plot. Prominent in this figure is the strong* 180° *peak associated with the "glory." Also shown are the primary and secondary rainbows peaks, indicated by the solid and dashed black lines, respectively. These are drawn at* 138° *and* 129° *as measured from the forward direction.*

will look at some of these in more detail when we discuss orbiting resonances.

Differential cross section: The glory and the rainbow

The gray features in Fig. 11.12 are a polar plot of a Mie calculation for a plane wave with wavelength 1 μm scattering from a water droplet with a 100 μm radius. The droplet is located at the convergence of the rays on the right hand side of the figure. The forward scattering is not shown here, but it is dominated by a large central diffraction peak flanked by a diffraction pattern very similar to that for an absorbing disk of the same cross sectional area. In the backward direction, while the resolution is not good enough to resolve individual interference peaks, there are several distinctive features. Directly backward is an enormous scattering peak that will create an optical glory as described next. Ranging from this backward scattering at 180° to the black lines in the diagram are a converging series of peaks.

These are the wave physics predictions for a rainbow with the black line drawn at 138° being the ray optics asymptote for the primary rainbow. Additionally, a weaker series of peaks, converging from the forward scattering direction (0°) to the dashed lines at 129°, describe a secondary rainbow. Beyond the primary rainbow scattering, the pattern may roughly be divided into two sections: a group of small peaks around the direct backward scattering that give way to significantly larger peaks leading up to the primary rainbow. Referring to the middle drawing of Fig. 11.9, the small peaks correspond to scattering of near-axial ray paths such as A and B. The larger peaks are a combination of contributions from both near-axial and near-grazing ray paths such as D and C–they result from the interference of two optical paths scattering into the same angle. To this point, the rainbow as described would be white. Actual rainbows observed in the sky are not white but rather show a color separation resulting from a slight dependence of the water dielectric constant, and thus the index of refraction, on light wavelength. Working against this is blurring due to diffraction which is significant even for the $100\,\mu$m radius droplet of the calculation. Indeed, the best rainbows are made from 1 mm or larger water drops. Finally, a casual observation of a rainbow through polarized sunglasses will show that rainbow light is strongly linearly polarized with the electric field vector tangent to the rainbow's arc. This is confirmed in the Mie calculation and may be understood from the ray picture of the middle drawing of Fig. 11.9 as follows: unpolarized light is transmitted almost entirely through the front of the droplet. However, because the incident angle of reflection at the back of the droplet is about 30° and near the Brewster's angle of 37° for the water→air system, P-polarized light is mostly transmitted but some S-polarized light (polarization perpendicular to the page) is reflected and continues on the path shown to form the rainbow.

The original account of the atmospheric glory is from 1735. Members of a French scientific expedition traveling in the Peruvian Andes observed their shadows on a ground level cloud. Each member of the expedition observed that his shadow's head was decorated by a multi-colored halo (**see Discussion 11.16**). Figure 11.13 shows a picture of a modern day glory. The location of the shadow defines the direction of incidence of the sun's rays with the glory-a clear result of

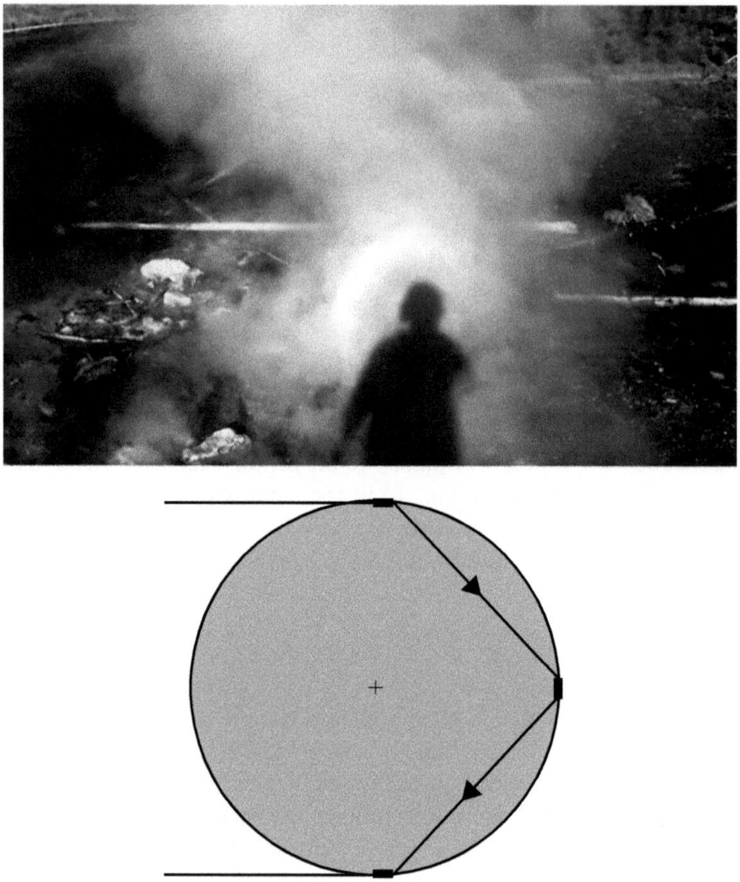

Fig. 11.13 *The glory. Top: Glory scattering is almost exactly backscattered (or forward scattered) light. Here the colored halo around the photographer is light backscattered off steam droplets formed in a hot spring. Credit: Brocken Inaglory (Own Work via Wikimedia Commons) Bottom: An improved ray optics-type picture includes diffraction of the rays on scattering to bridge gaps between ray segments. This is otherwise described as decaying surface waves and indicated by the black extended refraction area on the figure.*

strong backscattering of those rays from water droplets in the cloud. The exact origin of this backscattering, however, is subtle. Here we outline some general physical phenomena relevant to the problem. A more complete quantitative explanation is given by Nussenzveig.[27] The top drawing in Fig. 11.9 provides a geometrical optics description of the glory and as far as the term is used in more general scattering problems, this is an accurate picture. Yet, for a uniform dielectric sphere with an index of refraction less than $\sqrt{2}$, it is easily shown that there are no single bounce paths of the type shown. For water with index 1.33, the largest scattering angle for a single bounce non-axial ray ($b > 0$) is about 165° or 15° short of direct backscatter. In fact, the first $b > 0$ direct backward scattering path requires four internal reflections and the cumulative transmission losses for these yield rays much too weak to match up with observation. Van de Hulst[28] was first to describe the physics that was later shown to partially

[27] Nussenzveig, H. M., *op cit.*

[28] van de Hulst, H. C., *Light Scattering by Small Particles.* Wiley, N.Y (1957).

explain the phenomena. The basic idea is that we should include a diffraction broadening with the rays of the geometrical optics picture: the reflective bounce of a ray at the sphere's surface is not entirely point-like but will be broadened by diffraction. For our case, this diffraction will take the form of rapidly decaying surface waves. For the atmospheric glory, an improved ray-based picture is shown in the bottom of Fig. 11.13. The short, thick segments are surface waves, created by diffraction, which bridge the small gaps between the main ray segments. Since surface waves die off rapidly, smaller droplets with smaller gaps are somewhat better. Indeed, the best glories are observed from $\sim 25\ \mu$m diameter droplets and the phenomena is not observed in drops with $\beta > 1000$ or above $\sim 500\ \mu$m in diameter. Additionally, high-order orbiting type modes (as will be discussed in the next section) further enhance the backscattering.

Orbiting resonances and whispering gallery modes

In Fig. 11.14, we show a Mie calculation of the total cross section for a 5 μm radius sphere scattering light near the green part of the spectrum. In contrast to the earlier total cross section picture of Fig. 11.10, the size index, β, indicated along the x-axis at the top of the graph, is shown varying with the wavelength of the incident radiation rather than the size of the sphere. Imposed on the nominal twice-the-geometric cross section result are resonances. Some are so sharp that they are unresolved in the graph. Graphed in Fig. 11.4 in blue are the partial wave contributions from a couple of partial waves and these are clearly the sources of the resonances. As discussed before, a given partial wave, l, corresponds classically to scattering for an impact parameter $b = l\lambda/2\pi$. Working from the classical view, the

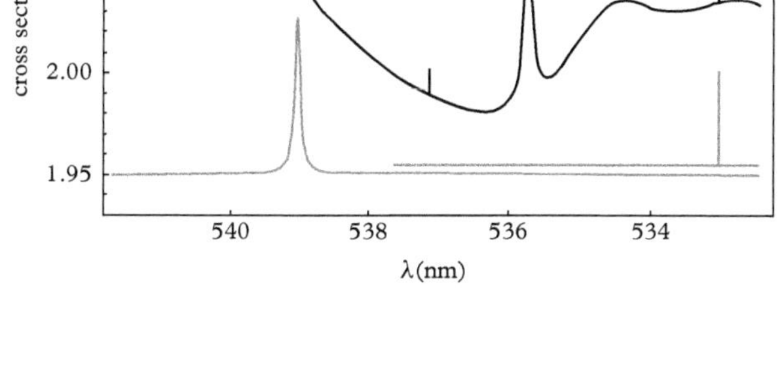

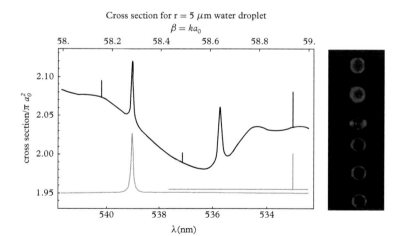

Fig. 11.14 *The wave theory equivalent of orbiting resonances, whispering gallery modes. Shown on the left is scattering of a part of the visible (green) spectrum off a water droplet. On the top x-axis, the size parameter is indicated. The total cross section is shown by the black curve. In the two lower curves, the origin of a couple of the resonances is identified with partial waves. The lowest curve is the partial wave of the magnetic-multipole-like contribution to the $l = 65$ partial wave, $\vec{M}_{65,1}$. The second, incredibly narrow, resonance is the electric-type contribution to the $l = 71$ partial wave, $\vec{N}_{71,1}$. The other peaks may be similarly associated with partial waves. In all of these cases, the partial wave index, l, is larger than the nominal size parameter, β. On the right: A spherical droplet can be used as an optical cavity.*

partial wave corresponding to the boundary of the sphere ($b = a_0$) is $l = 2\pi a_0/\lambda = \beta$, which, for this figure varies from 58 to 59, depending on wavelength.

Classically, the flux from higher partial waves do not hit the sphere. Yet the resonances in this figure are due to $l = 65$ and the incredibly sharp resonance, $l = 71$. Referring to the top drawing of Fig. 11.5, and considering instead a partial wave of value 71, the boundary of a sphere with $\beta = 59$ would lie to the left of the exponentially dying amplitude of the partial wave: physically, radiation of the partial wave would be very weakly but nonetheless coupled to the sphere. Contrariwise, while it is difficult to couple radiation into the sphere, once it is there, it is only weakly coupled out. Very narrow resonances in wavelength/frequency correspond, in the time domain, to photons rattling around for a long time in the sphere before finally making it out. Thus, that this flux classically would not even hit the sphere yet couples to it can be understood as a tunneling of the partial wave followed by, in the ray picture, shallow very near total reflections or, in the wave picture, surface waves. These are alternatively called morphology dependent resonances (MDRs) or whispering gallery mode resonances. The latter term is used because the behavior resembles the peculiar acoustics originally observed in St. Paul's Cathedral in London, where sounds on one side of the cathedral could be heard clearly on the other. The phenomena was originally explained by Rayleigh as due to surface waves traveling across the surface of the cathedral, continually being focused by the concave shape of the dome. The resonances can be incredibly sensitive to the details of the system and have found applications in sensing and high precision particle sizing.

An early demonstration of using these modes as a laser resonator is shown on the right of Fig. 11.14. Here, droplets of dye-doped ethanol are irradiated from left to right in the figure, by 532 nm laser light. Efficient coupling into a mode from the incident beam is only possible at the resonance of the mode. The orifice forming the droplets was piezo-controlled and allowed adjustment of the size of the droplets from 20 to 40 μm and, in turn, of the tuning of the mode frequencies. On generating droplets of a size for which a MDR occurred at 532 nm, lasing was observed. The figure shows the formation of droplets as they exit the orifice. They initially start out as blobs but by the lower frames have evolved into spheres with a well-defined radius. The gain curve of the dye is broad enough that for any sized droplet at least one resonance was in the red part of the gain spectrum and could produce gain and lasing.

Exercises

(11.1) The sodium D resonance occurs for light with wavelength 590 nm. Treating the transition as an electron on a spring, find the classical acceleration that would be imparted to an atom for resonant light of intensity I.

(11.2) Find the magnetic dipole cross section–both differential and total–for a perfectly conducting loop with radius $a_0 \ll \lambda$.

(11.3) Confirm that electric dipole radiation satisfies the optical theorem.

(11.4) Using the partial wave expansion, find the differential cross section for a perfectly conducting sphere of radius $a_0 = 2\lambda$. Compare with the dipole approximation.

(11.5) Compare the Born approximation for the total scattering cross section of a small dielectric sphere by (a) integrating the differential cross section and (b) using the imaginary part of the forward scattering.

(11.6) Solve the Mie scattering problem for a sphere with an index n for scalar waves. Assume continuity of the wave function and its perpendicular derivative at the sphere's boundary.

(11.7) Using the results of the previous problem, identify some sharp, scalar whispering gallery mode resonances.

(11.8) This and the next few problems will develop cylindrical symmetry analogs to the central potential and partial wave formalism of the chapter. Consider a long cylindrical target symmetric about the z-axis and extending infinitely along the z direction. Incident on the target is a plane wave, $\sim\exp(ikx)$. Write down the cylindrical analog of Eq. 11.3 and carry out the separation of variables of the Helmholtz equation in cylindrical coordinates to find a set of scalar basis functions suitable for expanding the incident and scattered waves.

(11.9) Expand $\sim\exp(ikx)$ using the basis functions you found in the previous problem.

(11.10) Using the results of the previous problems, carry out the cylindrical coordinate analog of the scalar partial wave expansion.

(11.11) Using the results of the previous problem, find the scattering cross section per unit length of a long, uniform dielectric cylinder oriented along the z-axis with radius a_0. Use $\exp(ikx)$ for the incident scalar wave.

(11.12) Numerically use the previous results and identify cylindrical whispering gallery modes.

(11.13) Find the cylindrical coordinate's analog of the Stratton $\vec{M}$ and $\vec{N}$ functions.

(11.14) Carry out the cylindrical coordinate analog of the vector partial wave expansion.

(11.15) Calculate c_l and d_l coefficients for the internal fields of the Mie solution.

(11.16) For real permittivity and $\mu_I = \mu_{II} = \mu_0$, show that the Mie solution obeys the optical theorem.

(11.17) Plot the fields within the sphere corresponding to the 4 lowest terms in the vector partial wave solutions of the Mie problem.

(11.18) Find the scattering in the backward direction for a whispering gallery mode resonance.

(11.19) Estimate how much polarization is to be expected in skylight scattered at angle θ. Assume that the light from the sun is unpolarized and work in the single scattering limit.

(11.20) Show that for $x = a_0/\lambda \ll 1$, the ratio of forward to backward scattering is:

$$1 + \frac{4}{15}x^2\frac{\left(n^2+4\right)\left(n^2+2\right)}{2n^2+3} + O\left(x^4\right)$$

11.6 Discussions

Discussion 11.1

This relation between the complex Poynting vector $\vec{S}_c$ and the time average of the real Poynting vector, $\langle\vec{S}\rangle = \langle Re\vec{E} \times Re\vec{H}\rangle$, is a general and useful relation for any quantity which is a product of two sinusoidally varying complex quantities such as $\vec{E}$ and $\vec{H}$. That is, for any observable real quantity, $\vec{C}$, which is the product of the real parts of two complex, sinusoidally varying quantities $\vec{A}$ and $\vec{B}$, we can easily find the time average of the observable by the relation $\langle\vec{C}\rangle = \frac{1}{2}Re\left(\vec{A} * \vec{B}^*\right)$ where the multiplication "*" can be a cross product, inner product, and so on. For products in which the two factors are the same, such as with energy density, it is not necessary to take the real part since the product is already completely real.

Discussion 11.2

We consider the damping of the same system that has been set into oscillation but is no longer driven. The decay of the oscillation will be determined by the radiation loss of energy from the system. We assume the system decays slowly in the sense that the amplitude changes fractionally by a very small amount over several cycles of oscillation. The equation of motion for the amplitude of the system is simply Eq. 11.18 without the driving term:

$$\ddot{x} = -\omega_0^2 x - \gamma\dot{x} \tag{11.164}$$

We assume a solution for this equation of the form, $x(t) = x_0 e^{-ist}$ and substitute into Eq. 11.164 to get

$$s^2 - \omega_0^2 = -i\gamma s \tag{11.165}$$

which, upon solving the quadratic equation for s and assuming very weak damping, that is, $s, \omega_0 \gg \gamma$, gives:

$$s \simeq \omega_0 - i\frac{\gamma}{2} \tag{11.166}$$

We write:

$$x = x_0 e^{-i\omega_0 t} e^{-(\gamma/2)t} \tag{11.167}$$

Thus, if we identify $\bar{x}(t)$ to be the slowly decaying amplitude of the electron's oscillation,

$$\bar{x}(t) = \bar{x}(0)e^{-(\gamma/2)t} \Leftrightarrow \frac{d\bar{x}}{dt} = -\frac{\gamma}{2}\bar{x} \tag{11.168}$$

As given by Eq. 11.6, the decay of the motion due to the radiation loss is

$$\frac{dE}{dt} = -\langle P\rangle_{ed} = -\frac{\omega_0^4 |\vec{p}|^2}{12\pi\varepsilon_0 c^3} \tag{11.169}$$

The energy in the system is given by the harmonic oscillator spring at its full extension when the kinetic energy is zero, $E = \frac{1}{2}k\bar{x}^2 = \frac{1}{2}m\omega_0^2\bar{x}^2$. Differentiating this with time and explicitly writing the dipole moment $|\vec{p}| = -e\bar{x}$ gives:

$$m\omega_0^2\bar{x}\frac{d\bar{x}}{dt} = -\frac{\omega_0^4 e^2\bar{x}^2}{12\pi\varepsilon_0 c^3} \Rightarrow \gamma = \frac{\omega_0^2 e^2}{6\pi\varepsilon_0 mc^3} \tag{11.170}$$

Discussion 11.3

Inserting Eq. 11.53 into the left-hand side of Eq. 11.52,

$$\oint_R \mathrm{Im}\left[(\varphi^*\nabla\varphi)\cdot\hat{r}\right] dS = R^2 \mathrm{Im} \oint_R d\phi d\theta \sin\theta \left\{\left(e^{-ikR\cos\theta} + \frac{e^{-ikR}}{R}f^*\right)\nabla\left(e^{ikR\cos\theta} + \frac{e^{ikR}}{R}f\right)\cdot\hat{r}\right\}$$

where $z = R\cos\theta$. The integrand in brackets is next written as

$$\left(e^{-ikR\cos\theta} + \frac{e^{-ikR}}{R}f^*\right)\left[ik\cos\theta e^{ikR\cos\theta} + \left(ik\frac{e^{ikR}}{R} - \frac{e^{ikR}}{R^2}\right)f\right]$$

which can be expanded to

$$ik\cos\theta + \left(ik\frac{e^{ik(R-R\cos\theta)}}{R} - \frac{e^{ik(R-R\cos\theta)}}{R^2}\right)f + \left(ik\cos\theta\frac{e^{-ik(R-R\cos\theta)}}{R}\right)f^* + \left(\frac{ik}{R^2} - \frac{1}{R^3}\right)|f^2|$$

Discussion 11.4

For $R \to \infty$, we expand about $\theta = 0$, keeping $\cos\theta$ out to 2nd order but keeping only the first non-zero terms of $f(\theta,\phi)$ and $\sin\theta$,

$$\begin{aligned} I_1 &= \mathrm{Im}\left[ikRe^{ikR}\int_0^{2\pi} d\phi \int_0^{\pi} f(\theta,\phi)\sin\theta\exp(-ikR\cos\theta)\,d\theta\right] \\ &\approx \mathrm{Im}\left[ikRe^{ikR}\int_0^{2\pi} d\phi \int_0^{\pi} f(0,\phi)\,\theta\exp\left[-ikR\left(1-\frac{\theta^2}{2}\right)\right]d\theta\right] \end{aligned}$$

Discussion 11.5

Generally, an integral of the form $\vec{I} = \int d^3x' \exp\left(-ik\hat{r}\cdot\vec{r'}\right) \nabla' \times \vec{F}\left(\vec{r'}\right)$ is equivalent to the same integral expression but with the substitution $\nabla' \to ik\hat{r}$. Proof: $\nabla' \times \vec{F}$ consists of combinations of terms of the form $\partial F_i/\partial x'_j$. We integrate by parts, over x'_j, one of these components:

$$\int d^3x' \exp\left(-ik\hat{r}\cdot\vec{r'}\right)\frac{\partial F_i}{\partial x'_j} = \int d^2x' \exp\left(-ik\hat{r}\cdot\vec{r'}\right) F_i\left(\vec{r'}\right)\Big|_{x'_j=-\infty}^{x'_j=+\infty} - \int d^3x' \exp\left(-ik\hat{r}\cdot\vec{r'}\right)\left(-ik\hat{r}_j F_i\right) \tag{11.171}$$

where the notation $\hat{r}_j = \left(x_j/r\right)$ indicates the jth cartesian component of the radial unit vector. The first, surface term is evaluated over the other variables for the j variable at infinity. As long is $F_i\left(\vec{r'}\right)$ is confined in space, this term vanishes leaving just the second term. Now, replace $\partial F_i/\partial x'_j$ in Eq. 11.171 with the kth component of $\nabla' \times \vec{F}$

$$\begin{aligned}\int d^3x' \exp\left(-ik\hat{r}\cdot\vec{r'}\right)\left(\nabla' \times \vec{F}\right)_k &= -\int d^3x' \exp\left(-ik\hat{r}\cdot\vec{r'}\right)\left(-ik\hat{r}_i F_j + ik\hat{r}_j F_i\right)\\ &= \int d^3x' \exp\left(-ik\hat{r}\cdot\vec{r'}\right)\left(ik\hat{r}_i F_j - ik\hat{r}_j F_i\right)\\ &\int d^3x' \exp\left(-ik\hat{r}\cdot\vec{r'}\right)\left(ik\hat{r} \times \vec{F}\right)_k\end{aligned}$$

so proving that $\nabla' \to ik\hat{r}$ for the jth and thus all the components of $\nabla' \times \vec{F}$.

Discussion 11.6

Using the relation $Y_{l0}(\theta) = \sqrt{\frac{2l+1}{4\pi}} P_l(\cos\theta)$, the ϕ integral in 11.99 simply gives a factor of 2π. We can rewrite Eqs. 11.98 and 11.99

$$\exp(ikr\cos\theta) = \sum_{l=0}^{\infty} \sqrt{\frac{2l+1}{4\pi}} A_l j_l(kr) P_l(\cos\theta) \tag{11.172}$$

and

$$A_l j_l(kr) = \sqrt{2\pi\frac{2l+1}{2}} \int_0^{\pi} \exp(ikr\cos\theta) P_l(\cos\theta) \sin\theta d\theta \tag{11.173}$$

where the Legendre polynomials, $P_l(\cos\theta)$, are a real, complete and orthogonal set on the interval $-1 \le \cos\theta \le 1$. Next, we note a useful integral expression for the spherical Bessel functions of the first kind (dlmf.nist.gov, eq. 10.54.2):

$$j_l(kr) = \frac{(-i)^l}{2} \int_0^{\pi} \exp(ikr\cos\theta) P_l(\cos\theta) \sin\theta d\theta \tag{11.174}$$

Comparing this result with Eq. 11.173, we obtain the coefficients A_l

$$A_l = i^l \sqrt{4\pi(2l+1)} \tag{11.175}$$

Discussion 11.7

We note that this treatment started with the assumption that the plane wave was directed along the z- (or polar) axis and that the angle θ at which we evaluate the wave was with respect to this direction. In general, however, the plane wave will be directed arbitrarily within a given coordinate system. We can generalize Eq. 11.172:

$$\exp\left(i\vec{k}\cdot\vec{r}\right) = \exp(ikr\cos\alpha) = \sum_{l=0}^{\infty} \sqrt{\frac{2l+1}{4\pi}} A_l j_l(kr) P_l(\cos\alpha) \tag{11.176}$$

where α is the angle between the direction of the wave's propagation, θ_k, ϕ_k, and the direction of observation, θ, ϕ. The addition theorem states that for any two directions, θ, ϕ and θ_k, ϕ_k, the Legendre polynomial of order l for the angle, α, between them is (dlmf.nist.gov, eq. 14.30.9):

$$P_l(\cos\alpha) = \frac{4\pi}{2l+1} \sum_{m=-l}^{l} Y_{lm}^*(\theta_k, \phi_k) Y_{lm}(\theta, \phi) \tag{11.177}$$

Substituting Eqs. 11.177 and 11.175 into 11.176, we thus obtain the general form

$$\exp\left(i\vec{k}\cdot\vec{r}\right) = 4\pi \sum_{l=0}^{\infty} \sum_{m=-l}^{l} i^l j_l(kr) Y_{lm}^*(\theta', \phi') Y_{lm}(\theta, \phi) \tag{11.178}$$

Discussion 11.8

Rewrite Eq. 11.114 as

$$\vec{M} = \nabla \times \psi\vec{C} = \nabla\psi \times \vec{C} = \psi\left(i\vec{k} \times \vec{C}\right)$$

since $\nabla \times \vec{C} = 0$, and

$$\nabla \times \vec{M} = \nabla \times \psi\left(i\vec{k} \times \vec{C}\right) = \psi\left[i\vec{k} \times \left(i\vec{k} \times \vec{C}\right)\right]$$

since $\nabla \times \left(i\vec{k} \times \vec{C}\right) = 0$. Taking one last curl, and noting that $\nabla \times \left[i\vec{k} \times \left(i\vec{k} \times \vec{C}\right)\right] = 0$,

$$\begin{aligned}\nabla \times \left(\nabla \times \vec{M}\right) &= i\vec{k} \times \left[i\vec{k} \times \psi\left(i\vec{k} \times \vec{C}\right)\right] = -k^2\left[\hat{k} \times \left(\hat{k} \times \vec{M}\right)\right] \\ &= -k^2\left[\left(\hat{k} \cdot \vec{M}\right)\hat{k} - \left(\hat{k} \cdot \hat{k}\right)\vec{M}\right] = k^2\vec{M}\end{aligned} \tag{11.179}$$

where we have used the vector identity $\vec{a} \times \left(\vec{b} \times \vec{c}\right) = (\vec{a} \cdot \vec{c})\vec{b} - \left(\vec{a} \cdot \vec{b}\right)\vec{c}$ and the fact that $\hat{k} \perp \vec{M}$. From Eq. 11.115 and $\nabla \cdot \vec{M} = 0$, we also have

$$\nabla \times \left(\nabla \times \vec{M}\right) = -\nabla^2\vec{M}$$

so $\vec{M}$ is a solution to the vector Helmholtz equation,

$$\nabla^2\vec{M} + k^2\vec{M} = 0$$

Discussion 11.9

Using the vector identities,

$$\begin{aligned}\nabla^2\left(\nabla \times \vec{A}\right) &= -\nabla \times \left[\nabla \times \left(\nabla \times \vec{A}\right)\right] = \nabla \times \nabla^2\vec{A} \\ (\nabla\psi \cdot \nabla)\vec{r} &= (\nabla\psi \cdot \hat{r})\,\hat{r} = \frac{\partial \psi}{\partial r}\hat{r} \\ \nabla \times \vec{r} &= 0\end{aligned}$$

it can be shown that the $\vec{M}$ and $\vec{N}$ vector functions are a suitable extension of the Stratton approach for spherical coordinates. Proof of first vector identity: The first equality is obtained in general since the divergence of a curl vanishes,

$$\begin{aligned}\nabla \times \left[\nabla \times \left(\nabla \times \vec{A}\right)\right] &= \nabla\left[\nabla \cdot \left(\nabla \times \vec{A}\right)\right] - \nabla^2\left(\nabla \times \vec{A}\right) \\ &= -\nabla^2\left(\nabla \times \vec{A}\right)\end{aligned}$$

And the second equality in the first vector identity is obtained in general by first writing,

$$\nabla \times \left(\nabla \times \vec{A}\right) = \nabla\left(\nabla \cdot \vec{A}\right) - \nabla^2 \vec{A}$$

and taking the curl

$$\begin{aligned}\nabla \times \left[\nabla \times \left(\nabla \times \vec{A}\right)\right] &= \nabla \times \left[\nabla\left(\nabla \cdot \vec{A}\right) - \nabla^2 \vec{A}\right] \\ &= -\nabla \times \nabla^2 \vec{A}\end{aligned}$$

where we have used the fact that the curl of a gradient vanishes.

Discussion 11.10

In the Appendix, the third vector spherical harmonic is defined as

$$\vec{\Phi}_{lm} = \frac{1}{\sqrt{l(l+1)}} \vec{r} \times \nabla Y_{lm}(\theta,\phi)$$

with $\psi_{lm}^{(a)} = z_l^{(a)}(kr)\, Y_{lm}(\theta,\phi)$, the curl in Eq. 11.121 can be expanded

$$\begin{aligned}\nabla \times \left(\vec{r}\psi_{lm}^{(a)}\right) &= \nabla \times \left(z_l^{(a)} Y_{lm}\right)\vec{r} = \nabla\left(z_l^{(a)} Y_{lm}\right) \times \vec{r} + \left(z_l^{(a)} Y_{lm}\right) \nabla \times \vec{r} \\ &= \nabla\left(z_l^{(a)} Y_{lm}\right) \times \vec{r} \\ &= \left[z_l^{(a)} \nabla Y_{lm} + Y_{lm} \nabla z_l^{(a)}\right] \times \vec{r} \\ &= \left[z_l^{(a)} \nabla Y_{lm} + Y_{lm} \frac{\partial z_l^{(a)}}{\partial r} \hat{r}\right] \times \vec{r} \\ &= z_l^{(a)} \nabla Y_{lm} \times \vec{r}\end{aligned}$$

so that Eq. 11.121 can be alternately expressed as

$$\vec{M}_{ml}^{(a)} = \frac{z_l^{(a)}(kr)}{\sqrt{l(l+1)}} \left(\vec{r} \times \nabla Y_{lm}(\theta,\phi)\right) = z_l^{(a)}(kr)\, \vec{\Phi}_{lm}$$

Discussion 11.11

From **Discussion 11.10**, M clearly has just $\hat{\theta}$ and $\hat{\phi}$ components:

$$\vec{M}_{lm} = -\frac{1}{\sqrt{l(l+1)}} z_l^{(a)} \nabla Y_{lm} \times \vec{r} = z_l^{(a)} \vec{\Phi}_{lm} \tag{11.180}$$

Let's evaluate this. In spherical coordinates,

$$\nabla f = \frac{\partial f}{\partial r}\hat{r} + \frac{1}{r}\frac{\partial f}{\partial \theta}\hat{\theta} + \frac{1}{r\sin\theta}\frac{\partial f}{\partial \phi}\hat{\phi} \tag{11.181}$$

$$\vec{M}_{lm} = z_l^{(a)} \frac{1}{\sqrt{l(l+1)}} \vec{r} \times \left(\frac{im}{r\sin\theta} Y_{lm}\hat{\phi} + \frac{1}{r}\frac{\partial Y_{lm}}{\partial \theta}\hat{\theta} \right) \tag{11.182}$$

To do the cross product, note:

$$\hat{r} \times \hat{\phi} = -\hat{\theta} \text{ and } \hat{r} \times \hat{\theta} = \hat{\phi} \tag{11.183}$$

$$\vec{M}_{lm} = z_l^{(a)} \frac{1}{\sqrt{l(l+1)}} \left(-\frac{im}{\sin\theta} Y_{lm}\hat{\theta} + \frac{\partial Y_{lm}}{\partial \theta}\hat{\phi} \right) \tag{11.184}$$

Next, referring to proof #3 in Problem A.2 of Appendix A

$$\begin{aligned}
\vec{N}_{lm} &= \tfrac{1}{k}\nabla \times \left(z_l^{(a)} \vec{\Phi}_{lm} \right) = -\tfrac{1}{k}\tfrac{\sqrt{l(l+1)}}{r} z_l^{(a)} \vec{Y}_{lm} - \tfrac{1}{k}\left(\tfrac{\partial z_l^{(a)}}{\partial r} + \tfrac{1}{r} z_l^{(a)} \right) \vec{\Psi}_{lm} \\
&= -\tfrac{\sqrt{l(l+1)}}{\rho} z_l^{(a)} Y_{lm}\hat{r} - \tfrac{1}{\rho}\tfrac{\partial}{\partial \rho}\left(\rho z_l^{(a)} \right) \tfrac{1}{\sqrt{l(l+1)}} \left(\tfrac{\partial Y_{lm}}{\partial \theta}\hat{\theta} + \tfrac{im}{\sin\theta} Y_{lm}\hat{\phi} \right) \\
&= -\tfrac{\sqrt{l(l+1)}}{\rho} z_l^{(a)} Y_{lm}\hat{r} - \tfrac{1}{\rho\sqrt{l(l+1)}}\tfrac{\partial}{\partial \rho}\left(\rho z_l^{(a)} \right) \left(\tfrac{\partial Y_{lm}}{\partial \theta}\hat{\theta} + \tfrac{im}{\sin\theta} Y_{lm}\hat{\phi} \right)
\end{aligned} \tag{11.185}$$

Since the $\vec{M}$ are radial functions times $\vec{\Phi}$-type spherical vector harmonics and the $\vec{N}$ are a linear combination of the $\vec{Y}$ and $\vec{\Psi}$ types, the orthogonality of these different types of spherical harmonics ensures the orthogonality of the $\vec{M}$'s relative to the $\vec{N}$'s when integrated over all solid angles.

Discussion 11.12

We know from Appendix A that vector spherical harmonics of differing types and/or differing lm are orthogonal over the function space of (θ, ϕ). We note, in particular, that the vector spherical harmonics we use are normalized such that:

$$\int \vec{\Phi}_{lm}^{*} \cdot G(r)\vec{\Phi}_{l'm'} d\Omega = G(r)\delta_{ll'}\delta_{mm'} \tag{11.186}$$

with these, therefore, the scalar product of $\vec{\Phi}^*_{lm}$ and Eq. 11.130,

$$\int \vec{\Phi}^*_{lm} \cdot \left[E_o \hat{e}^{\pm}_{\perp} \exp(ikr\cos\theta)\right] d\Omega = \int \vec{\Phi}^*_{lm} \cdot \sum_{l'm'} A^{\pm}_{l'm'} \vec{N}^{(1)}_{l'm'} d\Omega + \int \vec{\Phi}^*_{lm} \cdot \sum_{l'm'} A^{\pm}_{l'm'} \vec{M}^{(1)}_{l'm'} d\Omega$$

$$= \int d\Omega \vec{\Phi}^*_{lm} \cdot \sum_{l'm'} B^{\pm}_{l'm'} \left[-\frac{1}{k} \frac{\sqrt{l(l+1)}}{r} j^{(a)}_{l} \vec{Y}_{l'm'} - \frac{1}{k} \left(\frac{\partial j^{(a)}_{l'}}{\partial r} + \frac{1}{r} j^{(a)}_{l'} \right) \vec{\Psi}_{l'm'} \right]$$

$$+ \int \vec{\Phi}^*_{lm} \cdot \sum_{l'm'} \left[A^{\pm}_{l'm'} j_{l'}(kr) \right] \vec{\Phi}_{l'm'} d\Omega. \qquad (11.187)$$

Because of the orthogonality of the vector spherical harmonics, the first integral on the right-hand side is zero and this result is seen to reduce to Eq. 11.131.

Discussion 11.13

With the definition of $\vec{\Phi}_{lm}$ given in Appendix A as

$$\vec{\Phi}_{lm} = \frac{1}{\sqrt{l(l+1)}} \vec{r} \times \nabla Y_{lm}$$

and $\hat{\vec{L}} = \frac{1}{i}(\vec{r} \times \nabla)$, we see that

$$\vec{\Phi}^*_{lm} = \left[\frac{i}{\sqrt{l(l+1)}} \frac{1}{i} (\vec{r} \times \nabla) Y_{lm} \right]^* = \frac{-i}{\sqrt{l(l+1)}} \left(\hat{\vec{L}} Y_{lm} \right)^*$$

Expressing the $\hat{\vec{L}}$ operator as

$$\hat{\vec{L}} = \hat{L}_x \hat{x} + \hat{L}_y \hat{y} + \hat{L}_z \hat{z} = \frac{\left(\hat{L}_+ + \hat{L}_-\right)}{2} \hat{x} + \frac{\left(\hat{L}_+ - \hat{L}_-\right)}{2i} \hat{y} + \hat{L}_z \hat{z} \qquad (11.188)$$

and using the polarization unit vectors, $\hat{e}^{\pm}_{\perp}$, as in Eq. 11.127, the scalar product within the integral of Eq. 11.131 can be written as

$$\vec{\Phi}^*_{lm} \cdot \hat{e}^{\pm}_{\perp} = \frac{-i}{\sqrt{l(l+1)}} \left[\left(\frac{\left(\hat{L}_+ + \hat{L}_-\right) Y_{lm}}{2} \right)^* \hat{x} + \left(\frac{\left(\hat{L}_+ - \hat{L}_-\right) Y_{lm}}{2i} \right)^* \hat{y} + \left(\hat{L}_z Y_{lm}\right)^* \hat{z} \right] \cdot \frac{1}{\sqrt{2}} (\hat{x} \pm i\hat{y})$$

which, after taking the scalar products, simplifies to

$$\vec{\Phi}^*_{lm} \cdot \hat{e}^{\pm}_{\perp} = \frac{-i}{\sqrt{2l(l+1)}}\left[\left(\frac{(\widehat{L}_+ + \widehat{L}_-)\,Y_{lm}}{2}\right)^* \mp \left(\frac{(\widehat{L}_+ - \widehat{L}_-)\,Y_{lm}}{2}\right)^*\right] = \frac{-i\left(\widehat{L}_{\mp} Y_{lm}\right)^*}{\sqrt{2l(l+1)}}$$

so that Eq 11.131 becomes

$$\frac{-i}{\sqrt{2l(l+1)}} \int \left(\widehat{L}_{\mp} Y_{lm}\right)^* E_o \exp(ikr\cos\theta)d\Omega = A^{\pm}_{lm} j_l(kr)$$

Discussion 11.14

For the continuity of the electric field for $\vec{M}_{l,\pm 1}$ components (purely transverse contributions by magnetic multipoles) at the surface of the sphere, we write:

$$\begin{aligned} \vec{E}_{sph,lm} - \vec{E}_{scat,lm} &= \vec{E}_{inc,lm} \\ c^{\pm}_{lm}\vec{M}^{(1)}_{lm}(k_{II}a) - a^{\pm}_{lm}\vec{M}^{(3)}_{lm}(k_I a) &= \vec{M}^{(1)}_{l,\pm 1}(k_I a) \end{aligned}$$

where we have used Eqs. 11.142, 11.144, and 11.140 for the scattered field, the field inside the sphere and the incident field, respectively. The angular dependences of these are all the same so can be divided out reducing the equation to a matching of radial functions. Recalling the convention for the spherical Bessel function superscripts, this becomes:

$$c^{\pm}_{lm} j_l(k_{II}a) - a^{\pm}_{lm} h^{(1)}_l(k_I a) = j_l(k_I a) \tag{11.189}$$

For values of m other than ± 1, corresponding to right- and left-circular polarization, the plane wave expansion term, $j_l(k_I a)$, vanishes and since $j_l(k_{II}a) \neq h^{(1)}_l(k_I a)$, it must be that $c^{\pm}_{lm} = a^{\pm}_{lm} = 0$. Thus,

$$c^{\pm}_{l,\pm 1} j_l(k_{II}a) - a^{\pm}_{l,\pm 1} h^{(1)}_l(k_I a) = j_l(k_I a)$$

Using the symmetry relations between $\vec{M}$ and $\vec{N}$ functions for expansions of the electric and magnetic fields of a wave, Eqs. 11.118 and 11.119, the continuity of the $\vec{M}_{l,\pm 1}$ component for the magnetic field is:

$$\left(-\frac{i}{Z_{II}}\right) d^{\pm}_{l,\pm 1} j_l(k_{II}a) - \left(-\frac{i}{Z_I}\right) b^{\pm}_{l,\pm 1} h^{(1)}_l(k_I a) = \left(-\frac{i}{Z_I}\right) j_l(k_I a) \tag{11.190}$$

To find the corresponding expressions for the $\vec{N}$ functions, we note from Eq. 11.125 that their transverse components are given by:

$$\vec{N}_{lm\perp} = \frac{1}{\rho}\frac{\partial}{\partial\rho}\left(\rho z_l^{(a)}\right)\left[\frac{1}{\sqrt{l(l+1)}}\left(-\frac{\partial Y_{lm}}{\partial\theta}\hat{\theta} - \frac{im}{\sin\theta}Y_{lm}\hat{\phi}\right)\right] \tag{11.191}$$

where $\rho = kr$. For given l, m components, the interior, scattered, and incident fields will have in common the term in brackets. Thus the equations for the $\vec{N}_{lm}$ components for the electric fields are:

$$d_{l,\pm1}^{\pm}\frac{1}{k_{II}a}\frac{\partial}{\partial(k_{II}a)}\left[k_{II}a\cdot j_l(k_{II}a)\right] - b_{l,\pm1}^{\pm}\frac{1}{k_I a}\frac{\partial}{\partial(k_I a)}\left[k_I a\cdot h_l^{(1)}(k_I a)\right] = \frac{1}{k_I a}\frac{\partial}{\partial(k_I a)}\left[k_I a\cdot j_l(k_I a)\right] \tag{11.192}$$

Similarly, continuity of the transverse magnetic fields associated with the $\vec{N}_{lm}$ components requires:

$$\left(-\frac{i}{Z_{II}}\right)c_{l,\pm1}^{\pm}\frac{1}{k_{II}a}\frac{\partial}{\partial(k_{II}a)}\left[k_{II}a\cdot j_l(k_{II}a)\right] - \left(-\frac{i}{Z_I}\right)a_{l,\pm1}^{\pm}\frac{1}{k_I a}\frac{\partial}{\partial(k_I a)}\left[k_I a\cdot h_l^{(1)}(k_I a)\right]$$
$$= \left(-\frac{i}{Z_I}\right)\frac{1}{k_I a}\frac{\partial}{\partial(k_I a)}\left[k_I a\cdot j_l(k_I a)\right] \tag{11.193}$$

Discussion 11.15

We match the general scattering formalism outlined before to the scattering wave functions just found for scattering $\pm$ circularly polarized waves from a uniform dielectric sphere. The general scattering wave function for a σ^+ or or a σ^- circularly polarized incident wave may be written:

$$\vec{E}_{tot}^{\pm} = \vec{E}_{inc}^{\pm} + \vec{E}_{scat}^{\pm} = \hat{e}_{\perp}^{\pm}\exp(ikz) + \frac{e^{ikr}}{r}\vec{f}_{\pm}(\theta,\phi)$$
$$= \hat{e}_{\perp}^{\pm}\exp(ikz) + \frac{e^{ikr}}{r}\left[f_{\pm,\theta}(\theta,\phi)\hat{\theta}(\theta,\phi) + f_{\pm,\phi}(\theta,\phi)\hat{\phi}(\theta,\phi)\right] \tag{11.194}$$

In the brackets, we explicitly note that the θ and ϕ dependence of $\vec{f}$ includes specifying the directions of the θ and ϕ unit vectors. Writing this in terms of the Mie solutions:

$$\vec{E}_{tot} = \vec{E}_{inc} + \vec{E}_{scat} = \frac{E_0}{\sqrt{2}}\sum_l i^{l-1}\sqrt{4\pi(2l+1)}\left(\vec{M}_{l,\pm1}^{(1)}(k_I r) \pm \vec{N}_{l,\pm1}^{(1)}(k_I r)\right)$$
$$+ \frac{E_0}{\sqrt{2}}\sum_l i^{l-1}\sqrt{4\pi(2l+1)}\left(a_l\vec{M}_{l,\pm1}^{(3)} \pm b_l\vec{N}_{l,\pm1}^{(3)}\right) \tag{11.195}$$

Normalizing this by taking $E_0 = 1$ and equating the scattering wave function parts of Eqs. 11.194 and 11.195 with reference to Eqs. 11.124 and 11.125:

$$\frac{e^{ikr}}{r}\left[f_{\pm,\theta}(\theta,\phi)\hat{\theta}(\theta,\phi)+f_{\pm,\phi}(\theta,\phi)\hat{\phi}(\theta,\phi)\right]=\frac{1}{\sqrt{2}}\sum_l i^{l-1}\sqrt{4\pi(2l+1)}\left(a_l\vec{M}^{(3)}_{lm}\pm b_l\vec{N}^{(3)}_{elm}\right)$$

$$=\sqrt{2\pi}\sum_l i^{l-1}\sqrt{\frac{2l+1}{l(l+1)}}\left\{a_l\left(-h_l^{(1)}(k_Ir)\frac{i(\pm 1)}{\sin\theta}Y_{l,\pm 1}\hat{\theta}+h_l^{(1)}(k_Ir)\frac{\partial Y_{l,\pm 1}}{\partial\theta}\hat{\phi}\right)\right.$$

$$\left.\pm b_l\left(\frac{1}{k_Ir}\frac{\partial}{\partial k_Ir}\left[k_Irh_l^{(1)}(k_Ir)\right]\frac{\partial Y_{l,\pm 1}}{\partial\theta}\hat{\theta}+\frac{i(\pm 1)}{k_Ir\sin\theta}\frac{\partial}{\partial k_Ir}\left[k_Irh_l^{(1)}(k_Ir)\right]Y_{l,\pm 1}\hat{\phi}\right)\right\}$$

Recalling the $r\to\infty$ limit of the Hankel function:

$$h_l^{(1)}(k_Ir)\sim(-i)^{l+1}\frac{e^{+ik_Ir}}{k_Ir} \tag{11.196}$$

and separately equating θ and ϕ components:

$$f_{\pm,\theta}(\theta,\phi)=\frac{\pm i\sqrt{2\pi}}{k_I}\sum_l\sqrt{\frac{2l+1}{l(l+1)}}\left(a_l\frac{Y_{l,\pm 1}}{\sin\theta}-b_l\frac{\partial Y_{l,\pm 1}}{\partial\theta}\right)$$

$$f_{\pm,\phi}(\theta,\phi)=-\frac{\sqrt{2\pi}}{k_I}\sum_l\sqrt{\frac{2l+1}{l(l+1)}}\left(a_l\frac{\partial Y_{l,\pm 1}}{\partial\theta}-b_l\frac{Y_{l,\pm 1}}{\sin\theta}\right)$$

Discussion 11.16

A French scientific expedition, led by Pierre Bouguer, traveling in the Peruvian Andes for the purpose of measuring a degree of longitude provided a first account of atmospheric glory scattering. Bouguer describes

> A phenomenon which must be as old as the world, but which no one seems to have observed so far...A cloud that covered us dissolved itself and let through the rays of the rising sun...Then each of us saw his shadow projected upon the cloud...What seemed most remarkable to us was the appearance of a halo or glory about the head.

Atonio de Ulloa, a Spanish captain accompanying the expedition further explains

> The most surprising thing was that, of the six or seven people that were present, each one saw the phenomenon only around the shadow of his own head, and saw nothing around other people's heads.

Nowadays passengers on an airplane will frequently observe a glory about the shadow of the airplane on a lower lying cloud.

Diffraction and the Propagation of Light

12

- Geometric optics and the eikonal equation
- Kirchhoff's diffraction theory: Kirchhoff's integral theorem and boundary conditions; Rayleigh–Sommerfeld diffraction
- Fresnel approximation for the Kirchhoff integrals and Babinet's principle
- Fraunhofer (far-field) diffraction: diffraction by a rectangular and circular aperture
- Angular spectrum representation: Gaussian beams; Fourier optics (far-field) and tight focusing of fields
- The fields and modes of a tightly focused Gaussian beam: diffraction limits on microscopy

12.1 Diffraction

Sommerfeld defines diffraction as "any deviation of light waves from rectilinear paths which cannot be interpreted as reflection or refraction." By contrast with geometrical optics, diffraction constitutes those phenomena arising from the finite wavelength of light. Historically, the Italian Jesuit Francesco Maria Grimaldi coined the word "diffraction" and made the first accurate observations of the phenomena in the 1650s. That light is a wave phenomenon was suggested early on by Christian Huygens (1678). His idea, that a lightwave's propagation could be followed by viewing a wavefront as a series of microscopic secondary wave generators adding to make the advancing wave, is the essence of much that was to follow. However, Newton's geometrical optics (*Opticks*, 1704), a particle theory of light, was generally accepted as definitive until the dawn of the nineteenth century. At that time, Thomas Young, whose doctoral work was a mathematical theory of sound, argued by analogy that certain optical phenomena, specifically, the appearance of colors in stacked glass plates, could not be described

Electromagnetic Radiation. Richard Freeman, James King, Gregory Lafyatis, Oxford University Press (2019).
DOI: 10.1093/oso/9780198726500.001.0001

by a particle theory but were readily understood as a result of interference. Young performed his famous double slit–actually a double hole–experiment in 1803. But even at that the fundamental wave nature of light was slow in acceptance. The French Academy set the topic of diffraction as a subject for their 1817 prize. There, Augustin-Jean Fresnel offered a thesis that synthesized the ideas of Huygens and Young by making arbitrary but plausible assumptions of the phases and angular distribution of the waves resulting from Huygens secondary generators. Arguing to the contrary, Simeon Denis Poisson, a French mathematician and particle advocate, observed that Fresnel's theory made the clearly absurd prediction that for a point light source illuminating an on-axis opaque circular disk, there should be a bright spot in the center of its shadow. Francois Arago, who chaired the committee, actually performed the technically very challenging experiment and observed what today is alternatively known as "Arago's" or "Poisson's" spot. Fresnel was awarded the prize and the wave theory finally gained general acceptance. Nowadays, we may understand the diffraction of light as an early encounter with the uncertainty principle. Diffraction poses fundamental limits on the ability to resolve detail in venues ranging from optical microscopes to high energy particle accelerators. Diffraction is a universal feature of waves and most of the important results in optics–and for electromagnetic waves in general–do not depend on the details of Maxwell's equations but simply on the fact that they may be manipulated to produce wave equations for the fields.

The problems of interest for this chapter may be generically described as shown in Fig. 12.1. An electromagnetic wave originating at point P propagates through a region of constant index of refraction, usually vacuum, and interacts with a material system. The problem then frequently consists of describing the field downstream–at point P', for example. Depending on whether the material system's dimensions are large or small compared to the wavelength of the electromagnetic wave, the problems are usually labeled as diffraction or scattering, respectively. In the previous chapter we considered the latter case.

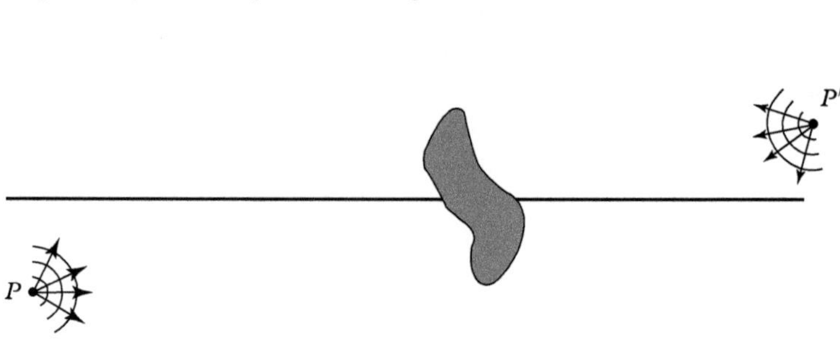

Fig. 12.1 *Generic diffraction problem: An electromagnetic signal originating at P propagates through vacuum or some constant index region, interacts with some material system and emerges, again propagating through vacuum. What is the net effect on this signal at point P′?*

Now, in this chapter, we are interested in the former case. Important "material systems" include screens with various shaped apertures, lenses, and optical systems. Most problems will involve the study of light near the axis defined by P and the center of the material system–the so-called "optical axis."

Recall that for a material whose permittivity and permeability are functions of position, a harmonic field obeys the generalized Helmholtz equation (Eq. 11.40),

$$\nabla^2 \vec{E} + n^2(\vec{r})\, k_0^2 \vec{E} = -\nabla\left(\vec{E}\cdot\nabla \ln \varepsilon_r\right) - i\omega\left(\nabla\mu \times \vec{H}\right) \tag{12.1}$$

where $n(\vec{r}) = c\sqrt{\varepsilon(\vec{r})\,\mu(\vec{r})}$ is the index of refraction and $k_0 \equiv \omega/c$ is the vacuum wavenumber. For this result, we made the assumption that the material is isotropic in the sense that $\varepsilon_r(\vec{r})$ and $\mu(\vec{r})$ are scalar, rather than tensor, quantities. For this case, only the right-hand side of this equation couples the different field components. However, for many problems the right-hand side essentially vanishes: at optical frequencies, most materials are non-magnetic and the second term on the right-hand side is negligible. In addition, the effects of the first right-hand side term's mixing of polarizations is significant only when the permittivity varies rapidly on the scale of the wavelength of light. This happens most notably at material boundaries where the mixing of polarizations is included and is most often treated by satisfying boundary conditions.[1] Thus, frequently the right-hand side is zero, and moreover, in what follows, the index of refraction is usually constant and most often approximately unity. Diffraction theory then just involves finding various ways to solve the Helmholtz equation.

Scalar diffraction theory is a way to solve the 1D Helmholtz equation where the "scalar" in question may be any component of the electric field in Eq. 12.1. Often referred to as Kirchhoff theory, scalar diffraction theory is based on the Helmholtz and Kirchhoff integral theorem. Fresnel and Fraunhofer diffraction are well-known examples.

Vector diffraction theory may mean several different things. First, a vector theory may be necessary for the coupling of polarization that occurs in birefringent materials. Alternatively, it is possible to extend Kirchhoff's scalar theory to vectors. However, the vector theory suffers most of the important problems of the scalar theory. Here, we will use the so-called "angular spectrum" representation to treat vector problems. Strictly speaking, this may be seen as a technique to simultaneously solve the three scalar Helmholtz equations of Eq. 12.1. This technique allows for the convenient imposition of boundary conditions that mix polarizations at the surfaces of optical elements.

[1] When there is no coupling of the field components, then the amplitude or rate of amplitude change of one polarization does not affect how another component changes.

12.2 Geometric optics and the eikonal equation

We begin by discussing what diffraction is not. Geometric optics, despite its limitations, frequently provides a useful visualization of light propagation. In terms of a wave theory, the concept of a ray may be defined by first finding wave surfaces of constant phase and then identifying rays with continuous lines perpendicular to these surfaces in the same way that electric field lines are derived from potential surfaces. We begin with the scalar version of Eq. 12.1 with the right-hand side taken to be zero,

$$\nabla^2 f(\vec{r}) = -n^2(\vec{r})\, k_0^2 f(\vec{r}) \tag{12.2}$$

where again, $f(\vec{r})$ is a single component of the field. Here, we allow for an index of refraction that may have a slow spatial variation. That is to say, it may have large variations only over distances much larger than the wavelength of the wave. Lens and mirror surfaces are important exceptions to this requirement in ray optics treatments of optical systems. But these may be consistently incorporated into this approach via boundary conditions on surfaces between regions that do satisfy this restriction. We consider a solution of the form

$$f(\vec{r}) = a(\vec{r})\, e^{ik_0\psi(\vec{r})} \tag{12.3}$$

In this and all of what follows, it is to be understood that the actual, physical value for a quantity is found by taking the real part of the complex expression. For ray optics to hold, $a(\vec{r})$ will be a slowly varying function of position. $\psi(\vec{r})$ is called the eikonal. It may be thought of as the local phase of the wave (it's actually proportional to the phase by the constant factor k_0) and its constant surfaces are those sought to define the rays. Substituting 12.3 into 12.2 (**see Discussion 12.1**)

$$\nabla^2 a + 2ik_0\nabla a\cdot\nabla\psi + ik_0 a\nabla^2\psi - ak_0^2(\nabla\psi)\cdot(\nabla\psi) = -ak_0^2 n^2(\vec{r}) \tag{12.4}$$

The real and imaginary parts of this equation must both be separately satisfied. For the left-hand side of this equation, the assumed slow variation of the amplitude term, $|\nabla a(\vec{r})| \ll k_0 a$, allows us to ignore the first term and keep the last.[2] The real part of the equation becomes:

$$(\nabla\psi)^2 = n^2(\vec{r}) \tag{12.5}$$

[2] The slow variation assumption

$$\frac{\delta a}{a} \ll \frac{\delta z}{\lambda}$$

where, if $\delta z \simeq \lambda$ then $\delta a/a \ll 1$ and the fractional variation of a on the scale of a wavelength is negligible,

$$|\nabla a(\vec{r})| \ll k_0 a$$

This is the "eikonal equation." For vacuum or constant index material, this result implies that light propagates along straight lines.[3]

12.3 Kirchhoff's diffraction theory

12.3.1 Kirchhoff's integral theorem

In 1882, in an effort to put the "Huygens–Fresnel" principle on a firmer theoretical footing, Kirchhoff derived his famous integral theorem that we will discuss presently. This theorem is the basis of most diffraction calculations. While Kirchhoff's work follows and is certainly in harmony with Maxwell's identification of optics as a branch of electromagnetism, the theorem relies only on the described phenomena obeying a wave equation and hence applies equally to sound.

Figure 12.2 shows the geometric arrangement used for Kirchhoff's integral theorem. This theorem expresses a solution, here a function $U(\vec{r})$, at a point P, of the Helmholtz equation in terms of the behavior of the function on some surface S that surrounds the point P. Again, $U(P)$ may be thought of as any single component of either the electric or magnetic fields. And since it is a complex function, the actual physical value of the amplitude quantity is found by taking the real part. Usually the experimentally accessible quantity is the intensity and this is proportional to the square of the field amplitude. We begin with a corollary of Gauss's theorem that relates a volume integral

[3] The gradient of ψ is normal to the surface of constant phase and thus defines the ray at that point. If the index is constant in space, the gradient is constant and the ray direction does not vary in space.

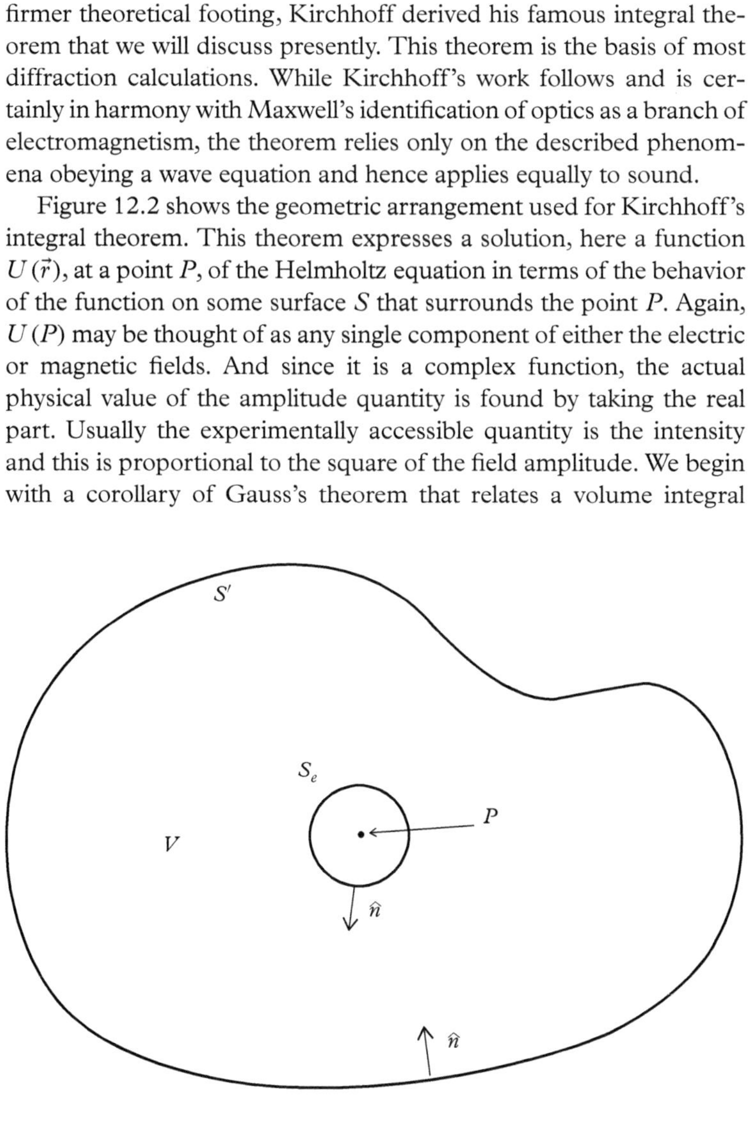

Fig. 12.2 *Geometry used for Kirchhoff's integral theorem.*

involving two functions, U and G, having continuous first-order and second-order derivatives, to an integral over the enclosing surface, S. For a volume V enclosed by this surface, this corollary, known also as Green's Theorem, gives:

$$\int_V dV\left(U\nabla^2 G - G\nabla^2 U\right) = -\oint_S dS\left(U\frac{\partial G}{\partial n} - G\frac{\partial U}{\partial n}\right) \tag{12.6}$$

where $\partial/\partial n$ indicates the inward differentiation along the normal to the surface.[4] Here, U will ultimately be the solution to the problem of interest, while G is an auxiliary function picked for its mathematical usefulness. Indeed, referring to Fig. 12.2, we take G to be the Green function for the Helmholtz equation with the boundary conditions of an outgoing spherical wave:

$$G(\vec{s}) = \frac{e^{iks}}{s} \tag{12.7}$$

where s is the distance from the point P to the position $\vec{r}$ within the volume or on the surface. Thus, if we exclude the point P from the volume V by surrounding it with the infinitesimal spherical surface S_ε of radius ε and assume there are no additional sources in V, then both U and G satisfy the homogeneous Helmholtz equation within V,

$$\left(\nabla^2 + k^2\right)U = 0 \quad \text{and} \quad \left(\nabla^2 + k^2\right)G = 0 \tag{12.8}$$

and the left-hand side of Eq. 12.6 vanishes so that:

$$\oint_S dS\left(U\frac{\partial G}{\partial n} - G\frac{\partial U}{\partial n}\right) = 0 \tag{12.9}$$

Because the surface S includes both S' and S_ε, Eq. 12.9 can be written more explicitly as,

$$\left\{\oint_{S'} dS + \oint_{S_\varepsilon} dS\right\}\left[U\frac{\partial}{\partial n}\left(\frac{e^{iks}}{s}\right) - \frac{e^{iks}}{s}\frac{\partial U}{\partial n}\right] = 0 \tag{12.10}$$

The integral over the infinitesimal sphere is evaluated for the limit $\varepsilon \to 0$ as follows:

$$\begin{aligned}
\oint_{S_\varepsilon} dS\left[U\frac{\partial}{\partial n}\left(\frac{e^{iks}}{s}\right) - \frac{e^{iks}}{s}\frac{\partial U}{\partial n}\right] &= \oint_{S_\varepsilon} dS\left[U\left(\frac{ik}{s} - \frac{1}{s^2}\right)e^{iks} - \frac{e^{iks}}{s}\frac{\partial U}{\partial n}\right] \\
&= \oint_{S_\varepsilon} d\Omega\varepsilon^2\left[U\left(\frac{ik}{\varepsilon} - \frac{1}{\varepsilon^2}\right) - \frac{1}{\varepsilon}\frac{\partial U}{\partial n}\right]e^{ik\varepsilon} \\
&= \oint_{S_\varepsilon} d\Omega\left[-U + \varepsilon\left(ikU - \frac{\partial U}{\partial n}\right)\right]e^{ik\varepsilon}
\end{aligned}$$

[4] Different authors use different conventions for the direction taken for this differentiation. We are following the notation and conventions of Born and Wolfe.

noting that both normals, $\hat{n}$ and $\hat{s}$, in this integral are "directly outward," into the volume within which the field propagates so that $\hat{n} = \hat{s}$. Next, taking the limit $\varepsilon \to 0$, so that the inner surface shrinks to zero around the point P, we obtain,

$$\oint_{S_\varepsilon} dS \left[U \frac{\partial}{\partial n} \left(\frac{e^{iks}}{s} \right) - \frac{e^{iks}}{s} \frac{\partial U}{\partial n} \right] \simeq -4\pi U(P)$$

Using this result in Eq. 12.10 gives the integral theorem of Helmholtz and Kirchhoff:

$$U(P) = \frac{1}{4\pi} \oint_{S'} dS \left[U \frac{\partial}{\partial n} \left(\frac{e^{iks}}{s} \right) - \frac{e^{iks}}{s} \frac{\partial U}{\partial n} \right] \tag{12.11}$$

The normal derivative in this equation can be expressed in terms of θ_{ns} defined as shown in Fig. 12.3, the angle between the inward pointing surface normal, $\hat{n}$, and $\vec{s}$, the outward pointing displacement between P and the surface element:

$$\frac{\partial}{\partial n} \left(\frac{e^{iks}}{s} \right) = \frac{\partial}{\partial s} \left(\frac{e^{iks}}{s} \right) (-\cos\theta_{ns}) = -\left(ik - \frac{1}{s} \right) \frac{e^{iks}}{s} \cos\theta_{ns}$$

$$\cong -ik \frac{e^{iks}}{s} \cos\theta_{ns} \tag{12.12}$$

where the last equality assumes the distance to the observation point is large compared to the optical wavelength.

12.3.2 Kirchhoff's diffraction theory: boundary conditions

The geometry shown in Fig. 12.3 is useful for discussing many diffraction results. Monochromatic light incident from the left onto an opaque planar screen with an aperture is observed downstream at the point, P. Often P will be on a second screen specifically for the observation. The dimensions of the aperture are assumed to be large compared to the wavelength of the incident light but otherwise may be taken as arbitrary. The aperture could be irregular or have multiple separated openings. The integral theorem (Eq. 12.11) is then used to find the value of the wave function at P. The surface used in the integral expression is divided into three parts: the screen aperture, that part of the screen other than the aperture, and the portion of a large sphere of radius R, "at infinity," that closes the surface. The integral theorem of Eq. 12.11 for this arrangement then provides:

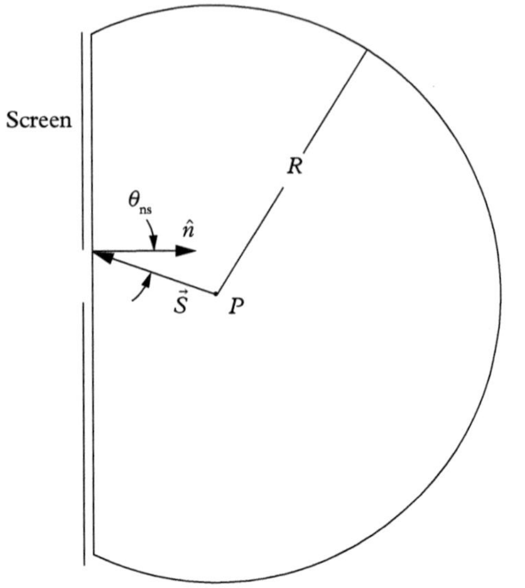

Fig. 12.3 *Geometry used for Kirchhoff diffraction. No assumptions are made concerning the shape of the aperture, however its dimensions will often be assumed small compared with the distance, s, to the observation point and also to the scale of transverse variations of the incident light source to the left of the screen. $\hat{n}$ is a unit vector for the screen.*

$$U(P) = \frac{1}{4\pi}\left\{\int_{aper} + \int_{screen} + \int_{R\to\infty}\right\} dS\left[U\frac{\partial}{\partial n}\left(\frac{e^{iks}}{s}\right) - \frac{e^{iks}}{s}\frac{\partial U}{\partial n}\right] \tag{12.13}$$

To evaluate the right-hand side, we need to estimate U and its normal derivative on the bounding surfaces. Kirchhoff suggested plausible boundary conditions that have been shown to produce remarkably accurate predictions. He argued that (1) as long as we look beyond a few wavelengths from the edges of the aperture, the wave function within the aperture should be well approximated by its incident form–its value in the absence of the screen–and that (2) on the downstream side of an opaque screen, moving away from the aperture, the wave function and its derivative should fall rapidly to zero. Summarizing, Kirchhoff's boundary conditions are:

$$\text{over the aperture:} \quad U = U^i, \ \frac{\partial U}{\partial n} = \frac{\partial U^i}{\partial n}$$

$$\text{over the rest of the screen:} \quad U = 0, \ \frac{\partial U}{\partial n} = 0$$

It remains to consider the integral on the sphere as $R \to \infty$. Using Eq. 12.12, valid for large R, and with $\theta_{ns} = 0$ and $\cos\theta_{ns} = 1$, the third surface integral of Eq. 12.13 can be written:

$$-\frac{1}{4\pi}\int d\Omega R^2 \frac{e^{ikR}}{R}\left[ikU(R) + \frac{\partial U(R)}{\partial n}\right] \tag{12.14}$$

Sufficiently far away, the field from a relatively local radiating source (i.e., U) also falls off as $\frac{e^{iks}}{s}$, so we can write 12.14 as,

$$-\frac{1}{4\pi}\int d\Omega R^2 \frac{e^{ikR}}{R}\left[ik\frac{e^{ikR}}{R} - \left(ik - \frac{1}{R}\right)\frac{e^{ikR}}{R}\right] = -\frac{1}{4\pi}\int d\Omega \frac{e^{2ikR}}{R} \tag{12.15}$$

which clearly vanishes as $R \to \infty$. Thus, only the surface integral over the aperture remains and these boundary conditions then serve to define Kirchhoff's diffraction theory:

$$U(P) = \frac{1}{4\pi}\int_{aper} dS\left[U^i \frac{\partial}{\partial n}\left(\frac{e^{iks}}{s}\right) - \frac{e^{iks}}{s}\frac{\partial U^i}{\partial n}\right] \tag{12.16}$$

Finally, we note that our previous assumption that U, as a radiating source, has the form of an outward spherical wave can be formalized by the Sommerfeld radiation condition. In reality, any given light signal will have originated at and be radiating outward from some point or region in space. Thus, the ideal (transversely) infinite plane wave cannot exist since its infinite radius of curvature would imply an infinite distance from its source and an equally infinitesimal amplitude. The Sommerfeld radiation condition rules out such plane wave solutions as well as the equally unrealistic solutions of incoming (advanced) spherical waves.[5]

12.3.3 Alternate boundary conditions: Rayleigh–Sommerfeld diffraction

Two criticisms of the Kirchhoff diffraction theory are that the boundary conditions overdetermine the solution to the problem and that they are inconsistent with the ultimate solution. To specify a unique solution to the Helmholtz equation in a source free region it is sufficient to specify, on a closed surface bounding the region, either the values of the solution on the surface, Dirichlet boundary conditions–or–the normal derivative of the function on the surface, Neumann boundary conditions. However, evaluation of the integral theorem expression requires both the value and the normal derivative of the function over the closed surface and so this is more information than is required for a unique solution. Morcover, it can be shown that if a solution to the Helmholtz equation has a value of zero and a normal derivative of zero over a finite area, then the function is necessarily zero everywhere, including the aperture. Thus the Kirchhoff solution is not even self-consistent.

[5] Generally, in three dimensions, the Sommerfeld radiation condition for a solution, $U(s)$, to the Helmholtz equation is written as

$$\lim_{s\to\infty} s\left(\frac{\partial U}{\partial s} - ikU\right) = 0$$

which, again noting that $\hat{n} = -\hat{s}$, is equivalent to an imposed requirement, in Eq. 12.14, of $\lim\left[R\left(ikU + \frac{\partial U}{\partial n}\right)\right] \to 0$ for $R \to \infty$.

The Rayleigh–Sommerfeld formulation of the diffraction problem addresses both of these problems. We can generalize the integral theorem, Eq. 12.11, by considering other auxiliary or Green functions. As long as U and G satisfy the Helmholtz equation and the resulting integrand decreases sufficiently rapidly at infinity, we can write the integral theorem, Eq. 12.13, for the geometry shown in Fig. 12.3 as:

$$U(P) = \frac{1}{4\pi} \int_{screen+aper} dS \left[U \frac{\partial G}{\partial n} - G \frac{\partial U}{\partial n} \right] \tag{12.17}$$

If Green functions could be found whose values or whose normal derivatives were zero over the entire plane of the screen, then we could pick out either the first or second term in brackets by using Dirichlet or Neumann boundary conditions, respectively. Such Green functions are readily found using the basic idea of the method of images. For example, referring to Fig. 12.4, we assume the screen to be an infinite plane and consider the function

$$G_{-}\left(\vec{s}, \vec{s}'\right) = \frac{e^{iks}}{s} - \frac{e^{iks'}}{s'} \tag{12.18}$$

where, as with the Green function, G, of Eq. 12.7, s is the distance from the point P to a position, $\vec{r}$, within the enclosed volume (right side of screen) or on the enclosing surface (screen + aperture). Now, however,

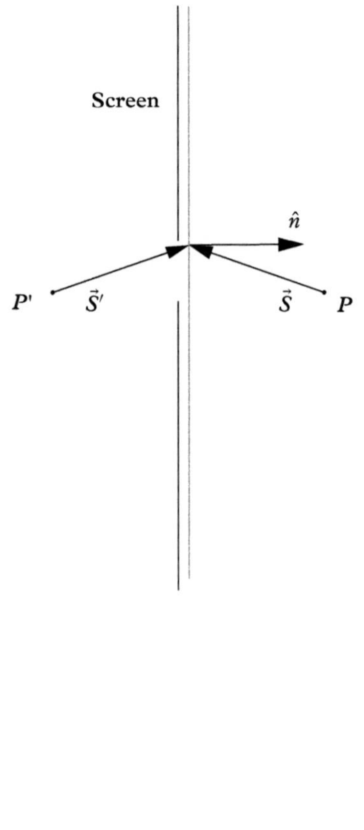

Fig. 12.4 *The images used to generate the Rayleigh–Sommerfeld Green functions.*

there is an additional term including a distance, s'. As depicted in the figure, s' is the distance from P', the image point of P, to $\vec{r}$. In the figure, $\vec{r}$ is located on the screen, at the top of the aperture. Like G, the Green function G_- is a solution to the Helmholtz equation defined at all points, apart from P, within the surface used in deriving the integral theorem (see Fig. 12.3). Note that the singularity of the image term is outside the integration surface and causes no problems for invoking Green's theorem. Since at any point on the screen $s = s'$, so G_- is identically zero and Eq. 12.17 becomes:

$$U_-(P) = \frac{1}{4\pi}\int_{screen+aper} dS\, U_- \frac{\partial G_-}{\partial n} \tag{12.19}$$

With U now approximated by its original, incident value within the aperture and by zero on the opaque part of the screen, the screen integral vanishes and we have,

$$U_-(P) = \frac{1}{4\pi}\int_{aper} dS\, U_i \frac{\partial G_-}{\partial n} \tag{12.20}$$

Thus, this diffraction integral is a solution to the Helmholtz equation simplified by vanishing Dirichlet boundary conditions (for G_- in the plane of the screen and for U_- on the opaque part of the screen). These approximations are at least consistent: with $\frac{\partial U}{\partial n}$ free to have a finite value on the opaque part of the screen, the Helmholtz solution is no longer necessarily zero everywhere. This is called the first Rayleigh–Sommerfeld solution.

In a similar vein,

$$G_+\left(\vec{s}, \vec{s}'\right) = \frac{e^{iks}}{s} + \frac{e^{iks'}}{s'} \tag{12.21}$$

is a solution to the Helmholtz equation that has a normal derivative, $\frac{\partial G_+}{\partial n}$, in the plane of the screen that is uniformly zero. With the normal derivative of U taken to be zero on the opaque part of the screen while retaining its original incident value within the aperture, the corresponding diffraction integral from Eq. 12.17 becomes:

$$U_+(P) = -\frac{1}{4\pi}\int_{aper} dS\, G_+ \frac{\partial U_i}{\partial n} \tag{12.22}$$

where U_+ is a solution for Neumann boundary conditions (for G_+ in the plane of the screen and for U_+ on the opaque part of the screen). U_+, so defined, is called the second Rayleigh–Sommerfeld solution.

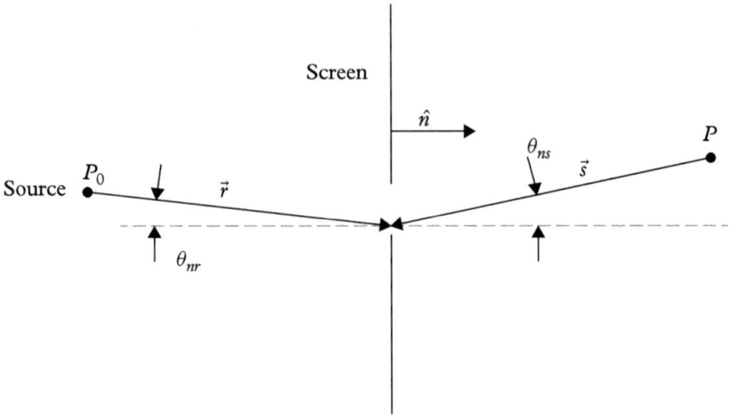

Fig. 12.5 *Light from a point source passes through a screen and is observed at point P.*

Summarizing: to this point, depending upon which of the three Green functions (Eqs. 12.7, 12.18, or 12.21) is used, there are three different choices for the downstream diffraction pattern. Note that the Kirchhoff Green function is the average of the two Rayleigh–Sommerfeld Green functions. Consequently, referring to Eq. 12.17, the Kirchhoff diffraction solution, $U(P)$, is just the average of the two Rayleigh–Sommerfeld solutions, U_+ and U_-. Detailed comparisons have concluded that the mathematical consistency of the Rayleigh–Sommerfeld formulations does not lead to better accuracy,[6] and that in regimes where any of them should be valid, it is observed that they all return similar estimates for the diffraction.

Equations 12.16, 12.19, and 12.22 seem to be very different integrands and that they should ever return similar results is not immediately clear. To understand this and see in general that the details of the boundary conditions are relatively unimportant as long as one is not observing the field near the boundaries, we consider a case originally discussed by Kirchhoff for which the incident illumination is provided by a point source radiating in front of the screen, as shown in Fig. 12.5.

The wave function and normal derivative for the wave incident on the screen are:

$$U^i(\vec{r}) = \frac{Ae^{ikr}}{r} \qquad \frac{\partial U^i}{\partial n} = \frac{Ae^{ikr}}{r}\left(ik - \frac{1}{r}\right)\cos\theta_{nr} \simeq \frac{Aike^{ikr}}{r}\cos\theta_{nr} \tag{12.23}$$

where $\vec{r}$ is the displacement between the source and where the field is being measured on the screen, θ_{nr} is the angle between that displacement and the normal to the screen, and we assume that the source-screen distance is much larger than the wavelength of the

[6] Wolf, E. and Marchand, E. W., Comparison of Kirchhoff and Rayleigh–Sommerfeld theories of diffraction at an aperture, *J. Opt. Soc. Am.*, 54, 587 (1964).

light. Kirchhoff's formula (Eq. 12.16) for the diffraction with these assumptions becomes:

$$U(P) = -\frac{1}{4\pi}\int_{aper} dS\left[\frac{Ae^{ikr}}{r}\frac{ike^{iks}}{s}\cos\theta_{ns} + \frac{Ae^{iks}}{s}\frac{ike^{ikr}}{r}\cos\theta_{nr}\right] \tag{12.24}$$

This result is known as the Fresnel–Kirchhoff diffraction formula:

$$U(P) = -\frac{iA}{2\lambda}\int_{aper} dS\frac{e^{ik(r+s)}}{rs}\left[\cos\theta_{nr} + \cos\theta_{ns}\right] \tag{12.25}$$

Note that this result is symmetrical in r and s. That is to say, the effect of a point source at P_0 observed at P is exactly the same as locating the point source at P and observing at P_0. This is known as the reciprocity theorem of Helmholtz. U_- and U_+ are readily found. Making the same assumptions on the positions of the source and observations:

$$U_-(P) = -\frac{iA}{2\lambda}\int_{aper} dS\frac{e^{ik(r+s)}}{rs}\left[2\cos\theta_{ns}\right] \tag{12.26}$$

$$U_+(P) = -\frac{iA}{2\lambda}\int_{aper} dS\frac{e^{ik(r+s)}}{rs}\left[2\cos\theta_{nr}\right] \tag{12.27}$$

These three expressions differ only in the bracketed terms. In many problems of interest, both the (effective) source and observation points are nearly on axis and at distances large compared with dimensions of the aperture. This is the paraxial limit. For these cases, the obliquity angles are small, their cosines are nearly constant and nearly equal to one over the aperture integration, and the three diffraction formulae will return similar results.

12.3.4 Babinet's principle

An interesting feature of diffraction is that complementary screens, that is, two screens for which the apertures in one are opaque regions in the other and vice versa, produce the same diffraction pattern for a given incident wave. See Fig. 12.6. The principle may be more precisely stated by referring back to Fig. 12.3 and considering the effects of three different screens. First consider the "no-screen-at-all" or "open" screen. Here, the Kirchhoff integral over the aperture (open part) must be evaluated over the entire vertical plane. This result, then, is equivalent to separately evaluating the integrals over the apertures of two complementary screens and then adding the results. For example, using the screens in Fig. 12.6 we have

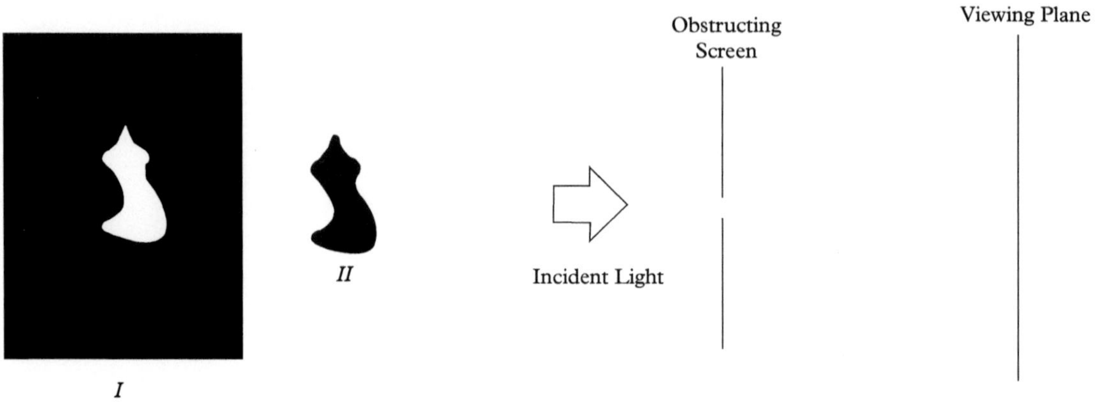

Fig. 12.6 *Optics for Babinet's principle. (a) On the left are two complementary screens. (b) On the right is an optics arrangement for observing the phenomena. Initially, the "open" screen is inserted, a beam of light–nominally a plane wave–is incident, and the Kirchhoffs integral evaluated over the screen plane returns whatever pattern the incident beam makes at the viewing plane. Babinet's principle observes that on the viewing plane, outside the incident beam's pattern–that is, in the diffraction region–the wave function is zero for the open screen case. And upon inserting an actual screen with apertures, a diffraction field is observed in this region. From this observation, Babinet's principle states that this diffraction field is identical in amplitude and opposite in phase for two complementary screens.*

$$U_{open}(P) = U_I(P) + U_{II}(P) \tag{12.28}$$

Now, consider an incident beam of finite width. In the open screen case, for any point P at the viewing plane and outside the incident beam's pattern, the wave amplitude is zero, $U_{open}(P) = 0$, and so clearly, $U_{II}(P) = -U_I(P)$. That is, the wave amplitudes from complementary screens will be equal in strength and 180° out of phase. Since the intensity is proportional to the square of the amplitude, the diffraction patterns formed by a given screen and its complement are identical. See Fig. 12.6. Note that Babinet's principle is frequently stated assuming an incident plane wave and since, strictly speaking, for an infinite plane wave in the open screen case the viewing plane would be uniformly illuminated, there is nowhere on the viewing plane for which $U_{open}(P) = 0$. However, for an experimental realization of the principle, it is only necessary that the light illuminate the vicinity of the obstructing target. Such is the case for a finite width beam. In this case, at the viewing screen, downstream from the apertures, the patterns are identical (only) in the diffraction region. This is illustrated and discussed in Fig. 12.6.

12.3.5 Fresnel approximation

Since there is no a priori reason to prefer any one of the three Kirchhoff and Rayleigh–Sommerfeld integral formulae, going forward we will use the first Rayleigh–Sommerfeld result, Eq. 12.20, because it is mathematically the simplest. Explicitly calculating the normal derivative of the $G_-(\vec{s},\vec{s}')$ Green function (Eq. 12.18) and assuming the observation distance is much larger than the light wavelength, we have:

$$U_-(P) = -\frac{i}{\lambda}\int_{aper} U^i \frac{e^{iks}}{s}\cos\theta_{ns}\, dS \qquad (12.29)$$

This lends itself to a Huygens construction interpretation where the integral is seen as a sum over a series of microscopic secondary wavelet generators, at the plane of the aperture, whose strength is $\frac{U^i}{\lambda}dS$. The $\cos\theta_{ns}$ term, referred to as the "obliquity factor," specifies the angular distribution of the secondary generator's wave and the $-i$ factor indicates the secondary wave leads the incident wave by 90°. In what follows, we will omit the "$-$" subscript.

In Fig. 12.7, the relevant quantities are defined. Writing the integral in Eq. 12.29 in Cartesian coordinates and using $\cos\theta_{ns} = z/s$ gives:

$$U_-(x,y,z) = -\frac{i}{\lambda}\int_{aper} dx_0\, dy_0\, U^i(x_0,y_0,0)\, z\frac{e^{iks}}{s^2} \qquad (12.30)$$

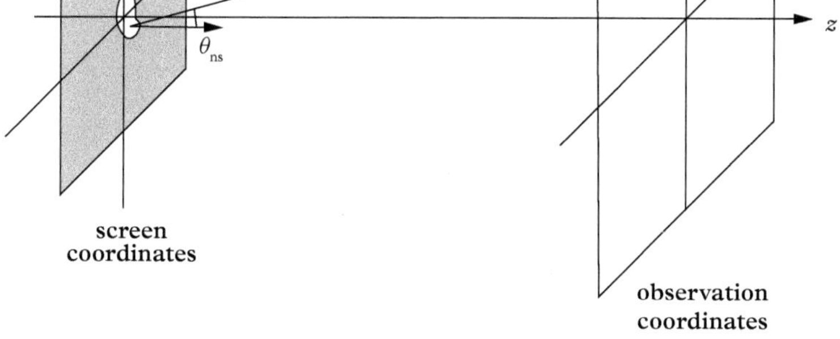

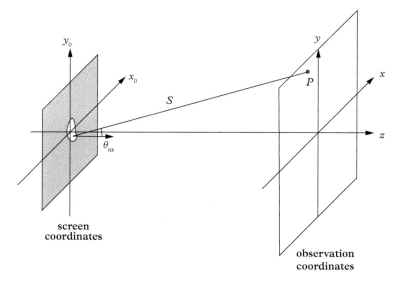

Fig. 12.7 *Defining the various quantities used to express the diffraction integral in Cartesian coordinates. Observations are carried out in a plane parallel to the screen defining the aperture, and both planes are perpendicular to the z- or optical axis. P is representative point for observation. The origin of coordinates is picked somewhere within the aperture–frequently symmetry will dictate this choice. The coordinates of the observation point, P are x, y, and z, and the corresponding coordinates of points within the aperture are* x_0, y_0, *and 0.*

where

$$s = \sqrt{z^2 + (x - x_0)^2 + (y - y_0)^2} \tag{12.31}$$

Many problems of interest involve downstream diffraction near the optical axis in which $x, y \ll z$. For such cases, the approximation $s \simeq z$ in the denominator of the integrand of Eq. 12.30 is adequate (see Fig. 12.7). For the exponential, more care is required. A Taylor expansion for the exponent:

$$iks \simeq ikz \left\{ 1 + \frac{1}{2} \frac{(x - x_0)^2 + (y - y_0)^2}{z^2} - \frac{1}{8} \left[\frac{(x - x_0)^2 + (y - y_0)^2}{z^2} \right]^2 + \ldots \right\} \tag{12.32}$$

The Fresnel approximation consists of keeping the first two terms in this expansion. To obtain reliable results, then, the contribution of the third term must be negligible. That is, it must contribute $\ll 2\pi$ to the exponential's phase. Identifying the transverse displacement from target to screen, $\rho \equiv \sqrt{(x - x_0)^2 + (y - y_0)^2}$, then allows this criterion for ignoring higher order terms in the Fresnel approximation to be written compactly as:

$$\frac{\rho^4}{z^3 \lambda} \ll 8 \tag{12.33}$$

Clearly, it is better satisfied for large z, far downstream, and for small ρ, near-axis or paraxial fields. Explicitly, the Fresnel approximation is:

$$U(x, y, z) = -\frac{ie^{ikz}}{\lambda z} \int dx_0\, dy_0\, U^i(x_0, y_0, 0) \times \exp\left[\frac{ik}{2z} \left((x - x_0)^2 + (y - y_0)^2 \right) \right] \tag{12.34}$$

This is a convolution integral of the form:

$$\int dx_0\, dy_0\, U^i(x_0, y_0)\, h(x - x_0, y - y_0) \tag{12.35}$$

with the convolution kernel

$$h(x, y) = -\frac{ie^{ikz}}{\lambda z} \exp\left[\frac{ik}{2z} \left(x^2 + y^2 \right) \right] \tag{12.36}$$

We will return to this viewpoint when we discuss the angular spectrum representation. Alternatively, we can expand the quadratics in the exponential and rewrite the Fresnel approximation:

$$U(x,y,z) = -\frac{ie^{ikz}}{\lambda z} e^{\frac{ik}{2z}(x^2+y^2)} \int_{aper} dx_0\, dy_0 \qquad (12.37)$$

$$\times \left\{ U^i(x_0,y_0,0) \exp\left[\frac{ik}{2z}\left(x_0^2+y_0^2\right)\right]\right\} \exp\left[-\frac{ik}{z}(xx_0+yy_0)\right]$$

Written this way, the integral appears as a 2D spatial Fourier transform of the quantity in braces, the incident field in the aperture multiplied by a complex exponential term quadratic in the aperture variables. Here, the two different transformation domains analogous to frequency and time, $\omega \Leftrightarrow t$, are the near-field and scaled far-field spatial regions, x_o, y_o and $(k/z)x$, $(k/z)y$, respectively.

12.3.6 Fraunhofer (far-field) diffraction

If the system further satisfies the condition $w^2 \ll z\lambda$ where w is the largest dimension of the diffracting aperture, the complex exponential of the bracketed term in Eq. 12.37 may be approximated as "1" for the integral over the aperture and the diffraction result simplifies to the Fraunhofer diffraction integral:

$$U(x,y,z) = -\frac{ie^{ikz}}{\lambda z} e^{\frac{ik}{2z}(x^2+y^2)} \int_{aper} dx_0\, dy_0\, U^i(x_0,y_0,0)$$

$$exp\left[-\frac{ik}{z}(xx_0+yy_0)\right] \qquad (12.38)$$

More formally, for a rectangular or circular aperture, the Fresnel number is defined:

$$N_F = \frac{w^2}{\lambda z} \qquad (12.39)$$

where w is the half-width of the longest dimension of a rectangular aperture or the radius of a circular aperture. For $N_F \ll 1$, the diffraction is called "far-field diffraction" and Eq. 12.38, the Fraunhofer diffraction approximation, is reliable. This equation gives the important interpretation that the Fraunhofer diffraction field in x,y is, to within some constant factors, the 2D Fourier transform of the initial incident field described by variables x_0, y_0. For analyzing macroscopic optical systems, the Fraunhofer approximation is usually appropriate and the condition that N_F is less than one can be satisfied for any system by going sufficiently far downstream in z. We next consider some examples.

Fraunhofer diffraction by a rectangular aperture

For a rectangular aperture of sides $2a$ and $2b$ illuminated by a normally incident plane wave, U^i is constant over the aperture and the Fraunhofer integral of Eq. 12.38 separates into two factors:

$$U(x,y,z) = C\int_{-a}^{a} dx_0 \exp\left[-i\frac{kx}{z}x_0\right]\int_{-b}^{b} dy_0 \exp\left[-i\frac{ky}{z}y_0\right]$$
$$= 4abC\frac{\sin k\frac{x}{z}a}{k\frac{x}{z}a}\frac{\sin k\frac{y}{z}b}{k\frac{y}{z}b} \qquad (12.40)$$

where $C = -\frac{ie^{ikz}}{\lambda z}\exp\frac{ik}{2z}\left(x^2+y^2\right)$.The intensity measured downstream on an observing screen at a distance z from the slit is given by the square of the amplitude:

$$I(x,y) \sim \left(\frac{\sin k\frac{x}{z}a}{k\frac{x}{z}a}\right)^2\left(\frac{\sin k\frac{y}{z}b}{k\frac{y}{z}b}\right)^2 \qquad (12.41)$$

This has nodes along the x and y directions for $k\frac{y}{z}b = m\pi$ and $k\frac{x}{z}a = n\pi$, respectively, for $m,n = \pm1,\pm2....$ The general shape of the two factors, a sinc function, is shown in Fig. 12.8.

Fraunhofer diffraction by a circular aperture

Next we consider a circular aperture of radius a, which is illuminated by a normally incident plane wave and observed on a screen located a distance z downstream. Polar coordinates are suitable for this problem. For (ρ_0,θ_0) taken as the polar coordinates within the aperture we have:

$$x_0 = \rho_0\cos\theta_0, \qquad y_0 = \rho_0\sin\theta_0 \qquad (12.42)$$

At the observation location, we use polar coordinates (ρ,θ) and define:

$$x = \rho\cos\theta \qquad y = \rho\sin\theta \qquad (12.43)$$

The Fraunhofer integral, Eq. 12.38, now becomes:

$$U(x,y,z) = C\int_0^a \rho_0\,d\rho_0\int_0^{2\pi} d\theta_0$$
$$\times\exp\left[-ik\frac{\rho}{z}\rho_0\left(\cos\theta_0\cos\theta+\sin\theta_0\sin\theta\right)\right]$$
$$= C\int_0^a \rho_0\,d\rho_0\int_0^{2\pi} d\theta_0\exp\left[-ik\frac{\rho}{z}\rho_0\cos\left(\theta_0-\theta\right)\right] \qquad (12.44)$$

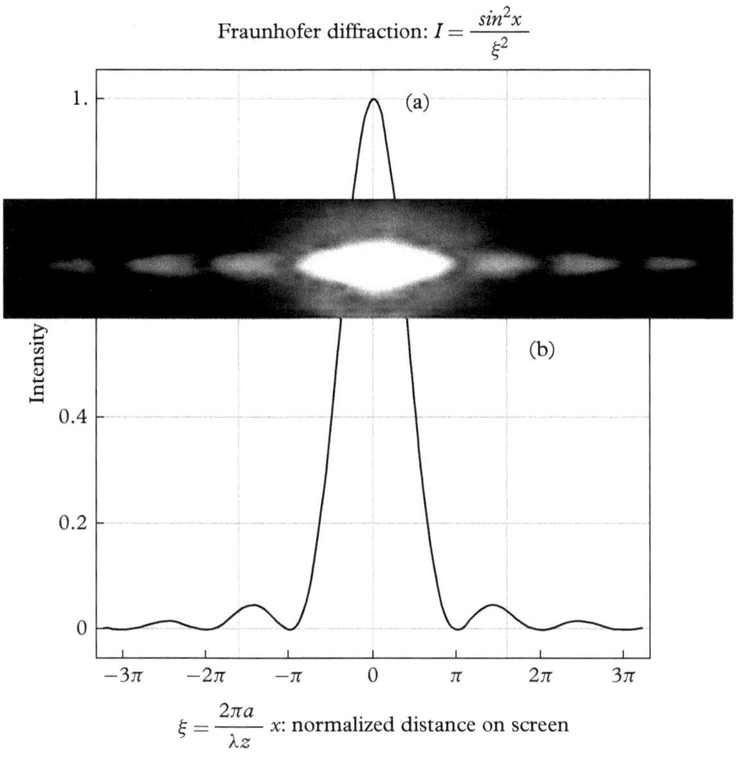

Fig. 12.8 *Fraunhofer diffraction of a rectangular slit. (a) The 1D intensity distribution for one of the factors in a rectangular diffraction corresponding to a side of length a. (b) a realization of this result showing the familiar single slit diffraction pattern. This picture is actually the diffraction pattern created by a laser hitting a human hair. By Babinet's principle, this pattern is the same as that of a slit. Babinet's principle assumes observation in a region where, with no screen, the downstream intensity is zero. It, therefore, does not apply in the center of this pattern where the original laser beam intensity is added to that of the diffraction pattern. Note the picture was intentionally overexposed to show weak outlying maxima.*

This integral may be expressed as a Bessel function using known analytic identities of those functions (**see Discussion 12.2**), resulting in

$$U(x,y,z) = \pi a^2 C \left(\frac{2\mathcal{J}_1\left(k\frac{\rho}{z}a\right)}{k\frac{\rho}{z}a} \right) \tag{12.45}$$

The intensity is then given by:

$$I(x,y,z) = |U(x,y,z)|^2 = \left(\frac{2\mathcal{J}_1\left(k\frac{\rho}{z}a\right)}{k\frac{\rho}{z}a} \right)^2 I_0 \tag{12.46}$$

This is called the Airy disk pattern and it is shown in Fig. 12.9.

12.3.7 Fresnel diffraction of rectangular slit: the near-field

For 1D single slit diffraction, there are again three length scales in the problem: the wavelength of light, the width of the slit, and the distance downstream at which the diffraction pattern is observed. Very

The Airy disk: $I = \dfrac{2\,\mathscr{J}_1^2\,(\xi)}{\xi^2}$

(a)

(b)

Intensity

$\xi = \dfrac{2\pi \cdot a}{\lambda \cdot z}\, \rho$: normalized radial distance

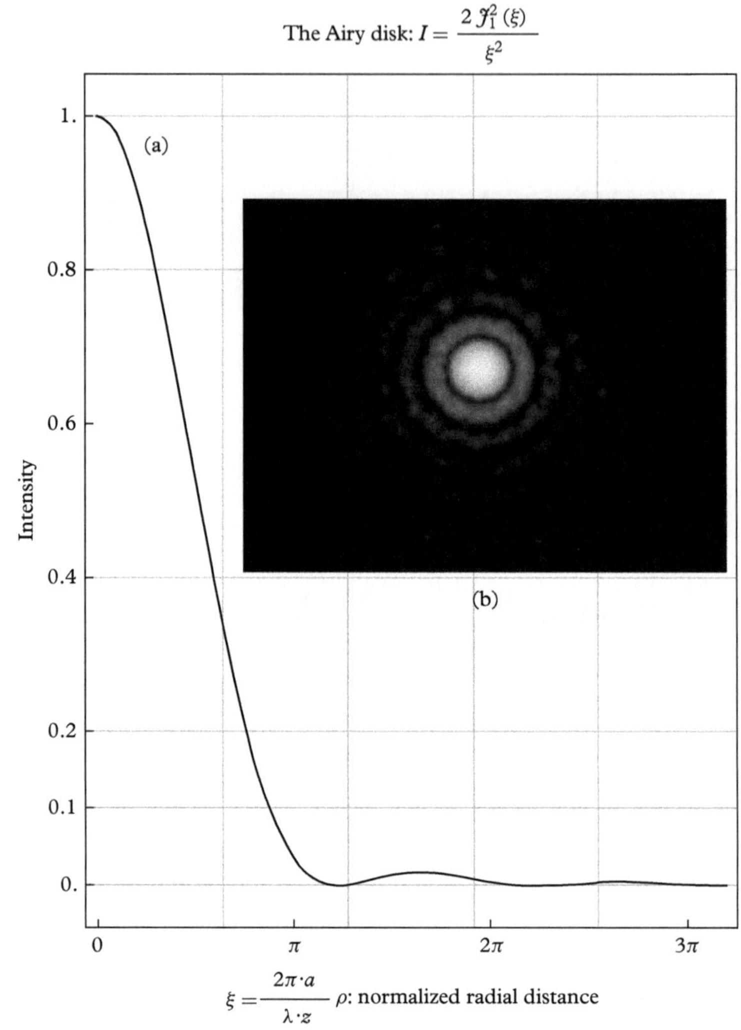

Fig. 12.9 *Diffraction from a circular aperture: (a) The Fraunhofer intensity for a circular aperture is an Airy function. (b) Photograph of an Airy diffraction pattern or "Airy disc."*

far downstream, Fraunhofer diffraction will always be observed. An observation for $N_F \gtrsim 1$ finds the "near-field diffraction," and here, as long as Eq. 12.33 is satisfied, the diffraction pattern is described by the Fresnel equation and its shape is determined by the Fresnel number of the arrangement. It is readily seen that the Fresnel pattern plotted against the "normalized" transverse distance, $X = x/\sqrt{\lambda z}$ is the same for all systems having the same Fresnel number. When Eq. 12.33 is violated–as it surely is for distances downstream that are smaller than the width of the slit–then the Fresnel result is unreliable and the exact integral, Eq. 12.30, provides the best estimate.

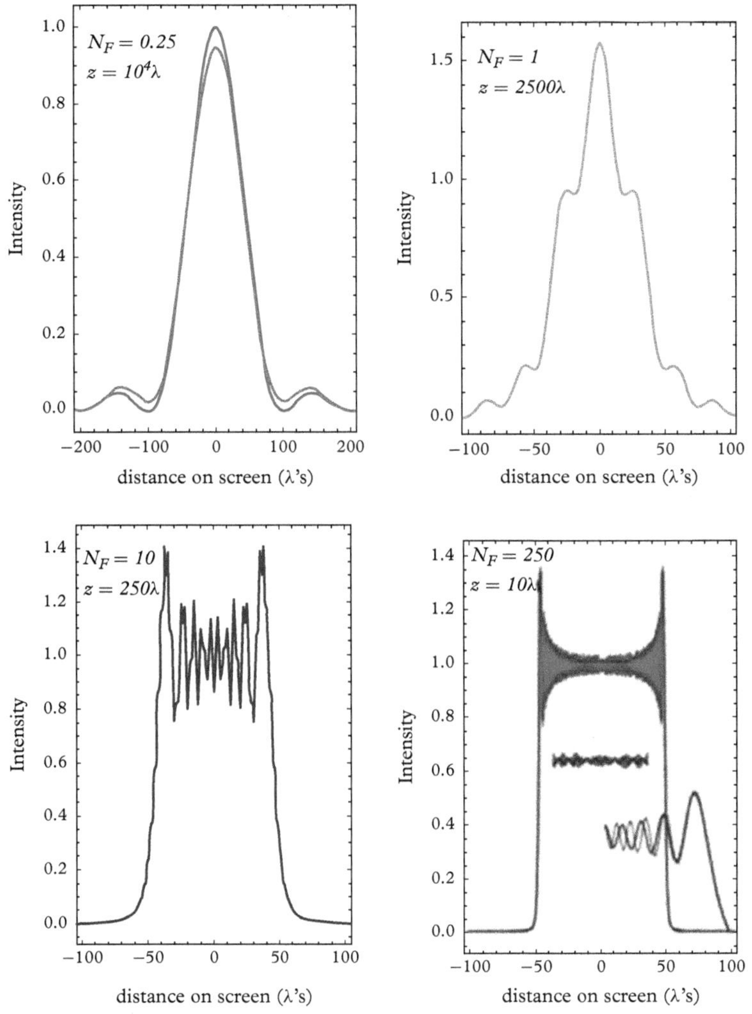

Fig. 12.10 *The intensity as a function of transverse distance for a plane wave on a slit for several distances downstream from the slit. The slit was taken to be 100 μm in the x direction and infinite in y. The blue curves show Fresnel diffraction calculations. On the upper leftmost plot, the red curve is the Fraunhofer result. For the bottom rightmost plot, the Fresnel criterion is violated. The red curve is the exact Sommerfeld integral result. The inset are comparisons for the center of the curve and the diffraction at the right edge of the slit.*

Figure 12.10 shows a series of calculations downstream from a slit illuminated by a plane wave. The results all scale with the wavelength. However, to get physical intuition, we consider a light wavelength of 1 μm in the near infrared part of the spectrum. The width of the slit for this study is 0.1 mm. $N_F = 1$, the nominal border between the near- and far-fields, occurs 2.5 mm downstream (upper right). At ~1 cm, $N_F = 1/4$ and the diffraction pattern is well represented by the Fraunhofer result (upper left). In the intermediate regime, the Fraunhofer formula is inaccurate and it is necessary to use the Fresnel expression. As the slit is approached, an examination of the x-integral in Eq. 12.34 reveals that the integrand near $x_0 = x$ is "stationary" as a function of x_0. That is, there is a minimum in the phase of the complex exponential and so its first derivative vanishes while over the rest of the integral, the rapidly varying phase is producing cancellation. Thus, near the slit the diffraction at x is dominated by the field within the slit that is at the same transverse location, and the shape of the slit becomes increasingly evident (lower left). For the lower right of Fig. 12.10, the condition of Eq. 12.33 is strongly violated. However, the Fresnel result qualitatively resembles the exact result: both reproduce the shape of the aperture.

12.4 The angular spectrum representation

The angular spectrum representation is a rigorous, vector theory for light propagation and diffraction.[7] The approach singles out a particular direction in space as "z." In applications, this could be the direction of the longitudinal axis of a microscope or lens system or the direction of a laser beam. Then, given an "input" field everywhere in a plane perpendicular to the axis, the angular spectrum representation provides the solution for electromagnetic fields at all other points in space by propagating the input field along the axis. We consider a harmonic (single frequency, ω) vector electric field:

$$\vec{E}(\vec{r},t) = Re\left[\vec{E}(x,y,z)\, e^{-i\omega t}\right] \tag{12.47}$$

where $\vec{E}(x,y,z)$ may be complex, indicating an additional phase. In a plane perpendicular to the defined z-axis, at point $z = z_i$, the 2D Fourier transform, $\vec{\mathscr{E}}\left(k_x,k_y;z_i\right)$, or the "angular spectrum," of this input field is found with respect to x and y. The transform and its inverse are given, respectively, by

$$\vec{\mathscr{E}}\left(k_x,k_y;z_i\right) = \frac{1}{2\pi}\iint_{-\infty}^{\infty} dxdy\, \vec{E}(x,y,z_i)\, e^{-i(k_x x+k_y y)} \tag{12.48}$$

$$\vec{E}(x,y,z_i) = \frac{1}{2\pi}\iint_{-\infty}^{\infty} dk_x dk_y\, \vec{\mathscr{E}}\left(k_x,k_y;z_i\right) e^{i(k_x x+k_y y)} \tag{12.49}$$

Since these are vector quantities, the transform and its inverse hold separately for the x and y components. For the region of the solution, the system is assumed to be linear, source free, and homogeneous. The field then must satisfy the Helmholtz equation:

[7] In this section, we are following the presentation and notation of Novotny, L. and Hecht, B., *Principles of Nano-optics*, 2nd edition, Cambridge University Press (2012).

$$\left(\nabla^2 + k^2\right)\vec{E}(\vec{r}) = 0 \tag{12.50}$$

where $k = n\frac{\omega}{c}$ and $n = \sqrt{\varepsilon/\varepsilon_0}$. Taking the 2D Fourier transform of the Helmholtz equation, Eq. 12.50 yields:

$$\mathscr{F}_{x,y}\left\{\left(\nabla^2 + k^2\right)\vec{E}(x,y,z)\right\}$$

$$= \frac{1}{2\pi}\iint_{-\infty}^{\infty} dxdy\left(\nabla^2 + k^2\right)\vec{E}(x,y,z)\,e^{-i(k_x x + k_y y)} = 0 \tag{12.51}$$

$$= \left(k^2 - k_x^2 - k_y^2\right)\vec{\mathscr{E}}\left(k_x, k_y; z\right) + \frac{\partial^2\vec{\mathscr{E}}\left(k_x, k_y; z\right)}{\partial z^2} = 0 \tag{12.52}$$

which is readily solved to give,

$$\vec{\mathscr{E}}\left(k_x, k_y; z\right) = \vec{\mathscr{E}}\left(k_x, k_y; z_i\right)\exp\left[\pm ik_z\left(z - z_i\right)\right] \tag{12.53}$$

where $k_z = \sqrt{k^2 - k_x^2 - k_y^2}$ can be real or imaginary. The former case gives regular propagating waves while imaginary values of k_z correspond to evanescent waves. We will discuss these shortly.

Now, generalizing Eq. 12.49 to any point along z,

$$\vec{E}(x,y,z) = \frac{1}{2\pi}\iint_{-\infty}^{\infty} dk_x dk_y\,\vec{\mathscr{E}}\left(k_x, k_y; z\right)e^{i(k_x x + k_y y)}$$

and substituting Eq. 12.53, with the input plane set to the origin of the z-axis, or $z_i = 0$, the electric field at any point in space is expressed in terms of its input plane values,

$$\vec{E}(x,y,z) = \frac{1}{2\pi}\iint_{-\infty}^{\infty} dk_x dk_y\,\vec{\mathscr{E}}\left(k_x, k_y; 0\right)e^{i(k_x x + k_y y \pm k_z z)} \tag{12.54}$$

This is called the *angular spectrum representation* of the field.

Equation 12.53, with $z_i = 0$, may be cast in terms of linear response theory as

$$\vec{\mathscr{E}}\left(k_x, k_y; z\right) = \mathfrak{h}\left(k_x, k_y; z\right)\vec{\mathscr{E}}\left(k_x, k_y; 0\right) \tag{12.55}$$

where $\mathfrak{h}\left(k_x, k_y; z\right)$ is called the "propagator in reciprocal space" or the "filter function" and $\vec{\mathscr{E}}\left(k_x, k_y; 0\right)$, the angular spectrum, is called the "input function." Explicitly,

$$\mathfrak{h}\left(k_x, k_y; z\right) = e^{\pm ik_z z} \tag{12.56}$$

The $\pm$ indicates that the propagation could be in either the positive or negative z direction and the argument of $\mathfrak{h}$ indicates that k_z is a function of k_x and k_y. The filter function describes the field at a point z due to a mixed-space point source at $z = 0$ and in that sense is directly related to the Green's function. With $k_z = \sqrt{k^2 - \left(k_x^2 + k_y^2\right)}$ real for $\left(k_x^2 + k_y^2\right) < k^2$, the propagator $\mathfrak{h}\left(k_x, k_y; z\right)$ is an oscillating function of z within a circle of radius k in the (k_x, k_y) plane. For features in the x–y plane at $z = z_i = 0$, with sufficiently high spatial frequency that $\left(k_x^2 + k_y^2\right) > k^2$, k_z is imaginary and the propagator is a real, decreasing exponential: $\mathfrak{h}\left(k_x, k_y; z\right) = e^{-|k_z|z}$. Physically, the corresponding signal/information is lost downstream. Said otherwise, the surviving image downstream has been low-pass filtered–the rapid transverse spatial variations quickly die out. This is our first example of a "spatial filter." In real space, only structures (input fields) with transverse features larger than

$$\Delta x \approx \lambda = \frac{2\pi}{k} = \frac{2\pi}{k_0 n} \tag{12.57}$$

can be accurately imaged. Clearly, resolution can be enhanced by using shorter wavelengths, but this strategy has its own problems. Generally, what is meant by "near-field imaging" is some scheme to retain the information in the high frequency components of the image.

The corresponding representation of all of this in real space is obtained by taking the 2D inverse Fourier transform of the 2D Fourier-transformed description of the field, Eq. 12.55,

$$\begin{aligned}
\mathscr{F}_{x,y}^{-1}\left\{\vec{\mathscr{E}}\left(k_x, k_y; z\right)\right\} &= \mathscr{F}_{x,y}^{-1}\left\{\mathfrak{h}\left(k_x, k_y; z\right)\vec{\mathscr{E}}\left(k_x, k_y; 0\right)\right\} \\
\vec{E}(x, y, z) &= \mathscr{F}_{x,y}^{-1}\left\{\mathfrak{h}\left(k_x, k_y; z\right)\right\} * \mathscr{F}_{x,y}^{-1}\left\{\vec{\mathscr{E}}\left(k_x, k_y; 0\right)\right\} \\
&= h(x, y; z) * \vec{E}(x, y; 0)
\end{aligned}$$

thus resulting in a convolution integral ($*$), in real space, of the source function, $\vec{E}(x, y; 0)$, and the real space propagator,

$$h(x, y; z) = \mathscr{F}_{x,y}^{-1}\left\{\mathfrak{h}\left(k_x, k_y; z\right)\right\} = \frac{1}{2\pi}\iint_{-\infty}^{\infty} e^{i\left[k_x x + k_y y \pm k_z z\right]} dk_x dk_y \tag{12.58}$$

Writing out the convolution explicitly,

$$\begin{aligned}
\vec{E}(x, y, z) = &\iint_{-\infty}^{\infty} \vec{E}\left(x', y'; 0\right) \\
&\times \left[\frac{1}{2\pi}\iint_{-\infty}^{\infty} e^{i\left[k_x(x-x') + k_y(y-y') \pm k_z z\right]} dk_x dk_y\right] dx' dy'
\end{aligned} \tag{12.59}$$

we note that $k_z = \sqrt{k^2 - \left(k_x^2 + k_y^2\right)}$ is generally a mathematically difficult-to-treat function of the other components. An important and useful special case is "paraxial" or mostly forward propagating optical fields, for which k_x, $k_y \ll k$, k_z, where k_z may be approximated as,

$$k_z = k\left(1 - \frac{k_x^2 + k_y^2}{k^2}\right)^{1/2} \approx k - \frac{k_x^2 + k_y^2}{2k} \tag{12.60}$$

12.4.1 Gaussian beams

As a first application of the angular spectrum formalism, we examine a linearly polarized Gaussian beam. We take the beam waist to have a radius, w_0, and to occur in the $z = 0$, input plane.

$$\vec{E}\left(x_0, y_0; 0\right) = \vec{E}_0 \exp\left[-\frac{x_0^2 + {y_0}^2}{w_0^2}\right] \tag{12.61}$$

Here, $\vec{E}_0$ is a real vector constant, the value of the field on axis at $z = 0$, which will shortly be further specialized to be *in the* x direction. Using Eq. 12.49, the 2D, $z = 0$ plane Fourier transform, the angular spectrum, is also a Gaussian function:

$$\begin{aligned}\vec{\mathscr{E}}\left(k_x, k_y; 0\right) &= \frac{1}{2\pi}\iint_{-\infty}^{\infty} \vec{E}_0 \exp\left[-\frac{x_0^2 + {y_0}^2}{w_0^2}\right] \\ &\quad \times \exp\left[-i\left(k_x x_0 + k_y y_0\right)\right] dx_0 dy_0 \\ &= \vec{E}_0 \frac{w_0^2}{2} \exp\left[-\left(k_x^2 + k_y^2\right)\frac{w_0^2}{4}\right]\end{aligned}$$

From Eq. 12.54, the angular spectrum representation of this field is then,

$$\begin{aligned}\vec{E}\left(x, y; z\right) &= \vec{E}_0 \frac{w_0^2}{4\pi}\iint_{-\infty}^{\infty} dk_x dk_y \\ &\quad \times \exp\left[-\left(k_x^2 + k_y^2\right)\frac{w_0^2}{4}\right] \exp\left[i\left(k_x x + k_y y + k_z z\right)\right]\end{aligned} \tag{12.62}$$

Evaluating this using the paraxial approximation of Eq. 12.60, the beam downstream is (**see Discussion 12.3**)

$$\vec{E}\left(x, y; z\right) \simeq \frac{\vec{E}_0 e^{ikz}}{1 + 2iz/kw_0^2} \exp\left[-\frac{\left(x^2 + y^2\right)}{w_0^2}\frac{1}{1 + 2iz/kw_0^2}\right] \tag{12.63}$$

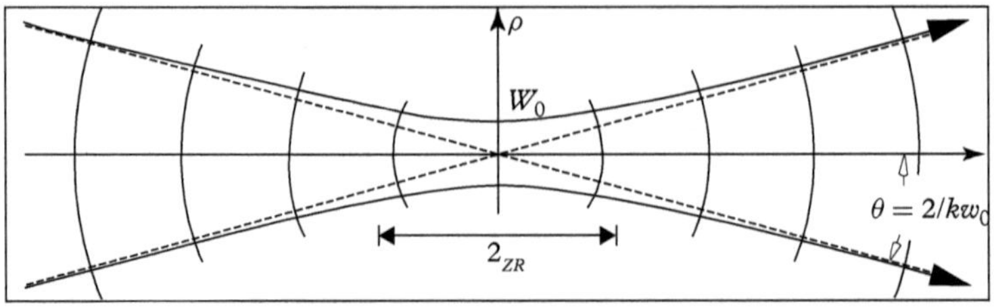

Fig. 12.11 *The Lowest order Hermite-Gaussian Beam,* $\vec{E}(\rho,z) \equiv \vec{E}^{H}_{00}(x,y,z)$*. As the beam propagates along the axis, its size varies as a hyperboloid,* $\pm w(z)$*. At* z_R*, the radius has increased by* $\sqrt{2}$ *from its size at the waist.*

This is the central result for describing Gaussian beams. We can rewrite this by defining quantities that can be used to empirically characterize a beam. See Fig. 12.11. Taking $\rho^2 = x^2 + y^2$ and defining the Rayleigh parameter, $z_R = \frac{kw_0^2}{2}$, this equation can be written in cylindrical coordinates as:

$$\vec{E}(\rho,z) = \frac{\vec{E}_0 e^{ikz}}{1+iz/z_R} \exp\left[-\frac{\rho^2}{w_0^2}\left(\frac{1}{1+iz/z_R}\right)\right] \tag{12.64}$$

The identifications:

$$w(z) = w_0\left(1+\frac{z^2}{z_R^2}\right)^{1/2} \qquad R(z) = z\left(1+\frac{z_R^2}{z^2}\right) \qquad \eta(z) = \arctan\left(\frac{z}{z_R}\right) \tag{12.65}$$

define the beam radius, wavefront radius of curvature, and phase correction. Substituting these into Eq. 12.64 and separating the amplitude dependence and the phase dependence (**see Discussion 12.4**):

$$\vec{E}(\rho,z) = \vec{E}_0 \frac{w_0}{w(z)} \exp\left[-\frac{\rho^2}{w^2(z)}\right] \exp\left[i\left(kz - \eta(z) + \frac{k\rho^2}{2R(z)}\right)\right] \tag{12.66}$$

The transverse radius of the beam for different positions along the beam's axis is conventionally defined to be the radius for which the intensity of the field is down from its on-axis maximum by a factor of $1/e^2$. For the Gaussian beam, this is $w(z)$. The "depth of focus" of a beam is described by the "Rayleigh Range", z_R, the longitudinal distance from the beam waist at which the area of the beam has doubled. Plotting the radius of the beam along the beam axis gives a hyperboloid, as shown in the figure, with asymptotes enclosing a divergence angle $\theta = 2/kw_0$. This is closely related to the numerical aperture used in microscopy: $NA = n\sin\theta \approx 2/kw_0 = \lambda/\pi w_0$ (where n is the index of refraction of the medium). Finally, there is an overall 180° phase shift in the temporal phase of the beam as it passes through the waist. This so-called "Gouy phase shift" results as $\eta(z) = \arctan \frac{z}{z_0}$ goes from $-90°$ for large negative z to $+90°$ for large positive z.

For a tightly focused beam we will require a more accurate description. Recall that small spatial features correspond to large transverse components of the wavenumber and the paraxial approximation, Eq. 12.60, breaks down. One might think about improving upon the paraxial approximation by taking further terms in the expansion but that is missing an important point: These large wavenumber components lead to evanescent waves and downstream, they simply are not there in a real beam. In fact, a true Gaussian beam cannot actually exist because there are always evanescent components that are improperly treated. The tighter the focus the greater the importance of the evanescent components and the further the departure of physical reality from an ideal Gaussian beam. Shortly, we will find the actual fields in the focal region of a tightly focused beam. First, however, we investigate some higher order modes. Derivatives of a solution to the homogeneous Helmholtz equation are also solutions to the equation. For example, the Hermite–Gaussian (HG) modes are given by:

$$\vec{E}^H_{nm}(x,y,z) = w_0^{n+m} \frac{\partial^n}{\partial x^n} \frac{\partial^m}{\partial y^m} \vec{E}^H_{00}(x,y,z) \tag{12.67}$$

Alternatively, the Laguerre–Gaussian (LG) modes have axial symmetry:

$$\vec{E}^L_{nm}(x,y,z) = w_0^{2n+m} \frac{\partial^n}{\partial z^n} \left(\frac{\partial^m}{\partial x^m} + i \frac{\partial^m}{\partial y^m} \right) \left\{ \vec{E}^L_{00}(x,y,z)\, e^{-ikz} \right\} \tag{12.68}$$

where the lowest order HG and LG beam modes, $\vec{E}^H_{00}$ and $\vec{E}^L_{00}$, are identified as the paraxial Gaussian solution of Eq. 12.66.

For the resulting fields, E_0 and w_0 are not necessarily the on-axis field and beam waist at the focus ($z = 0$), but rather they set the scales, respectively, for the near-focus field and focused mode length-scale near the focus. Figure 12.12 shows some low order Hermite–Gaussian modes. The first three are HG modes. Adding an x-polarized HG_{10} mode with a y-polarized HG_{10} of the same strength produces the radially polarized donut mode shown in the fourth figure. The origin is a node of the donut mode and is surrounded by steep field gradients.

Longitudinal fields

In free space, the only true TEM solution is an infinite plane wave. Gauss's law applied to a finite Gaussian beam indicates that there must be more than transverse components of the field. For example, if the transverse field component is x-polarized, there must be a longitudinal field component, E_z:

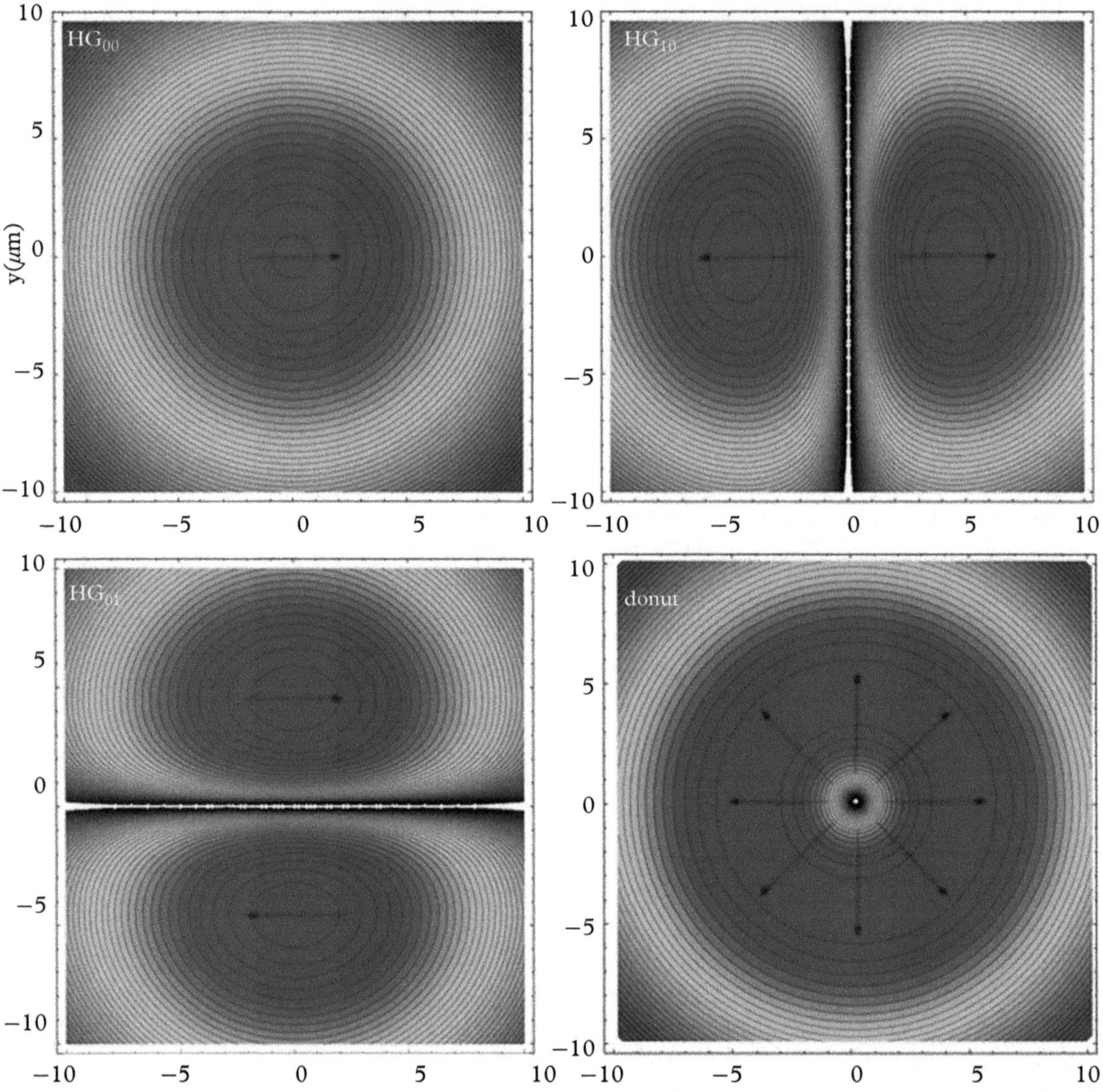

Fig. 12.12 *Intensity cross sections at $z = 0$ for low order Hermite–Gaussian modes. The arrows indicate the polarization direction. The* HG$_{00}$ *mode is the lowest order mode and is given by Eq. 12.66. Note the* HG$_{10}$ *mode has a node perpendicular to the polarization axis. The* HG$_{01}$ *has a node parallel to the polarization. The radial donut mode is found by adding two* HG$_{10}$ *modes, one x-polarized and the second y-polarized. The graphs are scaled logarithmically in the sense that a given contour indicates an intensity, $I \sim \left|\vec{E}^2\right|$, is down by a factor of 0.75 from the next higher up contour. The units for the axes may be taken as either "λ's" or, for the 1 μm beam used in the calculation, "μm's."*

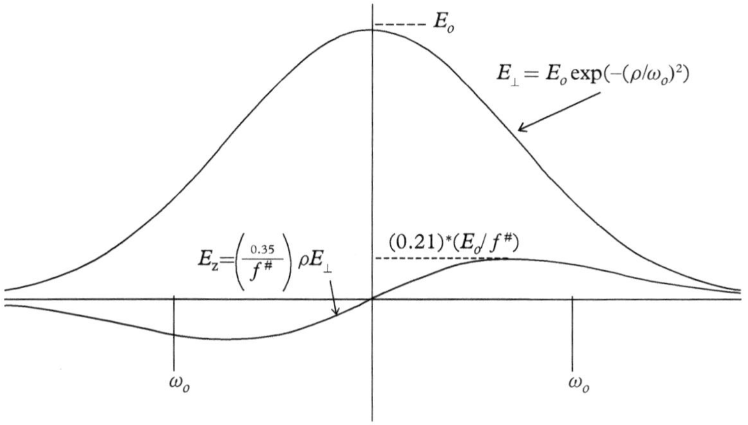

Fig. 12.13 *Relationship of the transverse field $E_\perp$ to the longitudinal z field E_z at the focus (z =0) of an electromagnetic wave with peak focused field of E_o. Here, $\rho^2 = (x^2 + y^2)$ and $f^\#$ is the f-number that is given in terms of the spot radius w_o by $f^\# = \frac{\pi w_o}{2\lambda}$. The plotted E_z field relative to the transverse field is for $f^\# = 1$. The peak E_z field in this case is 21% of the peak transverse field E_o, and out of phase by 90°.*

$$\frac{\partial E_x}{\partial x} + \frac{\partial E_z}{\partial z} = 0 \Rightarrow E_z = -\int \frac{\partial E_x(x,y,z)}{\partial x} dz \tag{12.69}$$

And so steep gradients in the transverse field lead to large longitudinal fields. Equation 12.63 for E_x may be differentiated with respect to x and integrated with respect to z to find the longitudinal field. In the focal plane:

$$\begin{aligned} E_z(x,y,0) &= -\int \frac{\partial}{\partial x}\left\{\frac{E_{0x}e^{ikz}}{1+2iz/kw_0^2}\exp\left[-\frac{(x^2+y^2)}{w_0^2}\frac{1}{1+2iz/kw_0^2}\right]\right\} dz \\ &= -i\left(\frac{2x}{kw_0^2}\right) E_x(x,y,0) \end{aligned} \tag{12.70}$$

From which we see that the longitudinal field (1) is zero on axis, (2) scales inversely with the area of the focal spot, and (3) leads the transverse field by 90°. See Fig. 12.13.

12.4.2 Fourier optics (far-field)

We next will examine the far-field regime using the angular spectrum representation. Specifically, starting with a particular localized field (object), how does the propagated field appear at a large distance? This will lead to Fourier optics. We start with the angular spectrum representation of the field, Eq. 12.54:

$$\vec{E}(x,y,z) = \frac{1}{2\pi}\iint_{-\infty}^{\infty} \vec{\mathscr{E}}\left(k_x,k_y;0\right) e^{i[k_x x + k_y y \pm k_z z]} dk_x dk_y \tag{12.71}$$

How does this look as $r \to \infty$? We define a direction vector: $\hat{s} = (s_x, s_y, s_z) = \left(\frac{x}{r}, \frac{y}{r}, \frac{z}{r}\right)$, where $r \to \infty$. Note the evanescent fields rapidly die out and do not contribute here. Therefore, we limit the transverse integration to a circle of radius k within which we get propagating wavevectors:

$$\vec{E}_\infty (s_x, s_y, s_z) = \frac{1}{2\pi} \iint_{(k_x^2+k_y^2 \le k^2)} \vec{\mathscr{E}} (k_x, k_y; 0) \exp\left[ikr\left(\frac{k_x}{k}s_x + \frac{k_y}{k}s_y \pm \frac{k_z}{k}s_z\right)\right] dk_x dk_y \quad (12.72)$$

In the $r \to \infty$ limit this integral can be evaluated by the use of the method of stationary phase, giving[8] (**see Discussions 12.5 and 12.6**):

$$\vec{E}_\infty (s_x, s_y, s_z) = -iks_z \vec{\mathscr{E}} (ks_x, ks_y; 0) \frac{e^{ikr}}{r} \quad (12.73)$$

In words, for a given field pattern about the origin, $\vec{E}(x, y, z)$, the far-field observed in the direction (s_x, s_y) is, to within some constants, found by taking the 2D Fourier transform of the field at the origin plane, $\vec{E}(x, y, 0)$, for the wave vector pointing in the observation direction, $\vec{\mathscr{E}} (ks_x, ks_y; 0)$. This is essentially the angular spectrum version of Fraunhofer diffraction. Ray optics may be used here: despite any complexities of the near-field radiation pattern, the far-field in the direction (s_x, s_y) is locally a plane wave and may be described by a bundle of rays pointing in that direction. In addition, we can (and will shortly) reverse the process by taking a far-field radiation pattern and using it to find the angular spectrum, $\vec{\mathscr{E}} (ks_x, ks_y; 0)$. This, in turn, can be used in Eq. 12.71 to find the field the near origin. In fact, solving Eq. 12.73 for $\vec{\mathscr{E}}$ and substituting this solution into Eq. 12.71 allows using the far-field limit to find the field at any location, to the extent that we can ignore evanescent components. This is the principal result of this section:

$$\vec{E}(x, y, z) = \frac{ire^{-ikr}}{2\pi} \iint_{(k_x^2+k_y^2 \le k^2)} \vec{E}_\infty (s_x, s_y, s_z) e^{i[k_x x + k_y y \pm k_z z]} \frac{1}{k_z} dk_x dk_y \quad (12.74)$$

where $s_i = k_i/k$.

12.4.3 Tight focusing of fields

Tightly focused fields are important in microscopy. In confocal microscopy a tightly focused beam is scanned over the sample being

[8] In the $r \to \infty$ limit, where the transverse extent of the source is negligible, light with a particular set of transverse wave vector components (k_x, k_y) will appear to originate from x_o, y_o, and travel at an angle with the z-axis of $\sqrt{(k_x^2 + k_y^2)}/k$. In this limit, then, we see that $k_x/k \to x/r$ and $k_y/k \to y/r$. In other words, at large r, the angular spectrum sorts itself out in the transverse direction, with k_x and k_y increasing along x and y. The projection in the far-field thus is seen as the Fourier transform of the source profile. This is proven explicitly in Discussion 12.5.

studied and the limit on the ability to resolve details of the sample is usually set by the size and shape of the beam near the focus. Similarly, optical data storage uses a tightly focused beam to locally alter a substrate, writing a zero or a one. The density of data that can be stored is ultimately limited by the geometry of the focused spot. Optical tweezers are used in both biological and cold atom applications where either cells or ultracold atoms are manipulated using a strongly focused electromagnetic field. Applications that use high intensity electromagnetic fields range from studying the nonlinear response of atomic and molecular systems to laser driven particle acceleration schemes. And attaining the required high intensity optical field usually involves the tight focusing of a laser. Here we will use the angular spectrum representation and, in particular, the results of the previous section to describe the focusing of a laser by a microscope objective. Specifically, we will find the near-focus optical field of a $\lambda = 1\ \mu$m linearly polarized laser focused by a state-of-the-art oil immersion objective having a numerical aperture (*NA*) of 1.4 and a focal length, f. The envisioned system is shown schematically in Fig. 12.14.

A monochromatic x-polarized plane wave in air along the z-axis is incident on an aplanatic lens. Aplanatic means that, for this case, the lens is free of spherical aberrations. In the ray picture, all rays parallel to the optical axis are refracted to a focal point, F. The numerical aperture of the lens is defined:

$$NA = n_2 \sin\theta_{\max} \tag{12.75}$$

where $\theta_{\max}$ is the maximum angle, measured from the optical axis, of the rays incident on the focus. That the NA is greater than one means

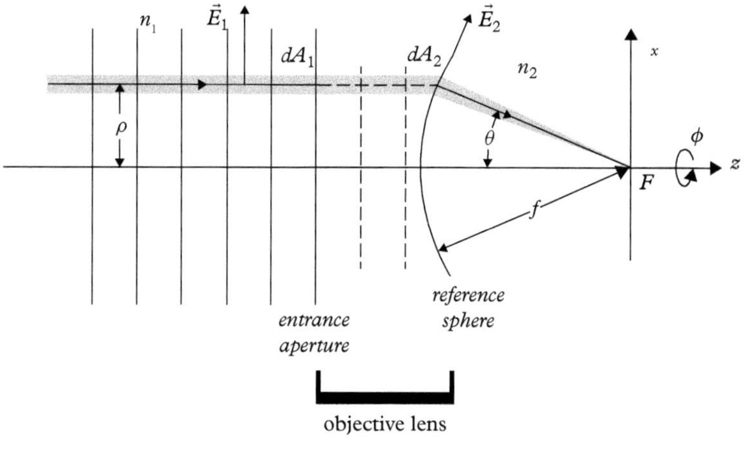

Fig. 12.14 *Tight focusing of a beam by an aplanatic objective lens. A plane wave is incident from the left. Shown are wavefronts and a pencil of rays around an explicitly shown central ray. The lens is described by an entrance aperture and a Gaussian reference sphere. The wavefront in the plane of the entrance aperture is mapped onto the reference sphere as described in the text.*

n_2 is greater than one. We will take $n_2 = 1.518$, the index of a common immersion oil used in microscopy. Referenced to the tightly focused image the lens creates, the field at the entrance aperture of the lens is the "far-field limit." In what follows, the field emerging from the lens, E_∞, will be found and then used in Eq. 12.74 to find the field in the region of the focus.

Description of the effect of an aplanatic lens

We follow Born and Wolf in modeling an aplanatic lens. An aplanatic objective lens takes rays parallel to the optical axis and directs them to the focus. Recalling that rays are perpendicular to the surfaces of constant phase–the wavefronts–the lens action may be mathematically implemented by taking the phase distribution at the plane of the entrance aperture of the objective lens and duplicating it on a spherical surface centered on the focus, a "Gaussian reference sphere." For the case of interest, here, both the plane at the entrance aperture and the Gaussian sphere are constant phase surfaces. Note that we are not emphasizing here the precise shapes of the optical surfaces of an aplanatic lens but only the effect on an incident phase surface. At any point on the reference sphere, the field is locally a plane wave perpendicular to the radius of the sphere at that point. The incident plane wave is most conveniently described in cylindrical coordinates while the converging wave after the lens is naturally described using spherical coordinates centered at F. The mapping of the field from the entrance aperture to the Gaussian sphere has the following three features (refer to Fig. 12.14):

(1) The sine rule: to give the correct magnifying power, the incident beam should be mapped onto a Gaussian sphere with radius $R = f$. The field at a radius ρ from the optic axis in the incident beam corresponds to the field at an angle θ on the Gaussian sphere where $\rho = f\,\sin\theta$. The angles ϕ are the same for the incident beam and its counterpart on the Gaussian sphere.

(2) The intensity law: the time-averaged power in a locally planar electromagnetic wave passing normal to a surface of area, dA, is:

$$dP = \frac{1}{2} Z_{\varepsilon\mu}^{-1} \, |E|^2 \, dA = \frac{1}{2}\sqrt{\frac{\varepsilon}{\mu}} \, |E|^2 \, dA \qquad (12.76)$$

As indicated by the grayed region in Fig. 12.14, the area of the pencil of rays in the incident beam is related to its corresponding area on the Gaussian sphere by $dA_1 = \cos\theta\, dA_2$. We assume all the materials are non-magnetic ($\mu_1 \simeq \mu_2 \simeq \mu_0$) and identify the indices of refraction as $n_i = \sqrt{\varepsilon_i/\varepsilon_0}$ where the ε_i are the material permittivities. By conservation of energy, then:[9]

$$|E_2| = |E_1|\sqrt{\frac{n_1}{n_2}\cos\theta} \qquad (12.77)$$

(3) Polarization transfer: assuming a purely transverse polarization vector (no z component in the incident beam), the polarization may be split up into a P component, in the plane of incidence (defined by the specific ray and the optical axis) and an S component, perpendicular to that plane. For the incident beam, these are the $\hat{i}_\rho$ and $\hat{i}_\phi$ directions and at the Gaussian sphere these are the $\hat{i}_\theta$ and $\hat{i}_\phi$ directions. Thus, in general, the incident beam's polarization unit vector may be written:

$$\hat{\varepsilon}(\rho,\phi) = a_\rho(\rho,\phi)\,\hat{i}_\rho(\phi) + a_\phi(\rho,\phi)\,\hat{i}_\phi(\phi)$$

where $a_\rho^2 + a_\theta^2 = 1$ and both the weighting and the direction of a unit vector depends on its location in the wavefront. For the present study, all the incident rays are x-polarized so $\hat{\varepsilon} = \hat{i}_x = \cos\phi\,\hat{i}_\rho - \sin\phi\,\hat{i}_\phi$. The polarization unit vector weights map to the spherical coordinates as $\hat{i}_\rho \Rightarrow \hat{i}_\theta$ and $\hat{i}_\phi \Rightarrow \hat{i}_\phi$. Thus the x-polarized beam at the reference sphere has polarization unit vector $\hat{\varepsilon}_\infty = \cos\phi\,\hat{i}_\theta - \sin\phi\,\hat{i}_\phi$.

We will use the convention that points near the lens, on the Gaussian sphere, will be indicated $(x_\infty, y_\infty, z_\infty) \Leftrightarrow (f,\theta,\phi)$ and points near the focus/origin are specified as $(x,y,z) \Leftrightarrow (\rho_0, \vartheta_0, z_0)$. For the incident beam, we assume an x- polarized Gaussian beam with its waist at the entrance aperture of the lens. The present vector diffraction theory is required to properly treat the features arising from the boundary conditions imposed on the fields as the beam passes through the lens. Putting features (1)–(3) all together, then, the total refracted field on the downstream side of the lens/Gaussian reference sphere is:[10]

$$\vec{E}_\infty(f,\theta,\phi) = t^{(p)}\sqrt{\frac{n_1}{n_2}}\cos\theta\left[\vec{E}_{inc}(f\sin\theta,\phi)\cdot\hat{i}_\rho\right]\hat{i}_\theta + t^{(s)}\sqrt{\frac{n_1}{n_2}}\cos\theta\left[\vec{E}_{inc}(f\sin\theta,\phi)\cdot\hat{i}_\phi\right]\hat{i}_\phi \qquad (12.78)$$

[9] Setting the powers in the incident and Gaussian sphere regions equal,

$$\begin{aligned} dP_2 &= dP_1 \\ \frac{1}{2}\sqrt{\frac{\varepsilon_2}{\mu_2}}|E_2|^2\, dA_2 &= \frac{1}{2}\sqrt{\frac{\varepsilon_1}{\mu_1}}|E_1|^2\, dA_1 \\ |E_2|^2\, dA_2 &= |E_1|^2\sqrt{\frac{\mu_2\varepsilon_1}{\varepsilon_2\mu_1}}\cos\theta\, dA_2 \\ |E_2|^2 &= |E_1|^2\,\frac{n_1}{n_2}\cos\theta \end{aligned}$$

where in the last step we used $\mu_1 \simeq \mu_2$ and $n_i = \sqrt{\varepsilon_i/\varepsilon_0}$.

[10] Note here that $\vec{E}_{inc}$ corresponds to $\vec{E}_1$ and $\vec{E}_\infty$ corresponds to $\vec{E}_2$ in Fig. 12.14.

where $t^{(p)}$ and $t^{(s)}$ are–possibly complex–transmission coefficients. This expression is general, but going forward, we consider the case where the transmission coefficients are unity. The strategy, here, is to next use Eq. 12.74 to find the field near the origin, $\vec{E}(x, y, z)$. To this end, it is useful to express the Cartesian variables, k_x, k_y, and k_z as the k-space spherical coordinates, k, θ_k, and ϕ_k. Using:

$$k_x = k\sin\theta_k\cos\phi_k \quad k_y = k\sin\theta_k\sin\phi_k \quad k_z = k\cos\theta_k \tag{12.79}$$

along with the Jacobian for the change of variables, $k_x, k_y \to \theta_k, \phi_k$, the 2D differential of Eq. 12.74 becomes:

$$\begin{aligned}\frac{1}{k_z}dk_x dk_y \Rightarrow &\frac{1}{k\cos\theta_k}\frac{\partial\left(k_x, k_y\right)}{\partial\left(\theta_k, \phi_k\right)} \\ &= \frac{1}{k\cos\theta_k}\left(k^2\cos\theta_k\sin\theta_k\cos\phi_k^2 + k^2\sin\theta_k\cos\theta_k\sin^2\phi_k\right) \\ &\times d\theta_k\, d\phi_k = k\sin\theta_k\, d\theta_k\, d\phi_k\end{aligned} \tag{12.80}$$

In addition, because of the cylindrical symmetry of the problem, the angular spectrum representation at the focus turns out to be more conveniently expressed using cylindrical (ρ_0, φ_0, z_0) rather than Cartesian coordinates. Thus in Eq. 12.74 we substitute:

$$x = \rho_0\cos\varphi_0 \quad y = \rho_0\sin\varphi_0 \quad z = z_0 \tag{12.81}$$

Putting the results of Eqs. 12.79, 12.80, and 12.81 together allows us to rewrite Eq. 12.74[11]

$$\begin{aligned}\vec{E}\left(\rho_0, \varphi_0, z_0\right) = &-\frac{ikfe^{-ikf}}{2\pi}\int_0^{\theta_{\max}}\int_0^{2\pi}\vec{E}_\infty\left(k, \theta_k, \phi_k\right)\exp\left(\pm ikz_0\cos\theta_k\right) \\ &\times \exp\left[ik\rho_0\sin\theta_k\cos\left(\phi_k - \varphi_0\right)\right]\sin\theta_k d\theta_k d\phi_k.\end{aligned} \tag{12.82}$$

Finally, the limit, $\theta_{\max} < \pi/2$, is determined by the size of the acceptance angle of the objective lens.

The focal field of a Gaussian beam

We are most interested here in studying beams with as tight a focus as possible. We assume a lens transmission of $t = 1$ for all angles and both polarizations. For incident fields, we consider the lowest few Gaussian modes. The beam waist of the input mode will be placed at the entrance aperture so that the incident wavefronts are planar surfaces and the paraxial approximation is excellent. The waist of the HG_{00} mode is given by Eq. 12.64 with $z = 0$:

$$\vec{E}_{00}^{inc} = \hat{i}_x E_0 e^{-\rho^2/w_0^2} \tag{12.83}$$

[11] On the second line of Eq. 12.82, we proceeded from Eq. 12.74 as

$$\begin{aligned}k_x x + k_y y &\to \left(k\sin\theta_k\cos\phi_k\right)\rho_0\cos\varphi_0 \\ &\quad + \left(k\sin\theta_k\sin\phi_k\right)\rho_0\sin\varphi_0 \\ &= k\rho_0\sin\theta_k\left[\left(\cos\phi_k\cos\varphi_0\right) + \left(\sin\phi_k\sin\varphi_0\right)\right] \\ &= k\rho_0\sin\theta_k\left[\cos\left(\phi_k - \varphi_0\right)\right]\end{aligned}$$

Also note that the minus sign in front of the right-hand side expression of Eq. 12.82 is because of the geometry: The integration over θ_k is actually around 180°, that is, referenced to the negative z-axis.

This is the field at the input to the lens, so it is not a tightly focused beam. Typically, to fill the entrance aperture of a high NA immersion lens, w_0 is on the order of a centimeter. As described by Eq. 12.78 in the previous section, this is mapped onto the Gaussian reference sphere for the lens:

$$\vec{E}^{\infty}_{00} = E_0 e^{-f^2 \sin^2\theta_k / w_0^2} \sqrt{\frac{n_1}{n_2}} \cos\theta_k \left(\cos\phi_k \hat{i}_{\theta_k} - \sin\phi_k \hat{i}_{\phi_k}\right) \tag{12.84}$$

Similarly, on the reference sphere, x-polarized HG_{10} and HG_{01} modes produce the fields,

$$\begin{aligned}\vec{E}^{\infty}_{10} = E_0 \left(\frac{2f}{w_0}\right) \sin\theta_k \cos\phi_k e^{-f^2 \sin^2\theta_k / w_0^2} \\ \sqrt{\frac{n_1}{n_2}} \cos\theta_k \left(\cos\phi_k \hat{i}_{\theta_k} - \sin\phi_k \hat{i}_{\phi_k}\right)\end{aligned} \tag{12.85}$$

and

$$\begin{aligned}\vec{E}^{\infty}_{01} = E_0 \left(\frac{2f}{w_0}\right) \sin\theta_k \sin\phi_k e^{-f^2 \sin^2\theta_k / w_0^2} \\ \sqrt{\frac{n_1}{n_2}} \cos\theta_k \left(\cos\phi_k \hat{i}_{\theta_k} - \sin\phi_k \hat{i}_{\phi_k}\right)\end{aligned} \tag{12.86}$$

The spherical polarization basis was useful in taking the incident field from the entrance aperture to the reference sphere. However, in carrying out the angular spectrum integral, Eq. 12.82, the directions of the basis vectors, $\hat{i}_{\theta_k}, \hat{i}_{\phi_k}$, themselves are functions of the integration variables. This means that, in that basis, the integral for even a single component of $\vec{E}_\infty$ is itself essentially a vector sum. This can be simplified by using a Cartesian basis for the polarization. Integrating each component is then simply a scalar-type integral: for example, the direction of $\hat{i}_x$ at a point P does not depend P's coordinates. The polarization unit vector on the reference sphere may be written as a column Cartesian vector following the discussion in the scattering chapter (see Eq. 11.158)

$$\cos\phi_k \hat{i}_{\theta_k} - \sin\phi_k \hat{i}_{\phi_k} = \left[\cos\phi_k \begin{pmatrix} \cos\phi_k \cos\theta_k \\ \sin\phi_k \cos\theta_k \\ -\sin\theta_k \end{pmatrix} - \sin\phi_k \begin{pmatrix} -\sin\phi_k \\ \cos\phi_k \\ 0 \end{pmatrix}\right] \tag{12.87}$$

Note how the bending of a light ray by the lens in the "θ" direction has introduced a field component in the "z" direction. See also Fig. 12.14. We define the beam "filling factor" as the ratio of the beam radius to the radius of the lens:

$$f_0 = \frac{w_0}{f \sin\theta_{\max}} \tag{12.88}$$

All modes contain the factor $f_w(\theta) = \exp\left[-f^2 \sin^2\theta / w_0^2\right]$; using Eq. 12.88, this may be rewritten as:

$$f_w(\theta) = \exp\left[-\frac{1}{f_0^2}\frac{\sin^2\theta}{\sin^2\theta_{\max}}\right] \tag{12.89}$$

and is identified a the "apodization function." It is like a pupil filter for the system. Summarizing these results, the field created near the focus by an objective lens of an x-polarized HG_{00} mode laser beam is:

$$\begin{aligned}\vec{E}_{00}(\rho_0, \varphi_0, z_0) = & -E_0\frac{ikfe^{-ikf}}{2\pi}\int_0^{\theta_{\max}}\int_0^{2\pi} f_w(\theta_k)\sqrt{\frac{n_1}{n_2}\cos\theta_k}\frac{1}{2}[\theta_k, \phi_k] \\ & \times \exp\left[ik\left(\rho_0\sin\theta_k\cos(\phi_k - \varphi) \pm z_0\cos\theta_k\right)\right] \\ & \times \sin\theta_k d\theta_k d\phi_k\end{aligned} \tag{12.90}$$

where the term $[\theta_k, \phi_k]$ represents the Cartesian vector,

$$[\theta_k, \phi_k] \equiv \begin{bmatrix} (1+\cos\theta_k) - (1-\cos\theta_k)\cos 2\phi_k \\ -(1-\cos\theta_k)\sin 2\phi_k \\ -2\cos\phi_k\sin\theta_k \end{bmatrix}$$

which has been obtained by collecting terms from Eq. 12.87. Again, to keep things straight, the positions of points near the focus are specified using cylindrical coordinates. However, the components of the electric field in the focal region are Cartesian. The ϕ integrals

$$\int_0^{2\pi} [\theta_k, \phi_k]\exp\left[ik\rho_0\sin\theta_k\cos(\phi_k - \varphi)\right]d\phi_k \tag{12.91}$$

may be evaluated analytically using the identities:

$$\begin{aligned}\textstyle\int_0^{2\pi}\cos n\phi e^{ix\cos(\phi-\varphi)}d\phi &= 2\pi(i)^n \mathcal{J}_n(x)\cos n\varphi \\ \textstyle\int_0^{2\pi}\sin n\phi e^{ix\cos(\phi-\varphi)}d\phi &= 2\pi(i)^n \mathcal{J}_n(x)\sin n\varphi\end{aligned} \tag{12.92}$$

where, by comparison with Eq. 12.91, we identify x as $k\rho_0\sin\theta_k$. This leaves a final integration over θ that must be evaluated numerically. Table (12.1) defines the numerical integrals required for the lowest few Hermite–Gaussian modes.

A tightly focused HG_{00} mode

The field in the focal region for an HG_{00}, Gaussian, incident beam is obtained using the identities 12.92 and Table 12.1 within the integral of Eq. 12.91:

Table 12.1 *Definitions of the numerical integrals needed for finding focused Hermite–Gaussian beams.*

$I_{00}=\int_0^{\theta_{\max}} f_w(\theta)\sqrt{\cos\theta}\sin\theta\,(1+\cos\theta)\,J_0(k\rho\sin\theta)\,e^{ikz\cos\theta}\,d\theta$
$I_{01}=\int_0^{\theta_{\max}} f_w(\theta)\sqrt{\cos\theta}\sin^2\theta\,J_1(k\rho\sin\theta)\,e^{ikz\cos\theta}\,d\theta$
$I_{02}=\int_0^{\theta_{\max}} f_w(\theta)\sqrt{\cos\theta}\sin\theta\,(1-\cos\theta)\,J_2(k\rho\sin\theta)\,e^{ikz\cos\theta}\,d\theta$
$I_{10}=\int_0^{\theta_{\max}} f_w(\theta)\sqrt{\cos\theta}\sin^3\theta\,J_0(k\rho\sin\theta)\,e^{ikz\cos\theta}\,d\theta$
$I_{11}=\int_0^{\theta_{\max}} f_w(\theta)\sqrt{\cos\theta}\sin^2\theta\,(1+3\cos\theta)\,J_1(k\rho\sin\theta)\,e^{ikz\cos\theta}\,d\theta$
$I_{12}=\int_0^{\theta_{\max}} f_w(\theta)\sqrt{\cos\theta}\sin^2\theta\,(1-\cos\theta)\,J_1(k\rho\sin\theta)\,e^{ikz\cos\theta}\,d\theta$
$I_{13}=\int_0^{\theta_{\max}} f_w(\theta)\sqrt{\cos\theta}\sin^3\theta\,J_2(k\rho\sin\theta)\,e^{ikz\cos\theta}\,d\theta$
$I_{14}=\int_0^{\theta_{\max}} f_w(\theta)\sqrt{\cos\theta}\sin^2\theta\,(1-\cos\theta)\,J_3(k\rho\sin\theta)\,e^{ikz\cos\theta}\,d\theta$

$$\vec{E}_{00}(\rho_0,\varphi_0,z_0)=-E_0\frac{ikfe^{-ikf}}{2}\sqrt{\frac{n_1}{n_2}}\begin{bmatrix} I_{00}+I_{02}\cos 2\varphi_0 \\ I_{02}\sin 2\varphi_0 \\ -2iI_{01}\cos\varphi_0 \end{bmatrix} \tag{12.93}$$

Figure 12.15 shows how the filling factor (see Eq. 12.88) affects the tightness of a focus for a high NA lens. Shown in the figure are profiles of $\left|\vec{E}^2\right|$ taken along the x-axis (solid lines) and y-axis (dashed lines) in the focal plane. For a small filling factor in which practically all of the incident beam passes through the entrance aperture, the focused beam is essentially a paraxial Gaussian, and Fig. 12.11 and Eq. 12.66 are reproduced at the focus. The limit of a large filling factor is an (infinite) plane wave incident in the entrance aperture and this produces an Airy disk image. Note that for this high numerical aperture, the shape of the focus takes an elliptical character at larger filling factors: the y-axis profile, that is, in the direction perpendicular to the incident polarization, is noticeably smaller than that taken along the x-axis. By contrast, scalar theory would predict these to be identical. Figure 12.16 shows, in more detail, the $f_0=1$ case where the beam size is well matched to the lens. Energy densities for the individual field components, $\left|\vec{E}^2{}_x\right|$, $\left|\vec{E}^2{}_y\right|$, and $\left|\vec{E}^2{}_z\right|$ in the focal region are shown. Scalar theory would predict the last two of these to be zero. For the longitudinal, z, field component, the maxima are on the x-axis, at $x=\pm 0.27\lambda$. The energy density of the z field component at these locations is 1/8th that of the maximum for the transverse field, whose maximum is at the origin. Figure 12.16 additionally shows contours of total field energy density along the direction of propagation for the x–z and y–z planes. Here again, the beam's polarization results in a departure from the radial symmetry of the Gaussian beam.

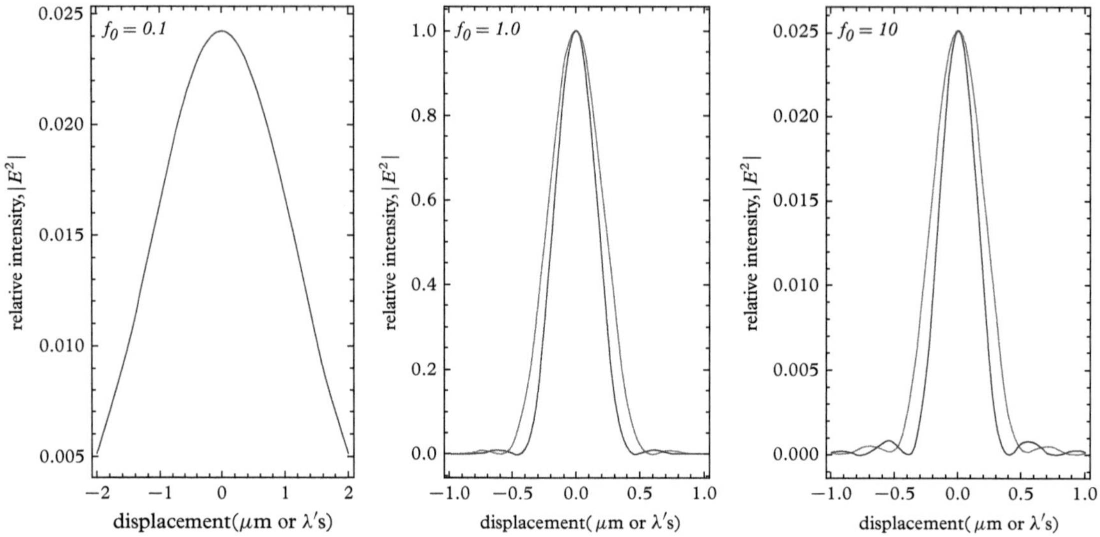

Fig. 12.15 *The effect of the filling factor on the sharpness of the focus for a Gaussian 00 beam focused by a lens with NA 1.4,* $n = 1.518$, *and* $\theta_{max} = 69^o$, *for filling factors of 0.1, 1.0, and 10. Shown are profiles of* $\left|\vec{E}^2\right|$ *taken along the x-axis (blue) and y-axis (red) in the focal plane for different filling factors of the lens aperture. The total power in the incident Gaussian beam is held constant for these figures and scaled such that the peak of the* $f_0 = 1$ *is 1. Note the differing scales for both axes.*

Higher order modes

Figure 12.12 shows x-polarized HG_{10} and HG_{01} input modes. On the Gaussian reference sphere, these are given by Eqs. 12.85 and 12.86, respectively. Proceeding as for the 00 mode, the fields at the focus are found to be:

$$\vec{E}_{10}(\rho,\varphi,z) = -E_0\frac{ikf^2e^{-ikf}}{2w_0}\sqrt{\frac{n_1}{n_2}}\begin{bmatrix} iI_{11}\cos\varphi + iI_{14}\cos 3\varphi \\ -iI_{12}\sin\varphi + iI_{14}\sin 3\varphi \\ -2iI_{10} + 2I_{13}\cos 2\varphi \end{bmatrix} \tag{12.94}$$

$$\vec{E}_{01}(\rho,\varphi,z) = -E_0\frac{ikf^2e^{-ikf}}{2w_0}\sqrt{\frac{n_1}{n_2}}\begin{bmatrix} i(I_{11}+2I_{12})\sin\varphi + iI_{14}\sin 3\varphi \\ -iI_{12}\cos\varphi - iI_{14}\cos 3\varphi \\ 2I_{13}\sin 2\varphi \end{bmatrix} \tag{12.95}$$

where the ϕ and θ integrals are defined, respectively, in the Eqs. of 12.92 and in Table 12.1.

Applications of higher order modes include atom trapping, optical tweezers, and particle acceleration. Frequently, the "donut modes"

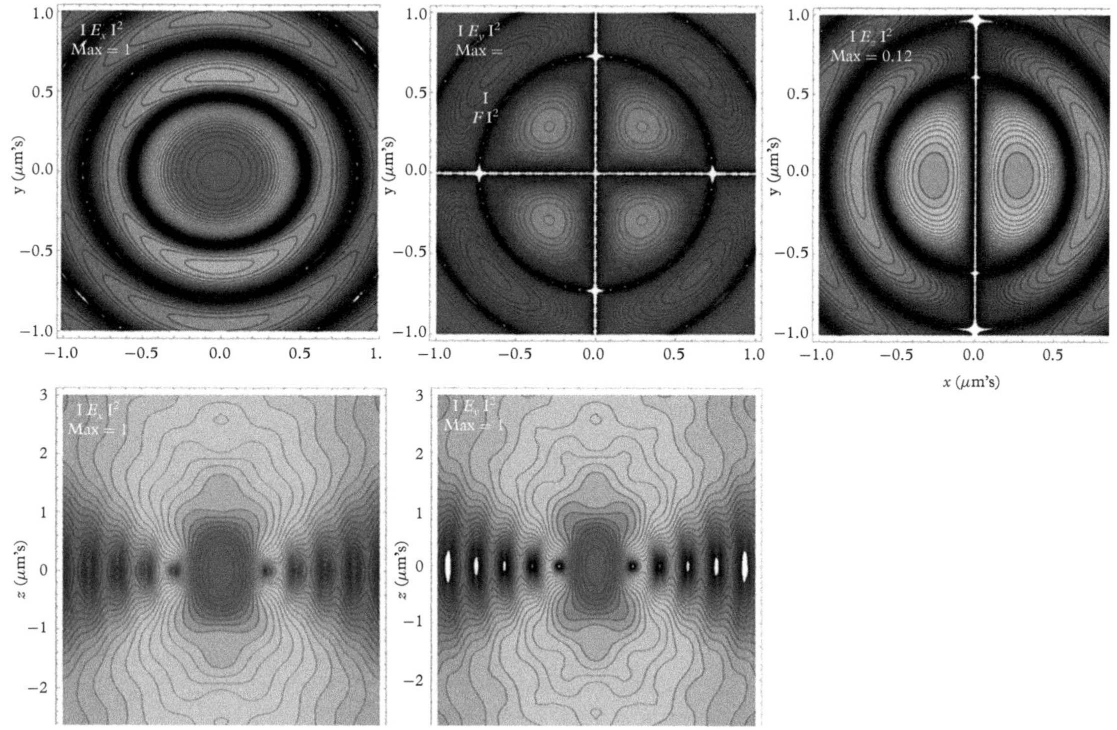

Fig. 12.16 *Electric field energy density in the focal region of Gaussian beam focused using a high NA immersion objective lens. The propagation direction is z and x is the polarization of the incident laser (before the lens). These are for NA = 1.4, n = 1.518, and a filling factor of 1. The top three maps show the energy in the individual field components,* $\left|\vec{E}_x^2\right|$, $\left|\vec{E}_y^2\right|$, *and* $\left|\vec{E}_z^2\right|$ *in the focal plane. Note especially the significant longitudinal,* E_z, *components near the focus. In the bottom maps, slices of* $|E_x|^2$ *(left) and* $|E_y|^2$ *(right) are taken to show the propagation of the fields along z. Differences between the x and y directions, not predicted in the Gaussian description, are readily seen. Careful examination of the* $\left|\vec{E}_x^2\right|$ *shows a similar effect: the contours are not perfectly circular. The contour scaling is logarithmic with successive contours differing by factors of 2. "Max" refers to the maximum energy density for a given figure relative to the total maximum electric field energy density in the beam at the origin. The color scaling is consistent among the figures. Again, the results scale such that the axis can be read as having either units of μm or λ's.*

created from linear combinations of HG modes are useful. Specifically, linearly polarized, radially polarized, and azimuthally polarized donut modes may be created by combining beams of two different input modes. Examples include:

$$\begin{aligned}
\text{x}-\text{polarized} &= \text{HG}_{10}\hat{i}_x + i\text{HG}_{01}\hat{i}_x \\
\text{radially polarized} &= \text{HG}_{10}\hat{i}_x + \text{HG}_{10}\hat{i}_y \\
\text{azimuthally polarized} &= -\text{HG}_{01}\hat{i}_x + \text{HG}_{01}\hat{i}_y
\end{aligned}$$

Here, for example, the notation $HG_{01}\hat{i}_y$ indicates a 01 Hermite-Gaussian mode linearly polarized in the y-direction. An "i" prefactor corresponds to a 90° phase shift for a particular beam. The radially polarized donut was introduced before and is shown in Fig. 12.12. One feature of the higher order modes relative to the 00 mode is that the derivatives are sharper and therefore the additional fields necessary to satisfy Gauss's law can be very large. This is very evident for large NA, large filling-factor beams that result in very tightly focused fields. The radial donut mode is especially interesting because when tightly focused it produces an extremely strong longitudinal field as we now discuss. The focal field for a radial donut mode input beam is found by adding Eq. 12.94, the focal field associated with the x-polarized HG_{10} input, to the same focal field rotated 90° about the z-axis (and therefore associated with a y-polarized HG_{10} input). This results in:

$$\vec{E}(\rho,\varphi,z) = -E_0\frac{ikf^2e^{-ikf}}{2w_0}\sqrt{\frac{n_1}{n_2}}\begin{bmatrix} i(I_{11}-I_{12})\cos\varphi \\ i(I_{11}-I_{12})\sin\varphi \\ -4I_{10} \end{bmatrix} \tag{12.96}$$

And, indeed, within the focal plane of the objective lens a radially polarized donut shaped field is produced. However, to satisfy Gauss's law, the steep field gradient on the inside of the donut requires a strong z component and the maximum of the longitudinal field, $|E_z|$, can be significantly stronger than the maximum of the radial field, $|E_\rho|$.

Figure 12.17 shows that in the focal plane of our NA = 1.4 optical system, the maximum energy density of the longitudinal field is over three times that of the radial field. Some applications such as particle acceleration schemes require a strong longitudinal field. These results also highlight the need, in some cases, for a vector diffraction theory: scalar diffraction theory would predict the longitudinal field to vanish.

12.4.4 Diffraction limits on microscopy

Diffraction presents fundamental limits on spatial resolution in microscopy. Important for most microscopy applications is the answer to the question "what is the minimum on the size of features the microscope can resolve?" A more precise way of framing the essence of this question is "how far apart must two point-source radiators be for them to be distinguished?" The resolution of a microscope is most often described using a "point spread function," or PSF. The PSF is a very general concept and for *any optical system* it is the signal pattern or the blurred spot observed at the end of an imaging system for a point source at the input. From the system's view of the angular frequency spectrum, the PSF is not a geometric point because of spatial filtering

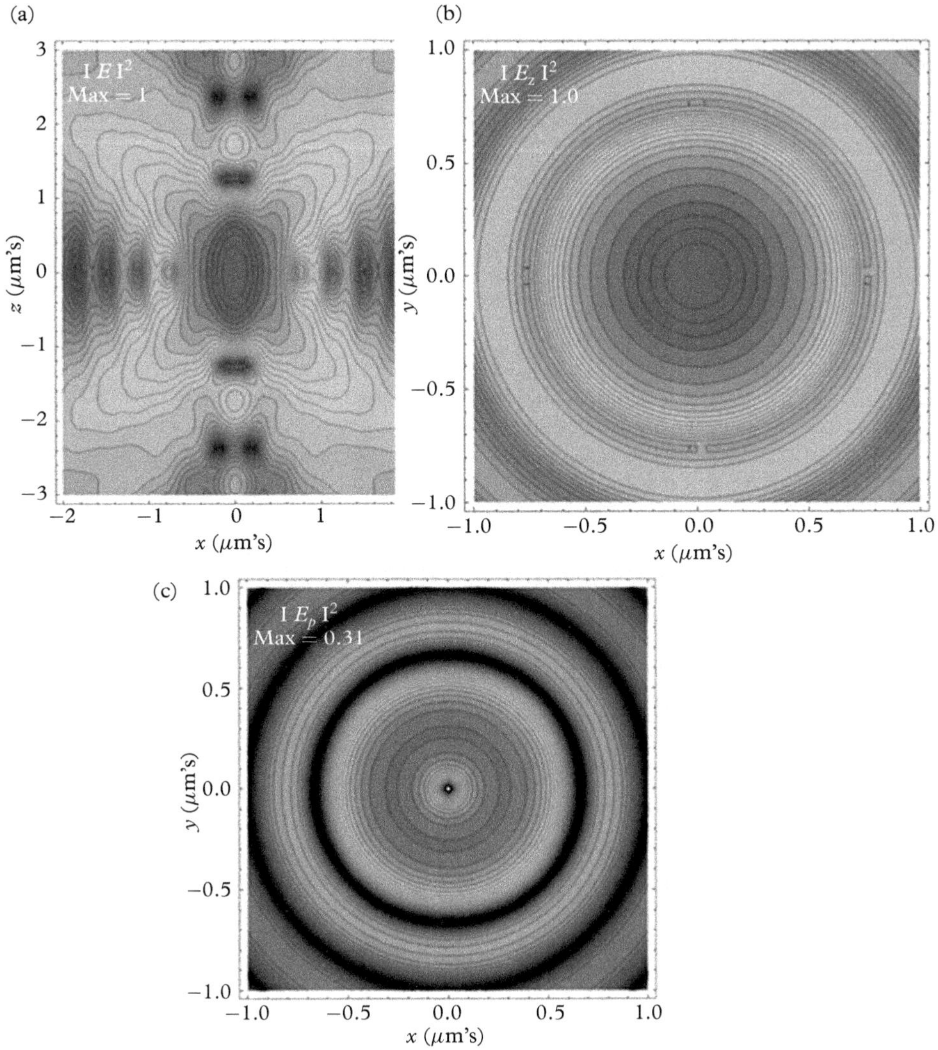

Fig. 12.17 $\left|\vec{E}^2\right|$ *in the focal region for a tightly focused beam from a radially polarized donut mode input. Here, NA=1.4,* $n = 1.518$*, and* $f_0 = 1$*. Map (a) is a plot showing contours of constant* $\left|\vec{E}^2\right|$ *along the axis. Successive contours represent factors of 2. Maps (b) and (c) are distributions for* $\left|\vec{E}_z^2\right|$, $\left|\vec{E}_\rho^2\right|$*, respectively.*

by the optical system. Not all of the Fourier components of the angular spectrum are captured by the system. Specifically, for regular, far-field optical microscopy, the evanescent components do not contribute to the image and even some of the propagating components are not captured. As a representative example, we will find the PSF of an

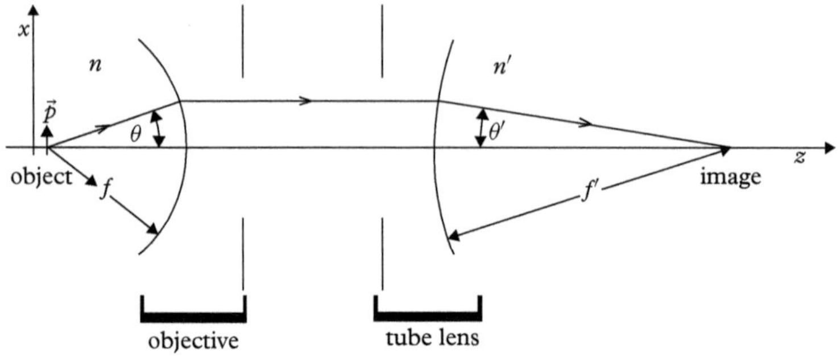

Fig. 12.18 *Schematic of microscope and definition of the quantities used to model the PSF.*

x-polarized dipole source of $\lambda = 1\ \mu$m light for the microscope shown schematically in Fig. 12.18. Most modern sophisticated microscopes use the infinity-corrected optics shown. Working from the left, in the geometrical optics picture: from a near-axis point source in its focal plane, an objective lens creates an image free of spherical aberration "at infinity." We will consider for this example calculation the NA = 1.4 objective lens of the previous section. We will choose its focal length such that the overall magnification of the system is 100×. Downstream, a "tube lens" with focal length f' focuses the rays onto a photosensitive surface located in its focal plane. Nowadays this surface is often a semiconductor sensor. If the object and image of the microscope are both in air, then in the geometrical optics limit an object dipole located in the objective's focal plane at a distance δ off axis creates a conjugate point source image a distance $\delta' = \frac{f'}{f}\delta$ from the optic axis.The magnification of the system, then, is $M = f'/f$. For our present case in which the object and possibly the image are embedded in a dielectric, the rays emerging from the objective lens are bent, by Snell's law, toward the optical axis by the factor $1/n$ and then they are bent away from the optical axis by $1/n'$ at the tube lens. The magnification thus becomes $M = \left(nf'\right) / \left(n'f\right)$. For most of what follows, we consider the on-axis dipole shown in the figure. The objective lens is the most technically sophisticated part of the microscope. With a typical focal length 2–5 mm and a numerical aperture >1, it requires a high degree of correction for both spherical and chromatic aberrations. Most present day tube lenses have a focal length of ~ 20 cm and, referring to the figure, since the required numerical aperture is down by the ratio of the focal lengths, a tube lens numerical aperture of 0.1 is adequate for most applications and the paraxial approximation for this part of the problem is well satisfied.

The microscope PSF may be found by proceeding from left to right in Fig. 12.18. We begin by using the multipole expansion results from previous chapters and finding the field on the Gaussian reference sphere of the objective lens. Next, we use the mapping, described earlier in the chapter, between the reference sphere and "entrance aperture" to find the field at the plane of the latter. Note that now the mapping is the reverse of the previous section and the entrance aperture for that problem is actually an exit aperture for this one. Now the propagation from the objective to the tube lens takes place in the geometrical optics regime. To a very good approximation, these rays are parallel to the optical axis and the effect of the propagation distance, Δz, between the two lens apertures is to add a constant phase factor, $\exp[ik\Delta z]$, to all points on the wavefront. Because the PSF is an intensity distribution, this has no effect on the final result and so we take the wavefront at the exit aperture of the objective equal to that of the input to the tube lens. Finding the final image–the PSF–in going from the entrance aperture to the focus of the tube lens is then essentially the problem of the previous section.

The radiation zone field of an oscillating dipole in a material with a relative dielectric constant is given by Eq. 4.40.

$$\vec{E}(\vec{r}) = \frac{\omega^2}{c^2}\frac{1}{4\pi\varepsilon_0}\frac{e^{ikr}}{r}\left[\left(\hat{i}_r \times \vec{p}\right)\times \hat{i}_r\right] \tag{12.97}$$

where $\vec{p}$ is the strength of the dipole, k is the wavevector in the dielectric in which the dipole is embedded, $\vec{r} = r\hat{i}_r$ is the displacement of the observation point from the dipole's position. For a dipole oscillating in the x-direction at the focus of the objective, the field on the Gaussian reference sphere in polar coordinates is[12]

$$\vec{E}_\infty(f,\theta_k,\phi_k) = \frac{\omega^2}{c^2}\frac{p}{4\pi\varepsilon_0}\frac{e^{ikf}}{f}\left(\cos\theta_k\cos\phi_k\hat{i}_\theta - \sin\phi_k\hat{i}_\phi\right) \tag{12.98}$$

This, then, is mapped to the objective's exit aperture as given by Eq. 12.77 (with $n_1 = 1$ and $n_2 = n$)

$$\vec{E}(\rho = f\sin\theta_k,\phi_k) = -\frac{\omega^2}{c^2}\frac{p}{4\pi\varepsilon_0}\frac{e^{ikf}}{f}\sqrt{\frac{n}{\cos\theta_k}}\left(\cos\theta_k\cos\phi_k\hat{i}_\rho - \sin\phi_k\hat{i}_\phi\right) \tag{12.99}$$

where the polarization change is carried out but we hold off on eliminating θ_k in favor of ρ. As described before, this is also taken to be the field at the entrance aperture of the tube lens. The field at the tube lens reference sphere is subsequently mapped, by Eq. 12.78, from the tube lens aperture as in the previous problem (now, with $n_1 = 1$ and $n_2 = n'$):

[12] Note that $\vec{p} = p\hat{x}$ and so,

$$\begin{aligned}\left[\left(\hat{i}_r \times \vec{p}\right)\times \hat{i}_r\right] &= \left(\hat{i}_r\cdot\hat{i}_r\right)\vec{p} - \left(\vec{p}\cdot\hat{i}_r\right)\hat{i}_r\\ &= p\left[\hat{x} - \left(\hat{x}\cdot\hat{i}_r\right)\hat{i}_r\right]\\ &= p\left[\left(\hat{x}\cdot\hat{i}_\theta\right)\hat{i}_\theta + \left(\hat{x}\cdot\hat{i}_\phi\right)\hat{i}_\phi\right]\\ &= p\left(\cos\theta_k\cos\phi_k\hat{i}_\theta - \sin\phi_k\hat{i}_\phi\right).\end{aligned}$$

$$\vec{E}'_{\infty}(f',\theta',\phi') = \frac{\omega^2}{c^2}\frac{p}{4\pi\varepsilon_0}\frac{e^{ikf}}{f}\sqrt{\frac{n}{\cos\theta_k}}\sqrt{\frac{\cos\theta'_{k'}}{n'}} \times \left(\cos\theta_k\cos\phi_k\hat{i}_{\theta'_{k'}} - \sin\phi_k\hat{i}_{\phi'_{k'}}\right) \tag{12.100}$$

This is, to within constants, the Fourier transform of the image we are after (see Eq. 12.73). The image field (in the focal region of the tube lens) is found by integrating this over $\theta'_{k'}$ and $\phi'_{k'}$, as in Eq. 12.90, to take the inverse transform. To carry out this vector integral, the field is expressed in Cartesian components as done previously

$$\begin{aligned}\vec{E}'(\rho'_0,\varphi'_0,z'_0) &= -p\frac{\omega^2}{c^2}\frac{ik'f'\exp\left[ikf - ik'f'\right]}{f\,8\pi^2\varepsilon_0}\sqrt{\frac{n}{n'}}\int_0^{\theta'_{\max}}\int_0^{2\pi}\sin\theta'_{k'}d\theta'_{k'}d\phi'_{k'}\sqrt{\frac{\cos\theta'_{k'}}{\cos\theta_k}} \\ &\times\left[\theta_k,\phi_k,\theta'_{k'},\phi'_{k'}\right]\exp\left[ik'\left(\rho'_0\sin\theta'_{k'}\cos\left(\phi'_{k'}-\varphi'_0\right) \pm z'_0\cos\theta'_{k'}\right)\right]\end{aligned}$$

where the term $\left[\theta_k,\phi_k,\theta'_{k'},\phi'_{k'}\right]$ represents the Cartesian vector,

$$\left[\theta_k,\phi_k,\theta'_{k'},\phi'_{k'}\right] = \left[\cos\theta_k\cos\phi_k\begin{pmatrix}\cos\phi'_{k'}\cos\theta'_{k'}\\ \sin\phi'_{k'}\cos\theta'_{k'}\\ -\sin\theta'_{k'}\end{pmatrix} - \sin\phi_k\begin{pmatrix}-\sin\phi'_{k'}\\ \cos\phi'_{k'}\\ 0\end{pmatrix}\right]$$

This integral is here expressed using both the unprimed angles of the objective lens and the primed variables of the tube lens. To evaluate the integral, it is necessary to eliminate one set. We will eliminate the primed variables in favor of the unprimed ones by noting that the field at the output of the objective lens maps directly into the input of the tube lens. Positions on the respective reference spheres are mapped: $\phi'_{k'} = \phi_k$. The "sine rules" relating the fields at the reference spheres and those at the planes of the respective output/input apertures give: $\rho = f\sin\theta_k = f'\sin\theta'_{k'}$. That is, in the previous equation:

$$\sin\theta'_{k'} = \frac{f}{f'}\sin\theta_k \qquad d\theta'_{k'} = \frac{f}{f'}\frac{\cos\theta_k}{\cos\theta'_{k'}}d\theta_k \simeq \frac{f}{f'}\cos\theta_k d\theta_k$$

which results from the assumption that $f \ll f'$, since,

$$\cos\theta'_{k'} = \left[1 - \left(\frac{f}{f'}\sin\theta_k\right)^2\right]^{1/2} \simeq 1 - \frac{1}{2}\left(\frac{f}{f'}\sin\theta_k\right)^2 \simeq 1$$

So the approximations are satisfied because the focal length of the tube lens is over an order of magnitude larger than that of the objective lens. For the cosine, the first approximation is used when it is part

Table 12.2 *Integrals for the point spread function of an x-directed oscillating electric dipole.*

$$\tilde{I}_{00}(\rho, z) = \int_0^{\theta_{\max}} d\theta \sqrt{\cos\theta}\sin\theta\,(1+\cos\theta)\,\mathcal{J}_0\left(k'\rho\sin\theta\,\tfrac{f}{f'}\right) \times \exp\left\{ik'z\left[1-\tfrac{1}{2}\left(\tfrac{f}{f'}\right)^2\sin^2\theta\right]\right\}$$

$$\tilde{I}_{02}(\rho, z) = \int_0^{\theta_{\max}} d\theta \sqrt{\cos\theta}\sin\theta\,(1-\cos\theta)\,\mathcal{J}_2\left(k'\rho\sin\theta\,\tfrac{f}{f'}\right) \times \exp\left\{ik'z\left[1-\tfrac{1}{2}\left(\tfrac{f}{f'}\right)^2\sin^2\theta\right]\right\}$$

of the phase of an exponent and the zeroth approximation is used otherwise. The evaluation of the several integrals proceeds similar to the previous tight focusing problem. The ϕ integrals may be evaluated analytically to give Bessel functions. The θ integrals must be evaluated numerically. The field in the image plane is found to be:

$$\vec{E}_{00}(\rho, \varphi, z) = -p\frac{\omega^2}{c^2}\frac{ik'f\exp\left[ikf - ik'f'\right]}{f'\,8\pi\varepsilon_0}\sqrt{\frac{n}{n'}}\begin{bmatrix}\tilde{I}_{00}+\tilde{I}_{02}\cos 2\varphi \\ \tilde{I}_{02}\sin 2\varphi \\ 0\end{bmatrix} \tag{12.101}$$

where the numerical integrals over θ are given in Table 12.2. Note the subscripts on the integrals give the order of the Bessel function and the larger the subscript, the higher the power of θ that enters into the $\theta-$numerical integral. The important parameters here are the numerical aperture, $NA = n\sin\theta_{\max}$, since it determines the upper integration limit, and the transverse magnification, $M = (nf')/(n'f)$. Figure 12.19 shows the PSF function measured along the x and y axis for the NA = 1.4, $M = 100$ system we are modeling. In scaling the axis by $1/M$, the figure shows the apparent size of the point source (in the object plane). For a small NA (the paraxial case) in the image plane ($z = 0$), the higher order integral can be ignored and the field is entirely along the polarization axis so small angle approximations are valid. The lowest order can thus be evaluated analytically (**see Discussion 12.7**)

$$\begin{aligned}\tilde{I}_{00}\left(\rho_0', z_0 = 0\right) &= \int_0^{\theta_{\max}} d\theta_k \sqrt{\cos\theta_k}\sin\theta_k\,(1+\cos\theta_k)\,\mathcal{J}_0\left(k'\rho_0'\sin\theta_k\frac{f}{f'}\right) \\ &\simeq 2\int_0^{\theta_{\max}} \theta_k\,\mathcal{J}_0\left(k'\rho_0'\theta_k\frac{f}{f'}\right)d\theta_k \\ &\simeq \mathcal{J}_1\left[2\pi\cdot\frac{NA\cdot\rho_0'}{M\lambda}\right]\cdot\frac{\lambda M\cdot NA}{\pi n^2\rho_0'}\end{aligned} \tag{12.102}$$

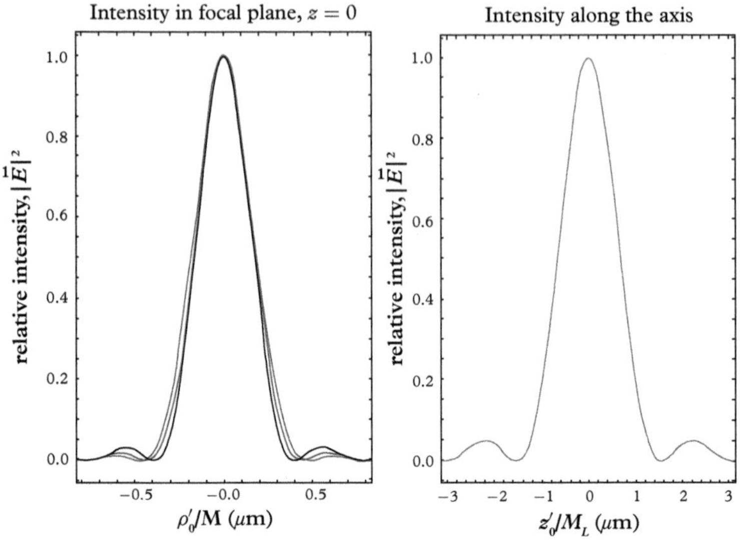

Fig. 12.19 *The point spread function. The left figure shows the PSF in the focal plane. The blue curve is the angular spectrum calculation along the x-axis, the black is the corresponding result along the y-axis. The red curve is the analytic result using the paraxial approximation, an Airy disk. By dividing out the magnification for the graph's x-axis, the curve represents the apparent size the optical system sees for a point source. The similar right hand curve is an angular spectrum result measured along the optical axis.*

Thus, the *paraxial point-spread function* for a dipole oriented along the x-axis (or y-axis) is an Airy disk:

$$\left|E\left(\rho_0', z=0\right)\right|^2 = \frac{\pi^4}{\varepsilon_0^2 n n'} \frac{p^2}{\lambda^6} \frac{NA^4}{M^2} \left[2\frac{\mathcal{J}_1\left(2\pi \tilde{\rho}_0'\right)}{2\pi \tilde{\rho}_0'}\right]^2, \quad \text{with } \tilde{\rho}_0' = \frac{NA \cdot \rho_0'}{M\lambda} \tag{12.103}$$

This result is shown in Fig. 12.19 for the NA = 1.4 system we are studying. One would expect the paraxial result to do poorly for a NA system, but in fact the agreement seen here with the full calculation is excellent.

The first zero of $\mathcal{J}_1(2\pi\tilde{\rho})$ occurs for $\tilde{\rho} = 0.6098$ or, in the image plane, $\Delta x = 0.6098\frac{M\lambda}{NA}$. This result is frequently used in specifying the resolution of a system. The range, Δz, about the image plane for which the axial point-spread function becomes zero is known as the *depth of field*. For the paraxial case:

$$\Delta z = 2n' \frac{M^2 \lambda}{NA^2} \tag{12.104}$$

for a typical microscope objective ($M = 100\times$ and NA $= 1.4$) imaging $\lambda = 1\ \mu$m light, $\Delta x \approx 43\mu m$ and $\Delta z \approx 10$ mm.

In discussing resolution, we could go back to the object plane and determine how close in the transverse plane, $\Delta r_{\|} = \sqrt{\Delta x^2 + \Delta y^2}$, two objects can get before they become indistinguishable in the image plane. Naturally, there is some arbitrariness in how to precisely define the minimum resolvable distance. Generally, of course, the narrower the point spread function, the better the resolution.

Alternatively, we might describe resolution in terms of the transverse spatial bandwidth, $\Delta k_{\|}$, of the system. For example, we can often describe our system in terms of the maximum transverse wavevector components $\Delta k_x, \Delta k_y$ that are captured and define $\Delta k_{\|} = \sqrt{\Delta k_x^2 + \Delta k_y^2}$. Then from Fourier transform theory (c.f. quantum mechanics):

$$\Delta r_{\|} \Delta k_{\|} \geq 1 \tag{12.105}$$

we can go on to call $\Delta r_{\|}$ "the minimum resolvable distance."

In far-field optics, we discard evanescent frequencies and thus the maximum possible transverse spatial bandwidth is $\Delta k_{\|} = 2k$. This includes positive and negative frequencies and suggests the best possible is:

$$\Delta r_{\|\min} = \frac{\lambda}{\pi n} \tag{12.106}$$

For our example problem, $\lambda = 1\ \mu$m, and this resolution limit is 210 nm. In contrast is the Abbe formulation of the Rayleigh criterion which requires that for two parallel dipoles to be resolved, the peak of one dipole image can be no closer than the first zero of the other dipole image. This leads to a limit of 435 nm in the example problem,

$$\Delta r_{\|\min} = 0.6089 \frac{\lambda}{NA} \tag{12.107}$$

Equation 12.105 allows a comprehensive approach to resolution. The equation suggests that there is no limit to the potential resolution at a given wavelength. However, getting extremely high resolution requires including evanescent wavevectors. This is what is done in near-field microscopy and this approach to resolution is suitable for considering those techniques. One of the ways to improve on the resolution limit is to use "prior knowledge." For example, if we *knew* that the objects we were looking at were perfect dipole sources we could model the images to be as shown in Fig. 12.19. That is, the only limit on finding the position of a single dipole is how well the available signal-to-noise allows fitting to the figure and in turn finding its center. If we knew we were observing two radiating point sources, this information could be used to deconvolve two close samples and do much better than either the limits of Eq. 12.106 or Eq. 12.107. Indeed, for a perfect signal-to-noise ratio, the two dipoles could be arbitrarily close and still be distinguished.

Exercises

(12.1) Reproduce Poisson's calculation showing that in the shadow of a disk, Arago's spot may be found. Find the spot's size and location as a function of disk size and light wavelength.

(12.2) A zone plate is sometimes considered as a generalization of the Arago spot geometry. It uses diffraction to focus a beam of light. Referring to Fig. 12.20, to achieve constructive interference from an incident plane wave, a zone plate consists of a series of concentric transparent and opaque rings with boundaries given by:

$$r_n = \sqrt{n\lambda f + \frac{n^2\lambda^2}{4}} \qquad (12.108)$$

where f is the focus of the zone plate and the n's are a set of consecutive positive integers. Find the intensity distribution of the focused spot as a function of λ and the n's used in the plate.

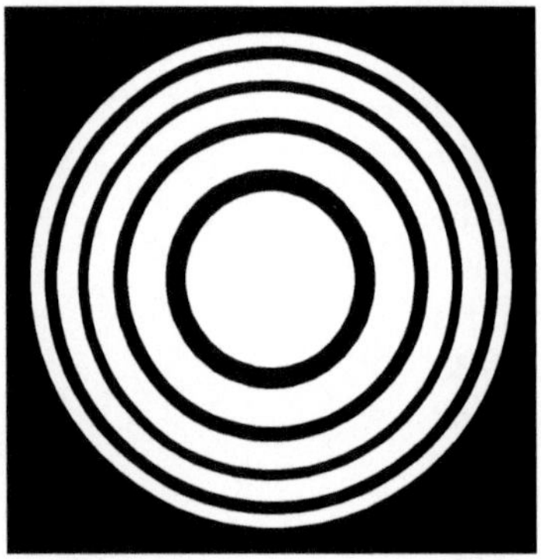

Fig. 12.20 *Zone plate.*

(12.3) A gradient index (or GRIN) lens, consists of a dielectric rod whose index varies as a function of radius:

$$N = N_0\left[1 - \frac{k}{2}r^2\right] \qquad (12.109)$$

where N_0 is the index at the center of the rod and k is the gradient constant. Using the eikonal equation, describe the propagation of rays down the GRIN rod from a point source on the rod's front surface. Discuss how a GRIN could be manufactured to collimate such a point source.

(12.4) The Talbot effect refers to interesting features in the near-field diffraction pattern of a transmission grating illuminated by a perpendicularly incident plane wave. See Fig. 12.21. The grating pattern repeats itself downstream from the grating, at the Talbot distance, given by:

$$Z_T = \frac{\lambda}{1 - \sqrt{1 - \lambda^2/a^2}} \approx \frac{2a^2}{\lambda} \qquad (12.110)$$

where a is the period of the grating. Moreover, downstream at integer fractions of z_T, other "gratings" appear in the diffraction pattern: at $\frac{1}{2}z_T$, a grating pattern is formed with a period half that of the original; at $\frac{1}{3}z_T$, a grating with a period one third the original, and so on. Using Fresnel diffraction theory find the intensity distribution across the plane, $z = z_T/3$.

(12.5) An ideal Bessel beam has a field of the form

$$\vec{E}(\rho,\phi,0) = \hat{i}A_0\mathcal{J}_n(k_\rho\rho)\exp(\pm in\phi) \qquad (12.111)$$

in the transverse plane containing the origin ($z = 0$). Here $\mathcal{J}_n$ is the Bessel function of the 1st kind of order n and with the radial wavenumber, $k_\rho < \omega/c$. Use the angular spectrum representation to find the downstream field and discuss in what sense such a beam is non-diffracting and the practical limitations of the "non-diffraction" for the $n = 0$ and $n = 1$ Bessel beams.

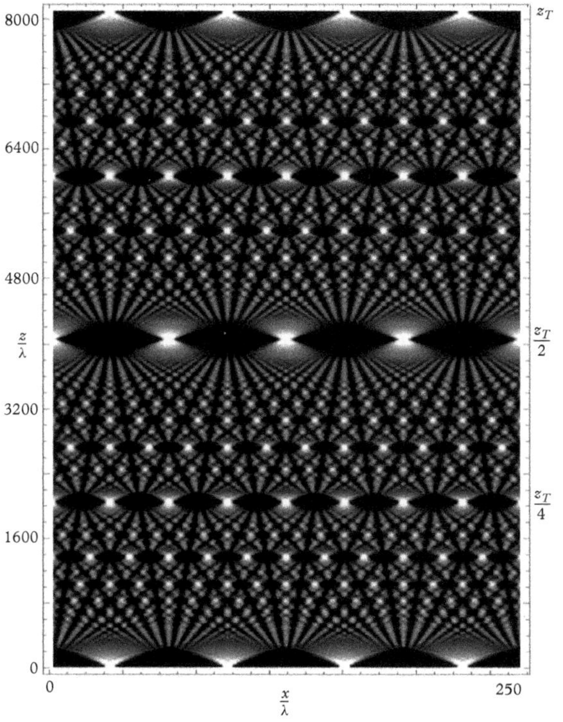

Fig. 12.21 *Optical Talbot effect. Credit: Ben Goodman (own work) via Wikimedia Commons.*

(12.6) A high numerical aperture cylinder lens can generate a tight line focus from an incident plane wave. Following the treatment of the tight focusing section (but assuming an infinitely wide beam and optics in the non-focusing dimension), find the E-field in the region of the focus for an NA=1 lens and a beam significantly overfilling the lens. This solution is used to generate "sheet of light" fields that assist in sectioning images for microscopy.

(12.7) Optical trapping or optical tweezers refer to the confinement of a polarizable particle at an electric field maxima. The polarization energy of a dielectric sphere in an optical field is given by $U = -\frac{1}{2}\alpha\left|\vec{E}\right|^2$ where α is the polarizability of the sphere and is given by:

$$\alpha = 4\pi\left(\frac{\varepsilon_2 - \varepsilon_1}{\varepsilon_2 + 2\varepsilon_1}\right)a_0^3 \tag{12.112}$$

Here ε_2 is the permittivity of the sphere, ε_1 is the permittivity of the medium in which it is embedded, and a_0 is the radius of the sphere. Consider a small polyethylene sphere ($n_2 = 1.51$) trapped in water ($n_1 = 1.33$) by a 1 watt, near-IR laser with $\lambda = 1064$ nm. Near the focus of the laser, the potential felt by the sphere is quadratic in both the transverse and longitudinal directions. Find the force constant, k, on a sphere small enough that the field across it may be taken as constant and its perturbation of the incident field is negligible. Find the dependence of k on sphere size and laser wavelength. And find the values of k_x, k_y, and k_z for a linearly x-polarized beam, a numerical aperture $NA = 1.1$, a filling factor $f_0 = 1$, and a sphere radius 0.01λ.

(12.8) Maxwell described the Maxwell "fish-eye" lens some time before he wrote down his equations. This lens is a sphere of radius R with an index of refraction that varies from its center to its surface as:

$$n(r) = \frac{2}{1 + (r/R)^2}$$

where r is the distance from the center of the sphere. Show, using the eikonal equation, that this lens images a point on the sphere's surface to the point opposite it.

(12.9) The eye lenses of all species have varying indices of refraction that have evolved to reduce both spherical and chromatic aberrations. As an example of the principle, we consider spherical, ball lenses.

a) Using ray tracing show that the focal length for paraxial parallel rays incident on a ball lens is given by

$$f = \frac{n_{rel}D}{4(n_{rel} - 1)} \tag{12.113}$$

where D is the diameter of the ball and $n_{rel} = n_{ball}/n_{med}$ is the relative index of the ball to the medium surrounding it and the focus is measured relative to the center of the sphere. For what follows, we consider a lens with $n_{rel} = 2$ for which paraxial rays are imaged to a point on the surface opposite their entrance. One problem with the ball lens is it suffers spherical aberration: incident rays far off axis are not imaged to the focal point.

b) Describe the refraction of parallel rays that are incident off axis by just under $D/2$– that is, rays approaching grazing incidence.

c) The Luneberg lens is a sphere that has a varying index of refraction given by:

$$n(r) = n_{med}\sqrt{2 - \left(\frac{2r}{D}\right)^2}$$

Show, using the eikonal equation, that the Luneberg lens images all parallel rays incident on a sphere to the point opposite their entrance. Note: the principle of the Luneberg lens is used in "radar optics" in the other direction, imaging a microwave point source into a collimated beam.

12.5 Discussions

Discussion 12.1

First consider the gradient of Eq. 12.3,

$$\begin{aligned}\nabla f &= \nabla a(\vec{r})\, e^{ik_0\psi(\vec{r})} = \nabla\{a(x,y,z)\exp[ik_0\psi(x,y,z)]\}\\ &= e^{ik_0\psi}\nabla a + e^{ik_0\psi} aik_0\nabla\psi\end{aligned}$$

and then the divergence of the result

$$\begin{aligned}\nabla^2 f &= \nabla^2 a(\vec{r})\, e^{ik_0\psi(\vec{r})} = \nabla\cdot\nabla a(\vec{r})\, e^{ik_0\psi(\vec{r})}\\ &= \nabla\cdot\left(e^{ik_0\psi}\nabla a + e^{ik_0\psi} aik_0\nabla\psi\right)\\ &= e^{ik_0\psi}\nabla^2 a + 2e^{ik_0\psi} ik_0\nabla\psi\cdot\nabla a + e^{ik_0\psi} a(ik_0)^2\nabla\psi\cdot\nabla\psi + e^{ik_0\psi} aik_0\nabla^2\psi\end{aligned}$$

Substituting this into Eq. 12.2, and dividing out the exponential

$$\nabla^2 a + 2ik_0\nabla\psi\cdot\nabla a + a(ik_0)^2(\nabla\psi)^2 + aik_0\nabla^2\psi = -ak_0^2 n^2$$

and rearranging the left-hand side,

$$\nabla^2 a + 2ik_0\nabla a\cdot\nabla\psi + ik_0 a\nabla^2\psi - ak_0^2(\nabla\psi)^2 = -ak_0^2 n^2$$

Discussion 12.2

An integral representation for Bessel functions :

$$\frac{i^{-n}}{2\pi}\int_0^{2\pi} e^{ix\cos\alpha}e^{in\alpha}\,d\alpha = \mathcal{J}_n(x) \tag{12.114}$$

and, specifically, for $n=0$,

$$\int_0^{2\pi} d\alpha \exp(ix\cos\alpha) = 2\pi\mathcal{J}_0(x)$$

permits the θ_0 integral of Eq. 12.44 to be carried out:

$$U(x,y,z) = 2\pi C\int_0^a \rho_0\,d\rho_0\mathcal{J}_0\left(k\frac{\rho}{z}\rho_0\right) \tag{12.115}$$

A recurrence relation for Bessel functions:

$$x^{n+1}\mathcal{J}_n(x) = \frac{d}{dx}\left[x^{n+1}\mathcal{J}_{n+1}(x)\right] \tag{12.116}$$

used for $n=0$ and integrated, gives the integration over ρ_0:

$$U(x,y,z) = \pi a^2 C\left(\frac{2\mathcal{J}_1\left(k\frac{\rho}{z}a\right)}{k\frac{\rho}{z}a}\right) \tag{12.117}$$

Discussion 12.3

Inserting the paraxial approximation of Eq. 12.60 into Eq. 12.62,

$$\begin{aligned}\vec{E}(x,y;z) &\simeq \vec{E}_0\frac{w_0^2}{4\pi}\iint_{-\infty}^{\infty} dk_x dk_y \exp\left[-\left(k_x^2+k_y^2\right)\frac{w_0^2}{4}\right]\exp\left[i\left(k_x x + k_y y + \left(k - \frac{k_x^2+k_y^2}{2k}\right)z\right)\right]\\ &\simeq \vec{E}_0\frac{w_0^2}{4\pi}e^{ikz}\iint_{-\infty}^{\infty} dk_x dk_y \exp\left[-\left(k_x^2+k_y^2\right)\frac{w_0^2}{4}\right]\exp\left[i\left(k_x x + k_y y - \left(\frac{k_x^2+k_y^2}{2k}\right)z\right)\right]\end{aligned}$$

Now considering the integral over k_x only, and rearranging,

$$\int_{-\infty}^{\infty}\exp\left\{-\left[\left(\frac{w_0^2}{4}+\frac{iz}{2k}\right)k_x^2 - ixk_x\right]\right\}dk_x$$

Using the general relation,

$$\int_{-\infty}^{\infty} \exp\left\{-\left[ax^2+bx\right]\right\} dx = \sqrt{\frac{\pi}{a}} \exp\left(\frac{b^2}{4a}\right)$$

and identifying

$$a \rightarrow \left(\frac{w_0^2}{4}+\frac{iz}{2k}\right) \quad \text{and} \quad b \rightarrow -ix$$

we obtain the solution for the integral over k_x,

$$\sqrt{\frac{4\pi}{w_0^2+2iz/k}} \exp\left(\frac{-x^2}{w_0^2+2iz/k}\right)$$

With the exact same form for the k_y integral solution, the full solution is,

$$\vec{E}(x,y;z) \simeq \frac{\vec{E}_0 e^{ikz}}{1+2iz/kw_0^2} \exp\left(\frac{-\left(x^2+y^2\right)}{w_0^2+2iz/k}\right)$$

Discussion 12.4

Writing

$$\frac{1}{1+iz/z_R} = \left[1+(z/z_R)^2\right]^{-1/2} \exp\left(-i\arctan\left(\frac{z}{z_R}\right)\right) = \frac{w_0}{w(z)} \exp(-i\eta(z))$$

the factor out front in Eq. 12.64 is rewritten,

$$\frac{\vec{E}_0 e^{ikz}}{1+iz/z_R} = \vec{E}_0 \frac{w_0}{w(z)} \exp\left[i\left(kz-\eta(z)\right)\right] \tag{12.118}$$

The same term in the exponential is next rewritten as

$$\frac{1}{1+iz/z_R} = \frac{1}{1+z^2/z_R^2} - i\frac{z/z_R}{1+z^2/z_R^2} = \frac{1}{1+z^2/z_R^2} - i\frac{z_R/z}{1+z_R^2/z^2}$$

Noting the identifications of Eq. 12.65, and using the Rayleigh parameter $z_R = \frac{kw_0^2}{2}$, this can be written,

$$\frac{1}{1+iz/z_R} = \left(\frac{w_0}{w(z)}\right)^2 - i\left(\frac{kw_0^2}{2R(z)}\right)$$

so that the exponential of Eq. 12.64 is written

$$\exp\left[-\frac{\rho^2}{w^2(z)} + i\frac{k\rho^2}{2R(z)}\right] \tag{12.119}$$

Combining Eqs. 12.118 and 12.119, we get,

$$\vec{E}(\rho,z) = \vec{E}_0\frac{w_0}{w(z)}\exp\left[-\frac{\rho^2}{w^2(z)}\right]\exp\left[i\left(kz - \eta(z) + \frac{k\rho^2}{2R(z)}\right)\right]$$

Discussion 12.5

Evaluating the far-field angular spectrum integral by the stationary phase method:

$$\vec{E}_\infty(s_x,s_y,s_z) = \frac{1}{2\pi}\iint_{(k_x^2+k_y^2\le k^2)} \vec{\mathscr{E}}(k_x,k_y;0)\exp\left[ikr\left(\frac{k_x}{k}s_x + \frac{k_y}{k}s_y \pm \frac{k_z}{k}s_z\right)\right]dk_xdk_y \tag{12.120}$$

Let's consider the signal in some fixed distant direction s_x, s_y:

$$s_x = \frac{x}{r} \quad s_y = \frac{y}{r} \quad s_z = \frac{z}{r} = \sqrt{1-s_x^2-s_y^2} \tag{12.121}$$

Now consider the exponential in Eq. 12.120. At a large distance, r, it is a rapidly varying function of k_x, k_y. For the term in brackets, unless its first partial derivatives with respect to k_x and k_y vanish as r gets very large, the oscillations of the exponential with respect to the integration variables become tighter and the cancellation of the integrand more complete. However, for those values of k_x and k_y in which the first derivatives of the phase vanish, the phase is stationary and we may get a finite contribution to the integrand even for $r\to\infty$. We will carry out a Taylor expansion to 2nd order about such a point. Define the phase function

$$\varphi(k_x,k_y) = r\left[k_xs_x + k_ys_y + k\left(1-\frac{k_x^2}{k^2}-\frac{k_y^2}{k^2}\right)^{1/2}\left(1-s_x^2-s_y^2\right)^{1/2}\right] \tag{12.122}$$

Its first derivatives are:

$$
\begin{aligned}
\frac{\partial\varphi}{\partial k_x} &= r\left[s_x - \frac{k_x}{k}\left(1-\frac{k_x^2}{k^2}-\frac{k_y^2}{k^2}\right)^{-1/2}\left(1-s_x^2-s_y^2\right)^{1/2}\right]\\
\frac{\partial\varphi}{\partial k_y} &= r\left[s_y - \frac{k_y}{k}\left(1-\frac{k_x^2}{k^2}-\frac{k_y^2}{k^2}\right)^{-1/2}\left(1-s_x^2-s_y^2\right)^{1/2}\right]
\end{aligned}
\tag{12.123}
$$

So that in the direction $(s_x, s_y s_z)$ these are zero and the exponent is "stationary" for the point in k space, $k_x = k_{x0} = s_x k$ and $k_y = k_{y0} = s_y k$. The phase function at this point is $\varphi\left(k_{x0}, k_{y0}\right) = kr$. The second derivatives at this point are:

$$
\begin{aligned}
\frac{\partial^2\varphi}{\partial k_x^2} &= r\left(1-s_x^2-s_y^2\right)^{1/2}\left[-\frac{1}{k}\left(1-\frac{k_x^2}{k^2}-\frac{k_y^2}{k^2}\right)^{-1/2}-\frac{k_x^2}{k^3}\left(1-\frac{k_x^2}{k^2}-\frac{k_y^2}{k^2}\right)^{-3/2}\right]\\
&= r\left[-\frac{1}{k}-\frac{k_x^2}{k^3}s_z^{-2}\right] = -\frac{r}{k}\left(1+s_x^2 s_z^{-2}\right)\\
\frac{\partial^2\varphi}{\partial k_y^2} &= r\left(1-s_x^2-s_y^2\right)^{1/2}\left[-\frac{1}{k}\left(1-\frac{k_x^2}{k^2}-\frac{k_y^2}{k^2}\right)^{-1/2}-\frac{k_y^2}{k^3}\left(1-\frac{k_x^2}{k^2}-\frac{k_y^2}{k^2}\right)^{-3/2}\right]\\
&= r\left[-\frac{1}{k}-\frac{k_y^2}{k^3}s_z^{-2}\right] = -\frac{r}{k}\left(1+s_y^2 s_z^{-2}\right)\\
\frac{\partial^2\varphi}{\partial k_y \partial k_x} &= r\left(1-s_x^2-s_y^2\right)^{1/2}\left[-\frac{k_x k_y}{k^3}\left(1-\frac{k_x^2}{k^2}-\frac{k_y^2}{k^2}\right)^{-3/2}\right]\\
&= r\left[-\frac{k_x k_y}{k^3}s_z^{-2}\right] = -\frac{r}{k}s_x s_y s_z^{-2}
\end{aligned}
$$

Discussion 12.6

Evaluating the far-field angular spectrum integral by the stationary phase method (continued): Where the final expressions for each of these derivatives is its value at the stationary point. For a nearby point in k space displaced δk_x, δk_y, the Taylor expansion for the phase is:

$$
\varphi = kr - \frac{1}{2}\frac{r}{k}\left(1+s_x^2 s_z^{-2}\right)\cdot\delta k_x^2 - \frac{1}{2}\frac{r}{k}\left(1+s_y^2 s_z^{-2}\right)\cdot\delta k_y^2 - \frac{r}{k}s_y s_x s_z^{-2}\cdot\delta k_x \delta k_y \tag{12.124}
$$

For $r \to \infty$ it is only the value of the function around that point that will matter since cancellation in the integrand occurs everywhere except near the stationary point. There, with the Taylor expansion for the phase inserted in Eq. 12.120, we find:

$$\vec{E}_{\infty}(x,y,z) = \frac{1}{2\pi}\vec{\mathscr{E}}\left(k_{x0},k_{y0};0\right)e^{ikr} \tag{12.125}$$

$$\times \int_{-\infty}^{+\infty} d(\delta k_x) \int_{-\infty}^{+\infty} d\left(\delta k_y\right) \exp\left[-i\frac{1}{2}\frac{r}{k}\left\{\left(1+s_x^2 s_z^{-2}\right)\cdot \delta k_x^2 + \left(1+s_y^2 s_z^{-2}\right)\cdot \delta k_y^2 + 2s_y s_x s_z^{-2}\cdot \delta k_x \delta k_y\right\}\right]$$

Here the k space integration was taken over $d\delta k_x d\delta k_y$, a simple shift. Since the rapidly oscillating phase kills the function for large displacements in k space from the stationary point, $\pm\infty$ may be taken for the integration limits. This integral may be evaluated in cylindrical k space coordinates, where the term in braces becomes:

$$\left(1+s_x^2 s_z^{-2}\right)\cdot \delta k_x^2 + \left(1+s_y^2 s_z^{-2}\right)\cdot \delta k_y^2 + 2s_y s_x s_z^{-2}\cdot \delta k_x \delta k_y = (\delta k)^2 + \frac{1}{s_z^2}\left(s_x \delta k_x + s_y \delta k_y\right)^2 \tag{12.126}$$

$$= (\delta k)^2 + \frac{1}{s_z^2}\left(\vec{s}_{\parallel}\cdot \delta\vec{k}\right)^2 \tag{12.127}$$

$$= (\delta k)^2 + \frac{1}{s_z^2}\left(\left|s_{\parallel}\right|\delta k\cos\phi_k\right)^2 \tag{12.128}$$

Here $(\delta k)^2 = (\delta k_x)^2 + \left(\delta k_y\right)^2$, $\vec{s}_{\parallel} \equiv \left(s_x, s_y\right)$ and ϕ_k is the angle between the observation direction and $\delta\vec{k} = \left(\delta k_x, \delta k_y\right)$. Thus Eq. 12.125 becomes $\vec{E}(x,y,0)$,

$$\vec{E}_{\infty}(x,y,z) = \frac{1}{2\pi}\vec{\mathscr{E}}\left(k_x,k_y;0\right)e^{ikr}\int_0^{2\pi} d\phi_k \int_0^{\infty}(\delta k)\, d(\delta k)\exp\left[-i\frac{1}{2}\frac{r}{k}(\delta k)^2\right]$$

$$\times \exp\left[-i\frac{1}{2}\frac{rs_{\parallel}^2}{ks_z^2}(\delta k\cos\phi_k)^2\right] \tag{12.129}$$

Using $1 = s_{\parallel}^2 + s_z^2$, the integral evaluates to $\frac{-2ik\pi s_z}{r}$ and in the limit $r \to \infty$ we find:

$$\vec{E}_{\infty}(x,y,z) = -iks_z\vec{\mathscr{E}}\left(k_{x0},k_{y0};0\right)\frac{e^{ikr}}{r} \tag{12.130}$$

Discussion 12.7

For the last equality, we use the Bessel function identity,

$$x^v \mathcal{J}_v(x) = \int x^v \mathcal{J}_{v-1}(x)\, dx$$

with $v = 1$, $x = k'\rho_0' f/f'\theta_k$ and $dx = \left(k'\rho_0' f/f'\right) d\theta_k$, we write,

$$k'\rho_0' \frac{f}{f'}\theta_k \mathcal{J}_1\left(k'\rho_0' \frac{f}{f'}\theta_k\right) = \int k'\rho_0' \frac{f}{f'}\theta_k \mathcal{J}_0\left(k'\rho_0' \frac{f}{f'}\theta_k\right) k'\rho_0' \frac{f}{f'} d\theta_k$$

or

$$\theta_k \mathcal{J}_1\left(k'\rho_0' \frac{f}{f'}\theta_k\right) = k'\rho_0' \frac{f}{f'} \int \theta_k \mathcal{J}_0\left(k'\rho_0' \frac{f}{f'}\theta_k\right) d\theta_k$$

or

$$\int \theta_k \mathcal{J}_0\left(k'\rho_0' \frac{f}{f'}\theta_k\right) d\theta_k = \frac{\theta_k}{k'\rho_0'} \frac{f'}{f} \mathcal{J}_1\left(k'\rho_0' \frac{f}{f'}\theta_k\right)$$

If we further identify $f/f' = n/\left(Mn'\right)$, $k' = 2\pi n'/\lambda$ and $n \sin\theta_k \simeq n\theta_k = NA$ (in the paraxial limit), then this is

$$\int \theta_k \mathcal{J}_0\left(k'\rho_0' \frac{f}{f'}\theta_k\right) d\theta_k \simeq \mathcal{J}_1\left[2\pi \cdot \frac{NA \cdot \rho_0'}{M\lambda}\right] \frac{\lambda}{2\pi n^2} \cdot \frac{NA}{\rho_0'} \cdot M$$

Radiation Fields in Constrained Environments

13

- The density of electromagnetic modes in space: restrictions due to boundary conditions
- Thermal radiation and the Planck spectrum
- Vacuum field energies and the force between two plates: the Casimir force
- The Einstein A and B coefficients and their properties under thermodynamic equilibrium
- Microwave cavities, wave guides, and their general properties; rectangular conducting waveguides, and transmission lines
- Optical waveguides: the ray optic picture. Planar waveguides, and variable index circular waveguides; single mode fibers and dispersion
- Photonic crystals: photonic band structures analogous to electronic bands

13.1 Constrained environments

Up to this point, we have considered mostly problems for which the frequency and direction of electromagnetic waves may be selected from a continuum of values. Here, we make the transition from analog to digital where for various problems the solution is constructed from a countable set of solutions to Maxwell's equations with discrete frequencies and/or wavevectors. This chapter is divided into two parts. In the first, all space is modeled as a large cavity in order to facilitate calculations. After all, arguing from physical grounds whether space is "very large" or in fact infinite would make little difference to most of the problems we have previously considered: for a large enough box, the two should converge. However, mathematically, the difference

Electromagnetic Radiation. Richard Freeman, James King, Gregory Lafyatis,

DOI: 10.1093/oso/9780198726500.001.0001

between working in a very large volume, "V," and "∞" is often a determining factor in the approach and tractability of a given problem. We begin by identifying, describing, and counting the normal modes of a large box for both periodic and conducting boundary conditions. These are then used to study blackbody radiation, the Casimir effect, and spontaneous emission by atoms and molecules. For all three of these, beyond solving Maxwell's equations, two additional pieces of physics are needed:

- **Quantization**–each mode of the radiation field behaves as a quantum mechanical harmonic oscillator.
- **Boltzmann distribution**–for a system in thermodynamic equilibrium, the probability ratio that the system is in state "a" with energy "E_a" to that of it being in state "b" with energy E_b is given by the Boltzmann factor: $\exp\left[-\left(E_a - E_b\right)/k_B T\right]$, where k_B is Boltzmann's constant.

The second part of the chapter treats various "real" cavities where we are interested in the modes themselves. We begin that part by treating microwave resonant cavities and waveguides. These are both of historical interest and provide useful visualizations for cultivating physical insights into what follows. The atomic beam clock, Fig. 13.1, represents a pivotal step in making extremely high precision measurements. The precision is largely derived via a two-slit type interference enabled by a "Ramsey microwave cavity," a resonant cavity that is large enough for the atomic beam to pass through at two well-separated interaction regions.

Moving on to dielectric waveguide structures, our treatment differs a little from other texts in the way we treat evanescent waves. Many texts identify "by hand" where a solution is expected to be evanescent and write the wave function as a decaying exponential, $\sim \exp\left[-\gamma x\right]$, where γ is real and positive. In contrast, we treat propagating waves and evanescent waves on equal footing, writing both in the form $\sim \exp\left[ikx\right]$. For an evanescent wave, the math will dictate that $k = i\gamma$ in the solution. This allows a consistent treatment of losses, metals, and perfect conductors simply by allowing the dielectric constant or index of refraction, $\varepsilon_r = n^2$, to be complex and the wavevectors describing the fields to be real, pure imaginary, or complex, depending on what the math dictates. Dielectric waveguides are staples of modern photonic technology. We examine the planar dielectric waveguide or "slab" first with a geometrical optics treatment and again by solving

Fig. 13.1 *Atomic beam clock (NIST). The cesium beam passes through resonant Ramsey microwave cavites in two locations: note the splitting of the microwaveguide (black arrow) delivering microwaves to the two cavities. Credit: NIST, public domain.*

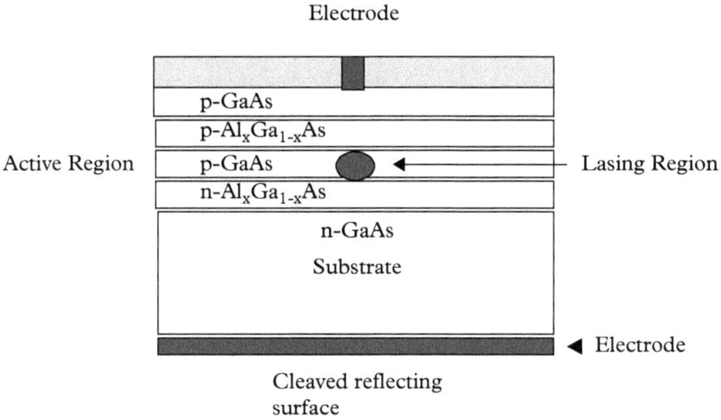

Fig. 13.2 *A planar waveguide is at the heart of many diode lasers. Shown is a AlGaAs laser in which the lasing mode is defined by a layer of GaAs, the lasing medium, sandwiched between mode-confining layers of $Al_xGA_{1-x}As$. The waveguide is sufficiently thin that only single TE and TM modes are active. The laser cavity is further defined by the cleaved semiconductor-air interfaces that reflect the TE mode slightly better than the TM mode and so lasing occurs in the former.*

Maxwell's equations for the propagating field. Figure 13.2 is a gain-guided diode-laser that is built around a symmetric planar waveguide. Strictly speaking, these may be considered 3D dielectric resonators in that the air-semiconductor interfaces at the ends have significant reflectivity that may be further enhanced or weakened by coating.

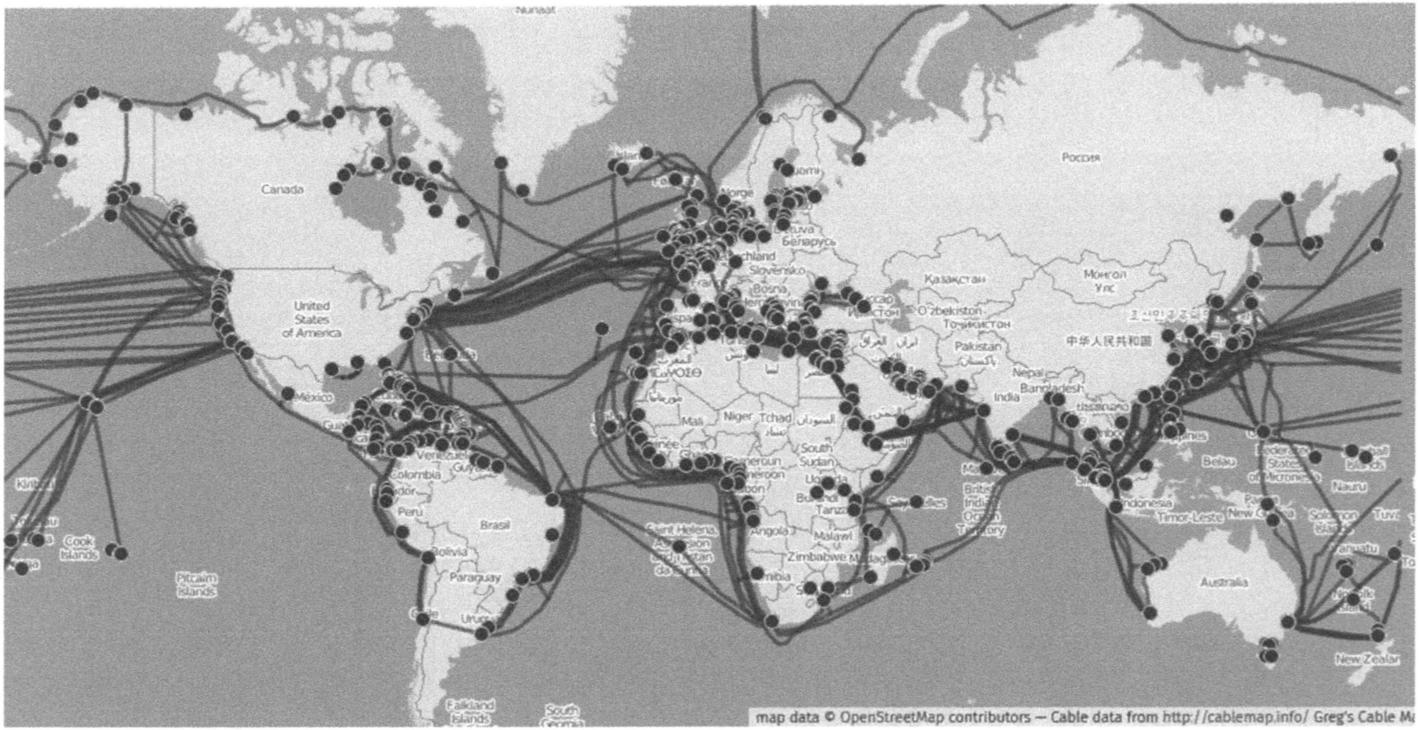

Fig. 13.3 *Transoceanic fiber optic cables as of July 2015. Credit: Openstreeetmap contributers: data by G. Mahlknecht (via Wikipedia Commons).*

Next, we consider cylindrical dielectric optical waveguides–optical fibers. The best of these have astonishingly small losses. Figure 13.3 shows the transoceanic cables that have been laid as of the July 2015. We conclude the chapter with a very brief examination of photonic crystals, a technology whose aim is to engineer modes of the radiation field to accomplish desired ends.

13.2 Mode counting: the density of electromagnetic modes in space

We begin with Maxwell's equations. It is most convenient to write these using the fields, $\vec{E}$ and $\vec{H}$. For harmonic fields with frequency ω, Maxwell's equations are:

$$\nabla \times \vec{H} = \vec{\mathcal{J}}_f - i\omega\varepsilon\vec{E} \tag{13.1}$$

$$\nabla \times \vec{E} = i\omega\mu\vec{H} \tag{13.2}$$

$$\nabla \cdot \vec{E} = \frac{\rho_f}{\varepsilon} \qquad \nabla \cdot \vec{H} = 0 \tag{13.3}$$

in which we have assumed that the permittivity and permeability, ε and μ, are spatially uniform within the medium. For many calculations it is useful to work with a discrete set of free electromagnetic wave solutions to Maxwell's equations in empty space ($\varepsilon = \varepsilon_o$, $\mu = \mu_o$). We approximate all space as a large, $L \times L \times L$ cubic box. The normal modes of the radiation field then are the wave solutions that satisfy the boundary conditions at the faces of the cube. Two sets of boundary conditions are popular: periodic and perfect conducting surfaces. The periodic boundary conditions can be satisfied by normal modes that are traveling waves with the requirement that a solution takes on the same values for any two points opposite one another on the two planes defining one of the dimensions, for example $\vec{E}(0,y,z,t) = \vec{E}(L,y,z,t)$. Alternatively, all surfaces of the box may be taken to be perfect conductors. This leads to standing wave solutions. For sufficiently large boxes these are, for all practical purposes, equivalent. We first consider the periodic boundary condition case. The wave equation, obtained from Eqs. 13.1–13.3, is separable, leading to solutions with periodic time dependence $\sim \exp(-i\omega t)$. The resulting spatial equation is also separable. The full solutions have the form:

$$\vec{E}(x,y,z,t) = \vec{E}_0 \exp\left[i\left(k_x x + k_y y + k_z z\right)\right]\exp(-i\omega t) \tag{13.4}$$

Here, $\vec{k} = (k_x, k_y, k_z)$ is the wavevector of the mode and specifies the direction of propagation and $\vec{E}_0 = (E_{0x}, E_{0y}, E_{0z})$ is a vector constant that gives the wave's polarization and may be used for normalization. In vacuum, the separation constants are related by $\omega = c\sqrt{k_x^2 + k_y^2 + k_z^2}$. To satisfy periodic boundary conditions, we require that for each wavevector component, $k_j L = 2\pi n_j$, where n_j represents an integer number of wavelengths that fit between the walls of the box in the jth direction. Thus, for each set of integers, we identify a wavevector $\vec{k} = 2\pi (n_x, n_y, n_z)/L$ that, when used in Eq. 13.4, will satisfy periodic boundary conditions. In what follows, we will use $k = |\vec{k}| = \omega/c$. With periodic boundary conditions, we allow n_x, n_y, n_z to range over the positive and negative integers and zero. To avoid double counting, we use only the positive square root when finding the frequency, ω. A negative square root solution is identical to the positive square root solution with a wavevector pointing in the opposite direction. From here on, we will omit the $\exp(-i\omega t)$ time dependence of the fields. Free electromagnetic plane waves, guided or unguided, are required by the transverse condition, $\vec{k} \cdot \vec{E} = 0$, to be transverse to their direction of propagation.[1] In general, this condition allows for two available modes, identified with the two independent polarizations, per allowed wavevector.

If, alternatively, the bounding box has perfectly conducting bounding surfaces, the normal modes again may be specified by a wavevector $\vec{k} = (k_x, k_y, k_z)$ and are given by superpositions of the traveling wave modes associated with $(\pm k_x, \pm k_y, \pm k_z)$. That is, standing waves can be seen as a pair of oppositely directed, equal wavelength traveling waves. Inside a perfect conductor the electric and magnetic fields must be zero or they will lead to infinite currents[2]. This requirement gives the conditions on the field components just outside the perfectly conducting surface:

(1) The components of electric field parallel to the surface must be zero ($E_\parallel = 0$).

(2) The component of magnetic field perpendicular to the surface must be zero ($H_\perp = 0$).

(3) The component of electric field perpendicular to the surface is allowed ($E_\perp \neq 0$). In this case, a surface charge density, Σ_S, is created that forces the perpendicular component to zero inside the conductor.

[1] Within free unbounded space the transverse demand of $\vec{k} \cdot \vec{E}_0 = 0$ is clearly satisfied, but within a bounded region such as a waveguide, the transverse condition of $\vec{k} \cdot \vec{E}_0 = 0$ does not require that the fields are transverse to the direction of the waveguide but only to the travelling wave which will be reflecting down the waveguide at specific skew angles associated with various modes.

[2] A finite static magnetic field may be "frozen" into the perfect conductor, but varying magnetic fields are not allowed.

(4) The components of magnetic field parallel to the surface are allowed ($H_{\|} \neq 0$). Here, a surface current density, $\vec{K}_S$, forces the parallel component of the magnetic field to zero inside of the conductor. The surface current is perpendicular to the field with $\left|\vec{K}_S\right| = \left|H_{\|}\right|$.

An electric field that satisfies these boundary conditions is given by:

$$\vec{E} = E_0\Big[\hat{x}\left(a_x \cos k_x x \sin k_y y \sin k_z z\right) + \hat{y}\left(a_y \sin k_x x \cos k_y y \sin k_z z\right) + \hat{z}\left(a_z \sin k_x x \sin k_y y \cos k_z z\right)\Big] \tag{13.5}$$

where $a_x^2 + a_y^2 + a_z^2 = 1$ and where, from condition 1, the electric field components parallel to bounding surfaces vanish and the allowed modes for conducting boundary conditions satisfy $\vec{k} = \left(k_x, k_y, k_z\right) = \left(n_x, n_y, n_z\right)\pi/L$. In analogy to E_{0x}, E_{0y} and E_{0z}, which specify a traveling wave's polarization, the a's here specify the polarizations of the standing waves. In particular, the general requirement of $\nabla \cdot \vec{E} = 0$ leads to the condition $k_x a_x + k_y a_y + k_z a_z = 0$, for which there are two independent solution sets, (a_x, a_y, a_z), and thus two polarizations or two modes for each allowed value of $\vec{k}$.[3]

The magnetic field of a mode is found from the electric field using Faraday's law, Eq. 13.2:

$$\vec{H} = \frac{E_0}{i\omega\mu_0}\Big[\hat{x}\left(a_z k_y - a_y k_z\right)\sin k_x x \cos k_y y \cos k_z z + \hat{y}\left(a_x k_z - a_z k_x\right)\cos k_x x \sin k_y y \cos k_z z + \hat{z}\left(a_y k_x - a_x k_y\right)\cos k_x x \cos k_y y \sin k_z z\Big] \tag{13.6}$$

By inspection, this field satisfies the boundary conditions for $\vec{H}$. With perfect conductor boundary conditions, because there is no directionality to standing waves, mode counting is now limited to the positive integer values of n_x, n_y, n_z. So it would appear, upon comparing the allowed $\vec{k}$ values here with those in the periodic boundary conditions discussion, that there are half as many modes for each dimension–or eight times fewer allowed modes out to a given k or ω. However, this is offset by the fact that the interval between allowed modes in

[3] Imposing the requirement $\nabla \cdot \vec{E} = 0$ on Eq. 13.5, we obtain,

$$\nabla \cdot \vec{E} = -E_0\Big[\left(k_x a_x \sin k_x x \sin k_y y \sin k_z z\right) + \left(k_y a_y \sin k_x x \sin k_y y \sin k_z z\right) + \left(k_z a_z \sin k_x x \sin k_y y \sin k_z z\right)\Big] = 0$$

And pulling out the common factor of $\sin k_x x \sin k_y y \sin k_z z$,

$$\nabla \cdot \vec{E} = -E_0\Big[k_x a_x + k_y a_y + k_z a_z\Big] \times \left(\sin k_x x \sin k_y y \sin k_z z\right) = 0$$

so that for $\nabla \cdot \vec{E} = 0$ to be satisfied everywhere in the cavity, it must be that

$$\left(k_x a_x + k_y a_y + k_z a_z\right) = \vec{k} \cdot \vec{a} = 0$$

which, like the transverse condition for traveling waves, ensures that a standing wave's electric field, $\vec{E}$, is everywhere perpendicular to its wavevector, $\vec{k}$.

k-space for the conducting boundaries is half of that for the periodic boundaries–π/L as compared to $2\pi/L$. Thus, the mode density per dimension has increased by a factor of two resulting in eight times more modes per k-space volume. Thus for this case, a complete set of modes is found by including only the positive (and zero) integer-valued triples.

We next find the mode density of the free electromagnetic field: the number of modes per unit frequency per unit volume. We will use the periodic boundary conditions although, as discussed, the results are essentially identical. We consider the number of modes, $\eta(\omega)$, whose frequency is equal to or less than some value ω or, equivalently, the number of modes with wavevector size equal to or smaller than $k = \omega/c$. In the continuum limit, $kL \gg 1$, this is equal to the number of integer triples (n_x, n_y, n_z) in a sphere of radius $n = kL/2\pi = L\omega/2\pi c$ times a factor of two for polarization, or

$$\eta(\omega) = 2 \times \frac{4}{3}\pi n^3 = 2 \times \frac{4}{3}\pi \left(\frac{L\omega}{2\pi c}\right)^3 \tag{13.7}$$

The mode density is the derivative of this divided by the volume of the cube:

$$\rho(\omega) = \frac{1}{L^3}\frac{d\eta(\omega)}{d\omega} = \frac{\omega^2}{\pi^2 c^3} \tag{13.8}$$

13.3 Thermal radiation

The electromagnetic field of a system that is in thermodynamic equilibrium is referred to as blackbody or cavity radiation. The energy density and spectrum of this field is found by summing over the contributions from the relevant modes in the radiation field. In classical statistical mechanics, the equipartition theorem states that each degree of freedom of a classical system with temperature T in thermodynamic equilibrium has, on average, $\frac{1}{2}k_B T$ of energy. Here k_B is the Boltzmann constant. A harmonic oscillator has two "quadratic degrees of freedom" corresponding to its kinetic and potential energies. Identifying each mode of the radiation field with an effective harmonic oscillator, the average classical energy of the fields of a mode is then $2 \times \frac{1}{2}k_B T$. The classical spectral energy density–energy per unit frequency per unit volume of space–then, is this average classical energy times the mode density of Eq. 13.8:

$$e(\omega) = \frac{\omega^2}{\pi^2 c^3} k_B T \tag{13.9}$$

This result is the Rayleigh–Jeans spectrum and is accurate for low frequencies. Going to high frequencies, however, this result becomes inconsistent with measurements. And furthermore, with an infinite number of high frequency modes, this expression predicts an infinite total thermal radiation energy per unit volume. This is referred to as the "ultraviolet catastrophe."

A resolution of this problem was originally put forth by Max Planck. His insight was that the energy of a mode of the radiation field did not vary continuously but rather was restricted to a set of discrete values, $0, \hbar\omega, 2\hbar\omega, 3\hbar\omega \ldots$ This is essentially treating each mode of the radiation field as a quantum harmonic oscillator with energies $\left(n+\frac{1}{2}\right)\hbar\omega$ in which the level number, n, of the quantum oscillator becomes the number of photons in the analogous quantum mode of the electromagnetic radiation field. The extra "$\frac{1}{2}\hbar\omega$" makes no difference in the blackbody radiation field discussion but will be central to Casimir forces. Such quantization modifies the average energy of a mode from the classical equipartition theorem value of $k_B T$. The ratio of the probability for finding n photons to that of finding $n-1$ photons in a mode is given by the Boltzmann factor, $\exp\left[-\hbar\omega/\left(kT_B\right)\right]$. The probability of finding n photons in a mode may therefore be written[4]

$$P(n) = \frac{\exp\left[-n\hbar\omega/\left(k_B T\right)\right]}{Z(\omega, T)} \tag{13.10}$$

where $Z(\omega, T) = 1/P(0)$ is a normalization factor that is identified as the partition function for the mode. Since the mode must have some number of photons, the probabilities sum to 1, giving[5]

$$\begin{aligned} Z(\omega, T) &= 1 + \exp\left(-\frac{\hbar\omega}{k_B T}\right) + \exp\left(-\frac{2\hbar\omega}{k_B T}\right) + \ldots \\ &= \frac{1}{1 - \exp\left[-\hbar\omega/k_B T\right]} \end{aligned} \tag{13.11}$$

Taking the derivative of Z with respect to $\chi = -1/\left(k_B T\right)$ and dividing by Z gives:

$$\begin{aligned} \frac{1}{Z}\frac{\partial Z(\omega, T)}{\partial \chi} &= \frac{\hbar\omega}{Z}\exp\left(\hbar\omega\chi\right) + \frac{2\hbar\omega}{Z}\exp\left(2\hbar\omega\chi\right) + \ldots \\ &= \hbar\omega P(1) + 2\hbar\omega P(2) + \ldots \end{aligned} \tag{13.12}$$

and so, term by term, this is the energy of an n-photon state times the probability of the mode being in that state. The sum is thus the mean energy of the mode:

[4] From the Boltzmann factor, the probability for finding n photons in a mode can be written,

$$P(n) = P(n-1)\exp\left(-\frac{\hbar\omega}{k_B T}\right)$$

and further expressing $P(n-1)$ in terms of $P(n-2)$, we write,

$$P(n) = P(n-2)\exp\left(-\frac{2\hbar\omega}{k_B T}\right)$$

so that extending this down to an expression in terms of $P(0)$,

$$P(n) = P(0)\exp\left(-\frac{n\hbar\omega}{k_B T}\right)$$

[5] Summing Eq. 13.10 over all n, and setting to unity,

$$\sum_n P(n) = 1 = \frac{1}{Z(\omega, T)}\sum_n \exp\left(-\frac{n\hbar\omega}{k_B T}\right)$$

or

$$\begin{aligned} Z(\omega, T) &= \sum_n \exp\left(-\frac{n\hbar\omega}{k_B T}\right) \\ &= 1 + \exp\left(-\frac{\hbar\omega}{k_B T}\right) + \exp\left(-\frac{2\hbar\omega}{k_B T}\right) + \ldots \end{aligned}$$

We then note the general exapansion

$$\frac{1}{1-x} = 1 + x + x^2 + x^3 + \ldots$$

and identify $x \to \exp\left[-\hbar\omega/k_B T\right]$ so

$$Z(\omega, T) = \frac{1}{1 - \exp\left[-\hbar\omega/k_B T\right]}$$

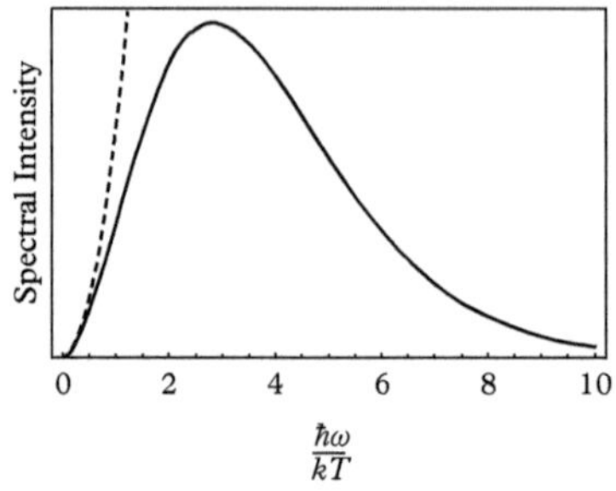

Fig. 13.4 *The Planck blackbody energy spectrum. The dashed curve is the Rayleigh–Jeans result.*

$$\langle E(\omega, T)\rangle = \frac{1}{Z}\frac{\partial Z(\omega, T)}{\partial \chi} = \frac{1}{Z}\frac{\partial}{\partial \chi}\left\{\frac{1}{1-\exp[\hbar\omega\chi]}\right\}$$
$$= \frac{\hbar\omega}{\exp[\hbar\omega/kT_B]-1} \tag{13.13}$$

And the spectral energy density is the mode density, Eq. 13.8, times this energy per mode:

$$e(\omega) = \rho(\omega)\langle E(\omega, T)\rangle = \frac{\hbar\omega^3}{\pi^2c^3}\frac{1}{\exp[\hbar\omega/(k_BT)]-1} \tag{13.14}$$

This is the Planck spectrum and is shown in Fig. 13.4. For low frequencies, the Rayleigh–Jeans result is reproduced. But, unlike the Rayleigh–Jeans result that rises monotonically with frequency, this spectrum has a maximum (at $\omega = 2.821k_BT/\hbar$). Additionally, the total frequency-integrated energy density–energy per unit volume–for the thermal radiation field is now finite and goes as the fourth power of the temperature:

$$\mathscr{E} = \frac{\pi^2k_B^4T^4}{15c^3\hbar^3} = aT^4 \tag{13.15}$$

The a in this equation is called the radiation constant and has the numerical value $a = 7.566\times10^{-16}\mathcal{J}\,m^{-3}\,K^{-4}$. Related to this is the familiar Stefan–Boltzmann constant, $\sigma = (c/4)\,a$, which specifies the power per unit area emitted by a blackbody radiator, $I = \sigma T^4$.

13.4 Casimir forces

Quantum harmonic oscillators have a zero point energy of $\frac{1}{2}\hbar\omega$ which may be understood as a consequence of the non-zero averages in the squares of the displacement and momentum quantum variables. Indeed, according to the Heisenberg uncertainty principle, the product of the position and momentum uncertainties has an absolute minimum value so that the sum of potential and kinetic energies, the total energy, must also be greater than zero. Similarly, the electric and magnetic fields of an electromagnetic mode are quantum variables and cannot be uniformly zero even in the lowest energy or "vacuum" states. Consequently, the squares of the uncertainties for a mode of the radiation field, $\vec{k}$, lead to an analogous zero point energy of $\frac{1}{2}\hbar\omega_{\vec{k}}$. While our understanding of vacuum state energies may be limited at this time, we may nevertheless extend the mode counting from above and discuss, quantitatively, an observable manifestation known as the Casimir force. Casimir and Polder originally derived the vacuum field force between an atom and a perfect conductor[6] and Casimir followed this shortly thereafter[7] with a derivation of the force between two perfectly conducting plates. We describe the latter here by first finding the energy in the vacuum field for the conducting box case considered earlier. The total vacuum field energy is equal to the sum of the zero point energies for all the modes of the system:

$$E = \sum_{\vec{k}} \frac{1}{2}\hbar\omega_{\vec{k}} \tag{13.16}$$

From the first section, for every volume $\Delta^3 k = \Delta k_x \Delta k_y \Delta k_z = (\pi/L)^3$ in $\vec{k}$-space, there are two modes of the radiation field. The inverse of this is the modes per volume or mode density in $\vec{k}$-space and we can take the product of the mode density and the corresponding zero point energy for the continuum limit of this equation:

$$\begin{aligned} E &= \int d^3k \cdot 2 \cdot \left(\frac{L}{\pi}\right)^3 \frac{1}{2}\hbar\omega_{\vec{k}} \\ &= \hbar V \left(\frac{1}{\pi}\right)^3 \iiint_0^\infty dk_z dk_y dk_x \cdot c\sqrt{k_x^2 + k_y^2 + k_z^2} \end{aligned} \tag{13.17}$$

where $V = L^3$ is the coordinate-space volume of the system and for the perfect conductor cube, the integral is taken over the positive octant of k-space. We note that with an infinite number of modes in their sums and integrals, Eqs. 13.16 and 13.17 predict an infinite vacuum energy within a finite volume. Figure 13.5, part (a), shows an infinitesimally thin, movable, perfectly conducting plate located at a position a above and $L - a$ below two perfectly conducting surfaces. If we take the zero potential energy configuration to be that of the original empty conducting cube without the plate, then the energy shift associated with conducting plate configuration is the Casimir potential. That is, the Casimir potential energy is the difference of the vacuum energies of the volume with and without the plate. We next address the case for which $a \ll L$.

[6] Casimir, H. B. G. and Polder D., The Influence of Retardation on the London-van der Walls Forces, *Phys. Rev.* 73, 360 (1948).

[7] Casimir, H. B. G. On the attraction between two perfectly conducting plates, *Proceedings of the Royal Netherlands Academy of Arts and Sciences*, 51, 793 (1948).

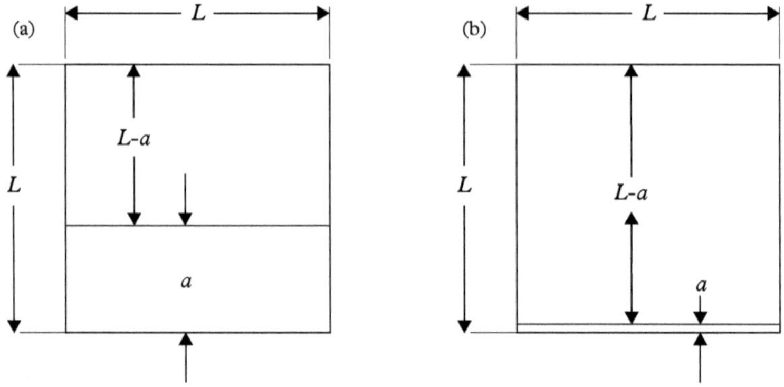

Fig. 13.5 *System for studying the Casimir force. (a) A perfectly conducting cubic box contains a movable plate* (blue). *(b) We find the potential energy of the system as the plate moves nearer the bottom plate of the cube.*

The difference of two infinite quantities is itself infinite so that the Casimir potential energy described is still infinite. This, however, may be tamed by noting the unphysical nature of taking the plate to be a perfect conductor. Specifically, waves at extremely high frequencies, x-rays and above, go right through without noticing the conductor: modes of sufficiently high energy would be unaffected by the presence of a thick gold or silver plate, for example. Thus, contributions from these modes are identical in the systems with and without the plate and so they exactly cancel upon taking the difference. Mathematically, we can treat this by defining a smooth function, $f(k/k_m)$, that goes to unity for low frequencies, vanishes for high frequencies and passes through 1/2 for $k = k_m$. It turns out that for most problems of interest the details of this function will not matter. Thus, we can write the Casimir potential as:

$$U(a) = \sum_{\vec{k}}^{a} f\left(\frac{k}{k_m}\right)\frac{1}{2}\hbar\omega_k + \sum_{\vec{k}}^{L-a} f\left(\frac{k}{k_m}\right)\frac{1}{2}\hbar\omega_k - \sum_{\vec{k}}^{\infty} f\left(\frac{k}{k_m}\right)\frac{1}{2}\hbar\omega_k \tag{13.18}$$

where "a," "$L-a$," and "∞" in these summations indicate that the sums, which are over all values of $\vec{k}$ in each case, are to be taken, respectively, in the two regions divided by the conducting plate and then within the undivided cube.

At first blush, this expression appears to give zero, since according to Eq. 13.17, the vacuum energy is proportional to volume and the summed volume of the two regions above and below the plate is equal to that of the undivided cube. However, when a is small, as in part (b) of Fig. 13.5, the continuum limit in the z direction cannot be used for the first summation. Indeed, the small region below the conducting plate must be examined more carefully. After rewriting the sums of

Eq. 13.18 in the general form of Eq. 13.17 and simplifying (**see Discussion 13.1**) we write:

$$U(a) = \hbar c\left(\frac{L}{\pi}\right)^2 \sum_{n_z=(0),1}^{\infty} \iint_0^{\infty} dk_y dk_x \cdot f\left(\frac{k}{k_m}\right)\sqrt{k_x^2+k_y^2+\left(\frac{n_z\pi}{a}\right)^2}$$
$$-\hbar c\left(\frac{L}{\pi}\right)^2 \int_0^{\infty} \frac{a}{\pi} dk_z \iint_0^{\infty} dk_y dk_x \cdot f\left(\frac{k}{k_m}\right)\sqrt{k_x^2+k_y^2+k_z^2} \tag{13.19}$$

where, in the first term, which is associated with the small, lower region below the plate, we have retained the continuum limit for k_x and k_y. The notation "(0)" under the summation sign indicates that a factor of 1/2 multiplies the $n_z = 0$ term because there is only one allowed polarization for those modes. We next evaluate the integrals over k_x and k_y in polar coordinates by defining $\kappa^2 = k_x^2 + k_y^2$:

$$U(a) = \hbar c\left(\frac{L}{\pi}\right)^2 \sum_{n_z=(0),1}^{\infty} \left(\frac{\pi}{2}\right)\int_0^{\infty} \kappa\, d\kappa \cdot f\left(\frac{\sqrt{\kappa^2+\left(\frac{n_z\pi}{a}\right)^2}}{k_m}\right)\sqrt{\kappa^2+\left(\frac{n_z\pi}{a}\right)^2}$$
$$-\hbar c\left(\frac{L}{\pi}\right)^2 \int_0^{\infty} \frac{a}{\pi} dk_z \left(\frac{\pi}{2}\right)\int_0^{\infty} \kappa\, d\kappa \cdot f\left(\frac{\sqrt{\kappa^2+k_z^2}}{k_m}\right)\sqrt{\kappa^2+k_z^2} \tag{13.20}$$

Thus, mathematically, the origin of the Casimir force is shown here as the difference, within the small lower region of Fig. 13.5, part (b), between the vacuum mode energy carefully summed over the actual modes–the first term–and the corresponding energy of the integral continuum approximation - the second term. If we note that the boundary condition requirement in z, $k_z = n_z\pi/a$, applies regardless of whether the k_z and n_z are discrete or continuous, then we can rewrite the second term in Eq. 13.20, the integral continuum approximation, as

$$-\hbar c\left(\frac{L}{\pi}\right)^2 \int_0^{\infty} dn_z \left(\frac{\pi}{2}\right)\int_0^{\infty} \kappa\, d\kappa \cdot f\left(\frac{\sqrt{\kappa^2+\left(\frac{n_z\pi}{a}\right)^2}}{k_m}\right)\sqrt{\kappa^2+\left(\frac{n_z\pi}{a}\right)^2}$$

If we now define the function common to both terms in Eq. 13.20,

$$g(k_z) = g\left(\frac{n_z\pi}{a}\right) = \frac{\pi}{2}\int_0^{\infty} \kappa\, d\kappa \cdot f\left(\frac{\sqrt{\kappa^2+\left(\frac{n_z\pi}{a}\right)^2}}{k_m}\right)\sqrt{\kappa^2+\left(\frac{n_z\pi}{a}\right)^2} \tag{13.21}$$

Equation 13.20 can be written more succinctly as

$$U(a) = \hbar c \left(\frac{L}{\pi}\right)^2 \left(\sum_{n_z=(0),1}^{\infty} g\left(\frac{n_z \pi}{a}\right) - \int_0^{\infty} dn_z g\left(\frac{n_z \pi}{a}\right) \right) \tag{13.22}$$

where it is understood that n_z is discrete under the summation and continuous under the integration. Looking closer at $g(k_z)$ with an eye toward simplification, we define $\kappa = \beta\pi/a$ to obtain[8]

$$g\left(\frac{n_z \pi}{a}\right) = \frac{\pi^4}{2a^3} \int_0^{\infty} \beta d\beta \cdot f\left(\frac{\pi\sqrt{\beta^2 + n_z^2}}{k_m a}\right) \sqrt{\beta^2 + n_z^2}$$

As a final simplification, we substitute the variable $\alpha = \beta^2 + n_z^2$ to obtain

$$g\left(\frac{n_z \pi}{a}\right) = \frac{\pi^4}{4a^3} \int_{n_z^2}^{\infty} d\alpha \cdot f\left(\frac{\pi \alpha^{1/2}}{k_m a}\right) \alpha^{1/2} = \frac{\pi^4}{4a^3} F(n_z) \tag{13.23}$$

thus defining the integral as the function $F(n_z)$. Written in terms of $F(n_z)$, Eq. 13.22,

$$U(a) = \hbar c \left(\frac{L}{\pi}\right)^2 \frac{\pi^4}{4a^3} \left(\sum_{n_z=(0),1}^{\infty} F(n_z) - \int_0^{\infty} dn_z F(n_z) \right) \tag{13.24}$$

Fortunately, there is a formula–the Euler–Maclaurin formula–which is directly applicable to this circumstance[9]. It gives the difference between the summed series and its integral approximation as:

$$\sum_{n_z=(0),1}^{\infty} F(n_z) - \int_0^{\infty} dn_z F(n_z) = -\frac{1}{12} F'(0) + \frac{1}{24 \times 30} F'''(0) + \cdots \tag{13.26}$$

where $F(n_z)$ must be a function capable of being differentiated several times for $n_z = 0$. From Eq. 13.23, the function $F(n_z)$ for this Euler–Maclaurin formula is

$$F(n_z) = \int_{n_z^2}^{\infty} d\alpha \cdot f\left(\frac{\pi \alpha^{1/2}}{k_m a}\right) \alpha^{1/2}$$

The derivatives required for the formula are:

$$F'(n_z) = -2n_z^2 f\left(\frac{\pi n_z^2}{k_m a}\right) \rightarrow F'(0) = 0 \quad \text{and} \quad F'''(0) = -4 \tag{13.27}$$

[8] Actually, $\kappa = n_\rho \pi / L$ would be more accurate since

$$\kappa^2 = k_x^2 + k_y^2 = \left(\frac{n_x \pi}{L}\right)^2 + \left(\frac{n_y \pi}{L}\right)^2 = \left(n_x^2 + n_y^2\right)\left(\frac{\pi}{L}\right)^2 = \left(\frac{n_\rho \pi}{L}\right)^2$$

but we are looking to pull out the π/a from the radicals in Eq. 13.21, so we use $\beta = an_\rho/L$.

[9] Generally, for a real or complex-valued function, $f(x)$, the Euler–Maclaurin formula gives the following relationship between the sum $S = f(m+1) + \ldots + f(n-1) + f(n)$ and the integral $I = \int_m^n f(x)\,dx$:

$$S - I = \sum_{k=1}^{p} \frac{B_k}{k!} \left[f^{(k-1)}(n) - f^{(k-1)}(m) \right] + R \tag{13.25}$$

where m and n are natural numbers $(0, 1, 2, 3\ldots)$, B_k is the kth Bernoulli number and R is a small remainder term.

the higher order derivatives all contain powers of $\frac{\pi}{k_m a}$ and contribute negligibly as long as $k_m a \gg 1$. The predicted Casimir potential energy per unit area of the plate is then:

$$\frac{U(a)}{L^2} = -\hbar c \frac{\pi^2}{24 \times 30} \cdot \frac{1}{a^3} \tag{13.28}$$

and the corresponding Casimir force is the negative derivative of this expression, $-\partial U/\partial a$. This may be interpreted as the zero-point pressure of electromagnetic waves. Good reflectors such as silver act as near ideal conductors/reflectors for wavelengths, λ_m, longer than a few hundreds of nanometers. So, for a choice of the cross-over function, f, with $k_m \simeq 2\pi/\lambda_m$, (see Eq. 13.18), the assumptions leading to the result for the potential energy are satisfied for distances, $a \gg \lambda_m$. Experimentally, the Casimir force was originally observed and measured by Lamoreaux in 1997[10] between two gold surfaces separated by distances ranging from 0.6 to 6 μm. Active research on Casimir forces has continued.[11]

13.5 Spontaneous emission: the Einstein *A* and *B* coefficients

An atom or molecule in an excited state isolated in empty space will, after some period of time, decay to a lower energy state, releasing a quantum of electromagnetic radiation, a photon, in the process. This process is called spontaneous emission and is the quantum analog of radiation from oscillating charges. A first principles derivation of this process is not possible in regular quantum mechanics but rather requires a quantizing of the field. That is, it requires a full quantum electrodynamics (QED) treatment. Einstein, however, by insisting on the consistency of electromagnetism and statistical mechanics with whatever the quantum predictions would eventually turn out to be, derived consistent results that can only be described as striking. Einstein considered two states of a system, an "atom," which he bathed in a radiation field but left otherwise isolated. If we label the lower energy state as "1," the higher energy state as "2," their energy difference as $E_0 = \hbar\omega_0$, and the spectral density of the radiation field in the vicinity of the atom as $e(\omega_0)$, then we can identify, as Einstein did, three processes connecting the states:

(1) Absorption, in which the atom in state 1 is excited by the field up to state 2 at a rate proportional, by the constant "B_{12}," to the spectral density: $R_{abs} = B_{12} e(\omega_0)$,

[10] Lamoreaux, S. K., Demonstration of the Casimir force in the 0.6 to 6 μm range, *Phys. Rev. Lett.* 78, 5, (1997).

[11] Klimchitskaya, G. L., Mohideen, U. and Mostepanenko, V. M., The Casimir force between real materials: Experiment and theory, *Reviews of Modern Physics* 81, 1827 (2009).

(2) Stimulated emission, for which an atom in state 2 is driven by the field down to state 1 (thus increasing the field energy by $E_0 = \hbar\omega_0$), at a rate proportional, by the constant "B_{21}," to the spectral density: $R_{stim} = B_{21} e(\omega_0)$,

(3) And spontaneous emission, for which an atom initially in the upper state, 2, decays to the lower state. This process, as described before, is taken to occur at a rate, $R_{spon} = A_{21}$, independent of the spectral density of the radiation field.

These Einstein A and B coefficients are understood to be properties of the atom alone, and not the field, although we will revisit this assumption next. In particular, the spectral density of the radiation field, $e(\omega_0)$, is only assumed to be isotropic, continuous and smooth for frequencies around ω_0. Therefore, it need not necessarily be thermal (blackbody), however, nor should it be sharply peaked around ω_0 as in a laser field. Generally, for a collection of atoms, some of which are in the two states of interest, the rate of change, for example, of the state 2 population due to transitions to and from state 1 by these processes is described by:

$$\frac{dN_2(2 \leftrightarrow 1)}{dt} = N_1 B_{12} e(\omega_0) - N_2 B_{21} e(\omega_0) - N_2 A_{21} \tag{13.29}$$

At this point we shall consider the special case of a system in thermodynamic equilibrium (TE). In TE, transition rates to and from state 2 are equal so the left-hand side of Eq. 13.29 is zero, $e(\omega_0)$ is given by the Planck distribution found in Section 13.3, and the ratio of the populations of the two states is given by the Boltzmann factor: $N_2/N_1 = \exp[-\hbar\omega_0/(k_B T)]$. We first consider the limit of extremely high temperatures for which the Boltzmann factor is unity, $N_2/N_1 = 1$, and the thermal radiation field is so incredibly large that it swamps the spontaneous emission rate in Eq. 13.29. For the system to maintain TE in this limit, it must be concluded that the B coefficients are equal:

$$B_{21} = B_{12}. \tag{13.30}$$

We next consider the extreme low temperature limit for which the radiation field in the vicinity of the resonance frequency is so small that the spontaneous emission term dominates the stimulated term. And in the Planck spectrum of Eq. 13.14, the exponential term in the denominator dominates. Equation 13.29 becomes:

$$N_2 A_{21} = N_1 B_{12} \frac{\hbar\omega_0^3}{\pi^2 c^3} \frac{1}{\exp[\hbar\omega_0/(k_B T)]} \tag{13.31}$$

with the ratio of the state populations given by the Boltzmann factor, the exponentials cancel and we obtain A_{21} in terms of $B_{12} = B_{21}$,

$$A_{21} = \frac{\hbar\omega_0^3}{\pi^2 c^3} B_{12} = \frac{\hbar\omega_0^3}{\pi^2 c^3} B_{21} \tag{13.32}$$

It is easily checked that with these definitions, the three rates satisfy thermodynamic equilibrium for any temperature. Working from what seem to be very different approaches, Eq. 13.30, the equality of the B coefficients, may be reproduced using the "semi-classical" approximation in regular, non-relativistic quantum mechanics,[12] while QED returns Eq. 13.32.

Two caveats are in order:

- Usually in atomic or molecular physics, coefficients are calculated for transitions between sets of degenerate states making up two atomic levels. If the degeneracies of the lower and upper levels are g_1 and g_2, respectively, the A and B coefficients averaged over two levels, satisfy:

$$g_2 B_{21} = g_1 B_{12} \quad \textit{and} \quad A_{21} = \frac{\hbar\omega_0^3}{\pi^2 c^3} B_{21} \tag{13.33}$$

- The various coefficients defined in this section assume a radiation field that is isotropic and the B's are actually averages over all directions in space. Quantal calculations frequently are carried out using plane waves whose polarization relative to the orientation of the atomic state is important.

The Einstein relations suggest some general qualitative insights into making coherent sources such as lasers and masers. A laser or maser starts with a population of excited state atoms and a few photons rattling around in a desired laser mode of the radiation field. Here, the desired "mode" is a real solution of Maxwell's equations for the laser or maser cavity of the device. Now the probability that a photon will be stimulated from one of the excited atoms by circulating photons is proportional to $N_2 B_{21}$. Competing processes include absorption removing photons from the mode, proportional to $N_1 B_{12}$ and spontaneous emission, proportional to $N_2 A_{21}$. Unlike in thermal equilibrium, where these processes conspire to produce no change in the state populations or radiation field, the condition for lasing requires a net radiation gain into the lasing mode and for this, Eq. 13.29 requires a population inversion, $N_2 > N_1$. Equation 13.32 shows the Einstein A coefficient for spontaneous emission increasing, relative to the B coefficient, by the cube of the frequency. Thus, with all other things

[12] Allen, L. and Eberly, J. H. *Optical Resonance and Two-Level Atoms,* Wiley, New York (1975).

being equal (which they are not necessarily), excited state losses for bluer transitions will tend to occur more via spontaneous emission than the desired stimulated emission. As a result, coherent sources at higher desired frequencies tend to be more difficult to produce.

It was claimed above that the A and B coefficients are related to atomic properties. This, however, is not entirely true[13] and the environment surrounding the atom–especially for confined systems–may be important, in two respects, to the determination of these coefficients. First, the excitation of an atom depends on its micro-environment. This is trivially true for atoms embedded in dielectrics or liquids: both the energies and the rates of transitions are so affected. More subtle shifts are found by tightly coupling otherwise free atoms to macroscopic systems such as the modes of a high finesse optical cavity.[14] Second, the "vacuum" an atom sees affects its spontaneous emission rate. Reducing the number of modes available for a spontaneously emitted photon reduces the corresponding A coefficient and increases the lifetime of the excited state. This effect has been observed[15] and an important goal for photonic crystal research is systematically controlling spontaneous emission by engineering the vacuum. Finally, a related question occasionally arises: Is it possible to "explain" spontaneous emission as the result of stimulated emission driven by the vacuum field? We leave it to the exercises to show that such an attempt falls short, by a factor of $1/2$, of accounting for the total spontaneous emission rate, A_{21}.

[13] Purcell, E. M., Spontaneous Emission Probabilities at Radio Frequencies, *Phys. Rev.* 69, 674 (1946).

[14] Bernardot, F., Nussenzveig, P., Brune, M., Raimond, J. M., and Haroche, S., Vacuum Rabi Splitting Observed on a Microscopic Atomic Sample in a Microwave Cavity, *Europhysics Lett.*, 17, 33 (1992).

[15] Hulet, R. G., Hilfer, E. S., and Kleppner, D., Inhibited spontaneous emission by a Rydberg Atom, *Phys. Rev. Lettr.* 20, 2137 (1985).

13.6 Microwave cavities

We next consider the case of a rectangular box with conducting surfaces. Such a "resonant cavity" is shown in Fig. 13.6.

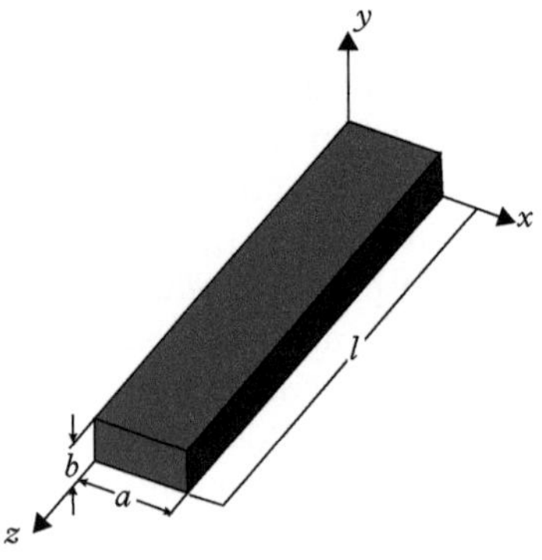

Fig. 13.6 *Rectangular micro-wave resonant cavity.*

In contrast to the previous discussion, we are now mostly interested in cavities having a small number of distinct modes for the frequency range of interest. While this discussion is general, the cavity in the figure has singled out the z dimension as being considerably larger than the other two. Such an elongated cavity is most relevant for modeling microwave cavities used in atomic beam clocks and also provides for a smooth transition to the microwave waveguide discussion in Section 13.7. The normal modes are a simple extension of the fields of Eqs. 13.5 and 13.6 where now the allowed wavevectors are given by:

$$\vec{k} = \left(k_x, k_y, k_z\right) = \left(\frac{n_x \pi}{a}, \frac{n_y \pi}{b}, \frac{n_z \pi}{l}\right) \tag{13.34}$$

in which n_x, n_y, and n_z are non-negative integers. Thus, only oscillating fields with frequencies satisfying:

$$\frac{\omega}{c} = k = \sqrt{k_x^2 + k_y^2 + k_z^2} \tag{13.35}$$

are resonant in the cavity. Usually there are two modes for each frequency. Convenient choices are modes with fields having either $E_z = 0$ or $B_z = 0$ for which $a_z = 0$ and $a_y k_x - a_x k_y = 0$, respectively, within Eqs. 13.5 and 13.6. If we specify modes by the integers (n_x, n_y, n_z), then the lowest frequency resonance for the cavity under consideration is given by $(0, 1, 1)$. This mode is non-degenerate with the fields given as:

$$\begin{aligned} \vec{E} &= E_0 \hat{x} \left(\sin k_y y \sin k_z z\right) \\ \vec{H} &= \frac{E_0}{i\omega\mu_0} \left[\hat{y}\left(k_z \sin k_y y \cos k_z z\right) - \hat{z}\left(k_y \cos k_y y \sin k_z z\right)\right] \end{aligned} \tag{13.36}$$

The most significant difference between ideal perfectly conducting waveguides and real waveguides is that the latter are lossy. The Q or quality factor of a resonance characterizes a cavity's losses. Specifically, Q is the ratio of the on-resonance electromagnetic energy stored in a cavity to the amount of energy lost from the cavity per oscillation period. Additionally, the finite Q of a resonance describes frequency broadening from the ideal of an infinitely sharp, infinite Q cavity resonance. Frequently, the reason for using a cavity is to provide a strong field by storing the energy from an input source and the Q of the resonance is central to describing this process. For metallic waveguides made from good conductors, the principal loss is from eddy currents driven in the waveguide walls. Specifically, referring to the discussion of boundary conditions in Section 13.2, any surface current density,

$\vec{K}_S$, driving a finite parallel magnetic field at a vacuum-wall interface must actually have some finite thickness due to the finite conductivity of the wall's metal. This current causes Ohmic losses in the wall. For good conductors, the fields in the vacuum are little-affected by the finite conductivity of the walls, so, in estimating the losses, we assume these fields are unchanged. In a good conductor, the displacement current contribution to Ampere's law is small compared to that of the conduction current and Maxwell's curl equations describing propagation into the conductor become:

$$\nabla\times\vec{H}=\vec{\jmath}=\sigma_f\vec{E}_c \qquad \nabla\times\vec{E}_c=i\omega\mu_0\vec{H} \tag{13.37}$$

where the σ_f is assumed to be real and $\vec{E}_c$ is the electric field in the metal due to its finite conductivity. Now, because derivatives of the fields taken normal to the surface are much larger than those parallel to the surface (the former are infinite for a perfect conductor), the latter can be ignored and the curl equations may be combined

$$-\nabla^2\vec{E}_c=-\frac{\partial^2\vec{E}_c}{\partial\xi^2}=i\omega\mu_0\sigma_f\vec{E}_c \tag{13.38}$$

where ξ refers to the coordinate in the direction perpendicular to the surface. This has the solution:

$$\vec{E}_c=\vec{E}_0\exp\left(-\frac{\xi}{\delta}\right)\exp\left(-i\frac{\xi}{\delta}\right) \tag{13.39}$$

from which we obtain the skin-depth of the conductor given by $\delta=\sqrt{2/\left(\omega\mu_0\sigma_f\right)}$. The total ohmic loss for a surface of area dA is:

$$P=dA\cdot\frac{1}{2}\int_0^\infty\vec{E}_c\cdot\vec{\jmath}^*\,d\xi=dA\frac{\sigma_f}{2}\int_0^\infty\left|\vec{E}_c\right|^2d\xi=dA\left|E_0^2\right|\frac{\sigma_f\delta}{4} \tag{13.40}$$

The total surface current density, $\vec{K}_S$, may be used to find $\vec{E}_0$:

$$\vec{K}_S=\int_0^\infty\vec{\jmath}d\xi=\sigma_f\vec{E}_0\int_0^\infty\exp\left(-\frac{\xi}{\delta}\right)\exp\left(-i\frac{\xi}{\delta}\right)d\xi=\frac{\vec{E}_0}{2}(1-i)\sigma_f\delta \tag{13.41}$$

Finally, using $\left|\vec{K}_S\right|=\left|H_\parallel\right|$ and Eq. 13.40, the eddy current loss for area dA is:

$$P=dA\frac{\left|H_\parallel\right|^2}{2\sigma_f\delta} \tag{13.42}$$

13.7 Microwave waveguides

We next examine how hollow metallic tubes can guide electromagnetic waves. We will begin by describing features common to all cylindrical waveguides (here, "cylindrical" refers to waveguides of arbitrary cross section- e.g., Fig. 13.7a). Specifically, we will consider the special utility of longitudinal components of the fields in describing waveguide modes. We next go on to examine the rectangular waveguide of Fig. 13.7, part (b), in more detail.

13.7.1 General features of waveguides

We take $+z$ to be the direction that the wave is to be guided and, for the general case of a cylindrical waveguide, we look for harmonic traveling wave solutions to Maxwell's equations with the form:

$$\begin{aligned}\vec{E}(\vec{r}) &= \vec{\mathscr{E}}(x,y)\exp(i\beta z) \\ \vec{H}(\vec{r}) &= \vec{\mathscr{H}}(x,y)\exp(i\beta z)\end{aligned} \tag{13.43}$$

where β, the "propagation constant" for the electromagnetic mode of the waveguide, is essentially the same quantity as k_z defined previously. Finding β is the first step in describing a waveguide mode. For these

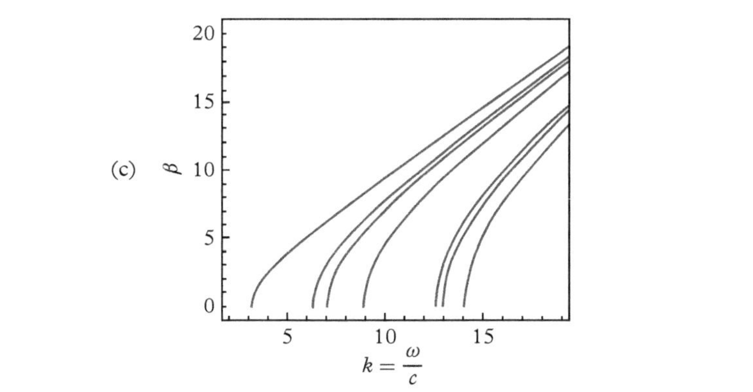

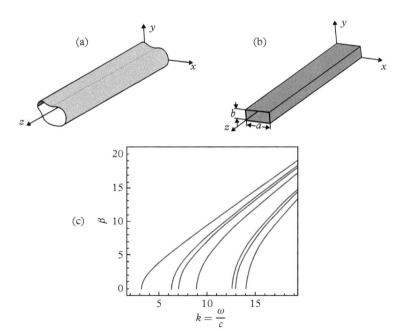

Fig. 13.7 *(a) General microwave wave guide; (b) rectangular waveguide; (c) the propagation constants for the lowest few modes of the rectangular waveguide of (b): leftmost is TE_{10} and then TE_{01}. The next curve describes the degenerate modes TE_{11} and TM_{11} and so on.*

types of wave functions, the gradient operator may be written:

$$\nabla = \hat{x}\frac{\partial}{\partial x} + \hat{y}\frac{\partial}{\partial y} + \hat{z}i\beta = \nabla_T + \hat{z}i\beta \tag{13.44}$$

Thus, the Helmholtz equation for the electric field becomes:

$$\nabla_T^2 \vec{\mathscr{E}} = -\left(k^2 - \beta^2\right)\vec{\mathscr{E}} \tag{13.45}$$

where the transverse gradient of $\mathscr{E}$ and $\hat{z}$ are orthogonal so the cross terms vanish. It is further useful to divide the field into transverse and longitudinal components:

$$\begin{aligned}\vec{\mathscr{E}} &= \vec{\mathscr{E}}_T + \vec{\mathscr{E}}_z \\ \vec{\mathscr{H}} &= \vec{\mathscr{H}}_T + \vec{\mathscr{H}}_z\end{aligned}$$

where, for example,

$$\begin{aligned}\vec{\mathscr{E}}_T &= \hat{z} \times \vec{\mathscr{E}} \times \hat{z} \\ \vec{\mathscr{E}}_z &= \hat{z}\mathscr{E}_z\end{aligned}$$

We now show the computationally very useful result that once the z field components, $\mathscr{E}_z$ and $\mathscr{H}_z$ are found, they may be used to construct the entire vector field. Upon multiplying each of the two free-space Maxwell's curl equations by $\hat{z}\times$, they become (**see Part (a) of Discussion 13.2**)

$$\begin{aligned}-i\beta\vec{\mathscr{E}}_T + \nabla_T\mathscr{E}_z &= i\omega\mu_0\hat{z} \times \vec{\mathscr{H}}_T \\ -i\beta\vec{\mathscr{H}}_T + \nabla_T\mathscr{H}_z &= -i\omega\varepsilon_0\hat{z} \times \vec{\mathscr{E}}_T\end{aligned} \tag{13.46}$$

And upon multiplying these equations by $\hat{z}\times$, the transverse field components may be solved in terms of the z components: (**see Part (b) of Discussion 13.2**):

$$\begin{aligned}\vec{\mathscr{E}}_T &= \frac{i}{\left(k^2 - \beta^2\right)}\left(\beta\nabla_T\mathscr{E}_z - \omega\mu_0\hat{z} \times \nabla_T\mathscr{H}_z\right) \\ \vec{\mathscr{H}}_T &= \frac{i}{\left(k^2 - \beta^2\right)}\left(\beta\nabla_T\mathscr{H}_z + \omega\varepsilon_0\hat{z} \times \nabla_T\mathscr{E}_z\right)\end{aligned} \tag{13.47}$$

For many (but not all) waveguides, it is possible to find solutions to Maxwell's equations and satisfy all boundary conditions for waves having either $\mathscr{E}_z = 0$ or $\mathscr{H}_z = 0$. These are referred to as transverse electric, *TE*, or transverse magnetic, *TM*, modes, respectively. For either type of mode, the transverse fields are related by:

$$\vec{\mathscr{E}}_T = -Z\left(\hat{z} \times \vec{\mathscr{H}}_T\right) \tag{13.48}$$

where for *TE* modes, $Z = \frac{k}{\beta}\sqrt{\frac{\mu_0}{\varepsilon_0}}$, and for *TM* modes, $Z = \frac{\beta}{k}\sqrt{\frac{\mu_0}{\varepsilon_0}}$.

13.7.2 Rectangular conducting waveguides

In Fig. 13.7(b), we show a rectangular waveguide with dimensions a and b. Shown is the most common case where $a = 2b$. Although we discuss here only the perfect conductor waveguides, losses, as described in the previous section, are readily included. Boundary conditions in x and y are the same and so the wavefunctions are similar to those of the conducting cavity. The z-dependence, however, is now that of a traveling wave rather than a standing wave. Equations 13.5 and 13.6 become:

$$\vec{\mathscr{E}} = E_0 \exp(i\beta z)\Big[\hat{x}a_x \cos k_x x \sin k_y y + \hat{y}a_y \sin k_x x \cos k_y y + \hat{z}a_z \sin k_x x \sin k_y y\Big] \tag{13.49}$$

and

$$\vec{\mathscr{H}} = \frac{E_0 \exp(i\beta z)}{i\omega\mu_0}\Big[\hat{x}\left(a_z k_y - a_y k_z\right) \sin k_x x \cos k_y y + \hat{y}\left(a_x k_z - a_z k_x\right) \cos k_x x \sin k_y y + \hat{z}\left(a_y k_x - a_x k_y\right) \cos k_x x \cos k_y y\Big] \tag{13.50}$$

Here, as before, $k_x = n_x\pi/a$ and $k_y = n_y\pi/b$, where n_x and n_y range over all of the non-negative integer combinations except $(n_x, n_y) = (0,0)$. Again, generally, for each pair of components, (k_x, k_y), there are two solutions and picking $a_z = 0$ or $a_y k_x - a_x k_y = 0$ gives these *TE* or *TM* solutions, respectively. To solve the wave equation, or the associated Helmoltz equation, the wavevector components must satisfy:

$$\frac{\omega^2}{c^2} = k^2 = k_x^2 + k_y^2 + \beta^2 \tag{13.51}$$

For the resonant cavity, this condition determined the resonant frequency of a mode. For waveguides, in order that β be real and yield a propagating solution, the frequency ω of the field must be greater than the minimum or "cut-off" frequency of a mode:

$$\omega \geq ck_{cutoff} = c\sqrt{k_x^2 + k_y^2} \tag{13.52}$$

The propagation constant, β, is shown in part (c) of Fig. 13.7 as a function of frequency for each mode. Here these results involve simply solving Eq. 13.51 for different modes. To generate the analogous figure for dielectric waveguides requires more work. The mode with the lowest cutoff frequency is often especially important for applications, as discussed next. In particular, it is frequently useful to use a waveguide with only a single allowed mode for the frequency of interest. For the rectangular, perfect conductor waveguide, this is the mode with $n_x = 1,\ n_y = 0$. Inspection of Eq. 13.49 shows that the electric field is completely in the y direction, so there is a TE_{10} mode. However, we see from Eq. 13.50 that the magnetic field has both x and z components so, as with all modes in which either n_x or n_y equals zero, there is no corresponding TM_{10} mode. Here, then, $\omega_{cutoff} = \pi c/a$ and, for waveguides such as this in which $a = 2b$, single mode operation holds for up to twice this frequency.

13.7.3 Transmission lines and coaxial cables: TEM modes

Equations 13.47 cannot be used for the important case, $\beta^2 = k^2$; here both fields are transverse and the modes are correspondingly designated "TEM." While it is true, then, that these TEM waves are not supported within the hollow waveguides discussed previously, it is possible to have TEM waves in a coaxial cable or a transmission line. We next examine these modes assuming perfectly conducting electrodes in a medium with the uniform, non-vacuum permittivity, ε. The magnetic permeability is still taken to be the vacuum value, $\mu = \mu_0$. Proceeding from Eqs. 13.46, the vanishing of the longitudinal field components gives:

$$\vec{\mathscr{E}}(x,y) = -Z\,\hat{z} \times \vec{\mathscr{H}} \tag{13.53}$$

$$\vec{\mathscr{H}}(x,y) = \frac{1}{Z}\,\hat{z} \times \vec{\mathscr{E}} \tag{13.54}$$

where Z is the wave impedance in the dielectric medium, $Z = \sqrt{\mu/\varepsilon}$. Thus, in the familiar way, the magnetic field is found immediately once the electric field is determined. In the medium, the divergence of the electric field and the z component of its curl both vanish. The latter follows from the transversality of the magnetic field ($B_z = 0$). Explicitly:

$$\frac{\partial \mathcal{E}_x(x,y)}{\partial x} + \frac{\partial \mathcal{E}_y(x,y)}{\partial y} = 0$$
$$\frac{\partial \mathcal{E}_x(x,y)}{\partial y} - \frac{\partial \mathcal{E}_y(x,y)}{\partial x} = 0$$

But these equations are identical to those for a 2D, source-free, static electric field problem. Recall that type of problem is most often solved by defining a scalar electrical potential, $V(x,y)$, that obeys the 2D Laplace equation. Conversely, we can say that any 2D electrical potential solution can serve as the basis of a TEM mode for the corresponding electrodynamic problem, with the one additional requirement that the fields go to zero sufficiently rapidly "at infinity." Electrical potential problems are typically posed by specifying voltages on all of the conductors (see Fig. 13.8). It is immediately seen, then, that the waveguides considered in the previous section can have no TEM modes: for a hollow conductor held at voltage V, the potential inside the conductor surface is also V and there are no fields. However, solutions do exist for systems with more than one conductor. Given the solution, $V(x,y)$, for a system of conductors such as in Fig. 13.8, parts (b), (c), and (d), $\vec{\mathcal{E}}_T(x,y)$ is obtained, in the usual way, by taking the gradient of the potential function. The magnetic field is found, in turn, using Eq. 13.54. The magnetic field distribution determines the current density distribution on the conductors (see Fig. 13.8e).

We carry out this program for the important example of the coaxial cable of Fig. 13.8, part (d). Here, we will take the radii of the inner and

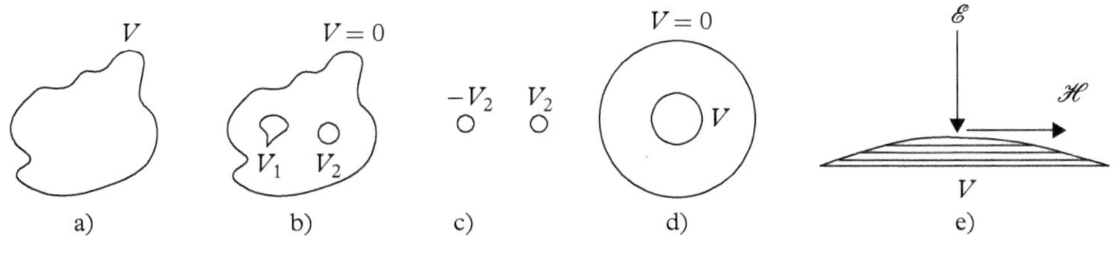

Fig. 13.8 *A transverse slice for various TEM solutions of transmission lines: (a) A hollow single conductor will not support TEM solutions, however (b) a general multi-conductor cable will. (c) The twin lead conductor is common. The corresponding solution to the Laplace equation must include the boundary condition that the fields go to 0 sufficiently rapidly at infinity. (d) The coax cable is the most widely used conductor of high frequency rf signals. (e) In the limit of perfect conductors, the electric field is everywhere perpendicular to a conductor's surface. For the transmission line problem, the magnetic field is used to find the surface current density of the electrode.*

outer shield conductors to be a and b, respectively. The static problem is well known. If the potentials on the conductors are as shown, then the electric field is in the radial direction with magnitude:

$$\mathscr{E}_\rho = \frac{V}{\ln(a/b)}\frac{1}{\rho} \tag{13.55}$$

From Eq. 13.54, the magnetic field at the inner conductor is axial with the corresponding current on that conductor:

$$\mathscr{H}_\theta = \frac{1}{Z}\frac{V}{\ln(a/b)}\frac{1}{a} \Longrightarrow I = \frac{2\pi V}{Z\ln(a/b)} \tag{13.56}$$

The characteristic impedance of the cable is defined as the ratio V/I. For a coaxial cable, this is found to be:

$$Z_{cable} = \frac{\ln(a/b)}{2\pi} Z \tag{13.57}$$

13.8 One-dimensional optical waveguides: the ray optic picture

Optical waveguides efficiently and compactly transport light with precise spatial control, effectively free of the diffraction effects that often adversely affect light transport by beams propagating in free space. In the rapidly developing industry of integrated optical circuits, optical waveguides are the functional equivalent of the conducting wire traces used in electronic circuits. Waveguides also serve as components in optical devices: such components include mixers, multiplexers, demodulators, interferometers, and more recently, chemical sensors. Communication traffic is increasingly carried by waveguides in the form of optical fibers for connections ranging from the intra-community to the intercontinental. Ray optics are a good starting place for discussing waveguide properties. Following rays through a guide gives an intuitive picture of the wave-guiding process and the ray picture can be pushed surprisingly far for quantitatively describing devices. We begin by considering 1D, planar-symmetry, dielectric waveguides. The "1D" refers to the effectively infinite extension of the dielectric layers in the directions parallel to their interfaces. A practical example of this type of waveguide is the "gain-guided" diode laser shown in Fig. 13.2. Basically, as shown in Fig. 13.9, a higher index, n_1, "core" is sandwiched between two lower index regions, n_2 and n_3, making up the waveguide's "cladding." For the most frequently

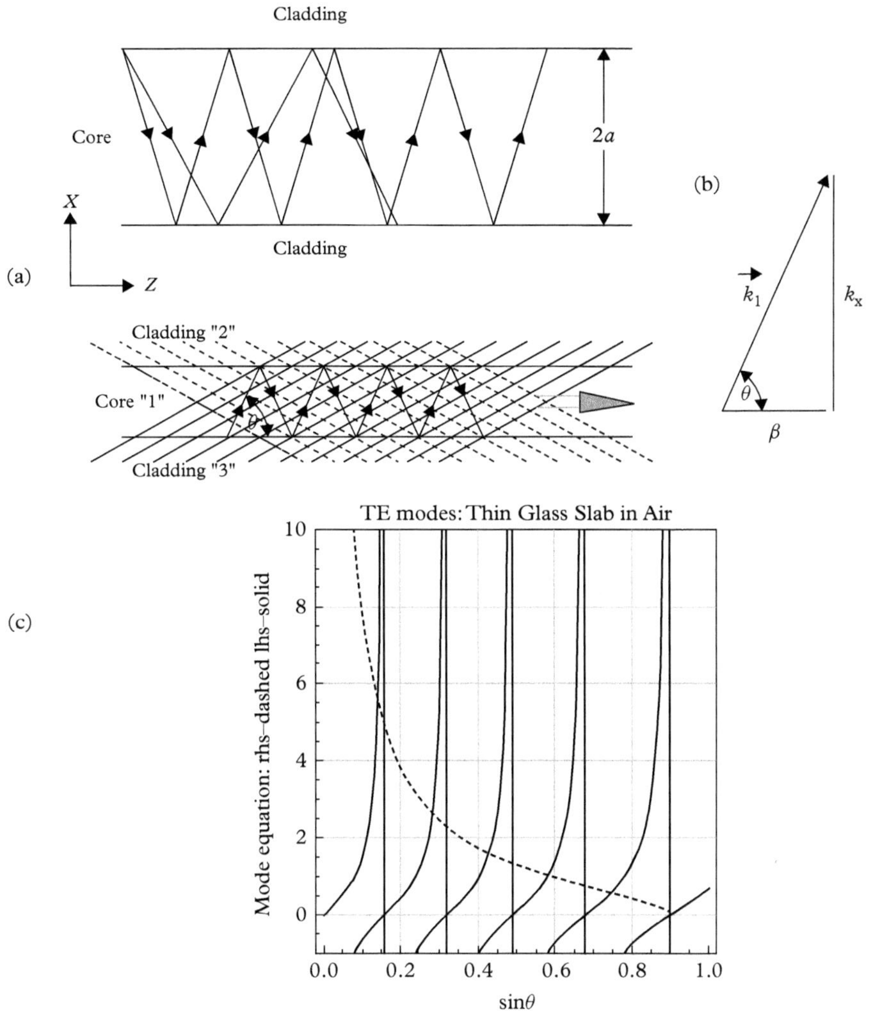

Fig. 13.9 *Ray optics for planar waveguides (a) Light propagation in a planar waveguide is understood as facilitated by total internal reflections at the core-cladding interfaces. (b) Shown are the parameters used to quantitatively describe single modes in a waveguide. (c) Graphical solution for waveguide modes. Propagating solutions are given by intersections of the dashed curve with the solid ones (the right- and left-hand sides of Eq. 13.62).*

encountered devices, the two cladding regions are often, but not always, the same material. Devices with same and different cladding are referred to, respectively, as symmetric and asymmetric planar waveguides. Here, we discuss the general case but use symmetric waveguides for quantitative examples. Waveguides are further categorized as either "step-index" or "graded-index." In the latter, the cladding is manufactured by treating–usually doping–regions near the surfaces so as to give an index of refraction profile that decreases progressively in the outward direction, reaching a minimum at the outer surface. A simple explanation of light guiding in a step index device is illustrated by two rays in Fig. 13.9(a): light rays propagate down the waveguide making total-internal-reflection bounces every time they hit a core-cladding interface. In the waveguide literature, a ray is described by the angle, θ, that it makes with the longitudinal axis of the waveguide. Note, from Fig 13.9 (b), that this "grazing" angle is the complement of the angle of incidence used in describing reflection and refraction along the waveguide surface. When we get to the wave picture, for which light propagation is described in terms of modes of the radiation field, that is, by solutions of the Helmholtz equation for a specific frequency, we will see that smaller angle, θ, rays correspond to lower order modes and higher order modes correspond to steeper angles up to a "cutoff" frequency. For a ray to propagate down the device, θ must be small enough for total internal reflection at the index-steps for both interfaces. That is, $\cos\theta$ must be greater than both n_2/n_1 and n_3/n_1 and cutoff occurs when this condition fails.

The ray picture of Fig. 13.9(a) is called the "zig-zag" model and for large multimode waveguides, the assumptions of ray optics are satisfied. The number of reflections occurring is striking: a typical communication fiber optic has a diameter of about 10 μm and a signal traveling 1 km will make about 10^{11} reflections. A recurring issue, the waveguide dispersion of a signal, is immediately evident from the figure. The two different paths shown have different lengths and thus as a signal propagates, the parts of the signal that are carried by the differently angled rays arrive at a point downstream at different times and so for large distances signals may become scrambled. The solution used for intercontinental optical connections is simply to use fibers for which only a single mode can propagate. A less effective but occasionally used alternative solution is offered by gradient index waveguides. Recall that the speed of a light wave goes inversely with index of refraction. By suitably engineering the index gradient, different paths can be made to have more nearly equal propagation times: the longer path of a large θ ray can be compensated for by having it go faster in the outer low index region of the effective cladding. Returning to Fig. 13.9, there are clearly two independent polarizations possible for a light ray–either with the electric field perpendicular to the page or in the page's plane. For the wave-guiding reflections shown, these correspond to S- and P-polarizations, respectively. When we get to waveguide modes, we will see that S-polarization corresponds to TE (transverse electric) modes whereas P-polarization gives TM (transverse magnetic) modes. For the latter, the zig-zag picture confirms that the electric field has a component in the z direction and so is not orthogonal to the overall z direction of propagation. Frequently, the loss in a beam as it propagates through a waveguide is dominated by surface interactions: most notably, scattering due to surface roughness. Ray optics correctly predicts these to be smaller for lower order modes both because there are fewer reflections and because near-grazing reflections are relatively insensitive to surface roughness.

Perhaps surprising–but very convenient–is that a ray optics picture can accurately describe individual waveguide modes and be used for waveguides whose width, shown as $2a$ in Fig. 13.9, is a fraction of an optical wavelength. This case is certainly not in keeping with the assumptions behind geometrical optics. As an example, consider Fig. 13.9(a). Here we have indicated a zig-zagging ray which now is to be identified with an individual mode. As shown in Fig. 13.9(b), each mode has an associated wavevector, $\vec{k}_1$ whose magnitude is $k_0 n_1 = 2\pi n_1/\lambda_0$ where n_1 is the core's refractive index and $\lambda_0 = 2\pi c/\omega$ is the free space wavelength. As found for microwave waveguides, central to describing a mode is the wavevector's projection in the direction of propagation, the propagation constant, β. One can thus speak of the mode's effective index of refraction, $n_m = \beta/k_0$ so that $v_p = c/n_m$ is the speed of a point of constant phase along the waveguide wall.[16]

In deriving ray optics from wave optics, we use a wave equation (discussed earlier in Section 12.2) solution having the form

$$f(\vec{r}, t) = a(\vec{r})\, e^{ik_0 \psi(\vec{r},t)} \tag{13.58}$$

The eikonal, ψ, allows a phase to be assigned along a ray. For a wave propagating in a uniform medium, $k_0\psi = \vec{k}\cdot\vec{r} - \omega t$. We drop the time dependence in what follows. To describe an allowed propagating mode, we insist that a complete cycle of the zig-zag–for example, from bottom to top to bottom–corresponds to the transverse contribution to the eikonal increasing by an integral multiple of 2π. To see why this works, we refer to the lower part of Fig. 13.9(a). The field in the core of the waveguide can be viewed as the superposition of two traveling waves, one moving generally upward to the right. Also drawn are the perpendiculars to these wavefronts, shown as zigzagging rays. For these traveling waves, ray optics is perfectly suited and the length of the zigs and zags correspond to the incremental phases. For a wave to propagate coherently down a waveguide, the guide must produce such a superposition in its core. In particular, in the x direction (tranverse), the waveguide must form a standing wave–or else the downstream waves will not add constructively. For a standing wave in the x dimension, the product of the waveguide's width, $2a$, and the transverse component of the wavevector, k_x, must equal an

[16] This should not be confused with $v_s = (\omega/k_1)\cos\theta = (c/n_1)\cos\theta$, the speed a signal propagates down the waveguide.

integral multiple of π. Or equivalently, the round-trip transverse phase of the eikonal must increase by an integral multiple of $2\pi/k_0$. In our notation:

$$k_0\delta\left(\psi_{rt}\right)_x = k_x \cdot 4a = k_1 \sin\theta \cdot 4a = m \cdot 2\pi \tag{13.59}$$

where higher order modes have increasing numbers of nodal surfaces along the x direction. Importantly, the reflections at the core-cladding interfaces add additional phase. These are readily found from the Fresnel coefficients (**see Discussion 13.3**). For total internal reflections these phases are referred to as Goos–Hanchen shifts. The shifts differ for S- and P-polarizations. For the former, corresponding to TE waveguide modes, at the 1–2 interface, we have:

$$k_0\delta\psi_{12} = -2\arctan\left(\frac{n_1^2 - n_2^2}{n_1^2\sin^2\theta} - 1\right)^{1/2} \tag{13.60}$$

The condition on the mode angle for propagation in the waveguide requires the total round trip phase change by an integral number of 2π:

$$k_0\delta\psi_{tot} = k_0\left(\delta\psi_{rt} + \delta\psi_{12} + \delta\psi_{13}\right) = m \cdot 2\pi$$
$$\Rightarrow ak_1\sin\theta - \arctan\left(\frac{n_1^2 - n_2^2}{n_1^2\sin^2\theta} - 1\right)^{1/2} = \frac{m\pi}{2} \tag{13.61}$$

where the second equation corresponds to a TE mode of a symmetric waveguide in which $\delta\psi_{12} = \delta\psi_{13}$. This may be conveniently rewritten:

$$\tan\left(ak_1\sin\theta - \frac{m\pi}{2}\right) = \left(\frac{n_1^2 - n_2^2}{n_1^2\sin^2\theta} - 1\right)^{1/2} \tag{13.62}$$

which, as a transcendental form, lends itself to a graphical solution. As an example of a planar waveguide, we consider a 2λ thick glass sheet ($n = 1.6$) in air. Figure 13.9(c) graphs both sides of the mode Eq. (13.62) and the propagating TE mode angles are then found as the intersections of two curves. In this case there are five propagating modes. The modes are of increasing order TE_0 to TE_4 going from left to right in the figure. From the form of the equation, and the nature of its graphical solution, it can be seen that there will always be a solution for the TE_0, regardless of how thin the waveguide becomes: the $m = 0$ solution on the left-hand side initially increases from zero to ∞ and the right-hand side is ∞ for $\sin\theta = 0$ and decreases to zero.

Similarly, it can be shown that there is also always a TM_0 mode for symmetric waveguides. For an infinitesimally thin symmetric waveguide, the propagating modes have nearly all of the energy residing in the cladding. In contrast, thin asymmetric waveguides may have no propagating modes.

13.8.1 The three-layer planar waveguide: the wave solutions of Maxwell's equations

Simply put, the rigorous treatment of waveguides using Maxwell's equations amounts to a boundary value problem matching solutions of Helmholtz equations in the core and claddings at the interfaces. For planar 1D waveguides this can all be done analytically (**see Discussion 13.4**). We consider the system from the previous section as shown in Fig. 13.9(a). Layers 2 and 3, the cladding layers, are assumed to extend to $x = +\infty$ and $x = -\infty$, respectively, and light is to be confined and guided in layer 1 along the positive z direction. For dielectric waveguides it is necessary that $n_1 > n_2$, n_3. We assume here a relative permeability of $\mu_r = 1$, which is the case for most materials at optical frequencies. In each layer, the solution of Maxwell's equations is the wave equation for that layer where the wave speed is reduced by the layer material's index of refraction, n_i. For convenience, in what follows, we work with the relative permittivity or dielectric constant, ε_r. In layer j, $\varepsilon_{rj} = n_j^2$. We will consider the general case, where ε_{rj} can be complex and thereby automatically include conduction currents (see Chapters 9 and 10). This will allow us to describe systems with losses- metals, and perfect conductors- for which $\varepsilon_r \to -\infty$ as $\omega \to 0$. Maxwell's equations in each medium yield wave equations:

$$\nabla^2 \vec{E}(\vec{r}, t) = \frac{\varepsilon_{rj}}{c^2} \frac{\partial^2 \vec{E}}{\partial t^2} \quad \text{and} \quad \nabla^2 \vec{H} = \frac{\varepsilon_{rj}}{c^2} \frac{\partial^2 \vec{H}}{\partial t^2} \tag{13.63}$$

Monochromatic wave solutions with time dependence $\exp(-i\omega t)$ reduce the wave equation to Helmholtz equations:

$$\nabla^2 \vec{E}(\vec{r}) = -k_j^2 \vec{E} \quad \text{and} \quad \nabla^2 \vec{H} = -k_j^2 \vec{H} \tag{13.64}$$

where $k_j = \sqrt{\varepsilon_j}\omega/c$ for each layer. We look at solutions propagating in z with the planar symmetry shown; that is, they are independent of y. (Note, however, they may have y components.) For this case, Eqs. 13.43 become:

$$\vec{E}(\vec{r}) = \vec{\mathscr{E}}(x) \exp(i\beta z) \quad \text{and} \quad \vec{H}(\vec{r}) = \vec{\mathscr{H}}(x) \exp(i\beta z) \tag{13.65}$$

Substituting this solution into the Helmholtz equation gives

$$\frac{\partial^2 \vec{\mathscr{E}}}{\partial x^2} = \left(\beta^2 - k_j^2\right) \vec{\mathscr{E}}(x) \quad \text{and} \quad \frac{\partial^2 \vec{\mathscr{H}}}{\partial x^2} = \left(\beta^2 - k_j^2\right) \vec{\mathscr{H}}(x) \qquad (13.66)$$

As before, all of the components of the fields may be determined from their z components. Here, we rewrite Eqs. 13.47 in terms of the reduced fields $\vec{\mathscr{E}}$ and $\vec{\mathscr{H}}$ noting that, for example, $\nabla_T \mathscr{E}_z(x) \to \hat{x}\frac{d}{dx}\mathscr{E}_z(x) \equiv \hat{x}\mathscr{E}_z'(x)$. Also, for each layer, the vacuum permittivity is replaced by the dielectric permittivity, $\varepsilon_0 \to \varepsilon_{rj}\varepsilon_0 \equiv \varepsilon_j$ and finally, we define $k_{jx} = \sqrt{k_j^2 - \beta^2}$ as the transverse, or $x-$directed, wavevector component. For the case of a planar waveguide, then, Eqs. 13.47 become:

$$\begin{aligned} \vec{\mathscr{E}}_T &= \frac{i}{k_{jx}^2}\left(\hat{x}\beta\mathscr{E}_z'(x) - \hat{y}\omega\mu_0\mathscr{H}_z'(x)\right) \\ \vec{\mathscr{H}}_T &= \frac{i}{k_{jx}^2}\left(\hat{x}\beta\mathscr{H}_z'(x) + \hat{y}\omega\varepsilon_j\mathscr{E}_z'(x)\right) \end{aligned} \qquad (13.67)$$

This naturally suggests the possibilities of pure *TE* and *TM* solutions with $\mathscr{H}_z$ non-zero and $\mathscr{E}_z$ zero for the *TE* case and vice-versa for *TM* solutions. For now, we will go forward by considering *TE* modes. For these, $\vec{\mathscr{E}}$ has only a y component and $\vec{\mathscr{H}}$ has an x component in addition to its longitudinal z component. Solutions in each layer consist of linear combinations of exponentials, $\exp(ik_{jx}x)$ and $\exp(-ik_{jx}x)$. In the core, for wave-guiding to work, ε_{r1} is dominantly real and $k_1^2 - \beta^2$ is positive and dominantly real so that the exponentials are essentially 1D traveling plane waves. Note the similarity of this wave picture with the zig-zag ray model. In the cladding regions, for ideal loss-free dielectrics, $k_j^2 - \beta^2$ is negative and the two wholly imaginary complex square roots, $\pm\sqrt{k_j^2 - \beta^2}$, describe solutions which exponentially decrease or increase with distance from the inferface. Only the first of these, the decreasing evanescent wave solution, makes for physically reasonable fields. We will indicate the root with the positive imaginary part by $\kappa_{jx} = \sqrt{k_j^2 - \beta^2} = i\sqrt{\beta^2 - k_j^2}$, where the Greek character is a reminder that we expect it to be an imaginary quantity. In summary, the *TE* solution for the z component of $\vec{\mathscr{H}}$ in a planar waveguide of width $2a$ is:

$$\mathscr{H}_z(x) = \begin{cases} A\exp(i\kappa_{2x}x) & x > a \\ B\exp(ik_{1x}x) + C\exp(-ik_{1x}x) & -a < x < a \\ D\exp(-i\kappa_{3x}x) & x < -a \end{cases} \qquad (13.68)$$

For the ideal dielectric waveguide, this is read as two plane waves sandwiched in the core–corresponding to the zigs and the zags–bounded by evanescent waves in the claddings. The problem now is one of matching boundary conditions to determine the constants, A, B, C, D, and, most importantly, β, which will in turn give all the k's and κ's. The boundary condition for $\mathscr{H}_z(x)$ is continuity at the interfaces. For example, at $x = a$

$$A\exp(i\kappa_{2x}a) = B\exp(ik_{1x}a) + C\exp(-ik_{1x}a) \tag{13.69}$$

The boundary condition for $\mathscr{H}_z'(x)$ in the TE solution is found from its relation to either $\mathscr{H}_x$ or $\mathscr{E}_y$ from Eq. 13.67, both of which are continuous at the 2–1 interface. Both boundary conditions are satisfied when:

$$A\frac{\exp(i\kappa_{2x}a)}{\kappa_{2x}} = B\frac{\exp(ik_{1x}a)}{k_{1x}} + C\frac{\exp(-ik_{1x}a)}{(-k_{1x})} \tag{13.70}$$

Similarly the 1–3 interface boundary conditions give two more equations. Counting variables, we now have expressed all of the boundary conditions with four equations and there are five unknowns, A, B, C, D, and β. Since the boundary conditions are all homogeneous linear equations in the first four of these, there is an overall normalization of the fields that, at this point, is free. It is convenient to combine the four boundary conditions in matrix form:

$$\begin{pmatrix} \exp(i\kappa_{2x}a) & -\exp(ik_{1x}a) & -\exp(-ik_{1x}a) & 0 \\ \frac{k_{1x}}{\kappa_{2x}}\exp(i\kappa_{2x}a) & -\exp(ik_{1x}a) & \exp(-ik_{1x}a) & 0 \\ 0 & \exp(-ik_{1x}a) & \exp(ik_{1x}a) & -\exp(-i\kappa_{3x}a) \\ 0 & \exp(-ik_{1x}a) & -\exp(ik_{1x}a) & -\frac{k_{1x}}{\kappa_{3x}}\exp(-i\kappa_{3x}a) \end{pmatrix} \begin{pmatrix} A \\ B \\ C \\ D \end{pmatrix} = 0 \tag{13.71}$$

where the first row of the matrix gives Eq. 13.69, the second row gives Eq. 13.70, and so on. Usually, the solution to this system of equations is $A = B = C = D = 0$. For there to be a non-trivial solution, the determinant of the matrix must equal zero. In general, this determinant is a lengthy expression, but for the symmetric waveguide, in which $\kappa_{3x} = \kappa_{2x}$, it gives a compact equation for TE modes[17]

$$2k_{1x}\kappa_{2x} - i\left(k_{1x}^2 + \kappa_{2x}^2\right)\tan(2k_{1x}a) = 0 \tag{13.72}$$

The equation may be solved for the remaining unknown, β. Graphically, we plot the magnitude of the left-hand side and each time it touches 0, there is an allowed mode. This is done in Fig. 13.10(a), part a, for the dielectric waveguide studied in the zigzag model: a $2a = 2\ \mu m$

[17] Specifically, for the symmetric waveguide case,

$$\det() = \frac{\exp(-2iak_{1x})}{k_{1x}^2} \times \left[(k_{1x} + \kappa_{2x})^2 - (k_{1x} - \kappa_{2x})^2\exp(4iak_{1x})\right]$$

and setting this to zero gives Eq. 13.72.

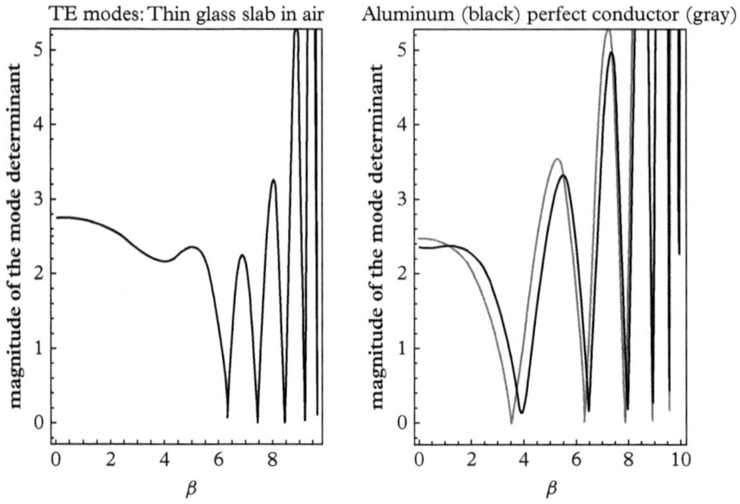

Fig. 13.10 *Results for planar waveguides. These figures show the left-hand side of Eq. 13.72. On the left is the glass-in-air dielectric waveguide studied in the previous section while on the right is the result for conducting claddings: perfect conducting (black) and aluminum (gray) coatings.*

glass core, sandwiched in air, guiding $\lambda = 1\ \mu$m waves. The five zeros correspond to the five curve intersections in Fig. 13.9(c). This time the lowest order mode is the rightmost in the figure, where β is largest. Even without referring to the ray optics picture, the angle associated with a mode may be defined by $\cos\theta = \beta/k_1$ and an effective refractive index for the mode, $n_m = \beta/k_0 = \beta\lambda_0/(2\pi)$.

The left-hand side of Fig. 13.10 shows the results for a planar waveguide with perfectly conducting cladding. The mode equation becomes $\tan(2k_{1x}a) = 0$ or $2k_{1x}a = m\pi$. This excludes the Goss–Hanchen effect and is exactly the condition of Eq. 13.59 that has a graphical interpretation shown in Fig. 13.9(b). The perfect conductor is compared, in the right-hand side of Fig. 13.10, with the result of a real waveguide made by coating the thin glass core with aluminum. Note that the minima for the aluminum coated waveguide are slightly shifted and do not reach all the way to zero. This is because the conduction losses in the aluminum coating lead to a decay in the wave going downstream in the waveguide: β becomes complex. For example, the minimum near $\beta = 6.4$ in the figure corresponds to a root in the complex plane for a propagation constant, $\beta = 6.432 + 0.0242i$. The imaginary part of the propagation constant corresponds to a decrease in power by a factor of $1/e^2$ for every $\delta = (1/0.0242)\lambda$ distance traveled. For the $\lambda = 1\mu$m of the calculation, this works out to a decay distance for the wave of about 42 μm. In contrast, for a communication quality fiber optic, this distance is several kilometers. Even the best real metal clad optical waveguides are woefully inferior to dielectric waveguides.

Finding detailed expressions for the modes is an exercise in solving linear equations. Solving the equations in 13.71 for a symmetric waveguide, we find (**see Discussion 13.5**):

$$B = \frac{A}{2}\left(1+\frac{k_{1x}}{\kappa_{2x}}\right)\exp\left(i\kappa_{2x}a - ik_{1x}a\right) \quad C = \frac{A}{2}\left(1-\frac{k_{1x}}{\kappa_{2x}}\right)\exp\left(i\kappa_{2x}a + ik_{1x}a\right)$$

$$\textit{and} \quad D = A\exp\left(-2ik_{1x}a\right) \qquad (13.73)$$

Qualitatively, there are a finite number of *TE* modes all with y-directed transverse electric fields and magnetic fields in the x–z plane. Higher order modes have increasing numbers of nodes in the core and increasingly steeper decays in the cladding regions. We are left with an overall normalization constant, A, which, for convenience, is frequently defined such that 1 watt of power flows per unit width, in y, of the waveguide. For a mode m, the power per unit width in y flowing through the waveguide is:

$$\frac{1}{2}\int_{-\infty}^{+\infty}\left(\vec{E}^m \times \vec{H}^{m*}\right)_z dx = -\frac{1}{2}\int_{-\infty}^{+\infty} E_y^m H_x^{m*}\,\hat{z}\,dx \qquad (13.74)$$

We consider the lossless case and describe the evanescent wave in the cladding by $\kappa_{2x} = i\gamma$ where γ is a positive real number. Combining Eqs. 13.67, 13.68, and 13.73 and carrying out the integrals in Eq. 13.74 yields (**see Discussion 13.6**):

$$A = \gamma k_{1x}\exp\left[\gamma a\right]\left[\frac{2\gamma}{\beta\omega\mu_0\left(1+a\gamma\right)\left(k_{1x}^2+\gamma^2\right)}\right]^{1/2} \qquad (13.75)$$

Similarly, it is possible to find the *TM* modes following a treatment parallel to this using $\mathscr{E}_z$ in place of $\mathscr{H}_z$.

13.8.2 Fiber optics: the step-index circular waveguide

We next turn to cylindrical waveguides which, in the optical regime, are optical fibers. The approach is similar to the previous section with a few new features that come about due to the cylindrical symmetry of the problem:

(1) Write vector Helmholtz equations for the core and cladding. Everywhere use "imaginary exponentials," or their cylindrical-coordinates analog, in the field expressions.

(2) Express the transverse fields in terms of the longitudinal, z components, as in Eq. 13.67.

(3) Apply boundary conditions to find solutions for the longitudinal components. In particular, express the boundary conditions as a matrix equation similar to Eq. 13.71.

(4) Find the propagation constants for the allowed modes by setting the determinant of the matrix to zero from the previous step.

We again allow the relative permittivity (dielectric constant) to be a complex quantity to provide a unified treatment of lossless and lossy dielectrics as well as metals and perfect conductors. Then, the expressions we find for lossless optical fibers equally well describe circular cylindrical metallic microwave waveguides. We consider a circular cross sectioned cylindrical waveguide that is infinitely long in $+z$, the propagation direction. The core, with dielectric constant $\varepsilon_{r1} = n_1^2$ and radius a is surrounded by a cladding with $\varepsilon_{r2} = n_2^2$. The cladding fields are evanescent and significant for only a very short distance from the interface. However, for the purpose of calculation, it is useful to model the cladding as extending to infinity. In each region, the harmonic solutions for the fields are solutions of the Helmholtz equation with the appropriate index of refraction. The Laplacian operator in cylindrical coordinates acting on a vector field sometimes creates mathematical difficulties because the radial and angle direction vectors are themselves functions of position. Fortunately, the z components of the field are well behaved in this respect. We look for solutions of the form:

$$\begin{aligned} \vec{E}(\vec{r}) &= \vec{\mathscr{E}}(\rho,\phi)\exp[i\beta z] \\ \vec{H}(\vec{r}) &= \vec{\mathscr{H}}(\rho,\phi)\exp[i\beta z] \end{aligned} \tag{13.76}$$

Time dependence could be included by adding factors of $\exp(-i\omega t)$. The transverse gradient in cylindrical coordinates is given by $\nabla_T = \hat{\rho}\frac{\partial}{\partial\rho} + \hat{\phi}\frac{1}{\rho}\frac{\partial}{\partial\phi}$ and the transverse field components are determined from the z components of the fields using Eq. 13.47:

$$\begin{aligned} \mathscr{E}_\rho &= \frac{i\beta}{k_j^2-\beta^2}\left(\frac{\partial}{\partial\rho}\mathscr{E}_z + \frac{\omega\mu_0}{\beta}\frac{\partial}{\rho\partial\phi}\mathscr{H}_z\right) \\ \mathscr{E}_\phi &= \frac{i\beta}{k_j^2-\beta^2}\left(\frac{\partial}{\rho\partial\phi}\mathscr{E}_z - \frac{\omega\mu_0}{\beta}\frac{\partial}{\partial\rho}\mathscr{H}_z\right) \\ \mathscr{H}_\rho &= \frac{i\beta}{k_j^2-\beta^2}\left(\frac{\partial}{\partial\rho}\mathscr{H}_z - \frac{\omega\varepsilon_j}{\beta}\frac{\partial}{\rho\partial\phi}\mathscr{E}_z\right) \\ \mathscr{H}_\phi &= \frac{i\beta}{k_j^2-\beta^2}\left(\frac{\partial}{\rho\partial\phi}\mathscr{H}_z + \frac{\omega\varepsilon_j}{\beta}\frac{\partial}{\partial\rho}\mathscr{E}_z\right) \end{aligned} \tag{13.77}$$

Here, $k_j = n_j\omega/c$ with $j = 1, 2$ for the core and cladding regions, respectively, and $\varepsilon_j = \varepsilon_{rj}\varepsilon_0$.[18] For the mode to be propagating and confined in the core, necessarily $k_1 > \beta > k_2$. Assuming the form of the solutions in Eq. 13.76, the Helmholtz equations for the reduced, transverse fields in cylindrical coordinates are:

$$\left[\frac{\partial^2}{\partial\rho^2} + \frac{1}{\rho}\frac{\partial}{\partial\rho} + \frac{1}{\rho^2}\frac{\partial^2}{\partial\phi^2} + \left(k_j^2 - \beta^2\right)\right]\left[\begin{array}{c}\mathscr{E}_z \\ \mathscr{H}_z\end{array}\right] = 0 \qquad (13.78)$$

These equations are separable in ρ and ϕ. Allowing $\mathscr{F}_z$ to stand in for either $\mathscr{E}_z$ or $\mathscr{H}_z$, we look for solutions of the form $\mathscr{F}_z(\rho, \phi) = R_{jz}(\rho)\,\Phi(\phi)$. Taking $-l^2$ as the separation constant, the angular equation is:

$$\frac{1}{\Phi}\frac{\partial^2\Phi}{\partial\phi^2} = -l^2 \qquad (13.79)$$

with solution,

$$\Phi(\phi) = \exp[\pm il\phi]\,. \qquad (13.80)$$

Here, l must be an integer for the wave function to be single valued and, mathematically at least, its role is very similar to that of orbital angular momentum in quantum mechanics. The ρ equation, then, is Bessel's equation (**see Discussion 13.7**):

$$\left[\frac{\partial^2}{\partial\rho^2} + \frac{1}{\rho}\frac{\partial}{\partial\rho} + \left(k_j^2 - \beta^2 - \frac{l^2}{\rho^2}\right)\right]R_{jz}(\rho) = 0 \qquad (13.81)$$

In the core region, $j = 1$ and $k_1^2 - \beta^2$ is positive and $k_t = \sqrt{k_\rho^2 + k_\phi^2} = \sqrt{k_1^2 - \beta^2}$ is the transverse component of the wavevector.[19] For the function to be finite at the origin, l must be non-negative and the radial solution must be a Bessel function of the first kind.[20] Thus, in the core:

$$\begin{aligned}\mathscr{E}_z(\rho, \phi) &= \mathscr{E}_0^{1,l}\mathcal{J}_l(k_t\rho)\exp[il\phi] \\ \mathscr{H}_z(\rho, \phi) &= \mathscr{H}_0^{1,l}\mathcal{J}_l(k_t\rho)\exp[il\phi]\end{aligned} \qquad (13.82)$$

where $\mathscr{E}_0^{1,l}$ and $\mathscr{H}_0^{1,l}$ are constants to be determined. In the cladding, $j = 2$ and $k_2^2 - \beta^2$ are negative for a loss-free dielectric. To work parallel to the previous treatment of planar waveguides, we define $\kappa_t = \sqrt{k_2^2 - \beta^2}$ and expect κ_t to be imaginary for a pure dielectric cladding. For an exponential fall off in the cladding region, the Hankel

[18] Note also, that in many texts Eqs. 13.77 are shown with minus signs. This difference is a result of their having a time dependence of $\exp(i\omega t)$ rather than our $\exp(-i\omega t)$.

[19] If we further identify l/ρ as the wavevector component in the ϕ direction, $l/\rho \equiv k_\phi$, then the wavevector component in the ρ direction is $k_\rho = \sqrt{k_1^2 - \beta^2 - (l/\rho)^2}$ and Bessel's equation becomes

$$\left[\frac{\partial^2}{\partial\rho^2} + \frac{1}{\rho}\frac{\partial}{\partial\rho} + k_\rho^2\right]R_{jz}(\rho) = 0$$

[20] If we were strictly following the planar waveguide pattern, we would write the core field as a sum of the two Hankel functions and then "discover" that to satisfy the boundary condition at the origin, it is necessary to take the linear combination of these leading to the corresponding $\mathcal{J}_l$.

function of the first kind, $H_l^{(1)}(\kappa_t\rho)$, is taken for the radial dependence. This is an exponentially decreasing function of distance for a positive imaginary argument, κ_t. Thus, in the cladding,

$$\begin{aligned}\mathscr{E}_z(\rho,\phi) &= \mathscr{E}_0^{2,l}H_l^{(1)}(\kappa_t\rho)\exp[il\phi]\\ \mathscr{H}_z(\rho,\phi) &= \mathscr{H}_0^{2,l}H_l^{(1)}(\kappa_t\rho)\exp[il\phi]\end{aligned} \tag{13.83}$$

with constants $\mathscr{E}_0^{2,l}$ and $\mathscr{H}_0^{2,l}$ to be determined. With the longitudinal (z) components in hand, we can now use Eq. 13.77 to express the transverse components of all the fields. For example, in the core, using Eqs. 13.82, the transverse components of Eqs. 13.77 are,

$$\begin{aligned}\mathscr{E}_\rho^{1,l} &= \frac{i\beta}{k_t^2}\left(\mathscr{E}_0^{1,l}k_t\mathcal{J}_l'(k_t\rho)+\mathscr{H}_0^{1,l}\frac{il\omega\mu_0}{\beta\rho}\mathcal{J}_l(k_t\rho)\right)\exp(il\phi)\\ \mathscr{E}_\phi^{1,l} &= \frac{i\beta}{k_t^2}\left(\mathscr{E}_0^{1,l}\frac{il}{\rho}\mathcal{J}_l(k_t\rho)-\mathscr{H}_0^{1,l}\frac{\omega\mu_0}{\beta}k_t\mathcal{J}_l'(k_t\rho)\right)\exp(il\phi)\\ \mathscr{H}_\rho^{1,l} &= \frac{i\beta}{k_t^2}\left(\mathscr{H}_0^{1,l}k_t\mathcal{J}_l'(k_t\rho)-\mathscr{E}_0^{1,l}\frac{il\omega\varepsilon_1}{\rho\beta}\mathcal{J}_l(k_t\rho)\right)\exp(il\phi)\\ \mathscr{H}_\phi^{1,l} &= \frac{i\beta}{k_t^2}\left(\mathscr{H}_0^{1,l}\frac{il}{\rho}\mathcal{J}_l(k_t\rho)_z+\mathscr{E}_0^{1,l}\frac{\omega\varepsilon_1}{\beta}k_t\mathcal{J}_l'(k_t\rho)\right)\exp(il\phi)\end{aligned} \tag{13.84}$$

where the primes indicate differentiation with respect to $k_t\rho$. The components in the cladding are obtained from these by taking $j=2$ everywhere and replacing $\mathcal{J}_l(k_t\rho)\to H_l^{(1)}(\kappa_t\rho)$. At the core-cladding interface, $\rho=a$ and the field components E_z, εE_ρ, H_z, and H_ρ must all be continuous, giving four equations. For example, the boundary condition for $E_z(\rho=a,\phi,z)$ yields the equation (referring to Eqs. 13.76, 13.82, and 13.83):

$$\mathscr{E}_0^{1,l}\mathcal{J}_l(k_ta)\exp[il\phi]\exp[i\beta z]-\mathscr{E}_0^{2,l}H_l^{(1)}(\kappa_ta)\exp[il\phi]\exp[i\beta z]=0 \tag{13.85}$$

Omitting the common exponentials for z and ϕ, these B.C. equations are written, respectively, as four rows of a matrix equation similar to that used in the planar waveguide example before (**see Discussion 13.8**):

$$\begin{pmatrix}\mathcal{J}_l(k_ta) & -H_l^{(1)}(\kappa_ta) & 0 & 0\\ \frac{i\varepsilon_1\beta}{k_t}\mathcal{J}_l'(k_ta) & \frac{-i\varepsilon_2\beta}{\kappa_t}H_l^{(1)\prime}(\kappa_ta) & \frac{-\varepsilon_1\omega\mu_0 l}{k_t^2a}\mathcal{J}_l(k_ta) & \frac{\varepsilon_2\omega\mu_0 l}{\kappa_t^2a}H_l^{(1)}(\kappa_ta)\\ 0 & 0 & \mathcal{J}_l(k_ta) & -H_l^{(1)}(\kappa_ta)\\ \frac{\varepsilon_1\omega l}{k_t^2a}\mathcal{J}_l(k_ta) & \frac{-\varepsilon_2\omega l}{\kappa_t^2a}H_l^{(1)}(\kappa_ta) & \frac{i\beta}{k_t}\mathcal{J}_l'(k_ta) & \frac{-i\beta}{\kappa_t}H_l^{(1)\prime}(\kappa_ta)\end{pmatrix}\begin{pmatrix}\mathscr{E}_0^{1,l}\\ \mathscr{E}_0^{2,l}\\ \mathscr{H}_0^{1,l}\\ \mathscr{H}_0^{2,l}\end{pmatrix}=0 \tag{13.86}$$

There are five unknowns at this point, β, $\mathscr{E}_0^{1,l}$, $\mathscr{H}_0^{1,l}$, $\mathscr{E}_0^{2,l}$, and $\mathscr{H}_0^{2,l}$. For there to be a non-trivial solution for the fields, the determinant of the matrix must be zero. Setting this determinant equal to zero, and dividing by $\left(a\beta\mathcal{J}_l H_l^{(1)}\right)^2$ gives the equation[21] for the propagation constant, β (**see Discussion 13.9**):

$$\left(\frac{\mathcal{J}_l'(k_t a)}{k_t a \mathcal{J}_l(k_t a)} - \frac{H_l^{(1)\prime}(\kappa_t a)}{\kappa_t a H_l^{(1)}(\kappa_t a)}\right)\left(\frac{k_1^2 \mathcal{J}_l'(k_t a)}{k_t a \mathcal{J}_l(k_t a)} - \frac{k_2^2 H_l^{(1)\prime}(\kappa_t a)}{\kappa_t a H_l^{(1)}(\kappa_t a)}\right)$$
$$= (\beta l)^2 \left[\frac{1}{(\kappa_t a)^2} - \frac{1}{(k_t a)^2}\right]^2 \tag{13.87}$$

The special case of $l = 0$ corresponds to the incident wave passing through the fiber center between each bounce. That is, with no azimuthal component ($k_\phi = 0$), k_t goes to k_ρ and the wavevector maintains the same plane of incidence, relative to the fiber, for each reflection. These are known as "meridional" modes. Mathematically, with $l = 0$, Eqs. 13.84 and 13.87 simplify considerably and the matrix elements in Eq. 13.86 coupling the z (longitudinal) fields–elements 3 and 4 in row 2 and elements 1 and 2 in row 4-vanish. With the right-hand side of Eq. 13.87 set to zero, one of the terms in parentheses on the right-hand side must be zero. For the first one equal to zero:

$$\frac{\mathcal{J}_0'(k_t a)}{k_t a \mathcal{J}_0(k_t a)} = \frac{H_0^{(1)\prime}(\kappa_t a)}{\kappa_t a H_0^{(1)}(\kappa_t a)} \tag{13.88}$$

Equations 13.84 for the core ($j = 1$), along with the equivalent equations for the cladding ($j = 2$), may be satisfied for $\mathscr{E}_0^{1,0} = \mathscr{E}_0^{2,0} = 0$. The resulting solution is a pure *TE* mode with non-vanishing field components H_ρ, H_z, and E_ϕ. The allowed propagation constants for these modes can be expressed graphically, in a way similar to the previous, by plotting both sides of Eq. 13.88 as functions of β, as shown in the left-hand side of Fig. 13.11. As β decreases, $\kappa_t = \sqrt{k_2^2 - \beta^2}$ and $k_t = \sqrt{k_1^2 - \beta^2}$ act as increasing scale factors for the functions $\mathcal{J}_0'$ and $H_0^{(1)\prime}$. Because of the oscillatory nature of $\mathcal{J}_0'$, there are multiple values of β at which Eq. 13.88 is satisfied. Physically, these "allowed modes" correspond to different numbers, m, of radial oscillations of the field, from $\rho = 0$ to $\rho = a$, as determined by $\mathcal{J}_0(k_t\rho)$. The number of these propagating $TE_{0,m}$ modes is finite and, in contrast with the planar waveguide, possibly zero.

We label these modes in order of descending propagation constant, from right to left on the graph, TE_{01}, TE_{02}, and so on. A similar

[21] The literature predominantly shows this equation in terms of the modified Bessel function of the second kind given as, $K_l(\gamma a)$, where $i\gamma = \kappa_t$, rather than with the Hankel function, $H_l^{(1)}(\kappa_t a) = H_l^{(1)}(i\gamma a)$, as we have written it. These two forms are easily converted using the relation, $H_l^{(1)}(ix) = \frac{2}{\pi}\frac{1}{i^{l+1}}K_l(x)$.

analysis is possible for the $l = 0$, pure TM modes–for which $\mathscr{H}_0^{1,0} = \mathscr{H}_0^{2,0} = 0$ and E_ρ, E_z, and H_ϕ are non-vanishing. These occur, alternatively, upon the vanishing of the second term on the left-hand side of Eq. 13.87.

13.8.3 Higher order modes, single mode fibers, and dispersion

For $l \neq 0$, we have the so-called "skew" rays or modes. For these, we first note that their skewness precludes the transverse modes, *TE* and *TM*. Mathematically, the longitudinal (z) components of the two fields are coupled in rows 2 and 4 of Eq. 13.86; in order to satisfy all four boundary conditions, these modes will all have longitudinal components for both $\vec{E}$ and $\vec{H}$. Second, because of the angular term, all modes are doubly degenerate: $\pm l$ give two independent solutions for the fields of a mode. Physically, this corresponds to the degenerate symmetry of rays skewed equally left or right. Eq. 13.87 may be recognized as being quadratic in $\mathcal{J}_l'/k_t a\mathcal{J}_l$ and we can solve that equation for this quantity:

$$\frac{\mathcal{J}_l'(k_t a)}{k_t a\mathcal{J}_l(k_t a)} = \left(\frac{k_2^2+k_1^2}{2k_1^2}\right)\frac{H_l^{(1)\prime}(\kappa_t a)}{\kappa_t a H_l^{(1)}(\kappa_t a)}$$
$$\pm\sqrt{\left(\frac{k_2^2-k_1^2}{2k_1^2}\right)^2\left(\frac{H_l^{(1)\prime}(\kappa_t a)}{\kappa_t a H_l^{(1)}(\kappa_t a)}\right)^2+\left(\frac{\beta l}{k_1}\right)^2\left(\frac{1}{(\kappa_t a)^2}-\frac{1}{(k_t a)^2}\right)^2} \tag{13.89}$$

For $l \neq 0$, we again get two types of modes given by the two roots of the quadratic. However, the modes are no longer pure transverse electric or magnetic–both fields have z components. The modes of the two solutions are designated *EH* or *HE* depending on which field has the dominant z contribution: if E_z makes the more significant contribution, the mode is labelled *EH* and modes with larger H_z are *HE* modes. The mode label has subscripts l, m, where l is the angular index and, as mentioned above, $m = 1, 2, 3...$ is the mode of a particular type in order of decreasing β or increasing number of wiggles in the transverse direction. Equation 13.89 may be solved graphically by again plotting the left- and right-hand sides. Alternatively, one can simply solve the equation numerically for the different modes.

Fibers used in communication are engineered to minimize losses and dispersion. Current state-of-the-art fibers are designed for operation at 1550 nm where the bulk attenuation of glass has a minimum.

Single mode fibers with attenuation less than 0.2 dB/km is now common. A representative fiber consists of a core of SiO_2 doped with several percent GeO surrounded by a cladding of SiO_2 that is often doped with fluoride. The relative indices are small for reasons discussed next. We consider, as a typical example, a $n_1 = 1.4475$ core surrounded by a $n_2 = 1.4440$ cladding. The right-hand side of Fig. 13.11 shows results for the allowed modes of this combination as a function of fiber diameter. We make the important observation that the HE_{11} mode has no cutoff–it propagates in fibers of any diameter. To minimize dispersion, transoceanic cables use fibers with only a single propagating mode. This is accomplished by designing a fiber with a suitably small core. By working with a cladding whose index is only slightly smaller than the core, as seen in the figure, the fiber diameter can be much larger than the wavelength of the light being carried and still be single mode.[22] Currently, the most commonly used single mode commercial fiber is called 9/125 that refers to a 9 μm diameter core surrounded by a 125 μm cladding. From the figure, 9 μm is just below the cutoff diameter for the second mode, TE_{02}. There are a couple of reasons for making the fiber diameter as large

[22] In such case, the closeness of the two indices results in a critical angle that restricts the incident ray to being nearly longitudinal, where β is only slightly less than k_1.

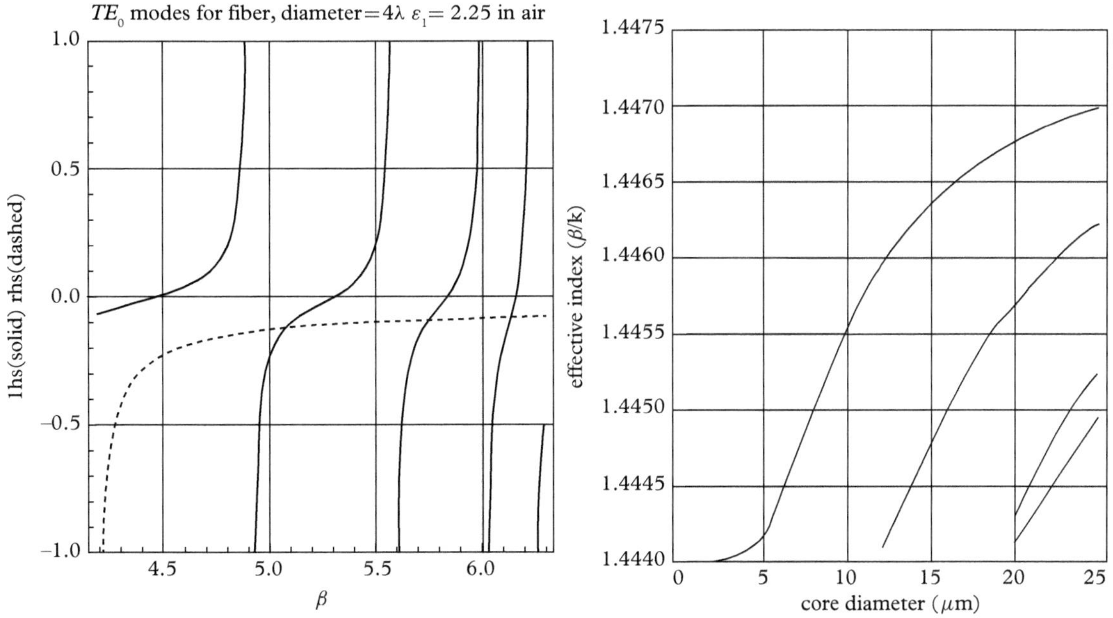

Fig. 13.11 *Left: Graphical determination of the propagation constants, β, for TE_{0m} modes, by plotting both sides of Eq. 13.88 as functions of β. The three TE modes occur for the curve crossings. Right: Propagation constants for several modes of a step index fiber plotted as a function of the core diameter. From left to right: HE_{11}, TE_{01}, TM_{01}, and HE_{21}.*

as possible. First, larger diameters lead to reduced waveguide losses from scattering at the core-cladding interface: in the zigzag ray picture, there are fewer bounces per unit length. A larger diameter core also allows more signal power to be coupled into the waveguide. More power is better because it allows for more attenuation in a fiber before the signal to noise requires a stage of amplification. However, the power in the fiber is limited by non-linearities in the wave propagation that result as the intensity in the core gets too high. Signal levels in modern transoceanic cables require an amplifier every 100 km or so.

The phase velocity, v_p, is not the same as the speed, v_s, a signal propagates down the waveguide. As described in Section 13.8, the phase velocity, along the z-direction, of a monochromatic wave propagating in a single mode of a waveguide is $v_p = \omega/\beta = \omega/(k_0 n_m) = c/n_m$. This motivates defining an "effective index" for the mode, $n_m(\omega) = \beta/k_0$. On the other hand, the signal speed along z for a light wave propagating in the waveguide is given by $v_s = (\omega/k_1)\cos\theta$. From $\beta = \sqrt{k_1^2 - k_t^2} = \sqrt{\varepsilon_1\mu_0\omega^2 - k_t^2}$, where k_t is a constant for a given mode, the effective index, $n_m(\omega) = \beta(\omega)/k_0$, is seen to be an increasing function of frequency. Thus, even a signal propagating in a single mode waveguide will get distorted via dispersion because of this variation in the effective index of refraction with frequency. But the effect is much less than for a multimode waveguide. And this is not the complete story: the right-hand side of Fig. 13.11 was generated assuming that the material indices of refraction for the core and cladding are spatially constant. This is generally not the case and so another source of dispersion is that of the materials. In a "dispersion-shifted fiber," however, these two sources of dispersion are engineered to cancel one another, allowing the fiber to transmit a larger bandwidth. Finally, we note here that the HE_{11} mode, with $l = \pm 1$, represents a polarization degree of freedom that is currently unexploited. The downside is that residual birefringence in the fiber can split these modes and appear as dispersion of a signal.

The most widely used multimode fibers are made of silica and labelled 50/125 and 62.5/125. This designation gives the diameters, in μm, of the fiber core and the cladding. The most commonly used wavelength bands are described by "telecom windows." The first window is from $\lambda = 800$–900 nm and the second is around $\lambda = 1.3$ μm where the material dispersion of the glass is typically the lowest. The third window is at $\lambda = 1.5$ μm where losses are lowest. Typically, between 200 and 800 modes are active and ray optics provides a suitable description for propagation in this regime. Rays passing through the origin correspond to $l = 0$ modes: only $\mathcal{J}_0$ Bessel functions are non-zero at the origin. While modal dispersion sharply limits the bandwidth and distances for communication, this effect can be greatly reduced by using graded index fibers. The basic idea is to fabricate the core of a fiber such that its index of refraction decreases with increasing ρ. Typically,

$$n(\rho) = n_1 \left[1 - 2\Delta \left(\frac{\rho}{a}\right)^g\right]^{1/2} \tag{13.90}$$

with the most common profile, referred to as a "parabolic profile," having $g = 2$. To treat profiles in ray optics, we recall the eikonal equation from Section 12.2:

$$(\nabla\psi)^2 = \frac{n^2(\rho)}{c^2}\omega^2 \tag{13.91}$$

In the geometric optics picture, the light of a ray is sped up as it travels significantly off axis with the net effect being a continuous lensing of the beam. Modal spreading of a signal in a well engineered graded index fiber can be reduced two to three orders of magnitude relative to a similar step index fiber.

13.9 Photonic crystals

The basic idea behind photonic crystals is to apply the same quantum mechanical and solid state physics formalisms and techniques used in describing electronic properties of materials–especially electronic band structure–to analogous systems for photons. Such systems generally consist of devices built up from periodic structures of varying dielectric constant and are referred to as "photonic" systems. A general difference between electronic and photonic systems is that where condensed matter physics concerns itself with materials whose unit cells and corresponding electronic de Broglie wavelengths are on the order of the size of an atom, a few tenths of a nanometer, photonic crystals will have unit cells on the scale of the wavelength of light–several hundreds of nanometers.[23] Fortunately, there are several technologies for controllably making structures with dielectric functions that vary periodically in 1D, 2D, or 3D over that length scale. While solutions to the 2D and 3D systems require numerical modeling, the 1D photonic crystal, which shows most of the basic ideas, can be solved analytically. We will now work through finding the "band structure" of such a device.

[23] Compare this definition of photonic crystals with the related field of "metamaterials" in which the atomic composition of materials is manipulated to affect the dielectric constant and index of refraction in novel ways.

We begin by establishing an optical analog to the time-independent Schrodinger equation of quantum mechanics. The Schrodinger equation presents a linear operator eigenvalue problem:

$$\hat{\mathscr{H}}\Psi_a = E_a\Psi_a \tag{13.92}$$

where the Hamiltonian operator has the form:

$$\hat{\mathscr{H}} = -\frac{\hbar^2}{2m}\nabla^2 + V \tag{13.93}$$

and the eigen-wavefunction, Ψ_a, typically has dimensions given by the spin of the system. Analogous to the electronics system, which consists of an electron wavefunction in a region of varying potential, $V(\vec{r})$, the photonic system consists of a radiation field in a region of varying dielectric function, $\varepsilon_r(\vec{r})$. The photonic analog to Schrodinger's equation is the Master equation for the magnetic field:

$$\hat{\Theta}\vec{H}_a = \left(\frac{\omega_a^2}{c^2}\right)\vec{H}_a \tag{13.94}$$

This convenient form holds only for the magnetic field within a system with a spatially varying dielectric function, $\varepsilon_r = \varepsilon_r(\vec{r})$, and a constant permeability, $\mu \simeq \mu_0$. We define the operator,

$$\hat{\Theta} = \nabla \times \left[\frac{1}{\varepsilon_r(\vec{r})}\nabla\times\right] \tag{13.95}$$

with eigenvalues, ω_a^2/c^2 and vector-valued eigenfunctions, $\vec{H}_a$.[24] The inner product is defined as usual:

$$\left\langle \vec{H}_1 | \vec{H}_2 \right\rangle = \int d^3r \vec{H}_1^*(\vec{r}) \cdot \vec{H}_2(\vec{r}) \tag{13.96}$$

The proof of orthogonality of two eigenfunctions having different eigenvalues follows nearly identically to that in quantum mechanics. We next consider a system with a periodically varying dielectric constant. For what follows, we will now outline a 1D case. Our system (Fig. 13.12) varies periodically in z with period R (R=2a in the figure) so that $\varepsilon_r(z) = \varepsilon_r(z+R)$. We define a translation operator, $\hat{T}_R$, that mathematically displaces the solution a distance R in z. Acting on an eigenfunction of the system, this operator yields $\hat{T}_R\vec{H}_a(\vec{r}) = \vec{H}_a(\vec{r} - R\hat{z})$. Thus, because the translated function must also be an eigenfunction of the system, with the same eigenvalue, we obtain,

[24] From the two macroscopic Maxwell's equations in the absence of free current or charge,

$$\nabla \times \vec{H} = \varepsilon\frac{\partial \vec{E}}{\partial t} = -i\omega\varepsilon\vec{E}$$

and $$\nabla \times \vec{E} = -\mu\frac{\partial \vec{H}}{\partial t} = i\omega\mu\vec{H}$$

we divide the first equation by ε, take the curl, and substitute the second equation to get,

$$\nabla \times \left(\frac{1}{\varepsilon}\nabla \times \vec{H}\right) = -i\omega\nabla \times \vec{E} = \omega^2\mu\vec{H}$$

Multiplying both sides by the vacuum permittivity and approximating $\mu \simeq \mu_0$, we get,

$$\nabla \times \left(\frac{1}{\varepsilon_r}\nabla \times \vec{H}\right) = \omega^2\varepsilon_0\mu_0\vec{H} = \left(\frac{\omega}{c}\right)^2\vec{H}$$

where $\varepsilon_r = \varepsilon/\varepsilon_0$. Thus, we obtain the form of the eigenvalue equation,

$$\nabla \times \left[\frac{1}{\varepsilon_r(\vec{r})}\nabla\times\right]\vec{H} = \hat{\Theta}\vec{H} = \left(\frac{\omega}{c}\right)^2\vec{H}$$

$$\hat{\Theta}\hat{T}_R\vec{H}_a(\vec{r}) = \left(\frac{\omega_a^2}{c^2}\right)\hat{T}_R\vec{H}_a(\vec{r}) = \hat{T}_R\hat{\Theta}\vec{H}_a(\vec{r}) \tag{13.97}$$

And since this is true for any eigenfunction of the system, the commutator of the two operators vanishes, $\left[\hat{T}_R, \hat{\Theta}\right] = 0$. If the eigenvalues of the system are all non-degenerate, then the translated function can differ from the original by at most a phase:

$$e^{i\alpha(\vec{r})}\vec{H}_a(\vec{r}) = \hat{T}_R\vec{H}_a(\vec{r}) \tag{13.98}$$

In solid-state physics, one of the principal results of electronic band structure theory is Bloch's theorem that, in the present context, states that the eigenfunctions of a periodic system consist of an exponential multiplied by a function, $\vec{u}$, with the periodicity of the system:

$$\vec{H}_a(\vec{r}) = e^{ik_a z}\vec{u}_{k_a}(\vec{r}) \tag{13.99}$$

where, for the present system, $\vec{u}_{k_a}(\vec{r}) = \vec{u}_{k_a}(\vec{r} - R\hat{z})$. Equation 13.99 defines the Bloch state. An important fact is that two solutions, with two different wave numbers, k_a and k'_a, which are related by $k'_a = k_a + mb$, where $b = 2\pi/R$, are identical from a physical point of view:

$$\vec{H}_{a'}(\vec{r}) = e^{i(k_a+mb)z}\vec{u}_{k'_a}(\vec{r}) = e^{ik_a z}\vec{u}_{k_a}(\vec{r}) = \vec{H}_a(\vec{r}) \tag{13.100}$$

Here, the new periodic function, $\vec{u}_{k'_a}(\vec{r})$, has "absorbed" the $\exp(imbz)$ factor. The translation in k-space, "mb," is a 1D example of a reciprocal lattice vector with "b" identified as a primitive reciprocal lattice vector. Bloch states and mode frequencies are conveniently labelled by subtracting a suitable reciprocal lattice vector, mb, from a large "k'_a" to identify the corresponding solution within the Brillouin zone–that region of k-space with $-b/2 \le k_a \le b/2$.

A simple ideal 1D photonic crystal is shown in Fig. 13.12. Such a configuration consists of an infinite set of non-magnetic ($\mu = \mu_0$) dielectric planar layers, each of thickness a, stacked in the z direction with successive layers alternating between indices of refraction $n_1^2 = \varepsilon_{r1}$ and $n_2^2 = \varepsilon_{r2}$. We identify the odd and even numbered layers with the subscripts $j = 1$ and 2, respectively. The "bands" for this "crystal" are the normal modes propagating in the direction perpendicular to the layers, z. We will use the same approach that we have followed previously for solving waveguide and optical fiber problems–the four step program of Section 13.8.2 that focuses on the z-component of the fields with a couple of important differences. We begin by writing wave equations for each layer, ν, of the waveguide:

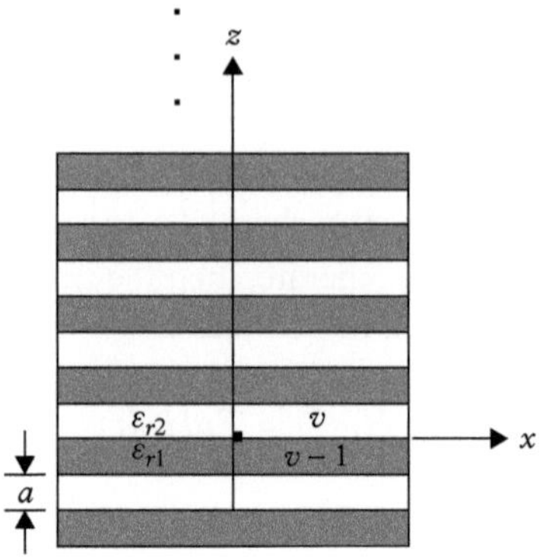

Fig. 13.12 *A 1D photonic crystal: A stack of layers with alternating dielectric function makes up a simple 1D photonic crystal.*

$$\nabla^2 \vec{E} + k_\nu^2 (z) \vec{E} = 0 \tag{13.101}$$

and similarly for $\vec{H}$. Here $k_\nu^2 = \varepsilon_{r\nu} (\omega/c)^2$ with $\varepsilon_{r\nu} = \varepsilon_{r1}$ or ε_{r2}, depending on the layer. We specify a solution by its wavevector, $\vec{k} = \left(k_x, k_{y,}, k_{\nu z}\right) = \left(\vec{k}_{\|}, k_{\nu z}\right)$, with solutions $\sim e^{ik_x x} e^{ik_y y} e^{ik_{\nu z}}$. For this system, in contrast to the above problem,

(1) To match boundary conditions over an entire *x*–*y* plane, the transverse exponentials, $e^{ik_x x} e^{ik_y y}$, will be the same for all waves.
(2) We use $k_{\nu z}$ instead of β since the "propagation constant" now is a function of the dielectric function of the layer given by $k_{\nu z} = \sqrt{k_\nu^2 - k_t^2}$ where $k_t = \sqrt{k_x^2 + k_y^2}$.
(3) Because of reflections at the interfaces, a second wave propagates in the $-z$ direction.

Summarizing, we are looking for solutions of the form:

$$\vec{E}^{\pm} (\vec{r}) = \vec{E_0}^{\pm} e^{ik_x x} e^{ik_y y} e^{\pm ik_z z} \quad \vec{H}^{\pm} (\vec{r}) = \vec{H_0}^{\pm} e^{ik_x x} e^{ik_y y} e^{\pm ik_z z} \tag{13.102}$$

As before, the transverse components may be obtained from the longitudinal ones (Eq. 13.47), The forms of these equations are a little different for the $+z$ and $-z$ directed waves:

$$\vec{E}_{\nu T}^{\pm} = \frac{i}{k_t^2} \left(\pm k_{\nu z} \nabla_T E_z^{\pm} - \omega \mu_0 \hat{z} \times \nabla_T H_z^{\pm}\right)$$
$$\vec{H}_{\nu T}^{\pm} = \frac{i}{k_t^2} \left(\pm k_{\nu z} \nabla_T H_z^{\pm} - \omega \varepsilon_\nu \hat{z} \times \nabla_T E_z^{\pm}\right) \tag{13.103}$$

We specialize now to look for a *TE* solution; this requires finding H_z. We write for each layer:

$$H_z^+(\vec{r}) = A_\nu e^{ik_x x} e^{ik_y y} e^{+ik_z z} \quad H_z^-(\vec{r}) = B_\nu e^{ik_x x} e^{ik_y y} e^{-ik_z z} \tag{13.104}$$

Thus, the problem has been reduced to finding the various A_ν's and B_ν's by applying boundary conditions. Referring to the figure, we consider the layers, $\nu-1$, ν, and $\nu+1$ which are conveniently located near the origin of coordinates.[25] Continuity of H_z at the two interfaces gives:

$$A_{\nu-1} + B_{\nu-1} = A_\nu + B_\nu \tag{13.105}$$

$$A_\nu e^{ik_{2z}a} + B_\nu e^{-ik_{2z}a} = A_{\nu+1} e^{ik_{1z}a} + B_{\nu+1} e^{-ik_{1z}a} \tag{13.106}$$

Inspection of Eq. 13.103 shows that this condition will also satisfy continuity of $\vec{E}$ across to interfaces. We get two additional independent equations by requiring continuity of $\vec{H}_T$:

$$k_{1z}(A_{\nu-1} - B_{\nu-1}) = k_{2z}(A_\nu - B_\nu) \tag{13.107}$$

$$k_{2z}\left(A_\nu e^{ik_{2z}a} - B_\nu e^{-ik_{2z}a}\right) = k_{1z}\left(A_{\nu+1} e^{ik_{1z}a} - B_{\nu+1} e^{-ik_{1z}a}\right) \tag{13.108}$$

We have four equations and seven unknowns: the A's and B's and k_t. The latter can be used to find k_{1z} and k_{2z}. At this point, we use the Bloch theorem, which shows that for an infinite periodic system, the wave solutions displaying the symmetry of the system have the form:

$$f(z + 2a) = \exp\left[ik_{Bloch} 2a\right] f(z) \tag{13.109}$$

where k_{Bloch}, the Bloch vector, is to be determined from the problem. For our purposes, we use this as an assumption and see if it works. For the present problem, the theorem implies that the coefficients for the $v-1$ and $v+1$ layers must be related by

$$A_{v+1} = A_{v-1} \exp\left[ik_{Bloch} 2a\right]$$

$$B_{v+1} = B_{v-1} \exp\left[ik_{Bloch} 2a\right]$$

which gives two more equations. As before, we move all terms to the left-hand side and write all six equations in matrix form:

[25] This example takes the x-axis to be the boundary between layer v-1 and layer v in Fig. 13.12.

$$\begin{pmatrix} 1 & 1 & -1 & -1 & 0 & 0 \\ 0 & 0 & e^{ik_{2z}a} & e^{-ik_{2z}a} & -e^{ik_{1z}a} & -e^{-ik_{1z}a} \\ k_{1z} & -k_{1z} & -k_{2z} & k_{2z} & 0 & 0 \\ 0 & 0 & k_{2z}e^{ik_{2z}a} & -k_{2z}e^{ik_{2z}a} & -k_{1z}e^{ik_{1z}a} & k_{1z}e^{-ik_{1z}a} \\ e^{i2k_{Bloch}a} & 0 & 0 & 0 & 1 & 0 \\ 0 & e^{i2k_{Bloch}a} & 0 & 0 & 0 & 1 \end{pmatrix} \begin{pmatrix} A_{\nu-1} \\ B_{\nu-1} \\ A_{\nu} \\ B_{\nu} \\ A_{\nu+1} \\ B_{\nu+1} \end{pmatrix}$$

Setting the determinant equal to zero gives the characteristic equation for k_{Bloch}:

$$\cos(2k_{Bloch}a) = \cos(k_{z1}a)\cos(k_{z2}a) - \frac{1}{2}\left(\frac{k_{z1}}{k_{z2}} - \frac{k_{z2}}{k_{z1}}\right) \times \sin(k_{z1}a)\sin(k_{z2}a) \tag{13.110}$$

Now the left-hand side of the equation necessarily lies between -1 and $+1$ and when the right-hand side is out of this range, the corresponding wave cannot propagate. By applying this criterion for different values of $k = \omega/c = 2\pi/\lambda_0$ and k_t, a band diagram may be generated. In Fig. 13.13, the bands are indicated in the light regions and propagation is not possible for wavevectors in the dark regions. From the graph, for a given frequency of light (the y-coordinate), a horizontal line at that level shows which directions of light (or cones of divergence) could propagate in the device. For those directions with dark regions on the graph, the crystal would act as a perfect

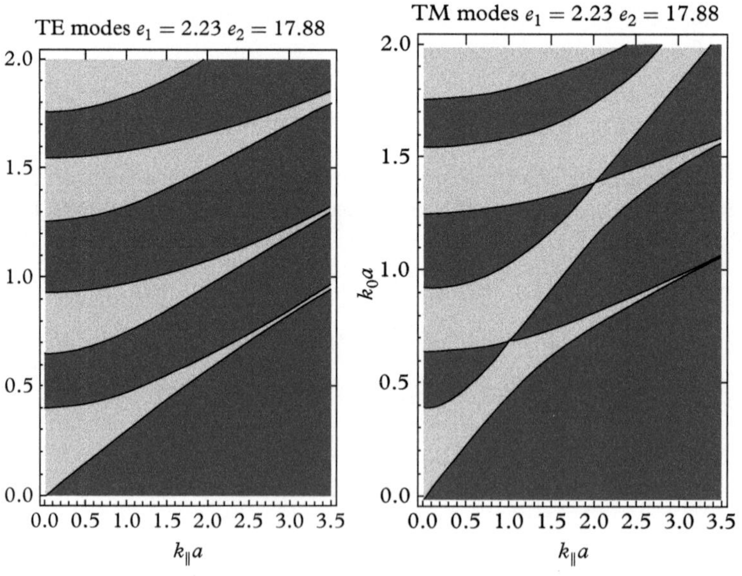

Fig. 13.13 *Band diagram for a 1D photonic crystal. The light swaths are the photonic bands for which light propagation is allowed.*

reflector. Similarly, TM modes are found in the same manner, but starting with $H_z = 0$ in the Eqs. of 13.10. And they ultimately satisfy the characteristic equation:

$$\cos(2k_{Bloch}a) = \cos(k_{z1}a)\cos(k_{z2}a) - \frac{1}{2}\left(\frac{\varepsilon_{r2}}{\varepsilon_{r1}}\frac{k_{z1}}{k_{z2}} - \frac{\varepsilon_{r1}}{\varepsilon_{r2}}\frac{k_{z2}}{k_{z1}}\right) \times \sin(k_{z1}a)\sin(k_{z2}a) \quad (13.111)$$

Exercises

(13.1) Repeat the Casimir force calculation for the case where there are significant thermal populations of cavity modes. What is the distance dependence of the resulting force? For a plate separation "d," at what temperature do thermal effects become important?

(13.2) Assuming each mode of the radiation field has a vacuum energy $\frac{1}{2}\hbar\omega_0$, find the stimulated emission rate of a two-state atom, whose states are separated by energy $\hbar\omega_0$, from its upper state to its lower state. Compare your result with the A coefficient.

(13.3) Estimate the loss, in dB per meter, for a coaxial cable made from a real metal with (small) resistivity ρ, in terms of the radii of the conductors, a and b. Assume that the frequency of interest is such that the skin depth for the wave is small compared to the thickness of the conductors. Note especially the frequency dependence of the loss.

(13.4) Integrate the Poynting vector over the cross section of a coaxial cable to find the power carried by the cable in terms of the voltage amplitude, V, and current amplitude, I.

(13.5) The TEM_{00} mode of a coaxial cable exists for all frequencies and has a signal transmission speed equal to that of the speed of light in the dielectric separating the conductors. Why? For higher frequencies, higher order modes exist and may be excited. Find the threshold and nature (TE or TM) and frequency for the next higher mode.

(13.6) A) The infinite series of LC filters shown in Fig. 13.14 is sometimes suggested as a lumped circuit model of a transmission line. Find the impedance of the circuit for the input indicated by the arrow.
B) Find the inductance per meter, L_{coax}, and capacitance per meter, C_{coax}, for a coaxial cable whose conductors have radii a and b. Show that the characteristic impedance of the cable is given by:

$$Z_{coax} = \sqrt{\frac{L_{coax}}{C_{coax}}} \quad (13.112)$$

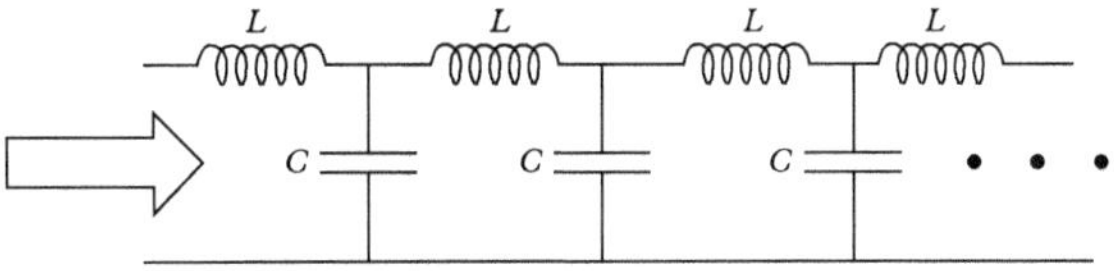

Fig. 13.14 *Infinite series of RC filters as a model for a transmission line.*

(13.7) A multimode optical fiber will have a much larger core than a single mode fiber. Typically, multimode fibers at the popular communications wavelength of 1550 nm will have a 62 μm diameter core whereas a single mode fiber is only 9 μm in diameter. While this allows the multimode fiber to carry much higher powers, dispersion is a problem. Since higher order modes tend to travel towards the edges of the core, this dispersion can be compensated for by grading the core–continuously reducing the index of refraction with increasing distance from the axis. Here we consider a similar problem for a planar waveguide: using geometric optics and the eikonal equation, design a symmetric planar waveguide graded (in both directions) from the mid-plane such that the propagation velocity is the same for both rays traveling along the mid-plane and rays zigzagging at angle α from the mid-plane.

(13.8) Starting from Maxwell's equations, derive the Master equation for the magnetic field, as discussed in Section 13.9, and, in particular, the form of the $\hat{\Theta}$ operator.

(13.9) Scale invariance has proved incredibly useful in advancing photonics by allowing reliable scaled microwave and rf models of devices ultimately intended to operate at optical frequencies. Show that, starting from a system with a varying dielectric function, $\varepsilon_r(\vec{r})$, which has been solved and found to have an eigenfunction $\vec{H}_a(\vec{r})$ with eigenfrequency ω_a, the spatially scaled system defined by $\varepsilon'_r(\vec{r}) = \varepsilon_r(\vec{r}/s)$ has an eigenfunction $\vec{H}'_a(\vec{r}) = \vec{H}_a(\vec{r}/s)$ with eigenfrequency $\omega'_a = \omega_a/s$.

13.10 Discussions

Discussion 13.1

With $\omega_k = c\sqrt{k_x^2 + k_y^2 + (n_z\pi/a)^2}$ for the 1st term (associated with the very thin slice of volume at the bottom of Fig. 13.5) and $\omega_k = c\sqrt{k_x^2 + k_y^2 + k_z^2}$ for the 2nd and 3rd terms, we write the sums of Eq. 13.18 in the general form of Eq. 13.17,

$$\sum_{\vec{k}}^{a} f\left(\frac{k}{k_m}\right)\frac{1}{2}\hbar\omega_k = \hbar c\left(\frac{L}{\pi}\right)^2 \sum_{n_z=(0),1}^{\infty} \iint_0^{\infty} dk_y dk_x \cdot f\left(\frac{k}{k_m}\right)\sqrt{k_x^2 + k_y^2 + \left(\frac{n_z\pi}{a}\right)^2}$$

$$\sum_{\vec{k}}^{L-a} f\left(\frac{k}{k_m}\right)\frac{1}{2}\hbar\omega_k = \hbar c\frac{(L^3 - L^2 a)}{\pi^3} \iiint_0^{\infty} dk_z dk_y dk_x \cdot f\left(\frac{k}{k_m}\right)\sqrt{k_x^2 + k_y^2 + k_z^2}$$

$$\sum_{\vec{k}}^{L} f\left(\frac{k}{k_m}\right)\frac{1}{2}\hbar\omega_k = \hbar c\left(\frac{L}{\pi}\right)^3 \iiint_0^{\infty} dk_z dk_y dk_x \cdot f\left(\frac{k}{k_m}\right)\sqrt{k_x^2 + k_y^2 + k_z^2}$$

Subtracting the 3rd term from the 2nd and adding the 1st yields,

$$\begin{aligned} U(a) = {} & \hbar c\left(\frac{L}{\pi}\right)^2 \sum_{n_z=(0),1}^{\infty} \iint_0^{\infty} dk_y dk_x \cdot f\left(\frac{k}{k_m}\right)\sqrt{k_x^2 + k_y^2 + \left(\frac{n_z\pi}{a}\right)^2} \\ & - \hbar c\left(\frac{L}{\pi}\right)^2 \int_0^{\infty} \frac{a}{\pi} dk_z \iint_0^{\infty} dk_y dk_x \cdot f\left(\frac{k}{k_m}\right)\sqrt{k_x^2 + k_y^2 + k_z^2} \end{aligned}$$

Discussion 13.2

Part(a) Take, for example, the curl equation expressing Faraday's Law multiplied by $\hat{z}\times$,

$$\hat{z}\times\left(\nabla\times\vec{\mathscr{E}}\exp(i\beta z)\right)=i\omega\mu_0\hat{z}\times\vec{\mathscr{H}}\exp(i\beta z)$$

The left-hand side with gradient operator written according to Eq. 13.44 is,

$$\hat{z}\times\left[\nabla_T\times\vec{\mathscr{E}}\exp(i\beta z)\right]+i\beta\left[\hat{z}\times\left(\hat{z}\times\vec{\mathscr{E}}\right)\right]\exp(i\beta z) \qquad (13.113)$$

Taking the transverse curl, the first term in square brackets is

$$\vec{V}=\left[\frac{\partial\mathscr{E}_z}{\partial y}\hat{x}-\frac{\partial\mathscr{E}_z}{\partial x}\hat{y}+\left(\frac{\partial\mathscr{E}_y}{\partial x}-\frac{\partial\mathscr{E}_x}{\partial y}\right)\hat{z}\right]\exp(i\beta z)$$

so that the first term in Eq. 13.113 is

$$\begin{aligned}\hat{z}\times\vec{V}&=\hat{z}\times\left[\nabla_T\times\vec{\mathscr{E}}\exp(i\beta z)\right]\\&=\left[\frac{\partial\mathscr{E}_z}{\partial x}\hat{x}+\frac{\partial\mathscr{E}_z}{\partial y}\hat{y}\right]\exp(i\beta z)=\nabla_T\mathscr{E}_z\exp(i\beta z)\end{aligned}$$

and with $\vec{\mathscr{E}}_T=\hat{z}\times\vec{\mathscr{E}}\times\hat{z}$ used in the second term, the curl equation can be written as

$$\nabla_T\mathscr{E}_z-i\beta\vec{\mathscr{E}}_T=i\omega\mu_0\hat{z}\times\vec{\mathscr{H}}_T$$

Part (b) Continuing with the example of Faraday's Law, we now further multiply the result for part (a) by $\hat{z}\times$

$$\begin{aligned}\hat{z}\times\nabla_T\mathscr{E}_z-i\beta\hat{z}\times\vec{\mathscr{E}}_T&=i\omega\mu_0\hat{z}\times\left(\hat{z}\times\vec{\mathscr{H}}_T\right)\\&=-i\omega\mu_0\vec{\mathscr{H}}_T\end{aligned}$$

or

$$\vec{\mathscr{H}}_T=\frac{i}{\omega\mu_0}\left[\hat{z}\times\nabla_T\mathscr{E}_z-i\beta\hat{z}\times\vec{\mathscr{E}}_T\right]$$

Next, solving the second equation in Eq. 13.46 for $\hat{z}\times\vec{\mathscr{E}}_T$ and substituting, we get

$$\vec{\mathscr{H}}_T = \frac{i}{\omega\mu_0}\left[\hat{z}\times\nabla_T\mathscr{E}_z - \frac{i\beta^2}{\omega\varepsilon_0}\vec{\mathscr{H}}_T + \frac{\beta}{\omega\varepsilon_0}\nabla_T\mathscr{H}_z\right]$$

$$= \frac{i}{\omega\mu_0}\left(\hat{z}\times\nabla_T\mathscr{E}_z\right) + \frac{\beta^2}{k^2}\vec{\mathscr{H}}_T + \frac{i\beta}{k^2}\nabla_T\mathscr{H}_z$$

where we have used $k^2 = \varepsilon_0\mu_0\omega^2$. Pulling the $\vec{\mathscr{H}}_T$ term to the left-hand side,

$$\frac{k^2-\beta^2}{k^2}\vec{\mathscr{H}}_T = \frac{i}{\omega\mu_0}\left(\hat{z}\times\nabla_T\mathscr{E}_z\right) + \frac{i\beta}{k^2}\nabla_T\mathscr{H}_z$$

or

$$\vec{\mathscr{H}}_T = \frac{ik^2}{\omega\mu_0\left(k^2-\beta^2\right)}\left(\hat{z}\times\nabla_T\mathscr{E}_z\right) + \frac{i\beta}{\left(k^2-\beta^2\right)}\nabla_T\mathscr{H}_z$$

or

$$\vec{\mathscr{H}}_T = \frac{i}{\left(k^2-\beta^2\right)}\left(\beta\nabla_T\mathscr{H}_z + \omega\varepsilon_0\hat{z}\times\nabla_T\mathscr{E}_z\right)$$

Discussion 13.3

Equations 9.66 and 9.70 in Chapter 9 give the Fresnel coefficients for the reflected electric fields from S- and P-polarized electromagnetic waves in a medium of index n incident onto the surface of a second medium with index n' where θ_i is the angle of the incident wave with respect to the normal of the surface

$$\left(\frac{E_r}{E_0}\right)_S = \left(\frac{E''}{E}\right)_S = \frac{n\cos\vartheta_i - \sqrt{(n')^2 - n^2\sin^2\vartheta_i}}{n\cos\vartheta_i + \sqrt{(n')^2 - n^2\sin^2\vartheta_i}} \tag{13.114}$$

$$\left(\frac{E_r}{E_0}\right)_P = \left(\frac{E''}{E}\right)_P = \frac{n\sqrt{(n')^2 - n^2\sin^2\vartheta_i} - (n')^2\cos\vartheta_i}{(n')^2\cos\vartheta_i + n\sqrt{(n')^2 - n^2\sin^2\vartheta_i}} \tag{13.115}$$

For total internal reflection to occur, $n \sin\theta \geq n'$ and therefore the radicals in these expressions are imaginary numbers. These equations are of the form:

$$\frac{E_r}{E_0} = \exp(i\delta\psi k_0) = \frac{a+ib}{a-ib} = \frac{Z}{Z^*} = \frac{|Z|\exp(i\phi)}{|Z^*|\exp(-i\phi)} = \exp(2i\phi) \tag{13.116}$$

Thus, $\phi = k_0\delta\psi/2$ and for S-polarization we have:

$$\tan\left(\frac{k_0\delta\psi_s}{2}\right) = -\frac{\sqrt{n^2 sin^2\vartheta_i - (n')^2}}{n\, cos\,\vartheta_i} \tag{13.117}$$

Noting that the waveguide angle θ is measured with respect to the surface (not the normal), we have $\vartheta_i = 90 - \theta$ and so:

$$k_0\delta\psi_s = -2\arctan\left(\frac{n^2 sin^2\vartheta_i - (n')^2}{n^2 cos^2\vartheta_i}\right)^{1/2} = -2\arctan\left(\frac{n^2 - (n')^2}{n^2\sin^2\theta} - 1\right)^{1/2} \tag{13.118}$$

Similarly for P-polarization:

$$k_0\delta\psi_p = -2\arctan\left[\frac{n^2\left(n^2\cos^2\theta - n'^2\right)}{n'^4\sin^2\theta}\right]^{1/2} \tag{13.119}$$

Discussion 13.4

Why not treat rectangular dielectric waveguides following a treatment similar to the microwave metal waveguides? A careful look at that treatment turns up the fact that the fields in the corners of the waveguide are omitted from consideration. For the microwave waveguide, even for the finite conductor case, practically all of the fields are very near the surface and, relative to the other dimensions of the waveguide, the "corner effects" will change the fields for the real waveguide negligibly. For dielectric waveguides, frequently a significant fraction of the wave travels in the cladding and in turn, the corner regions are much more important. Modeling a dielectric waveguide using the "corner neglecting" approximation frequently does provide adequate estimates for the field. That procedure is sometimes called the Marcatili solution. See Marcatili, E. A. J. Dielectric rectangular waveguide and directional coupler for integrated optics, *Bell System Technical Journal* p. 2071 (1969).

Discussion 13.5

First two rows of Eq. 13.71 are:

$$A\exp(i\kappa_{2x}a) = B\exp(ik_{1x}a) + C\exp(-ik_{1x}a)$$
$$\frac{k_{1x}}{\kappa_{2x}}A\exp(i\kappa_{2x}a) = B\exp(ik_{1x}a) - C\exp(-ik_{1x}a)$$

When added and subtracted, they yield,

$$\Rightarrow B = \frac{A}{2}\left(1 + \frac{k_{1x}}{\kappa_{2x}}\right)\exp(i\kappa_{2x}a - ik_{1x}a)$$
$$\Rightarrow C = \frac{A}{2}\left(1 - \frac{k_{1x}}{\kappa_{2x}}\right)\exp(i\kappa_{2x}a + ik_{1x}a)$$

Noting that this is being carried out for the symmetric case in which $\kappa_{3x} = \kappa_{2x}$, the third and fourth rows of Eq. 13.71 are:

$$D\exp(i\kappa_{2x}a) = B\exp(-ik_{1x}a) + C\exp(+ik_{1x}a)$$
$$\frac{k_{1x}}{\kappa_{2x}}D\exp(i\kappa_{2x}a) = B\exp(-ik_{1x}a) - C\exp(ik_{1x}a)$$

When added they yield,

$$\Rightarrow B = \frac{D}{2}\left(1 + \frac{k_{1x}}{\kappa_{2x}}\right)\exp(i\kappa_{2x}a + ik_{1x}a)$$

and comparing this with the other equation for B,

$$\Rightarrow D = A\exp(-2ik_{1x}a)$$

Discussion 13.6

For a TE mode the power per unit width, in y, of the waveguide, Eq. 13.67 provides:

$$\mathscr{E}_y = -\frac{i}{k_{jx}^2}\omega\mu_0\frac{d}{dx}\mathscr{H}_z(x)$$
$$\mathscr{H}_x^* = -\frac{i}{k_{jx}^2}\beta\frac{d}{dx}\mathscr{H}_z^*(x) \qquad (13.120)$$

$$\Rightarrow P = \frac{1}{2}\frac{\omega\mu_0}{\left|k_{jx}^4\right|}\beta\int_{-\infty}^{\infty} dx\left|\frac{d\mathscr{H}_z(x)}{dx}\right|^2$$

First, by symmetry, both claddings contribute equally:

$$P_{2+3} = \frac{\omega\mu_0}{\left|\gamma^2\right|}\int_a^{\infty} dx \exp\left[-2\gamma x\right] = \frac{\omega\mu_0}{2\left|\gamma^3\right|}\exp\left[-2\gamma a\right] \tag{13.121}$$

The core integral is significantly more difficult. First the fields in the core:

$$\begin{aligned} \mathscr{E}_y &= -\tfrac{1}{k_{1x}}\omega\mu_0\left[B\exp[+ik_{1x}x] - B^*\exp[-ik_{1x}x]\right] = -\tfrac{2i}{k_{1x}}\omega\mu_0\left[ImB\exp[+ik_{1x}x]\right] \\ \mathscr{H}_x^* &= -\tfrac{i}{k_{1x}^2}\beta\left[Bik_{1x}\exp[+ik_{1x}x] - B^*ik_{1x}\exp[-ik_{1x}x]\right]^* = \tfrac{2i}{k_{1x}}\beta\left[ImB\exp[+ik_{1x}x]\right] \end{aligned} \tag{13.122}$$

Thus we have:

$$\begin{aligned} P_1 &= \frac{1}{2}\frac{\omega\mu_0}{\left|k_{1x}^4\right|}\beta\int_{-a}^{a} dx\left|\frac{d}{dx}\left(B\exp\left(ik_{1x}x\right) + B^*\exp\left(-ik_{1x}x\right)\right)\right|^2 = 2\frac{\omega\mu_0}{\left|k_{1x}^2\right|}\beta\int_{-a}^{a} dx\left|Im\left(B\exp\left(ik_{1x}x\right)\right)\right|^2 \\ &= \frac{A^2}{2}\frac{\omega\mu_0}{\left|k_{1x}^2\right|}\beta\exp\left[-2\gamma a\right]\int_{-a}^{a} dx\left|Im\left(\left(1 - \frac{ik}{\gamma}\right)\exp[ik_{1x}x - ika]\right)\right|^2 \\ &= \frac{A^2}{2}\frac{\omega\mu_0}{\left|k_{1x}^2\right|}\beta\exp\left[-2\gamma a\right]\int_{-a}^{a} dx\left|\left(-\frac{k}{\gamma}\cos[kx - ka] + \sin[kx - ka]\right)\right|^2 \\ &= \frac{A^2}{2}\frac{\omega\mu_0}{\left|k_{1x}^2\right|}\beta\exp\left[-2\gamma a\right]\left\{\frac{2k\gamma + 4a\gamma^2 k_{1x}^2 + 4ak_{1x}^3 - 2\gamma k_{1x}\cos\left[4ak_{1x}\right] + \left(k_{1x}^2 - \gamma^2\right)\sin\left(4ak_{1x}\right)}{4\gamma^2 k_{1x}}\right\} \end{aligned} \tag{13.123}$$

(NB integral is carried out in Mathematica) From the side note on the determinant (#17), we find that an equivalent way of writing the propagation constant equation is:

$$\begin{aligned} &\left[\cos\left(4ak_{1x}\right) + i\sin\left(4akxl1x\right)\right]\left[k_{1x}^2 - \gamma^2 - 2ik\gamma\right] - \left(k_{1x}^2 - \gamma^2 + 2ik_{1x}\gamma\right) = 0 \\ &\left(k_{1x}^2 - \gamma^2\right)\sin\left(4ak_{1x}\right) - 2k_{1x}\gamma\cos\left(4ak_{1x}\right) - 2k_{1x}\gamma = 0 \\ &\left(\gamma^2 - k_{1x}^2\right)\cos 4ak_{1x} - 2k_{1x}\gamma\sin\left(4ak_{1x}\right) + k_{1x}^2 - \gamma^2 = 0 \end{aligned} \tag{13.124}$$

where we have separated out the imaginary and real parts in the lower two lines. Using the middle line to clean up the last expression in the previous equation:

$$P_1 = A^2\frac{\omega\mu_0}{2}\beta\exp\left[-2\gamma a\right]\left\{\frac{\gamma + a\gamma^2 + ak_{1x}^2}{\gamma^2 k_{1x}^2}\right\} \tag{13.125}$$

Finally, collecting all the terms and solving for A:

$$P_{tot} = A^2 \frac{\omega\mu_0}{2\gamma} \beta \exp[-2\gamma a] \left[\frac{k_{1x}^2 + \gamma^2 + a\gamma^3 + a\gamma k_{1x}^2}{\gamma^2 k_{1x}^2} \right] = 1$$
$$A = \gamma k_{1x} \exp[\gamma a] \left[\frac{2\gamma}{\beta\omega\mu_0 (1 + a\gamma)(k_{1x}^2 + \gamma^2)} \right]^{1/2} \tag{13.126}$$

Discussion 13.7

The Bessel functions that we consider here are Bessel functions of the first and second kind, $\mathcal{J}_l(x)$ and $Y_l(x)$ and the two Hankel functions, $H_l^{(1)}(x)$ and $H_l^{(2)}(x)$. These are all solutions of the Bessel differential equation and regularly turn up in problems with cylindrical symmetry. The $\mathcal{J}_l$ and Y_l may roughly be thought of as analogous to the cosine and sine functions whereas the Hankel functions are analogous to oscillating exponentials in the following sense: The Hankel functions are related to these first two Bessel functions:

$$H_l^{(1)}(x) = \mathcal{J}_l(x) + iY_l(x) \tag{13.127}$$
$$H_l^{(2)}(x) = \mathcal{J}_l(x) - iY_l(x) \tag{13.128}$$

The asymptotic forms of the first two near zero are:

$$\lim_{x\to 0} \mathcal{J}_l(x) = \frac{x^l}{2^l l!} \tag{13.129}$$

$$\lim_{x\to 0} Y_l(x) = \begin{cases} \frac{2}{\pi} \ln\left[\left(\frac{1}{2}x\right) + \gamma\right] & \text{for } l = 0 \\ -\frac{\Gamma(l)}{\pi}\left(\frac{2}{x}\right)^l & \text{for } l > 0 \end{cases} \tag{13.130}$$

Only the $\mathcal{J}$'s are finite at the origin. The asymptotic forms of the Hankel functions are:

$$\lim_{x\to\infty} H_l^{(1)}(x) = (-i)^{l+1} \frac{\exp(ix)}{x} \tag{13.131}$$

$$\lim_{x\to\infty} H_l^{(2)}(x) = (i)^{l+1} \frac{\exp(-ix)}{x} \tag{13.132}$$

Thus, when an imaginary argument (with a positive coefficient multiplying the i) is used, solutions are returned that are either exponentially decaying or exponentially growing with distance. The former describes the evanescent behavior of the field of a circular cylindrical waveguide.

Discussion 13.8

The second row of the matrix, Eq. 13.86, is generated from the boundary condition, $\varepsilon_1 E_\rho^{1,l} = \varepsilon_2 E_\rho^{2,l}$ or $\varepsilon_1 \mathscr{E}_\rho^{1,l} = \varepsilon_2 \mathscr{E}_\rho^{2,l}$, at $\rho = a$. Specifically, from the first equation in Eq. 13.84,

$$\mathscr{E}_\rho^{1,l} = \frac{i\beta}{k_t^2}\left(\mathscr{E}_0^{1,l} k_t \mathcal{J}_l'(k_t a) + \mathscr{H}_0^{1,l} \frac{il\omega\mu_0}{\beta a} \mathcal{J}_l(k_t a)\right) \exp(il\phi)$$

and on the other side of the boundary, in the cladding material, we get the same form but with $k_t \to \kappa_t$ and $\mathcal{J}_l(k_t a) \to H_l^{(1)}(\kappa_t a)$

$$\mathscr{E}_\rho^{2,l} = \frac{i\beta}{\kappa_t^2}\left(\mathscr{E}_0^{2,l} \kappa_t H_l^{(1)\prime}(\kappa_t a) + \mathscr{H}_0^{2,l} \frac{il\omega\mu_0}{\beta a} H_l^{(1)}(\kappa_t a)\right) \exp(il\phi)$$

so the associated boundary condition equation is

$$\frac{i\varepsilon_1\beta}{k_t^2}\left(\mathscr{E}_0^{1,l} k_t \mathcal{J}_l'(k_t a) + \mathscr{H}_0^{1,l} \frac{il\omega\mu_0}{\beta a} \mathcal{J}_l(k_t a)\right) - \frac{i\varepsilon_2\beta}{\kappa_t^2}\left(\mathscr{E}_0^{2,l} \kappa_t H_l^{(1)\prime}(\kappa_t a) + \mathscr{H}_0^{2,l} \frac{il\omega\mu_0}{\beta a} H_l^{(1)}(\kappa_t a)\right) = 0$$

or in terms of the second row of the matrix of Eq. 13.86,

$$\left[\frac{i\varepsilon_1\beta}{k_t}\mathcal{J}_l'(k_t a)\right]\mathscr{E}_0^{1,l} - \left[\frac{i\varepsilon_2\beta}{\kappa_t} H_l^{(1)\prime}(\kappa_t a)\right]\mathscr{E}_0^{2,l} - \left[\frac{\varepsilon_1\omega\mu_0 l}{k_t^2 a}\mathcal{J}_l(k_t a)\right]\mathscr{H}_0^{1,l} + \left[\frac{\varepsilon_2\omega\mu_0 l}{\kappa_t^2 a} H_l^{(1)}(\kappa_t a)\right]\mathscr{H}_0^{2,l} = 0$$

The fourth row is similarly generated from the condition $\mathscr{H}_\rho^{1,l} = \mathscr{H}_\rho^{2,l}$, at $\rho = a$:

$$\frac{i\beta}{k_t^2}\left(\mathscr{H}_0^{1,l} k_t \mathcal{J}_l'(k_t a) - \mathscr{E}_0^{1,l} \frac{il\omega\varepsilon_1}{a\beta} \mathcal{J}_l(k_t a)\right) - \frac{i\beta}{\kappa_t^2}\left(\mathscr{H}_0^{2,l} \kappa_t H_l^{(1)\prime}(\kappa_t a) - \mathscr{E}_0^{2,l} \frac{il\omega\varepsilon_2}{a\beta} H_l^{(1)}(\kappa_t a)\right) = 0$$

or in terms of the fourth row of the matrix of Eq. 13.86,

$$\left[\frac{\varepsilon_1\omega l}{k_t^2 a}\mathcal{J}_l(k_t a)\right]\mathscr{E}_0^{1,l} - \left[\frac{\varepsilon_2\omega l}{\kappa_t^2 a} H_l^{(1)}(\kappa_t a)\right]\mathscr{E}_0^{2,l} + \left[\frac{i\beta}{k_t}\mathcal{J}_l'(k_t a)\right]\mathscr{H}_0^{1,l} - \left[\frac{i\beta}{\kappa_t} H_l^{(1)\prime}(\kappa_t a)\right]\mathscr{H}_0^{2,l} = 0$$

Discussion 13.9

Setting the determinant to zero

$$\mathcal{J}_l\left\{\frac{\varepsilon_2\beta}{\kappa_t}H_l^{(1)\prime}\left(H_l^{(1)}\frac{\beta}{k_t}\mathcal{J}_l'-\mathcal{J}_l\frac{\beta}{\kappa_t}H_l^{(1)\prime}\right)-\frac{\varepsilon_1\omega\mu_0 l}{k_t^2 a}\mathcal{J}_l\left(H_l^{(1)}\frac{\varepsilon_2\omega l}{\kappa_t^2 a}H_l^{(1)}\right)+\frac{\varepsilon_2\omega\mu_0 l}{\kappa_t^2 a}H_l^{(1)}\left(\mathcal{J}_l\frac{\varepsilon_2\omega l}{\kappa_t^2 a}H_l^{(1)}\right)\right\}$$

$$+H_l^{(1)}\left\{\frac{\varepsilon_1\omega\mu_0 l}{k_t^2 a}\mathcal{J}_l\left(H_l^{(1)}\frac{\varepsilon_1\omega l}{k_t^2 a}\mathcal{J}_l\right)-\frac{\varepsilon_2\omega\mu_0 l}{\kappa_t^2 a}H_l^{(1)}\left(\mathcal{J}_l\frac{\varepsilon_1\omega l}{k_t^2 a}\mathcal{J}_l\right)-\frac{\varepsilon_1\beta}{k_t}\mathcal{J}_l'\left(H_l^{(1)}\frac{\beta}{k_t}\mathcal{J}_l'-\mathcal{J}_l\frac{\beta}{\kappa_t}H_l^{(1)\prime}\right)\right\}=0$$

the first and last terms combine and the second through fifth terms combine to the following,

$$\beta^2\left(\frac{1}{k_t}\mathcal{J}_l'H_l^{(1)}-\frac{1}{\kappa_t}\mathcal{J}_lH_l^{(1)\prime}\right)\left(\frac{\varepsilon_2}{\kappa_t}\mathcal{J}_lH_l^{(1)\prime}-\frac{\varepsilon_1}{k_t}\mathcal{J}_l'H_l^{(1)}\right)$$

$$+\left[-\frac{\varepsilon_1\mu_0}{k_t^2}\left(\frac{\varepsilon_2}{\kappa_t^2}\right)+\frac{\varepsilon_2\mu_0}{\kappa_t^2}\left(\frac{\varepsilon_2}{\kappa_t^2}\right)+\frac{\varepsilon_1\mu_0}{k_t^2}\left(\frac{\varepsilon_1}{k_t^2}\right)-\frac{\varepsilon_2\mu_0}{\kappa_t^2}\left(\frac{\varepsilon_1}{k_t^2}\right)\right]\left(\mathcal{J}_lH_l^{(1)}\frac{\omega l}{a}\right)^2=0$$

Focusing on the second, larger term, we can simplify it to,

$$\beta^2()()+\mu_0\left(\frac{\varepsilon_2}{\kappa_t^2}-\frac{\varepsilon_1}{k_t^2}\right)\left(\frac{\varepsilon_2}{\kappa_t^2}-\frac{\varepsilon_1}{k_t^2}\right)\left(\mathcal{J}_lH_l^{(1)}\frac{\omega l}{a}\right)^2=0$$

or

$$\beta^2()()+\frac{1}{\mu_0}\left(\frac{\varepsilon_2\mu_0\omega^2}{\kappa_t^2}-\frac{\varepsilon_1\mu_0\omega^2}{k_t^2}\right)\left(\frac{\varepsilon_2\mu_0\omega^2}{\kappa_t^2}-\frac{\varepsilon_1\mu_0\omega^2}{k_t^2}\right)\left(\mathcal{J}_lH_l^{(1)}\frac{l}{a\omega}\right)^2=0$$

and since $\varepsilon_j\mu_0\omega^2=k_j^2$

$$\beta^2()()+\frac{1}{\mu_0}\left(\frac{k_2^2}{\kappa_t^2}-\frac{k_1^2}{k_t^2}\right)\left(\frac{k_2^2}{\kappa_t^2}-\frac{k_1^2}{k_t^2}\right)\left(\mathcal{J}_lH_l^{(1)}\frac{l}{a\omega}\right)^2=0$$

But we note here that

$$\frac{k_2^2}{\kappa_t^2}-\frac{k_1^2}{k_t^2}=\frac{\kappa_t^2+\beta^2}{\kappa_t^2}-\frac{k_t^2+\beta^2}{k_t^2}=1+\frac{\beta^2}{\kappa_t^2}-1-\frac{\beta^2}{k_t^2}=\beta^2\left(\frac{1}{\kappa_t^2}-\frac{1}{k_t^2}\right)$$

so we have

$$\beta^2\left(\frac{1}{k_t}\mathcal{J}_l' H_l^{(1)} - \frac{1}{\kappa_t}\mathcal{J}_l H_l^{(1)\prime}\right)\left(\frac{\varepsilon_2}{\kappa_t}\mathcal{J}_l H_l^{(1)\prime} - \frac{\varepsilon_1}{k_t}\mathcal{J}_l' H_l^{(1)}\right) + \frac{\beta^4}{\mu_0}\left(\frac{1}{\kappa_t^2} - \frac{1}{k_t^2}\right)^2\left(\mathcal{J}_l H_l^{(1)}\frac{l}{a\omega}\right)^2 = 0$$

dividing by $\left[a\beta\mathcal{J}_l H_l^{(1)}\right]^2$, we obtain

$$\left(\frac{\mathcal{J}_l'}{k_t a\mathcal{J}_l} - \frac{H_l^{(1)\prime}}{\kappa_t a H_l^{(1)}}\right)\left(\frac{\varepsilon_1\mathcal{J}_l'}{k_t a\mathcal{J}_l} - \frac{\varepsilon_2 H_l^{(1)\prime}}{\kappa_t a H_l^{(1)}}\right) - \frac{1}{\mu_0\omega^2}(\beta l)^2\left(\frac{1}{(\kappa_t a)^2} - \frac{1}{(k_t a)^2}\right)^2 = 0$$

or

$$\left(\frac{\mathcal{J}_l'}{k_t a\mathcal{J}_l} - \frac{H_l^{(1)\prime}}{\kappa_t a H_l^{(1)}}\right)\left(\frac{\varepsilon_1\mu_0\omega^2\mathcal{J}_l'}{k_t a\mathcal{J}_l} - \frac{\varepsilon_2\mu_0\omega^2 H_l^{(1)\prime}}{\kappa_t a H_l^{(1)}}\right) = (\beta l)^2\left(\frac{1}{(\kappa_t a)^2} - \frac{1}{(k_t a)^2}\right)^2$$

and finally,

$$\left(\frac{\mathcal{J}_l'}{k_t a\mathcal{J}_l} - \frac{H_l^{(1)\prime}}{\kappa_t a H_l^{(1)}}\right)\left(\frac{k_1^2\mathcal{J}_l'}{k_t a\mathcal{J}_l} - \frac{k_2^2 H_l^{(1)\prime}}{\kappa_t a H_l^{(1)}}\right) = (\beta l)^2\left(\frac{1}{(\kappa_t a)^2} - \frac{1}{(k_t a)^2}\right)^2$$

Appendix A
Vector Multipole Expansion of the Fields

In Chapter 4, we derived the electric and magnetic dipole and quadrupole moment radiation fields from an arbitrary distribution of charge and current. Specifically, we considered a simple far-field ($R \gg \lambda$) Taylor expansion of the vector potential. With the additional approximation that the source sizes were much less than the wavelength of any emitted radiation ($kr \ll 1$), this expansion converged in the first few terms. From these terms we found expressions for the fields related to the first few charge moments of the source. This method is limited, however, because it becomes increasingly more difficult to continue the expansion beyond the quadrupole terms. In the more general case for which expansion of the charge and current distributions are not adequately described by the first few moments of the source expansion, it would be preferable to have a method to obtain exact terms, without approximations, of any order, quadrupole and higher, which are as easy to compute as the dipole terms. In fact, there is a method to describe a vector multipole field in such an exact manner: Vector Spherical Harmonics (VSH).

A.1 Vector spherical harmonics

In general, any arbitrary three-dimensional scalar field can be expressed as an expansion in scalar spherical harmonics, $Y_{lm}(\theta,\phi)$ with radially-dependent coefficients. It would then seem natural to extend this to vector fields by simply applying the same expansion to each of the three vector field components. This sort of expansion of a vector field, however, turns out to be of little use because such terms in vector operations involving the del ∇ operator do not reproduce a function proportional to the Y_{lm} and so the separation of angular dependence (in the form of the Y_{lm}'s) and radial dependence in vector equations is not preserved.

There is, however, a type of complete, orthogonal set of angle-dependent vector fields, based on the spherical harmonics, with which we can, given the correct radially dependent coefficients, construct any arbitrary vector field and that, in vector operations involving the del operator, reproduce functions proportional to the Y_{lm} and finally that, like the spherical harmonics, exhibit definite parity. There have been constructed multiple related forms of these so-called "vector spherical harmonics (VSH)" (References[1]). In this work, we will adopt the normalized version of Barrera's form. Based on the Y_{lm}, these VSH are defined as:

$$\vec{Y}_{lm} = Y_{lm}(\theta,\phi)\,\hat{r} \tag{A.1}$$

$$\vec{\Psi}_{lm} = \frac{r}{\sqrt{l(l+1)}}\nabla Y_{lm}(\theta,\phi) \tag{A.2}$$

$$\vec{\Phi}_{lm} = \frac{1}{\sqrt{l(l+1)}}\vec{r}\times\nabla Y_{lm}(\theta,\phi) \tag{A.3}$$

Thus, an arbitrary vector field , $\vec{V}(\vec{r},\omega)$, can be represented by a sum, over l and m, of pure multipole vector fields, $\vec{V}_{lm}(\vec{r},\omega)$. That is,

$$\vec{V}(\vec{r},\omega) = \sum_{l,m} \vec{V}_{lm}(\vec{r},\omega) \tag{A.4}$$

and each pure multipole is, in turn, composed of the three orthonormal VSH in the order of l, m:

$$\vec{V}_{lm}(\vec{r},\omega) = V^{r}_{lm}(r)\,\vec{Y}_{lm}(\theta,\phi) + V^{(1)}_{lm}(r)\,\vec{\Psi}_{lm}(\theta,\phi) + V^{(2)}_{lm}(r)\,\vec{\Phi}_{lm}(\theta,\phi) \tag{A.5}$$

where the coefficients result from the orthonormality of the VSH,

$$\begin{aligned} V^{r}_{lm}(r) &= \int \vec{Y}^{*}_{lm}(\theta,\phi)\cdot\vec{V}(r,\theta,\phi)\,d\Omega \\ V^{(1)}_{lm}(r) &= \int \vec{\Psi}^{*}_{lm}(\theta,\phi)\cdot\vec{V}(r,\theta,\phi)\,d\Omega \\ V^{(2)}_{lm}(r) &= \int \vec{\Phi}^{*}_{lm}(\theta,\phi)\cdot\vec{V}(r,\theta,\phi)\,d\Omega \end{aligned}$$

A.1.1 VSH expansion of general radiation fields

In the case of the electric and magnetic vector fields, $\vec{E}(\vec{r},\omega)$ and $\vec{B}(\vec{r},\omega)$, which represent a single frequency component of the radiation field, the pure multipole fields, $\vec{E}_{lm}(\vec{r},\omega)$ and $\vec{B}_{lm}(\vec{r},\omega)$, can each be further subdivided into two components associated with the lm^{th} electric (E) and magnetic (M) multipoles (moments) of the source distribution. Thus, for example, in the case of the electric vector field expansion:

$$\vec{E}(\vec{r},\omega) = \sum_{l,m}\vec{E}_{lm}(\vec{r},\omega) = \sum_{l,m}\left(\vec{E}^{E}_{lm}(\vec{r},\omega) + \vec{E}^{M}_{lm}(\vec{r},\omega)\right) \tag{A.6}$$

and for a single lm term expressed in the components of the VSH,

$$\begin{aligned} \vec{E}_{lm}(\vec{r},\omega) &= \vec{E}^{E}_{lm}(\vec{r},\omega) + \vec{E}^{M}_{lm}(\vec{r},\omega) \\ &= \left(E^{rE}_{lm}(r) + E^{rM}_{lm}(r)\right)\vec{Y}_{lm}(\theta,\phi) + \left(E^{(1)E}_{lm}(r) + E^{(1)M}_{lm}(r)\right)\vec{\Psi}_{lm}(\theta,\phi) \\ &\quad + \left(E^{(2)E}_{lm}(r) + E^{(2)M}_{lm}(r)\right)\vec{\Phi}_{lm}(\theta,\phi) \end{aligned} \tag{A.7}$$

with an equivalent representation for the magnetic vector field, $\vec{B}(\vec{r},\omega)$.

A.2 Multipole expansion of electromagnetic radiation

A.2.1 Non-homogeneous field wave equations

We will now attempt a multipole expansion of electromagnetic radiation emitted from an arbitrary, time-varying charge and current distribution *near the origin* in terms of the VSH defined in Section A.1. Starting with the non-homogenous wave equations for the $\vec{E}$ and $\vec{B}$ fields with given charge and current densities, $\rho(\vec{r},t)$ and $\vec{\mathcal{J}}(\vec{r},t)$:

$$\left(\nabla^2 - \frac{1}{c^2}\frac{\partial^2}{\partial t^2}\right)\vec{E}(\vec{r},t) = \frac{1}{\varepsilon_o}\left(\nabla\rho(\vec{r},t) + \frac{1}{c^2}\frac{\partial\vec{\mathcal{J}}(\vec{r},t)}{\partial t}\right) \tag{A.8}$$

$$\left(\nabla^2 - \frac{1}{c^2}\frac{\partial^2}{\partial t^2}\right)\vec{B}(\vec{r},t) = -\mu_o\nabla\times\vec{\mathcal{J}}(\vec{r},t) \tag{A.9}$$

we take the Fourier Transform with respect to time (or equivalently, assume sources that vary sinusoidally in time with frequency, ω) and obtain (Eqs. 3.18 and 3.19),

$$\left(\nabla^2 + k^2\right)\vec{E}(\vec{r},\omega) = \frac{1}{\varepsilon_o}\nabla\rho(\vec{r},\omega) - \frac{i\omega}{\varepsilon_o c^2}\vec{\mathcal{J}}(\vec{r},\omega) \tag{A.10}$$

$$\left(\nabla^2 + k^2\right)\vec{B}(\vec{r},\omega) = -\mu_o\nabla\times\vec{\mathcal{J}}(\vec{r},\omega) \tag{A.11}$$

Next, we can use the general vector identity for the vector Laplace operator, $\nabla^2\vec{V} = \nabla\left(\nabla\cdot\vec{V}\right) - \nabla\times\left(\nabla\times\vec{V}\right)$, and the two Maxwell's divergence equations, $\nabla\cdot\vec{E} = \rho/\varepsilon_o$ and $\nabla\cdot\vec{B} = 0$, to re-express these equations as,

$$\left[\nabla\times(\nabla\times) - k^2\right]\vec{E} = i\omega\mu_o\vec{\mathcal{J}} \tag{A.12}$$

$$\left[\nabla\times(\nabla\times) - k^2\right]\vec{B} = \mu_o\left(\nabla\times\vec{\mathcal{J}}\right) \tag{A.13}$$

A.2.2 VSH expansion of the field wave equations

Expressing the electric field and current density within Eq. A.12 in terms of their VSH expansions (as in Eqs. A.4 and A.5), we get,

$$\left[\nabla\times(\nabla\times) - k^2\right]\sum_{l,m}\left(E^r_{lm}\vec{Y}_{lm} + E^{(1)}_{lm}\vec{\Psi}_{lm} + E^{(2)}_{lm}\vec{\Phi}_{lm}\right)$$

$$= i\omega\mu_o\sum_{l,m}\left(\mathcal{J}^r_{lm}\vec{Y}_{lm} + \mathcal{J}^{(1)}_{lm}\vec{\Psi}_{lm} + \mathcal{J}^{(2)}_{lm}\vec{\Phi}_{lm}\right) \tag{A.14}$$

with a similar expansion for the magnetic field. As mentioned, the VSH are a quite useful vector field basis set mainly owing to their advantageous properties associated with vector operations involving the "del" (nabla symbol) operator. For example, for curl operations on the VSH, the angular dependence of the resulting field remains in terms of the VSH. Specifically, (**Problem set**[2]),

$$\nabla\times\left(F(r)\vec{Y}_{lm}\right) = -\left(\frac{\sqrt{l(l+1)}}{r}F(r)\right)\vec{\Phi}_{lm} \tag{A.15}$$

$$\nabla \times \left(F(r)\vec{\Psi}_{lm}\right) = \left(\frac{1}{r}\frac{d}{dr}rF(r)\right)\vec{\Phi}_{lm} \tag{A.16}$$

$$\nabla \times \left(F(r)\vec{\Phi}_{lm}\right) = -\left(\frac{\sqrt{l(l+1)}}{r}F(r)\right)\vec{Y}_{lm} - \left(\frac{1}{r}\frac{d}{dr}rF(r)\right)\vec{\Psi}_{lm} \tag{A.17}$$

We now show that direct substitution of these relations into A.14 gives a full VSH expansion of A.12. If the VSH-expanded electric field is understood to be,

$$\vec{E} = \sum_{l,m}\left(E^{r}_{lm}\vec{Y}_{lm} + E^{(1)}_{lm}\vec{\Psi}_{lm} + E^{(2)}_{lm}\vec{\Phi}_{lm}\right) \tag{A.18}$$

then substitution of the relations A.15–A.17 into the curl of this VSH-expanded electric field yields

$$\nabla \times \vec{E} = \sum_{l,m}\left(-\frac{\sqrt{l(l+1)}}{r}E^{(2)}_{lm}\vec{Y}_{lm} - \frac{1}{r}\frac{d}{dr}rE^{(2)}_{lm}\vec{\Psi}_{lm} + \left(\frac{1}{r}\frac{d}{dr}rE^{(1)}_{lm} - \frac{\sqrt{l(l+1)}}{r}E^{r}_{lm}\right)\vec{\Phi}_{lm}\right) \tag{A.19}$$

Taking the curl of this,

$$\nabla \times \left(\nabla \times \vec{E}\right) = \nabla \times \sum_{l,m}\left(-\frac{\sqrt{l(l+1)}}{r}E^{(2)}_{lm}\vec{Y}_{lm} - \frac{1}{r}\frac{d}{dr}rE^{(2)}_{lm}\vec{\Psi}_{lm} + \left(\frac{1}{r}\frac{d}{dr}rE^{(1)}_{lm} - \frac{\sqrt{l(l+1)}}{r}E^{r}_{lm}\right)\vec{\Phi}_{lm}\right)$$

and again applying the relations A.15–A.17, we rearrange to get,

$$\begin{aligned}\nabla \times \left(\nabla \times \vec{E}\right) = -\sum_{l,m}\Bigg[&\frac{\sqrt{l(l+1)}}{r^2}\left(\frac{d}{dr}rE^{(1)}_{lm} - \sqrt{l(l+1)}E^{r}_{lm}\right)\vec{Y}_{lm} \\ &+\left(\frac{1}{r}\frac{d^2}{dr^2}rE^{(1)}_{lm} - \frac{\sqrt{l(l+1)}}{r}\frac{d}{dr}E^{r}_{lm}\right)\vec{\Psi}_{lm} + \left(\frac{1}{r}\frac{d^2}{dr^2}rE^{(2)}_{lm} - \frac{l(l+1)}{r^2}E^{(2)}_{lm}\right)\vec{\Phi}_{lm}\Bigg]\end{aligned}$$

and, substituting into A.14, we finally obtain

$$\begin{aligned}&\sum_{l,m}\left[\frac{\sqrt{l(l+1)}}{r^2}\left(\frac{d}{dr}rE^{(1)}_{lm} - \sqrt{l(l+1)}E^{r}_{lm}\right) + k^2E^{r}_{lm}\right]\vec{Y}_{lm} \\ &\quad+\sum_{l,m}\left[\frac{1}{r}\frac{d^2}{dr^2}rE^{(1)}_{lm} - \frac{\sqrt{l(l+1)}}{r}\frac{d}{dr}E^{r}_{lm} + k^2E^{(1)}_{lm}\right]\vec{\Psi}_{lm} \\ &\quad+\sum_{l,m}\left[\frac{1}{r}\frac{d^2}{dr^2}rE^{(2)}_{lm} - \frac{l(l+1)}{r^2}E^{(2)}_{lm} + k^2E^{(2)}_{lm}\right]\vec{\Phi}_{lm} \\ &= -i\omega\mu_0\sum_{l,m}\left(\mathcal{J}^{r}_{lm}\vec{Y}_{lm} + \mathcal{J}^{(1)}_{lm}\vec{\Psi}_{lm} + \mathcal{J}^{(2)}_{lm}\vec{\Phi}_{lm}\right)\end{aligned} \tag{A.20}$$

Because the $\vec{Y}_{lm}$, $\vec{\Psi}_{lm}$, and $\vec{\Phi}_{lm}$ are mutually orthogonal, Eq. A.20 can be separated into three sums of radial equations over *lm*:

$$\sum_{l,m}\left[\frac{\sqrt{l(l+1)}}{r^2}\left(\frac{d}{dr}rE_{lm}^{(1)}-\sqrt{l(l+1)}E_{lm}^{r}\right)+k^2E_{lm}^{r}\right]\vec{Y}_{lm}=-i\omega\mu_o\sum_{l,m}\mathcal{J}_{lm}^{r}\vec{Y}_{lm} \tag{A.21}$$

$$\sum_{l,m}\left[\left(\frac{1}{r^2}\frac{d}{dr}\left(r^2\frac{d}{dr}E_{lm}^{(1)}\right)-\frac{\sqrt{l(l+1)}}{r}\frac{d}{dr}E_{lm}^{r}\right)+k^2E_{lm}^{(1)}\right]\vec{\Psi}_{lm}=-i\omega\mu_o\sum_{l,m}\mathcal{J}_{lm}^{(1)}\vec{\Psi}_{lm} \tag{A.22}$$

$$\sum_{l,m}\left[\left(\frac{1}{r^2}\frac{d}{dr}\left(r^2\frac{d}{dr}E_{lm}^{(2)}\right)-\frac{l(l+1)}{r^2}E_{lm}^{(2)}\right)+k^2E_{lm}^{(2)}\right]\vec{\Phi}_{lm}=-i\omega\mu_o\sum_{l,m}\mathcal{J}_{lm}^{(2)}\vec{\Phi}_{lm} \tag{A.23}$$

A.2.3 Parity considerations

Not only can the electric and magnetic fields of radiation be expanded into VSH multipole terms, as demonstrated previously, but they can each be further subdivided, as expressed in Eq. A.7, into two types of multipole radiation: one type originating from the electric multipole moments and the other originating from the magnetic multipole moments. Furthermore, for a given multipole term in an electric or magnetic field expansion, the two types will have opposite parities, each determined by the order l and whether they originated from an electric or magnetic multipole. Specifically, for radiation originating from an electric multipole of order lm, the electric vector field has a parity of $(-1)^{l+1}$ and the magnetic field has a parity of $(-1)^l$. And for radiation originating from a magnetic multipole of order lm, the parities of the fields are exactly reversed. This differing parity is consistent with the relationship of these two terms, $\vec{E}_{lm}^{E}=\frac{ic}{k}\nabla\times\vec{B}_{lm}^{E}$, and the fact that the curl of a vector field with definite parity has the opposite parity. These results for the electric and magnetic multipole fields have important consequences in the following development because the VSH that will represent them also have well defined parities. The parities of the VSH are:

$$\begin{aligned}\hat{P}\vec{Y}_{lm}(\theta,\phi)&=(-1)^{l+1}\,\vec{Y}_{lm}(\theta,\phi)\\ \hat{P}\vec{\Psi}_{lm}(\theta,\phi)&=(-1)^{l+1}\,\vec{\Psi}_{lm}(\theta,\phi)\\ \hat{P}\vec{\Phi}_{lm}(\theta,\phi)&=(-1)^{l}\,\vec{\Phi}_{lm}(\theta,\phi)\end{aligned} \tag{A.24}$$

A comparison of these VSH parities with those of the multipole fields is revealing. For example, in the full expansion of $\vec{E}$, as defined in Eqs. A.6 and A.7, we know that the $E_{lm}^{(2)E}(r)$ term must vanish because any contribution to the electric field from an electric multipole of order lm necessarily has a parity of $(-1)^{l+1}$ and cannot therefore be represented by $\vec{\Phi}_{lm}$, which as has a parity of $(-1)^l$. That is to say, $E_{lm}^{(2)}=E_{lm}^{(2)M}+E_{lm}^{(2)E}\rightarrow E_{lm}^{(2)M}$. By the same argument, the terms $E_{lm}^{rM}(r)$ and $E_{lm}^{(1)M}(r)$ also vanish.

If we consider contributions to the electric vector field from multipoles of a single order lm and if we eliminate the vanished coefficients, the Eqs. A.21–A.23 reduce to:

$$\left[\frac{\sqrt{l(l+1)}}{r^2}\left(\frac{d}{dr}rE_{lm}^{(1)E}-\sqrt{l(l+1)}E_{lm}^{rE}\right)+k^2E_{lm}^{rE}\right]\vec{Y}_{lm}=-i\omega\mu_o\mathcal{J}_{lm}^{r}\vec{Y}_{lm} \tag{A.25}$$

$$\left[\frac{1}{r^2}\frac{d}{dr}\left(r^2\frac{d}{dr}E_{lm}^{(1)E}\right)-\frac{\sqrt{l(l+1)}}{r}\frac{d}{dr}E_{lm}^{rE}+k^2E_{lm}^{(1)E}\right]\vec{\Psi}_{lm}=-i\omega\mu_o\mathcal{J}_{lm}^{(1)}\vec{\Psi}_{lm} \tag{A.26}$$

$$\left[\frac{1}{r^2}\frac{d}{dr}\left(r^2\frac{d}{dr}E_{lm}^{(2)M}\right)-\frac{l(l+1)}{r^2}E_{lm}^{(2)M}+k^2E_{lm}^{(2)M}\right]\vec{\Phi}_{lm}=-i\omega\mu_o\mathcal{J}_{lm}^{(2)}\vec{\Phi}_{lm} \tag{A.27}$$

for which we now see from Eq. A.27 that the $\vec{\Phi}_{lm}$ components of the vector field $\vec{\mathcal{J}}$ are associated with the magnetic (M) multipole contributions to the electric field and from Eqs. A.25 and A.26, that the $\vec{Y}_{lm}$ and $\vec{\Psi}_{lm}$ components of $\vec{\mathcal{J}}$ are associated with the electric (E) multipole contributions to the electric field. Now, it is clear that the angular dependence of these equations cancels out and we are left with three non-homogeneous ordinary linear differential equations (ODE's) with solutions $E_{lm}^{(1)E}(r), E_{lm}^{rE}(r)$, and $E_{lm}^{(2)M}(r)$. It is the last one, Eq. A.27, however, which remains uncoupled from the others and is thus more easily solved. We find that, as usual, there is a set of equations, similar to Eqs. A.25–A.27, for the magnetic vector field in which the $\vec{\Phi}_{lm}$ components of the source field (this time, $\nabla\times\vec{\mathcal{J}}$) are associated with the electric multipole contributions to the magnetic field.

A.2.4 Multipole expansion in a source-free region

For points outside the source distribution, Eq. A.27 becomes the homogeneous ODE,

$$\left(\frac{1}{r^2}\frac{d}{dr}\left(r^2\frac{d}{dr}\right)-\frac{l(l+1)}{r^2}+k^2\right)E_{lm}^{(2)M}=0 \tag{A.28}$$

The two main linearly independent solutions, of independent variable kr, to these equations are known as the spherical Bessel and Neumann functions. These functions, respectively labeled $j_l(kr)$ and $y_l(kr)$, have the general form of combinations of sine and cosine functions multiplied by power series in inverse powers of kr. A second pair of linearly independent solutions, known as the spherical Hankel functions, h_l^1 and h_l^2, are constructed as linear combinations of $j_l(kr)$ and $y_l(kr)$. Specifically, $h_l^{1,2}=j_l(kr)\pm iy(kr)$. The general form of the spherical Hankel functions is a complex exponential multiplied by a complex power series in inverse powers of kr. The first few ($l=0,1,2$) of these three types of functions are given as:

$$\begin{aligned}
&j_0=\frac{\sin(kr)}{kr} && j_1=\frac{\sin(kr)}{(kr)^2}-\frac{\cos(kr)}{kr} && j_2=\left(\frac{3}{(kr)^3}-\frac{1}{kr}\right)\sin(kr)-\frac{3\cos(kr)}{(kr)^2}\\
&y_0=-\frac{\cos(kr)}{kr} && y_1=-\frac{\cos(kr)}{(kr)^2}-\frac{\sin(kr)}{kr} && y_2=-\left(\frac{3}{(kr)^3}-\frac{1}{kr}\right)\cos(kr)-\frac{3\sin(kr)}{(kr)^2}\\
&h_0^1=\frac{-ie^{ikr}}{kr} && h_1^1=-\frac{e^{ikr}}{kr}\left(1+\frac{i}{kr}\right) && h_2^1=\frac{ie^{ikr}}{kr}\left(1+\frac{3i}{kr}-\frac{3}{(kr)^2}\right)
\end{aligned} \tag{A.29}$$

where the $h_l^2(kr)$, which are not shown, are the complex conjugates of the $h_l^1(kr)$. The spherical Bessel and Neumann functions can be seen as analogs to the sine and cosine functions while the spherical Hankel functions, as linear combinations of $j_l(kr)$ and $y_l(kr)$, are analogous to the complex exponentials.

Because their complex exponential form is better suited for the description of spherical waves, the spherical Hankel functions are used and, in particular, appealing to the condition of causality, we choose those functions, $h_l^1\propto e^{ikr}/kr$, which, when combined with a sinusoidal time dependence of $e^{-i\omega t}$, result in causally correct outward expanding waves (as opposed to $h_l^2\propto e^{-ikr}/kr$ that correspond to inward converging waves). Thus we see the general form of the solution $E_{lm}^{(2)M}$ is:

$$E_{lm}^{(2)M}(r) = a_{lm}^{M} h_l^1(kr) \tag{A.30}$$

And, indeed, with the vanishing of the terms E_{lm}^{rM} and $E_{lm}^{(1)M}$, it is seen from Eq. A.7 that the lm^{th} magnetic multipole contribution to the electric field expansion is due solely to $E_{lm}^{(2)M}$ so that

$$\vec{E}_{lm}^{M}(\vec{r},\omega) = E_{lm}^{(2)M}(r)\,\vec{\Phi}_{lm}(\theta,\phi) = a_{lm}^{M} h_l^1(kr)\,\vec{\Phi}_{lm}(\theta,\phi) \tag{A.31}$$

With $\vec{E}_{lm}^{M}(\vec{r},\omega)$ in hand, we can immediately find the associated magnetic field due to the lm^{th} magnetic multipole using $\vec{B}_{lm}^{M}(\vec{r},\omega) = -\frac{i}{ck}\nabla\times\vec{E}_{lm}^{M}(\vec{r},\omega)$. Thus, with Eq. A.17, this trick, when applied to the similarly found $\vec{B}_{lm}^{E}(\vec{r},\omega)$, eliminates the need to uncouple and solve the homogeneous versions of Eqs. A.25 and A.26 for the terms $E_{lm}^{(1)E}$ and E_{lm}^{rE}. The electric and magnetic fields of the electric multipole radiation can be found in a similar way. So, with all the components, we can now state the general solution to the sinusoidally time-dependent, *free-space* Maxwell's equations in terms of a VSH multipole expansion of the electric and magnetic fields:

$$\begin{aligned}
\vec{E}(\vec{r},\omega) &= \sum_{lm}\left[\vec{E}_{lm}^{E} + \vec{E}_{lm}^{M}\right] = \sum_{lm}\left[\frac{ic}{k}\nabla\times a_{lm}^{E} h_l^1(kr)\,\vec{\Phi}_{lm}(\theta,\phi) + a_{lm}^{M} h_l^1(kr)\,\vec{\Phi}_{lm}(\theta,\phi)\right] \\
\vec{B}(\vec{r},\omega) &= \sum_{lm}\left[\vec{B}_{lm}^{E} + \vec{B}_{lm}^{M}\right] = \sum_{lm}\left[a_{lm}^{E} h_l^1(kr)\,\vec{\Phi}_{lm}(\theta,\phi) - \frac{i}{ck}\nabla\times a_{lm}^{M} h_l^1(kr)\,\vec{\Phi}_{lm}(\theta,\phi)\right]
\end{aligned} \tag{A.32}$$

In summary, the method of solution in a source-free region of space for all field components of a given order lm is to first solve the uncoupled ODE's associated with the $\vec{\Phi}_{lm}$ equations to obtain $\vec{E}_{lm}^{M} = E_{lm}^{(2)M}\vec{\Phi}_{lm}$ and $\vec{B}_{lm}^{E} = B_{lm}^{(2)E}\vec{\Phi}_{lm}$ and then use the free space Maxwell' equations in the forms $\vec{B} = -\frac{i}{ck}\nabla\times\vec{E}$ and $\vec{E} = \frac{ic}{k}\nabla\times\vec{B}$, to obtain $\vec{B}_{lm}^{M} = B_{lm}^{rM}\vec{Y}_{lm} + B_{lm}^{(1)M}\vec{\Psi}_{lm}$ and $\vec{E}_{lm}^{E} = E_{lm}^{rE}\vec{Y}_{lm} + E_{lm}^{(1)E}\vec{\Psi}_{lm}$.

A.3 Multipole radiation: energy and angular momentum

A.3.1 Energy density and the Poynting vector

Electromagnetic radiation emitted from a source of time varying charges and currents carries with it energy, momentum, and angular momentum. Because the fields exist as a continuum we must deal with the densities of these quantities. The time-averaged energy per unit volume per frequency component within electromagnetic radiation is given by,

$$\langle\varepsilon\rangle = \frac{\varepsilon_o}{4}\vec{E}\cdot\vec{E}^* + \frac{1}{4\mu_o}\vec{B}\cdot\vec{B}^* \tag{A.33}$$

In Section A.4.2, we will approximate the VSH expansion, Eq. A.32, in the radiation or "far" zone as Eq. A.58. Here, we use this result within Eq. A.33 to consider the energy density in the far zone contributed by a single magnetic multipole moment (l,m)

$$\begin{aligned}\left\langle\varepsilon_{lm}^{M}\right\rangle &= \frac{\varepsilon_o}{4}\vec{E}_{lm}^{M}\cdot\vec{E}_{lm}^{*M}+\frac{1}{4\mu_o}\vec{B}_{lm}^{M}\cdot\vec{B}_{lm}^{*M}\\ &= \frac{-\varepsilon_o}{4}\left(ia_{lm}^{M}\frac{e^{i\left(kr-\frac{\pi}{2}l\right)}}{kr}\vec{\Phi}_{lm}\right)\cdot\left(ia_{lm}^{*M}\frac{e^{-i\left(kr-\frac{\pi}{2}l\right)}}{kr}\vec{\Phi}_{lm}^{*}\right)\\ &\quad+\frac{1}{4\mu_o}\left(\frac{1}{ck}\nabla\times a_{lm}^{M}\frac{e^{i\left(kr-\frac{\pi}{2}l\right)}}{kr}\vec{\Phi}_{lm}\right)\cdot\left(\frac{1}{ck}\nabla\times a_{lm}^{*M}\frac{e^{-i\left(kr-\frac{\pi}{2}l\right)}}{kr}\vec{\Phi}_{lm}^{*}\right)\Bigg]\end{aligned}\tag{A.34}$$

Applying Eq. A.17, the second term, the time-averaged magnetic field energy density from the lm^{th} magnetic multipole moment, is rewritten as,

$$\begin{aligned}\frac{1}{4\mu_o}\vec{B}_{lm}^{M}\cdot\vec{B}_{lm}^{*M} &= \frac{1}{4\mu_o}\left[\left(a_{lm}^{M}\sqrt{l(l+1)}\frac{e^{i\left(kr-\frac{\pi}{2}l\right)}}{c(kr)^2}\vec{Y}_{lm}\right)\cdot\left(a_{lm}^{*M}\sqrt{l(l+1)}\frac{e^{-i\left(kr-\frac{\pi}{2}l\right)}}{c(kr)^2}\vec{Y}_{lm}^{*}\right)\right.\\ &\quad\left.+\left(ia_{lm}^{M}\frac{e^{i\left(kr-\frac{\pi}{2}l\right)}}{ckr}\vec{\Psi}_{lm}\right)\cdot\left(-ia_{lm}^{*M}\frac{e^{-i\left(kr-\frac{\pi}{2}l\right)}}{ckr}\vec{\Psi}_{lm}^{*}\right)\right]\end{aligned}$$

where the cross terms, $\vec{Y}_{lm}\cdot\vec{\Psi}_{lm}^{*}$ and $\vec{\Psi}_{lm}\cdot\vec{Y}_{lm}^{*}$, have vanished due to simple vector orthogonality. This second term then simplifies to,

$$\begin{aligned}\frac{|a_{lm}^{M}|^2}{4\mu_o c^2}\left[\frac{l(l+1)}{(kr)^4}\vec{Y}_{lm}\cdot\vec{Y}_{lm}^{*}+\frac{1}{(kr)^2}\vec{\Psi}_{lm}\cdot\vec{\Psi}_{lm}^{*}\right] &\cong \frac{|a_{lm}^{M}|^2}{4\mu_o c^2(kr)^2}\vec{\Psi}_{lm}\cdot\vec{\Psi}_{lm}^{*}\\ &= \frac{\varepsilon_o|a_{lm}^{M}|^2}{4(kr)^2}\vec{\Psi}_{lm}\cdot\vec{\Psi}_{lm}^{*}\end{aligned}$$

where we have recognized the radiation zone condition, $kr \gg l$, and omitted the (negligible) term due to the $1/r^2$ radial contribution to the magnetic field. Noting that, as expected in the radiation zone, the $\vec{E}$ and $\vec{B}$ field contributions to the energy density are equal in amplitude, we combine the two terms and Eq. A.34 simplifies to,

$$\left\langle\varepsilon_{lm}^{M}\right\rangle \cong \frac{\varepsilon_o|a_{lm}^{M}|^2}{4(kr)^2}\left(\vec{\Phi}_{lm}\cdot\vec{\Phi}_{lm}^{*}+\vec{\Psi}_{lm}\cdot\vec{\Psi}_{lm}^{*}\right)$$

Furthermore, with $\vec{r}\perp\nabla Y_{lm}$, we have $|\vec{\Phi}_{lm}| = |\vec{r}\times\nabla Y_{lm}| = r|\nabla Y_{lm}| = |\vec{\Psi}_{lm}|$, and so $|\vec{\Psi}_{lm}|^2 = |\vec{\Phi}_{lm}|^2$. The energy density can then be written,

$$\left\langle\varepsilon_{lm}^{M}\right\rangle \cong \frac{\varepsilon_o|a_{lm}^{M}|^2}{2(kr)^2}\left(\vec{\Psi}_{lm}\cdot\vec{\Psi}_{lm}^{*}\right)$$

Multiplying by c yields the time-averaged Poynting vector magnitude, $\left|\left\langle\vec{s}_{lm}^{M}\right\rangle\right|$, of a single magnetic multipole moment,

$$\left|\left\langle\vec{s}_{lm}^{M}\right\rangle\right| = c\left\langle\varepsilon_{lm}^{M}\right\rangle = \frac{\varepsilon_o c|a_{lm}^{M}|^2}{2(kr)^2}\left(\vec{\Psi}_{lm}\cdot\vec{\Psi}_{lm}^{*}\right)$$

where the time-averaged Poynting vector or energy-flux density is more commonly expressed as

$$\langle \vec{s} \rangle = \frac{1}{2\mu_o} Re\left(\vec{E} \times \vec{B}^*\right) \textbf{(Problem set}^3\textbf{)}.$$

The time-averaged energy, $\langle dU \rangle$, in a spherical shell between r and $r + dr$ due to this lm^{th} magnetic multipole moment is then given as,

$$\left\langle dU_{lm}^M \right\rangle = \frac{\varepsilon_o |a_{lm}^M|^2 dr}{2k^2} \int \left(\vec{\Psi}_{lm} \cdot \vec{\Psi}_{lm}^*\right) d\Omega = \frac{\varepsilon_o |a_{lm}^M|^2 dr}{2k^2} \qquad \text{(A.35)}$$

where we see that for any spherical shell of thickness dr, regardless of radius, the energy content is the same. In other words, energy is conserved as the pure harmonic spherical wave expands outward. Following an identical procedure, we find the lm^{th} electric multipole moment contribution of energy to a spherical shell of thickness dr in the far zone to be of the same form with $|a_{lm}^M|^2 \rightarrow |a_{lm}^E|^2$. Finally, if we consider both magnetic and electric moment contributions to the multipole fields and sum over all multipoles, the total time-averaged energy in a spherical shell of thickness dr is,

$$\langle dU \rangle = \sum_{lm} \left(\left\langle d\varepsilon_{lm}^M \right\rangle + \left\langle d\varepsilon_{lm}^E \right\rangle \right) = \sum_{lm} \frac{\varepsilon_o}{2k^2} \left(|a_{lm}^M|^2 + |a_{lm}^E|^2 \right) dr$$

A.3.2 Momentum density and angular momentum density

Through consideration of the Lorentz forces on charges and currents within an electromagnetic field, along with the law of momentum conservation, we can identify the time-averaged momentum density of a radiation field as,

$$\left\langle \vec{p} \right\rangle = \frac{1}{2\mu_o c^2} Re\left(\vec{E} \times \vec{B}^*\right) = \frac{\varepsilon_o}{2} Re\left(\vec{E} \times \vec{B}^*\right) \qquad \text{(A.36)}$$

which is related to the Poynting vector by $\left\langle \vec{p} \right\rangle = \langle \vec{s} \rangle / c^2$. In the far zone ($kr \gg 1$), the radiation is essentially expanding radially outward from a source approximated to be at the origin and so the momentum carried by the fields should be directed radially outward. From Eq. A.36 and the fact that the fields are dominantly transverse in the far zone, we see that this is indeed the case.

The radiation fields also carry angular momentum. The time-averaged angular momentum density $\langle \vec{g} \rangle$ is, as expected,

$$\langle \vec{g} \rangle = \vec{r} \times \left\langle \vec{p} \right\rangle = \frac{\varepsilon_o}{2} Re\left[\vec{r} \times \left(\vec{E} \times \vec{B}^*\right)\right] \qquad \text{(A.37)}$$

At first glance it would seem that in the far zone the field angular momentum would vanish because $\vec{E}$ and $\vec{B}$ are transverse and so the momentum $\left\langle \vec{p} \right\rangle$ is directed outward along $\vec{r}$. This is not the case, however, because the far zone fields in fact do have components, albeit small, directed radially. Because the lm^{th} contribution of angular momentum density in the far zone results from the interplay of magnetic multipole fields of order (l, m), $(l, m-1)$, and $(l, m+1)$, we cannot come to a proper description of the total field angular momentum by considering the contribution from a single magnetic (or electric) multipole moment (l, m) as we did with the field energy but rather we must consider a superposition of magnetic (or electric) multipoles of different m but all with the same l. So, considering the angular momentum density in the far zone contributed by all $2m+1$ magnetic multipole moments of order l,

$$\left\langle \vec{g}_l^M \right\rangle = \frac{\varepsilon_o}{2} Re\left[\vec{r} \times \left(\vec{E}_l^M \times \vec{B}_l^{*M}\right)\right] \tag{A.38}$$

where $\vec{E}_l^M = \sum_m \vec{E}_{lm}^M$ and $\vec{B}_l^{*M} = \sum_m \vec{B}_{lm}^{*M}$. Using the identity $\vec{a} \times \left(\vec{b} \times \vec{c}\right) = (\vec{a} \cdot \vec{c})\,\vec{b} - \left(\vec{a} \cdot \vec{b}\right)\vec{c}$ and the fact that $\vec{E}_l^M$ is purely transverse, $\langle \vec{g}_l^M \rangle$ becomes,

$$\left\langle \vec{g}_l^M \right\rangle = \frac{\varepsilon_o}{2} Re\left[\left(\vec{r} \cdot \vec{B}_l^{*M}\right)\vec{E}_l^M\right]$$

With a general relation between the electric and magnetic fields of radiation, $\vec{B} = \frac{1}{ikc}\nabla \times \vec{E}$, we next substitute for $\vec{B}_l^{*M}$ to get

$$\left\langle \vec{g}_l^M \right\rangle = \frac{\varepsilon_o}{2kc} Re\left[\frac{1}{i}\vec{r} \cdot \left(\nabla \times \vec{E}_l^{*M}\right)\vec{E}_l^M\right]$$

Combining the vector identity $\vec{a} \cdot \left(\nabla \times \vec{b}\right) = (\vec{a} \times \nabla) \cdot \vec{b}$ with the definition of the angular momentum operator $\vec{L} = \frac{1}{i}(\vec{r} \times \nabla)$ then yields

$$\left\langle \vec{g}_l^M \right\rangle = \frac{\varepsilon_o}{2kc} Re\left[\left(\vec{L} \cdot \vec{E}_l^{*M}\right)\vec{E}_l^M\right] = \frac{\varepsilon_o}{2kc} Re\left[\left(\vec{L} \cdot \sum_m \vec{E}_{lm}^{*M}\right)\sum_{m'} \vec{E}_{lm'}^M\right]$$

For the far zone, according to Eq. A.58, this can be written explicitly as,

$$\left\langle \vec{g}_l^M \right\rangle = \frac{-\varepsilon_o}{2ck} Re\left[\left(\vec{L} \cdot \sum_{m'} i a_{lm'}^{*M} \frac{e^{-i\left(kr - \frac{\pi}{2}l\right)}}{kr} \vec{\Phi}_{lm'}^*\right)\sum_m i a_{lm}^M \frac{e^{i\left(kr - \frac{\pi}{2}l\right)}}{kr} \vec{\Phi}_{lm}\right]$$

Noting that $\vec{L}$ operates only on functions of angle and that the exponentials cancel, this simplifies to

$$\left\langle \vec{g}_l^M \right\rangle = \frac{\varepsilon_o}{2ck^3r^2} \sum_{m,m'} Re\left[a_{lm'}^{*M} a_{lm}^M \left(\vec{L} \cdot \vec{\Phi}_{lm'}^*\right)\vec{\Phi}_{lm}\right]$$

Again using the definition of the angular momentum operator, this time to express the $\vec{\Phi}_{lm}$ of Eq. A.3 as $\vec{\Phi}_{lm} = \frac{i}{\sqrt{l(l+1)}}\vec{L}Y_{lm}$, we get

$$\left\langle \vec{g}_l^M \right\rangle = \frac{\varepsilon_o}{2ck^3r^2} \frac{1}{l(l+1)} \sum_{m,m'} Re\left[a_{lm'}^{*M} a_{lm}^M \left(L^2 Y_{lm'}^*\right)\vec{L}Y_{lm}\right]$$

but $L^2 Y^*_{lm} = l(l+1)\, Y^*_{lm}$, so finally Eq. A.38 is re-expressed as,

$$\left\langle \vec{g}^M_l \right\rangle = \frac{\varepsilon_o}{2ck^3r^2} \sum_{m,m'} Re\left[a^{*M}_{lm'} a^M_{lm} Y^*_{lm'} \vec{L} Y_{lm} \right] \tag{A.39}$$

The $\vec{L}$ operator can be expressed in Cartesian coordinates as

$$\vec{L} = L_x \hat{x} + L_y \hat{y} + L_z \hat{z} = \frac{(L_+ + L_-)}{2} \hat{x} + \frac{(L_+ - L_-)}{2i} \hat{y} + L_z \hat{z} \tag{A.40}$$

where L_+ and L_-, known, respectively, as the raising and lowering operators, act on the spherical harmonics to effectively "raise" or "lower" them by one unit of m with the introduction of an l, m dependent coefficient. That is,

$$L_\pm Y_{lm} = \left[l^2 + l - \left(m^2 \pm m \right) \right]^{\frac{1}{2}} Y_{l,m\pm 1}$$

while the z component of the operator yields,

$$L_z Y_{lm} = m Y_{lm}$$

Using Eq. A.40 within Eq. A.39 with the abbreviated notation $\{m^\pm\} = \left[l^2 + l - \left(m^2 \pm m \right) \right]^{\frac{1}{2}}$ it is straightforward to obtain the Cartesian components of $\langle \vec{g}^M_l \rangle$,

$$\left\langle \vec{g}^M_l \right\rangle_x = \frac{\varepsilon_o}{4\omega k^2 r^2} \sum_{m,m'} Re\left[a^{*M}_{lm'} a^M_{lm} Y^*_{lm'} \left(\{m^+\} Y_{l,m+1} + \{m^-\} Y_{l,m-1} \right) \right]$$

$$\left\langle \vec{g}^M_l \right\rangle_y = \frac{\varepsilon_o}{4\omega k^2 r^2} \sum_{m,m'} Im\left[a^{*M}_{lm'} a^M_{lm} Y^*_{lm'} \left(\{m^+\} Y_{l,m+1} - \{m^-\} Y_{l,m-1} \right) \right]$$

$$\left\langle \vec{g}^M_l \right\rangle_z = \frac{\varepsilon_o}{2\omega k^2 r^2} \sum_{m,m'} Re\left[a^{*M}_{lm'} a^M_{lm} Y^*_{lm'} m Y_{lm} \right]$$

where for the $\hat{y}$ component, we have used $Re(-iZ) = Im(Z)$ in which Z is a complex function. With these results we see that the Cartesian components of the lth magnetic multipole contribution to the field angular momentum density for a given position (r, θ, ϕ) are proportional to sums of $Y_{lm}(\theta, \phi)$ pair products. If, in addition, we want to know the angular momentum, dG, in a spherical shell between r and $r + dr$ due to the lth magnetic multipole moment, we integrate over 4π steradians, taking advantage of the orthonormality of the spherical harmonics,

$$\left\langle d\vec{G}^M_l \right\rangle_x = \int_\Omega \left\langle \vec{g}^M_l \right\rangle_x d\Omega r^2 dr = \frac{\varepsilon_o dr}{4\omega k^2} \sum_m Re\left[a^{*M}_{l,m+1} a^M_{lm} \{m^+\} + a^{*M}_{l,m-1} a^M_{lm} \{m^-\} \right] \tag{A.41}$$

$$\left\langle d\vec{G}_l^M \right\rangle_y = \int_\Omega \left\langle \vec{g}_l^M \right\rangle_y d\Omega r^2 dr = \frac{\varepsilon_o dr}{4\omega k^2} \sum_m Im\left[a_{l,m+1}^{*M} a_{lm}^M \{m^+\} - a_{l,m-1}^{*M} a_{lm}^M \{m^-\}\right] \tag{A.42}$$

$$\left\langle d\vec{G}_l^M \right\rangle_z = \int_\Omega \left\langle \vec{g}_l^M \right\rangle_z d\Omega r^2 dr = \frac{\varepsilon_o dr}{2\omega k^2} \sum_m m|a_{lm}^M|^2 \tag{A.43}$$

Now we see what was pointed out earlier, namely that the lm^{th} contribution of angular momentum results from the interplay of magnetic multipole fields of order $(l,m),(l,m-1)$, and $(l,m+1)$, at least for the x and y components. For the z component, the situation is simpler in that, just as with the field energy, this component of the field angular momentum can be parsed according to individual magnetic multipole moments (l,m). In fact, a comparison of the lm^{th} magnetic multipole time-averaged contributions of energy (Eq. A.35) and z component of angular momentum (Eq. A.43) reveals the relation,

$$\left\langle d\vec{G}_{lm}^M \right\rangle_z = \frac{m}{\omega} \left\langle dU_{lm}^M \right\rangle$$

Because this treatment can be carried out for electric multipoles with identical results, this relationship between energy and angular momentum is general for monochromatic radiation. Although this is a purely classical result, it can be interpreted quantum mechanically as a general statement that for each quanta of radiation energy, $\hbar\omega$, in radiation from a multipole of order lm, there exists an amount $\hbar m$ of z component angular momentum.

A.4 Multipole fields from vector harmonic expansion

A.4.1 Multipole expansion including sources

Ultimately, the fields result from the sources and more precisely, the lm^{th} multipole radiation fields originate with the lm^{th} electric and magnetic multipole moments of the source. It seems natural, then, that the multipole expansion of the radiation be expressible in terms of the multipole moments (multipoles) of the source. Continuing our treatment of the electric part of the radiation, with the understanding the magnetic part can be treated identically, a description of the field can be obtained by first solving the decoupled inhomogeneous differential equations associated with the $\vec{\Phi}_{lm}$, Eq. A.27, for $E_{lm}^{(2)M}$. For a given lm we have,

$$\left(\frac{1}{r_o^2}\frac{d}{dr_o}\left(r_o^2\frac{d}{dr_o}\right) - \frac{l(l+1)}{r_o^2} + k^2\right) E_{lm}^{(2)M} = -i\omega\mu_o \mathcal{J}_{lm}^{(2)} \tag{A.44}$$

where $\mathcal{J}_{lm}^{(2)}$ is the $\vec{\Phi}_{lm}{}^{th}$ component of the source current density and we now explicitly associate r with the source and r_o with the observer. Noting that the VSH are an orthonormal and complete set, $\mathcal{J}_{lm}^{(2)}$ is given by

$$\mathcal{J}_{lm}^{(2)}(r) = \int \vec{\Phi}_{lm}^*(\theta,\phi) \cdot \vec{\mathcal{J}}(r,\theta,\phi)\, d\Omega \tag{A.45}$$

For the magnetic field, through an identical treatment starting with Eq. A.13, we again obtain three sets of inhomogeneous radial equations, in $\vec{Y}_{lm}$, $\vec{\Psi}_{lm}$, and $\vec{\Phi}_{lm}$, similar to A.25–A.27. And again, with the $\vec{\Phi}_{lm}$ equations uncoupled, we get

$$\left(\frac{1}{r_o}\frac{d^2}{dr_o^2}r_o - \frac{1}{r_o^2}l(l+1) + k^2\right)B_{lm}^{(2)E} = -\mu_o\left(\nabla\times\vec{\mathcal{J}}\right)_{lm}^{(2)} \tag{A.46}$$

where the $\vec{\Phi}_{lm}$ th component of the curl of the source current density, $\left(\nabla\times\vec{\mathcal{J}}\right)_{lm}^{(2)}$, is similarly given by

$$\left(\nabla\times\vec{\mathcal{J}}\right)_{lm}^{(2)} = \int \vec{\Phi}_{lm}^{*}(\theta,\phi)\cdot\nabla\times\vec{\mathcal{J}}(r,\theta,\phi)\,d\Omega \tag{A.47}$$

For points outside the source distribution, away from the origin, the Eq. A.44 reduce to homogeneous equations such as Eq. A.28 and, as we have shown, the fields are then given in general by A.32. However, in the present case we would like to describe the fields at all points in space. We want the solution to not only represent outgoing waves at large distances but also to describe the fields within the source distribution at and near the origin. In particular, the solutions are required to be finite at the origin. Now, as can be seen from the Eq. A.29, the $y_l(kr)$, and thus the $h_l^{(1,2)}(kr)$, all asymptote to infinity as $r\to 0$ and so fail to satisfy this requirement. In fact, the only homogeneous solutions to A.27 that are finite at the origin are the $j_l(kr)$. This implies that a proper solution to satisfy both outgoing waves at large distances and finiteness at the origin should somehow involve the product $j_l(kr)\,h_l^1(kr)$. With this in mind, consider the Green function equation associated with Eq. A.44:

$$\left(\frac{1}{r_o^2}\frac{d}{dr_o}\left(r_o^2\frac{d}{dr_o}\right) - \frac{l(l+1)}{r_o^2} + k^2\right)G(r_o,r) = -\frac{1}{r_o^2}\delta(r_o-r) \tag{A.48}$$

where the Green function (GF), which is interpreted as the response at r_o to an element of current density at r, can be expressed in terms of the solutions of the associated homogeneous differential equation. It is found to be:

$$G(r_o,r) = ikj_l(kr_<)\,h_l^{(1)}(kr_>) \tag{A.49}$$

with $r_<$ $(r_>)$ representing the smaller (larger) of r_o and r. Using this GF as a weighting function for the lm^{th} magnetic multipole contribution to the electric part of the radiation field (i.e., for the $i\omega\mu_o\mathcal{J}_{lm}^{(2)}$ inhomogeneity in Eq. A.44), we obtain $E_{lm}^{(2)M}(r_o)$:

$$E_{lm}^{(2)M}(r_o) = i\omega\mu_o\int G(r_o,r)\,r^2\mathcal{J}_{lm}^{(2)}(r)\,dr \tag{A.50}$$

where the r^2 results from having to solve the GF in the Sturm–Liouville form which is r_o^2 times Eq. A.48. Plugging the Green Function of Eq. A.49 into Eq. A.50 and noting that $r_> = r_o$ and $r_< = r$ for the solution in the region outside the source, we get the explicit statement,

$$E_{lm}^{(2)M}(r_o) = -\omega k \mu_o \left[\int j_l(kr)\, r^2 \mathcal{J}_{lm}^{(2)}(r)\, dr \right] h_l^{(1)}(kr_o)$$

from which, through comparison with A.30, we can identify the amplitude, a_{lm}^M, of the lm^{th} order magnetic multipole radiation,

$$a_{lm}^M = -\omega k \mu_o \int j_l(kr)\, r^2 \mathcal{J}_{lm}^{(2)}(r)\, dr$$

substituting the representation in Eq. A.45 for $\mathcal{J}_{lm}^{(2)}$, we obtain a workable expression for the coefficient a_{lm}^M,

$$\begin{aligned} a_{lm}^M &= -\omega k \mu_o \int \int j_l(kr) \left[\vec{\Phi}_{lm}^*(\theta,\phi) \cdot \vec{\mathcal{J}}(r,\theta,\phi) \right] r^2 d\Omega dr \\ &= -\frac{k^2}{\varepsilon_o c} \int j_l(kr) \left[\vec{\Phi}_{lm}^*(\theta,\phi) \cdot \vec{\mathcal{J}}(r,\theta,\phi) \right] dV \end{aligned} \tag{A.51}$$

And finally, a similar treatment for the magnetic field yields a_{lm}^E,

$$a_{lm}^E = \frac{ik}{\varepsilon_o c^2} \int j_l(kr) \left[\vec{\Phi}_{lm}^*(\theta,\phi) \cdot \nabla \times \vec{\mathcal{J}}(r,\theta,\phi) \right] dV \tag{A.52}$$

and so connecting the radiation with its sources has provided us with expressions for the coefficients, a_{lm}^M (and a_{lm}^E), of the magnetic (and electric) multipole radiation. Insertion of these terms into Eq. A.32 then gives the solution for the VSH multipole expansion of the fields in the presence of sources.

So, we have again derived electric and magnetic multipole moments of the radiation fields from an arbitrary distribution of charge and current. This time, however, unlike the simpler Taylor expansion of Chapter 4, no approximations regarding the relative sizes of the source, evaluation point radius, and radiation wavelength have yet been made. Thus, the expressions for the fields in terms of the multipole moments of the source, given by Eq. A.32 along with Eqs. A.51 and A.52, are exact in all orders and correct in all zones. Indeed, here we have a method of obtaining exact terms, without approximations, of any order, which are relatively easy to compute.

For wavelengths long compared to source dimensions ($kr \ll 1$)

With Eqs. A.32, A.51 and A.52 as starting points, we can proceed to consider simplifications to specific cases. For the common case in which the source size is much smaller than any of the radiated wavelengths, or $kr \ll 1$, such as occurs in atomic and nuclear systems of charges, $j_l(kr)$ can be replaced by its asymptotic form (see Arfken and Weber)[4],

$$j_l(kr) \simeq \frac{(kr)^l}{(2l+1)!!} \tag{A.53}$$

where $(2l+1)!! = [(2l+1)(2l-1)(2l-3)\ldots 5 \cdot 3 \cdot 1]$. Using this asymptotic form and the explicit form of $\vec{\Phi}_{lm}(\theta,\phi)$,

$$\vec{\Phi}_{lm}(\theta,\phi) = \frac{1}{\sqrt{l(l+1)}}\vec{r}\times\nabla Y_{lm}(\theta,\phi) = \frac{1}{\sqrt{l(l+1)}}i\vec{L}Y_{lm}(\theta,\phi)$$

where $\vec{L} = \frac{1}{i}(\vec{r}\times\nabla)$ is essentially the angular momentum operator of quantum mechanics encountered in Section A.3.2, Eq. A.51 can be expressed as:

$$a_{lm}^{M} \simeq \frac{-k^{l+2}}{\varepsilon_o c(2l+1)!!\sqrt{l(l+1)}}\int r^l\left[\left(i\vec{L}Y_{lm}(\theta,\phi)\right)^*\cdot\vec{\jmath}(r,\theta,\phi)\right]dV \tag{A.54}$$

Now, since $\vec{L}$ is Hermitian, $\left(\vec{L}Y_{lm}\right)^* = Y_{lm}^*\vec{L}^* = Y_{lm}^*\vec{L}$ and we can write,

$$a_{lm}^{M} \simeq \frac{ik^{l+2}}{\varepsilon_o c(2l+1)!!\sqrt{l(l+1)}}\int r^l\left[Y_{lm}^*(\theta,\phi)\vec{L}\cdot\vec{\jmath}(r,\theta,\phi)\right]dV$$

And using the vector identity $\vec{L}\cdot\vec{\jmath} = i\nabla\cdot\left(\vec{r}\times\vec{\jmath}\right)$, we get (**Problem set**[5]),

$$a_{lm}^{M} \simeq \frac{-k^{l+2}}{\varepsilon_o c(2l+1)!!\sqrt{l(l+1)}}\int r^l Y_{lm}^*(\theta,\phi)\nabla\cdot\left(\vec{r}\times\vec{\jmath}\right)dV$$

or

$$a_{lm}^{M} \simeq \frac{1}{\varepsilon_o(2l+1)!!}\left(\frac{l+1}{l}\right)^{\frac{1}{2}}k^{l+2}m_{lm} \tag{A.55}$$

where the $m_{lm} = \frac{-1}{c(l+1)}\int r^l Y_{lm}^*(\theta,\phi)\nabla\cdot(\vec{r}\times\vec{\jmath})dV$ are the magnetic multipole moments. Again, a similar approximation can be carried out for $kr \ll 1$ with the electric multipole radiation coefficient a_{lm}^{E} as given by Eq. A.52 resulting in (**Problem set**[6])

$$a_{lm}^{E} \simeq \frac{-1}{\varepsilon_o c(2l+1)!!}\left(\frac{l+1}{l}\right)^{\frac{1}{2}}k^{l+2}q_{lm} \tag{A.56}$$

where the $q_{lm} = \int r^l Y_{lm}^*(\theta,\phi)\rho(r,\theta,\phi)dV$ are the electric multipole moments introduced and shown, in Discussion 2.3 of Chapter 2, as linear combinations of the Taylor-expanded multipole moment components. In particular, we note the relations between the three $l = 1$ electric multipole moments $\left(q_{1,1}, q_{1,0}, q_{1,-1}\right)$ and the electric dipole moment, $\vec{p} = \int\vec{r}\rho(\vec{r})dV$ (**Problem set**[7]):

$$q_{1,1} = \mp\sqrt{\frac{3}{8\pi}}\left(p_x \mp ip_y\right)$$

$$q_{1,0} = \sqrt{\frac{3}{4\pi}}p_z \tag{A.57}$$

Similar relations exist between the three $l=1$ magnetic multipole moments $(m_{1,1}, m_{1,0}, m_{1,-1})$ and magnetic dipole moment, $\vec{\mathrm{m}} = \frac{1}{2}\int \vec{r} \times \vec{J} dV$ (**Problem set**[8]).

A.4.2 The small source approximation: near and far zones

In the interest of completeness, we would like to connect the exact spherical multipole expansion results expressed in Eqs. A.32, A.51 and A.52 of this chapter with the near and far zone approximations to the radiation fields obtained in Chapter 4. Recall that in that chapter the Taylor expansion of the vector potential in Cartesian coordinates was taken in the limit of source dimensions that are small relative to both the wavelength ($kr \ll 1$) and the radius of the evaluation point ($r \ll r_o$), the combination of which we are calling the "small source" approximation. Having noted that the resulting small source expressions, though approximated, were nonetheless valid everywhere, they were then further approximated in the limits of the near or "static" ($kr_o \ll 1$) and far or "radiation" ($kr_o \gg 1$) zones. To reproduce these results from the exact spherical multipole expansion terms of this chapter, we must therefore consider the same limits and approximations. At the end of Section A.4.1, we obtained the approximate forms of the multipole coefficients a_{lm}^M and a_{lm}^E (Eqs. A.55 and A.56) in the limit of $kr \ll 1$, the so-called "long wavelength" limit. With the remaining necessary approximation of $r \ll r_o$ assumed in what follows, we can proceed in the small source approximation within which we can express the spherical multipole terms in the near and far zones for comparison with the Cartesian terms.

Fields in the radiation or far zone ($kr_o \gg 1$)

As discussed in Chapter 2, the radiation zone is characterized by $r_o \gg \lambda$ (or equivalently $kr_o \gg 1$). Later, in Chapter 3, we found the dominant fields in this zone to be transverse and to scale as r_o^{-1}. It is therefore expected that the electric and magnetic fields obtained in Eq. A.32 will behave identically in this zone. To get some insight into the far zone behavior of these fields first notice from the original radial ODE Eq. A.28 that for $k^2 \gg l(l+1)/r_o^2$ or equivalently $kr_o \gg l$, the equation approximates a homogeneous radial Helmholtz equation. The solutions must then correspondingly transform from the spherical Bessel functions to the Helmholtz equation solutions as $kr_o \to kr_o \gg l$. And they do: noting that the outgoing solution to the Helmholtz equation is given by $\psi(r_o) \propto e^{ikr_o}/r_o$ and recalling that the general outgoing spherical Hankel function solutions to Eq. A.28, $h_l^1(kr_o)$, have the form of e^{ikr_o}/r_o multiplied by a complex power series in inverse powers of kr_o, it is no surprise that these spherical Hankel functions all asymptote to e^{ikr_o}/kr_o, as $r_o \to \infty$. Specifically, as shown in Arfken and Weber,[9] the forms of the spherical Bessel and Neumann functions in the limit of $kr_o \gg l$ are as follows:

$$j_l(kr_o) \sim \frac{1}{kr_o}\sin\left(kr_o - \frac{\pi}{2}l\right)$$

$$y_l(kr_o) \sim \frac{-1}{kr_o}\cos\left(kr_o - \frac{\pi}{2}l\right)$$

and the form of the spherical outgoing (1) and incoming (2) Hankel functions in this limit follow since, by definition,

$$h_l^{1,2}(kr_o) = j_l(kr_o) \pm i y_l(kr_o)$$
$$= \mp \frac{i}{kr_o}\left[\cos\left(kr_o - \frac{\pi}{2}l\right) \pm i \sin\left(kr_o - \frac{\pi}{2}l\right)\right]$$
$$= \mp i \frac{e^{\pm i\left(kr_o - \frac{\pi}{2}l\right)}}{kr_o}$$

Using this approximation for $h_l^1(kr_o)$ in the limit of $kr_o \gg l$, and noting the l dependent phase factor $-ie^{-i\frac{\pi}{2}l}$, we can approximate the VSH multipole expansion of Eq. A.32 in the radiation zone as,

$$\vec{E}_{kr_o \gg l}(\vec{r}_o, \omega) = \sum_{lm}\left[\vec{E}_{lm}^E + \vec{E}_{lm}^M\right]$$
$$= \sum_{lm}\left[\frac{c}{k}\nabla \times a_{lm}^E \frac{e^{i\left(kr_o - \frac{\pi}{2}l\right)}}{kr_o}\vec{\Phi}_{lm} - i a_{lm}^M \frac{e^{i\left(kr_o - \frac{\pi}{2}l\right)}}{kr_o}\vec{\Phi}_{lm}\right]$$
$$\vec{B}_{kr_o \gg l}(\vec{r}_o, \omega) = \sum_{lm}\left[\vec{B}_{lm}^E + \vec{B}_{lm}^M\right]$$
$$= \sum_{lm}\left[-i a_{lm}^E \frac{e^{i\left(kr_o - \frac{\pi}{2}l\right)}}{kr_o}\vec{\Phi}_{lm} - \frac{1}{ck}\nabla \times a_{lm}^M \frac{e^{i\left(kr_o - \frac{\pi}{2}l\right)}}{kr_o}\vec{\Phi}_{lm}\right] \quad \text{(A.58)}$$

So we see immediately that the purely transverse VSH terms $(f(r_o)\vec{\Phi}_{lm})$ fall off as $1/r_o$. On the other hand, it can be shown with the use of Eq. A.17 that the $\nabla \times f(r_o)\vec{\Phi}_{lm}$ terms contain both radial and transverse field components that fall off as $1/r_o^2$ and $1/r_o$, respectively (**Problem set**[10]). This behavior is consistent with the transverse, $1/r_o$ dependent "decoupled" fields that dominate in the radiation zone and the radial, $1/r_o^2$ dependent "coupled" fields that dominate in the near zone both of which were first discussed in Section 3.2. In other words, in this expansion approximated for the radiation zone, the transverse fields dominate and when these complex exponential $1/r_o$ terms are combined with the sinusoidal time dependence $e^{-i\omega t}$, outgoing transverse spherical waves result. We note again that the purely transverse VSH terms $(f(r_o)\vec{\Phi}_{lm})$ represent the magnetic/electric fields contributed by electric/magnetic multipoles.

Because we are in the long-wavelength $(kr \ll 1)$ limit, we can use the approximate forms of the multipole coefficients given by Eqs. A.55 and A.56 in Eq. A.58. And because these approximations $(kr \ll 1$ and $kr_o \gg 1)$ now amount to being in the small source, radiation zone regime, Eq. A.58 should yield field terms that are linear combinations of the associated Cartesian multipole field term components. To show this, we start by considering the electric multipole contributions to the $\vec{B}$ field in Eq. A.58

$$\sum_{lm}\vec{B}_{lm}^E = -i\sum_{lm} a_{lm}^E \frac{e^{i\left(kr_o - \frac{\pi}{2}l\right)}}{kr_o}\vec{\Phi}_{lm}$$

After expressing $\vec{\Phi}_{lm}$ explicitly as given in Eq. A.3 and using the long-wavelength limit form of a_{lm}^E as given in Eq. A.56 this becomes,

$$\sum_{lm}\vec{B}^{E}_{lm} = -i\sum_{lm}\left(\frac{-1}{\varepsilon_o c(2l+1)!!}\left(\frac{l+1}{l}\right)^{\frac{1}{2}}k^{l+2}q_{lm}\right)\frac{e^{i(kr_o-\frac{\pi}{2}l)}}{kr_o}\left(\frac{1}{\sqrt{l(l+1)}}\vec{r}_o\times\nabla Y_{lm}\right)$$

where it is seen that lowest order multipole contribution corresponds to $l=1$. This $l=1$ term can be simplified to,

$$\vec{B}^{E}_{1,m} = \frac{k^2 e^{ikr_o}}{3\varepsilon_o c}q_{1,m}\left(\hat{r}_o\times\nabla Y_{1,m}\right) \tag{A.59}$$

where $m=1,0,-1$ and we have used $ie^{-i\frac{\pi}{2}}=1$. It can further be shown that (**Problem set**[11])

$$\hat{r}_o\times\nabla Y_{1\pm1} = i\sqrt{\frac{3}{8\pi}}\frac{1}{r_o}\left[\cos\theta\hat{x}\pm i\cos\theta\hat{y}-(\sin\theta\cos\phi\pm i\sin\theta\sin\phi)\hat{z}\right] \tag{A.60}$$

$$\hat{r}_o\times\nabla Y_{10} = \sqrt{\frac{3}{4\pi}}\frac{1}{r_o}\left(\sin\theta\sin\phi\hat{x}-\sin\theta\cos\phi\hat{y}\right) \tag{A.61}$$

Considering the $m=0$ term first, substitution of Eqs. A.61 and A.57 into A.59 yields

$$\vec{B}^{E}_{1,0} = \frac{k^2 e^{ikr_o}}{3\varepsilon_o c}\left(\sqrt{\frac{3}{4\pi}}p_z\right)\left(\sqrt{\frac{3}{4\pi}}\frac{1}{r_o}\left(\sin\theta\sin\phi\hat{x}-\sin\theta\cos\phi\hat{y}\right)\right) \tag{A.62}$$

If we note that $\hat{\phi}=-\sin\phi\hat{x}+\cos\phi\hat{y}$ and $\mu_o c=(\varepsilon_o c)^{-1}$, this becomes

$$\vec{B}^{E}_{1,0} = -ck^2\frac{\mu_o}{4\pi}\frac{e^{ikr_o}}{r_o}p_z\sin\theta\hat{\phi}$$

which tells us that the $l=1$, $m=0$ electric multipole contribution to the magnetic field in the small source, radiation zone limit, is purely azimuthal. Furthermore, for the specific case in which an electric dipole moment has only a $\hat{z}$ component ($\vec{p}=p_z\hat{z}$), this can be written as,

$$\vec{B}^{E}_{1,0} = ck^2\frac{\mu_o}{4\pi}\frac{e^{ikr_o}}{r_o}\left(\hat{r}_o\times\vec{p}\right) = \hat{r}_o\times ck^2\frac{\mu_o}{4\pi}\frac{e^{ikr_o}}{r_o}\vec{p}$$

For the more general case in which the electric dipole moment has extension in all three coordinates, we substitute Eqs. A.60 along with the q_{11} and q_{1-1} terms of A.57 into A.59 to get

$$\begin{aligned}\vec{B}^{E}_{1\pm1} = \frac{ik^2}{8\pi\varepsilon_o c}\frac{e^{ikr_o}}{r_o}\Big[&\mp\cos\theta\left(p_x\hat{x}+p_y\hat{y}\right)\pm\sin\theta\left(\cos\phi p_x+\sin\phi p_y\right)\hat{z}\\ &-i\left(\cos\theta p_x\hat{y}-\cos\theta p_y\hat{x}\right)-i\left(\sin\theta\cos\phi p_y-\sin\theta\sin\phi p_x\right)\hat{z}\Big]\end{aligned} \tag{A.63}$$

where we note here that $\vec{B}^{E}_{10}$ is associated with p_z, and $\vec{B}^{E}_{1\pm1}$ are associated with p_x and p_y. If we now consider the sum of $\vec{B}^{E}_{1,1}$ and $\vec{B}^{E}_{1,-1}$, the real terms in the brackets cancel and we get

$$\vec{B}^E_{1,1} + \vec{B}^E_{1-1} = \frac{k^2}{4\pi\varepsilon_o c}\frac{e^{ikr_o}}{r_o}\left[\left(\cos\theta p_x\hat{y} - \cos\theta p_y\hat{x}\right) + \left(\sin\theta\cos\phi p_y - \sin\theta\sin\phi p_x\right)\hat{z}\right]$$
$$= ck^2\frac{\mu_o}{4\pi}\frac{e^{ikr_o}}{r_o}\left[\left(\cos\theta p_x\hat{y} - \cos\theta p_y\hat{x}\right) + \left(\hat{r}_o \times \vec{p}\right)_z\hat{z}\right]$$

Finally, combining this sum with $\vec{B}^E_{1,0}$ of Eq. A.62,

$$\vec{B}^E_{1,1} + \vec{B}^E_{1-1} + \vec{B}^E_{1,0} = ck^2\frac{\mu_o}{4\pi}\frac{e^{ikr_o}}{r_o}\left[\left(\sin\theta\sin\phi p_z - \cos\theta p_y\right)\hat{x} + \left(\cos\theta p_x - \sin\theta\cos\phi p_z\right)\hat{y} + \left(\hat{r}_o \times \vec{p}\right)_z\hat{z}\right]$$
$$= ck^2\frac{\mu_o}{4\pi}\frac{e^{ikr_o}}{r_o}\left[\left(\hat{r}_o \times \vec{p}\right)_x\hat{x} + \left(\hat{r}_o \times \vec{p}\right)_y\hat{y} + \left(\hat{r}_o \times \vec{p}\right)_z\hat{z}\right]$$
$$= ck^2\frac{\mu_o}{4\pi}\frac{e^{ikr_o}}{r_o}\left(\hat{r}_o \times \vec{p}\right) = \left[\vec{B}\right]_{ed} \tag{A.64}$$

which is exactly the magnetic field contribution from the electric dipole (zeroth order) term, Eq. 4.32 in the Cartesian multipole expansion of Chapter 4. It is the connection between the $l = 1$ spherical electric multipole and the (zeroth order) Cartesian electric dipole contributions to the magnetic field in the small source, radiation zone limit. The associated electric field naturally follows and is given by

$$\vec{E}^E_{1,1} + \vec{E}^E_{1,-1} + \vec{E}^E_{1,0} = \frac{ic}{k}\nabla \times \left(\vec{B}^E_{1,1} + \vec{B}^E_{1-1} + \vec{B}^E_{1,0}\right)$$
$$= -c\hat{r}_o \times \left(\vec{B}^E_{1,1} + \vec{B}^E_{1-1} + \vec{B}^E_{1,0}\right)$$
$$= -\hat{r}_o \times k^2\frac{1}{4\pi\varepsilon_o}\frac{e^{ikr_o}}{r_o}\left(\hat{r}_o \times \vec{p}\right) = \left[\vec{E}\right]_{ed}$$

where we have used the operator relation $(\nabla\times) \to \left(ik\hat{r}_o\times\right)$ that can be used for single wavenumber harmonic vector fields. This, then, completes the connection of the $l = 1$ electric multipole fields to electric dipole fields and shows they are the same fields just parsed differently. What about the other half of the $l = 1$ contribution, the magnetic multipole contribution? It can be shown in an analogous way that the (first order) magnetic dipole fields of Eq. 4.48 are similarly connected to the $l = 1$ magnetic multipole fields. Indeed, unlike for the Cartesian expansion in which, for example, the first order represents both dipole and quadrupole terms, the various l orders in the spherical expansion represent similar terms: $l = 1$ represents dipole terms, $l = 2$ represents quadrupole terms, and so on.

Fields in the static or near zone ($kr_o \ll 1$)

The near zone is defined by $r_o \ll \lambda$ (or $kr_o \ll 1$). We found in Chapter 4 that fields in this zone, in the limit of "small sources" ($r \ll r_o$ and $r \ll \lambda$), are a mixture of radial and transverse components and can be characterized as classic static fields which are oscillating at $e^{-i\omega t}$. We expect, therefore, that the exact electric and magnetic fields obtained in Eq. A.32, when subject to the same approximations, will behave identically.

Implementing the near zone ($kr_o \ll 1$) approximations of the spherical Bessel and Neumann functions[12],

$$j_l(kr_o) \simeq \frac{(kr_o)^l}{(2l+1)!!}$$

$$y_l(kr_o) \simeq -\frac{(2l-1)!!}{(kr_o)^{l+1}}$$

and noting that under the condition $kr_o \ll 1$, we have $j_l(kr_o) \ll y(kr_o)$ and so the associated Hankel functions, $h_l^{1,2}(kr_o) \simeq j_l(kr_o) \pm iy_l(kr_o)$, reduce to

$$h_l^{1,2}(kr_o) \simeq \mp i\frac{(2l-1)!!}{(kr_o)^{l+1}}$$

and we can replace $h_l^1(kr_o)$ in Eq. A.32 by the asymptotic form

$$h_l^1(kr_o) = -i\frac{(2l-1)!!}{(kr_o)^{l+1}} \tag{A.65}$$

This results in the near zone approximation of the VSH multipole expansion,

$$\vec{E}_{kr_o\ll 1} = \sum_{lm}\left[\frac{c}{k}\nabla\times a_{lm}^E\left(\frac{(2l-1)!!}{(kr_o)^{l+1}}\right)\vec{\Phi}_{lm} - ia_{lm}^M\left(\frac{(2l-1)!!}{(kr_o)^{l+1}}\right)\vec{\Phi}_{lm}\right]$$

$$\vec{B}_{kr_o\ll 1} = \sum_{lm}\left[-ia_{lm}^E\left(\frac{(2l-1)!!}{(kr_o)^{l+1}}\right)\vec{\Phi}_{lm} - \frac{1}{ck}\nabla\times a_{lm}^M\left(\frac{(2l-1)!!}{(kr_o)^{l+1}}\right)\vec{\Phi}_{lm}\right] \tag{A.66}$$

where we first note that, just as for static multipole fields, the terms fall off in the near zone as $1/r_o^{l+1}$ and $1/r_o^{l+2}$ (see discussion and note after Eq. A.58). That is, the fields fall off at rates depending on the multipole order. Because we are in the long-wavelength ($kr \ll 1$) limit, we can use the approximate forms of the multipole coefficients given by Eqs. A.55 and A.56 in Eq. A.66. Note also that the combined limits of long-wavelength ($kr \ll 1$) and near zone ($kr_o \ll 1$) do not specify the size of r relative to r_o and so we must additionally assume ($r \ll r_o$) for the small source approximation. In these limits, Eq. A.58 should again yield field terms that are linear combinations of the associated Cartesian multipole field term components. To show this, just as with the radiation zone, we start by considering the electric multipole contributions to the $\vec{B}$ field in Eq. A.66

$$\sum_{lm}\vec{B}_{lm}^E = -i\sum_{lm}a_{lm}^E\frac{(2l-1)!!}{(kr_o)^{l+1}}\vec{\Phi}_{lm}$$

Expressing $\vec{\Phi}_{lm}$ as in Eq. A.3 and a_{lm}^E as in Eq. A.56,

$$\sum_{lm}\vec{B}_{lm}^E = -i\sum_{lm}\left(\frac{-1}{\varepsilon_o c(2l+1)!!}\left(\frac{l+1}{l}\right)^{\frac{1}{2}}k^{l+2}q_{lm}\right)\frac{(2l-1)!!}{(kr_o)^{l+1}}\left(\frac{1}{\sqrt{l(l+1)}}\vec{r}_o\times\nabla Y_{lm}\right)$$

As before, consider the $l=1$; $m=1,0,-1$ terms that reduce to

$$\vec{B}^E_{1,m} = \frac{ik}{3\varepsilon_o c r_o} q_{1,m} \left(\hat{r}_o \times \nabla Y_{1,m}\right) \tag{A.67}$$

For the $m=0$ term, substitution of Eqs. A.61 and A.57 into A.67, along with the replacements $\hat{\phi} = -\sin\phi\hat{x} + \cos\phi\hat{y}$ and $\mu_o c = (\varepsilon_o c)^{-1}$, yields

$$\vec{B}^E_{1,0} = -ick\frac{\mu_o}{4\pi}\frac{1}{r_o^2} p_z \sin\theta\hat{\phi} \tag{A.68}$$

Previously, we noted that in the radiation zone the $l=1$, $m=0$ electric multipole contribution to the magnetic field was purely azimuthal. Now we see that to be true in the near zone as well. This is consistent with the fact observed in Chapter 4 that $\vec{B}$ is transverse to the direction of the field point for all values of r_o. And again, in the specific case in which an electric dipole moment has only a $\hat{z}$ component $\left(\vec{p} = p_z\hat{z}\right)$, Eq. A.68 can be written as,

$$\vec{B}^E_{1,0} = ick\frac{\mu_o}{4\pi}\frac{1}{r_o^2}\left(\hat{r}_o \times \vec{p}\right) = \hat{r}_o \times ick\frac{\mu_o}{4\pi}\frac{1}{r_o^2}\vec{p}$$

which, apart from the additional factor of $e^{ikr_o} \sim 1$, is exactly the magnetic field contribution from the electric dipole term, Eq. 4.31, in the Cartesian multipole expansion of Chapter 4. For the more general case in which the electric dipole moment has extension in all three coordinates, we continue identically as with the far field case: we substitute Eqs. A.60 along with the q_{11} and q_{1-1} terms of A.57 into A.67. Through a comparison of $\vec{B}^E_{1,m}$ in the near and far zones (Eqs. A.67 and A.59), we see that with a substitution of $\frac{i}{r_o}$ for ke^{ikr_o} in the radiation zone result for $\vec{B}^E_{1\pm1}$ (Eq. A.63), we can obtain the near zone result for $\vec{B}^E_{1\pm1}$

$$\begin{aligned}\vec{B}^E_{1\pm1} = -\frac{k}{8\pi\varepsilon_o c}\frac{1}{r_o^2}\Big[&\mp\cos\theta\left(p_x\hat{x} + p_y\hat{y}\right) \pm \sin\theta\left(\cos\phi p_x + \sin\phi p_y\right)\hat{z} \\ &- i\left(\cos\theta p_x\hat{y} - \cos\theta p_y\hat{x}\right) - i\left(\sin\theta\cos\phi p_y - \sin\theta\sin\phi p_x\right)\hat{z}\Big]\end{aligned} \tag{A.69}$$

and finally, combining $\vec{B}^E_{1,1}$, $\vec{B}^E_{1-1}$, and $\vec{B}^E_{1,0}$ as before, we get

$$\begin{aligned}\vec{B}^E_{1,1} + \vec{B}^E_{1-1} + \vec{B}^E_{1,0} &= ick\frac{\mu_o}{4\pi}\frac{1}{r_o^2}\left[\left(\sin\theta\sin\phi p_z - \cos\theta p_y\right)\hat{x} + \left(\cos\theta p_x - \sin\theta\cos\phi p_z\right)\hat{y} + \left(\hat{r}_o \times \vec{p}\right)_z\hat{z}\right] \\ &= ick\frac{\mu_o}{4\pi}\frac{1}{r_o^2}\left[\left(\hat{r}_o \times \vec{p}\right)_x\hat{x} + \left(\hat{r}_o \times \vec{p}\right)_y\hat{y} + \left(\hat{r}_o \times \vec{p}\right)_z\hat{z}\right] \\ &= ick\frac{\mu_o}{4\pi}\frac{1}{r_o^2}\left(\hat{r}_o \times \vec{p}\right) = \left[\vec{B}\right]_{ed}\end{aligned} \tag{A.70}$$

but now the connection is between the $l=1$ spherical electric multipole and the (zeroth order) electric dipole in the near zone limit. The associated spherical-Cartesian multipole relations for electric field follow by

$$\vec{E}^E_{1,1} + \vec{E}^E_{1,-1} + \vec{E}^E_{1,0} = \frac{ic}{k}\nabla \times \left(\vec{B}^E_{1,1} + \vec{B}^E_{1-1} + \vec{B}^E_{1,0}\right)$$
$$= -c\hat{r}_o \times \left(\vec{B}^E_{1,1} + \vec{B}^E_{1-1} + \vec{B}^E_{1,0}\right)$$

which can be shown to be equivalent to the electric field from the electric dipole of Eq. (4.38). Likewise, it can be shown in an analogous way that the near field term $(1/r_o^2)$ in the magnetic dipole expression of Eq. (4.48) is connected to the $l = 1$ magnetic multipole field.

Exercises

[1] Various related forms of Vector Spherical Harmonics have been presented by the following: Morse and Feshbach, H. *Methods of Theoretical Physics, Part II,* New York: McGraw-Hill, 1898–1901 (1953). Hill, E. L. The theory of vector spherical harmonics, *Am. J. Phys.* 22, 211–214 (1954). Blatt J. M. and Weisskopf, V. F. *Theoretical Nuclear Physics, 2nd edition,* New York: Wiley (1978). Barrera, R. G., Estévez, G. A., and Giraldo, J. Vector spherical harmonics and their application to magnetostatics, *Eur. J. Phys.* 6, 287–294 (1985). Weinberg, E. J. *Monopole vector spherical harmonics, Phys. Rev. D.* 49, 1086–1092 (1994).

[2] **Problem:** Prove Eqs. A.15, A.16, and A.17:

Solution: Noting that the (normalized) VSH are expressed in spherical coordinates as:

$$\vec{Y}_{lm}(\theta,\phi) = Y_{lm}(\theta,\phi)\,\hat{r}$$
$$\vec{\Psi}_{lm}(\theta,\phi) = \frac{1}{\sqrt{l(l+1)}}\left(\frac{\partial Y_{lm}}{\partial\theta}\hat{\theta} + \frac{1}{\sin\theta}\frac{\partial Y_{lm}}{\partial\phi}\hat{\phi}\right)$$
$$\vec{\Phi}_{lm}(\theta,\phi) = \frac{1}{\sqrt{l(l+1)}}\left(-\frac{1}{\sin\theta}\frac{\partial Y_{lm}}{\partial\phi}\hat{\theta} + \frac{\partial Y_{lm}}{\partial\theta}\hat{\phi}\right)$$

Proof #1)

$$\nabla \times \left(F(r)\,\vec{Y}_{lm}\right) = \nabla F(r) \times Y_{lm}\hat{r} + F(r)\left(\nabla \times Y_{lm}\hat{r}\right) = F(r)\left(\nabla \times Y_{lm}\hat{r}\right)$$
$$= F(r)\left[\frac{1}{r\sin\theta}\frac{\partial Y_{lm}}{\partial\phi}\hat{\theta} - \frac{1}{r}\frac{\partial Y_{lm}}{\partial\theta}\hat{\phi}\right]$$

but,

$$\sqrt{l(l+1)}\vec{\Phi}_{lm}(\theta,\phi) = \vec{r} \times \nabla Y_{lm} = \begin{pmatrix} \hat{r} & \hat{\theta} & \hat{\phi} \\ r & 0 & 0 \\ 0 & \frac{1}{r}\frac{\partial Y_{lm}}{\partial\theta} & \frac{1}{r\sin\theta}\frac{\partial Y_{lm}}{\partial\phi} \end{pmatrix} = -\left[\frac{1}{\sin\theta}\frac{\partial Y_{lm}}{\partial\phi}\hat{\theta} - \frac{\partial Y_{lm}}{\partial\theta}\hat{\phi}\right]$$

so

$$\nabla \times \left(F(r)\,\vec{Y}_{lm}\right) = -\frac{F(r)}{r}\left(\vec{r} \times \nabla Y_{lm}\right) = -\left(\frac{\sqrt{l(l+1)}}{r}F(r)\right)\vec{\Phi}_{lm}$$

Proof #2)

$$\sqrt{l(l+1)}\left(\nabla\times\left(F(r)\vec{\Psi}_{lm}\right)\right)=\nabla\times\left(F(r)r\nabla Y_{lm}\right)=\nabla\left(rF(r)\right)\times\nabla Y_{lm}+rF(r)\left(\nabla\times\nabla Y_{lm}\right)$$
$$=\nabla\left(rF(r)\right)\times\nabla Y_{lm}=\begin{pmatrix}\hat{r} & \hat{\theta} & \hat{\phi}\\ \frac{d}{dr}\left(rF(r)\right) & 0 & 0\\ 0 & \frac{1}{r}\frac{\partial Y_{lm}}{\partial\theta} & \frac{1}{r\sin\theta}\frac{\partial Y_{lm}}{\partial\phi}\end{pmatrix}$$
$$=-\frac{1}{r}\frac{d}{dr}\left(rF(r)\right)\left[\frac{1}{\sin\theta}\frac{\partial Y_{lm}}{\partial\phi}\hat{\theta}-\frac{\partial Y_{lm}}{\partial\theta}\hat{\phi}\right]=\frac{1}{r}\frac{d}{dr}\left(rF(r)\right)\left[\vec{r}\times\nabla Y_{lm}\right]$$

or, dividing by $\sqrt{l(l+1)}$,

$$\nabla\times\left(F(r)\vec{\Psi}_{lm}\right)=\frac{1}{r}\frac{d}{dr}\left(rF(r)\right)\left[\frac{\vec{r}\times\nabla Y_{lm}}{\sqrt{l(l+1)}}\right]=\left(\frac{1}{r}\frac{d}{dr}rF(r)\right)\vec{\Phi}_{lm}$$

Proof #3)

$$\sqrt{l(l+1)}\left(\nabla\times\left(F(r)\vec{\Phi}_{lm}\right)\right)=\nabla\times\left(F(r)\vec{r}\times\nabla Y_{lm}\right)$$
$$=\nabla\left(rF(r)\right)\times\left(\hat{r}\times\nabla Y_{lm}\right)+rF(r)\left[\nabla\times\left(\hat{r}\times\nabla Y_{lm}\right)\right]\tag{A.71}$$

with,

$$\left(\hat{r}\times\nabla Y_{lm}\right)=\hat{r}\times\left[\frac{1}{r}\frac{\partial Y_{lm}}{\partial\theta}\hat{\theta}+\frac{1}{r\sin\theta}\frac{\partial Y_{lm}}{\partial\phi}\hat{\phi}\right]=\frac{1}{r}\frac{\partial Y_{lm}}{\partial\theta}\hat{\phi}-\frac{1}{r\sin\theta}\frac{\partial Y_{lm}}{\partial\phi}\hat{\theta}$$

so the first term on the right side of Eq. A.71 can be expressed,

$$\frac{d}{dr}\left(rF(r)\right)\hat{r}\times\left[\frac{1}{r}\frac{\partial Y_{lm}}{\partial\theta}\hat{\phi}-\frac{1}{r\sin\theta}\frac{\partial Y_{lm}}{\partial\phi}\hat{\theta}\right]=-\frac{1}{r}\frac{d}{dr}\left(rF(r)\right)\left[\frac{\partial Y_{lm}}{\partial\theta}\hat{\theta}+\frac{1}{\sin\theta}\frac{\partial Y_{lm}}{\partial\phi}\hat{\phi}\right]$$
$$=-\frac{1}{r}\frac{d}{dr}\left(rF(r)\right)\sqrt{l(l+1)}\vec{\Psi}_{lm}\tag{A.72}$$

Now, considering the second term on the right side of Eq. A.71, first note,

$$\nabla\times\left(\hat{r}\times\nabla Y_{lm}\right)=\nabla\times\left[\frac{1}{r}\frac{\partial Y_{lm}}{\partial\theta}\hat{\phi}-\frac{1}{r\sin}\frac{\partial Y_{lm}}{\partial\phi}\hat{\theta}\right]$$
$$=\frac{1}{r\sin\theta}\left[\frac{\partial}{\partial\theta}\left(\frac{\sin\theta}{r}\frac{\partial Y_{lm}}{\partial\theta}\right)+\frac{\partial}{\partial\phi}\left(\frac{1}{r\sin\theta}\frac{\partial Y_{lm}}{\partial\phi}\right)\right]\hat{r}$$
$$-\frac{1}{r}\frac{\partial}{\partial r}\left(\frac{\partial Y_{lm}}{\partial\theta}\right)\hat{\theta}-\frac{1}{r}\frac{\partial}{\partial r}\left(\frac{1}{\sin\theta}\frac{\partial Y_{lm}}{\partial\phi}\right)\hat{\phi}$$

but the $\hat{\theta}$ and $\hat{\phi}$ terms vanish so the second term on the right-hand side of Eq. A.71 is

$$rF(r)\left[\nabla\times\left(\hat{r}\times\nabla Y_{lm}\right)\right] = rF(r)\left[\frac{1}{r^2\sin\theta}\frac{\partial}{\partial\theta}\left(\sin\theta\frac{\partial}{\partial\theta}\right)+\frac{1}{r^2\sin^2\theta}\frac{\partial^2}{\partial\phi^2}\right]Y_{lm}\hat{r}$$
$$= rF(r)\nabla^2 Y_{lm}\hat{r} = -\frac{l(l+1)}{r}F(r)\vec{Y}_{lm} \qquad \text{(A.73)}$$

So, combining Eqs. A.71, A.72 and A.73, and dividing by $\sqrt{l(l+1)}$, we get

$$\nabla\times\left(F(r)\vec{\Phi}_{lm}\right) = -\left(\frac{\sqrt{l(l+1)}}{r}F(r)\right)\vec{Y}_{lm} - \left(\frac{1}{r}\frac{d}{dr}rF(r)\right)\vec{\Psi}_{lm}$$

[3] **Problem:** Show explicitly the relationship between the time-averaged Poynting vector and energy density for radiation from a single magnetic multipole moment in the far zone:

Solution: The time averaged Poynting vector divided by c is $\left\langle\vec{S}\right\rangle\frac{1}{c} = \frac{1}{2\mu_0 c}Re\left(\vec{E}\times\vec{B}^*\right)$. For a single magnetic multipole moment in the far zone this becomes,

$$\left\langle\vec{S}^M_{lm}\right\rangle\frac{1}{c} = \frac{1}{2\mu_0 c}Re\left(\vec{E}^M_{lm}\times\vec{B}^{*M}_{lm}\right)$$
$$= \frac{1}{2\mu_0 c}Re\left(ia^M_{lm}\frac{e^{i(kr-\frac{\pi}{2}l)}}{kr}\vec{\Phi}_{lm}\times ia^{*M}_{lm}\frac{e^{-i(kr-\frac{\pi}{2}l)}}{ckr}\vec{\Psi}^*_{lm}\right)$$
$$= \frac{-\varepsilon_0|a^M_{lm}|^2}{2(kr)^2}\left(\vec{\Phi}_{lm}\times\vec{\Psi}^*_{lm}\right)$$

where we have again, as in the previous footnote, omitted the (negligible) term due to the $1/r^2$ radial contribution to the magnetic field. Now,

$$\vec{\Phi}_{lm}\times\vec{\Psi}^*_{lm} = \frac{1}{l(l+1)}\left(\vec{r}\times\nabla Y_{lm}\right)\times\left(r\nabla Y^*_{lm}\right) = -\frac{r^2}{l(l+1)}\left[\nabla Y^*_{lm}\times\left(\hat{r}\times\nabla Y_{lm}\right)\right]$$

using the vector identity $\vec{a}\times\left(\vec{b}\times\vec{c}\right) = (\vec{a}\cdot\vec{c})\vec{b} - \left(\vec{a}\cdot\vec{b}\right)\vec{c}$,

$$\vec{\Phi}_{lm}\times\vec{\Psi}^*_{lm} = -\frac{r^2}{l(l+1)}\left[\left(\nabla Y^*_{lm}\cdot\nabla Y_{lm}\right)\hat{r} - \left(\nabla Y^*_{lm}\cdot\hat{r}\right)\nabla Y_{lm}\right] = -\left(\vec{\Psi}^*_{lm}\cdot\vec{\Psi}_{lm}\right)\hat{r}$$

since $\nabla Y^*_{lm}\perp\hat{r}$. Plugging this in, we get the result

$$\left\langle\vec{S}^M_{lm}\right\rangle\frac{1}{c} = \frac{\varepsilon_0|a^M_{lm}|^2}{2(kr)^2}\left(\vec{\Psi}^*_{lm}\cdot\vec{\Psi}_{lm}\right)\hat{r}$$

or

$$\left|\left\langle\vec{S}^M_{lm}\right\rangle\right|\frac{1}{c}=\frac{\varepsilon_o|a^M_{lm}|^2}{2(kr)^2}\left(\vec{\Psi}^*_{lm}\cdot\vec{\Psi}_{lm}\right)=\varepsilon^M_{lm}$$

and so we confirm that, indeed, the magnitude of the time-averaged Poynting vector divided by c is equal to the energy density for the case of a single magnetic multipole moment in the far zone.

[4] Arfken, G. and Weber, H. *Mathematical Methods for Physicists, 4th edition,* Academic Press (1995), p. 681.

[5] **Problem:** Show that $\vec{L}\cdot\vec{\jmath}=i\nabla\cdot\left(\vec{r}\times\vec{\jmath}\right)$.

Solution:

$$\begin{aligned}
i\vec{L}\cdot\vec{\jmath} &= (\vec{r}\times\nabla)\cdot\vec{\jmath}\\
&=\left[\vec{r}\times\left(\frac{1}{r^2}\frac{\partial}{\partial r}r\hat{r}+\frac{1}{r\sin\theta}\frac{\partial}{\partial\theta}(\sin\theta)\hat{\theta}+\frac{1}{r\sin\theta}\frac{\partial}{\partial\phi}\hat{\phi}\right)\right]\cdot\vec{\jmath}\\
&=\left[-\frac{1}{\sin\theta}\frac{\partial}{\partial\phi}\hat{\theta}+\frac{1}{\sin\theta}\frac{\partial}{\partial\theta}(\sin\theta)\hat{\phi}\right]\cdot\left(\jmath_r\hat{r}+\jmath_\theta\hat{\theta}+\jmath_\phi\hat{\phi}\right)\\
&=\left[\frac{1}{\sin\theta}\frac{\partial}{\partial\theta}\left(\jmath_\phi\sin\theta\right)-\frac{1}{\sin\theta}\frac{\partial\jmath_\theta}{\partial\phi}\right]\\
&=\nabla\cdot\left(r\jmath_\phi\hat{\theta}-r\jmath_\theta\hat{\phi}\right)\\
&=-\nabla\cdot\left(\vec{r}\times\vec{\jmath}\right)
\end{aligned}$$

[6] **Problem:** Carry out a similar approximation with the electric multipole radiation coefficient a^E_{lm} as given by Eq. A.52

Solution: Given Eq. A.52,

$$a^E_{lm}=\frac{ik}{\varepsilon_o c^2}\int j_l(kr)\left[\vec{\Phi}^*_{lm}(\theta,\phi)\cdot\nabla\times\vec{\jmath}(r,\theta,\phi)\right]dV$$

As with a^M_{lm} for $kr\ll 1$, we first replace $j_l(kr)$ with its asymptotic form in Eq. A.53, express $\vec{\Phi}^*_{lm}$ as $[l(l+1)]^{-1/2}\left(i\vec{L}Y_{lm}\right)^*$, and use the Hermitian property of $\vec{L}$ to get

$$a^E_{lm}\simeq\frac{k^{l+1}}{\varepsilon_o c^2(2l+1)!!\sqrt{l(l+1)}}\int r^l Y^*_{lm}(\theta,\phi)\left(\vec{L}\cdot\nabla\times\vec{\jmath}\right)dV \tag{A.74}$$

Now using the vector identity $\hat{L}\cdot\left(\nabla\times\vec{\jmath}\right)=-i\left[(\vec{r}\cdot\nabla+2)\left(\nabla\cdot\vec{\jmath}\right)-\nabla^2\left(\vec{r}\cdot\vec{\jmath}\right)\right]$, the integral becomes,

$$\int r^l Y^*_{lm}(\theta,\phi)\left[\omega(\vec{r}\cdot\nabla)\rho+2\omega\rho+i\nabla^2\left(\vec{r}\cdot\vec{\jmath}\right)\right]dV \tag{A.75}$$

where we have used $\nabla \cdot \vec{\jmath} = -\frac{\partial \rho}{\partial t} = i\omega\rho$. Now $(\vec{r} \cdot \nabla)\rho = r\frac{\partial \rho}{\partial r}$, so the first component of this integral can be expressed as,

$$1^{st}component = \omega \int r^l Y^*_{lm}(\theta,\phi)(\vec{r} \cdot \nabla)\rho dV = \omega \int r^l Y^*_{lm}(\theta,\phi) r \frac{\partial \rho}{\partial r} r^2 dr d\Omega$$

where $dV = r^2 dr d\Omega$. First separating the angular and radial integrals, we then perform a radial integration by parts,

$$\begin{aligned} 1^{st}component &= \omega \int_\Omega Y^*_{lm} \left(\int_0^\infty r^{l+3} \frac{\partial \rho}{\partial r} dr \right) d\Omega \\ &= \omega \int_\Omega Y^*_{lm} \left(\left[r^{l+3}\rho \right]_0^\infty - (l+3) \int_0^\infty r^{l+2} \rho dr \right) d\Omega \end{aligned}$$

but we are assuming a finitely extended source, ρ, so the first term vanishes leaving.

$$1^{st}component = -(l+3)\omega \int r^l Y^*_{lm}(\theta,\phi)\rho dV$$

while the second component in the integral in Eq. A.75 can be written as,

$$2^{nd}component = 2\omega \int r^l Y^*_{lm}(\theta,\phi)\rho dV$$

For the third component of this integral, we use Green's theorem,

$$3^{rd}component = i \int r^l Y^*_{lm}(\theta,\phi)\nabla^2\left(\vec{r} \cdot \vec{\jmath}\right) dV$$

Let $\phi = r^l Y^*_{lm}$ and $\psi = \vec{r} \cdot \vec{\jmath}$, then Green's theorem says,

$$\int_V \left(\phi \nabla^2 \psi - \psi \nabla^2 \phi \right) dV = \oint_S \left(\phi \frac{\partial \psi}{\partial n} - \psi \frac{\partial \phi}{\partial n} \right) da$$

But, again, we note that the sources (ρ and $\vec{\jmath}$) are assumed finite so as the surface of integration (S) goes to infinity, both ψ and $\frac{\partial \psi}{\partial n}$ vanish. We then have,

$$\int_V \phi \nabla^2 \psi dV = \int_V \psi \nabla^2 \phi dV$$

or

$$\int_V r^l Y^*_{lm} \nabla^2 \left(\vec{r} \cdot \vec{\jmath}\right) dV = \int_V \left(\vec{r} \cdot \vec{\jmath}\right) \nabla^2 \left(r^l Y^*_{lm}\right) dV$$

and thus we can write the third component as

$$3^{rd} component = i \int \left(\vec{r}\cdot\vec{\jmath}\right) \nabla^2 \left(r^l Y_{lm}^* (\theta,\phi)\right) dV$$

but notice that $\nabla^2 \left(r^l Y_{lm}^*\right) = Y_{lm}^* \frac{1}{r}\frac{\partial^2}{\partial r^2} r\left(r^l\right) + r^l \nabla^2 Y_{lm}^* = Y_{lm}^* \frac{l(l+1)}{r^2} r^l - r^l \frac{l(l+1)}{r^2} Y_{lm}^* = 0$ so the third term vanishes and the integral of Eq. A.75 becomes

$$1^{st} component + 2^{nd} component = -(l+1)\,\omega \int r^l Y_{lm}^* (\theta,\phi)\, \rho(r,\theta,\phi)\, dV$$

and substituting this into Eq. A.74,

$$\begin{aligned} a_{lm}^E &\simeq \frac{-\omega k^{l+1}(l+1)}{\varepsilon_0 c^2 (2l+1)!!\sqrt{l(l+1)}} \int r^l Y_{lm}^* (\theta,\phi)\, \rho(r,\theta,\phi)\, dV \\ &\simeq \frac{-1}{\varepsilon_0 c (2l+1)!!} \left(\frac{l+1}{l}\right)^{\frac{1}{2}} k^{l+2} \int r^l Y_{lm}^* (\theta,\phi)\, \rho(r,\theta,\phi)\, dV \end{aligned}$$

7 **Problem:** Using $Y_{1,\pm1}$ and $Y_{1,0}$, prove the relations of Eq. A.57.

Solution: The relations between the three $l = 1$ electric multipole moments and the electric dipole moment are obtained as:

$$\begin{aligned} q_{1,\pm1} &= \int r Y_{1,\pm1}^* (\theta,\phi)\, \rho(r,\theta,\phi)\, dV = \mp\sqrt{\frac{3}{8\pi}} \int r \sin\theta\, e^{\mp i\phi} \rho dV \\ &= \mp\sqrt{\frac{3}{8\pi}} \left(\int r\sin\theta\cos\phi\rho dV \mp i \int r\sin\theta\sin\phi\rho dV\right) \\ &= \mp\sqrt{\frac{3}{8\pi}} \left(\int x\rho dV \mp i \int y\rho dV\right) = \mp\sqrt{\frac{3}{8\pi}} \left(p_x \mp i p_y\right) \end{aligned}$$

and

$$\begin{aligned} q_{1,0} = \int r Y_{1,0}^* (\theta,\phi)\, \rho(r,\theta,\phi)\, dV &= \sqrt{\frac{3}{4\pi}} \int r\cos\theta\rho dV \quad . \\ &= \sqrt{\frac{3}{4\pi}} \int z\rho dV = \sqrt{\frac{3}{4\pi}} p_z \end{aligned}$$

where we have recalled that the electric dipole moment is $\vec{p} = \int \vec{r}\rho(\vec{r})\, dV$.

8 **Problem:** With the general form of $m_{lm} = \frac{-1}{c(l+1)} \int r^l Y_{lm}^* (\theta,\phi)\, \nabla\cdot(\vec{r}\times\vec{\jmath}) dV$ for the spherical magnetic multipole moments and the general form of $\vec{m} = \frac{1}{2}\int \vec{r}\times\vec{\jmath}$ for the Cartesian magnetic dipole moment, show the relations between the three $l = 1$ magnetic multipole moments and the magnetic dipole moment.

Solution:

$$m_{1,\pm 1} = -\frac{1}{2c}\int rY^*_{1,\pm 1}(\theta,\phi)\,\nabla\cdot(\vec{r}\times\vec{\mathcal{J}})dV = \pm\sqrt{\frac{3}{8\pi c^2}}\frac{1}{2}\int r\sin\theta e^{\mp i\phi}\nabla\cdot(\vec{r}\times\vec{\mathcal{J}})dV$$
$$= \pm\sqrt{\frac{3}{8\pi c^2}}\frac{1}{2}\left(\int r\sin\theta\cos\phi\nabla\cdot(\vec{r}\times\vec{\mathcal{J}})dV \mp i\int r\sin\theta\sin\phi\nabla\cdot(\vec{r}\times\vec{\mathcal{J}})dV\right)$$
$$= \pm\sqrt{\frac{3}{8\pi c^2}}\frac{1}{2}\left(\int x\nabla\cdot(\vec{r}\times\vec{\mathcal{J}})dV \mp i\int y\nabla\cdot(\vec{r}\times\vec{\mathcal{J}})dV\right)$$

using integration by parts on the two terms yields

$$m_{1,\pm 1} = \mp\sqrt{\frac{3}{8\pi c^2}}\frac{1}{2}\left(\int(\vec{r}\times\vec{\mathcal{J}})_x dV \mp i\int(\vec{r}\times\vec{\mathcal{J}})_y dV\right) = \mp\sqrt{\frac{3}{8\pi c^2}}\left(\mathsf{m}_x \mp i\mathsf{m}_y\right)$$

and likewise

$$m_{1,0} = -\frac{1}{2c}\int rY^*_{1,0}(\theta,\phi)\,\nabla\cdot(\vec{r}\times\vec{\mathcal{J}})dV = -\sqrt{\frac{3}{4\pi c^2}}\frac{1}{2}\int r\cos\theta\nabla\cdot(\vec{r}\times\vec{\mathcal{J}})dV$$
$$= -\sqrt{\frac{3}{4\pi c^2}}\frac{1}{2}\int z\nabla\cdot(\vec{r}\times\vec{\mathcal{J}})dV$$
$$= \sqrt{\frac{3}{4\pi c^2}}\frac{1}{2}\int(\vec{r}\times\vec{\mathcal{J}})_z dV = \sqrt{\frac{3}{4\pi c^2}}\mathsf{m}_z$$

[9] Arfken and Weber, p. 682.

[10] **Problem:** With the use of Eq. A.17, show that the $\nabla\times f(r_o)\,\vec{\Phi}_{lm}$ terms in Eq. A.58 contain both radial and transverse field components that fall off as $1/r_o^2$ and $1/r_o$, respectively.

Solution: Starting with the curl terms in Eq. A.58,

$$\vec{E}^E_{lm} = \frac{c}{k}\nabla\times a^E_{lm}\frac{e^{i\left(kr_o-\frac{\pi}{2}l\right)}}{kr_o}\vec{\Phi}_{lm}$$
$$\vec{B}^M_{lm} = -\frac{1}{ck}\nabla\times a^M_{lm}\frac{e^{i\left(kr_o-\frac{\pi}{2}l\right)}}{kr_o}\vec{\Phi}_{lm}$$

applying Eq. A.17,

$$\nabla\times\left(F(r_o)\,\vec{\Phi}_{lm}\right) = -\left(\frac{\sqrt{l(l+1)}}{r_o}F(r_o)\right)\vec{Y}_{lm} - \left(\frac{1}{r_o}\frac{d}{dr_o}r_oF(r_o)\right)\vec{\Psi}_{lm}$$

and noting that,

$$\frac{d}{dr_o} r_o F(r_o) = \frac{d}{dr_o} r_o \left(\frac{e^{i(kr_o - \frac{\pi}{2}l)}}{kr_o} \right) = ie^{i(kr_o - \frac{\pi}{2}l)}$$

reveals the radial $1/r_o^2$ and transverse $1/r_o$ electric field terms (from the electric multipoles)

$$\vec{E}_{lm}^E = -a_{lm}^E c\sqrt{l(l+1)} \frac{e^{i(kr_o - \frac{\pi}{2}l)}}{(kr_o)^2} \vec{Y}_{lm} - ia_{lm}^E c \frac{e^{i(kr_o - \frac{\pi}{2}l)}}{kr_o} \vec{\Psi}_{lm}$$

and the radial $1/r_o^2$ and transverse $1/r_o$ magnetic field terms (from the magnetic multipoles)

$$\vec{B}_{lm}^M = a_{lm}^M \sqrt{l(l+1)} \frac{e^{i(kr_o - \frac{\pi}{2}l)}}{c(kr_o)^2} \vec{Y}_{lm} + ia_{lm}^M \frac{e^{i(kr_o - \frac{\pi}{2}l)}}{ckr_o} \vec{\Psi}_{lm}$$

[11] **Problem:** Derive the relations of Eqs. A. 60 and A. 61.
Solution: The $l = 1$ spherical harmonics are

$$Y_{1\pm 1} = \mp\sqrt{\frac{3}{8\pi}} \sin\theta e^{\pm i\phi}$$
$$Y_{10} = \sqrt{\frac{3}{4\pi}} \cos\theta$$

and the gradients are

$$\nabla Y_{1\pm 1} = \mp\sqrt{\frac{3}{8\pi}} \frac{1}{r_o} \left[\cos\theta e^{\pm i\phi} \hat{\theta} \pm ie^{\pm i\phi} \hat{\phi} \right]$$
$$\nabla Y_{10} = -\sqrt{\frac{3}{4\pi}} \frac{1}{r_o} \sin\theta \hat{\theta}$$

and thc cross product of $\hat{r}_o$ with the gradients (non-normalized $\vec{\Phi}_{1,m}$),

$$\hat{r}_o \times \nabla Y_{1\pm 1} = \sqrt{\frac{3}{8\pi}} \frac{e^{\pm i\phi}}{r_o} \left[i\hat{\theta} \mp \cos\theta \hat{\phi} \right]$$
$$\hat{r}_o \times \nabla Y_{10} = -\sqrt{\frac{3}{4\pi}} \frac{1}{r_o} \sin\theta \hat{\phi}$$

Noting the relationships of $\hat{\theta}$ and $\hat{\phi}$ to $\hat{x}$, $\hat{y}$ and $\hat{z}$

$$\hat{\theta} = \cos\theta \cos\phi \hat{x} + \cos\theta \sin\phi \hat{y} - \sin\theta \hat{z}$$
$$\hat{\phi} = -\sin\phi \hat{x} + \cos\phi \hat{y}$$

these $\vec{\Phi}_{1,m}$ terms can be expressed

$$\hat{r}_o \times \nabla Y_{1\pm1} = \sqrt{\frac{3}{8\pi}}\frac{e^{\pm i\phi}}{r_o}\left[i\left(\cos\theta\cos\phi\hat{x} + \cos\theta\sin\phi\hat{y} - \sin\theta\hat{z}\right) \pm \cos\theta\left(\sin\phi\hat{x} - \cos\phi\hat{y}\right)\right]$$

$$\hat{r}_o \times \nabla Y_{10} = \sqrt{\frac{3}{4\pi}}\frac{1}{r_o}\left(\sin\theta\sin\phi\hat{x} - \sin\theta\cos\phi\hat{y}\right)$$

and after some simple algebraic manipulations of the $\hat{r}_o \times \nabla Y_{1\pm1}$ terms, we have

$$\hat{r}_o \times \nabla Y_{1\pm1} = i\sqrt{\frac{3}{8\pi}}\frac{1}{r_o}\left[\cos\theta\hat{x} \pm i\cos\theta\hat{y} - (\sin\theta\cos\phi \pm i\sin\theta\sin\phi)\,\hat{z}\right]$$

$$\hat{r}_o \times \nabla Y_{10} = \sqrt{\frac{3}{4\pi}}\frac{1}{r_o}\left(\sin\theta\sin\phi\hat{x} - \sin\theta\cos\phi\hat{y}\right)$$

[12] Arfken and Weber, p. 681.

References

Chapter 1

[5] Jackson, J.D., "Classical Electrodynamics", 3rd edition, Wiley, NY., NY 1999
[9] Panofsky, W.K.H. and Phillips, M., "Classical Electricity and Magnetism" 2nd edition, Addison-Wesley, 1962
[6] Purcell, E.M., and Morin, D.J. "Electricity and Magnetism", 3rd edition, Cambridge University Press, 2013
[24] Shintake, Nuc Inst. And Methods in Physics Research, **507**, 189 (2003)
[25] The Feynman Lectures on Physics, Feynman, R.P., Leighton, R.B., Sands, M., **1** chapter 28, eq., 28.3, Addison-Wesley, 1964

Chapter 3

[6] McDonald K.T. Am. J. Physics, **65**, 1074 (1997)
[14] Griffiths, D.J., "Introduction to Electrodynamics" 3rd edition, Prentice-Hall (1981)

Chapter 4

[1] Souza, R. M, et al., Am J. Physics **77** 67 (2009)

Chapter 5

[3] Darrigol, O., Seminare Poincare **1** 1–22 (2005)
[4] Poincare, H., Revue de Mataphysique et de Morale, **6** 1–13 (1898)
[32] Evenson, K.M., et al., Physical Review Letter, **29** (19) 1346 (1972)

Chapter 6

[2] Panofsky, W.K.H. and Phillips, M., "Classical Electricity and Magnetism" 2nd edition, Addison-Wesley, 1962
[14] Pollock, H.C., Am. J. Physics, **51** 278 (1983)
[17] Panofsky, W.K.H. and Phillips, M., "Classical Electricity and Magnetism" 2nd edition, Addison-Wesley, 1962

Chapter 7

[5] Goldstein, H., "Classical Mechanics", Chapter 2, Addison-Wesley, 1951
[17] Goldstein, H., Poole, C.P., Salko, J.L., "Classical Mechanics" 3rd edition, Addison-Wesley 2001

Chapter 8

[2] Panofsky, W.K.H. and Phillips, M., "Classical Electricity and Magnetism" 2nd edition, Addison-Wesley, 1962
[11] Griffiths, et al., Am. J. Physics **78** 391 (2010)
[12] Landau, L.D. and Lifshitz, E.M. "The Classical Theory of Fields" Pergamon, Oxford, Sec 75 (1971)
[15] Panofsky, W.K.H. and Phillips, M., "Classical Electricity and Magnetism" 2nd edition, Pgs 309–310, Addison-Wesley, 1962
[18] Schwinger, J., Foundations of Physics, **13**, 373 (1983)

Chapter 9

[14] Forsmann, A., et al. Phys Rev. E **58** R1248 (1998); Widmann, K., Phys. Rev. Lett **92** 125002 (2004); Y.Ping, et al., Phys. Rev. Lett **96** 255003 (2006)

Chapter 10

[6] Milchberg, H.M., et al., Phys. Rev. Lett **61** 2364 (1988); Lee, Y.T. and Moore, R.M., Physics of Fluids 1958, **27** (1984)

Chapter 11

[1] Tesersky, V., App. Opt. **3** 1150 (1964)
[3] Strutt, J.W. (3rd Baron Rayleigh) "The Theory of Sound", MacMillan and Co. (1896)
[4] Jackson, J.D., "Classical Electrodynamics", 3rd edition, (eq. 4.56) Wiley, NY., NY 1999
[7] Cai, W., Shalaev, V. "Optical Metamaterials", Springer, NY (2010)
[15] Newton, R., "Scattering Theory of Waves and Particles", McGraw-Hill, Inc, NY (1966)
[22] Stratton, J.A., "Electromagnetic Theory", McGraw-Hill, NY (1941)
[26] Nussenzweig, H.M., "Diffraction Effects in Semiclassical Scattering", Cambridge University Press (1992)

[27] Nussenzweig, H.M., "Diffraction Effects in Semiclassical Scattering", Cambridge University Press (1992)
[28] van de Hlst, H.C., "Light Scattering by Small Particles", Wiley, N.Y. (1957)

Chapter 12

[5] Wolf, E., and Marchand, E.W. J. Opt. Soc. Am. **54**, 587 (1964)

Chapter 13

[6] Casimir, H.B.G. and Polder, D., Phys. Rev. **73** 360 (1948)
[7] Casimir, H.B.G., Proc. Royal Netherlands Academy of Scienes, **51** 793 (1948)
[10] Phys. Rev. Lett, **78** 5 (1997)
[11] Klimchitskaya, G.L., et al., Rev. mod. Phys., **81** 1827 (2009)
[12] Allen, L., and Eberley, J.H. "Optical Resonances and Two-level Atoms, Wiley, N.Y. (1975)
[13] Purcell, E.M., Phys.Rev. **69** 674 (1946)
[14] Berbardot, F., et al, Europhysics Lett. **17** 33 (1992)
[15] Hulet, R.G., et al., Phys. Rev. Lett. **20** 2137 (1985)

Index